T0328951

SECOND EDITION

BIOMARKERS OF KIDNEY DISEASE

Academic Press is an imprint of Elsevier
125 London Wall, London EC2Y 5AS, United Kingdom
525 B Street, Suite 1800, San Diego, CA 92101-4495, United States
50 Hampshire Street, 5th Floor, Cambridge, MA 02139, United States
The Boulevard, Langford Lane, Kidlington, Oxford OX5 1GB, United Kingdom

Notices
Knowledge and best practice in this field are constantly changing. As new research and experience broaden our understanding, changes in research methods, professional practices, or medical treatment may become necessary.

Practitioners and researchers must always rely on their own experience and knowledge in evaluating and using any information, methods, compounds, or experiments described herein. In using such information or methods they should be mindful of their own safety and the safety of others, including parties for whom they have a professional responsibility.

To the fullest extent of the law, neither the Publisher nor the authors, contributors, or editors, assume any liability for any injury and/or damage to persons or property as a matter of products liability, negligence or otherwise, or from any use or operation of any methods, products, instructions, or ideas contained in the material herein.

Library of Congress Cataloging-in-Publication Data
A catalog record for this book is available from the Library of Congress

British Library Cataloguing-in-Publication Data
A catalogue record for this book is available from the British Library

ISBN: 978-0-12-803014-1

For information on all Academic Press publications
visit our website at https://www.elsevier.com/

Publisher: Mica Haley
Acquisition Editor: Tari Broderick
Editorial Project Manager: Lisa Eppich
Production Project Manager: Karen East and Kirsty Halterman
Designer: Mark Rogers

Typeset by Thomson Digital

SECOND EDITION

BIOMARKERS OF KIDNEY DISEASE

Edited by

CHARLES L. EDELSTEIN, MD, PhD

Division of Renal Diseases and Hypertension
University of Colorado Denver
Aurora, CO, United States

Amsterdam • Boston • Heidelberg • London
New York • Oxford• Paris • San Diego
San Francisco • Singapore • Sydney • Tokyo
Academic Press is an imprint of Elsevier

CONTENTS

LIST OF CONTRIBUTORS

J.M. Arthur, MD, PhD
Division of Nephrology, University of Arkansas for Medical Sciences and Central Arkansas Veterans Healthcare System, Little Rock, AR, United States

R.E. Banks, PhD
Biomedical Proteomics, Clinical and Biomedical Proteomics Group, Leeds Institute of Cancer and Pathology, St James's University Hospital, Leeds, United Kingdom

M.R. Bennett, PhD
Division of Nephrology and Hypertension, Cincinnati Children's Hospital Medical Center, University of Cincinnati, College of Medicine, Cincinnati, OH, United States

U. Christians, MD, PhD
iC42 Clinical Research and Development, Department of Anesthesiology, University of Colorado, Anschutz Medical Campus, Aurora, CO, United States

P. Devarajan, MD, FAAP
Division of Nephrology and Hypertension, Cincinnati Children's Hospital Medical Center, University of Cincinnati, College of Medicine, Cincinnati, OH, United States

C.L. Edelstein, MD, PhD
Division of Renal Diseases and Hypertension, University of Colorado Denver, Aurora, CO, United States

E. Elnagar, MBBS, MPH
Division of Nephrology, University of Arkansas for Medical Sciences and Central Arkansas Veterans Healthcare System, Little Rock, AR, United States

Z.H. Endre, BScMed, MBBS, PhD, RACP, FASN
Department of Nephrology, Prince of Wales Hospital and Clinical School, University of New South Wales, Sydney, NSW; School of Medicine, University of Queensland, Brisbane, QLD, Australia; Department of Medicine, University of Otago, Christchurch, New Zealand

S. Faubel, MD
Medicine, Division of Renal Diseases and Hypertension, University of Colorado Denver, Veteran Affairs Medical Center, Denver, CO, United States

G. Fick-Brosnahan, MD
Division of Renal Diseases and Hypertension, Anschutz Medical Campus, Aurora, CO, United States

A. Grubb, MD, PhD
Department of Clinical Chemistry and Pharmacology, University Hospital, Lund University, Lund, Sweden

S. Jain, PhD
University of Colorado, Aurora, CO, United States

A. Jani, MD
University of Colorado, Aurora; Denver Veteran Affairs Medical Center, Denver, CO, United States

N. Karakala, MD
Division of Nephrology, University of Arkansas for Medical Sciences and Central Arkansas Veterans Healthcare System, Little Rock, AR, United States

S.A. Karumanchi, MD
Department of Medicine, Obstetrics and Gynecology, Beth Israel Deaconess Medical Center and Harvard Medical School, Boston, MA, United States

J. Klawitter, PhD
iC42 Clinical Research and Development, Department of Anesthesiology, University of Colorado, Anschutz Medical Campus, Aurora, CO, United States

J. Klawitter, PhD
iC42 Clinical Research and Development, Department of Anesthesiology, University of Colorado, Anschutz Medical Campus, Aurora, CO, United States

J. Klepacki, PhD
iC42 Clinical Research and Development, Department of Anesthesiology, University of Colorado, Anschutz Medical Campus, Aurora, CO, United States

J.B. Klein, MD, PhD
Robley Rex Veterans Administration Medical Center, Louisville, KY, United States

M.L. Merchant, PhD
Division of Nephrology and Hypertension, Department of Medicine, University of Louisville, Louisville, KY, United States

C.R. Parikh, MD, PhD
Program of Applied Translational Research, Department of Medicine, Yale University, New Haven; Veterans Affairs Medical Center, West Haven, CT, United States

H. Thiessen Philbrook, MMath, AStat
Program of Applied Translational Research, Department of Medicine, Yale University, New Haven; Veterans Affairs Medical Center, West Haven, CT, United States

B.Y. Reed, PhD
Division of Renal Diseases and Hypertension, Anschutz Medical Campus, Aurora, CO, United States

N.S. Vasudev, MD, PhD
Medical Oncology, Clinical and Biomedical Proteomics Group, Leeds Institute of Cancer and Pathology, St James's University Hospital, Leeds, United Kingdom

R.J. Walker, MBChB, MD, FRACP, FASN, FAHA
Department of Nephrology, Dunedin Hospital and University of Otago, Dunedin, New Zealand

PREFACE

Developing and defining biomarkers of kidney diseases that can be used for early diagnosis, assessment of severity, assessment of short- and long-term prognosis and risk-stratification is extremely important for the practicing physician. Biomarkers can help physicians in determining the timely prevention, severity, more effective treatment, prognosis and response to therapy of disease. Biomarkers of disease are a fertile area of research for scientists.

During the last 6 years since the first edition of the book, there has continued to be exponential growth in research on biomarkers of kidney diseases and as a result, we can now bring preclinical studies to the bedside and diagnose certain kidney diseases at earlier stages than was possible with conventional tests. One of the most important advances has been NephroCheck, the first FDA-approved biomarker of acute kidney injury (AKI). NephroCheck uses a combination of urinary insulin-like growth factor-binding protein-7 (IGF-BP7) and tissue inhibitor of metalloproteinases-2 (TIMP2) and with its approval, early diagnosis and treatment of kidney diseases has now become a reality in clinical practice.

The second edition of the book provides an update of biomarkers of kidney diseases that are of particular importance to the practicing physician while remaining the most comprehensive work published on this crucial topic. New chapters include "Biomarkers of Extra-Renal Complications of AKI," "Diagnostic and Prognostic Biomarkers in Autosomal Dominant Polycystic Kidney Disease," and "Biomarkers of Cardiovascular Risk in Chronic Kidney Disease." In addition, the second edition expands coverage of certain diseases, including AKI, CKD, kidney transplant rejection, delayed kidney allograft function, polycystic kidney disease, renal cell cancer, glomerular disease, diabetic nephropathy, and preeclampsia.

Successful biomarker candidates are now being advanced as tools for personalized and predictive approaches to kidney disease. Prasad Devarajan provides a brief review of how novel biomarkers are discovered and validated, and what the general characteristics of an ideal biomarker are.

For the physician interpreting or planning biomarker studies, Chirag R. Parikh and Heather Thiessen Philbrook, both experts in the field, discuss traditional and emerging statistical methods for evaluating the prediction performance of diagnostic biomarkers.

Proteomic and metabolomic profiling of body fluids and tissues has great potential to advance our understanding of kidney diseases and drug effects, to advance clinical diagnostics and to be an important tool in the individualization of treatment. Dr. Uwe Christians, who has state-of-the-art laboratories at the University of Colorado for biomarker discovery, has updated his comprehensive chapter on the use of metabolomics and proteomics in kidney diseases with the most exciting studies in the field in the last 6 years.

BUN and serum creatinine are not very sensitive and specific markers of kidney function in AKI as they are influenced by many renal and non-renal factors independent of kidney function. Charles L. Edelstein reviews the new biomarkers for the diagnosis and prognosis of AKI that have been discovered over the last 6 years including newly FDA-approved biomarkers. Dr. Alkesh Jani, a transplant nephrologist, has updated the chapter on biomarkers for the early diagnosis of delayed kidney graft function, kidney rejection, and polyoma virus infection.

Clinical and experimental data indicate that AKI contributes to distant organ injury. Thus, the high mortality of AKI may be due to deleterious systemic effects of AKI. In a new addition to the book, Dr. Sarah Faubel discusses the inflammatory and pulmonary complications of AKI as well as their potential biomarkers.

We are fortunate to have Dr. Grubb, who helped isolate and sequence the "mysterious protein" cystatin C that was discovered in the urine in 1961, write the chapter on cystatin C as a biomarker in kidney diseases. The updated chapter includes the role of cystatin C in identifying the novel "Shrunken Pore Syndrome."

Determining prognosis for individual patients with renal cell cancer is important to allow targeting of high-risk patients for trials of adjuvant therapy and more intensive follow-up. The current field of renal cancer biomarkers is comprehensively reviewed by Dr. Roz E. Banks and Dr. Naveen S. Vasudev.

Diabetic nephropathy and glomerulonephritis are the commonest causes of ESRD in the USA. Dr. Jon B. Klein and colleagues update the evolving role that proteomics has played in expanding our understanding of the natural history of diabetic nephropathy. The most promising candidate biomarkers for the early diagnosis, early prediction of flares and prediction of outcome in patients with glomerulonephritis like membranous GN, FSGS, and IgA nephropathy are reviewed by Dr. John M. Arthur and colleagues.

In an exciting new addition to the book, Zoltan H. Endre and Robert J. Walker review traditional markers of kidney disease, traditional markers of cardiovascular disease, and novel markers of kidney damage as markers of cardiovascular risk in subjects with CKD.

Autosomal dominant polycystic kidney disease (ADPKD) is the commonest hereditary kidney disease. Drs. Berenice Y. Reed and Godela Fick-Brosnahan, well-known researchers in ADPKD, have written a unique new addition to the book. A prognostic biomarker predicting the disease course at an early age would be helpful for patient counseling, selecting those patients most likely to benefit from an intervention and could serve as a surrogate endpoint in clinical trials testing new therapeutic interventions in ADPKD. Total kidney volume is qualified as a biomarker by the FDA for ADPKD Trials.

Preeclampsia can be a devastating disease and is a leading cause of maternal and perinatal morbidity and mortality. Dr. S. Ananth Karumanchi, a world expert on this topic, has updated his chapter to include new angiogenic factors, placental protein-13 (PP-13), and combinations of these and other parameters with Doppler analysis that hold promise for future predictive testing for preeclampsia.

The advances in our knowledge of biomarkers of kidney disease continue to grow and I believe that the use of novel biomarkers of kidney disease has become a reality in clinical practice. It is my pleasure and privilege to edit the second edition of a book written by distinguished authors that continue to contribute to the exciting advances in our knowledge of biomarkers of kidney disease.

Charles L. Edelstein

CHAPTER ONE

Characteristics of an Ideal Biomarker of Kidney Diseases

M.R. Bennett, PhD and P. Devarajan, MD, FAAP
Division of Nephrology and Hypertension, Cincinnati Children's Hospital Medical Center, University of Cincinnati, College of Medicine, Cincinnati, OH, United States

Contents

THE DISCOVERY OF BIOMARKERS

The quest for biomarkers is as old as medicine itself. From the earliest days of diagnostic medicine in ancient Egypt, to the misguided science of phrenology (the belief that skull measurements could predict personality traits), to the powerful discoveries of modern science, we have been searching for measurable biologic cues that will give us an insight into the physiologic workings of the human organism. In its simplest definition, a biomarker is anything that can be measured to extract information about a biologic state or process. The NIH Biomarkers Definitions Working Group has defined a biologic marker (biomarker) as "A characteristic that is objectively measured and evaluated as an indicator of normal biologic processes, pathogenic processes, or pharmacologic responses to a therapeutic intervention [1]."

Biomarkers appear in every form. Body temperature, in the form of a fever, can signal infection. Blood pressure and cholesterol levels can predict cardiovascular risk. Tracking biomarkers, such as, height and weight can give clues about normal human growth and development. Such general biomarkers have been used for decades or even centuries and have remained powerful tools for tracking general biologic activity. However, the era of personalized medicine is well on us. Ushered in by the remarkable genomic

Biomarkers of Kidney Disease. http://dx.doi.org/10.1016/B978-0-12-803014-1.00001-7

Table 1.1 Phases of biomarker discovery, translation, and validation

Phase	Terminology	Action steps
Phase 1	Preclinical discovery	• Discover biomarkers in tissues or body fluids • Confirm and prioritize promising candidates
Phase 2	Assay development	• Develop and optimize clinically useful assay • Test on existing samples of established disease
Phase 3	Retrospective study	• Test biomarker in completed clinical trial • Test if biomarker detects the disease early • Evaluate sensitivity, specificity and receiver operating characteristic (ROC)
Phase 4	Prospective screening	• Use biomarker to screen population • Identify extent and characteristics of disease • Identify false-referral rate
Phase 5	Disease control	• Determine impact of screening on reducing disease burden

Source: Adapted from States DJ, Omenn GS, Blackwell TW, Fermin D, Eng J, Speicher DW, Hanash SM. Challenges in deriving high-confidence protein identifications from data gathered by a HUPO plasma proteome collaborative study. Nat Biotechnol 2006;24(3):333–8 [7].

and proteomic advances in our understanding of health and disease, personalized medicine promises a more precise determination of disease predisposition, diagnosis, and prognosis, earlier preventive and therapeutic interventions, a more efficient drug development process, and a safer and more fiscally responsive approach toward medicine. Biomarkers are the essential tools for the implementation of personalized medicine. The quest for the advancement of personalized medicine pushes us further and further into the realm of molecular medicine to discover biomarkers with increasing sensitivity and specificity. For most of our history, biomarker discovery has relied on the intimate knowledge of the pathophysiology of the diseases being studied. Biologic substances, which we knew were related to a disease state, were investigated to see if they could serve as diagnostic markers, provide a target for therapy, or lend further insight into the etiology of the disease. While this can be tedious, and relies heavily on prior knowledge of the disease mechanism, this hypothesis-driven method of research almost always provides useful scientific results, whether positive or negative.

The biomarker-development process has typically been divided into five phases, as shown in Table 1.1. The preclinical discovery phase requires high-quality, well-characterized tissue or body fluid samples from carefully chosen animal or human models of the disease under investigation. In the last 20 years, the ready availability of powerful tools that scan both the genome and the proteome of an organism have revolutionized and greatly accelerated biomarker discovery. Transcriptome profiling, using complementary DNA (cDNA) microarrays that can measure the entire complement of messenger RNA (mRNA) in a given sample type, has yielded a number of promising biomarkers of kidney disease, as well as, novel disease mechanisms in many fields [2–4]. This approach can be combined with other techniques, such as laser capture, microdissection, to target specific areas of a diseased tissue to give mechanistic clues that was not possible just a decade ago. Even with this level of specificity, these techniques can yield a daunting array of data that must be sifted through for relevance. A shortcoming of transcriptomic profiling approaches is that it cannot be performed directly in biologic fluids. Another problem with this approach is that ultimately the mRNA does not always reflect protein levels or activity, which must be further confirmed at the protein level prior to larger validation studies. Despite these limitations, transcriptome-profiling studies have been extensively utilized to study models of acute kidney injury (AKI) [5]. A metaanalysis of gene-expression profiles from 150 distinct microarray experiments from 21 different models of AKI identified several upregulated genes previously known to be associated with AKI [6]. The most consistently and most highly upregulated gene has been neutrophil gelatinase-associated lipocalin (NGAL), whose protein product has now successfully passed through the preclinical, assay development, and clinical testing stages of the biomarker-development process.

In the last 5 years, deep sequencing techniques, such as, RNAseq have supplanted microarrays as the preferred transcriptomic "shotgun" method for biomarker discovery, though it is not without limitations in terms of clinical utility. RNAseq uses deep sequencing technologies to sequence the RNA in a given sample as opposed to hybridizing mRNA onto a known cDNA array [8]. This gives a more precise measurement of the level of transcripts and sequence variations [8]. The difficulties with this technology lies not only at the bioinformatic level—as there needs to be the ability to deal with massive amounts of data and narrow them down to a usable format—but also at a cost level. The deeper the sequencing, the more expensive it is to run, and that limits the utility in a clinical environment to large institutions that can afford the specialized equipment, but that also have the bioinformatic capabilities to interpret the resulting profiles.

Proteomic approaches move a step beyond genomic studies and screen the actual proteins and peptides present in a sample. This approach allows one to go beyond simple translation of mRNA into protein and allows a look into protein regulation, posttranslational modifications (such as, glycosylation and methylation), and even disease-specific fragmentation. There are a number of proteomic approaches including gel electrophoresis and modern mass spectometry techniques, such as, matrix assisted laser desorption ionization time of flight (MALDI-TOF) mass spectrometry and isobaric tag for relative and absolute quantitation (iTRAQ). iTRAQ is a labeling technique for relative quantitation and identification of differentially expressed proteins. These iTRAQ methods, which originated from the isotope-codes affinity tag (ICAT) approach reported by Gygi et al. [9], have the added advantages of using labeling chemistry targeted at primary amines (rather than sulfhydryl groups) and the ability to simultaneously measure relative quantities of proteins under multiple conditions [10]. This method is surpassing other forms of mass spectrometry because of the advantage of readily being able to identify differentially expressed proteins without further isolation steps. These techniques are capable of identifying and quantifying proteins and peptides in exceedingly large numbers [11]. The urinary proteome itself is quite large, with laboratories having identified over 1500 proteins to date [12,13]. The blood proteome is even larger, with over 3000 nonredundant proteins identified in the plasma alone [7,14,15]. Adding the proteome of the cellular component of blood will yield thousands more. To this end we have entered what has been termed an "open loop" [16] or an unbiased approach to biomarker discovery. This is in stark contrast to the hypothesis-driven approach of our past. With such a vast pool of potential biomarkers from readily available, noninvasive sources one must take care to plan and design the proper experimental approach to ensure parsimony. Despite the potential complexities of proteomics, these approaches can identify and quantify human proteins that change rapidly and specifically with pathophysiologic changes in the individual and are therefore indispensable to predictive and personalized medicine [5]. A recent analysis of the most commonly reported upregulated urinary proteins in AKI determined by proteomic profiling studies have identified NGAL, albumin, α1-microglobulin, β2-microglobulin, α1-antitrypsin, and IGFBP-7 [5], many of which have now been further validated as early biomarkers of AKI [17,18].

CHARACTERISTICS OF AN IDEAL BIOMARKER

Prior to beginning the search for biomarkers of renal disease, one has to ask, what are the ideal characteristics of a renal biomarker? To be certain, what constitutes an ideal biomarker is highly dependent on the disease you are investigating. However, certain universal characteristics are important for any biomarker: (1) they should be noninvasive, easily measured, inexpensive, and produce rapid results; (2) they should be from readily available sources, such as, blood or urine; (3) they should have a high sensitivity, allowing early detection, and no overlap in values between diseased patients and healthy controls; (4) they should have a high specificity, being greatly upregulated (or downregulated) specifically in the diseased samples, and unaffected by comorbid conditions; (5) biomarker levels should vary rapidly to reflect disease severity and in response to treatment; (6) biomarker levels should aid in risk stratification and possess prognostic value in terms of hard clinical outcomes; and (7) biomarkers should be biologically plausible and provide insight into the underlying disease mechanism [1,19].

Of course, very few biomarkers will meet all of the characteristics of an ideal marker, but let us discuss these characteristics in a little more detail. First, a biomarker should be noninvasive. For example, many chronic kidney diseases (CKDs) present with a range of proteinuria. Currently the preferred method for differentiating nephrotic syndrome-producing CKDs, such as, focal segmental glomerulosclerosis, membranous nephropathy, or minimal change disease is an invasive biopsy. In addition to the health risks, these procedures cause undue anxiety, especially in pediatric populations. While typically a safe procedure, there are associated risks, especially for those patients with contraindications to percutaneous renal biopsies who must elect for an "open," or operative renal biopsy. A recent study found major (cardiac arrest, stroke, and sepsis) and minor (wound infection, pneumonia, arrhythmia, postoperative retroperitoneal bleed, and deep vein thrombosis) complication rates of 6.1 and 27% in a group of 115 open biopsy patients from 1991 to 2006 [20]. While these are relatively rare occurrences, they illustrate the need for less invasive diagnostic procedures.

Regarding the source of biomarkers, the most readily available ones are urine and blood. These are substances obtained in the normal care of a patient, easily collected at the bedside, and associated with little to no health risks to the patient. Each source has desirable and negative characteristics. Urine is an excellent source of biomarkers produced in the kidney [21] and thus may give a better mechanistic insight into specific renal pathologies. Urine is less complex than serum and thus is easier to screen for potential biomarkers. Collection of urine is easy enough and it can be readily employed in home-testing kits. The handling of urine, however, greatly influences the stability of its proteins and measurements should be made immediately after collection or the urine should be promptly frozen at −80°C to avoid degradation [22]. Finally, urinary biomarker studies typically adjust for urine creatinine to account for differences in urine concentration due to hydration status and medications, such as diuretics. However, the utility of urine creatinine in biomarker correction has been questioned due to its variable excretion throughout the day and its dependence on normal renal function. Serum or plasma can also be a good source of biomarkers and is even available in anuric patients. Serum is less prone to bacterial contamination than urine and is considered to be more stable. Serum biomarkers, however, are more likely to represent a systemic response to disease, rather than an organ-specific response. A real problem with serum as a source of biomarkers lies in the discovery phase. Serum has a wide range of protein concentrations across several orders of magnitude, with a small number of proteins (such as albumin) accounting for a large percentage of the volume. This can be compared to trying to spot a single strand of cotton in a large tapestry. The more abundant proteins simply overwhelm the signal of those in less abundance. While there exist assays to remove these high-abundance proteins from serum, many potential biomarkers have, for example, been shown to bind to albumin. Thus, when you deplete the albumin, the rest of the tapestry unravels with it and may result in a loss of proteins relevant to the disease.

The sensitivity and specificity of a biomarker go hand-in-hand. The ROC curve is a binary classification test, based on the sensitivity and specificity of a biomarker at certain cutoff points. ROC curves are often used to determine the clinical diagnostic value of a marker [19,23]. The area under the ROC curve (AUC) is a common statistic derived from ROC curves. An AUC of 1.0 represents a perfect biomarker, while an AUC of 0.5 is a result that is no better than expected by chance. An AUC of 0.75 or greater is generally considered a good biomarker, while an AUC of 0.90 is considered as an excellent biomarker [19]. However, even a sensitive biomarker, with what experimentally would be considered an excellent specificity of 90%, would still yield a false positive rate of 10%, which may be unacceptably high for clinical use as a stand-alone marker [18]. As a result, the best approach clinically may be to find multiple biomarkers that can be combined as part of a panel to achieve an even higher specificity.

Lack of specificity and slow response to alterations in disease severity or treatment are primary reasons why serum creatinine is an unsatisfactory biomarker for renal disease, especially in cases of AKI. First, serum-creatinine levels change with factors unrelated to renal disease, such as, age, gender, diet, muscle mass, muscle metabolism, race, strenuous exercise, and hydration status. Creatinine levels are also influenced by certain drugs [24,25]. Furthermore, in AKI, serum creatinine is not a real-time indicator of kidney function because the patients are not in steady state; so rises in serum creatinine occur long after the inciting renal injury is sustained. In fact, serum-creatinine concentrations may not change until approximately 50% of kidney function has been lost. This makes serum creatinine a poor diagnostic marker for AKI, as treatments need to be administered soon after injury to be effective. Animal studies have shown that treatments that can prevent or alleviate AKI need to begin well before the serum-creatinine levels begin to rise [21,26,27]. As so many variables affect creatinine levels, it also lacks precision in assessing disease progression or risk stratification. Finally, it is well known that significant renal disease, such as fibrosis, can exist with little or no change in creatinine because of the renal reserve or enhanced tubular secretion of creatinine [28,29]. Despite having few of outlined characteristics of an ideal biomarker, serum creatinine remains in widespread use as an indicator of renal function and until recently was the only FDA-approved diagnostic marker of AKI. The problems with creatinine have been evident for over 35 years [29], yet until the last decade little progress had been made in the search for replacement markers that will aid in earlier, more accurate, and specific diagnosis of renal disease.

BIOMARKERS IN AKI

AKI is a serious clinical problem and is increasing in incidence, lacks satisfactory therapeutic options, and presents an enormous financial burden to society. Conservative estimates have placed the annual healthcare expenditures attributable to hospital-acquired AKI at greater than $10 billion in the United States alone [30,31]. AKI is a major side effect of other medical procedures and can result from insults ranging from ischemia reperfusion injury following cardiopulmonary bypass surgery or renal transplant to damage from nephrotoxic agents, such as, contrast agents, aminoglycosides, and cisplatin. Although many new insights into the mechanisms of AKI have been achieved and novel interventions in animal models have shown promise, translational efforts in humans have been disappointing. There are many plausible reasons for this lack of success, among them is a paucity of early diagnostic markers of AKI leading to delayed initiation of therapy and an incomplete pathophysiologic understanding of the disease process [19].

Table 1.2 RIFLE criteria (ADQI)

Stage	Serum creatinine criteria	Glomerular filtration rate (GFR) criteria	Urine output criteria (mL/kg per h)
R = risk for renal dysfunction	Increase in serum creatinine ≥ 1.5× baseline	Decrease in GFR ≥ 25%	<0.5 for 6 h
I = injury to the kidney	Increase in serum creatinine ≥ 2.0× baseline	Decrease in GFR ≥ 50%	<0.5 for 12 h
F = failure of kidney function	Increase in serum creatinine ≥ 3.0× baseline OR serum creatinine ≥ 4.0 mg/dL in the setting of an acute rise ≥ 0.5 mg/dL	Decrease in GFR ≥ 75%	<0.3 for 24 h OR anuria for 12 h
L = loss of kidney function	Persistent failure > 4 weeks		
E = ESRD	Persistent failure > 3 months		

Source: Adapted from Chertow GM, Burdick E, Honour M, Bonventre JV, Bates DW. Acute kidney injury, mortality, length of stay, and costs in hospitalized patients. J Am Soc Nephrol 2005;16(11):3365–70 [30].

Another major hindrance to the successful implementation of new therapies is the lack of a consensus definition of AKI (previously known as acute renal failure). In fact, the Acute Dialysis Quality Initiative (ADQI) workgroup found that over 30 definitions for acute renal failure were used in the literature. The definitions varied from a 25% increase over baseline serum creatinine to the need for dialysis [32]. The term AKI was introduced around 2005 and was proposed to better account for the diverse spectrum of molecular, biochemical, and structural processes that characterize the AKI syndrome [33]. In order to classify AKI better, the Risk–Injury–Failure–Loss–End Stage Renal Disease (ESRD) or the RIFLE classification system (Table 1.2) was developed [34]. The first three classes represent degrees of injury and the last two are outcome measures. This system has shown to correlate well with mortality rates [35]. In order to further refine the definition of AKI, the Acute Kidney Injury Network (AKIN) was created, which proposed a modified version of the RIFLE classification, known as the AKIN criteria. The AKIN criteria define AKI as an abrupt (within 48 h) reduction in kidney function as measured by an absolute increase in serum creatinine ≥ 0.3 mg/dL, a percentage increase in serum creatinine ≥ 50%,

Table 1.3 Comparison of the RIFLE criteria with the AKIN staging criteria

RIFLE stage	RIFLE criteria	AKIN stage	AKIN criteria
R	≥150% increase in serum creatinine OR >25% GFR decrease	I	≥150% OR ≥ 0.3 mg/dL increase in serum creatinine
I	≥200% increase in serum creatinine OR >50% GFR decrease	II	>200% increase in serum creatinine
F	≥300% increase in serum creatinine OR serum creatinine of ≥ 4.0 mg/dL in setting of increase ≥ 0.5 mg/dL, OR > 75% GFR decrease	III	>300% increase in serum creatinine OR serum creatinine of ≥ 4.0 mg/dL in setting of increase ≥ 0.5 mg/dL

Note: The urine output criteria are the same for both RIFLE and AKIN.

or documented oliguria (<0.5 mL/kg per h) for more than 6 h [36]. Minor modifications of the RIFLE criteria (Table 1.3) include broadening the "risk" category of RIFLE to include an increase in serum creatinine of at least 0.3 mg/dL in order to increase the sensitivity of RIFLE for detecting AKI at an earlier time point. In addition, the AKIN criteria sets a window on the first documentation of any criteria to 48 h and categorizes patients in the "failure" category of RIFLE if they are treated with renal replacement therapy, regardless of either changes in creatinine or urine output. Finally, AKIN replaces the three levels of severity R, I, and F with stage I, II, and III [37]. The nonprofit group Kidney Disease: Improving Global Outcomes (KDIGO) was founded in 2003 to "improve the care and outcomes of kidney disease patients worldwide through promoting coordination, collaboration, and integration of initiatives to develop and implement clinical practice guidelines" [38]. This group has worked to develop consensus recommendations and consolidate previous definitions of AKI. The resulting updated definition of AKI is similar to the AKIN definition and is shown in Table 1.4. Both the definition and staging of AKI now include a 0.3 mg/dL serum-creatinine increase criterion that is specifically applicable to pediatric AKI. Modifications also allow for a child with an eGFR <35 mL/min per 1.73 m^2 to be included in Stage 3, in contrast with the adult criterion of ≥4 mg/dL serum creatinine, which would be unusual in infants and young children. Extension of the diagnostic timeframe for a serum-creatinine rise to 7 days allows for capture of subjects with late-onset AKI. It is recommended that the KDIGO AKI definition and staging be used to

Table 1.4 The KDIGO AKI criteria

Stage	Serum creatinine	Urine output (mL/kg per h)
1	1.5–1.9× baseline, OR ≥ 0.3 mg/dL (≥26.5 μmol/L) increase	<0.5 for 6–12 h
2	1.0–2.9× baseline	<0.5 for ≥ 12 h
3	3.0× baseline, OR SCr ≥ 4.0 mg/dL (≥353.6 μmol/L), OR Initiation of renal replacement therapy, OR Estimated (eGFR) < 35 mL/min per 1.73 m^2 (<18 years)	<0.3 for ≥ 24 h, OR Anuria for ≥ 12 h

Source: Adapted from Eknoyan G, Lameire N, Barsoum R, Eckardt KU, Levin A, Levin N, Locatelli F, MacLeod A, Vanholder R, Walker R, Wang H. The burden of kidney disease: improving global outcomes. Kidney Int 2004;66(4):1310–4 [38].

guide clinical care and as a standardized inclusion and outcome measure in AKI studies.

Many conventional markers of kidney function have suffered from a lack of specificity and poor standardized assays. The insensitivity of these measurements, such as, casts and fractional secretion of sodium, make them poor candidates for the early detection of AKI. As mentioned, creatinine is an unreliable marker of acute changes in kidney function due to its slow response time and the fact that many variables can alter creatinine levels [39]. The failure of two pivotal clinical trials on promising new interventions in AKI, human insulin-like growth factor 1 and anaritide, is at least partly attributable to the lack of early biomarkers for AKI [40,41]. Despite these and other potential advances in clinical care and groundbreaking research into the mechanisms of AKI, it remains a devastating clinical condition and studies suggest its incidence may be increasing [42–44]. AKI has been reported to complicate up to 7% of all hospital admissions [45,46] and as high as 25% of intensive care-unit admissions [47]. The prognosis of AKI is has remained quite poor over the past 50 years with a mortality rate of 40–80% in the intensive care setting [27,48]. Identification of novel AKI biomarkers has been designated as a top priority by the American Society of Nephrology and the concept of developing a new collection of tools for earlier diagnosis of disease states is a prominent feature in the National Institutes of Health road map or biomedical research [33,49].

Besides establishing the early diagnosis, biomarkers are needed for several other purposes in AKI (summarized in Table 1.5). Thus, biomarkers are needed for (1) pinpointing the location of primary injury (proximal tubule, distal tubule, interstitium, or vasculature); (2) determining the duration of kidney

Table 1.5 Areas of need for biomarkers in AKI

Biomarkers are needed to determine:
1. Location of injury
2. Duration of AKI
3. AKI subtypes
4. AKI etiologies
5. Differentiation from other forms of acute kidney disease
6. Risk stratification and prognostication
7. Defining course of AKI
8. Monitoring response to interventions

failure (AKI, CKD, or "acute-on-chronic" kidney disease); (3) discerning AKI subtypes (prerenal, intrinsic renal, or postrenal); (4) identifying AKI etiologies (ischemia, toxins, sepsis, or a combination); (5) differentiating AKI from other forms of acute kidney disease (urinary tract infection, glomerulonephritis, or interstitial nephritis); (6) risk stratification and prognostication (duration and severity of AKI, need for renal replacement therapy, length of hospital stay, and mortality); (7) defining the course of AKI; and (8) monitoring the response to AKI interventions [19]. Biomarkers are also needed for use as surrogate endpoints in clinical trials evaluating potential therapeutics for AKI. Surrogate markers are precise measurements that can accurately correlate with a clinical endpoint [1]. Surrogate endpoints can expedite clinical trials evaluating the safety and efficacy of new drug applications. If the intervention has the desired effect on the surrogate endpoint, then further evaluations are warranted to directly address the effect of the intervention on the appropriate clinical endpoint. This linking of the surrogate endpoint to the clinical endpoint is referred to as validation and is an essential step in the biomarker-discovery process.

With respect to the desirable characteristics of AKI biomarkers, the most important remain those that are clinically applicable and can lead to an early diagnosis and treatment of AKI. Other important properties of clinically relevant biomarkers of AKI are similar in concept to the properties of ideal biomarkers in general. Specific characteristics should include (1) measurements from noninvasive sources, such as blood or urine; (2) easy to perform either at bedside or in a standard clinical laboratory; (3) measurements should be reliable and have a rapid turnaround time; (4) they should be sensitive for early detection and have a wide dynamic range of values with cut offs to allow for risk stratification; (5) they should be highly specific, and ideally allow for AKI subtype classification; and (6) they should be inexpensive to allow for broad global use.

Several promising candidates for clinical use as biomarkers in AKI are under intense contemporary study and some have already been approved for clinical use in many parts of the world. Many of these biomarkers will be discussed in more detail in other chapters, but we will offer a brief description of the major candidates. Perhaps the most widely applicable marker found to date for the early diagnosis of AKI is NGAL. NGAL was discovered by cDNA-microarray analysis to be induced very early following ischemic or nephrotoxic injury and the protein is easily detectable in urine and plasma soon after AKI [4,50–54], but is also (mildly) elevated in patients with urinary tract infections, as well as, those with preexisting CKD. Another emerging candidate for inclusion in the AKI panel of biomarkers is the proinflammatory cytokine interleukin-18 (IL-18), which is induced in the proximal tubule after AKI. IL-18 does not appear to be significantly affected by CKD or UTIs [55–60]. Serum cystatin C is another candidate for inclusion in the AKI panel. Cystatin C is produced in the blood and is filtered by the glomerulus, then completely reabsorbed by the proximal tubules, and is not normally excreted in the urine [61]. Serum cystatin C is primarily a sensitive marker of glomerular filtration-rate reduction and not that of kidney injury, but it has been shown to predict AKI earlier than serum creatinine in the intensive care setting [26]. Kidney injury molecule-1 is a transmembrane protein upregulated in dedifferentiated proximal tubule cells after ischemic or nephrotoxic injury, but not expressed in normal kidney [62]. Kidney injury molecule-1 detects AKI later than NGAL or IL-18, (e.g., 12–24 h vs. 2–6 h post-CPB, respectively [61]), but shows promise in differentiating between subtypes of AKI. More recently, the combination of tissue inhibitor of metalloproteinase 2 (TIMP-2) and insulin-like growth factor-binding protein 7 (IGFBP7) for the prediction of moderate to severe risk of AKI in critically ill patients has garnered a lot of attention [63]. Both TIMP-2 and IGFBP7 are markers of G1 cell-cycle arrest and are elevated in the initial phases of tubular cell injury following insults ranging from ischemia to inflammation [64–66]. A clinical test for this marker combination was recently approved for use by FDA, but enough time has yet to elapse to determine the effect of these markers on patient outcomes.

BIOMARKERS IN CKD

CKD is a devastating illness that has reached epidemic proportions and continues to increase in incidence at an alarming rate. Estimates place the prevalence of CKD in the general population at 10–13% [67]. It is estimated that the medicare costs in the fee for service (Medicare FFS) population are $49 billion per year for those patients in stage 2–4 CKD [68]. For those patients progressing to ESRD, the mortality levels even exceed those of most malignancies [69]. Even those with mild CKD have increased risk of premature death when compared to the general population, mainly due to associated cardiovascular disease [70]. CKD is a complex disease that often affects multiple organ systems and often coexists with numerous associated conditions, such as, cardiovascular disease, diabetes mellitus, lupus, and chronic inflammation. In many cases these conditions (especially cardiovascular disease) are independently associated with CKD, implying a vicious circle in which cardiovascular disease can lead to CKD, which worsens cardiovascular disease down the line.

The "gold standard" measurement for CKD is the "true" GFR as tracked by 24-h urine isotope clearance. This method is quite expensive and not always practical in the clinical setting. A commonly used clinical surrogate for nuclear GFR is serum-creatinine clearance. However, as noted previously, the accuracy of serum creatinine is greatly affected by a number of patient-dependent and -independent variables. Additionally, serum creatinine may fall to one-third of its normal level in advanced kidney disease, unrelated to its renal clearance [61]. Even serial 24-h creatinine measurements fail to determine risk progression in approximately 20% of CKD patients [71]. Even when accurate, 24-h creatinine clearance fails to offer reliable prognosis of CKD progression. KDIGO CKD working group recommends using calculated GFR, Modification of Diet in Renal Disease [72,73], and Chronic Kidney Disease Epidemiology Collaboration equations [73]. The Modification of Diet in Renal Disease formula estimates GFR adjusted for body-surface area, with age and gender as variables [74]. The Chronic Kidney Disease Epidemiology Collaboration equation, was developed to create a more precise formula when actual GFR is > 60 mL/min per 1.73 m^2 [75–77]. CKD is defined by the presence of kidney damage or GFR less than 60 mL/min per 1.73 m^2 for 3 months or greater, regardless of cause. However, significant increases in cardiovascular disease risk occur at more subtle loss of kidney function (a GFR of approximately 75 mL/min per 1.73 m^2) [70] so it is inherently important that CKD be caught in its earliest stages when possible.

Proteinuria is another useful marker of progressive functional decline in renal function. Proteinuria has been shown to directly represent kidney damage and higher levels of proteinuria correlate well with a more rapid progression of kidney disease [78]. Proteinuria is the earliest known marker

Table 1.6 KDIGO classification of CKD according to GFR category

GFR category	GFR (mL/min per 1.73 m^2)	Terms
G1	≥90	Normal or high
G2	60–89	Mildly decreased
G3a	45–59	Mildly to moderately decreased
G3b	30–44	Moderately to severely decreased
G4	15–29	Severely decreased
G5	<15	Kidney failure

Source: Adapted from Kidney Disease: Improving Global Outcomes. Definition and classification of CKD. Kidney Int Suppl 2013;3:19–62 [Chapter 1] [81].

of kidney damage in glomerular diseases, diabetes, and hypertension and is the most common marker of kidney damage in the adult population. However, proteinuria has certain limitations. Proteinuria may occur long after the renal injury has occurred and it is not always present in many types of renal disease [61]. Treatments, such as lowering urinary protein excretion using renin–angiotensin system blockade and controlling hypertension, can reduce CKD progression rates [79]. CKD is often not caught until shortly before the onset of symptomatic kidney failure, so it is typically too late to prevent many adverse outcomes [80]. At this point, early diagnosis would entail routine testing of asymptomatic individuals in at-risk categories for the development of CKD and allow for determining staging and appropriate treatment options for those individuals identified as having renal disease. KDIGO has recently developed updated CKD-staging guidelines, expanding from 5 to 6 stages based on GFR to predict outcomes (Table 1.6) [81].

The need for biomarkers that can aid in diagnosing, distinguishing subtypes, and prognosticating the severity of CKD and associated conditions are greatly needed, as the risk factors of this population are different from those of the general population. The search for biomarkers of CKD, especially those for the early diagnosis, is more difficult than that of AKI because the timing and nature of the insult is harder to pin down. With AKI, it is easier to pick a population undergoing a procedure, such as, contrast administration or cardiopulmonary bypass, where the timing and nature of the insult can be tightly controlled and measurements can be made in scheduled periods before and after the potential injury has occurred. With pediatric populations, it is often the case that you can control for many comorbid conditions, such as, prolonged cardiovascular disease, effects of obesity, and lifestyle that can effect renal function and isolate the AKI incident from other variables that might influence potential biomarker levels. Such is not the case with CKD. Individuals with acquired or even hereditary forms of CKD can go years without the knowledge of their condition until it becomes severe enough to adversely affect the general health of the individual. As many comorbid conditions are likely to exist, the results of biomarker studies on these individuals may be subject to high-individual variability and be difficult to interpret and subsequently reproduce.

As is the case with AKI, biomarkers are needed in many areas of CKD, including the following: (1) determining the site of predominant kidney damage (e.g., glomerular and tubular); (2) providing insight into disease mechanism; (3) prognostication of disease progression (e.g., if it is determined that an individual is more likely to progress to ESRD, more aggressive treatments may be employed); (4) subtype classification and ability to direct the course of treatment (e.g., distinguishing progressive focal segmental glomerular sclerosis, which is normally resistant to steroid treatment, from minimal change disease, which is nonprogressive and typically sensitive to steroid treatment); (5) determining risk of complications from comorbid conditions, such as cardiovascular disease; and (6) more sensitive and reliable surrogate measurements for the estimation of GFR. In terms of characteristics of clinically applicable biomarkers for CKD, they should be nearly identical to those discussed for AKI (Table 1.5).

For CKD, due to its complexity and coexistence with other conditions, it is even more unlikely that any one marker can be found to possess all of the ideal characteristics of a biomarker. Ongoing research has produced some promising candidates for possible inclusion in a panel of biomarkers for CKD. In addition to its place as a promising AKI biomarker, NGAL has been shown to be a potential marker for CKD severity and progression [82]. Likewise, cystatin C is a promising marker of GFR in both AKI and CKD [83]. It should be noted that more studies are needed to determine if cystatin C is truly a better marker of GFR than serum creatinine. Asymmetric dimethylarginine is a nitric oxide synthase inhibitor and a marker of endothelial function. Increases in ADMA levels are predictive of CKD progression rates and are a risk factor for mortality in ESRD patients [84–86]. Liver-type fatty acid binding protein is expressed in the proximal tubule of the kidney and its elevation has been shown to predict progression in CKD [87]. Larger longitudinal studies are needed to determine the utility of these and the other biomarkers mentioned in predicting CKD progression in multiple etiologies.

CONCLUSIONS AND FUTURE DIRECTIONS

In the last 2 decades, the ready availability of powerful methods of genomics and proteomics have revolutionized and greatly accelerated biomarker discovery in kidney disease. Successful candidates are now being advanced as tools for personalized and predictive approaches to kidney disease. As a prime example, NGAL as an AKI biomarker has successfully passed through the preclinical, assay development, and initial clinical testing stages of the biomarker-development process. It has now entered the prospective screening stage, facilitated by the development of commercial tools for the measurement of NGAL on large populations across different laboratories. Similarly, targeted proteomics have identified the combination of TIMP-2 and IGFBP7 for prediction of moderate to severe risk of AKI, and clinical test for this marker combination was recently approved for use by FDA. The widespread availability of a panel of validated AKI biomarkers will further revolutionize renal and critical care in the not-too-distant future [88].

REFERENCES

[1] Biomarkers Definitions Working Group. Biomarkers and surrogate endpoints: preferred definitions and conceptual framework. Clin Pharmacol Ther 2001;69(3):89–95.

[2] Supavekin S, et al. Differential gene expression following early renal ischemia/reperfusion. Kidney Int 2003;63(5):1714–24.

[3] Devarajan P, et al. Gene expression in early ischemic renal injury: clues towards pathogenesis, biomarker discovery, and novel therapeutics. Mol Genet Metab 2003;80(4):365–76.

[4] Mishra J, et al. Identification of neutrophil gelatinase-associated lipocalin as a novel early urinary biomarker for ischemic renal injury. J Am Soc Nephrol 2003;14(10):2534–43.

[5] Devarajan P. Genomic and proteomic characterization of acute kidney injury. Nephron 2015;131:85–91.

[6] Grigoryev DN, et al. Meta-analysis of moldecular response of kidney to ischemia reperfusion injury for the identification of new candidate genes. BMC Nephrol 2013;14:231.

[7] States DJ, et al. Challenges in deriving high-confidence protein identifications from data gathered by a HUPO plasma proteome collaborative study. Nat Biotechnol 2006;24(3):333–8.

[8] Wang Z, Gerstein M, Snyder M. RNA-Seq: a revolutionary tool for transcriptomics. Nat Rev Genet 2009;10(1):57–63.

[9] Gygi SP, et al. Quantitative analysis of complex protein mixtures using isotope-coded affinity tags. Nat Biotechnol 1999;17(10):994–9.

[10] Ross PL, et al. Multiplexed protein quantitation in *Saccharomyces cerevisiae* using amine-reactive isobaric tagging reagents. Mol Cell Proteomics 2004;3(12):1154–69.

[11] Knepper MA. Proteomics and the kidney. J Am Soc Nephrol 2002;13(5):1398–408.

[12] Thongboonkerd V, et al. Proteomic analysis of normal human urinary proteins isolated by acetone precipitation or ultracentrifugation. Kidney Int 2002;62(4):1461–9.

[13] Adachi J, et al. The human urinary proteome contains more than 1500 proteins, including a large proportion of membrane proteins. Genome Biol 2006;7(9):R80.

[14] Omenn GS. Exploring the human plasma proteome. Proteomics 2005;5(13):3223–5.
[15] Omenn GS, et al. Overview of the HUPO Plasma Proteome Project: results from the pilot phase with 35 collaborating laboratories and multiple analytical groups, generating a core dataset of 3020 proteins and a publicly-available database. Proteomics 2005;5(13):3226–45.
[16] Knepper MA. Common sense approaches to urinary biomarker study design. J Am Soc Nephrol 2009;20(6):1175–8.
[17] Devarajan P, et al. Proteomic identification of early biomarkers of acute kidney injury after cardiac surgery in children. Am J Kidney Dis 2010;56:632–42.
[18] Ho J, et al. Mass spectrometry-based proteomic analysis of urine in acute kidney injury following cardiopulmonary bypass: a nested case-control study. Am J Kidney Dis 2009;53:584–95.
[19] Devarajan P. Proteomics for biomarker discovery in acute kidney injury. Semin Nephrol 2007;27(6):637–51.
[20] Stec AA, et al. Open renal biopsy: comorbidities and complications in a contemporary series. BJU Int 2009.
[21] Hewitt SM, Dear J, Star RA. Discovery of protein biomarkers for renal diseases. J Am Soc Nephrol 2004;15(7):1677–89.
[22] Schuh MP, et al. Long term stability of biomarkers of acute kidney injury in children. Am J Kidney Dis 2015;67(1):56–61.
[23] Zweig MH, Campbell G. Receiver-operating characteristic (ROC) plots: a fundamental evaluation tool in clinical medicine. Clin Chem 1993;39(4):561–77.
[24] Letellier G, Desjarlais F. Analytical interference of drugs in clinical chemistry: II—The interference of three cephalosporins with the determination of serum creatinine concentration by the Jaffe reaction. Clin Biochem 1985;18(6):352–6.
[25] Weber JA, van Zanten AP. Interferences in current methods for measurements of creatinine. Clin Chem 1991;37(5):695–700.
[26] Herget-Rosenthal S, et al. Early detection of acute renal failure by serum cystatin C. Kidney Int 2004;66(3):1115–22.
[27] Devarajan P. Update on mechanisms of ischemic acute kidney injury. J Am Soc Nephrol 2006;17(6):1503–20.
[28] Branten AJ, Vervoort G, Wetzels JF. Serum creatinine is a poor marker of GFR in nephrotic syndrome. Nephrol Dial Transplant 2005;20(4):707–11.
[29] Carrie BJ, et al. Creatinine: an inadequate filtration marker in glomerular diseases. Am J Med 1980;69(2):177–82.
[30] Chertow GM, et al. Acute kidney injury, mortality, length of stay, and costs in hospitalized patients. J Am Soc Nephrol 2005;16(11):3365–70.
[31] Hoste EA, Schurgers M. Epidemiology of acute kidney injury: how big is the problem? Crit Care Med 2008;36(4 Suppl.):S146–51.
[32] Kellum JA, et al. Developing a consensus classification system for acute renal failure. Curr Opin Crit Care 2002;8(6):509–14.
[33] American Society of Nephrology. American Society of Nephrology Renal Research Report. J Am Soc Nephrol 2005;16(7):1886–903.
[34] Bellomo R, et al. Acute renal failure—definition, outcome measures, animal models, fluid therapy and information technology needs: the Second International Consensus Conference of the Acute Dialysis Quality Initiative (ADQI) Group. Crit Care 2004;8(4):R204–12.
[35] Ricci Z, Cruz D, Ronco C. The RIFLE criteria and mortality in acute kidney injury: a systematic review. Kidney Int 2008;73(5):538–46.
[36] Mehta RL, et al. Acute Kidney Injury Network: report of an initiative to improve outcomes in acute kidney injury. Crit Care 2007;11(2):pR31.
[37] Soni SS, et al. Early diagnosis of acute kidney injury: the promise of novel biomarkers. Blood Purif 2009;28(3):165–74.

[38] Eknoyan G, et al. The burden of kidney disease: improving global outcomes. Kidney Int 2004;66(4):1310–4.
[39] Bellomo R, Kellum JA, Ronco C. Defining acute renal failure: physiological principles. Intensive Care Med 2004;30(1):33–7.
[40] Allgren RL, et al. Anaritide in acute tubular necrosis. Auriculin Anaritide Acute Renal Failure Study Group. N Engl J Med 1997;336(12):828–34.
[41] Hirschberg R, et al. Multicenter clinical trial of recombinant human insulin-like growth factor I in patients with acute renal failure. Kidney Int 1999;55(6):2423–32.
[42] Waikar SS, et al. Declining mortality in patients with acute renal failure, 1988 to 2002. J Am Soc Nephrol 2006;17(4):1143–50.
[43] Xue JL, et al. Incidence and mortality of acute renal failure in medicare beneficiaries, 1992 to 2001. J Am Soc Nephrol 2006;17(4):1135–42.
[44] Ympa YP, et al. Has mortality from acute renal failure decreased? A systematic review of the literature. Am J Med 2005;118(8):827–32.
[45] Chertow GM, et al. Guided medication dosing for inpatients with renal insufficiency. JAMA 2001;286(22):2839–44.
[46] Liangos O, et al. Epidemiology and outcomes of acute renal failure in hospitalized patients: a national survey. Clin J Am Soc Nephrol 2006;1(1):43–51.
[47] de Mendonca A, et al. Acute renal failure in the ICU: risk factors and outcome evaluated by the SOFA score. Intensive Care Med 2000;26(7):915–21.
[48] Vaidya VS, Ferguson MA, Bonventre JV. Biomarkers of acute kidney injury. Annu Rev Pharmacol Toxicol 2008;48:463–93.
[49] Zerhouni E. Medicine. The NIH Roadmap. Science 2003;302(5642):63–72.
[50] Mishra J, et al. Neutrophil gelatinase-associated lipocalin (NGAL) as a biomarker for acute renal injury after cardiac surgery. Lancet 2005;365(9466):1231–8.
[51] Mishra J, Kidney NGAL, et al. is a novel early marker of acute injury following transplantation. Pediatr Nephrol 2006;21(6):856–63.
[52] Mishra J, et al. Neutrophil gelatinase-associated lipocalin: a novel early urinary biomarker for cisplatin nephrotoxicity. Am J Nephrol 2004;24(3):307–15.
[53] Mishra J, et al. Amelioration of ischemic acute renal injury by neutrophil gelatinase-associated lipocalin. J Am Soc Nephrol 2004;15(12):3073–82.
[54] Mori K, et al. Endocytic delivery of lipocalin-siderophore-iron complex rescues the kidney from ischemia-reperfusion injury. J Clin Invest 2005;115(3):610–21.
[55] Hall IE, et al. IL-18 and urinary NGAL predict dialysis and graft recovery after kidney transplantation. J Am Soc Nephrol 2009;21(1):189–97.
[56] Parikh CR, et al. Urine IL-18 is an early diagnostic marker for acute kidney injury and predicts mortality in the intensive care unit. J Am Soc Nephrol 2005;16(10):3046–52.
[57] Parikh CR, et al. Urinary interleukin-18 is a marker of human acute tubular necrosis. Am J Kidney Dis 2004;43(3):405–14.
[58] Parikh CR, et al. Urine NGAL and IL-18 are predictive biomarkers for delayed graft function following kidney transplantation. Am J Transplant 2006;6(7):1639–45.
[59] Parikh CR, et al. Urinary IL-18 is an early predictive biomarker of acute kidney injury after cardiac surgery. Kidney Int 2006;70(1):199–203.
[60] Washburn KK, et al. Urinary interleukin-18 is an acute kidney injury biomarker in critically ill children. Nephrol Dial Transplant 2008;23(2):566–72.
[61] Nickolas TL, Barasch J, Devarajan P. Biomarkers in acute and chronic kidney disease. Curr Opin Nephrol Hypertens 2008;17(2):127–32.
[62] Zhang Z, Humphreys BD, Bonventre JV. Shedding of the urinary biomarker kidney injury molecule-1 (KIM-1) is regulated by MAP kinases and juxtamembrane region. J Am Soc Nephrol 2007;18(10):2704–14.
[63] Gocze I, et al. Urinary biomarkers TIMP-2 and IGFBP7 early predict acute kidney injury after major surgery. PLoS One 2015;10(3):pe0120863.

[64] Seo DW, et al. Shp-1 mediates the antiproliferative activity of tissue inhibitor of metalloproteinase-2 in human microvascular endothelial cells. J Biol Chem 2006;281(6):3711–21.
[65] Yang QH, et al. Acute renal failure during sepsis: potential role of cell cycle regulation. J Infect 2009;58(6):459–64.
[66] Kashani K, et al. Discovery and validation of cell cycle arrest biomarkers in human acute kidney injury. Crit Care 2013;17(1):pR25.
[67] Coresh J, et al. Prevalence of chronic kidney disease in the United States. JAMA 2007;298(17):2038–47.
[68] Honeycutt AA, et al. Medical costs of CKD in the medicare population. J Am Soc Nephrol 2013;24(9):1478–83.
[69] Kovesdy CP, Kalantar-Zadeh K. Review article: biomarkers of clinical outcomes in advanced chronic kidney disease. Nephrology 2009;14(4):408–15.
[70] Stenvinkel P, et al. Emerging biomarkers for evaluating cardiovascular risk in the chronic kidney disease patient: how do new pieces fit into the uremic puzzle? Clin J Am Soc Nephrol 2008;3(2):505–21.
[71] Coresh J, et al. Prevalence of chronic kidney disease and decreased kidney function in the adult US population: Third National Health and Nutrition Examination Survey. Am J Kidney Dis 2003;41(1):1–12.
[72] Levey AS, et al. A more accurate method to estimate glomerular filtration rate from serum creatinine: a new prediction equation. Modification of Diet in Renal Disease Study Group. Ann Intern Med 1999;130(6):461–70.
[73] Levey AS, Stevens LA. Estimating GFR using the CKD Epidemiology Collaboration (CKD-EPI) creatinine equation: more accurate GFR estimates, lower CKD prevalence estimates, and better risk predictions. Am J Kidney Dis 2010;55(4):622–7.
[74] Levey AS, et al. Expressing the Modification of Diet in Renal Disease Study equation for estimating glomerular filtration rate with standardized serum creatinine values. Clin Chem 2007;53(4):766–72.
[75] Horio M, et al. Performance of the Japanese glomerular filtration rate equation based on standardized serum cystatin C in potential kidney donors. Transplant Proc 2014;46(2):314–7.
[76] Horio M, et al. Modification of the CKD epidemiology collaboration (CKD-EPI) equation for Japanese: accuracy and use for population estimates. Am J Kidney Dis 2010;56(1):32–8.
[77] White SL, et al. Comparison of the prevalence and mortality risk of CKD in Australia using the CKD Epidemiology Collaboration (CKD-EPI) and Modification of Diet in Renal Disease (MDRD) Study GFR estimating equations: the AusDiab (Australian Diabetes, Obesity and Lifestyle) Study. Am J Kidney Dis 2010;55(4):660–70.
[78] Zandi-Nejad K, et al. Why is proteinuria an ominous biomarker of progressive kidney disease? Kidney Int Suppl 2004;(92):S76–89.
[79] de Zeeuw D, et al. Renal risk and renoprotection among ethnic groups with type 2 diabetic nephropathy: a post hoc analysis of RENAAL. Kidney Int 2006;69(9):1675–82.
[80] Kinchen KS, et al. The timing of specialist evaluation in chronic kidney disease and mortality. Ann Intern Med 2002;137(6):479–86.
[81] Kidney Disease: Improving Global Outcomes. Definition and classification of CKD. Kidney Int Suppl 2013;3:19–62 [Chapter 1]
[82] Bolignano D, et al. Neutrophil gelatinase-associated lipocalin (NGAL) and progression of chronic kidney disease. Clin J Am Soc Nephrol 2009;4(2):337–44.
[83] Zahran A, El-Husseini A, Shoker A. Can cystatin C replace creatinine to estimate glomerular filtration rate? A literature review. Am J Nephrol 2007;27(2):197–205.
[84] Ravani P, et al. Asymmetrical dimethylarginine predicts progression to dialysis and death in patients with chronic kidney disease: a competing risks modeling approach. J Am Soc Nephrol 2005;16(8):2449–55.

[85] Fliser D, et al. Asymmetric dimethylarginine and progression of chronic kidney disease: the mild to moderate kidney disease study. J Am Soc Nephrol 2005;16(8):2456–61.
[86] Kronenberg F. Emerging risk factors and markers of chronic kidney disease progression. Nat Rev Nephrol 2009;5(12):677–89.
[87] Kamijo A, et al. Urinary liver-type fatty acid binding protein as a useful biomarker in chronic kidney disease. Mol Cell Biochem 2006;284(1-2):175–82.
[88] Devarajan P, Murray P. Biomarkers in acute kidney injury: are we ready for prime time? Nephron Clin Pract 2014;127(1–4):176–9.

CHAPTER TWO

Statistical Considerations in Analysis and Interpretation of Biomarker Studies

C.R. Parikh, MD, PhD*, and H. Thiessen Philbrook, MMath, AStat*,****

*Program of Applied Translational Research, Department of Medicine, Yale University, New Haven, CT, United States

**Veterans Affairs Medical Center, West Haven, CT, United States

Contents

INTRODUCTION

Biomarkers can be broadly defined as biologic parameters, which objectively can be measured and evaluated as indicators of normal biologic processes, pathogenic processes, or pharmacologic responses to therapeutic interventions. The development of biomarkers into clinical applications can be categorized into three broad phases: biomarker discovery, the evaluation of biomarker-prediction performance, and the impact of using biomarkers in clinical care (Table 2.1) [1]. Each phase requires unique statistical considerations and tailored study design to accurately evaluate research objectives. There are several

Biomarkers of Kidney Disease. http://dx.doi.org/10.1016/B978-0-12-803014-1.00002-9

Table 2.1 Research phases of biomarker development

Phase I	Biomarker discovery • Identify candidate biomarker for disease of interest
Phase II	Biomarker-prediction performance • Establish that the biomarker can discriminate between diseased and nondiseased • Determine if biomarker precedes current methods of diagnosis • Develop combinations or cut-offs that can be validated in larger studies
Phase III	Biomarker use in clinical care • Assess the impact and additive benefit of the integration of the biomarker into clinical care • Determine cost effectiveness and improvement in outcomes

resources available for each phase of biomarker development, pertaining to study design, statistical analysis, and sample-size calculations [2–4].

In this chapter, we will focus on human studies evaluating the prediction performance of diagnostic biomarkers. We will use examples of biomarkers in acute kidney injury, to highlight concepts of the prediction performance of biomarkers. The methodology and framework described here can easily be extended for research and the development of biomarkers in other clinical settings. Statistical metrics required for the assessment of predictive performance and utility of biomarkers differ from the classical methods used in epidemiology or therapeutic research [5–8]. In biomarker development, we are focused on classification or discrimination [e.g., true positive rates (TPR) and false positive rates (FPR)], rather than measures of association (e.g., odds ratio and relative risks).

At the end of the biomarker-discovery phase, we assume that a candidate biomarker is promising for a disease of interest and can be measured by reliable methods or assays. During the second phase of biomarker development, we want to establish that the biomarker can: (1) discriminate between diseased and nondiseased patients earlier than the current clinical standard, (2) explore covariates associated with the biomarker, and (3) validate the biomarker-screening criteria and the combination of biomarkers, if applicable. In some cases, identifying the screen-positive criteria and the combination of biomarkers also may be completed in this second phase. Several studies are usually required to complete the second stage of biomarker development. The final phase will examine the impact of biomarker usage in clinical care [1,9–11].

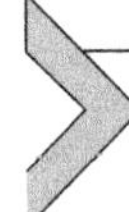

PLANNING A STUDY

Research Objectives for Assessing Biomarker Performance

In planning a research study assessing biomarker performance, a well-defined research question relevant to the phase of biomarker development is required and ample consideration should be given to defining the target population and ensuring that the data elements (both clinical information and sample-processing details) will be collected in sufficient detail. The minimum biomarker-performance level must be specified in advance and this will drive sample-size requirements. The determination of a minimal required performance level should be based on the current clinical standards and consequences of potential misclassification by the biomarker.

The first research aim is to determine if the biomarker can adequately discriminate between diseased and nondiseased patients [1,11]. Generally, this can be completed in a retrospective study, using samples from an existing sample biorepository, where the biomarker is measured at approximately the same time that the disease is diagnosed. The time-dependent discriminatory ability of the biomarker will not be assessed in this study.

The next research aim is to evaluate whether the biomarker can discriminate between diseased and nondiseased patients earlier than the current clinical standard [1,11]. For this, a prospective study is required, where specimens are taken at several time points prior to the clinical diagnosis. The biomarker then can be measured in all patients or in a subgroup of patients (nested case–control study). To reduce bias, a patient's clinical information should be blinded when the specimen is assayed. To assess whether the biomarker can discriminate between diseased and nondiseased patients earlier than current clinical care, time-dependent receiver operating characteristic (ROC) curves or ROC regression should be used [2,12].

If the biomarker will be measured only in a subgroup of patients, consideration should be given to how the sample should be selected and if matching should be implemented [1]. The use of matching will introduce additional complexities in the analytic methods, such as accounting for the matching factors. As the nondiseased patients are no longer a representative sample of the target population, it will not be possible to evaluate the influence of matching factors with the biomarker, and the interpretation of false positives will change [1].

Explore Covariates That May Affect Biomarker Values

An important step, often overlooked when evaluating the prediction performance of a biomarker, is to determine the existence of factors that influence a biomarker's prediction performance that are unrelated to the outcome of interest [13]. It is important to explore such factors by examining the distribution of the biomarker in the nondiseased patients. Factors to consider may be related to patient demographics (e.g., age, race, and gender), clinical parameters (e.g., protein in urine, oliguria, and chronic kidney disease), sample-processing details (e.g., collection time, freezing time, and length of storage), or study sample (e.g., study center). If there are factors identified relating to the biomarker in the nondiseased patients, then diagnostic accuracy can be assessed separately (e.g., determining biomarker in adults and children separately or between high- and low body-mass index), or an adjusted ROC-curve analysis can be completed [13]. An adjusted ROC-curve analysis is analogous to covariate adjustment in studies of association.

Potential Sources of Biases

A frequent criticism of biomarker discovery and validation is irreproducible results [4,10], which may be due to several sources of bias. Two concepts discussed in risk-model development can also be applied to biomarker identification and discovery: model selection bias and resubstitution bias [14]. The first occurs when many biomarkers (or models) are assessed but only the best result is reported. Resubstitution bias refers to the situation where a model is fit and performance is evaluated in the same data. This is frequently seen where the identification and validation of the biomarker was completed in the same data. If the identification of the screen-positive criteria or the combination of biomarkers is to be completed in the same study as the evaluation of the biomarker's performance, methods need to be implemented to reduce resubstitution bias. The most straightforward method is to split the data into two sets: a derivation and a validation dataset [1,5,10]. Alternatively, bootstrapping or crossvalidation methods could be applied [1,5,10] but they require more advanced statistical techniques. Rigorous reporting of methods is needed to properly assess potential biases and interpret published findings [14,15].

METRICS FOR PREDICTION PERFORMANCE

The analytic methods required for a study depend on the research question and the study design. In this second phase of biomarker development, we are focused on evaluating the prediction performance of the

Table 2.2 Biomarker predictions by disease status

		True disease state	
		Diseased	**Nondiseased**
Biomarker test	Positive (diseased)	TP	FP
	Negative (nondiseased)	FN	TN

TP, true positive; FN, false negative, FP, false positive; TN, true negative.
True positive rate (TPR) = Sensitivity = TP/(TP + FN).
False positive rate (FPR) = 1−Specificity = FP/(FP + TN).

biomarker. In general, for both the retrospective and the prospective studies described earlier, we recommend quantifying the prediction performance with TPR, FPR, and ROC curves. In medical literature, these rates are also referred to as Sensitivity (TPR) and Specificity (1−FPR).

True Positive Rate and False Positive Rate

If we compare the prediction of the biomarker to the true disease status, the results can be categorized as a true positive (TP), a false positive (FP), a true negative (TN), or a false negative (FN) (Table 2.2). A TP result occurs when the biomarker correctly classifies the patient as a diseased patient, and similarly, a TN result occurs when the biomarker correctly classifies the patient as a nondiseased patient. A FP or a FN occurs when the biomarker incorrectly classifies a nondiseased patient as a diseased patient, or a diseased patient as a nondiseased patient, respectively. The TPR is the proportion of diseased patients that the biomarker correctly classified as diseased patients and the FPR the proportion of nondiseased patients that the biomarker incorrectly classified as diseased patients. The range of possible values for both the TPR and FPR is between 0 and 1. A good biomarker has a high TPR and a low FPR.

For studies of prediction, TPR and FPR should be used instead of an odds ratio. Prediction performance can differ even if the odds ratio remains the same [6,8].

ROC Curve

The ROC curve provides a complete description of the biomarker's prediction performance. It is a single curve plotted on a graph with the FPR on the horizontal axis and the TPR on the vertical axis (Fig. 2.1). The curve is a plot of the prediction performance (FPR and TPR) of the biomarker as the screen-positive criteria changes. ROC curves can guide the selection of screen-positive criteria [2,16]. Biomarkers with ROC curves closer to the

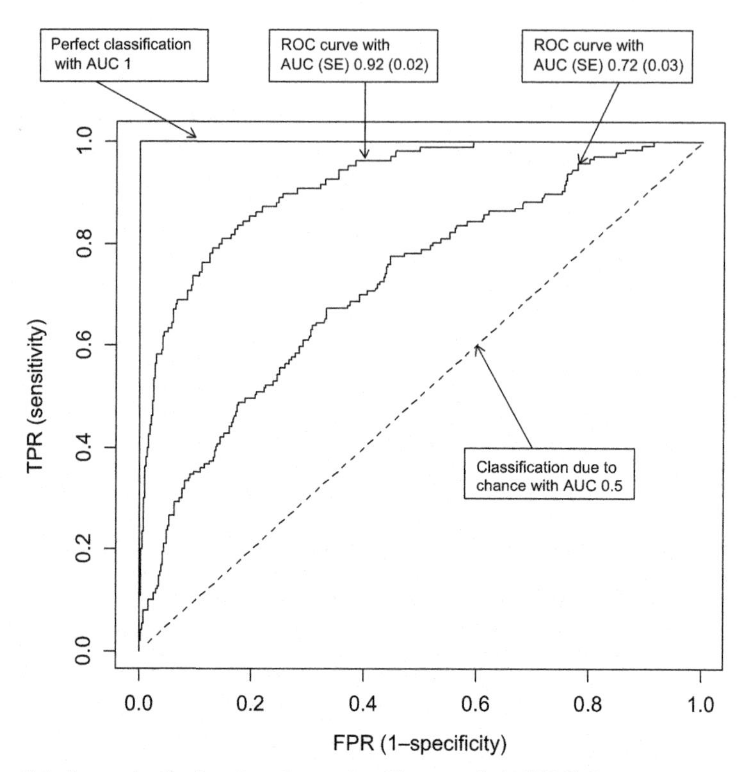

Figure 2.1 ***Example of a Receiver Operating Characteristic (ROC) Curve.***

top left-hand corner have better prediction performance. A perfect biomarker that accurately discriminates all diseased and nondiseased patients would have an ROC curve along the left side of the graph and along the top of the graph. A diagonal line is included on ROC curves to demonstrate the performance of a biomarker purely due to chance. If the entire ROC curve lies below the diagonal line, this indicates that the distribution of the biomarker is opposite of usual convention (e.g., lower values of the biomarker are associated with diseased patients) [2,17]. In such situations, transform the biomarker data so that they follow usual convention (e.g., biomarker values multiplied by negative one) and recreate the ROC curve. The top left-hand corner of the ROC curve corresponds to the biomarker level where specificity and sensitivity are optimized.

Area Under the Curve

If it is difficult to produce an ROC curve or if there are too many biomarkers to compare with ROC curves, summary indices of ROC curves are frequently used. The area under the curve (AUC) is probably the most widely-used summary index. The AUC ranges from 0.5 (the area under the diagonal line representing discrimination based on random chance) to 1 (the area of the entire square representing perfect discrimination). The AUC can be interpreted as the probability of the biomarker value being higher in a diseased patient compared with a nondiseased patient, if the diseased and nondiseased patients are randomly chosen [2,17]. However, applying this interpretation in a clinical context is cumbersome as patients do not present as pairs of randomly selected cases and controls.

ROC curves and AUC can be calculated using most statistical software packages. The area under the curve can be estimated by the c-index (usually calculated by the trapezoidal rule) or by the Mann–Whitney U statistic [18]. The trapezoidal rule and U-statistic are nearly identical when the biomarker is continuous [19] but if the biomarker only has a few distinct values (5 or 6), the trapezoidal rule systematically underestimates the true area [20].

Optimal Prediction Threshold

Another summary index frequently reported is the set of FPR and TPR that corresponds to a particular screening threshold. Often, the optimal prediction threshold is defined as the cut point with the maximum difference between the TPR and FPR [e.g., the Youden Index is calculated as max(TPR − FPR) or equivalently max(sensitivity + specificity − 1)]. This definition may not be the optimal threshold, depending on the clinical context. For example, for a biomarker to be accepted in clinical practice, it must have a better prediction performance than the existing test, which has a FPR of 10%. Thus, the optimal threshold in this scenario would be defined as the maximum TPR for an FPR of at least 5%.

Partial Area Under the Curve

In some contexts, it might be of interest to summarize the prediction performance of the biomarker based on more than one screening threshold but less than the full range of FPR values. The partial area under the curve can be used to describe the prediction performance within a range of FPR values. For example, certain settings may require very low FPR values (e.g., ≤ 0.05); therefore only the AUC between FPR values of 0 and 0.05 would be of interest. There are other summary indices that have been proposed for this measure but are not discussed here [2,17].

SAMPLE-SIZE CALCULATIONS

Sample-size calculations should be linked to the statistical methods used in the analysis. Margaret Pepe has developed a rigorous methodology for sample-size calculations [2]. Here, we will provide examples of sample sizes for a continuous biomarker based on TPR and FPR using Pepe's methodology.

For the sample-size calculation, we will determine if the TPR is above some minimally acceptable value for a given minimally acceptable FPR. The following assumptions are required for the calculations: significance level, power level, disease event rate, and the ratio of the variability of the biomarker in diseased and nondiseased patients. We assume the variances of the biomarker in diseased and nondiseased patients are equal, in order to provide us with the largest sample sizes. In addition, we assume a significance level of 5% and 80% power.

For example, suppose we are evaluating a new continuous biomarker for a disease with an event rate of 7%. The largest acceptable false positive rate is 5% (corresponds to specificity of 95%) and at that rate, the biomarker must have a true positive rate of at least 5% (TPR null), in order to be considered a useful biomarker. It is expected that the biomarker will have a TPR of at least 10% (TPR alternative). Given these assumptions, 3300 patients are required (231 diseased patients and 3069 nondiseased patients). A smaller sample size is required with a higher event rate and larger effect sizes (Fig. 2.2). Thus, 190 patients will be required for a biomarker with expected TPR rate of 40%, where the event rate of disease is 20%.

EVALUATING INCREMENTAL VALUE

In settings where existing clinical measures or risk scores are being used for outcome prediction, we need to quantify the additional value the biomarker adds above and beyond those existing current measures [7,8]. Here we have described several metrics to quantify incremental value for the biomarker and have assumed we are evaluating a biomarker compared to an existing clinical risk score.

Prior to quantifying the incremental value of a biomarker, it is important to ensure the underlying clinical risk model is well calibrated [7]. With a well calibrated model, the first step is determine if the biomarker is associated with the outcome of interest after adjustment for the existing clinical risk factors. This can be completed with a multivariable significance test

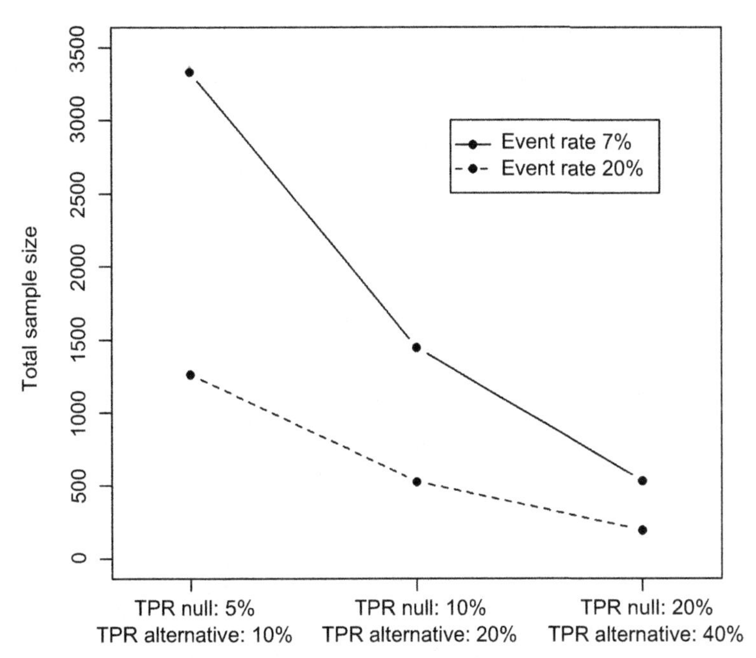

Figure 2.2 ***Sample Size Required for a False Positive Rate (FPR) of 5%.***

[21,22] (Table 2.3). It is then important to further examine the incremental value of the biomarker by looking at measures of discrimination (ΔAUC, ΔTPR, and ΔFPR) or reclassification (NRI and IDI) (Table 2.3) [23–26].

Change in Discrimination

The AUC is a common method to summarize the discriminatory ability of a biomarker with a dichotomous outcome. To quantify the incremental value, the ΔAUC is calculated as the difference in AUCs when a biomarker is added to a risk-prediction model. For a given risk threshold, the ΔTPR can be calculated as the change in proportion of cases correctly classified and similarly the ΔFPR is the change in proportion of controls incorrectly classified.

Net Reclassification Index and Integrated Discrimination Improvement

NRI and IDI are two newer metrics based on the concept of reclassification tables [23–26]. This applies to the situation where two models are being compared. For example, comparing the original risk model to the risk model with the biomarker added. If a diseased patient moves "up" in risk

Table 2.3 Summary of measures of incremental value

Statistical metric	Description
Multivariable significance test	Demonstrates if the biomarker is independent of existing covariates and has some incremental value but does not determine if the incremental value is clinically meaningful
ΔAUC	Provides a single summary measure comparing AUCs between two prediction models; the interpretation is not clinically relevant
ΔTPR, ΔFPR, or NRI (2-way)	Provides a direct comparison of performance prediction but the selected risk thresholds may not be well defined in the clinical setting
NRI 3-way categorical	Examines the changes in risk categories for cases and controls before and after the addition of a biomarker to a prediction model; the clinical implication of the changes in risk categories is not taken into consideration and the metric depends on the threshold levels selected for the risk categories
NRI (continuous)	Examines the changes in the predicted probabilities between cases and controls after the addition of a biomarker to a prediction model; the interpretation is problematic as there are no guidelines for meaningful magnitude and it is not directly linked to clinical use
IDI	Calculated as the difference in discrimination slopes; the metric is sensitive to differences in event rates
Relative IDI	Ratio of IDI and the discrimination slope in the baseline clinical model; the relative scale improves the interpretability but there is no defined range for meaningful improvement

NRI, net reclassification index; IDI, integrated discrimination improvement.

classification in the new model this is seen as an improvement in classification and any "downward movement" is considered as a worse reclassification. NRI is the sum of two differences $NRI_{event} + NRI_{nonevent}$ where each is the difference in the proportion of individuals with improved minus worsened reclassification. For example, the NRI_{event} is calculated as the proportion of individuals moving up minus the proportion of individuals moving down for individuals with the outcome (events or cases). It has been suggested to report the separate the differences separately (NRI_{event} and $NRI_{nonevent}$) instead of the overall NRI [27,28]. The "upward" and "downward" movement can be defined as movement between specified risk categories (NRI 2-way or 3-way) or by any change in the predicted probabilities (NRI continuous).

In the case when NRI is calculated using two risk categories the NRI_{event} and $NRI_{nonevent}$ are equivalent to ΔTPR and ΔFPR, respectively. The IDI does not use risk categories and instead examines the actual change in the predicted probabilities for each individual. It is the difference in discrimination slopes or difference of average probabilities between events and non-events [29].

SUMMARY

Biomarker development is a phased program and requires several years to develop a biomarker for clinical use. For each phase of biomarkoer development, it is important to customize the study design, statistical analysis, and sample-size calculation, to evaluate the clearly defined research objective. Biomarker-prediction performance should be quantified with appropriate metrics, such as, TPR, FPR, and ROC curves. Having a clear understanding of the research methodology and research goals can improve efficiency for successful biomarkers and prevent wastage of resources and effort on failed biomarkers.

REFERENCES

[1] Pepe MS, Feng Z, Janes H, Bossuyt PM, Potter JD. Pivotal evaluation of the accuracy of a biomarker used for classification or prediction: standards for study design. J Natl Cancer Inst 2008;100(20):1432–8.
[2] Pepe MS. The statistical evaluation of medical tests for classification and prediction. Oxford: Oxford University Press; 2003.
[3] Baker SG, Kramer BS, Srivastava S. Markers for early detection of cancer: statistical guidelines for nested case-control studies. BMC Med Res Methodol 2002;2:4.
[4] Feng Z, Prentice R, Srivastava S. Research issues and strategies for genomic and proteomic biomarker discovery and validation: a statistical perspective. Pharmacogenomics 2004;5(6):709–19.
[5] Ransohoff DF. How to improve reliability and efficiency of research about molecular markers: roles of phases, guidelines, and study design. J Clin Epidemiol 2007;60(12):1205–19.
[6] Pepe MS, Janes H, Longton G, Leisenring W, Newcomb P. Limitations of the odds ratio in gauging the performance of a diagnostic, prognostic, or screening marker. Am J Epidemiol 2004;159(9):882–90.
[7] Kerr KF, Meisner A, Thiessen-Philbrook H, Coca SG, Parikh CR. Developing risk prediction models for kidney injury and assessing incremental value for novel biomarkers. Clin J Am Soc Nephrol 2014;9(8):1488–96.
[8] Parikh CR, Thiessen-Philbrook H. Key concepts and limitations of statistical methods for evaluating biomarkers of kidney disease. J Am Soc Nephrol 2014;25(8):1621–9.
[9] Parikh CR, Garg AX. Acute kidney injury: better biomarkers and beyond. Kidney Int 2008;73(7):801–3.
[10] Baker SG, Kramer BS, McIntosh M, Patterson BH, Shyr Y, Skates S. Evaluating markers for the early detection of cancer: overview of study designs and methods. Clin Trials 2006;3(1):43–56.

[11] Pepe MS, Etzioni R, Feng Z, Potter JD, Thompson ML, Thornquist M, et al. Phases of biomarker development for early detection of cancer. J Natl Cancer Inst 2001;93(14):1054–61.
[12] Pepe MS. Evaluating technologies for classification and prediction in medicine. Stat Med 2005;24(24):3687–96.
[13] Janes H, Pepe MS. Adjusting for covariates in studies of diagnostic, screening, or prognostic markers: an old concept in a new setting. Am J Epidemiol 2008;168(1):89–97.
[14] Kerr KF, Meisner A, Thiessen-Philbrook H, Coca SG, Parikh CR. RiGoR: reporting guidelines to address common sources of bias in risk model development. Biomark Res 2015;3(1):2.
[15] Meisner A, Kerr KF, Thiessen-Philbrook H, Coca SG, Parikh CR. Methodological issues in current practice may lead to bias in the development of biomarker combinations for predicting acute kidney injury. Kidney Int 2015;89(2):429–38.
[16] Baker SG. The central role of receiver operating characteristic (ROC) curves in evaluating tests for the early detection of cancer. J Natl Cancer Inst 2003;95(7):511–5.
[17] Krzanowski WJ, Hand DJ. ROC curves for continuous data. Boca Raton, FL: Chapman & Hall/CRC; 2009.
[18] Pepe MS, Cai T, Longton G. Combining predictors for classification using the area under the receiver operating characteristic curve. Biometrics 2006;62(1):221–9.
[19] Hanley JA, McNeil BJ. The meaning and use of the area under a receiver operating characteristic (ROC) curve. Radiology 1982;143(1):29–36.
[20] DeLong ER, DeLong DM, Clarke-Pearson DL. Comparing the areas under two or more correlated receiver operating characteristic curves: a nonparametric approach. Biometrics 1988;44(3):837–45.
[21] Demler OV, Pencina MJ, D'Agostino RB Sr. Misuse of DeLong test to compare AUCs for nested models. Stat Med 2012;31(23):2577–87.
[22] Pepe MS, Kerr KF, Longton G, Wang Z. Testing for improvement in prediction model performance. Stat Med 2013;32(9):1467–82.
[23] Pencina MJ, D'Agostino RB Sr, Demler OV. Novel metrics for evaluating improvement in discrimination: net reclassification and integrated discrimination improvement for normal variables and nested models. Stat Med 2012;31(2):101–13.
[24] Pencina MJ, D'Agostino RB Sr, D'Agostino RB Jr, Vasan RS. Evaluating the added predictive ability of a new marker: from area under the ROC curve to reclassification and beyond. Stat Med 2008;27(2):157–72.
[25] Pencina MJ, D'Agostino RB Sr, Steyerberg EW. Extensions of net reclassification improvement calculations to measure usefulness of new biomarkers. Stat Med 2011;30(1):11–21.
[26] Pencina MJ, D'Agostino RB, Pencina KM, Janssens AC, Greenland P. Interpreting incremental value of markers added to risk prediction models. Am J Epidemiol 2012;176(6):473–81.
[27] Kerr KF, Bansal A, Pepe MS. Further insight into the incremental value of new markers: the interpretation of performance measures and the importance of clinical context. Am J Epidemiol 2012;176(6):482–7.
[28] Kerr KF, Wang Z, Janes H, McClelland RL, Psaty BM, Pepe MS. Net reclassification indices for evaluating risk prediction instruments: a critical review. Epidemiology 2014;25(1):114–21.
[29] Pencina MJ, D'Agostino RB Sr, D'Agostino RB Jr, Vasan RS. Evaluating the added predictive ability of a new marker: from area under the ROC curve to reclassification and beyond. Stat Med 2008;27(2):157–72.

The Role of Metabolomics in the Study of Kidney Diseases and in the Development of Diagnostic Tools

U. Christians, MD, PhD, J. Klawitter, PhD, J. Klepacki, PhD and J. Klawitter, PhD
iC42 Clinical Research and Development, Department of Anesthesiology, University of Colorado, Anschutz Medical Campus, Aurora, CO, United States

Contents

Biomarkers of Kidney Disease. http://dx.doi.org/10.1016/B978-0-12-803014-1.00003-0

INTRODUCTION

Although functionally and physically separate entities within the body, all of the body's cells are in constant communication with the various fluid compartments of the body. Cell metabolites, peptides, and proteins are in constant flux, being alternatively released from cells or taken up by cells from body fluids via a variety of mechanisms, such as normal excretion, transmembrane diffusion, or transport, and during the death process when cells release all of their contents. Thus, at least to a certain extent, the biochemical and protein-based changes, which are occurring within cells and organs, are reflected in body fluids.

It was already recognized in ancient Greece that changes in tissue and biological fluids were observed to be coincident with the development of pathology, and thus were capable of serving as indicators of a given disease processes. To that end, the so-called urine charts were developed and have widely been used since the Middle Ages [1]. The developments in chemistry in the late 18th century and the emergence of analytical methods, albeit simple, provided the basis for the first clinical chemistry diagnostic tools used in nephrology. In 1795 the nitric acid test for proteinuria was described and only a few decades later, already more than 100 organic and inorganic compounds in urine were known [2]. Technological advances in nuclear magnetic resonance (NMR), mass spectrometry (MS), and chemometrics (biostatistical pattern recognition methods) have opened up new opportunities in biochemistry by introducing metabolomics as an approach to study metabolism and its regulation in response to drugs, disease, genetic, and environmental factors [1]. In general, metabolomics-based strategies have been developed and employed to:

1. identify unknown molecular mechanisms,
2. discover molecular markers that can be used for drug discovery, preclinical, and clinical drug development, and
3. develop diagnostic tools.

Definitions

Metabonomics has been defined as "the quantitative measurement of the multiparametric metabolic response of living systems to pathophysiological stimuli or genetic modification" [3]. There are numerous and often conflicting uses of the terms metabonomics and metabolomics in the literature and both words have been used interchangeably. The definitions of metabolomics, metabonomics, and other related terms are listed—in Table 3.1. As in most cases it is in fact metabolic profiling that is being performed in body fluids or in specific organs (in this case, the kidney), the term metabolomics will be used here for the sake of simplicity.

Table 3.1 Terms and definitions [4,5]

Metabolome	A quantitative descriptor of all endogenous low-molecular weight components in a biological sample, such as urine or plasma; each cell type and biological fluid has a characteristic set of metabolites that reflects the organism under a particular set of environmental conditions and that fluctuates according to physiological demands; the metabolome can be divided into the primary metabolome (as controlled by the host genome) and the cometabolome (dependent on the microbiome)
Cometabolome	Metabolites that can only be formed by the integrated biochemical actions of more than one genome, such as the gut microbial metabolism of a mammalian metabolite or vice versa
Metabonome	Theoretical combinations, sums, and products of the interactions of multiple metabolomes (primary, symbiotic, parasitic, environmental, and cometabolic) in complex systems
Metabolomics	The comprehensive quantitative analysis of all the metabolites of an organism or a specific biological sample
Metabonomics	The quantitative measurement over time of the metabolic responses of an individual or population to a disease, drug treatment, or other challenge
Microbiolome	The consortium of microorganisms, bacteria, protozoa, and fungi that live commensally or symbiotically with a host
Xenometabolome	Characteristic profile of nonendogenous compounds, such as drugs, their metabolites, and their excipients, dietary components, herbal medicines, and environmental exposure

A biomarker is defined as "a characteristic that is objectively measured and evaluated as an indicator of normal biological processes, pathogenic processes, or pharmacological responses to therapeutic intervention" [6]. On the basis of this definition, biomarkers have been in use since the emergence of clinical diagnostics and include a whole host of procedures, ranging from the mundane, such as measurement of clinical signs and symptoms (blood pressure readings or temperature assessment), to the slightly more sophisticated analysis seen in ECG tracings, to the various refined examinations available, including imaging technologies, such as CT or magnetic resonance imaging (MRI), and ultimately extending to the most modern technologies, such as high-throughput gene arrays [7]. As metabolomics is based on technologies that directly or indirectly assess molecular mechanisms, the more focused term "molecular marker" will be used here instead of the broader term "biomarker." A molecular marker can consist of the measurement of a single molecular entity but it can also be a set of several molecular entities, as in a molecular pattern or fingerprint.

Why are Metabolomics-Based Molecular Markers Expected to be More Sensitive and Specific Than Currently Established Markers Used in Nephrology?

As of today, clinical laboratory diagnostics is usually based on a limited set of molecular markers, often only one parameter that is closely correlated with a functional aspect of the organ in question or with a specific disease process. However, there is not and there will never be a single molecular entity that captures the function of the kidney in all its complexity. Although the limitations of the currently most widely used molecular markers for the detection of acute and chronic kidney injury, such as proteinuria, creatinine in serum, and blood urea nitrogen, are well known and have often been discussed, these diagnostic markers remain the standard of care. All of these markers are less than optimal, in large part because they focus on the later stages of kidney injury when therapeutic interventions may be less effective and less likely to result in complete reversal of the injury [8]. In essence, these are often indicators of irreversible or only partially reversible kidney injury. Moreover, these markers provide little information about the causation or location of the said injury.

Modern analytical technologies allow for the identification of patterns that confer significantly more information than the measurement of a single parameter, much as a bar code contains more information than a single number. Well-qualified molecular marker patterns will yield more detailed and mechanistically relevant results than the measurements currently available, ultimately translating into good specificity. The better the specificity of a molecular marker pattern; the greater the reduction in nonspecific background noise. Reduced background noise can be expected to result in better sensitivity, and thereby an enhanced ability to recognize a disease process, while it remains early in the making.

While, for example, creatinine concentrations in serum typically need to increase by 20% before such an increase is considered clinically significant, several signals in a pattern revealing smaller changes in a certain direction may be sufficient to draw reliable conclusions on the basis of their being congruous [9]. Moreover, a molecular marker that is composed of several qualified parameters that describe and measure different aspects of kidney function will convey more comprehensive diagnostic information and thereby reduce the risk of overlooking disease processes or drug effects that may have been subtly indicated but considered insignificant when only a single parameter marker is used.

Metabolomic-Based Molecular Markers Versus Protein and Genomic Markers: Advantages and Challenges

Genomics, proteomics, and metabolomics when taken together as a whole, provide a comprehensive framework, also referred to as systems biology, that describes the biochemical function of an organism and its response to challenges. Genomic and phenotypic molecular markers, including proteins and metabolites, have been differentiated. The genotype of a patient defines the risk or probability of reacting to a disease, drug, or environmental challenge in a certain way and is static. The phenotype more closely reflects clinical reality at any given moment. In recent years, gene arrays have extensively been used for not only molecular marker discovery, but also in drug development and the identification of molecular mechanisms. One of the reasons gene arrays are considered so desirable is the availability of standardized high-throughput technologies, while the analytical technologies used for metabolomics are not yet as mature. Unfortunately, it cannot be assumed that changes of mRNA concentrations, also known as the transcriptome, translate directly into corresponding changes in the number of functional proteins. Accordingly, it cannot be assumed that changes in the transcriptome are necessarily associated with changes in signal transduction and cell biochemistry. Therefore, downstream confirmation by analyzing protein concentrations and/or metabolites is usually required [10]. However, the changes of a protein concentration may also not necessarily translate into changes in cell biochemistry and function, as protein concentration is not always correlated with activity. Reasons include changes in translational modifications, reaction with oxygen radicals, and allosteric and competitive regulation by substrates, products, and other inhibitors and activators. Pathophysiological changes and histological damage is in most cases directly caused by changes in cell metabolism. Thus, metabolomics typically is more closely associated with a disease process or drug effect than proteins, mRNA, or genes [11,12].

While transcriptomics and proteomics strictly detect endogenous changes, the metabolome communicates with the environment and is an open system [12]. The exact number of metabolites varies at any given time. Metabolic profiles include endogenous and exogenous chemical entities including peptides, amino acids, nucleic acids, carbohydrates, organic acids, vitamins, hormones, drugs, drug metabolites, drug excipients, food additives, phytochemicals, toxins, and other chemicals ingested or synthesized by a cell or organism. The metabolome can also be influenced by environment, gut flora and its metabolites, diet, and general activities and responses, such as stress, hormones, physical injury, and exercise [4] (Fig. 3.1). In comparison

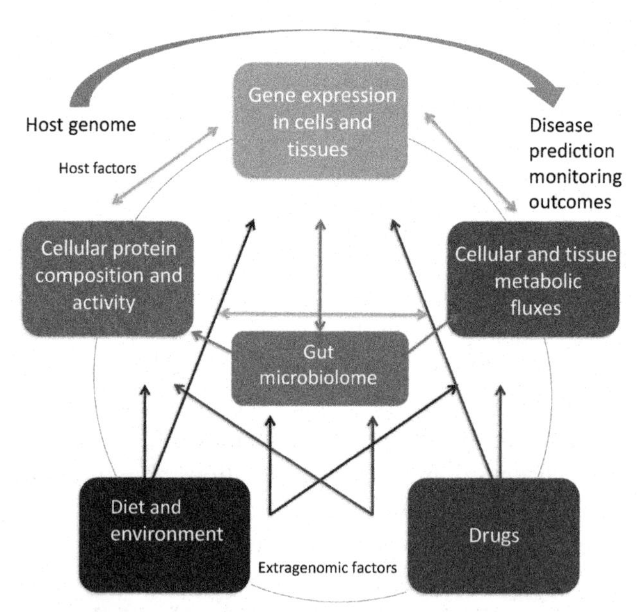

Figure 3.1 Interactions between the mammalian system, the microbial metabolome, diet and environment. ***(Please also see Nicholson JK, Wilson ID. Understanding global systems biology: metabonomics andthe continuum of metabolism. Nature Rev Drug Discov 2003;2:668–76.)***

to the larger proteins and mRNA, small molecules, such as metabolites can distribute quickly all over the body. This will result in comprehensive information and a rather complete picture of the complex interactions of an organism's metabolism and its interactions with the microbiome [13], diet [14], environment, and other exogenous factors, but deconvolution of this information can be challenging [4]. The food metabolome, for example, has been estimated to consist of more than 25,000 compounds [14].

Most endogenous metabolites are tied to specific biochemical pathways, such as glycolysis, Krebs cycle, lipid or amino acid metabolism, signaling pathways, such as transmitters and hormones, and specific pathobiochemical processes, such as oxidative stress. Thus, changes in specific metabolite patterns may reflect changes in pathways and processes [7,15]. Psychogios et al. [16]. found 4229 confirmed and highly probable human serum metabolites, 3247 of which were glycerolipids and phospholipids. A similar study of the urine metabolome suggested that there are at least 3079 detectable metabolites in human urine, of which 1350 have been quantified [17]. Interestingly, 866 of the detectable metabolites in urine are lipids. Nevertheless, the overall size of the human metabolome is unknown [18].

Lipids are a specific group of cell metabolites and the term lipidomics has been used to describe the comprehensive identification and quantification of all lipid molecular species in a biological system [19]. Lipids are loosely defined as biological compounds that are generally hydrophobic in nature and soluble in organic solvents. Lipids are membrane components, mediators in cell signaling, and are utilized as fuel and energy storage [20]. Tens and hundreds of thousands of lipid molecular species at the attomolar and nanomolar level per milligram protein in cells have been predicted [21]. Their distinct solubility properties may require separate extraction and analysis in more targeted high-performance liquid chromatography–mass spectrometry (LC–MS) bioanalytical strategies [19,21], but also nontargeted, direct infusion shotgun lipidomics bioanalytical approaches have successfully been used [21].

The metabolome is considered the most predictive phenotype and holds the promise to extensively contribute to the understanding of phenotypic changes as an organism's answer to disease, genetic changes, and nutritional, toxicological, environmental, and pharmacological influences [4]. Another advantage of metabolomics is that in contrast to genes and proteins, metabolites are often tissue- and species-independent. This facilitates translation of molecular markers strategies from bench-to-bedside or vice versa [14], which is advantageous for drug development and molecular marker qualification (vide infra). Also, while it may take hours, days, and sometimes weeks for protein and mRNA expression to change in response to a challenge, metabolic responses can often be measured within seconds or minutes [4].

METABOLIC MAPPING OF THE KIDNEY

The kidney has a wide range of biochemical, physiological, and endocrine functions including, but not limited to: the regulation of blood pressure, fluid volume and systemic electrolyte concentrations, the elimination of waste products, the recovery of desired substrates from urine, the metabolism of endogenous compounds and xenobiotics, and the synthesis of hormones, such as erythropoietin, renin, and 1,25-hydroxy vitamin D_3 [22]. The kidney consists of the following major regions (listed from outside in): the cortex, the outer and inner medulla, and the papilla. All of the separate regions of the kidney are associated with unique functionalities and face dissimilar metabolic challenges. These challenges are in part driven by differences in osmolarity and oxygen tension faced by the various areas of the kidney. Accordingly, enzymes, transporters, and other proteins are differentially distributed across the different regions according to the needs of the various functional anatomical structures of the kidney. The consequences of this are region-specific differences in cell metabolism. Thus, the majority of the kidney's drug metabolizing enzymes, and enzymes involved in the detoxification of radicals are located in the proximal tubule. Regions in an environment with high osmolarity, such as the Loop of Henle and the collecting ducts, are rich in osmolyte transporters and glycolytic enzymes due to the high-energy requirements in such an environment. For a more comprehensive summary of the distribution of enzymes and transporters in different regions of the nephron, please refer to Niemann and Serkova [22] and Burckhardt and Burckhardt [23].

It has been shown that the region-specific differences in metabolite distribution across the kidney can readily be detected through metabolic profiling of kidney tissue samples, such as by high-resolution magic angle spinning proton NMR spectroscopy (^{1}H-NMR) [22] and MS imaging [24]. The following is a brief summary of the regional differences in cell metabolism and metabolite patterns. For a more in-depth discussion of renal metabolite distribution patterns relative to kidney metabolomics, please refer to Niemann and Serkova [22]. However, it must be noted that most information about metabolite distribution in the kidney is based on studies of the rat kidney and that it is not clear to which extent this can be translated to the human kidney.

Cortex

The cortex is characterized by a relatively high expression of mitochondrial oxidative enzymes, as well as Krebs cycle enzymes, while it shows relatively low concentrations of enzymes associated with anaerobic glycolysis, such as phosphofructokinase and lactate dehydrogenase. Cortical nephrons contain rather high concentrations of free amino acids, organic acids, choline, glucose, and trimethyl amine-*N*-oxide (TMAO), as well as high concentrations of triglycerides and phospholipids. It has been speculated that this may be related to the significant need for membrane turnover and maintenance generated by the abundance of transporters in the proximal *tubuli* [22].

Medulla

The oxygen tension in the inner medulla is significantly lower than in the cortex and in contrast to the cortex that mainly relies on mitochondrial oxidation to fulfill its energy requirements, cells in the medulla rely on both mitochondrial and glycolytic pathways. Accordingly, glucose, lactate, and hydroxybutyrate play a much more important role in the energy metabolism in the medulla than in cortex cells. With increasing osmolarity in the inner medulla, the intracellular concentrations of osmolytes, such as betaine, taurine, sorbitol, glycerophosphocholine, and myoinositol also increase.

Papilla

The papilla is characterized by a low density of mitochondria and seems to mainly rely on anaerobic energy metabolism, which is reflected by the metabolite patterns. In addition, papilla cells are characterized by high concentrations of osmolytes, such as betaine, myoinositol, sorbitol, taurine, and glycerophosphocholine.

Due to the aforementioned differences in functionalities and metabolism, the disturbance of a specific segment of the nephron will lead to characteristic changes in urine metabolite patterns (see also section "Metabolomics in Renal Research: Kidney Function, Disease, and Injury Markers"). Urinary metabolic profiling therefore presents the possibility of allowing for not only the sensitive detection of disturbances in kidney metabolism, function, and extent of injury, but concurrently presents the potential to establish localization of the injury. With that depth of understanding into the process at hand, metabolic profiling offers an insight into the underlying mechanism of injury that we cannot currently generate through any of the noninvasive means at use today.

It has been established that the following processes affect urine metabolite patterns: filtration, active secretion and absorption, transport and synthesis of osmolytes, exchange of cell metabolites with urine, oxidative stress, and release of cell contents during injury [22]. With a basic understanding of the processes at work and the differing metabolic profiles within the regions of the kidney, conclusions can be drawn regarding the location of damage based on the altered levels of metabolites. Thus, an increase of TMAO serves as a marker for medullary injury. Glutaric acid and adipic acid are markers of mitochondrial dysfunction. Glucosuria is a marker of proximal tubular dysfunction. Decreases in citrate, α-ketoglutarate, and succinate concentrations are rather specific markers for mitochondrial dysfunction in the proximal tubule. This is due to the fact that only proximal tubule cells possess the ability to compensate for inhibition of their mitochondrial Krebs cycle by importing Krebs cycle intermediates from the urine via the sodium–dicarboxylate symporter, NaDC3 [23]. Increased concentrations of dimethyl amine, sorbitol, and myoinositol are indicators of papillary damage [22].

NONTARGETED AND TARGETED METABOLOMICS

Although by definition true metabolomics assays are nontargeted [25], nontargeted and targeted assays have been differentiated in the literature. The goal of a nontargeted assay is to capture as much information as possible [26]. As the goal is the nonbiased detection of unknowns, these are semiquantitative at best and are minimally, if at all, validated. In contrast to nontargeted assays, targeted assays measure one or several well-defined compounds, can be validated, and are quantitative. Although the quality of the results is much better understood, these assays are limited in terms of their ability to detect unknown effects and are only used when the target of a drug or disease process is at least partially understood. Although the classification of metabolomics assays into targeted and nontargeted assays is tempting, it is an oversimplification as many assays are somewhere in between. Such assays are here referred to as semitargeted assays.

If scarce previous information is available, a nontargeted assay–based discovery strategy is usually a potent first step. Thus when considered broadly, nontargeted assays are hypothesis-generating strategies whose results usually require follow-up with more targeted, quantitative approaches. The major problem with nontargeted assays is the false-positive result. Due to the large number of analytes detected in relationship to the number of samples, signals may be picked up that are random and have no relationship to a disease or drug effect. In contrast, the main problem of targeted assays is the false negative result. The potential that an effect is missed because it could not be captured by the limited amount of metabolites included in the assay is a significant concern. Hence, there is value in combining targeted and nontargeted assays. In such a way, the targeted assays may be employed to test a hypothesis and the nontargeted approach is utilized to ensure that no important information is missed [27].

Metabolic fingerprinting describes the least targeted analysis of the metabolome by examination of metabolite patterns in different experimental groups with the subsequent classification of these patterns into a "fingerprint" [28,29]. Samples can be classified if the metabolite fingerprints differ between groups, allowing for sample clustering. In most cases, ^{1}H-NMR– and MS-based assays are used for metabolic fingerprinting. In ^{1}H-NMR–based assays, the chemical shift and "area under the peak," and in MS-based assays, the mass-to-charge ratios (m/z) and the signal intensities are used to describe a specific fingerprint. If separation steps, such as gas chromatography or HPLC, are used to separate compounds before detection, retention times provide additional information for indexing metabolites. Fingerprinting methods benefit from added resolution, such as 2D-NMR [30], 2D-GC–MS [31], ultrahigh performance liquid chromatography (U-HPLC), 2D-HPLC, and high-resolution MS [32,33]. The fingerprint of the analyzed sample is then exported for sample classification using multivariate analysis [34–36]. Fingerprinting is solely based on pattern analysis and comparison, the metabolites underlying the signals or peaks are not further identified. Therefore, at this stage, not much mechanistic information is gained and the resulting molecular markers cannot be validated or qualified. To generate the mechanistic data desired, it is necessary to first pursue statistical analysis and identification of the differences between samples from control and treatment groups or healthy controls and disease groups. Once that data has been established, the metabolites of interest should be identified by database search or further structural identification using analytical technologies, such as homo- and heteronuclear 2D-NMR. A representative workflow is shown in Fig. 3.2 (please also see Refs. [25,26,36]).

In comparison to completely nonbiased fingerprinting strategies, semitargeted technologies screen for a multitude of key compounds in specific metabolic pathways, such as amino acids, fatty acids, phospholipids, high-energy phosphates, and nitric oxide (NO)–synthesis pathway. Several assays can be used to screen for changes in known compounds across a range of biochemical pathways. These "multianalyte" assays typically capture 5–50 compounds and are quantitative or at least semiquantitative [27]. Although these assays may be

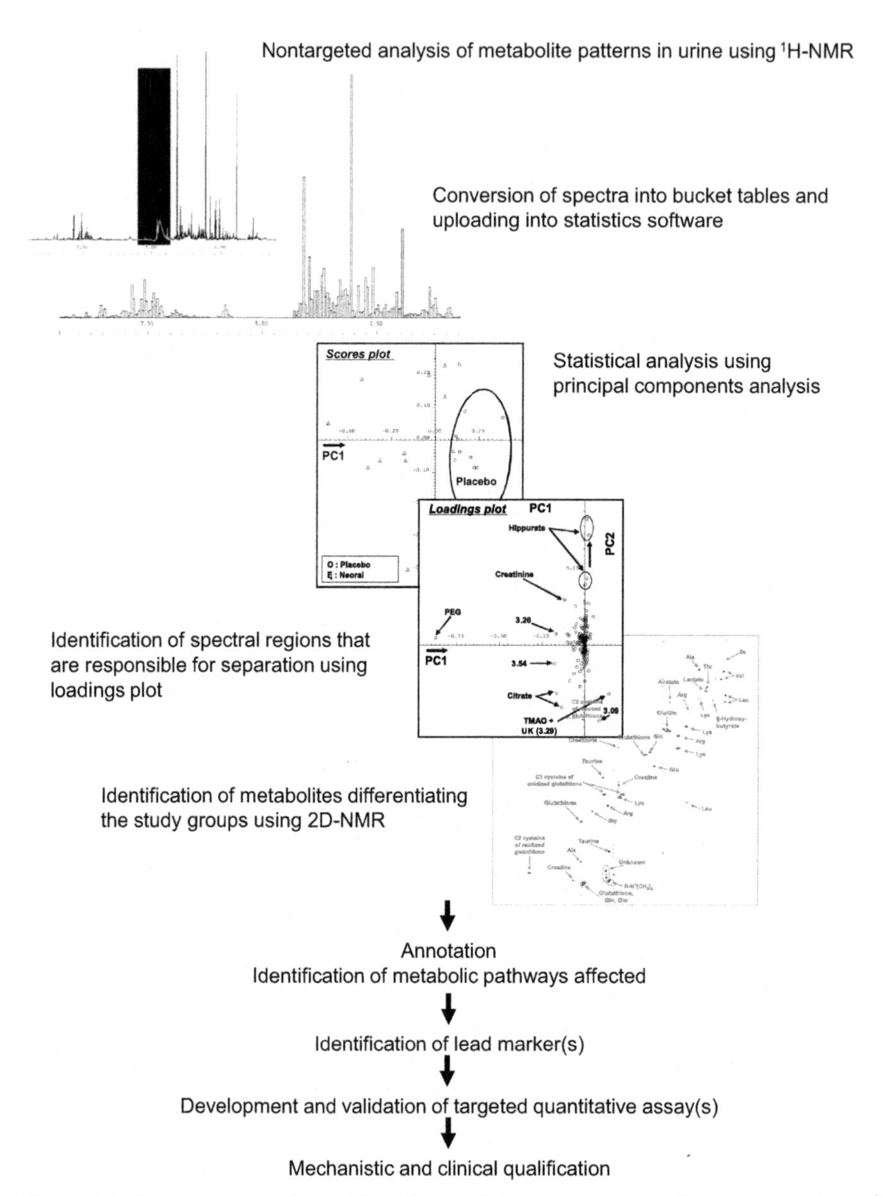

Figure 3.2 ***Representative Flow of Metabolite Marker Discovery and Development.*** The workflow of a non-targeted metabolome analysis as used in a cross-over, two-period clinical study to compare the effect of a single oral 5 mg/kg cyclosporine dose (Neoral, Novartis, Basel, Switzerland) to placebo (Neoral formulation without cyclosporine) on the kidney in thirteen healthy individuals is shown.[41] Metabolome profiling started with the acquisition of a set of ^{1}H-NMR spectra in urine. The spectra were then reduced to histograms ("binning") which represent the area under the curve in a certain spectral region. This created an ensemble of XY-tables (spectral region versus integral), the so-called bucket tables. The spectra were analyzed using a principal components analysis (PCA) and partial least squares fit analysis (PLS) (AMIX software, Bruker, Rheinstetten,

associated with a much lower throughput than straight forward fingerprinting, this type of semitargeted discovery strategy can give a rather complete picture. Despite the fact that the analytical workload and effort is higher than for finger printing approaches, a semitargeted strategy avoids some of the analytical and statistical uncertainties associated with completely nontargeted data sets. Specifically, semitargeted assays generate higher quality data, are more quantitative, and are able to minimize interference and false positives. Another advantage of the semiquantitative approach is that it provides the ability to assess drug or disease effects within a broader range of already known compounds. Thus, within one or several assays, first mechanistic information will already be available. An example of such an assay is that described by Yuan et al. [37], which is based on hydrophilic interaction liquid chromatography (HILIC) with positive/negative ion switching to analyze 258 metabolites (289 MS/MS ion transitions) in a single 15-min LC–MS/MS run. In addition, commercial semitargeted metabolomics assay solutions, such as the Biocrates Life Sciences Absolute IDQ P180 kit [38] and the Sciex Lipidyzer [39] have become available.

It is important to note that at the moment "global" or "nonbiased" analysis of all metabolites is only a theoretical concept [27]. In reality, the available analytical technologies allow for only partial analysis depending on their biophysical principles and the chemicophysical properties of the analytes, which reach from strong ions to extremely lipophilic compounds. Although nontargeted metabolomics assays are usually also considered nonbiased, it has to be kept in mind that the use of different analytical methods will introduce bias simply because of the chemical and physical properties of the different compounds, a potentially wide range in concentrations, and differences in stability of the analytes [27]. For example, if a GC–MS assay is used for metabolic profiling, only compounds that can be derivatized, go into the gas phase, and can sufficiently be ionized. ^{1}H-NMR is not a very sensitive technology; metabolite concentrations are often the limiting factor. The different methods used for nontargeted metabolomics are compared in Table 3.2. In most cases, the combination of different metabolomics technologies results

Germany). In the PCA, the principal components are constructed in such a way that the first explains most of the variance in the ensemble, the second explains the second most, and so on. The clustering analysis of the scores plots, the PC_1 versus the PC_2, was used to determine if groups of spectra differed from each other. Thus, hidden phenomena that were not obvious from the usual spectral dimension could be discovered. The spectral regions that caused the separation were identified in the loading plots, which form the link back to the spectral dimension. The compounds under the signals that were responsible for the separation of the effects of drug and placebo identified using of 2D-NMR.

Table 3.2 Comparison of technologies used for nontargeted metabolic profiling [41]

	Number of metabolites	Sensitivity	Quantitation	Sample prep	Metabolite ID	Comments
GC–MS	+	+++	++	+	+++	Requirement for derivatization excludes compounds that do not react, sample preparation with derivatization can be extensive, low throughput with run times typically between 20 and 60 min, large databases for metabolite identification based on fragmentation patterns, such as the NIST database are available
CE–MS	++	++	++	+	+	Relatively extensive sample preparation, limited software and databases, rather low throughput
LC–MS/TOF LC–orbitrap MS LC–FT–ICR–MS LC–QTRAP	+++	+++	+	+++	+	Very sensitive, detects the most metabolites (>1000), ion suppression in the electrospray source limits quantitation, poor separation and resolution of peaks, as well as relatively poor reproducibility, limited body of software and databases, sample preparation can be automated, low throughput with HPLC runtimes typically between 20 and 60 min, shorter when U-HPLC is used

MALDI–MS	+++	+++	+	+++	+	Ion suppression limits quantitation, little sample preparation required, can be used for metabolite mapping or imaging of tissue slices, rather high throughput
Infusion–nanospray–high-resolution MS	+++	+++	++	++	+	Infusion in combination with nanospray sources mostly eliminates the ion suppression problems observed with high-flow electrospray sources, more extensive sample preparation required than for HPLC or U-HPLC–MS assays, low throughput with 10–40 min infusion times
"Shotgun," sample effusion and atmospheric sample introduction EESI–MS DESI–MS DART–MS	+++	+++	+	+	+	In general less quantitative and suffer from ion suppression in complex biological samples, virtually no sample preparation required, relatively high throughput
NMR	+	+	+++	++	+++	Quantitative, nondestructive, low sensitivity, very robust technology, good metabolite capabilities identification using 2D NMR and databases

CE–MS, Capillary electrophoresis–mass spectrometry; DART, direct analysis in real-time mass spectrometry; DESI, desorption electrospray atmospheric ionization–mass spectrometry; EESI–MS, extractive electrospray ionization–mass spectrometry; FT–ICR–MS, Fourier transformation ion cyclotron mass spectrometers; GC–MS, gas chromatography–mass spectrometry; HPLC, high-performance liquid chromatography; LC, liquid chromatography; MALDI–MS, matrix-assisted laser desorption ionization–mass spectrometry; MS, mass spectrometry; NMR, nuclear magnetic resonance spectroscopy; Prep, preparation; QTOF, quadrupole–linear ion trap mass spectrometry; TOF, time-of-flight mass spectrometry; U-HPLC, ultrahigh performance liquid chromatography.

in some overlap but will also give significant additional information. It was found that LC–MS–based profiling on two different MS systems (QTOF and QTRAP) using the same chromatographic eluate resulted in a similar statistical separation of the study groups; however, this separation was not based on the same metabolites, most likely due to the different sensitivities and specificities of the two instruments to detect specific metabolites [25,40].

After the molecular marker(s) of interest has (have) been identified, the next step is to establish targeted and validated assays that are capable of quantifying these specific compounds with acceptable total imprecision and sensitivity. In many cases targeted quantitative assays have been described in the literature or are even established in clinical routine laboratories.

THE SAMPLE

In general, metabolomics studies utilize biofluids, cells, or tissues [42]. Cells and tissues can be extracted before analysis, but as NMR spectroscopy is a nondestructive technology, they can also be perfused and thereby preserved inside an NMR magnet. This is an attractive approach for the study of the time dependency of effects following exposure to a challenge, as metabolic changes can be assessed continuously and in real time. The perfusion of intact organs or tissue slices is also referred to as "ex vivo" experiments. The sample is the key to metabolomics analysis and its quality. A wide variety of sample collection and processing procedures among metabolomics study exists [43]. The lack of consensus of how samples should be collected, treated, handled, and stored may be leading to spurious molecular markers being reported, as well as general lack of reproducibility between laboratories [43]. It should also be noted that the character and quality of a sample is already determined during the study design phase. There are factors that cannot be controlled for clinical metabolomics studies, such as metabotype and gut flora. But on the other hand, the study protocol can control factors that may influence metabolite patterns, such as age, gender, ethnicity, body mass index, diet, medications, physical activity, diseases and disease status, sample collection method, timing of sample collection, sample storage, and sample preparation [43].

Tissues

Tissues samples can come from animal experiments or patient biopsies. The challenge with the collection of tissues is that as soon as the sample is collected, secondary to the hypoxia incurred as a result of the collection process, there are almost instantaneous metabolic changes occurring. These alterations in metabolism make it difficult to exclude "after the fact" changes and artifacts from the primary process of interest. To that end, effort has been made to find methods that will immediately arrest metabolic processes during sample collection. This is referred to as quenching [44]. Common approaches are freeze clamping with lower temperature receptacles, immediate freezing in liquid nitrogen, and acidic protein precipitation with perchloric or nitric acid. If the latter method is used, stability of the compounds of interest has to be assured.

Biofluids

A major limitation of genomics approaches is that in most cases clinical diagnostics based on gene chips or arrays will require a biopsy, while phenotypic molecular markers, such as metabolites and proteins can be monitored in body fluids. In nephrology, urine is an attractive matrix, as it can be considered a proximal matrix that can noninvasively be collected in large quantities, and is nonviscous and contains fewer proteins and lipids than blood, plasma, or serum [43]. In contrast to measuring molecular markers in blood, plasma, or serum that reflect changes in the systemic compartment, a "proximal" fluid is defined as a biofluid closer to, or in direct contact with, the site of disease or drug effect [45]. Proximal fluids are local sinks for metabolites, proteins, or peptides secreted, shed or leaked from diseased tissue. Once in the systemic circulation, these get quickly diluted and eventually mixed with metabolites, proteins, and peptides from other sources, which may complicate location of an injury. Unlike in serum and plasma, where metabolites from a specific organ are usually diluted, metabolites in urine are concentrated by the kidneys.

First-void urine or spot urine samples are commonly used for metabolomics analyses [46]. Based on NMR profiling of morning and afternoon urine samples, it was found that the effect of diurnal variation on healthy human urine samples is insignificant and that potential differences may rather be caused by diet [47]. Recent detailed recommendations favor the collection of midstream urine [43,48]. Said samples are less contaminated with epithelial cells and bacteria from the urinary tract [43]. Nevertheless, first-void urine samples have been reported to be less influenced by life style factors, such as diet, physical activity, and stress [49]. However, collection of first-void urine samples may be more challenging due to poor patient compliance [46].

Sample integrity is defined as stability of the analyte(s) in the biological matrix throughout variable environments spanning from sample collection, storage, shipping, and further storage up to the last sample analysis [50]. The analytical results and the conclusions drawn from the results can only be valid if the sample that reaches the laboratory is of sufficient quality (the so-called "garbage in, garbage out" principle). In most cases, the typical quality control measures taken during analysis will not catch samples of poor quality. Moreover, preanalytical sample treatment of urine samples may alter the original metabolic profiles [51] and the time period from sample collection until the sample reaches the analytical laboratory is often poorly controlled and validated. Thus, method development and validation for molecular markers will have to start with the moment the samples are collected [50] and take into account sampling devices and tubes [35,52,53]. Considerations regarding sample tubes should include potential interferences due to compounds leaking from the tubes [54,55] and blood coagulants. EDTA is usually the best choice when it comes to the prevention of clotting of the matrix during long-term storage; however, EDTA has the potential to interfere with ^{1}H-NMR analyses [48]. Standard measures employed for the stabilization of biofluids for metabolomics analysis span a broad range of methods, including the following: maintaining samples on ice, flash freezing in liquid nitrogen, addition of preservatives (such as sodium azide and antioxidants for makers that are prone to concentration changes due to autoxidation), and immediate extraction [43,48]. Moreover, some urine analytes are sensitive to light and the use of amber vials may be appropriate [56]. Based on recent recommendations, urine samples should be aliquoted within 2 h from the time of collection [48]. They should be kept refrigerated at +4°C and should not be frozen to avoid cell breakage before processing [51]. The addition of sodium azide (0.01–0.1%) to limit residual bacterial growth and enzyme activity may be considered [43]. Before aliquotation and long-term storage, centrifugation of urine samples at 1000–3000*g* and filtration through a 0.22-μm filter to remove cells and other particles is recommended [48,51]. For long-term storage, samples should be stored at −80°C or in liquid nitrogen [48,51]. Blood, serum, and plasma samples stored for 1 year and urine samples stored for 9 months under such conditions have been shown to maintain stability [57].

ANALYTICAL TECHNOLOGIES

NMR Spectroscopy

The first quantitative analysis of a mixture of small organic compounds by ^{1}H-NMR was reported in 1963 [58]. Henceforward, NMR-based chemical shift imaging technologies have extensively been used for monitoring metabolic changes in vitro, ex vivo, and noninvasively in vivo. NMR is nondestructive, highly discriminatory, and can quantify compounds in rather crude samples without the requirement for extensive sample cleanup [43]. Sample preparation and the setup of NMR experiments are described in detail by Emwas et al. [43,48] and Beckonert et al. [57] Urine samples can be used without further sample preparation; however, strong salt and pH variations between urine samples can lead to NMR signal shifts [43,57]. Blood, plasma, and serum samples require extraction, as otherwise broad macromolecule peaks may interfere with the signals of low-molecular weight molecules [59]. Different deproteinization methods have been compared. Acetonitrile precipitation at physiological pH was found to result in the resolution and detection of the most low-molecular weight molecule signals [60]. A problem with one-dimensional (1D) ^{1}H-NMR spectra, especially in urine, is the substantial spectral overlap [48,61]. Single metabolites often give several signals in the spectra. For example, urine NMR spectra may consist of more than 2000 detectable peaks corresponding to only approximately 200 metabolites [17]. Water signals are also a problem and must be suppressed [48].

Sensitivity is another limiting factor of NMR spectroscopy–based metabolomics. Often metabolite concentrations in the range of 1–10 μmol/L are required for detection and quantification by NMR. High-field NMR spectroscopy [62] and cryoprobes can improve sensitivity with detection limits in the nanomolar per liter range [63,64]. Still, in comparison to MS-based methods that are several orders of magnitude more sensitive [63,64], rather large sample volumes or numbers of cells (often >3 million) are required [44]. The sensitivity depends on the natural abundance of the *nucleus* studied (^{1}H, ^{31}P, or ^{13}C) and the potential concentration of isotopes that a cell culture, animal, or human has been exposed to. ^{13}C-NMR provides a greater spectral range in comparison to ^{1}H-NMR (200 ppm vs. 15 ppm) and has less spectral overlap of peaks; nevertheless, the low natural abundance of ^{13}C of 1.1% and low-gyromagnetic ratio limits its sensitivity [43,62].

Most metabolomics NMR studies have been based on recording 1D–^{1}H-NMR spectra, but, as aforementioned, the resulting spectra are usually complex with many overlapping peaks. This can be improved by separation of the hydrophilic and hydrophobic components in a sample by dual-step extraction before NMR analysis [65], as well as by using NMR pulse sequences [30,43,48,62], such as J-resolved, homonuclear correlated spectroscopy (COSY), ^{1}H-^{1}H total correlated spectroscopy (TOCSY), ^{1}H-^{13}C heteronuclear single quantum correlation (HSQC), nuclear Overhauser effect spectroscopy (NOESY), and flip angle adjustable 1D NOESY (FLIPSY) [64,66]. 2D–J resolved spectra are attractive as they simplify the spectra due to increased resolution in comparison to 1D spectra, and metabolites can be quantified even if they are present in concentration 10- to 100-fold lower than the major components. However, one drawback is that the integrals of the method are strongly influenced by T_2 relaxation during the long T_1 evolution period and hence only relative quantification of metabolites is possible [67]. Notwithstanding their advantages, the drawbacks of 2D-NMR methods have been the lack of speed and reproducible quantification. Recent advances in 2D-NMR techniques and data processing, such as 2D ^{1}H-INADEQUATE in combination with spare sampling/nonlinear sampling and ultrafast 2D-NMR spectroscopy, have addressed these disadvantages [48,62,64]. ^{15}N and ^{13}C isotope tagging in combination with ^{1}H-^{15}N HSQC and ^{1}H-^{13}C HSQC have shown good accuracy and precision of 2D-NMR metabolomics analysis of human plasma [68].

One of the strengths of NMR spectroscopy is that it allows tracking of stable isotope metabolites labeled at a specific atomic position, also referred to as isotopomer analysis [64]. There are two approaches to NMR-based isotopomer analysis, direct observation of the nucleus of interest or indirect detection via attached spectator spin, usually the proton [64]. While direct detection has the advantage that all sites can be observed, indirect detection is often more sensitive [64]. Stable isotope enrichment patterns of metabolites allow for reconstruction of metabolic networks, as well as flux modeling (please also see section "Identification of Disease and Pharmacodynamic Molecular Mechanisms" further).

Magic angle spinning allows for analysis of intact tissue in a nondestructive manner. The sample is spun by a rotor at 3–6 kHz at an angle of 54.7 degree relative to the magnetic field resulting in high-resolution, liquid-like NMR spectra [69]. This technology can be used for the analysis of tissue biopsies and thus can directly be compared to histopathological findings using the same tissue sample [68]. Sensitivity can further be enhanced by the use of microcoils (magic angle coil spinning, MACS) [70,71].

Standardized protocols for NMR-based metabolomics analysis in urine samples have been proposed [43,48].

Mass Spectrometry

A wide variety of different MS-based metabolomics strategies are available, all with their own specific strengths and limitations depending on the analytes of interest and the purpose and goals of a metabolomics project. MS-based metabolomics approaches can vary in terms of their chromatographic separation, ionization technique, and the type of mass spectrometer [72].

Gas chromatography–mass spectrometry (GC–MS) is still considered the gold standard in metabolite detection and quantification [73]. It is also an established clinical technology to detect inborn metabolic errors in newborns [74]. GC–MS is the oldest and most robust coupling between a separation technique and MS. Moreover, it is also the most reproducible in terms of retention times and mass spectra [73]. GC–MS has good sensitivity, peak resolution, reproducibility, and robustness [46]. It allows for the separation of isomers and does not suffer as much from matrix effects, such as ion suppression, as LC–MS [73]. Depending on the analyte, detection limits are typically in the picomolar and nanomolar range [72,73]. Drawbacks are the rather long-run times (usually between 20 and 60 min), the requirement for extensive sample preparation including derivatization, and the limitation to volatile compounds [73]. It has also been observed that derivatization may cause artifacts; for example, silylation can convert arginine into ornithine [46,73]. While the number of metabolites that can be identified in one GC–MS run is usually between 100 and 300, deconvolution software can increase this number to 1000 [44]. A representative GC–MS ion chromatogram of a urine samples from a healthy volunteer is shown in Fig. 3.3. Another strategy to increase the number of metabolites that can be differentiated in a sample is the use of GC/GC–TOF–MS [75–77]. In addition to the differentiation of more metabolites, approximately 1200, spectral purity is better than using 1D GC–MS. This improves spectral deconvolution and the reliability of peak identification [78].

Compared with GC/MS, a major advantage of LC–atmospheric pressure ionization MS is that samples usually do not require derivatization and metabolites with a larger range of physicochemical properties can be detected. LC–MS assays that are able to detect more than 2000 metabolites in 1 run have been described [79]. The number and type of metabolites detected depends not only on the extraction procedure, but also on the ionization technology used. It has been suggested that true global metabolomics requires multiple polarities (positive and negative mode) and ionization technologies to address the inherent metabolite diversity, and therefore the complexity in and of metabolomics studies [80]. Switching between positive and negative ionization mode, which is possible with most mass spectrometers nowadays, results in more comprehensive metabolite patterns because some analytes are only detectable in the positive, while others are only detectable in the negative mode [72]. In fact, it was estimated that more than 90% of the ions detected in the positive ionization mode in human blood plasma had no corresponding signals in the negative ionization

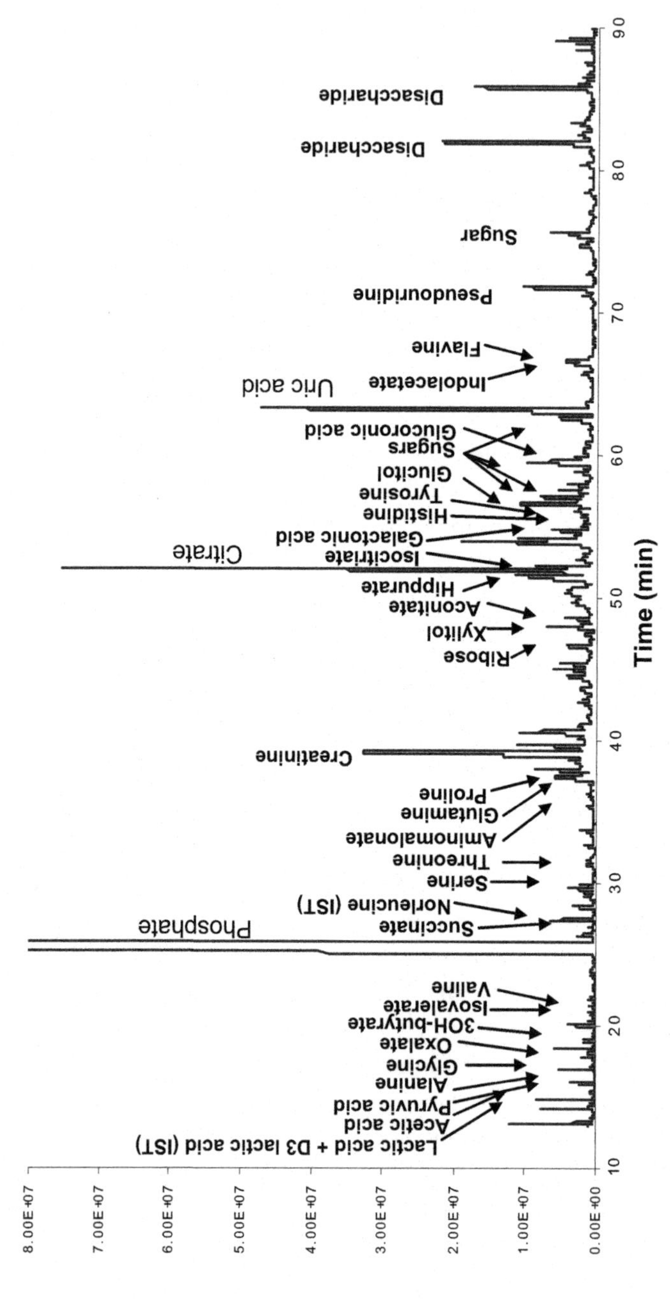

Figure 3.3 *A representative GC–MS ion chromatogram of a urine samples from a healthy volunteer.*

mode [80]. A major drawback of MS ionization technologies is that they require ionization in a fluid or matrix. As a result, ion suppression and/or ion enhancement may be caused by the interaction of multiple analytes that are present in the ionization source at the same time [81]. This means that the MS signal of an analyte is not only dependent on the concentration of the analyte itself, but potentially also dependent on the concentration and physicochemical properties of other compounds that are ionized simultaneously.

Direct injection MS is an attractive concept as it allows for high-sample throughput. It may involve direct injection using an HPLC system without column separation or direct infusion of the sample into the ionization source using a syringe pump and detection of the different metabolites solely based on their mass-to-charge ratio or high-resolution mass spectra [34,53,72]. Direct injection assays have also been referred to as "shotgun" metabolomics or lipidomics [21]. In most cases, electrospray ionization has been used and shotgun metabolomics benefits greatly from the use of high-resolution mass spectrometers [34,72]. In comparison to MS-based metabolomics assays combined with chromatographic separation, shotgun metabolomics has several drawbacks including the inability to distinguish between adduct and product ions, to differentiate isomers and their high susceptibility to matrix effects, such as ion suppression and enhancement, negatively affecting the ability to quantify [72]. At the moment nanospray ionization seems the most viable direct injection MS strategy for high-complexity samples. Nano electrospray ionization liquid chromatography is performed at flow rates of approximately 200 nL/min. This produces small, submicron-sized droplets requiring less evaporation and a greater ability to focus the resulting ions into the analyzer, thereby increasing sensitivity and ultimately offering a greater dynamic range [82]. Nanoelectrospray ionization also reduces the risk of ion suppression commonly associated with electrospray ionization. Infusion chips have successfully been coupled to nanospray electrospray sources for metabolomic profiling in highly diluted samples [83].

As an alternative to chromatographic separation, sample effusion and atmospheric sample introduction methods have become available [67,72]. These include extractive electrospray ionization (EESI)–MS, desorption electrospray atmospheric ionization (DESI)–MS, and direct analysis in real time (DART)–MS [67,72]. A commonality that these methods share is that they use very little or no sample preparation, but all suffer from the limited ability to quantify metabolites due to ion suppression when complex biological matrices are analyzed. In brief, EESI–MS uses two colliding spray sources for ionization and introduction into the mass spectrometer, DESI–MS involves a charged and nebulized solvent directed toward the sample, and DART–MS uses a stream of excited metastable helium gas and hot nitrogen to ionize the analytes. Due to matrix suppression issues for low-molecular weight molecules, matrix-assisted laser desorption/ionization (MALDI) applications have been limited [82]. Desorption ionization on porous silicon (DIOS) seems to have more potential for metabolomics studies as it allows for the detection of small molecules in both positive and negative mode, with little background interference [82].

The most frequently used detectors for LC–MS–based metabolomics are triple-stage quadrupole, time-of-flight (TOF), linear ion traps, and ultrahigh resolution detectors, such as Fourier transformation ion cyclotron mass spectrometers (FT–ICR–MS) and orbitraps [72,82]. The linear TOF mass analyzer is the simplest mass analyzer, with virtually unlimited mass range, whereas the TOF reflectron has mass range up to mass/charge ratios of approximately 10,000. TOF instruments offer high-resolution, fast-scanning capabilities (milliseconds), and accuracy in the order of 3–5 parts per million (ppm). Combination of the TOF analyzer with a quadrupole (QTOF) allows for fragmentation of a metabolite, thus rendering additional information. Ion traps allow for the isolation of a specific ion species and all others are ejected from the trap. The isolated ions can subsequently be further fragmented (MS^n). However, a limitation is that the ratio between precursor mass-to-charge ratio and the lowest trapped fragment ion is $\sim$0.3 (the "one-third rule"). Further limitations of 3D ion traps are their inability to perform high-sensitivity, triple quadrupole-type, precursor ion scanning and neutral loss scanning experiments [80]. The dynamic range is also limited due to space charge effects when too many ions are in the trap, which diminishes the performance of the ion trap. Linear ion traps have advantages over the 3D trap. A larger analyzer volume results in a greater dynamic range and an improved range of quantitative analysis.

While multiple-pass TOF–MS instruments typically have a resolving power around 40,000 (mass accuracy < 5 ppm), FT–ICR–MS and orbitraps offer a higher resolution of 1,000,000 (<1 ppm) and 150,000 (1–5 ppm), respectively [72]. This is usually sufficient to allow for specific structural identification of larger molecules based on the exact molecular mass alone. In addition, hybrid instruments, such as iontrap–FT–ICR–MS transformation and iontrap–orbitrap mass spectrometers are available.

The coupling of HPLC to a mass spectrometer allows for the separation of complex mixtures with the aforementioned advantages in comparison to a shotgun approach. Both normal phase and reverse phase HPLC have been used for metabolomics including lipidomics; nevertheless, most assays have been based on C_8 and C_{18} reversed phase HPLC columns [72]. Normal phase and reversed phase HILIC–MS has emerged as an efficient separation strategy for predominantly polar metabolites, such as those in urine [37,72,84]. In comparison to HPLC, which typically uses columns with 3–5 μm particle size, U-HPLC uses column packing particles of less than 2 μm in size. The larger interaction surface of the stationary phase results in a faster and more efficient resolution, albeit at the price of 2- to 3-fold higher back pressure requiring specific U-HPLC equipment. It has been estimated that U-HPLC, in comparison to HPLC, increases the number of detectable metabolites by more than 20% [85]. Detailed U-HPLC–based protocols for global metabolic profiling in tissues and urine have been published [86,87].

Capillary electrophoresis in combination with MS (CE–MS) has not been widely applied for global metabolic profiling. Its analytical system stability is inferior to GC–MS and LC–MS, often due to capillary wall modification following sorption of macromolecules present in biological samples [26]. Nevertheless, it has been proven to be a valuable tool for the targeted analysis of polar metabolites [26].

Other Technologies for Metabolic Profiling

Other technologies that are used for metabolomics are Raman and infrared spectroscopy [53]. It has been shown that the analysis of the same sample by a combination of multiple technologies, such as GC–MS, NMR, and LC–MS will result in a far more complete picture than each of these technologies used alone [64,88]. The number of "shared" compounds identified by one method *versus* another is often less than 50% and indeed may be as low as 20% [44].

Imaging Technologies

Metabolic profiling is also possible in vivo using MRI spectroscopy. In past animal models, radiofrequency coils have successfully been used to study kidney metabolism [89,90]. There has been an increased interest in MRI and image guided spectroscopy for in vivo assessment of metabolite patterns. This has been driven by new technology and software solutions that facilitate their use, improve sensitivity, and allow for circumvention of motion-mediated artifacts [64].

MS imaging in combination with MALDI was introduced in 1997 [91]. MS imaging can detect labeled and label-free endogenous compounds and xenobiotics across a wide mass range, including drugs, metabolites, lipids, and proteins with good spatial resolution (typically 10–100 μm depending on the instrumentation) directly on thin sections cut from fresh frozen tissue specimens [24,92,93]. Tissues are usually cut at 10–20 μm. This is comparable to the thickness of mammalian cells and the majority of cells are cut open, which allows the MALDI matrix to cocrystallize with the cell contents. The role of the MALDI matrix is to absorb the energy of pulsed, localized laser spots, which then leads to the explosive desorption of molecules into the gas phase, usually without causing degradation [24,93]. For an overview of MALDI matrices and workflows for imaging, please see Lalowski et al. [24] and Cobice et al. [93] The majority of MALDI imaging instruments are based on TOF mass spectrometers, although iontraps, FT–ICR–MS, and orbitraps have also been used [93,94]. In addition to MALDI, SIMS–MS and DESI–MS have also successfully been used for imaging [95].

More recent developments include matrix-free MALDI imaging, improved software, faster scan rates, and resolution with <1-μm spot sizes [96–98]. Thus, the combination of DESI–MS and electroincision techniques ("iKnife") allows for real-time tissue typing during surgery [96,99]. This involves real-time comparison of the mass spectra against a database of similarly acquired reference spectra in healthy and diseased tissues [96]. These developments have the potential to revolutionize the field of histology [96].

Chemometrics and Databases

Chemometrics is defined as the application of mathematical and statistical methods to chemistry [98–100]. Chemometric or nonquantitative metabolomics does not require the initial identification of compounds. It is solely based on spectral patterns and intensities. Chemometric analyses are necessary to develop statistical pattern recognition models, achieve optimal characterization of the samples, and detect molecular markers from diverse, highly dimensional omics datasets [97]. The spectra are statistically compared, clustered, and/or correlated, and used to make diagnoses, identify phenotypes, and/or to draw conclusions [35,99]. Common approaches to analyze metabolomics data sets are summarized by Alonso et al. [36] and in Table 3.3. Lists of tools available for metabolomics spectral processing and data analysis are provided by Theodoridis et al. [25] and Alonso et al. [36]

In quantitative metabolomics, metabolites are identified before statistical analysis is carried out. Databases are important tools for metabolite identification [19]. The Human Metabolome Database is the metabolomic

Table 3.3 Analysis of metabolomic data sets

Quality control and quality assurance	This may include the acceptance and rejection of data or whole data sets based on predefined acceptance criteria. Data are also checked for completeness, integrity, correctness, and queries are resolved. Once a "clean" database exists, the database is locked and data can be analyzed. Conclusions drawn from data can only be as good as the quality of the data itself. This is especially important if data are generated by multiple laboratories and entered at different sites.
Spectral preprocessing	This may include baseline correction, noise filtering, peak detection, peak alignment, peak integration, normalization, and deconvolution.
Data processing	This may include normalization of data–data transforms, background reduction, missing value corrections, data binning, and data scaling to emphasize smaller concentration metabolites.
Data reduction	This may include limiting data analysis to a specific region of interest, removal of data of poor quality, removal of data that is outside the analytical limits or that cannot consistently be replicated, and exclusion of outliers.
Unsupervised data analysis	This may include principal component analysis, multiple component analysis, independent component analysis and their subtypes, hierachical cluster analysis, nonlinear mapping, *k*-means clustering, and self-organizing maps.
Supervised data analysis	In contrast to unsupervised methods, supervised methods will require a training dataset or require that the classes of the samples are already known. Examples of such methods include Fisher discriminant analysis, SIMCA, artificial and polynominal neuronal networks, partial least square discriminate analysis, and support vector machines.
Quantification	NMR is inherently quantitative. For GC–MS and LC–MS assays a calibration strategy is required.
Statistical comparison	This may include univariate and multivariate statistics, correlation and regression analysis, ANOVA, or MANOVA, and calculation of coefficients of variance. In general, statistical comparison is used in combination with quantitative metabolomics.

(*Continued*)

Table 3.3 Analysis of metabolomic data sets (*cont.*)

Annotation	The results are put into context with existing knowledge about molecular interaction networks, such as metabolic pathways and signaling pathways. The metabolite and/or protein changes that are indicated by nontargeted discovery technologies may be complex and may present surrogate markers of complex molecular interactions. Current manual curation processes will take far too long to complete the annotations of even just the most important model organisms, and they will never be sufficient for completing the annotation of all currently available metabolome, proteome, and genome interactions [36]. Computational strategies are required that include molecular pathway and network analysis tools [101,102], computational systems biology approaches [103], as well as knowledge-based systems that combine reading, reasoning, and reporting methods to facilitate analysis of experimental data, such as the Hanalyzer software [104].
Biological interpretation	This may include pathway analyses, such as overrepresentation analysis, quantitative enrichment analysis, and single-sample profiling, as well as correlation-based network analysis; potentially followed by integration of metabolomics with genomics and other omics data. Software suites, such as KEGGarray, are capable of integrating data from transcriptomics, proteomics, and metabolomics studies [105].

Depending on the nature of the data and the goal of the study, some or all of these steps are required. For a more detailed review, please see Refs. [35,36]. ANOVA, Analysis of variance; MANOVA, multiple analyses of variance; SIMCA, soft independent modeling of class analogy.

equivalent of GenBank. It is web accessible and provides reference NMR and mass spectra, metabolite disease associations, metabolic pathway data, and reference metabolite concentrations for hundreds of human metabolites from several biofluids [35,105–107]. An overview table of available spectral databases for metabolite identification is provided by Alonso et al. [36]

There are two basic principles in MS that allow for the structural identification of a molecule: fragmentation patterns and exact molecular mass. Traditionally, due to the early use of GC–MS and the lack of high-resolution mass spectrometers, MS libraries have identified molecules based on low-resolution mass, fragmentation patterns, and retention times.

NMR

NMR, GC–MS, and LC–MS spectra contain hundreds and thousands of peaks. The identification of the individual metabolites that are underlying these peaks is a challenge and usually involves fitting the spectrum of the mixture to a set of individual reference spectra [108]. If successful, this yields information about the identity of the metabolites, as well as information about their relative and even absolute concentrations.

Substantial information is already available from 1D–^{1}H-NMR spectra, including but not limited to: chemical shifts, signal multiplicities, homonuclear (^{1}H–^{1}H) coupling constants, heteronuclear coupling constants (typically ^{14}N–^{1}H or ^{31}P–^{1}H), the first- or second-order nature of the signal, the half bandwidth of the signal, and the stability and integrity of the signal [109]. Spectral overlap is an inherent problem in the analysis of complex metabolomics 1D–^{1}H-NMR spectra and thus an increased interest in 2D-NMR for metabolomics allowing for easier identification of molecular markers has emerged during the past decade [110]. This development has been facilitated by the development of nonuniform sampling and spatially encoded ultrafast methods of 2D-NMR data acquisition, which may replace 1D–^{1}H-NMR spectroscopy as the standard NMR metabolomics approach in the future [109,110].

Profiling of NMR spectra is a complex pattern recognition problem and is often accomplished by a trained expert. Nevertheless, this process is slow, may lead to inconsistent results, may negatively affect reproducibility, and is prone to investigator bias and errors [108]. Algorithms to analyze NMR spectra of complex mixtures, such as Statistical Correlation Spectroscopy (STOCSY) and Subset Optimization by Reference Matching (STORM) have been developed [109]. Automated spectral profiling software solutions are also emerging [108]. However, because the NMR shifts are affected by factors, such as pH and interaction with other metabolites, as well as with matrix compounds, standardized sample processing procedures are required [43,48,108].

Most metabolites that can be observed by NMR spectroscopy are known [109]. Relevant information and reference spectra of at least some of those are available in databases, such as the HMDB [107], the BioMagResBank (BMRB) [111] and the Birmingham Metabolite Library (BML) [112]. The HMDB is the largest repository of NMR data on human metabolites and, as of 2015, contained information of 41,993 metabolites, of which 1,381 had experimental NMR data, totaling 3,186 NMR spectra [109]. This includes 2D ^{13}C, ^{1}H HSQC spectra that provide a good starting point for metabolite identification [109]. But even databases as large as the HMDB are still incomplete.

GC–MS

Automated analysis software and extensive databases, such as the National Institute of Standards and Technology (NIST) database and the Automated Mass Spectral Deconvolution and Identification System (AMDIS) are available [113,114]. The 2014 version of the NIST library contains 276,248 GC–MS spectra of 242,466 unique compounds and now also contains MS/MS and high-resolution spectra [113]. Although these numbers seem extensive, this database contains only a relatively small amount of endogenous compounds [46]. AMDIS provides deconvolution, quality matching using advanced spectral matching algorithms, adjacent peak deconvolution, and background subtraction, as well as retention index comparison [113,114]. For an overview table of other spectral preprocessing and deconvolution software packages and databases used for analysis of GC–MS spectra, please see Mastrangelo et al. [73]. Various algorithms and software solutions for automatically matching GC–MS mass spectra to those in libraries, such as the NIST database, have been developed [115].

LC–MS

Although atmospheric pressure ionization MS/MS techniques have been used for many years, research libraries of production mass spectra have only reluctantly been created as the mass spectral patterns are less reproducible among instruments from different manufacturers than GC–MS electron impact ionization (EI) mass spectra [116]. The fragmentation patterns depend on a large number of factors, many of which are not properly understood, such as ion source designs, ion source potentials, fragmentation gases, and mobile phase effects. In the meantime, several atmospheric pressure ionization–MS libraries, such as the HMDB [107], MetLin databases [117], and, as a representative example of a more specialized database, LIPID MAPS [118] have been compiled and successfully utilized [36]. The use of HPLC retention parameters is complicated by the variety of column stationary phases available and the infinite number of mobile phase combinations that can be used to provide suitable separations. HPLC retention times may also be influenced by column age and column load. Despite such difficulties, retention parameters have been included in LC–MS databases and this area has importance in metabolic profiling by LC/MS. The quality of metabolite annotation increases and the false discovery rate decreases with the resolution of the mass spectra and the quality of the database [35,107]. Moreover, MS metabolomics databases are often queried based on the neutral mass value using an appropriate tolerance window. The peak *m/z* value of an unknown metabolite, depending on chemical nature and ionization mode, can lead to multiple plausible neutral molecular masses that can represent different ionization adducts (such as H^+, Na^+, K^+, and others) and thus can lead to false-positive results [36]. The correct identification of metabolites based on untargeted LC–MS/MS analysis can be improved by purity of spectra [119] and the recording of high-resolution spectra. As aforementioned, the more targeted an LC–MS/MS metabolomics assay is, the more reliable is the metabolite identification; however, the more limited is the bandwidth of metabolic information. It has also been determined that today's databases are not capable of comprehensively retrieving all known metabolites [18,116].

Normalization of Urine Data

Urine presents a very attractive matrix for the metabolomics-based studies as discussed previously, yet a variety of difficulties must be overcome to gain meaningful information regarding urinary metabolite markers. One initial difficulty that must be surmounted, is normalizing urinary samples for differences in dilution. Normalization is used to identify and remove sources of systematic variation between sample profiles due to factors that are irrelevant with regard to biological processes, such as sample dilution, to ensure that spectra are comparable across runs and across related sample sets [120]. Up to 15-fold changes in urine volume are commonly observed under normal conditions, resulting in significant variation in dilution of metabolite concentrations across samples [120,121].

Normalization methods for metabolomics analysis in urine and can be categorized into curative pre- and postacquisition methods. Preacquisition methods involve normalization by dilution or reconstitution before sample analysis. Postacquisition methods adjust the results, such as spectra or concentrations, after analysis of the urine samples [120,121]. Urine is commonly normalized based on total urinary volume, creatinine, osmolality, or mass spectral total usable signal (MSTUS) [122]. MSTUS is the total intensity of reproducible peaks common to all samples [122].

The most common method of normalization is based on the creatinine concentrations in urine samples postacquisition [121,123]. The assumption that creatinine is an acceptable surrogate marker for dilutional leveling may be correct as long as creatinine clearance is normal. However, the urine creatinine concentration is a function of glomerular filtration, tubular excretion, gender and age, and may be affected by creatinine release from other sources, such as muscle. Urinary molecular marker concentrations may be misleading when based on creatinine levels in patients with disease processes or drug effects that alter release and handling of creatinine by the kidney [121,123]. Thus, normalization based on the more robust urinary cystatin C concentrations has been proposed for clinical samples [123]. Variability observed in targeted and nontargeted urine metabolomics analyses was highly dependent on the normalization strategy, which had a substantial impact on the quantitative and statistical results [122,124]. A study found that among the aforementioned normalization strategies, osmolality or MSTUS provide the best results for postacquisition normalization, but suggested that the use of two different normalization methods may be more robust [122]. Moreover, it was shown that several postacquisition normalization strategies applied to serially diluted urine samples after LC–MS/MS metabolomics analysis failed to correct for variations in urine concentrations. This is due to signal saturation or ion suppression in the most concentrated samples and default detection of some metabolites in the most diluted samples [125], suggesting that preacquisition normalization may be beneficial. This is supported by a study that showed the different normalization strategies tested, preacquisition normalization to specific gravity gave the best results [121].

In ^{1}H-NMR–based metabolomics studies, postacquisition scaling based on MSTUS represents the standard approach [126–128]. However, integral normalization may not always be the best strategy for metabolomics studies. Massive amounts of single metabolites in samples may significantly hamper the normalization based on integrals, which yields incorrectly scaled spectra. In the case of ^{1}H-NMR–based metabolite profiles, probabilistic quotient normalization was found to work best [126]. Probabilistic quotient normalization involves calculation of probable dilution factors by analyzing the distribution of the quotients of the amplitudes within a test spectrum, and comparison with a reference spectrum. Zhang et al. [128] showed that peak-picked and logarithm-transformed ^{1}H-NMR spectra are preferred. Signal processing and statistical analysis steps seemed not to be independent. While variance stabilizing transformation worked best in conjunction with principal component analysis, constant normalization seemed more appropriate for analysis using *t*-test. Overall, given the fact that this has significant impact on the results, there is still surprisingly little consensus on this critical issue. It seems that choice of the appropriate normalization procedures is dependent on context, analytical technology, and statistical algorithms [128].

Validation of Analytical Assays, Quality Control, and Standardization

To allow for interlaboratory comparison and exchange of metabolomics results in databases, a move toward consolidation is critical [25,129]. This requires standardization of assays and sample collection procedures, the assessment of assay performance, quality control, and the development of quality assurance strategies [130,131]. Development of standardization protocols is especially desirable for nontargeted metabolomics fingerprints, where data analysis in terms of biological endpoints is carried out prior to metabolite identification [129]. First steps toward standardization are the Metabolomics Standards Initiative (MSI) and the EU Coordination of Standards in Metabolomics (COSMOS) initiative. MSI, which followed earlier work by the Standard Metabolic Reporting Structure initiative and the Architecture for Metabolomics Consortium, focuses on reporting standards when submitting data for publication and to data repositories [132]. MetaboLights is the first general purpose database in metabolomics and adheres to MSI standards for metadata reporting [133–135]. The goal of the Framework Programme 7 EU Initiative COSMOS is to develop a robust data infrastructure and exchange standards for metabolomics data and metadata [135,136]. This initiative includes the support of workflows for a broad range of metabolomics applications.

The successful translation of a molecular marker into a viable clinical diagnostic test requires the availability of a robust, precise, simple, and sensitive assay that can be automated and is reasonably high throughput [50,137,138].

The key to any bioanalytical method development is the validation of every step, including sample handling and storage, to ensure accurate and reproducible results [138]. Nontargeted metabolomics is no exception. Nevertheless, the challenges are different than those for targeted assays [138,139]. Several guidelines for the development and validation of bioanalytical assays have been developed [140–145]. The following are generally considered the fundamental parameters of a validation [142,143]:

- accuracy
- precision
- selectivity
- sensitivity
- reproducibility
- stability

Method validation should demonstrate that a particular assay is "reliable for the intended application," and thus, the rigor of method depends on the purpose [139,143,146]. The challenge with nontargeted metabolomics assays is that there is no blank matrix. Endogenous compounds are ubiquitous and as these assays capture unknown metabolites, only a limited number of reference materials are available [138,139]. This means that accuracy, selectivity, and sensitivity cannot often be determined. Nevertheless, a "fit-for-purpose validation" of a nontargeted metabolomics assay may include within- and interbatch imprecision [147–149], linearity (by sample dilution), reproducibility, and stabilities [sample benchtop stability, short- and long-term storage stability [150], freeze–thaw cycle stability, and extracted sample (autosampler) stability] [138,139,150].

Quality control protocols for nontargeted metabolomics assays have been developed [148,149]. These are based upon quality controls samples that are used to monitor precision and stability of the assay [149]. Such quality controls can be pooled samples generated by mixing aliquots from study samples [148]. It has been recommended to include a quality control sample as often as every 10 study samples [149]. Statistical analysis of these quality control samples consists of data quality assessment, multivariate analysis, and comparison of exported peak table data (for more detail please see Gika et al. [149]). The addition of isotope-labeled internal standards of key metabolites spread over the mass/charge and HPLC retention time range may be considered for LC–MS/MS–based metabolomics assays. The study design should not only consider precision and stability monitoring, but should also address important issues, such as source cleaning for MS assays. The rather nonselective sample preparation procedures for these assays often consist of a simple protein precipitation step and produce rather "dirty" extracts. The first samples in an LC–MS/MS batch tend to show more variability, so it may be beneficial to run a few quality control samples first before measuring the first study samples [149]. Moreover, cross-validation between laboratories using the same assay can be carried out. The United States National Institute of Standards and Technology (NIST) offers standardized human plasma reference material for metabolomics analysis [151].

Targeted metabolomics assays can be validated and, as in many cases reference materials are available, accuracy and sensitivity can be established. It is often challenging finding an appropriate blank matrix for assay validation and preparing calibrators and quality control samples for an assay measuring endogenous compounds. Solutions may include charcoal stripping, diluted matrices, the use of corresponding matrices from other species, or artificial surrogate matrices. In the case that blank matrices are not available, samples from healthy individuals or animals can be utilized depending on the species relevant for molecular marker testing, preferably with low concentrations of the compound of interest. These samples are then enriched with the reference compounds and the endogenous signal is later subtracted. In principle, targeted multianalyte metabolomics assays can be validated following regulatory guidelines [140–143]. However, it has to be realized that current regulatory guidelines have been written mostly with the quantification of single-drug compounds in mind and may be too rigid for targeted multianalyte metabolomics assays [131,138]. The challenge with such multianalyte assays is that several compounds with different physicochemical properties are quantified simultaneously. Thus it is not possible to optimize the assay for each compound to the extent that is possible for analysis of single compounds. Moreover, the larger the number of simultaneously measured compounds, the higher the statistical probability that one accidentally fails to meet the standard acceptance criteria for quality controls samples of ±15% [131]. Wider acceptance limits for molecular markers of ±25% and even ±30% may be considered [131,138,143].

To compare data across different experiments and among different laboratories, standardization is critical [43,48,152,153]. This includes study design [152,153], sample collection, labeling, handling and storage [43,48,152–155], assay harmonization and cross-validation, and as already mentioned earlier, standardized reporting structures for metabolomics data [132–136,156,157].

The comparison of results among different analytical laboratories and datasets has shown that metabolic profiling using ^{1}H-NMR spectroscopy is surprisingly robust [57,158,159], and that most variability could be assigned to sample handling rather than ^{1}H-NMR analysis. One of the reasons is that NMR spectroscopy setups are fairly similar across laboratories. This is very different with LC–MS/MS instruments and metabolomics assays; therefore standardization is much more difficult [152,153]. As aforementioned, in an experiment to compare the results generated with a triple-stage, quadrupole-linear iontrap and a QTOF MS/MS system, rat urine samples were separated by U-HPLC and then the flow was split equally. Both streams of eluent were simultaneously directed to the inlets of the two mass spectrometers [40]. After statistical analysis, the data generated by both instruments differentiated the treatment from the control samples, but this separation was based on a different set of metabolite ions. Another study compared the results of urine amino acid patterns in kidney transplant patients with different glomerular filtration rates (GFRs) in both a semitargeted metabolomics and a targeted amino acid LC–MS/MS assay [160]. Although there was an overlap between the results of the targeted and nontargeted metabolomics assays, there were also substantial inconsistencies, with the nontargeted assay resulting in more "hits" than the targeted assay. Without further verification of the hits detected by the nontargeted discovery assay, this would have led to different interpretation of the results [160].

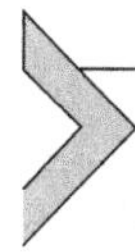

METABOLIC MOLECULAR MARKER DISCOVERY AND DEVELOPMENT

The goal of a molecular marker development is to take the marker from discovery to a status where it becomes an accepted clinical diagnostic tool and/or outcomes marker for clinical drug development [161–163]. The increasing importance of biomarkers has been recognized by drug agencies. So the 2012 FDA Safety and Innovation Act included a provision to advance the use of molecular markers in drug development and regulation [161]. In general there are two regulatory pathways through which the FDA reviews and approves biomarkers (Fig. 3.4). If the biomarker is used for drug development, it will be part of the review and approval process that leads to a new drug application (NDA) [161]. If a biomarker is not linked to a specific drug development and is of general clinical interest or for use in multiple drug development programs, such as a novel kidney function marker, stakeholders in said biomarkers including industry or public consortia, disease-specific foundations, and health research organizations can submit through the FDA Biomarker Qualification Program (BFQ) [161]. For further details regarding the biomarker development and submission process, please see Amur et al. [161], Goodsaid and Mattes [164], and applicable FDA guidance [165].

There are two keys to regulatory biomarker approval: qualification and bioanalytical validation [162]. Qualification and validation have sometimes been used interchangeably in the literature, but in a regulatory sense they are two different concepts. Validation focuses on the reliability and performance characteristics of the analytical assay used to measure molecular markers [166,167]. On the other hand, qualification has been defined as the "conclusion that within the stated context of use, a biomarker can be relied upon to have a specific interpretation and application in drug development and regulatory review" [161] and as "a graded, fit-for-purpose evidentiary process linking a biomarker with biology and clinical endpoints" [168].

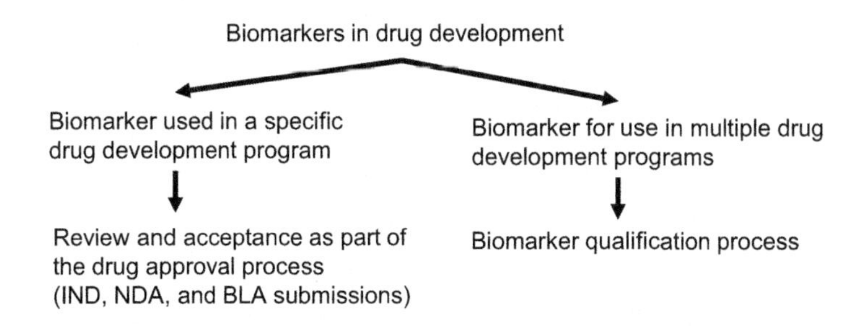

Type of biomarker	Context of use
Diagnostic	• Patient selection
Prognostic	• Patient stratification • Enrichment of trials with patients likely to have disease
Predictive	• Stratification • Enrichment: inclusion criteria • Enrichment: companion diagnostics
Response (pharmacodynamics)	• Pharmacodynamic marker as an indicator of intended drug activity • Efficacy response biomarker as surrogate for clinical endpoint • Safety biomarker for monitoring adverse effects

Figure 3.4 ***FDA regulatory pathways for review and approval of biomarkers.*** *BLA*, Biologics license application; *IND*, investigational new drug application; *NDA*, new drug application.

As the latter definition indicates, there are two key aspects to the qualification of a molecular marker:

1. To mechanistically link the molecular marker to the biochemical process underlying a disease or drug effect.
2. To establish a link between the molecular marker and clinical outcomes.

The most important first step of a molecular marker qualification is a clear understanding of what the molecular marker will be used for in a scientific, preclinical, clinical, and regulatory context. This has a critical impact on the extent and depth of the required work. There are three basic strategies that can be used for establishing a mechanistic link between the molecular marker and the biochemical process underlying a disease or drug effect:

1. leverage of preexisting knowledge,
2. biostatistical strategies, and
3. experiments identifying the underlying molecular mechanisms leading to changes in the molecular marker.

In most cases, a thorough literature analysis and/or data-mining approach will provide substantial information. The next step is a gap analysis that provides the basis for a qualification plan and then will map out which further in vitro and in vivo studies will be required. Experiments supporting a mechanistic qualification strategy may include, but are not limited to the assessment of dose dependency, time dependency, gene knockouts and knockdowns, and gene silencing. Among the three mechanistic qualification strategies, the weakest is solely relying on biostatistical evaluations, such as algorithms that are available in several current molecular marker discovery software packages. It must be kept in mind that most biostatistical methods establish associations and correlations, which may suggest cause–effect relationships but rarely prove them. Establishing cause–effect relationships between a drug or disease effect and a molecular marker is the core purpose of a robust mechanistic qualification strategy. Even if a molecular marker is discovered during clinical trials, this does not by any means establish that this marker is clinically relevant.

The next part of a molecular marker qualification is to show that a molecular marker is associated with the target disease process or drug effect in humans. In addition to sensitivity and specificity, a rigorous clinical qualification should also include the assessment of time and dose dependency. The extent and rigor of these studies ("evidentiary standards") will depend on the goal of the molecular marker qualification [161,168]. If a molecular marker is used as a clinical diagnostic tool or to support regulatory claims, studies must go beyond simply proof of concept in terms of statistical power considerations, documentation, monitoring, and regulatory compliance. Receiver operating characteristic (ROC) curves for the definition of sensitivity and specificity [36,169,170] are basic metrics to assess molecular marker performance [171]. In general, area under the ROC curves (AUC_{ROC}) ≤ 0.5 are considered not useful and indicate that the molecular marker cannot discriminate between treatment or disease and the control group [36,169,170]. While in the ROC analyses of preclinical animal studies, histology is often used as the gold standard endpoint; established outcome parameters are used in clinical trials. It is important for the quality of the ROC analysis that the reference outcome parameters are precise and nonbiased [171].

When appropriately qualified, molecular markers can support primary outcomes in a number of different ways. They may help to understand and monitor mechanisms of toxicity, drug–drug interactions, disease–drug interactions, and the effects of genotypes, gender, and age [162]. Molecular markers can also be used to stratify patient populations and guide subgroup analyses in such a manner as to bridge safety and efficacy data between different populations [161]. This becomes increasingly important when drugs are considered for use in pediatric populations, where it is more difficult to run the appropriate clinical trials. As such, the utilization of molecular markers with known correlations between adult and pediatric populations could provide an added measure of security when making this bridge.

The current status of regulatory biomarker approval, as well as an overview of biomarkers approved over the recent years by the FDA and European Medicines Agency (EMA) are provided by Amur et al. [163].

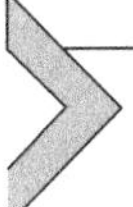

METABOLOMICS IN RENAL RESEARCH: KIDNEY FUNCTION, DISEASE, AND INJURY MARKERS

As aforementioned, one of the challenges in nephrology today is the limited set of established clinical diagnostic markers that are not very specific, are rather insensitive, and detect a disease process or negative drug effect at a later stage when the injury often cannot be fully reversed [172,173].

Historically, molecular markers have been established empirically, sometimes throughout years and decades of use in clinical practice and drug development. Incremental numbers of publications during their period of utilization have established their validity, as well as their limitations. Due to the absence of clear rules and guidelines, the qualification of a biomarker was mostly accepted based on consensus among clinicians/scientists, and between clinicians/scientists and regulatory agencies. It is reasonable to assume that many of today's established clinical markers would not meet the acceptance criteria and standards that are required by regulatory agencies and scientific consensus today in terms of qualification, sensitivity, and specificity.

Metabolomics holds the promise to serve as a potent tool to discover and develop new diagnostic strategies and to develop into a specific and sensitive diagnostic tool itself. In addition, metabolomics strategies can help to better understand the molecular mechanisms of disease processes and drug toxicities. This knowledge can be leveraged to develop new therapeutic approaches and better and safer drugs.

Identification of Disease and Pharmacodynamic Molecular Mechanisms

The problem with targeted research approaches to assess molecular mechanisms is that some information must already exist that allows for generation of a hypothesis. Another limitation is that the approach itself will bias the results; one will only find what one is looking for. It is often not possible to completely understand the results in context of the complex cause–effect relationships, correlations, and interactions of the biochemistry and signal transduction pathways of a cell, an organ, or an organism. Metabolomics alone, and even more so in combination with proteomics and genomics, is a hypothesis generator that when combined with molecular, cellular, and pharmacological techniques provides a framework for understanding molecular mechanisms [19]. These are critical tools for the mechanistic qualification of molecular markers. It also has to be taken into account that in most cases during today's drug development flow, a molecular target is identified and then often combinatorial chemistry compound libraries are screened to identify suitable molecules that interact with the target. This means that the mechanism of action is known almost from the beginning. However, toxicities are usually detected for the first time during preclinical animal toxicology studies or even later during clinical development. There is significant value in identifying the toxicodynamic mechanism for risk assessment, to evaluate if the toxicodynamic mechanism is linked to the pharmacodynamic target, and/or to identify molecular marker strategies for toxicodynamic monitoring during the preclinical and clinical phases of drug development.

The identification of the unknown molecular marker will have to incorporate a nonbiased, nontargeted screening strategy to generate a hypothesis that guides subsequent targeted studies and/or to ensure that no important unexpected effects are overlooked.

A powerful metabolomics tool to identify unknown molecular mechanisms is the assessment of fluxes in the metabolic network of a cell, organ, or organism [18,174]. This strategy is also termed "metabolic flux analysis (MFA)" or "fluxomics" [175]. It provides a true dynamic picture of the phenotype as it captures the metabolome in its functional interactions with the environment and the genome and provides a link [176]. MFA integrates in vivo measurements of metabolic fluxes with stoichiometric network models to allow the determination of absolute flux through large networks of the central carbon metabolism [175]. Although several methods for flux quantification are available, the most reliable strategies are still based on isotope-labeled precursors of metabolic pathways, mostly using ^{2}H- and ^{13}C-labeled substrates [18,174]. Depending on the metabolic pathway, the ^{2}H and ^{13}C atoms of the precursor are incorporated into the newly formed downstream metabolites in distinct numbers and specific positions. Each metabolite may have several isotope isomers, meaning molecules of the same metabolite with distinct labeling states, also known as isotopomers. Isotopomer distribution is assessed by metabolomics platform strategies, most importantly ^{13}C-NMR, GC–MS, isotope ratio mass spectrometers, and high-resolution mass spectrometers. The analysis of fluxes and the effects of disease and drugs on these using tracer-based metabolomic data requires a prior knowledge of the possible distribution of a tracer within the metabolic network [177]. But there are challenges. The size of the studied metabolic network should be restricted, otherwise too many alternative formation pathways will confound data interpretation [18]. As of today, most fluxome analyses have focused on the central carbon metabolism [175]. The number of usable labeled substrates is limited. Substrates that are formed by several alternative pathways may potentially dilute and confuse the analysis. The most widely used substrates are 1-^{13}C-, 1,2-^{13}C-, and uniformly labeled U-^{13}C glucose [18,177]. Software packages for the calculation and interpretation of fluxes, such as ^{13}C-FLUX2 and INCA have been developed [178–180]. Metabolic flux analysis and visualization workflows and software are discussed in more detail by Nöh et al. [181]

The following two examples illustrate how metabolomics can be used to gain further insights into disease mechanisms. Urine samples from patients with Fanconi's syndrome, healthy volunteers, and patients with tubular proteinuria were compared to assess the downstream molecular mechanisms in three genetic types of renal Fanconi's syndrome: Dent's disease, Loewe's syndrome, and autosomal dominant idiopathic forms. This was performed using a combined proteomics and metabolomics approach [182]. Like the protein patterns, cluster analysis grouped Loewe's and Dent's metabolomes together, whereas the autosomal dominant idiopathic forms and urines from ifosfamide-treated patients clustered together. The differences in the urine metabolomes were mainly due to different amino acid patterns (increased concentrations of basic and neutral, but not of branched amino acids in the case of Loewe's and Dent's disease) and differences in *N*-methyl nicotinic acid suggesting the involvement of cation transporters in the proximal tubule [182]. Taylor et al. [183] used a GC–TOF–MS–based metabolomics approach to partially qualify a juvenile mouse polycystic kidney disease model. Before there was serological evidence of kidney dysfunction, there were already marked changes in the urine metabolome. Functional score analysis, and the Kyoto Encyclopedia of Genes and Genomes (KEGG) pathway database suggested significant early changes in the purine and galactose metabolism pathways. The study also revealed several candidate molecular markers in urine, most notably allantoic acid and adenosine [183], which still need to be qualified in patients [184].

Drug Development and Nephrotoxicity

Today's most important diseases, including many kidney diseases, have a metabolic basis. In other words, they are caused by or are associated with metabolic changes [185]. Hence it is not surprising that many of the most effective and successful drugs are enzyme inhibitors. As of today, thousands of metabolite-inspired inhibitors or antimetabolites have been discovered. Moreover, if it is found that a disease is caused by metabolites, the reduction or increase of its intake may already provide a simple therapeutic option, often in the form of nutritional supplements or specific diets [185]. Metabolomics is also a potential strategy to reduce costly failures of drug candidates during the clinical stages of development by detecting toxic effects during the preclinical development. Metabolomics strategies to assess drug toxicity have been developed as early as in the 1980s with a focus mainly on hepato- and nephrotoxicity. Since then a large knowledge base has been developed [185–188]. Metabolomics approaches are useful to assess: [185,189]

- the target organ or region of toxicity,
- the biochemical mechanism contributing to toxicity,
- molecular marker profiles of nephrotoxicity in plasma or serum and urine, and
- the time course of nephrotoxicity, its dose dependency, and its recovery.

These efforts cumulated in the Consortium for Metabonomic Toxicology (COMET), a consortium of five major pharmaceutical companies and the Imperial College of London. The goal of the COMET study was to build expert systems and predictive models of target organ toxicity based upon ^{1}H-NMR spectra mainly observing renal and hepatic toxins. COMET led to the creation of a database of 35,000 NMR spectra with conventional histopathology data on mice and rats for 147 model toxins [28,42,185–189]. Based on the COMET database, Ebbels et al. [188] conducted an analysis including 12,935 NMR spectra from 1,652 rats that had received 80 different treatments to build a modeling system for toxicity prediction. Where predictions could be made, there was an error rate of 8%. The sensitivities to liver and kidney toxicity were 67% and 41%, respectively, whereas the corresponding specificities were 77% and 100%, respectively. In some cases, it was not possible to make predictions because of interference by drug-related metabolite signals (18%), an inconsistent histopathological or urinary response (11%), genuine class overlap (8%), or lack of similarity to any other treatment (2%). This study constituted the largest validation of the metabolomics approach to preclinical nephrotoxicity and in vivo drug toxicity screening. It confirmed earlier observations that the pattern changes of urinary metabolites can be used with good sensitivity and specificity to identify the nephrotoxic potential of compounds [188].

In a representative metabolomics-based nephrotoxicity study, rats were dosed with the nephrotoxins gentamicin, cisplatin, or tobramycin. A combined nontargeted GC–MS– and LC–MS–based metabolomics analysis showed that increases in urinary concentrations of polyamines and amino acids could be detected after the first dose before any histopathological changes occurred [190]. Upon prolonged exposure, a progressive loss of amino acids in urine were observed with a concomitant decrease of amino acid and nucleoside concentrations in the kidney tissue. A nephrotoxicity prediction model based on urinary concentrations of branched amino acids distinguished samples from rats treated with the nephrotoxins from vehicle controls with 70, 93, and 100% accuracy after 1, 5 ,and 28 days of treatment, respectively [190].

Metabolomics can also determine the site of nephrotoxicity (cortex or medulla) [191–193]. The following metabolite signatures in urine have been associated with injury to specific regions of the kidney: [189]

- Proximal straight tubules (via D-serine): increase of lactate, phenylalanine, tryptophan, tyrosine, and valine.
- Proximal convolute tubules (via gentamycin): increase of glucose; reduction of TMAO, xanthurenic acid, and kynurenic acid.
- Cortical injury (via mercuric chloride): increased glucose, alanine, valine, lactate, and hippurate; decreased citrate, succinate, and oxoglutarate.
- Papilla and medulla (via bromoethanamide): increase of glutaric acid, creatine, and adipic acid; reduction of citrate, succinate, oxoglutarate, and TMAO.

The changes of urine metabolite patterns found in several key nephrotoxicity studies in the rat are summarized in Table 3.4. The table interestingly indicates that under the experimental conditions used in these studies, the changes in urine metabolite patterns caused by these toxins or drugs is largely determined by the region of injury rather than by specific drug effects. It is also important to observe that results differed among studies using the same toxins. This may be explained by the use of different instrumentation, analytical strategies, different doses, time of sample collection relative to drug administration, length of treatment, the rat strains studied, diet, and the environment at the study site.

Table 3.4 Effects of selected nephrotoxin on metabolite patterns in the rat

Toxin	Location of injury (histology)	Crea	Citr	Succ	2-Oxo	Gluc	Hipp	TMAO	Tau	AA	Acet	Lac	Refs
Adriamycin	Glomerulus	↑	↓	—	↑↓	—	—	—	↑	—	—	—	[194]
Puromycin aminonucleoside	Glomerulus and proximal tubule	↑	↓	—	↓	↑	—	↑	↑	↑	↑	—	[186]
Sodium chromate	S1 proximal tubule	—	↓	—	↓	↑	↓	—	—	—	—	—	[194]
D-Serine[a]	S1 proximal tubule	↓	—	—	—	—	—	—	—	↑	—	↑	[195]
Gentamycin[b]	S1/S2 proximal tubule	←→	↓	↓	↓	↑	↓	↓	—	↑	↑	↑	[196]
DCVHC	S2/S3 proximal tubule	↓	↓	↓	↓	↑	↓	—	—	↑	↑	—	[194]
DCVC	S2/S3 proximal tubule	—	↑	↑	—	—	—	—	—	—	—	—	[194]
Hexachlorobutadiene	S3 proximal tubule	↓	↓	↓	↓	↑	↓	—	—	↑	↑	—	[194]
Mercuric chloride	S3 proximal tubule	↑	↓	↓	↓	↑	↓	↓	↑	↑	↑	↑	[197]
Mercuric chloride[c]	S3 proximal tubule	—	↓	↑	↓	↑	↓	↓	—	↑	↑	↑	[198]
Mercuric chloride	S3 proximal tubule	↓	↓	↑	↑	↑	↑↓	—	—	—	↑	—	[199]
p-Aminophenol	S3 proximal tubule	↓	↓	↓	↓	↑	↓	—	—	↑	↑	—	[186]
TCTFP	S3 proximal tubule	↓	↓	↓	↓	↑	↓	—	—	↑	↑	—	[194]
Uranyl nitrate	S3 proximal tubule	↓	↓	↓	↓	↑	↓	—	—	↑	↑	—	[200]
Cyclosporine[c,e]	S3 proximal tubule	↓	↓	←→	←→	↑	↓	—	—	—	↑	—	[201]
Ochratoxin[c]	S3 proximal tubule	↑	↓	—	↑↓	↑	—	←→	—	↑	—	↑	[202]
Cisplatin	S3 proximal tubule	—	—	—	↑	↑	—	↓	—	—	—	—	[203]
Cyclosporine[c,d,f]	Proximal tubule and medulla	—	—	↑	—	↑	—	↓	—	↑	↑	↑	[204]
Gentamycin[c,g]	Proximal tubule and medulla	—	↑	—	↑	—	—	↓	—	—	—	—	[205]
Gentamycin[c,h]	Proximal tubule and medulla	—	↑	—	—	↑	—	↓	—	—	—	↑	[205]

Doxorubicin	Medulla	↑	↓	↓	↓	↑	↓	↑	↑	↑	↑	—	[206]
Thioacetamide	Medulla	↓	↓	↓	↓	↑	↓	↑	↑	↑	↑	↑	[194]
2-Bromoethanamine	Medulla/papilla	—	↓	↓	↓	—	↓	↓	↑	↑	↑	↑	[207]
2-Bromoethanamine	Medulla/papilla	—	—	↑↓	—	—	—	↑↓	—	—	—	—	[194]
2-Chloroethanamine	Medulla/papilla	—	—	↑↓	—	—	—	↑↓	—	—	—	—	[186]

If not mentioned otherwise, metabolite patterns were assessed using ^{1}H-NMR spectroscopy–based metabolomics. Due to the large number of metabolites that can be captured with metabolomics technologies, only a selection of major metabolites can be shown here. For more details, please see the original references. Also the list of studies referenced and toxins that has been tested cannot be considered complete. AA, Amino acids; acet, acetate; crea, creatinine; citr, citrate; gluc, D-glucose; hipp, hippurate; lac, lactate; 2-oxo, 2-oxoglutarate; refs, references; succ, succinate; tau, taurine; TMAO, trimethyl amine-*N*-oxide.

[a] LC–MS–based metabolomics.

[b] GC–MS– and ^{1}H-NMR–based metabolomics.

[c] LC–MS– and ^{1}H-NMR–based metabolomics.

[d] GC–MS–, LC–MS–, and ^{1}H-NMR–based metabolomics.

[e] 10 mg/kg over 28 days.

[f] High dose of 45 mg/kg for 9 days.

[g] On day 3.

[h] On day 8–9.

Wishart [185] proposes a metabolomics-based drug discovery and development strategy, which is an attractive concept and has promising feasibility data. Nevertheless, as of today, metabolomics has not become a widely accepted drug development tool in the industry and for regulatory submissions. Instead an FDA- and EMA-approved panel of protein kidney injury molecular markers is gaining acceptance [208,209].

Acute Kidney Injury

As described earlier, metabolomics has extensively been used for studying acute drug–induced kidney injury in animal models. Nevertheless, there are still few clinical studies [210].

Beger et al. [211] studied the metabolic changes in serial urine samples in 40 children undergoing cardiopulmonary bypass. Twenty-one of these children developed acute kidney injury (AKI) defined as an increase of creatinine concentrations in serum 50% or greater from baseline after 48–72 h. The urine metabolite patterns were analyzed using U-HPLC–TOF–MS in the negative ionization mode. The urine metabolomes of children developing AKI were distinct, and further analysis showed that the dopamine metabolite, homovanillic acid sulfate, was a major molecular marker indicating AKI in this patient population. Using a cut-off value of 24 ng/μL at 12 h after surgery, a sensitivity of 90% and a specificity of 95% was found [211]. In a prospective targeted metabolomics study, Ujike-Omori et al. [212] assessed a potential association between urinary prostanoids and the development of AKI in 93 adult intensive care unit patients. Based on ROC analysis, the urinary 2,3-dinor-6-oxo-prostaglandin $F_{1\alpha}$/creatinine and 11-dehydro-thromboxane B2/creatinine ratios exhibited the best diagnostic and predictive performance in terms of subsequent onset of AKI and poor outcomes [212]. In a pilot study, Sun et al. [213] compared U-HPLC–TOF–MS serum metabolomics profiles in 17 hospitalized patients with newly diagnosed AKI to those of 13 age-matched subjects with normal kidney function. In the AKI patients, they found higher serum concentrations of acyl carnitines and amino acids [methionine, homocysteine, pyroglutamate, asymmetric dimethylarginine (ADMA), and phenylalanine] and lower serum concentrations of arginine and several lysophosphatidyl cholines. Increases in homocysteine and ADMA are also discussed as molecular markers of cardiovascular renal disease, and acyl carnitines are markers of defective fatty acid oxidation [213].

A list of metabolites reported as associated with renal disorders in metabolomics studies is shown by Barrios et al. [214]

Chronic Kidney Disease

Chronic kidney disease (CKD) is a heterogenous class of renal disorders, which all lead to the characteristic pathological hallmarks of interstitial fibrosis and inflammation [215]. Here clinical studies are discussed that evaluated the association of metabolomics profiles with CKD without specifically focusing on the underlying disease.

Several studies have explored the use of metabolomics profiling to identify individuals who are at risk to develop CKD in the following years. Based on 2 clinical trials with altogether 3995 samples, Goek et al. [216] studied the association between serum metabolite patterns and estimated GFRs using a metabolomics kit [38] in combination with tandem MS that measured 151 metabolites. In addition, 22,650 metabolite ratios were calculated. Of these, 22 metabolites and 516 ratios were statistically significantly associated with estimated GFRs. Acyl carnitines, especially glutaryl carnintine, showed the best correlation among single metabolites. Nevertheless, the ratio serine/glutaryl carnitine outperformed glutaryl carnitine alone. Rhee et al. [217] analyzed plasma from 1434 participants in the Framingham study using 3 distinct LC–MS/MS–based metabolomics assays. Over the next 8 years, 123 of these study subjects developed CKD. Nine metabolites were found to predict CKD, among which of most interest were citrulline and choline as markers of renal metabolism, and kynureic acid as marker of renal secretion. It was concluded that the combination of metabolomics profiles and clinical data may improve the prediction as to whether or not an individual will develop CKD. Nevertheless, the study also showed that the metabolomics profiles alone were insufficient as a predictor [217].

Yu et al. [218] studied the serum metabolomics profiles of 1921 African American individuals enrolled in the Atherosclerotic Risk in Communities (ARIC) study, including 204 CKD events with a median follow up of 19.6 years. Samples were analyzed using both untargeted GC–MS and LC–MS/MS. Forty named and 34 unnamed metabolites were found associated with estimated GFRs. Among these, 5-oxoproline and 1,5-anhydroglucitol were considered candidate risk factors for the development of CKD in the study population [218]. In a metabolome-wide association study of kidney function in the general population based on 2899 individuals, metabolomics profiles in serum were assessed using LC–MS/MS and GC–MS assays. The results showed that C-mannosyl tryptophan, pseudouridine, and O-sulfo-L-tyrosine were highly correlated with the annual change in estimated GFR, as well as the incidence of CKD, which occurred in 95 cases [219].

Another set of studies assessed metabolomic differences in healthy individuals and patients with different stages of CKD. Posada-Ayala et al. [220] took a two-stage approach. First ^{1}H-NMR was used to discover metabolites in urine that distinguished between CKD and healthy individuals. Seven such metabolites were found: 5-oxoproline, glutamate, guanidoacetate, α-phenyl acetyl glutamine, taurine, citrate, and TMAO. For these metabolites a targeted LC–MS/MS assay was developed. This panel could detect CKD with 80% sensitivity and 86% specificity [220]. A drawback of this study is the relatively small number of patients. Kobayashi et al. [221] identified 10 plasma metabolites associated with CKD and measured those with a targeted, quantitative LC–MS assay in 69 patients with different CKD stages. Based on the results, a multivariate regression equation was constructed that was found to predict CKD stage with 81.3% accuracy [221]. In plasma samples from patients with CKD stages 2, 3, and 4 (n = 10/stage), Shah et al. [222] assessed metabolite profiles using a combination of three U-HPLC–MS/MS and GC–MS assays. The plasma metabolite patterns between CKD stages were indeed found different. Stage 3 differed from stage 2 in 62 metabolites, stage 4 from stage 2 in 111 metabolites, and stage 4 from stage 3 in 11 metabolites [222]. In another study, Duranton et al. [223] collected samples from 77 patients focusing on amino acid patterns in plasma and urine. The amino acid profiles were assessed using a metabolomics kit [38] in combination with LC–MS/MS. Citrulline, ADMA, as well as tyrosine/phenylalanine and valine/glycine ratios were able to differentiate between different CKD stages [223].

Two recent comprehensive reviews by Zhao [224] and Breit and Weinberger [225] summarized urine, serum, and plasma metabolites that have been identified as potential molecular markers in CKD metabolomics studies. Overall, these metabolites include intermediates of dimethylarginine metabolism, tryptophan metabolism, NO synthesis, acyl carnitines, and oxidative stress markers [225]. In addition to the identification of single metabolites associated with GFRs, the value of multiparametric metabolite molecular marker panels has been explored. Breit and Weinberger [225] found that an aggregate molecular marker including symmetric dimethylarginine (SDMA), arginine, kynureine, tryptophan, and others when summarized into a weighted score, outperformed individual markers and ratios. This was suggested by better ROC curves for the differentiation between stage 3 and 4 CKD [225]. In their clinical study including 49 patients with different stages of CKD, Nkuipou-Kenfack et al. [226] assessed the performance of plasma and urine metabolites, as well as urinary peptides, all of which individually were found to statistically significantly correlate with estimated GFRs, summarized into 3 aggregate classifiers (plasma: $n = 17$ metabolites, urine: $n = 13$ metabolites, and urine: $n = 46$ peptides). Both metabolite classifiers (plasma and urine) and the peptide classifier (urine) were statistically significantly inversely correlated with estimated GFRs. However, combination of metabolite and peptide classifiers did not further improve the correlation with estimated GFRs [226].

Autosomal Polycystic Kidney Disease

Autosomal polycystic kidney disease (ADPKD) is one of the leading hereditary causes of kidney disease. It has become evident that estimated GFR is a poor predictor of ADPKD progression. However, early diagnosis and monitoring of ADPKD progression to start intervention at the best time point can be expected to slow the advancement toward end-stage renal disease [184]. The effect of the development of ADPKD on metabolite patterns in plasma, urine, and kidney tissues has been studied in mice and rat models [227–229]. In serum samples from 110 patients with different estimated GFRs and total kidney volumes, Klawitter et al. [230] studied the metabolite patterns using a portfolio of targeted LC–MS/MS assays in comparison with healthy individuals. ADPKD patients with eGFR > 60 mL/min per 1.73 m^2 showed higher levels of 5- and 12/15-lipoxygenase (LOX) and cyclooxygenase-generated hydroxy-octadecadienoic acids (9-HODE and 13-HODE) and HETEs (8-HETE, 11-HETE, 12-HETE, and 15-HETE) than healthy individuals. Serum concentrations of 9-HODE, and 13-HODE, as well as cytochrome P450–generated arachidonic acid metabolite, 20-HETE, correlated with estimated GFR and total kidney volume [230]. Moreover, ADPKD patients with eGFR > 60 mL/min per 1.73 m^2 showed higher serum concentrations of the cardiovascular disease risk markers ADMA and SDMA, homocysteine, and S-adenosyl homocysteine compared with the healthy controls. In addition, serum concentrations of prostaglandins (PG), including the oxidative stress marker 8-isoprostane, as well as $PGF_{2\alpha}$, PGD_2, and PGE_2 were markedly elevated in patients with ADPKD compared with healthy controls [231]. In a study in 91 pediatric ADPKD patients, plasma and urine samples were collected at baseline, 18 months and 36 months from 91 patients [232]. A targeted LC–MS/MS metabolomics strategy was used to assess the effect of treatment with the HMG-CoA reductase inhibitor, pravastatin, on disease progression. Changes in plasma concentrations of proinflammatory and oxidative stress markers, such as 9-HODE, 13-HODE, and 15-HETE over 3 years were significantly different between the placebo control and pravastatin-treated groups, with the pravastatin group showing a slower increase of these molecular markers. Urinary 8-HETE, 9-HETE, and 11-HETE concentrations were positively associated with the changes in height-corrected total kidney volume [232].

Diabetic Nephropathy

Diabetic nephropathy is caused by the chronic negative effects of diabetes on the renal microvasculature. Early diagnosis and management is critically important [233]. The effect of different stages of diabetic nephropathy on serum, plasma, and urine metabolite patterns has extensively been studied [234–249]. The associated metabolites that changed with the different stages of diabetic nephropathy were members of the following metabolic pathways: lipid metabolism (unsaturated and saturated fatty acids, carnitines, and phospholipids), amino acid metabolism, urea cycle, NO synthesis, nucleotide metabolism, and the Krebs cycle [233]. Several of these studies were designed to assess the value of metabolomics markers to predict and monitor progression of diabetic nephropathy [250]. Lipid markers, such as butenoylcarnitine, nonesterified fatty acids, phospholipids, very long ceramides, as well as various amino acids, such as histidine, glutamine, tyrosine, and tryptophan especially seem to be potential markers for the classification of patients with a high risk for progression [233,250]. Barrios et al. [214] and Zhang et al. [233] reviewed the status of metabolomics molecular marker research in diabetic nephropathy in detail.

Kidney Cancer

Renal cell carcinoma is generally asymptomatic at an early stage and is often discovered in an advanced state, when already metastatic [184]. Noninvasive, sensitive, and specific diagnostic tools will facilitate identifying renal cell carcinoma at an earlier stage [251]. In addition to ex vivo metabolomic analysis of body fluids or biopsies using NMR spectroscopy or MS, noninvasive MRI/magnetic resonance spectroscopy imaging (MRSI) and PET have been used to assess the metabolism of tumors in vivo. Metabolomics for the diagnosis of tumor and monitoring treatment is a promising concept as tumor metabolism markedly differs from the metabolism of normal cells. In addition to cancer tissue–specific metabolite changes, major differences include the energy metabolism with an upregulation of glycolysis and inhibition of the mitochondrial oxidation and the mitochondrial Krebs cycle, the so-called Warburg effect [252] that can be promoted by p53, hypoxia-inducible factor-1, c-Myc, Akt, and the mTOR pathway [253]. Choline metabolism is modulated by growth factor signaling, cytokines, oncogene activation, and chemical carcinogenesis [254,255]. This has been confirmed in metabolomics studies based on renal cell carcinoma tissue. The results could be explained by increased glucose uptake and increased fluxes through the glycolytic and pentose phosphate pathways, which was associated with an increase in upstream glycolytic intermediates, such as glucose-6-phosphate and fructose-6-phosphate, and a reduction in downstream intermediates, such as 3-phosphoglycerate, 2-phosphoglycerate, and phosphoenolpyruvate [256,257]. This is accompanied by the downregulation of the Krebs cycle, mitochondrial oxidative metabolism, and β-oxidation. The latter was found to be associated with increased concentrations of acylcarnitines. The tryptophan pathway was upregulated with higher concentrations of kynurenine and quinolinate in cancer than in normal renal tissue, while the other arm of tryptophan metabolism resulting in serotonin and indolacetate was downregulated [258]. In their MS-based metabolomics study of 138 human clear cell renal cell carcinoma/normal tissue pairs, Hakimi et al. [259] compiled integrated metabolic maps. This study also assessed the relationship between metabolic gene expression and metabolite patterns. Interestingly, this analysis revealed lack of linear correlation between transcriptomics and metabolomics. A large number of metabolic pathways exhibited reduced levels of gene expression, while metabolite levels were increased. This observation was particularly striking for glutathione metabolism [259].

A study in a mouse xenograft model suggested that serum metabolomics analysis is a more accurate surrogate for metabolome changes in renal cancer tissue than urine [260]. Nevertheless, as of today, most clinical metabolomics studies have focused on urine. To a certain extent the changes in urine metabolite patterns in patients with renal cell carcinoma reflect tumor cell metabolism including corresponding changes in glycolytic, energy, and amino acid metabolism, such as tryptophan metabolism [261]. Moreover, increased acyl carnitine concentrations in urine could differentiate patients with renal cell cancer from normal individuals, as well as different cancer status and grade [262]. Kim et al. [263] assessed the utility of urine metabolome profiling to detect renal cell carcinoma. In a clinical study, urine from 50 patients with renal cell carcinoma were collected and compared with urine samples from 13 healthy individuals. The urine metabolites were profiled using hydrophilic interaction chromatography–electrospray–linear iontrap MS. Mass spectra in the positive and negative mode were recorded. Data was analyzed using cluster, principal components, differential, and variance components analyses. Urine samples of patients with renal cell carcinoma could be differentiated from healthy individuals. Interestingly, it was found that these metabolome differences persisted after the tumor was removed. Although this study provided statistical proof-of-concept, it can only be considered a first step, as the metabolites responsible for the separation were not identified [263]. In a pilot study, Kind et al. [264] compared the urine metabolite patterns from six patients with clear cell renal cell carcinoma with those of the urine of six randomly selected healthy individuals using three independent analytical techniques: HILIC–LC–MS, reversed phase U-HPLC–MS, and GC–TOF–MS. The combination of these techniques covered a large part of the urine metabolome by enabling the detection of both lipophilic and hydrophilic metabolites. The results were analyzed by a feature selection algorithm with subsequent univariate analysis of variance and a multivariate partial least squares approach. From more than 2000 mass spectral features detected in the urine, several significant components were detected that enabled discrimination between urine samples from patients with renal cell carcinoma and controls despite the relatively small sample size. A feature selection process condensed the significant features to less than 30 components in each of the data sets. However, none of these metabolites were identified [264].

In serum samples from 33 patients with different stages of renal cell carcinoma and 25 healthy controls, Lin et al. [265] compared metabolite patterns after analysis with two LC–MS/MS assays, one based on HILIC and the other on reversed phase chromatography. The combined data of the two assays correctly differentiated patients with renal cell carcinoma and healthy controls with 100% sensitivity and specificity and correctly clustered patients with renal cell carcinoma according to the cancer stage. The metabolic changes in serum metabolite patterns from patients with renal cell carcinoma indicated disturbances of phospholipid catabolism, sphingolipid, phenylalanine, tryptophan, cholesterol, and arachidonic acid metabolism, as well as fatty acid β-oxidation and transport [265]. In another study, ^{1}H-NMR spectra in plasma samples from 32 renal cell carcinoma patients and 13 controls were recorded and analyzed using multivariate statistical techniques [266]. Different concentrations of LDL/VLDL, *N*-acetyl glycoproteins, lactate, and choline were observed between the groups of plasma samples. The authors concluded that the combination of ^{1}H-NMR spectroscopy and principal component analysis is a potential tool in cancer diagnosis; however, the major molecular markers identified, such as lipoproteins and choline, are not unique to renal cell carcinoma but may also be the result of other malignancies [266].

Renal biopsies have mainly been studied using magic angle spinning ^{1}H-NMR spectroscopy. As in the case of the analysis of biofluids and tissue homogenates, sample collection, preparations, and metabolic stabilities during the NMR experiment can cause artifacts and affect spectral quality [267]. In a study by Moka et al. [268], paired biopsy samples from the same kidney, one from a region of the renal cell carcinoma, the other from the a unchanged cortex region, were collected and immediately stored at −70°C. Samples weighing 80 mg were used for magic angle spinning NMR experiments including 1D–^{1}H-NMR spectroscopy, and 2D–J resolved, TOCSY, and ^{1}H-^{13}C HMQC experiments. The differences included a significant accumulation of lipids, as well as higher concentrations of *N*-acetyl neuramic acid, *N*-acetyl glucosamine, and of various amino acids in the samples from the tumor than in the samples from the unchanged cortex. These results were confirmed by a later study using a similar paired-sample study design and magic angle spinning ^{1}H-NMR spectroscopy [269]. Unsupervised and supervised statistical procedures distinguished between renal cell carcinoma samples and healthy cortex samples with 100% accuracy. This study also included a sample from a renal metastasis of a primary lung cell carcinoma and a sample of a renal collecting duct tumor. Both samples could clearly be differentiated from the cortical tumors based on their metabolite patterns [269]. A more recent study compared 1D–magic angle spinning ^{1}H-NMR spectra of clear cell and papillary renal cell carcinomas in comparison to normal renal cortex and papilla tissues, respectively [270]. The spectra of human normal cortex and medulla showed the presence of differently distributed organic osmolytes as markers of a physiological renal condition. As found in the earlier studies, the marked decrease or disappearance of osmolytes and the high-lipid content was typical for clear cell renal cell carcinoma tissues, while papillary renal cell carcinoma were characterized by the absence of lipids and very high amounts of taurine [270].

Metabolomics has extensively been used to assess the biochemical response of tumors during exposure to anticancer drugs [254,271]. Studies have mainly focused on cell and animal models. As aforementioned, there are also several metabolomics studies assessing the nephrotoxic effects of cancer drugs and the associated metabolome changes in plasma and urine (please also see Table 3.4). Relatively little data about the metabolic effects of renal cancer treatment is available, which is of interest to monitor treatment success. In serum samples from 121 patients with metastatic renal cell carcinoma, Jobard et al. [272] assessed metabolomic profiles using ^{1}H-NMR spectroscopy. Patients had received the mTOR inhibitor temsirolimus and bevacizumab in combination, and were compared to patients receiving standard therapy with either sunitinib alone or bevacizumab and interferon-α in combination. Serum samples were collected at baseline before treatment began and after 2 weeks and 5–6 weeks. Temsirolimus treatment resulted in a more rapid change of serum metabolite patterns consistent with mTOR inhibition and changes in renal cell carcinoma metabolism. Serum plasma concentrations that changed after 5–6 weeks of temsirolimus treatment included lipids, LDL, VLDL, end products of β-oxidation, glucose, and glutamine [272].

Kidney Transplantation

Although current immunosuppressive protocols have dramatically decreased acute rejection episodes after kidney transplantation, there has been only a relatively small improvement in long-term graft survival after kidney transplantation over the last 2 decades [273,274].

As of today, serum creatinine concentrations are routinely used as a clinical marker for monitoring function of kidney allografts [275]. Once an elevation in serum creatinine concentrations is detected, a biopsy is then procured to differentiate between the possible diagnoses. A Banff-graded, two-core allograft biopsy remains the gold standard with which all novel diagnostic tools must be compared. However, even biopsies will not necessarily allow for conclusive diagnosis of the etiology of the observed histopathological changes with sufficient confidence. Lesions, such as interstitial fibrosis and tubular atrophy, as well as glomerular injury are nonspecific responses to injury. Antibody-mediated endothelial activation, calcineurin inhibitor toxicity, recurrent disease, chronic inflammation, innate immune mechanisms, as well as diabetes mellitus and hypertension have all been invoked as potential etiologies [276,277]. Unfortunately, as aforementioned, serum creatinine is neither a specific nor a sensitive molecular marker. There is evidence that up to 30% of grafts with stable creatinine may have chronic/subclinical rejection [277]. The key to reducing chronic renal allograft dysfunction is early detection [278]. A common strategy to reduce the prevalence and severity of renal allograft dysfunction has been minimizing or discontinuing the doses of calcineurin inhibitors during long-term maintenance immunosuppression [275]. This is often performed without foreknowledge of which factors are contributing to chronic allograft dysfunction in any individual kidney transplant patient and without guidance by an appropriate diagnostic strategy. Overall, this frequently results in a reduction of the immunosuppressive efficacy of the drug regimen and creates a dilemma. As mentioned previously, another major factor contributing to renal allograft dysfunction is allograft immune response. Treatment to avoid damage by immunological responses requires enhanced immunosuppressive drug regimens. There is currently no noninvasive diagnostic tool available that allows for differentiating between renal allograft dysfunction due to alloimmune response or immunosuppressant toxicity [278]. The concept of monitoring biochemical changes and of detecting disease processes and immunosuppressant toxicity before significant histological or pathophysiological damage occurs and while said process is still potentially reversible, is attractive. Metabolomics can be applied toward: [7,278,279]

- assessment of transplant kidney quality before and during cold storage,
- monitoring of ischemia/reperfusion injury,
- toxicodynamic drug monitoring of immunosuppressants and individualization of immunosuppressive and other drug regimens, and
- detection of acute and chronic alloimmune reactions.

Organ Quality, Organ Storage, and Ischemia Reperfusion Injury of Kidney Transplants

Injury of a kidney transplant in the donor (in the case of cadaveric donors) during explantation, machine perfusion, cold ischemic storage, and reperfusion after transplantation may affect the extent of damage by oxidative stress, inflammation, and alloantigen-dependent factors. All of these damages may have a negative effect on outcomes of a kidney transplant. Mostly in animal models, ^{1}H-NMR–based metabolomics has been used to study the effects of donor treatment, explantation techniques, conditions during cold storage, machine perfusion, cold storage times, and of ischemia/reperfusion, as well as pharmacological prophylaxis against ischemia/reperfusion injury [280–290].

Bon et al. studied the time-dependent changes of metabolite patterns in explanted pig kidneys during hypothermic machine perfusion using ^{1}H-NMR spectroscopy [289]. During perfusion, the concentrations of lactate, choline, and of the amino acids valine, glycine, and glutamine increased in the perfusate with time, while the concentration of glutathione significantly decreased. There was a significant association between the concentration of these metabolites and function recovery after transplantation [289]. In a similar study, the metabolite patterns in perfusates from 26 transplanted cadaveric kidneys after hypothermic machine perfusion were assessed using ^{1}H-NMR [290]. Out of these, 73% of the kidneys had immediate and 27%

had delayed graft function. In the corresponding perfusates, concentrations of glucose were lower in those from kidneys with delayed graft function. Other differences between perfusate metabolite concentrations from kidneys with delayed and immediate graft function included inosine, leucine, and gluconate concentrations [290].

In rat kidney transplants, it was shown that after ischemia/reperfusion in kidney tissues, polyunsaturated fatty acids were decreased and allantoin, a known marker of oxidative stress, was increased [288]. At the same time, blood concentrations of TMAO and allantoin were significantly increased. Interestingly, no statistically significant changes in serum creatinine concentrations were found [288]. In 20 renal transplant recipients, HPLC was used to measure whole blood and plasma concentrations of adenosine triphosphate, adenosine monophosphate, guanosine, inosine, hypoxanthine, xanthine, uric acid, and uridine [291]. Hypoxanthine and xanthine concentrations were found increased in the renal allograft vein after reperfusion as compared with peripheral vein during the pre- and postreperfusion periods. The results suggested that differences in hypoxanthine and xanthine concentrations between renal and peripheral veins reflect metabolic alterations in renal tissue [291]. Overall, these studies indicate that metabolomics may be a valuable tool to study and monitor transplant kidney quality, cold ischemia conditions, and ischemia/reperfusion injury to assess the effect of pharmacological prophylaxis and interventions and to study the correlation with outcomes [279,289].

Immunosuppressant Nephrotoxicity

Although immunosuppressants have made organ transplantation possible, immunosuppressive drug regimens have serious side effects that not only may damage the transplant kidney, but may also limit patient survival. These include but are not limited to, an increased prevalence of cardiovascular disease, diabetes, neurotoxicity, cancer, and nephrotoxicity [276]. Nephrotoxicity of immunosuppressants is also a relevant problem for transplant patients who have received organs other than kidneys, and it has been shown that in these patients the development of immunosuppressant nephrotoxicity also negatively affects long-term outcomes [292]. Pharmacokinetic therapeutic drug monitoring and blood level–guided dosing of immunosuppressants is a common clinical practice; however, this strategy does not seem sufficient to prevent chronic nephrotoxicity [9]. Therefore, the concept of pharmacokinetic and toxicodynamic immunosuppressive drug monitoring seems attractive [293].

Several studies have focused on the effects of immunosuppressants alone and in combination on kidney tissue and the metabolite patterns in blood and urine (see also Table 3.4). Most of these studies have been purely descriptive and showed urine metabolite pattern changes typical for primary proximal tubular injury. But, a series of systematic studies also included mechanistic qualification to link urine and blood metabolome changes with toxicodynamic mechanisms of immunosuppressive drugs [9,201,294–297]. After treatment of rats with calcineurin inhibitors and their combination with sirolimus or mycophenolic acid (mycophenolate mofetil) for 28 days, GFRs were significantly reduced. The changes of metabolite patterns in urine were associated with a combination of changes in glomerular filtration, changes in secretion/absorption by tubule cells, and changes in kidney cell metabolism [201]. Based on these results, a combinatorial metabolite marker for monitoring immunosuppressant-induced kidney dysfunction in rats treated with calcineurin inhibitors was proposed [9,84]. Markers of glomerular filtration (creatinine), reabsorption (glucose), tubule cell metabolism (citrate, oxoglutarate, and lactate), active secretion and kidney amino acylase activity (hippurate), as well as oxidative stress (isoprostanes), and the release of metabolites that are protective against the protein-precipitating effect of uric acid (TMAO). An association between immunosuppressant-induced changes in kidney metabolism and urine metabolite patterns was confirmed by proteomics studies that were conducted to mechanistically explain and qualify the urinary metabolite pattern changes [296]. The changes in expression of specific enzymes compared to untreated controls explained several of the changes in metabolite patterns observed in urine. The extent of changes in GFRs after 28 days was predicted by the extent of metabolite pattern changes in urine after 6 days, even though GFRs at that time were not different from baseline, and histological changes were not detectable [201]. In this study after 6 days of treatment, urine metabolite patterns were similar to those reported for agents causing oxidative damage, while pattern changes after 28 days were typical for agents that cause S3 tubular damage [201]. These results matched the histologies showing specific damage of the proximal *tubuli*. After 28 days, there was also histological damage to *glomeruli*. These studies suggested the following mechanism causing the characteristic changes in urine metabolite patterns: calcineurin inhibitors directly and/or indirectly (via endothelial dysfunction) derail mitochondrial oxidation causing oxygen radical formation, inhibition of Krebs cycle, and decline of energy production. In a clinical open label, placebo-controlled cross-over study, the time-dependent toxicodynamic effects of a single-oral cyclosporine dose (5 mg/kg) on the kidney was assessed in 13 healthy individuals [294]. In plasma and urine samples, concentrations of 15-F_{2t}-isoprostane using LC–MS/MS and metabolite profiles using ^{1}H-NMR spectroscopy were analyzed. The increase in urinary 15-F_{2t}-isoprostane observed 4 h after administration of cyclosporine indicated an increase in oxidative stress. In average, 15-F_{2t}-isoprostaglandin concentrations were 2.9-fold higher after cyclosporine than after placebo. Unsupervised metabolome analysis using principal components analysis and partial least square fit analysis revealed significant changes in urine metabolites typically associated with negative effects on proximal tubulus cells. The major metabolites that differed between the 4-h urine samples after cyclosporine and the placebo were citrate, hippurate, lactate, TMAO, creatinine, and phenylalanine (please see Fig. 3.2). This indicated that analysis of urinary metabolites was a sensitive enough marker for detection of the effects of a single cyclosporine dose shortly after drug administration and that the results in rats translate into at least healthy humans. Creatinine concentrations in serum remained unchanged [294]. A decrease in citrate concentrations in urine during treatment with immunosuppressants had also been reported by others [298].

The results of the study by Klawitter et al. [201] also suggested that changes in urine metabolite patterns reflected the negative effects of immunosuppressants on kidneys with better sensitivity and specificity than metabolite changes in blood. Other studies have concluded that immunosuppressants alone and in combination lead to marked changes of metabolite patterns in the blood of rats [299] and transplant patients [300–302].

Alloimmune Reactions

Foxall et al. studied the changes in urine metabolite spectra early after transplantation [303]. In this study no patient showed clinical or histopathological evidence of cyclosporine nephrotoxicity. Urine samples were collected daily for 14 days from 33 patients who underwent primary renal allograft transplantation, and were analyzed by 500 and/or 600 MHz ^{1}H-NMR spectroscopy. The NMR spectra of urine from patients with immediate functioning grafts were similar with respect to their patterns of amino acids, organic acids, and organic amines, whereas the patients with delayed or nonfunctioning grafts showed significantly different metabolite excretion patterns. In longitudinal studies on individual patients, there were increased urinary levels of TMAO, dimethylamine, lactate, acetate, succinate, glycine, and alanine during episodes of graft dysfunction. However, only the urinary concentration of TMAO was statistically significantly higher in the urine collected from patients during episodes of graft dysfunction (410 ± 102 μM of TMAO/mM creatinine) than in patients with good graft function (91 ± 18 μM of TMAO/mM creatinine) or healthy control subjects (100 ± 50 μM of TMAO/mM creatinine). These findings suggested that early graft dysfunction is associated with damage to the renal medulla, which causes the release of TMAO into the urine from the damaged renal medullary cells [303]. Urine and plasma samples from 39 patients, who underwent renal transplantation, were analyzed by ^{1}H-NMR spectroscopy [304]. In this study, Le Moyec et al. found that the most relevant ^{1}H-NMR signals for evaluating renal function after transplantation were those arising from citrate, TMAO, alanine, and lactate when compared to creatinine. The respective variations of these metabolites in urine were associated with cyclosporine toxicity and rejection [304]. Knoflach and Binswanger [305] reported that hippuric acid concentrations in plasma may be a sensitive and early marker of acute allograft rejection, but may also be a marker for the response to antirejection treatment. The classification of urine metabolite spectra from 33 kidney transplant patients with normal histology and from 35 patients with rejection, as confirmed by Banff-graded protocol biopsies taken shortly after the collection of the urine samples, resulted in 96.3% sensitivity and 93.1% specificity when samples were analyzed using ^{1}H-NMR spectroscopy and in 96.2% specificity and 88.9% specificity when samples were analyzed with infrared spectroscopy [306]. Wang et al. analyzed the metabolite patterns of 15 midstream urine samples from patients with acute cellular rejection of a kidney allograft and 24 urine samples from 8 patients without evidence of rejection using MALDI–FTMS [307]. Seven molecules with mass/charge ratios between 278 and 424 were identified that differentiated the two sets of urine samples with 100% specificity. However, the molecular structures of these molecules were not further identified [307].

In a more recent clinical study, Blydt-Hansen [308] evaluated the potential value of urinary metabolomics for the noninvasive diagnosis of T-cell mediated rejection in pediatric kidney transplant patients. Quantitative LC–MS/MS was used to measure 134 metabolites in 277 samples from 57 patients with corresponding kidney biopsies. Partial least squares discriminant analysis identified distinct classifiers for T-cell mediated rejections, which significantly correlated with Banff classification scores. The most important metabolites contributing to said T-cell mediated rejection discriminant scores were phosphatidylcholines and the amino acids kynureine, proline, sarcosine, methionine sulfoxide, threonine, glutamine, phenylalanine, and alanine. While the latter study focused on urine, Zhao et al. [309] assessed the serum metabolome in 11 subjects with acute graft rejection. The results were compared to the metabolite patterns in 16 serum samples from patients with nonacute graft rejection. Samples were analyzed using a nontargeted metabolomics approach based on a combination of HILIC and reversed phase chromatography U-HPLC–TOF–MS. Metabolites that allowed for discrimination of acute and nonacute graft rejection included creatinine, kynurenine, uric acid, polyunsaturated fatty acids, phosphatidyl cholines, sphingomyelines, and lysophasphatidylcholines. Moreover, Klepacki et al. [310] used a targeted quantitative and validated LC–MS/MS assay and showed that the transmethylation pathway intermediates, S-adenosyl methionine and S-adenosyl homocysteine, were significantly elevated in plasma of kidney transplant patients compared to healthy individuals, and increased prior to and during acute rejection episodes. The same group also showed that amino acid patterns in plasma of de novo kidney transplant patients, especially the concentrations of hydroxyproline and L-methyl histidine, were significantly correlated with estimated GFRs when analyzed using a targeted and nontargeted LC–MS/MS metabolomics approach [161]. The results in this study were comparable to those previously described for CKD patients [224].

Urine metabolomics has also been used to study the graft recovery process after kidney transplantation [84,311,312]. After ^{1}H-NMR analysis of at least 9 urine samples from each of the 15 kidney transplant patients enrolled in the study, principal component analysis could differentiate 3 stages after kidney transplantation: initial dysfunction, recovery of functions, and stable function during follow-up. In another pilot study based on a targeted metabolomics, LC–MS/MS assay of urine samples longitudinally collected from 9 pediatric kidney transplant patients and 36 healthy controls, it was shown that glucose, sorbitol, and TMAO concentrations were significantly higher in transplant patients early after transplantation, but that urinary metabolite patterns matched those of healthy controls 1 month after transplantation [84].

Bonneau et al. [313] summarized the metabolite changes associated with kidney transplantation discussed earlier in a comprehensive overview table. Although metabolomics seems to be a promising concept to provide reliable indications of transplant kidney function, injury, and immunosuppressant toxicity, metabolomics-based diagnostic and monitoring strategies have not yet been implemented into clinical practice [293,314].

Metabolomics in the Comparison of Immunosuppressive Drug Regimens

Metabolomics strategies have also been used to compare immunosuppressive drug regimens in animal studies and clinical trials.

Enhancement of calcineurin inhibitor nephrotoxicity by sirolimus is limiting the clinical use of this drug combination. Bohra et al. [315] compared the dose-dependent effects of the structurally related everolimus and sirolimus alone and in combination with cyclosporine on the rat kidney over 28 days of oral treatment with different dose combinations. ^{1}H-NMR and GC–MS were used to assess metabolite patterns in urine. The combination of cyclosporine with sirolimus led to higher urinary glucose concentrations and decreased levels of urinary Krebs cycle metabolites when compared to controls, suggesting that cyclosporine + sirolimus negatively impacted proximal tubule metabolism. Unsupervised principal component analysis of ^{1}H-NMR spectra distinguished unique urine metabolite patterns of rats treated with cyclosporine + sirolimus from those treated with cyclosporine + everolimus and the controls. Sirolimus, but not everolimus, blood concentrations were inversely correlated with urine Krebs cycle metabolite concentrations, indicating distinct effects of these structurally related drugs in combination with cyclosporine on the kidney [315].

The immunosuppressive calcineurin inhibitors tacrolimus and cyclosporine have similar toxicology profiles including nephrotoxicity. Kim et al. [312] compared the serum metabolomes of 27 de novo kidney transplant patients receiving a cyclosporine-based drug regimen with those of 30 de novo kidney transplant patients receiving a tacrolimus-based immunosuppressive drug regimen. All patients also received mycophenolate mofetil and corticosteroids. Serum metabolite patterns were assessed using ^{1}H-NMR spectroscopy. The study showed that serum metabolite patterns of cyclosporine- and tacrolimus-treated patients differed in lipid, glucose, hypoxanthine, lactate, succinate, and taurine concentrations. On the other hand, TMAO concentrations potentially associated with kidney graft dysfunction were not different during the observation period of 6 months after transplantation [312]. In a clinical multicenter trial, plasma and urine metabolomes between de novo kidney transplant patients in the two treatment arms: everolimus + low-dose tacrolimus and mycophenolate mofetil + standard dose tacrolimus, were compared [316]. In this molecular marker substudy, samples were collected longitudinally (baseline, 1, 2, 4, and 6 months after transplantation) from 120 patients evenly distributed between the treatment arms. A targeted/nontargeted metabolomics approach based on a set of quantitative multianalyte LC–MS/MS assays was used. There were no significant differences in any of the more than 600 evaluated metabolites in plasma or urine between the two treatment arms, including known inflammation molecular markers, plasma vascular endothelial dysfunction markers, amino acid profiles, oxidative stress markers, fatty acid patterns in enriched blood cells, and urine metabolite kidney function markers, as well as plasma and urine metabolomics profiles. The only exception were metabolic effects directly associated with the mycophenolic acid and everolimus mechanisms of action [316].

Urine as a Matrix for Nonrenal Disease and Injury

As most small molecules can pass the glomerular membrane and are often concentrated by the kidney, systemic metabolite changes in blood are reflected in urine, at least to a certain extent. Urine metabolite patterns can be affected by the gut microbiolome, diet, and other factors [15] (Fig. 3.1). This also explains why several metabolomics studies reported kidney disease–dependent gut microbiolome metabolite concentration changes in urine. Analysis of the urine metabolome has been used to study the effects of diet on human biochemistry [317] and the biochemistry of term and preterm neonates [318–322]. The monitoring of inborn metabolic errors is an established clinical procedure that is based on the profiling of urine metabolites [67,321].

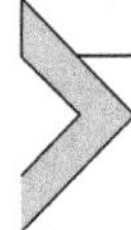

METABOLOMICS AS A CLINICAL DIAGNOSTIC TOOL IN NEPHROLOGY

Challenges

While metabolic marker discovery is relatively easy and fast, the process of actually making them useful and clinical implementation is much slower, trickier, and requires substantial resources [323]. It has also to be taken into account that, although untargeted metabolomics is considered to still be in its infancy, there is evidence that the complexity of even the cellular metabolome exceeds that expected based on classical biochemical pathways [26]. It appears that the metabolome is much larger than anticipated and in several metabolomics studies the number of "unknown" metabolites exceeded that of known metabolites of interest. Molecular markers can be potentially misleading if they correlate with a disease process or drug effect, but there is no cause–effect relationship. Hence they do not necessarily reflect mechanistically relevant changes. It has to be considered that many biostatistical methods used for molecular marker discovery are based on chemometric algorithms, which establish statistical associations but hardly ever establish cause–effect relationships. Other typical problems with these procedures and study designs include, but are not limited to the following points.

- Due to the complexity of the analyses, studies often have relatively small numbers of observations, while hundreds and thousands of parameters are measured and compared.
- The quality of the data entered into such analyses is often poorly controlled, if at all. This is especially relevant if data has been collected and analyzed over a period of time and/or by different sites and laboratories.
- As the metabolome is an open system, differences in diet, exercise, and environment need to be taken into account as potential confounders and, if not appropriately controlled, may increase the risk of false positives even further.
- The lack of a community-wide, consensus-based, human- and machine-interpretable language for describing their phenotypes and genomic and environmental contexts has been identified as a major obstacle.

Computational and mathematical techniques, such as multivariate analysis or machine learning, can find differences or clusters of differences that distinguish members of one group from the other, at least for a specific sample set. Given enough variables these algorithms may discriminate groups by chance, with sometimes misleading impressive statistical significance [323].

It is impossible to decide if a molecular marker is a valid surrogate of a disease or drug effect if the pathophysiological and pathobiochemical links are not understood [323]. The key to establishing this link is mechanistic qualification. A widely recognized problem is the translation of molecular markers from animal models to humans and vice versa. Fortunately, this is less of a problem with metabolites than with proteins or genes. Unlike genes and proteins, metabolites are often tissue- and species-independent [41].

A molecular marker solely based on a metabolic fingerprint without any attempt to understand the underlying molecular changes cannot be mechanistically qualified and thus may not meet current regulatory standards. Another problem with fingerprinting is that validation of such assays following current laboratory standard practices and regulatory requirements is difficult [324]. Therefore, it is reasonable to assume that such a strategy is a powerful discovery tool but, at least for the next few years, will remain clinically irrelevant.

In terms of molecular marker discovery, qualification, and determination of sensitivity and specificity, it is critical to consider the time dependency of biochemical changes. While the genome is static, the proteome and metabolome are in constant flux [325]. In the later stages of a kidney injury, the biochemical signature often remains unchanged, but during the earlier stages, cell and organ biochemistry may change quickly as the injury progresses. This may include compensatory mechanisms, the onset of secondary reactions, such as oxygen radical formation and damage, changes in cell function and regulation, and the onset of additional systemic processes, such as immune reactions and inflammation. Different stages during the development of a biochemical injury may be characterized by different sets of metabolite markers, and thus time dependency and its underlying mechanistic dynamics need to be understood. A good example is the aforementioned study by Klawitter et al. [201] that showed that cyclosporine caused rat urine metabolite changes consistent with oxidative stress of the kidney during the first 6 days, the primary mechanism through which cyclosporine causes nephrotoxicity. But after 28 days of cyclosporine exposure, the urine metabolites had shifted to a pattern typical for S3 tubular damage. Accordingly, the correct timing of sample collection is critical for the success of molecular marker development and the later use of a specific molecular marker in clinical trials and as a clinical diagnostic tool. Although there is great potential in understanding such time-dependent metabolic changes and patterns, this important aspect has often not been taken into account and systematically been explored in terms of prediction, early detection and monitoring disease progression, and treatment response and organ recovery.

Interestingly, most original publications in this field are limited to describing the metabolite pattern changes and a fair number of these manuscripts conclude that the discovered metabolite changes may be useful as diagnostic markers. However, with a very few exceptions, investigators have followed up on their results and have taken steps toward mechanistic, clinical qualification, and implementation. As suggested by Table 3.4., the urinary metabolite pattern changes associated with proximal tubular and medullary kidney damage are fairly consistent and as discussed in section "Metabolomics in Renal Research: Kidney Function, Disease and Injury Markers." Several specific metabolites are consistently found as potential markers across metabolomics studies of kidney diseases (please see also Breit and Weinberger [226]). Although similar urinary metabolite pattern changes have been described over and over again, to the best of our knowledge, never has a systematic approach been made to explain why this is the case and to explore the mechanistic reasons for these changes, nor has there been an attempt to further develop these targeted metabolomics markers or marker panels toward clinical implementation.

Even fundamental issues, such as normalization of metabolite concentrations to compensate for differences of dilution in urine samples, have not systematically been approached. There is consensus that the use of creatinine concentrations in urine for this purpose can be misleading in certain cases and even though this is a critical and fundamental problem, normalization based on creatinine remains the clinical standard.

Overall and in comparison to urinary protein kidney dysfunction markers [209,210], there has been very little systematic effort yet to develop metabolite molecular markers into clinical diagnostic tools. Most of the work has focused on the development of metabolomics strategies for preclinical drug development. This is reflected by the literature with the majority of data published in animal models and yet only a relatively few publications in humans. The advantages, opportunities, and risks of metabolomics as a clinical diagnostic tool or for the development of diagnostic tools were discussed by Miller [326] and are summarized in Table 3.5.

It has been argued that while transcriptomics and proteomics are important research tools, metabolic profiling will offer the greatest impact on the field of personalized health and as an outcomes parameter [5] (Fig. 3.5). One reason is that metabolomics reflects best, the interaction between phenotype and environment (Fig. 3.1).

Metabolomics allows for a global view of an individual's metabolome and its interactions with the microbiolome, the environment, drugs, and disease agents. Profiling the whole metabolome and to extract relevant information using chemometrics has been referred to as a "top-down" approach. Although intriguing, the amount of information generated by modern nontargeted screening technologies is often clinically irrelevant, impractical, and can only be meaningful if it assists in drawing clear and valid conclusions. Using truly nontargeted screening technologies in clinical decision making is not yet feasible, mostly because of the complexity of the data generated and, as discussed earlier, the lack of algorithms to convert this information into robust and meaningful clinical information. Another problem is that most of the hundreds and thousands of data points generated are not relevant to a specific disease or drug effect. Instead of conveying additional information, they only cause random statistical noise including false-positive results and may mask valid information. However, while nontargeted omics technologies are mostly hypothesis-generating technologies, this information is valuable to develop new targeted diagnostic strategies and tools [327].

Table 3.5 Strengths, weaknesses, opportunities, and threats of metabolomics as a clinical diagnostic tool and as a tool to discover and develop new diagnostic strategies and in drug development [326]

Strengths	**Weaknesses**
Comprehensive profile of the entire metabolome hypothesis generation and identification of unknown molecular mechanisms	Lack of databases with a comprehensive information for metabolite identification; many unknown metabolites are still not identified
Small molecules in the metabolome are the ultimate manifestation of cellular genomic and proteomic signaling	Lack of software for automated identification and quantitation; currently available chemometric approaches are still based on many assumptions and the analysis is vulnerable to false-positive results
Analytical high-throughput screening technologies, such as NMR spectroscopy and MS, for metabolite measurement are already in place	Minor and potentially toxicologically important metabolites may be overlooked
Development of molecular markers of effect not just exposure	Metabolic cause of toxicity and consequence of damage can be difficult to distinguish
Ability to define normal ranges	Dealing with data where "normal" encompasses a wide range
Due to the fast response, early and time-dependent changes can be monitored	Current analytical technologies are probably unable to cover the complete metabolome
Opportunities	**Threats**
Predictive molecular markers of metabolomic disruption, drug effects, and the development of disease states	Will the technology live up to its promise and result in deliverables?
Metabolomic profiles are an open system and can assess and predict interactions between humans and their environment	What to do with the data?
New mechanistic insights from discovery-driven research	Technological ability to detect metabolomic changes but what do changes mean?
Use of genetically modified models of disease to understand metabolic mechanisms	What is a "normal" metabolome for humans?
Use of high-throughput systems to rapidly generate vast amounts of data	When is the "normal" metabolome perturbed in a manner that is consequential?
Technology complementary to and integrated with genomics and proteomics (→ systems biology)	How will regulatory agencies use the data?
Personalized "susceptibility" index	How to implement metabolomics in clinical laboratories and how to control quality?

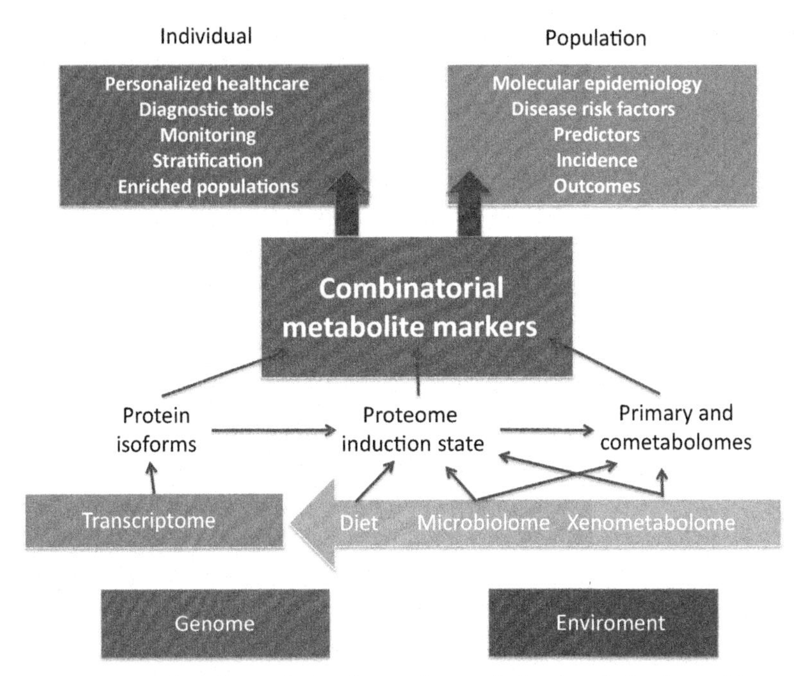

Figure 3.5 The role of metabolomics and metabolomics-derived combinatorial metabolite markers for individualized medicine and molecular epidemiology.

Outlook: Metabolomics-Based Phenotyping and Precision Medicine

The fundamental concept of personalized medicine is the combination of established clinical–pathological indices with state-of-the art molecular profiling to devise predictive, prognostic, diagnostic, therapeutic, and clinical monitoring strategies that precisely meet each individual patient's requirements, also referred to as "precision medicine" [328].

As already discussed earlier, it is reasonable to expect that metabolomics will play a key role in precision medicine, mainly for two reasons. The "metabotypes" of individuals can be analyzed in biofluids that can easily and minimally invasively be collected. Moreover, metabolomics is an interface between the individual and the environment, diet, lifestyle, and gut microbes [329] (Fig. 3.5). This has also been referred to as "phenomics" [330]. While the cell metabolome is determined by the gene–protein–metabolite interactions, the phenome is an aggregate of the cell metabolome as reflected in the body fluids, microbiolome, and xenobiolome. Hence it constitutes an interface between the gene–environment interactions.

Deconvolution of the macro- and microenvironmental contributions in phenomics may resolve the subtleties that explain the basis for why humans of the same relevant genotype exposed to similar environmental conditions respond differently, for example, to explain why an individual patient responds to a nephrotoxin, while another exposed to exactly the same dose over the same time period does not. Thus phenomics has applications in both population-based disease-risk investigations, can solve problems related to personalized healthcare, and allows for patient risk and treatment response stratification [329,330].

As of today there are approximately 190 clinically approved metabolite molecular markers in the United States, most of which require individual assays [186]. In comparison, metabolomics offers more molecular marker options as all of today's approved metabolite molecular markers combined in a single analytical run. In addition, clinical implementation of metabolomics assays is a realistic option, as the required instrumentation (LC–MS/MS) is nowadays robust, quantitative, flexible, and often already available in many clinical laboratories [186]. LC–MS/MS assays can be validated and automated. Nevertheless, successful comprehensive clinical implementation of metabolomics, metabotyping, or phenomics as tools of precision medicine will first require a shift toward a deeper understanding of disease based on molecular biology and will also inevitably require reclassification of disease states incorporating this knowledge [328]. Moreover, precision medicine will require handling of multiparametric data, as well as proficiency in interpreting complex data [328]. Taking full advantage of these new technologies and concepts will require training and education of medical professionals to deal with the anticipated complexity and volume of new information, as well as effective clinical decision support tools and new educational models.

Metabolomics, metabotyping, and phenomics are promising concepts that can provide a critical component of precision medicine strategies and have the potential to revolutionize nephrology and medicine in general. Yet, as discussed in this chapter, it still has a long way to go until successful comprehensive clinical implementation and acceptance [329,330].

REFERENCES

[1] Nicholson JK, Lindon JC. Metabonomics. Nature 2008;455:1054–6.
[2] Neild GH, Foxall PJD, Lindon JC, et al. Uroscopy in the 21st century: high-field NMR spectroscopy. Nephrol Dial Transplant 1997;12:404–17.
[3] Nicholson JK, Lindon LC, Holmes E. "Metabonomics": understanding the metabolic response of living systems to pathophysiological stimuli via multivariate statistical analysis of biological NMR spectroscopic data. Xenobiotica 1999;11:1181–9.

[4] Holmes E, Wilson ID, Nicholson JK. Metabolic phenotyping in health and disease. Cell 2008;134:714–7.
[5] Nicholson JK. Global systems biology, personalized medicine and molecular epidemiology. Mol Syst Biol 2006;2:52.
[6] Biomarkers Definition Working Group. Biomarkers and surrogate endpoints: preferred definitions and conceptual framework. Clin Pharmacol Ther 2001;69:89–95.
[7] Wishart DS. Metabolomics: the principles and potential applications to transplantation. Am J Transplant 2005;5:2814–20.
[8] Rosner MH. Urinary biomarkers for the detection of renal injury. Adv Clin Chem 2009;49:73–97.
[9] Christians U, Klawitter J, Bendrick-Peart J, et al. Toxicodynamic therapeutic drug monitoring of immunosuppressants: promises, reality and challenges. Ther Drug Monit 2008;30:151–8.
[10] Devarajan P. Emerging urinary biomarkers in the diagnosis of acute kidney injury. Expert Opin Med Diagn 2008;2:387–98.
[11] Schnackenberg LK. Global metabolic profiling and its role in systems biology to advance personalized medicine in the 21st century. Expert Rev Mol Diagn 2007;7:247–59.
[12] Everett JR. Pharmacometabonomics in humans: a new tool for personalized medicine. Pharmacogenomics 2015;16:737–54.
[13] Wikoff WR, Anfora AT, Liu J, Schultz PG, Lesley SA, Peters EC, Siuzdak G. Metabolomics analysis reveals large effects of gut microflora on mammalian blood metabolites. Proc Natl Acad Sci USA 2009;106:3698–703.
[14] Scalbert A, Brennan L, Manach C, Andres-Lacueva C, Dragsted LO, Draper J, Rappaport SM, van der Hooft JJ, Wishart DS. The food metabolome: a window over dietary exposure. Am J Clin Nutr 2014;99:1286–308.
[15] Nicholson JK, Wilson ID. Understanding global systems biology: metabonomics and the continuum of metabolism. Nature Rev Drug Discov 2003;2:668–76.
[16] Psychogios N, Hau DD, Peng J, Guo AC, Mandal R, Bouatra S, Sinelnikov I, Krishnamurthy R, Eisner R, Gautam B, Young N, Xia J, Knox C, Dong E, Huang P, Hollander Z, Pedersen TL, Smith SR, Bamforth F, Greiner R, McManus B, Newman JW, Goodfriend T, Wishart DS. The human serum metabolome. PLoS One 2011;6:e16957.
[17] Bouatra S, Aziat F, Mandal R, Guo AC, Wilson MR, Knox C, Zjorndahl TC, Krishnamurthy R, Saleem F, Liu P, Dame ZT, Poelzer J, Huynh J, Yallou FS, Psychogios N, Dong E, Bogumil R, Roehring C, Wishart DS. The human urine metabolome. PLoS One 2013;8:e73076.
[18] Zamboni N, Saghatelian A, Patti GJ. Defining the metabolome: size, flux, and regulation. Mol Cell 2015;58:699–706.
[19] Griffiths WJ, Wang Y. Mass spectrometry: from proteomics to metabolomics and lipidomics. Chem Soc Rev 2009;38:1882–96.
[20] German JB, Gillies LA, Smilowitz JT, et al. Lipidomics and lipid profiling in metabolomics. Curr Opin Lipidol 2007;18:66–71.
[21] Wang M, Wang C, Han RH, Han X. Novel advances in shotgun lipidomics for biology and medicine. Prog Lipid Res 2016;61:83–108.
[22] Niemann CU, Serkova NJ. Biochemical mechanisms of nephrotoxicity: application for metabolomics. Expert Opin Drug Metab Toxicol 2007;3:527–44.
[23] Burckhardt BC, Burckhardt G. Transport of organic anions across the basolateral membrane of proximal tubule cells. Rev Physiol Biochem Pharmacol 2003;146:95–158.
[24] Lalowski M, Magni F, Mainini V, Monogioudi E, Gotsopoulos A, Soliymani R, Chinello C, Baumann M. Imaging mass spectrometry: a new tool for kidney disease investigations. Nephrol Dial Transplant 2013;28:1648–56.
[25] Theodoridis GA, Gika HG, Want EJ, Wilson ID. Liquid chromatography-mass spectrometry based global metabolite profiling: a review. Anal Chim Acta 2012;711:7–16.

[26] Patti GJ, Yanes O, Siuzdak G. Innovation: Metabolomics: the apogee of the omics trilogy. Nat Rev Mol Cell Biol 2012;13:263–9.
[27] Christians U, Klawitter J, Hornberger A, Klawitter J. How unbiased is non-targeted metabolomics and is targeted pathway screening the solution? Curr Pharm Biotechnol 2011;12:1053–66.
[28] Coen M, Holmes E, Lindon JC, Nicholson JK. NMR-based metabolic profiling and metabonomic approaches to problems in molecular toxicology. Chem Res Toxicol 2008;21:9–27.
[29] Oldiges M, Lütz S, Pflug S, Schroer K, Stein N, Wiendahl C. Metabolomics: current state and evolving methodologies and tools. Appl Microbiol Biotechnol 2007;76:495–511.
[30] Xia J, Bjorndahl TC, Tang P, Wishart DS. MetaboMiner—semi-automated identification of metabolites from 2D NMR spectra of complex biofluids. BMC Bioinformatics 2008;28:507.
[31] Almstetter MF, Appel IJ, Gruber MA, et al. Integrative normalization and comparative analysis for metabolic fingerprinting by comprehensive two-dimensional gas chromatography-time-of-flight mass spectrometry. Anal Chem 2009;81:5731–9.
[32] Nicholson JK, Connelly J, Lindon JC, Holmes E. Metabonomics: a platform for studying drug toxicity and gene function. Nat Rev Drug Discov 2002;1:153–61.
[33] Matsuda F, Shinbo Y, Oikawa A, Hirai MY, Fiehn O, Kanaya S, Saito K. Assessment of metabolome annotation quality: a method for evaluating the false discovery rate of elemental composition searches. PLoS One 2009;16:e7490.
[34] Dettmer K, Aronov PA, Hammock BD. Mass spectrometry-based metabolomics. Mass Spectrom Rev 2007;26:51–78.
[35] Wishart DS. Computational approaches to metabolomics. Methods Mol Biol 2010;593:283–313.
[36] Alonso A, Marsal S, Julià A. Analytical methods in untargeted metabolomics: state of the art in 2015. Front Bioeng Biotechnol 2015;3:23.
[37] Yuan M, Breitkopf SB, Yang X, Asara JM. A positive/negative ion-switching, targeted mass spectrometry-based metabolomics platform for bodily fluids, cells, and fresh and fixed tissue. Nat Protoc 2012;7:872–81.
[38] Koal T, Klavins K, Seppi D, Kemmler G, Humpel C. Sphingomyelin SM(d18:1/18:0) is significantly enhanced in cerebrospinal fluid samples dichotomized by pathological β-amyloid, tau and phospho-tau-181 levels. J Alzheimers Dis 2014;44:1193–201.
[39] Available from: http://sciex.com/lipidyzer
[40] Gika HG, Theodoridis GA, Earll M, Snyder RW, Sumner SJ, Wilson ID. Does the mass spectrometer define the marker? A comparison of global metabolite profiling data generated simultaneously via UPLC-MS on two different mass spectrometers. Anal Chem 2010;82:8226–34.
[41] Xu EY, Schaefer WH, Xu Q. Metabolomics in pharmaceutical research and development: metabolites, mechanisms and pathways. Curr Opin Drug Discov Devel 2009;12:40–52.
[42] Lindon JC, Holmes E, Nicholson JK. Metabonomics techniques and applications to pharmaceutical research & development. Pharm Res 2006;23:1075–88.
[43] Emwas AH, Luchinat C, Turano P, Tenori L, Roy R, Salek RM, Ryan D, Merzaban JS, Kaddurah-Daouk R, Zeri AC, Nagana Gowda GA, Raftery D, Wang Y, Brennan L, Wishart DS. Standardizing the experimental conditions for using urine in NMR-based metabolomic studies with a particular focus on diagnostic studies: a review. Metabolomics 2015;11:872–94.
[44] Khoo SHG, Al-Rubeal M. Metabolomics as a complementary tool in cell culture. Biotechnol Appl Biochem 2007;47:71–84.
[45] Rifai N, Gillette MA, Carr SA. Protein biomarker discovery and validation: the long and uncertain path to clinical utility. Nat Biotechnol 2006;24:971–83.

[46] Pasikanti KK, Ho PC, Chan EC. Gas chromatography/mass spectrometry in metabolic profiling of biological fluids. J Chromatogr B Analyt Technol Biomed Life Sci 2008;871:202–11.
[47] Slupsky CM, Rankin KN, Wagner J, et al. Investigations of the effects of gender, diurnal variation, and age in human urinary metabolomic profiles. Anal Chem 2007; 79:6995–7004.
[48] Emwas AH, Roy R, McKay RT, Ryan D, Brennan L, Tenori L, Luchinat C, Gao X, Zeri AC, Gowda GA, Raftery D, Steinbeck C, Salek RM, Wishart DS. Recommendations and standardization of biomarker quantification using NMR-based metabolomics with particular focus on urinary analysis. J Proteome Res 2016;15:360–73.
[49] Lenz EM, Bright J, Wilson ID, Morgan SR, Nash AF. A ^{1}H NMR-based metabonomic study of urine and plasma samples obtained from healthy human subjects. J Pharm Biomed Anal 2003;33:1103–15.
[50] Lee JW, Weiner RS, Sailstad JM, et al. Method validation and measurement of biomarkers in nonclinical and clinical samples in drug development: a conference report. Pharm Res 2005;22:499–511.
[51] Bernini P, Bertini I, Luchinat C, Nincheri P, Staderini S, Turano P. Standard operating procedures for pre-analytical handling of blood and urine for metabolomic studies and biobanks. J Biomol NMR 2011;49:231–43.
[52] Issaq HJ, Van QN, Waybright TJ, Muschik GM, Veenstra TD. Analytical and statistical approaches to metabolomics research. J Sep Sci 2009;32:2183–99.
[53] Dunn WB, Bailey NJ, Johnson HE. Measuring the metabolome: current analytical technologies. Analyst 2005;130:606–25.
[54] Drake SK, Bowen RAR, Remaley AT, Hortin GL. Potential interferences from blood collection tubes in mass spectrometric analyses of serum polypeptides. Clin Chem 2004;50:2398–401.
[55] Bowen RA, Hortin GL, Csako G, Otañez OH, Remaley AT. Impact of blood collection devices on clinical chemistry assays. Clin Biochem 2010;43:4–25.
[56] Delanghe J, Speeckaert M. Preanalytical requirements of urinalysis. Biochem Med 2014;24:89–104.
[57] Beckonert O, Keun HC, Ebbels TMD, et al. Metabolic profiling, metabolomic and metabonomic procedures for NMR spectroscopy of urine, plasma, serum and tissue extracts. Nat Protoc 2007;2:2692–703.
[58] Hollis DP. Quantitative analysis of aspirin, phenacetin, and caffeine mixtures by nuclear magnetic resonance spectrometry. Anal Chem 1963;35:1682–4.
[59] Daykin CA, Foxall PJD, Connor SC, et al. The comparison of plasma deproteinization methods for the detection of low-molecular weight metabolites by ^{1}H nuclear magnetic resonance spectroscopy. Anal Biochem 2002;304:220–30.
[60] Lindon JC, Holmes E, Bollard ME, et al. Metabonomics technologies and their applications in physiological monitoring, drug safety assessment and disease diagnosis. Biomarkers 2004;9:1–31.
[61] Giraudeau P. Quantitative 2D liquid-state NMR. Magn Reson Chem 2014;52:259–72.
[62] Reo NV. NMR-based metabolomics. Drug Chem Toxicol 2002;25:375–82.
[63] Pan Z, Raftery D. Comparing and combining NMR spectroscopy and mass spectrometry in metabolomics. Anal Bioanal Chem 2007;387:525–7.
[64] Fan TW, Lane AN. Applications of NMR spectroscopy to systems biochemistry. Prog Nucl Magn Reson Spectrosc 2016;92–93:18–53.
[65] Serkova N, Fuller TF, Klawitter J, et al. ^{1}H-NMR-based metabolic signatures of mild and severe ischemia/reperfusion injury in rat kidney transplants. Kidney Int 2005;67:1142–51.
[66] Lauridsen M, Maher AD, Keun H, et al. Application of the FLIPSY pulse sequence for increased sensitivity in ^{1}H NMR-based metabolic profiling studies. Anal Chem 2008;80:3365–71.

[67] Gowda GA, Zhang S, Gu H, et al. Metabolomics-based methods for early disease diagnostics. Expert Rev Mol Diagn 2008;8:617–33.
[68] Gowda GA, Tayyari F, Ye T, Suryani Y, Wei S, Shanaiah N, Raftery D. Quantitative analysis of blood plasma metabolites using isotope enhanced NMR methods. Anal Chem 2010;82:8983–90.
[69] Bathen TF, Sitter B, Sjøbakk TE, Tessem MB, Gribbestad IS. Magnetic resonance metabolomics of intact tissue: a biotechnological tool in cancer diagnostics and treatment evaluation. Cancer Res 2010;70:6692–6.
[70] Takeda K. Microcoils and microsamples in solid-state NMR. Solid State Nucl Magn Reson 2012;47–48:1–9.
[71] Wong A, Li X, Sakellariou D. Refined magic-angle coil spinning resonator for nanoliter NMR spectroscopy: enhanced spectral resolution. Anal Chem 2013;85:2021–6.
[72] Lei Z, Huhman DV, Sumner LW. Mass spectrometry strategies in metabolomics. J Biol Chem 2011;286:25435–42.
[73] Mastrangelo A, Ferrarini A, Rey-Stolle F, García A, Barbas C. From sample treatment to biomarker discovery: A tutorial for untargeted metabolomics based on GC-(EI)-Q-MS. Anal Chim Acta 2015;900:21–35.
[74] Schnackenberg K, Beger RD. Monitoring the health to disease continuum with global metabolic profiling and systems biology. Pharmacogenomics 2006;7:1077–86.
[75] Pierce KM, Hoggard JC, Mohler RE, Synovec RE. Recent advancements in comprehensive two-dimensional separations with chemometrics. J Chromatogr A 2008;1184: 341–52.
[76] Almstetter MF, Oefner PJ, Dettmer K. Comprehensive two-dimensional gas chromatography in metabolomics. Anal Bioanal Chem 2012;402:1993–2013.
[77] Marney LC, Hoggard JC, Skogerboe KJ, Synovec RE. Methods of discovery-based and targeted metabolite analysis by comprehensive two-dimensional gas chromatography with time-of-flight mass spectrometry detection. Methods Mol Biol 2014;1198:83–97.
[78] Adahchour M, Beens J, Brinkman UA. Recent developments in the application of comprehensive two-dimensional gas chromatography. J Chromatogr A 2008;1186: 67–108.
[79] Want EJ, O'Maille G, Smith CA, et al. Solvent-dependent metabolite distribution, clustering, and protein extraction for serum profiling with mass spectrometry. Anal Chem 2006;78:743–52.
[80] Nordström A, Want E, Northen T, Lehtiö J, Siuzdak G. Multiple ionization mass spectrometry strategy used to reveal the complexity of metabolomics. Anal Chem 2008;80:421–9.
[81] Annesley TM. Ion suppression in mass spectrometry. Clin Chem 2007;49:1041–4.
[82] Want EJ, Nordström A, Morita H, Siuzdak G. From exogenous to endogenous: the inevitable imprint of mass spectrometry in metabolomics. J Proteome Res 2007;6: 459–68.
[83] Boernsen KO, Gatzek S, Imbert G. Controlled protein precipitation in combination with chip-based nanospray infusion mass spectrometry. An approach for metabolomics profiling of plasma. Anal Chem 2005;77:7255–64.
[84] Klepacki J, Klawitter J, Klawitter J, Thurman JM, Christians U. A high-performance liquid chromatography-tandem mass spectrometry-based targeted metabolomics kidney dysfunction marker panel in human urine. Clin Chim Acta 2015;446:43–53.
[85] Nordström A, O'Maille G, Qin C, Siuzdak G. Nonlinear data alignment for UPLC-MS and HPLC-MS based metabolomics: quantitative analysis of endogenous and exogenous metabolites in human serum. Anal Chem 2006;78:3289–95.
[86] Want EJ, Masson P, Michopoulos F, Wilson ID, Theodoridis G, Plumb RS, et al. Global metabolic profiling of animal and human tissues via UPLC–MS. Nat Protoc 2013;8:17–32.

[87] Want EJ, Wilson ID, Gika H, Theodoridis G, Plumb RS, Shockcor J, et al. Global metabolic profiling procedures for urine using UPLC–MS. Nat Protoc 2010;5: 1005–18.
[88] Van der Werf MJ, Overkamp KM, Muilwijk B, Coulier L, Hankemeier T. Microbial metabolomics: toward a platform with full metabolic coverage. Anal Biochem 2007;370:17–25.
[89] Gordon RE, Hanley PE, Shaw D, et al. Localization of metabolites in animals using ^{31}P topical magnetic resonance. Nature 1980;287:736–8.
[90] Koretsky AP, Wang S, Murphy-Boesch J, et al. ^{31}P NMR spectroscopy of rat organs, in situ, using chronically implanted radiofrequency coils. Proc Natl Acad Sci USA 1983;80:7491–5.
[91] Crecelius AC, Schubert US, von Eggeling F. MALDI mass spectrometric imaging meets "omics": recent advances in the fruitful marriage. Analyst 2015;140:5806–20.
[92] Caprioli RM, Farmer TB, Gile J. Molecular imaging of biological samples: localization of peptides and proteins using MALDI-TOF-MS. Anal Chem 1997;69:4751–60.
[93] Cobice DF, Goodwin RJ, Andren PE, Nilsson A, Mackay CL, Andrew R. Future technology insight: mass spectrometry imaging as a tool in drug research and development. Br J Pharmacol 2015;172:3266–83.
[94] Wishart DS. Emerging applications of metabolomics in drug discovery and precision medicine. Nat Rev Drug Discov 2016;15(7):473–84.
[95] Trim PJ, Snel MF. Small molecule MALDI MS imaging: current technologies and future challenges. Methods 2016;104:127–41.
[96] Murray KK, Seneviratne CA, Ghorai S. High resolution laser mass spectrometry bioimaging. Methods 2016;104:118–26.
[97] Sekuła J, Nizioł J, Rode W, Ruman T. Gold nanoparticle-enhanced target (AuNPET) as universal solution for laser desorption/ionization mass spectrometry analysis and imaging of low molecular weight compounds. Anal Chim Acta 2015;875:61–72.
[98] Gessel MM, Norris JL, Caprioli RM. MALDI imaging mass spectrometry: spatialmolecular analysis to enable a new age of discovery. J Proteomics 2014;107:71–82.
[99] Balog J, Sasi-Szabó L, Kinross J, Lewis MR, Muirhead LJ, Veselkov K, Mirnezami R, Dezső B, Damjanovich L, Darzi A, Nicholson JK, Takáts Z. Intraoperative tissue identification using rapid evaporative ionization mass spectrometry. Sci Transl Med 2013;5. 194ra93.
[100] Deming SN. Chemometrics: an overview. Clin Chem 1986;32:1702–6.
[101] Ganter B, Zidek N, Hewitt PR, Müller D, Vladimirova A. Pathway analysis tools and toxicogenomics reference databases for risk assessment. Pharmacogenomics 2008;9:35–54.
[102] Wheelock CE, Wheelock AM, Kawashima S, et al. Systems biology approaches and pathway tools for investigating cardiovascular disease. Mol Biosyst 2009;5:588–602.
[103] Materi W, Wishart DS. Computational systems biology in drug discovery and development: methods and applications. Drug Discov Today 2007;12:295–303.
[104] Leach SM, Tipney H, Feng W, et al. Biomedical discovery acceleration, with applications to craniofacial development. PLoS Comput Biol 2009;5:e1000215.
[105] Wishart DS, Knox C, Guo AC, et al. HMDB: a knowledgebase for the human metabolome. Nucleic Acids Res 2009;37(Database issue):D603–10.
[106] Wishart DS, Mandal R, Stanislaus A, Ramirez-Gaona M. Cancer metabolomics and the human metabolome database. Metabolites 2016;6(1):10.
[107] Wishart DS, Jewison T, Guo AC, Wilson M, Knox C, Liu Y, Djoumbou Y, Mandal R, Aziat F, Dong E, Bouatra S, Sinelnikov I, Arndt D, Xia J, Liu P, Yallou F, Bjorndahl T, Perez-Pineiro R, Eisner R, Allen F, Neveu V, Greiner R, Scalbert A. HMDB 3.0—The Human Metabolome Database in 2013. Nucleic Acids Res 2013;41(Database issue):D801–7.

[108] Ravanbakhsh S, Liu P, Bjorndahl TC, Mandal R, Grant JR, Wilson M, Eisner R, Sinelnikov I, Hu X, Luchinat C, Greiner R, Wishart DS. Accurate, fully-automated NMR spectral profiling for metabolomics. PLoS One 2015;10:e0124219.
[109] Dona AC, Kyriakides M, Scott F, Shephard EA, Varshavi D, Veselkov K, Everett JR. A guide to the identification of metabolites in NMR-based metabonomics/metabolomics experiments. Comput Struct Biotechnol J 2016;14:135–53.
[110] Guennec AL, Giraudeau P, Caldarelli S. Evaluation of fast 2D NMR for metabolomics. Anal Chem 2014;86:5946–54.
[111] Ulrich EL, Akutsu H, Doreleijers JF, Harano Y, Ioannidis YE, Lin J, Livny M, Mading S, Maziuk D, Miller Z, Nakatani E, Schulte CF, Tolmie DE, Kent Wenger R, Yao H, Markley JL. BioMagResBank. Nucleic Acids Res 2008;36(Database issue):D402–8.
[112] Ludwig C, Easton JM, Lodi A, Tiziani S, Manzoor SE, Southam AD, Byrne JJ, Bishop LM, He S, Arvanitis TN, Günther UL, Viant MR. Birmingham Metabolite Library: a publicly accessible database of 1-D H-1 and 2-D H-1 J-resolved NMR spectra of authentic metabolite standards (BML-NMR). Metabolomics 2012;8:8–18.
[113] NIST14 Spectral Library. Available from: http://nistmassspeclibrary.com/
[114] Halket JM, Przyborowska A, Stein SE, Mallard WG, Down S, Chalmers RA. Deconvolution gas chromatography/mass spectrometry of urinary organic acids—potential for pattern recognition and automated identification of metabolic disorders. Rapid Commun Mass Spectrom 1999;13:279–84.
[115] Koo I, Kim S, Shi B, Lorkiewicz P, Song M, McClain C, Zhang X. EIder: A compound identification tool for gas chromatography mass spectrometry data. J Chromatogr A 2016;1448:107–14.
[116] Kind T, Scholz M, Fiehn O. How large is the metabolome? A critical analysis of data exchange practices in chemistry. PLoS One 2009;4:e5440.
[117] Tautenhahn R, Cho K, Uritboonthai W, Zhu Z, Patti GJ, Siuzdak G. An accelerated workflow for untargeted metabolomics using the METLIN database. Nat Biotechnol 2012;30:826–8.
[118] Lipid maps. Available from: http://www.lipidmaps.org
[119] Nikolskiy I, Mahieu NG, Chen YJ, Tautenhahn R, Patti GJ. An untargeted metabolomic workflow to improve structural characterization of metabolites. Anal Chem 2013;85:7713–9.
[120] Veselkov KA, Vingara LK, Masson P, Robinette SL, Want E, Li JV, Barton RH, Boursier-Neyret C, Walther B, Ebbels TM, Pelczer I, Holmes E, Lindon JC, Nicholson JK. Optimized preprocessing of ultra-performance liquid chromatography/mass spectrometry urinary metabolic profiles for improved information recovery. Anal Chem 2011;83:5864–72.
[121] Edmands WM, Ferrari P, Scalbert A. Normalization to specific gravity prior to analysis improves information recovery from high resolution mass spectrometry metabolomic profiles of human urine. Anal Chem 2014;86:10925–31.
[122] Warrack BM, Hnatyshyn S, Ott KH, Reily MD, Sanders M, Zhang H, Drexler DM. Normalization strategies for metabonomic analysis of urine samples. J Chromatogr B 2009;877:547–52.
[123] Conti M, Moutereau S, Esmilaire L, et al. Should kidney tubular markers be adjusted for urine creatinine? The example of urinary cystatin C. Clin Chem Lab Med 2009;47:1553–6.
[124] Schnackenberg LK, Sun J, Espandiari P, Holland RD, Hanig J, Beger RD. Metabonomics evaluations of age-related changes in urinary compositions of male Sprague Dawley rats and effects of data normalization methods on statistical and quantitative analysis. BMC Bioinformatics 2007;8(Suppl. 7):S3.
[125] Chen Y, Shen G, Zhang R, He J, Zhang Y, Xu J, Yang W, Chen X, Song Y, Abliz Z. Combination of injection volume calibration by creatinine and MS signals' normalization to overcome urine variability in LC-MS-based metabolomics studies. Anal Chem 2013;85:7659–65.

[126] Dieterle F, Ross A, Schlotterbeck G, Senn H. Probabilistic quotient normalization as robust method to account for dilution of complex biological mixtures. Application in ^{1}H NMR metabonomics. Anal Chem 2006;78:4281–90.

[127] Craig A, Cloarec O, Holmes E, Nicholson JK, Lindon JC. Scaling and normalization effects in NMR spectroscopic metabonomic data sets. Anal Chem 2006;78:2262–7.

[128] Zhang S, Zheng C, Lanza IR, Nair KS, Raftery D, Vitek O. Interdependence of signal processing and analysis of urine ^{1}H NMR spectra for metabolic profiling. Anal Chem 2009;81:6080–8.

[129] Lindon JC, Nicholson JK, Holmes E, Darzi AW. Future visions for clinical metabolic phenotyping: prospects and challenges. In: Holmes E, Nicholson JK, Darzi AW, Lindon JC, editors. Metabolic phenotyping in personalized and public health care. Tokyo: Academic Press; 2016. p. 369–88.

[130] Beisken S, Eiden M, Salek RM. Getting the right answers: understanding metabolomics challenges. Expert Rev Mol Diagn 2015;15:97–109.

[131] Christians U, Klepacki J, Shokati T, Klawitter J, Klawitter J. Mass spectrometry-based multiplexing for the analysis of biomarkers in drug development and clinical diagnostics—how much is too much? Microchem J 2012;105:32–8.

[132] Salek RM, Steinbeck C, Viant MR, Goodacre R, Dunn WB. The role of reporting standards for metabolite annotation and identification in metabolomics studies. Gigascience 2013;2:13.

[133] Kale NS, Haug K, Conesa P, Jayseelan K, Moreno P, Rocca-Serra P, Nainala VC, Spicer RA, Williams M, Li X, Salek RM, Griffin JL, Steinbeck C, MetaboLights: An open-access database repository for metabolomics data. Curr Protoc Bioinformatics 2016;53. 14.13.1–14.13.18.

[134] Salek RM, Haug K, Conesa P, Hastings J, Williams M, Mahendraker T, Maguire E, González-Beltrán AN, Rocca-Serra P, Sansone SA, Steinbeck C. The MetaboLights repository: curation challenges in metabolomics. Database 2013;2013. bat029.

[135] Salek RM, Haug K, Steinbeck C. Dissemination of metabolomics results: role of MetaboLights and COSMOS. Gigascience 2013;2:8.

[136] Salek RM, Neumann S, Schober D, Hummel J, Billiau K, Kopka J, Correa E, Reijmers T, Rosato A, Tenori L, Turano P, Marin S, Deborde C, Jacob D, Rolin D, Dartigues B, Conesa P, Haug K, Rocca-Serra P, O'Hagan S, Hao J, van Vliet M, Sysi-Aho M, Ludwig C, Bouwman J, Cascante M, Ebbels T, Griffin JL, Moing A, Nikolski M, Oresic M, Sansone SA, Viant MR, Goodacre R, Günther UL, Hankemeier T, Luchinat C, Walther D, Steinbeck C. COordination of Standards in MetabOlomicS (COSMOS): facilitating integrated metabolomics data access. Metabolomics 2015;11:1587–97.

[137] Dancey JE, Dobbin KK, Groshen S, Jessup JM, Hruszkewycz AH, Koehler M, Parchment R, Ratain MJ, Shankar LK, Stadler WM, True LD, Gravell A, Grever MR. Biomarkers Task Force of the NCI Investigational Drug Steering Committee. Guidelines for the development and incorporation of biomarker studies in early clinical trials of novel agents. Clin Cancer Res 2010;16:1745–55.

[138] Naz S, Vallejo M, Garcia A, Barbas C. Method validation strategies involved in non-targeted metabolomics. J Chromatogr A 2014;1353:99–105.

[139] Chau CH, Rixe O, McLeod H, Figg WD. Validation of analytic methods for biomarkers used in drug development. Clin Cancer Res 2008;14:5967–76.

[140] Clarke W, editor. Liquid chromatography-mass spectrometry methods. Approved guideline C62-A. Wayne, PA: Clinical Laboratory Standards Institute; 2014.

[141] Sargent M, editor. Guide to achieving reliable quantitative LC-MS measurements. Teddington: RSC Analytical Methods Committee; 2013.

[142] US Department of Health and Human Services, Food and Drug Administration, Center for Drug Evaluation and Research and Center for Veterinary Medicine. Guidance for the industry. Bioanalytical method validation. 2001. Available from: http://www.fda.gov/downloads/Drugs/GuidanceComplianceRegulatoryInformation/Guidances/UCM070107.pdf

[143] US Department of Health and Human Services, Food and Drug Administration. Bioanalytical method validation. 2013. Available from: http://www.fda.gov/downloads/drugs/guidancecomplianceregulatoryinformation/guidances/ucm368107.pdf
[144] Clinical Laboratory and Standard Institute. Available from: www.clsi.org
[145] Guideline IH. Validation of analytical procedures: text and methodology. Q2 (R1). 2005. Available from: http://www.ich.org/fileadmin/Public_Web_Site/ICH_Products/Guidelines/Quality/Q2_R1/Step4/Q2_R1__Guideline.pdf
[146] Wagner JA, Williams SA, Webster CJ. Biomarkers and surrogate end points for fit-for-purpose development and regulatory evaluation of new drugs. Clin Pharmacol Ther 2007;81:104–7.
[147] Gika HG, Theodoridis GA, Earll M, Wilson ID. A QC approach to the determination of day-to-day reproducibility and robustness of LC-MS methods for global metabolite profiling in metabonomics/metabolomics. Bioanalysis 2012;4:2239–47.
[148] Dunn WB, Wilson ID, Nicholls AW, Broadhurst D. The importance of experimental design and QC samples in large-scale and MS-driven untargeted metabolomic studies of humans. Bioanalysis 2012;4:2249–64.
[149] Gika HG, Zisi C, Theodoridis G, Wilson ID. Protocol for quality control in metabolic profiling of biological fluids by U(H)PLC-MS. J Chromatogr B 2016;1008:15–25.
[150] Gika HG, Theodoridis GA, Wilson ID. Liquid chromatography and ultra-performance liquid chromatography-mass spectrometry fingerprinting of human urine: sample stability under different handling and storage conditions for metabonomics studies. J Chromatogr A 2008;1189:314–22.
[151] United States National Institute of Standards and Technology (NIST). Available from: http://srm1950.nist.gov/
[152] Dunn WB, Broadhurst D, Begley P, Zelena E, Francis-McIntyre S, Anderson N, Brown M, Knowles JD, Halsall A, Haselden JN, Nicholls AW, Wilson ID, Kell DB, Goodacre R. Human Serum Metabolome (HUSERMET) Consortium. Procedures for large-scale metabolic profiling of serum and plasma using gas chromatography and liquid chromatography coupled to mass spectrometry. Nat Protoc 2011;6:1060–83.
[153] Gika HG, Wilson ID, Theodoridis GA. LC-MS-based holistic metabolic profiling. Problems, limitations, advantages, and future perspectives. J Chromatogr B 2014;966:1–6.
[154] Morrison N, Cochrane G, Faruque N, et al. Concept of sample in OMICS technology. OMICS 2006;10:127–37.
[155] Gika H, Theodoridis G. Sample preparation prior to the LC-MS-based metabolomics/metabonomics of blood-derived samples. Bioanalysis 2011;3:1647–61.
[156] The Standard Metabolic Reporting Structures Working Group. Summary recommendations for standardization and reporting of metabolic analyses. Nat Biotechnol 2005;23:833–8.
[157] Castle AL, Fiehn O, Kaddurah-Daouk R, Lindon JC. Metabolomics standards workshop and the development of international standards for reporting metabolomics experimental results. Brief Bioinform 2006;7:159–62.
[158] Keun HC, Ebbels TMD, Antti H, et al. Analytical reproducibility in ^{1}H NMR-based metabonomic urinalysis. Chem Res Toxicol 2002;15:1380–6.
[159] Dumas ME, Maibaum EC, Teague C, et al. Assessment of the analytical reproducibility of ^{1}H NMR spectroscopy based metabonomics for large-scale epidemiological research: the INTERMAP study. Anal Chem 2006;78:2199–208.
[160] Klepacki J, Klawitter J, Klawitter J, Karimpour-Fard A, Thurman J, Ingle G, Patel D, Christians U. Amino acids in a targeted versus a non-targeted metabolomics LC-MS/MS assay. Are the results consistent? Clin Biochem 2016. [Epub ahead of print].
[161] Amur S, LaVange L, Zineh I, Buckman-Garner S, Woodcock J. Biomarker qualification: toward a multiple stakeholder framework for biomarker development, regulatory acceptance, and utilization. Clin Pharmacol Ther 2015;98:34–46.

[162] Zhao X, Modur V, Carayannopoulos LN, Laterza OF. Biomarkers in pharmaceutical research. Clin Chem 2015;61:1343–53.
[163] Lavezzari G, Womack AW. Industry perspectives on biomarker qualification. Clin Pharmacol Ther 2016;99:208–13.
[164] Goodsaid F, Mattes WB. The path from biomarker discovery to regulatory qualification. 1st ed. Oxford, UK: Elsevier; 2013.
[165] US Department of Health and Human Services, Food and Drug Administration. Center for Drug Evaluation and Research. Guidance for Industry. E16 Biomarkers related to drug or biotechnology product development: context, structure, and format of qualification submissions. 2011. Available from: http://www.fda.gov/downloads/drugs/guidancecomplianceregulatoryinformation/guidances/ucm267449.pdf
[166] Burckart GJ, Amur S, Goodsaid FM, et al. Qualification of biomarkers for drug development in organ transplantation. Am J Transplant 2008;8:267–70.
[167] Müller PY, Dieterle F. Tissue-specific, noninvasive toxicity biomarkers: translation from preclinical safety assessment to clinical safety monitoring. Expert Opin Drug Metab Toxicol 2009;5:1023–38.
[168] Lesko LJ, Atkinson AJ Jr. Use of biomarkers and surrogate endpoints in drug development and regulatory decision making: criteria, validation, strategies. Annu Rev Pharmacol Toxicol 2001;41:347–66.
[169] Fawcett T. Introduction to ROC analysis. Pattern Recogn Lett 2006;27:861–74.
[170] Xia J, Broadhurst DI, Wilson M, Wishart DS. Translational biomarker discovery in clinical metabolomics: an introductory tutorial. Metabolomics 2013;9:280–99.
[171] Goodsaid FM, Frueh FW, Mattes W. Strategic paths for biomarker qualification. Toxicology 2008;245:219–23.
[172] Berl T. American Society of Nephrology Renal Research Report. J Am Soc Nephrol 2005;16:1886–903.
[173] Slocum JL, Heung M, Pennathur S. Marking renal injury: can we move beyond serum creatinine? Transl Res 2012;159:277–89.
[174] Cascante M, Boros LG, Comin-Anduix B, de Atauri P, Centelles JJ, Lee PW. Metabolic control analysis in drug discovery and disease. Nat Biotechnol 2002;20:243–9.
[175] Winter G, Krömer JO. Fluxomics—connecting 'omics analysis and phenotypes. Environ Microbiol 2013;15:1901–16.
[176] Zamboni N. ^{13}C metabolic flux analysis in complex systems. Curr Opin Biotechnol 2011;22:103–8.
[177] Martin G, Chauvin MF, Dugelay S, Bavarel G. Non-steady state model applicable to NMR studies for calculating flux rates in glycolysis, gluconeogenesis, and citric acid cycle. J Biol Chem 1994;42:26034–9.
[178] Wiechert W, Möllney M, Petersen S, de Graaf AA. A universal framework for ^{13}C metabolic flux analysis. Metab Eng 2001;3:265–83.
[179] Weitzel M, Nöh K, Dalman T, Niedenführ S, Stute B, Wiechert W. ^{13}CFLUX2—high-performance software suite for $^{(13)}$C-metabolic flux analysis. Bioinformatics 2013;29:143–5.
[180] Young JD. INCA: a computational platform for isotopically non-stationary metabolic flux analysis. Bioinformatics 2014;30:1333–5.
[181] Nöh K, Droste P, Wiechert W. Visual workflows for ^{13}C-metabolic flux analysis. Bioinformatics 2015;31:346–54.
[182] Vilasi A, Cutillas PR, Maher AD, et al. Combined proteomic and metabonomic studies in three genetic forms of the renal Fanconi syndrome. Am J Physiol Renal Physiol 2007;293:F456–67.
[183] Taylor SL, Gant S, Bukanov NO, et al. A metabolomics approach using juvenile cystic mice to identify urinary biomarkers and altered pathways in polycystic kidney disease. Am J Physiol Renal Physiol 2010;298:F909–22.

[184] Weiss RH, Kim K. Metabolomics in the study of kidney diseases. Nat Rev Nephrol 2011;8:22–33.
[185] Wishart DS. Emerging applications of metabolomics in drug discovery and precision medicine. Nat Rev Drug Discov 2016. [Epub ahead of print].
[186] Shockcor JP, Holmes E. Metabonomic applications in toxicity screening and disease diagnostics. Curr Topics Med Cem 2002;2:35–51.
[187] van Ravenzwaay B, Herold M, Kamp H, Kapp MD, Fabian E, Looser R, Krennrich G, Mellert W, Prokoudine A, Strauss V, Walk T, Wiemer J, Metabolomics: a tool for early detection of toxicological effects and an opportunity for biology based grouping of chemicals-from QSAR to QBAR. Mutat Res 2012;746:144–50.
[188] Ebbels TM, Keun HC, Beckonert OP, Bollard ME, Lindon JC, Holmes E, Nicholson JK. Prediction and classification of drug toxicity using probabilistic modeling of temporal metabolic data: the consortium on metabonomic toxicology screening approach. J Proteome Res 2007;6:4407–22.
[189] Wishart DS. Application of metabolomics in drug discovery and development. Drugs R D 2008;9:307–22.
[190] Boudnock KJ, Mitchell MW, Nemet L, et al. Discovery of metabolomics biomarkers for early detection of nephrotoxicity. Toxicol Pathol 2009;37:280–92.
[191] Gartland KPR, Bonner FW, Nicholson JK. Investigations into the biochemical effects of region-specific nehprotoxins. Mol Pharmacol 1989;35:242–50.
[192] Anthony ML, Rose VS, Nicholson JK, Lindon JC. Classification of toxin-induced changes in ^{1}H-NMR spectra of urine using an artificial neural network. J Pharm Biomed Anal 1995;13:205–11.
[193] Anthony ML, Sweatman BC, Beddell CR, Lindon LC, Nicholson JK. Pattern recognition classification of the site of nephrotoxicity based in metabolic data derived from proton nuclear magnetic resonance spectra of urine. Mol Pharmacol 1994;48:199–211.
[194] Holmes E, Nicholls AW, Lindon JC, et al. Development of a model for classification of toxin-induced lesions using ^{1}H NMR spectroscopy of urine combined with pattern recognition. NMR Biomed 1998;11:235–44.
[195] Williams RE, Major H, Lock EA, Lenz EM, Wilson ID. D-Serine nephrotoxicity: a HPLC-TOF/MS-based metabonomics approach. Toxicology 2005;207:179–209.
[196] Sieber M, Hoffmann D, Adler M, et al. Comparative analysis of novel noninvasive renal biomarkers and metabonomic changes in a rat model of gentamycin nephrotoxicity. Toxicol Sci 2009;109:336–49.
[197] Holmes E, Bonner FW, Sweatman BC, et al. Nuclear magnetic resonance spectroscopy and pattern recognition analysis of the biochemical processes associated with the progression of and recovery from nephrotoxic lesions in the rat induced by mercury(II) chloride and 2-bromoethanamine. Mol Pharmacol 1992;42:922–30.
[198] Lenz EM, Bright J, Knight R, Wilson ID, Major H. A metabonomic investigation of the biochemical effects of mercuric chloride in the rat using ^{1}H-NMR and HPLC-TOF/MS: time dependent changes in the urinary profile of endogenous metabolites as a result of nephrotoxicity. Analyst 2004;129:535–41.
[199] Nicholson JK, Timbrell JA, Sadler PJ. Proton NMR spectra of urine as indicators of renal damage. Mercury-induced nephrotoxicity in rats. Mol Pharmacol 1985;27:644–51.
[200] Anthony ML, Gartland KP, Beddell CR, Lindon JC, Nicholson JK. Studies of the biochemical toxicology of uranyl nitrate in the rat. Arch Toxicol 1994;68:43–53.
[201] Klawitter J, Bendrick-Peart J, Rudolph B, et al. Urine metabolites reflect time-dependent effects of cyclosporine and sirolimus on rat kidney function. Chem Res Toxicol 2009;22:118–28.
[202] Sieber M, Wagner S, Rached E, et al. Metabonomic study of ochratoxin A toxicity in rats after repeat administration: phenotypic anchoring enhances the ability for biomarker discovery. Chem Res Toxicol 2009;22:1221–31.

[203] Portilla D, Li S, Nagothu KK, et al. Metabolomic study of cisplatin-induced nephrotoxicity. Kidney Int 2006;69:2194–204.
[204] Lenz EM, Bright J, Knight R, Wilson ID, Major H. Cyclosporin A-induced changes in endogenous metabolites in rat urine: a metabonomic investigation using high field 1H NMR spectroscopy, HPLC-TOF/MS and chemometrics. J Pharm Biomed Anal 2004;35:599–608.
[205] Lenz EM, Bright J, Knight R, et al. Metabonomics with ^{1}H-NMR spectroscopy and liquid chromatography-mass spectrometry applied to the investigation of metabolic changes caused by gentamycin-induced nephrotoxicity in the rat. Biomarkers 2005;10:173–87.
[206] Park JC, Hong YS, Kim YJ, et al. A metabonomic study on the biochemical effects of doxorubicin in rats using ^{1}H-NMR spectroscopy. J Toxicol Environ Health A 2009;72:374–84.
[207] Holmes E, Caddick S, Lindon JC, et al. ^{1}H and ^{2}H NMR spectroscopic studies on the metabolism and biochemical effects of 2-bromoethanamine in the rat. Biochem Pharmacol 1995;49:1349–59.
[208] Dieterle F, Sistare F, Goodsaid F, et al. Renal biomarker qualification submission: a dialog between the FDA-EMEA and Predictive Safety Testing Consortium. Nat Biotechnol 2010;28:455–62.
[209] Brott DA, Furlong ST, Adler SH, Hainer JW, Arani RB, Pinches M, Rossing P, Chaturvedi N. DIRECT Programme Steering Committee. Characterization of renal biomarkers for use in clinical trials: effect of preanalytical processing and qualification using samples from subjects with diabetes. Drug Des Devel Ther 2015;9:3191–8.
[210] He JC, Chuang PY, Ma'ayan A, Iyengar R. Systems biology of kidney diseases. Kidney Int 2012;81:22–39.
[211] Beger RD, Holland RD, Sun J, Schnackenberg LK, Moore PC, Dent CL, Devarajan P, Portilla D. Metabonomics of acute kidney injury in children after cardiac surgery. Pediatr Nephrol 2008;23:977–84.
[212] Ujike-Omori H, Maeshima Y, Kinomura M, Tanabe K, Mori K, Watatani H, Hinamoto N, Sugiyama H, Sakai Y, Morimatsu H, Makino H. The urinary levels of prostanoid metabolites predict acute kidney injury in heterogeneous adult Japanese ICU patients: a prospective observational study. Clin Exp Nephrol 2015;19:1024–36.
[213] Sun J, Shannon M, Ando Y, Schnackenberg LK, Khan NA, Portilla D, Beger RD. Serum metabolomic profiles from patients with acute kidney injury: a pilot study. J Chromatogr B 2012;893–894:107–13.
[214] Barrios C, Spector TD, Menni C. Blood, urine and faecal metabolite profiles in the study of adult renal disease. Arch Biochem Biophys 2016;589:81–92.
[215] Cisek K, Krochmal M, Klein J, Mischak H. The application of multi-omics and systems biology to identify therapeutic targets in chronic kidney disease. Nephrol Dial Transplant 2015; gfv364.
[216] Goek ON, Döring A, Gieger C, Heier M, Koenig W, Prehn C, Römisch-Margl W, Wang-Sattler R, Illig T, Suhre K, Sekula P, Zhai G, Adamski J, Köttgen A, Meisinger C. Serum metabolite concentrations and decreased GFR in the general population. Am J Kidney Dis 2012;60:197–206.
[217] Rhee EP, Clish CB, Ghorbani A, Larson MG, Elmariah S, McCabe E, Yang Q, Cheng S, Pierce K, Deik A, Souza AL, Farrell L, Domos C, Yeh RW, Palacios I, Rosenfield K, Vasan RS, Florez JC, Wang TJ, Fox CS, Gerszten RE. A combined epidemiologic and metabolomic approach improves CKD prediction. J Am Soc Nephrol 2013;24: 1330–8.
[218] Yu B, Zheng Y, Nettleton JA, Alexander D, Coresh J, Boerwinkle E. Serum metabolomic profiling and incident CKD among African Americans. Clin J Am Soc Nephrol 2014;9:1410–7.

[219] Sekula P, Goek ON, Quaye L, Barrios C, Levey AS, Römisch-Margl W, Menni C, Yet I, Gieger C, Inker LA, Adamski J, Gronwald W, Illig T, Dettmer K, Krumsiek J, Oefner PJ, Valdes AM, Meisinger C, Coresh J, Spector TD, Mohney RP, Suhre K, Kastenmüller G, Köttgen A. A metabolome-wide association study of kidney function and disease in the general population. J Am Soc Nephrol 2016;27:1175–88.
[220] Posada-Ayala M, Zubiri I, Martin-Lorenzo M, Sanz-Maroto A, Molero D, Gonzalez-Calero L, Fernandez-Fernandez B, de la Cuesta F, Laborde CM, Barderas MG, Ortiz A, Vivanco F, Alvarez-Llamas G. Identification of a urine metabolomic signature in patients with advanced-stage chronic kidney disease. Kidney Int 2014;85:103–11.
[221] Kobayashi T, Yoshida T, Fujisawa T, Matsumura Y, Ozawa T, Yanai H, Iwasawa A, Kamachi T, Fujiwara K, Kohno M, Tanaka N. A metabolomics-based approach for predicting stages of chronic kidney disease. Biochem Biophys Res Commun 2014;445:412–6.
[222] Shah VO, Townsend RR, Feldman HI, Pappan KL, Kensicki E, Vander Jagt DL. Plasma metabolomic profiles in different stages of CKD. Clin J Am Soc Nephrol 2013;8:363–70.
[223] Duranton F, Lundin U, Gayrard N, Mischak H, Aparicio M, Mourad G, Daurès JP, Weinberger KM, Argilés A. Plasma and urinary amino acid metabolomic profiling in patients with different levels of kidney function. Clin J Am Soc Nephrol 2014;9:37–45.
[224] Zhao YY. Metabolomics in chronic kidney disease. Clin Chim Acta 2013;422:59–69.
[225] Breit M, Weinberger KM. Metabolic biomarkers for chronic kidney disease. Arch Biochem Biophys 2016;589:62–80.
[226] Nkuipou-Kenfack E, Duranton F, Gayrard N, Argilés À, Lundin U, Weinberger KM, Dakna M, Delles C, Mullen W, Husi H, Klein J, Koeck T, Zürbig P, Mischak H. Assessment of metabolomic and proteomic biomarkers in detection and prognosis of progression of renal function in chronic kidney disease. PLoS One 2014;9:e96955.
[227] Toyohara T, Suzuki T, Akiyama Y, Yoshihara D, Takeuchi Y, Mishima E, Kikuchi K, Suzuki C, Tanemoto M, Ito S, Nagao S, Soga T, Abe T. Metabolomic profiling of the autosomal dominant polycystic kidney disease rat model. Clin Exp Nephrol 2011;15:676–87.
[228] Taylor SL, Ganti S, Bukanov NO, Chapman A, Fiehn O, Osier M, Kim K, Weiss RH. A metabolomics approach using juvenile cystic mice to identify urinary biomarkers and altered pathways in polycystic kidney disease. Am J Physiol Renal Physiol 2010;298:F909–22.
[229] Klawitter J, Zafar I, Klawitter J, Pennington AT, Klepacki J, Gitomer BY, Schrier RW, Christians U, Edelstein CL. Effects of lovastatin treatment on the metabolic distributions in the Han:SPRD rat model of polycystic kidney disease. BMC Nephrol 2013;14:165.
[230] Klawitter J, Reed-Gitomer BY, McFann K, Pennington A, Klawitter J, Abebe KZ, Klepacki J, Cadnapaphornchai MA, Brosnahan G, Chonchol M, Christians U, Schrier RW. Endothelial dysfunction and oxidative stress in polycystic kidney disease. Am J Physiol Renal Physiol 2014;307:F1198–206.
[231] Klawitter J, Klawitter J, McFann K, Pennington AT, Abebe KZ, Brosnahan G, Cadnapaphornchai MA, Chonchol M, Gitomer B, Christians U, Schrier RW. Bioactive lipid mediators in polycystic kidney disease. J Lipid Res 2014;55:1139–49.
[232] Klawitter J, McFann K, Pennington AT, Wang W, Klawitter J, Christians U, Schrier RW, Gitomer B, Cadnapaphornchai MA. Pravastatin therapy and biomarker changes in children and young adults with autosomal dominant polycystic kidney disease. Clin J Am Soc Nephrol 2015;10:1534–41.
[233] Zhang Y, Zhang S, Wang G. Metabolomic biomarkers in diabetic kidney diseases—a systematic review. J Diabetes Complications 2015;29:1345–51.

[234] Han LD, Xia JF, Liang QL, Wang Y, Wang YM, Hu P, Li P, Luo GA. Plasma esterified and non-esterified fatty acids metabolic profiling using gas chromatography-mass spectrometry and its application in the study of diabetic mellitus and diabetic nephropathy. Anal Chim Acta 2011;689:85–91.
[235] Ng DP, Salim A, Liu Y, Zou L, Xu FG, Huang S, Leong H, Ong CN. A metabolomics study of low estimated GFR in non-proteinuric type 2 diabetes mellitus. Diabetologia 2012;55:499–508.
[236] Mäkinen VP, Tynkkynen T, Soininen P, Forsblom C, Peltola T, Kangas AJ, Groop PH, Ala-Korpela M. Sphingomyelin is associated with kidney disease in type 1 diabetes (The FinnDiane Study). Metabolomics 2012;8:369–75.
[237] Mäkinen VP, Tynkkynen T, Soininen P, Peltola T, Kangas AJ, Forsblom C, Thorn LM, Kaski K, Laatikainen R, Ala-Korpela M, Groop PH. Metabolic diversity of progressive kidney disease in 325 patients with type 1 diabetes (the FinnDiane Study). J Proteome Res 2012;11:1782–90.
[238] van der Kloet FM, Tempels FW, Ismail N, van der Heijden R, Kasper PT, Rojas-Cherto M, van Doorn R, Spijksma G, Koek M, van der Greef J, Mäkinen VP, Forsblom C, Holthöfer H, Groop PH, Reijmers TH, Hankemeier T. Discovery of early-stage biomarkers for diabetic kidney disease using MS-based metabolomics (FinnDiane study). Metabolomics 2012;8:109–19.
[239] Fiehn O, Garvey WT, Newman JW, Lok KH, Hoppel CL, Adams SH. Plasma metabolomics profiles reflective of glucose homeostasis in non-diabetic and type 2 diabetic obese African-American women. PLoS One 2010;5:e15234.
[240] Hirayama A, Nakashima E, Sugimoto M, Akiyama S, Sato W, Maruyama S, Matsuo S, Tomita M, Yuzawa Y, Soga T. Metabolic profiling reveals new serum biomarkers for differentiating diabetic nephropathy. Anal Bioanal Chem 2012;404:3101–9.
[241] Sharma K, Karl B, Mathew AV, Gangoiti JA, Wassel CL, Saito R, Pu M, Sharma S, You YH, Wang L, Diamond-Stanic M, Lindenmeyer MT, Forsblom C, Wu W, Ix JH, Ideker T, Kopp JB, Nigam SK, Cohen CD, Groop PH, Barshop BA, Natarajan L, Nyhan WL, Naviaux RK. Metabolomics reveals signature of mitochondrial dysfunction in diabetic kidney disease. J Am Soc Nephrol 2013;24:1901–12.
[242] Sirolli V, Rossi C, Di Castelnuovo A, Felaco P, Amoroso L, Zucchelli M, Ciavardelli D, Di Ilio C, Sacchetta P, Bernardini S, Arduini A, Bonomini M, Urbani A. Toward personalized hemodialysis by low molecular weight amino-containing compounds: future perspective of patient metabolic fingerprint. Blood Transfus 2012;10(Suppl. 2):s78–88.
[243] Xia JF, Hu P, Liang QL, Zou TT, Wang YM, Luo GA. Correlations of creatine and six related pyrimidine metabolites and diabetic nephropathy in Chinese type 2 diabetic patients. Clin Biochem 2010;43:957–62.
[244] Xia JF, Liang QL, Liang XP, Wang YM, Hu P, Li P, Luo GA. Ultraviolet and tandem mass spectrometry for simultaneous quantification of 21 pivotal metabolites in plasma from patients with diabetic nephropathy. J Chromatogr B 2009;877:1930–6.
[245] Zhang J, Yan L, Chen W, Lin L, Song X, Yan X, Hang W, Huang B. Metabonomics research of diabetic nephropathy and type 2 diabetes mellitus based on UPLC-oaTOF-MS system. Anal Chim Acta 2009;650:16–22.
[246] Zhang X, Wang Y, Hao F, Zhou X, Han X, Tang H, Ji L. Human serum metabonomic analysis reveals progression axes for glucose intolerance and insulin resistance statuses. J Proteome Res 2009;8:5188–95.
[247] Niewczas MA, Sirich TL, Mathew AV, Skupien J, Mohney RP, Warram JH, Smiles A, Huang X, Walker W, Byun J, Karoly ED, Kensicki EM, Berry GT, Bonventre JV, Pennathur S, Meyer TW, Krolewski AS. Uremic solutes and risk of end-stage renal disease in type 2 diabetes: metabolomic study. Kidney Int 2014;85:1214–24.

[248] Pena MJ, Lambers Heerspink HJ, Hellemons ME, Friedrich T, Dallmann G, Lajer M, Bakker SJ, Gansevoort RT, Rossing P, de Zeeuw D, Roscioni SS. Urine and plasma metabolites predict the development of diabetic nephropathy in individuals with Type 2 diabetes mellitus. Diabet Med 2014;31:1138–47.
[249] Klein RL, Hammad SM, Baker NL, Hunt KJ, Al Gadban MM, Cleary PA, Virella G, Lopes-Virella MF. DCCT/EDIC Research Group. Decreased plasma levels of select very long chain ceramide species are associated with the development of nephropathy in type 1 diabetes. Metabolism 2014;63:1287–95.
[250] Pena MJ, de Zeeuw D, Mischak H, Jankowski J, Oberbauer R, Woloszczuk W, Benner J, Dallmann G, Mayer B, Mayer G, Rossing P, Lambers Heerspink HJ. Prognostic clinical and molecular biomarkers of renal disease in type 2 diabetes. Nephrol Dial Transplant 2015;30(Suppl. 4):iv86–95.
[251] Ganti S, Weiss RH. Urine metabolomics for kidney cancer detection and biomarker discovery. Urol Oncol 2011;29:551–7.
[252] Warburg O. On the origin of cancer cells. Science 1956;123:309–14.
[253] Laplante M, Sabatini DM. mTOR signaling in growth control and disease. Cell 2012;149:274–93.
[254] Serkova NJ, Spratlin JL, Eckhardt SG. NMR-based metabolomics: translational application and treatment of cancer. Curr Opin Mol Ther 2007;9:572–85.
[255] Aboud OA, Weiss RH. New opportunities from the cancer metabolome. Clin Chem 2013;59:138–46.
[256] Ngo TC, Wood CG, Karam JA. Biomarkers of renal cell carcinoma. Urol Oncol 2014;32:243–51.
[257] Lucarelli G, Galleggiante V, Rutigliano M, Sanguedolce F, Cagiano S, Bufo P, Lastilla G, Maiorano E, Ribatti D, Giglio A, Serino G, Vavallo A, Bettocchi C, Selvaggi FP, Battaglia M, Ditonno P. Metabolomic profile of glycolysis and the pentose phosphate pathway identifies the central role of glucose-6-phosphate dehydrogenase in clear cell-renal cell carcinoma. Oncotarget 2015;6:13371–86.
[258] Wettersten HI, Hakimi AA, Morin D, Bianchi C, Johnstone ME, Donohoe DR, Trott JF, Aboud OA, Stirdivant S, Neri B, Wolfert R, Stewart B, Perego R, Hsieh JJ, Weiss RH. Grade-dependent metabolic reprogramming in kidney cancer revealed by combined proteomics and metabolomics analysis. Cancer Res 2015;75: 2541–52.
[259] Hakimi AA, Reznik E, Lee CH, Creighton CJ, Brannon AR, Luna A, Aksoy BA, Liu EM, Shen R, Lee W, Chen Y, Stirdivant SM, Russo P, Chen YB, Tickoo SK, Reuter VE, Cheng EH, Sander C, Hsieh JJ. An integrated metabolic atlas of clear cell renal cell carcinoma. Cancer Cell 2016;29:104–16.
[260] Ganti S, Taylor SL, Abu Aboud O, Yang J, Evans C, Osier MV, Alexander DC, Kim K, Weiss RH. Kidney tumor biomarkers revealed by simultaneous multiple matrix metabolomics analysis. Cancer Res 2012;72:3471–9.
[261] Kim K, Taylor SL, Ganti S, Guo L, Osier MV, Weiss RH. Urine metabolomics analysis identifies potential biomarkers and pathogenic pathways in kidney cancer. OMICS 2011;15:293–303.
[262] Ganti S, Taylor SL, Kim K, Hoppel CL, Guo L, Yang J, Evans C, Weiss RH. Urinary acylcarnitines are altered in human kidney cancer. Int J Cancer 2012;130:2791–800.
[263] Kim K, Aronov P, Zakharin SO, et al. Urine metabolomics analysis for kidney cancer detection and biomarker discovery. Mol Cell Proteomics 2009;8(3):558–70.
[264] Kind T, Tolstikov V, Fiehn O, Weiss RH. A comprehensive urinary metabolomic approach for identifying kidney cancer. Anal Biochem 2007;363:185–95.
[265] Lin L, Huang Z, Gao Y, Yan X, Xing J, Hang W. LC-MS based serum metabonomic analysis for renal cell carcinoma diagnosis, staging, and biomarker discovery. J Proteome Res 2011;10:1396–405.

[266] Zira AN, Theocharis SE, Mitropoulos D, Migdalis V, Mikros E. ^{1}H NMR metabonomic analysis in renal cell carcinoma: a possible diagnostic tool. J Proteome Res 2010;9:4038–44.
[267] Waters NJ, Garrod S, Farrant RD, et al. High-resolution magic angle spinning ^{1}H NMR spectroscopy of intact liver and kidney: optimization of sample preparation procedures and biochemical stability of tissue during spectral acquisition. Anal Biochem 2000;282:16–23.
[268] Moka D, Vorreuther R, Schicha H, et al. Biochemical classification of kidney carcinoma biopsy samples using magic-angle-spinning ^{1}H nuclear magnetic resonance spectroscopy. J Pharm Biomed Anal 1998;17:125–32.
[269] Tate RA, Foxall PJD, Holmes E, et al. Distinction between normal and renal cell carcinoma kidney cortical biopsy samples using pattern recognition of ^{1}H magic angle spinning (MAS) NMR spectra. NMR Biomed 2000;13:64–71.
[270] Righi V, Mucci A, Schenetti L, et al. Ex vivo HR-MAS magnetic resonance spectroscopy of normal and malignant human renal tissues. Anticancer Res 2007;27: 3195–204.
[271] Chung YL, Griffith JR. Using metabolomics to monitor anticancer drugs. Ernst Schering Found Symp Proc 2008;4:55–78.
[272] Jobard E, Blanc E, Négrier S, Escudier B, Gravis G, Chevreau C, Elena-Herrmann B, Trédan O. A serum metabolomic fingerprint of bevacizumab and temsirolimus combination as first-line treatment of metastatic renal cell carcinoma. Br J Cancer 2015;113:1148–57.
[273] Womer KL, Kaplan B. Recent developments in kidney transplantation—a critical assessment. Am J Transplant 2009;9:1265–71.
[274] Matas AJ, Smith JM, Skeans MA, Thompson B, Gustafson SK, Stewart DE, Cherikh WS, Wainright JL, Boyle G, Snyder JJ, Israni AK, Kasiske BL. OPTN/SRTR 2013 Annual Data Report: kidney. Am J Transplant 2015;15(Suppl. 2):1–34.
[275] Nankivell BJ, Kuypers DR. Diagnosis and prevention of chronic kidney allograft loss. Lancet 2011;378:1428.
[276] Nankivell BJ, P'Ng CH, O'Connell PJ, Chapman JR. Calcineurin inhibitor nephrotoxicity through the lens of longitudinal histology: comparison of cyclosporine and tacrolimus eras. Transplantation 2016;100(8):1723–31.
[277] Chapman JR, O'Connell PJ, Nankivell BJ. Chronic renal allograft dysfunction. J Am Soc Nephrol 2005;16:3015–26.
[278] Bohra R, Klepacki J, Klawitter J, Klawitter J, Thurman JM, Christians U. Proteomics and metabolomics in renal transplantation—quo vadis? Transpl Int 2013;26: 225–41.
[279] Wishart DS. Metabolomics: a complementary tool in renal transplantation. Contrib Nephrol 2008;160:76–87.
[280] Fuller TF, Serkova N, Neimann CU, Freise CE. Influence of donor pretreatment with N-acetylcysteine on ischemia/reperfusion injury in rat kidney grafts. J Urol 2004;171:1296–300.
[281] Hauet T, Gibelin H, Godart C, Eugene M, Carretier M. Kidney retrieval conditions influence damage to renal medulla: evaluation by proton nuclear magnetic resonance (NMR) spectroscopy. Clin Chem Lab Med 2000;38:1085–92.
[282] Hauet T, Baumert H, Gibelin H, et al. Noninvasive monitoring of citrate, acetate, lactate, and renal medullary osmolyte excretion in urine as biomarkers of exposure to ischemic reperfusion injury. Cryobiology 2000;41:280–91.
[283] Gibelin H, Eugene M, Hebrard W, Henry C, Carretier M, Hauet T. A new approach to the evaluation of liver graft function by nuclear magnetic resonance spectroscopy. A comparative study between Euro-Collins and University of Wisconsin solutions. Clin Chem Lab Med 2000;38:1133–6.

[284] Hauet T, Baumert H, Gibelin H, Godart C, Carretier M, Eugene M. Citrate, acetate and renal medullary osmolyte excretion in urine as predictor of renal changes after cold ischaemia and transplantation. Clin Chem Lab Med 2000;38:1093–8.
[285] Hauet T, Gibelin H, Richer JP, Godart C, Eugene M, Carretier M. Influence of retrieval conditions on renal medulla injury: evaluation by proton NMR spectroscopy in an isolated perfused pig kidney model. J Surg Res 2000;93:1–8.
[286] Hauet T, Goujon JM, Tallineau C, Carretier M, Eugene M. Early evaluation of renal reperfusion injury after prolonged cold storage using proton nuclear magnetic resonance spectroscopy. Br J Surg 1999;86:1401–9.
[287] Schmitz V, Klawitter J, Bendrick-Peart J, et al. Graft flushing with histidine-tryptophane-ketoglutarate (HTK) followed by extended cold preservation in University of Wisconsin (UW) solution in a rat kidney transplantation model- An improved preservation protocol? Eur J Surg Res 2006;38:388–98.
[288] Serkova N, Fuller TF, Klawitter J, Freise CE, Niemann CU. ^{1}H-NMR-based metabolic signatures of mild and severe ischemia/reperfusion injury in rat kidney transplants. Kidney Int 2005;67:1142–57.
[289] Bon D, Billault C, Thuillier R, Hebrard W, Boildieu N, Celhay O, Irani J, Seguin F, Hauet T. Analysis of perfusates during hypothermic machine perfusion by NMR spectroscopy: a potential tool for predicting kidney graft outcome. Transplantation 2014;97:810–6.
[290] Guy AJ, Nath J, Cobbold M, Ludwig C, Tennant DA, Inston NG, Ready AR. Metabolomic analysis of perfusate during hypothermic machine perfusion of human cadaveric kidneys. Transplantation 2015;99:754–9.
[291] Domański L, Safranow K, Ostrowski M, Pawlik A, Olszewska M, Dutkiewicz G, Ciechanowski K. Oxypurine and purine nucleoside concentrations in renal vein of allograft are potential markers of energy status of renal tissue. Arch Med Res 2007;38:240–6.
[292] Ojo AO. Renal disease in recipients of nonrenal solid organ transplantation. Semin Nephrol 2007;27:498–507.
[293] Brunet M, Shipkova M, van Gelder T, Wieland E, Sommerer C, Budde K, Haufroid V, Christians U, López-Hoyos M, Barten MJ, Bergan S, Picard N, Millán López O, Marquet P, Hesselink DA, Noceti O, Pawinski T, Wallemacq P, Oellerich M. Barcelona consensus on biomarker-based immunosuppressive drugs management in solid organ transplantation. Ther Drug Monit 2016;38(Suppl. 1):S1–S20.
[294] Klawitter J, Haschke M, Kahle C, et al. Toxicodynamic effects of ciclosporin are reflected by metabolite profiles in the urine of healthy individuals after a single dose. Br J Clin Pharmacol 2010;70:241–51.
[295] Schmitz V, Klawitter J, Bendrick-Peart J, Schoening W, Puhl G, Haschke M, Klawitter J, Consoer J, Rivard CJ, Chan L, Tran ZV, Leibfritz D, Christians U. Metabolic profiles in urine reflect nephrotoxicity of sirolimus and cyclosporine following rat kidney transplantation. Nephron 2009;111:e80–91.
[296] Klawitter J, Klawitter J, Kushner E, Jonscher KR, Bendrick-Peart J, Leibfritz D, Christians U, Schmitz V. Association of immunosuppressant-induced protein changes in the rat kidney with changes in urine metabolite patterns: A proteo-metabonomic study. J Proteome Res 2010;9:865–75.
[297] Klawitter J, Klawitter J, Schmitz V, Shokati T, Epshtein E, Thurman JM, Christians U. Mycophenolate mofetil enhances the negative effects of sirolimus and tacrolimus on rat kidney cell metabolism. PLoS One 2014;9:e86202.
[298] Stapenhorst L, Sassen L, Beck B, Laube N, Hesse A, Hoppe B. Hypocitrateuria as a risk factor for nephrocalcinosis after kidney transplantation. Pediatr Nephrol 2005;20:652–6.

[299] Serkova NJ, Christians U. Biomarkers for toxicodynamic monitoring of immunosuppressants: NMR-based quantitative metabonomics of the blood. Ther Drug Monit 2005;20:652–6.
[300] Kanaby M, Akcay A, Huddam B, et al. Influence of cyclosporine and tacrolimus on serum uric acid levels in stable kidney transplant recipients. Transplant Proc 2005;37:3119–20.
[301] Perico N, Codreanu I, Caruso M, Remuzzi G. Hypoeruricemia in kidney transplantation. Contrib Nephrol 2005;147:124–31.
[302] Armstrong KA, Johnson DW, Campbell SB, Isbel NM, Hawley CM. Does uric acid have a pathogeneric role in graft dysfunction and hypertension in renal transplant patients? Transplantation 2005;80:1565–71.
[303] Foxall PJ, Mellotte GJ, Bending MR, Lindon JC, Nicholson JK. NMR spectroscopy as a novel approach to the monitoring of renal transplant function. Kidney Int 1993;43:234–45.
[304] Le Moyec L, Pruna A, Eugène M, Bedrossian J, Idatte JM, Huneau JF, Tomé D. Proton nuclear magnetic resonance spectroscopy of urine and plasma in renal transplantation follow-up. Nephron 1993;65:433–9.
[305] Knoflach A, Binswanger U. Serum hippuric acid concentration in renal allograft rejection, ureter obstruction, and tubular necrosis. Transpl Int 1994;7:17–21.
[306] Rush D, Somorjai R, Deslauriers R, Shaw A, Jeffery J, Nickerson P. Subclinical rejection—a potential surrogate marker for chronic rejection—may be diagnosed by protocol biopsy or urine spectroscopy. Ann Transplant 2000;5:44–9.
[307] Wang NJ, Zhou Y, Zhu TY, Wang X, Guo YL. Prediction of acute cellular renal allograft rejection by urinary metabolomics using MALDI-FTMS. J Proteome Res 2008;7:3597–601.
[308] Blydt-Hansen TD, Sharma A, Gibson IW, Mandal R, Wishart DS. Urinary metabolomics for noninvasive detection of borderline and acute T cell-mediated rejection in children after kidney transplantation. Am J Transplant 2014;14:2339–49.
[309] Zhao X, Chen J, Ye L, Xu G. Serum metabolomics study of the acute graft rejection in human renal transplantation based on liquid chromatography-mass spectrometry. J Proteome Res 2014;13:2659–67.
[310] Klepacki J, Brunner N, Schmitz V, Klawitter J, Christians U, Klawitter J. Development and validation of an LC-MS/MS assay for the quantification of the trans-methylation pathway intermediates S-adenosylmethionine and S-adenosylhomocysteine in human plasma. Clin Chim Acta 2013;421:91–7.
[311] Calderisi M, Vivi A, Mlynarz P, Tassin M, Banasik M, Dawiskiba T, Carmellini M. Using metabolomics to monitor kidney transplantation patients by means of clustering to spot anomalous patient behavior. Transplant Proc 2013;45:1511–5.
[312] Kim CD, Kim EY, Yoo H, Lee JW, Ryu DH, Noh DW, Park SH, Kim YL, Hwang GS, Kwon TH. Metabonomic analysis of serum metabolites in kidney transplant recipients with cyclosporine A- or tacrolimus-based immunosuppression. Transplantation 2010;90:748–56.
[313] Bonneau E, Tétreault N, Robitaille R, Boucher A, De Guire V, Metabolomics: Perspectives on potential biomarkers in organ transplantation and immunosuppressant toxicity. Clin Biochem 2016;49:377–84.
[314] Christians U, Klawitter J, Klawitter J. Biomarkers in transplantation—proteomics and metabolomics. Ther Drug Monit 2016;38:S70–4.
[315] Bohra R, Schöning W, Klawitter J, Brunner N, Schmitz V, Shokati T, Lawrence R, Arbelaez MF, Schniedewind B, Christians U, Klawitter J. Everolimus and sirolimus in combination with cyclosporine have different effects on renal metabolism in the rat. PLoS One 2012;7:e48063.

[316] Klepacki J, Klawitter J, Klawitter J, Karimpour-fard A, Anderson E, Ingle G, Patel D, Johnson K, Cibrik D, Christians U. A Comprehensive biomarker study to compare tacrolimus and mycophenolic acid versus half-dose tacrolimus and everolimus in de novo kidney transplant patients in the Novartis US92 Study. Am J Transplant 2016;16(Suppl. 3):535.
[317] Legido-Quigley C, Stella C, Perez-Jimenez F, Lopez-Miranda J, Ordovas J, Powell J, van-der-Ouderaa F, Ware L, Lindon JC, Nicholson JK, Holmes E. Liquid chromatography-mass spectrometry methods for urinary biomarker detection in metabonomic studies with application to nutritional studies. Biomed Chromatogr 2010;24:737–43.
[318] Foxall PJD, Bewley S, Neild G, Rodeck CH, Nicholson JK. Analysis of fetal and neonatal urine using proton nuclear magnetic resonance spectroscopy. Arch Dis Child 1995;73:F153–7.
[319] Trump S, Laudi S, Unruh N, Goelz R, Leibfritz D. ^{1}H-NMR metabolic profiling of human neonatal urine. Magn Reson Mater Phy 2006;19:305–12.
[320] Fanos V, Van den Anker J, Noto A, Mussap M, Atzori L. Metabolomics in neonatology: fact or fiction? Semin Fetal Neonatal Med 2013;18:3–12.
[321] Mussap M, Antonucci R, Noto A, Fanos V. The role of metabolomics in neonatal and pediatric laboratory medicine. Clin Chim Acta 2013;426:127–38.
[322] Mussap M, Noto A, Fanos V, Van Den Anker JN. Emerging biomarkers and metabolomics for assessing toxic nephropathy and acute kidney injury (AKI) in neonatology. Biomed Res Int 2014;2014:602526.
[323] Baker M. In biomarkers we trust? Nat Biotechnol 2005;23:297–304.
[324] Shipkova M, López OM, Picard N, Noceti O, Sommerer C, Christians U, Wieland E. Analytical aspects of the implementation of biomarkers in clinical transplantation. Ther Drug Monit 2016;38(Suppl. 1):S80–92.
[325] Billelo JA. The agony and ecstasy of "omic" technologies in drug development. Curr Mol Med 2005;5:39–52.
[326] Miller MG. Environmental metabolomics: a SWOT analysis (strengths, weaknesses, opportunities and threats). J Proteome Res 2007;6:540–5.
[327] Heijne WH, Kienhuis AS, van Ommen B, Stierum RH, Groten JP. Systems toxicology: applications of toxicogenomics, transcriptomics, proteomics and metabolomics in toxicology. Expert Rev Proteomics 2005;2:767–80.
[328] Mirnezami R, Nicholson J, Darzi A. Preparing for precision medicine. N Engl J Med 2012;366:489–91.
[329] Nicholson JK, Holmes E, Kinross JM, Darzi AW, Takats Z, Lindon JC. Metabolic phenotyping in clinical and surgical environments. Nature 2012;491(7424):384–92.
[330] Chitayat S, Rudan JF. Phenome centers and global harmonization. In: Holmes E, Nicholson JK, Darzi AW, Lindon JC, editors. Metabolic phenotyping in personalized and public health care. Tokyo: Academic Press; 2016. p. 291–315.

CHAPTER FOUR

The Role of Proteomics in the Study of Kidney Diseases and in the Development of Diagnostic Tools

U. Christians, MD, PhD, J. Klawitter, PhD, J. Klepacki, PhD and J. Klawitter, PhD
iC42 Clinical Research and Development, Department of Anesthesiology, University of Colorado, Anschutz Medical Campus, Aurora, CO, United States

Contents

Biomarkers of Kidney Disease. http://dx.doi.org/10.1016/B978-0-12-803014-1.00004-2

INTRODUCTION

Most of the physiologic functions within a cell, a tissue, an organ, and an organism are mediated by proteins and, hence, proteins are of substantial interest as clinical diagnostic molecular markers [1]. Proteins are the functional output of genes [2]. While the genome is static, the proteome is dynamic, or constantly in flux, and changes in response to external and internal stimuli. Changes in gene expression patterns ("transcriptomics") are neither complete nor accurate surrogate markers of protein concentrations, their structures, or activities [3]. Concentrations and activities of proteins are controlled by processes affecting gene expression such as transcription, mRNA splicing and mRNA stability, processes regulating activity, such as protein folding and posttranslational modifications, allosteric interactions with substrates, products, inhibitors and activators, and processes inactivating proteins through covalent binding or breakdown. In particular, posttranslational modifications often play a critical role in the regulation of the activity of a protein. After translation, most proteins are modified through the addition of carbohydrates, phosphates, cholesterol synthesis pathways, intermediates, and other molecules. Posttranslational modifications are not encoded by genes.

The proteome is defined as the expressed protein and peptide complement of a cell, organ or organism, including all isoforms and posttranslational variants. While an organism possesses a single genome, it possesses multiple proteomes depending on the cell compartment, type of cell, type of tissue, and organ. Proteomes undergo constant temporal changes. Changes can occur within minutes, hours, and sometimes days if regulated via translation, but can also occur within seconds at the functional level. This may involve mechanisms, such as phosphorylation, substrate and cosubstrate interactions, allosteric inhibition and activation, reaction with radicals, and proteolytic cleavage. The term peptidome has been used for the peptide subset of the proteome [4].

Proteomics has been defined as the "systematic analysis of proteins for their identity, quantity, and function" [5]. Thus, the term proteomics summarizes the procedures required for analysis of a proteome. While typical protein analysis involves the assessment of an individual protein, proteomics investigate populations of proteins rather than a single protein [1].

Until the 1990s, enzymatic or chemical evaluation, such as Edman degradation, of highly purified proteins constituted the mainstream methods for the determination of amino acid sequences of polypeptides and proteins.

Protein profiling started with the introduction of two-dimension gel electrophoresis in 1975 [6]. However, it was not until the introduction of mass spectrometry, the availability of protein databases, search algorithms, and other informatics procedures during the last 20 years that identification of proteins cut from 2D-gels became routine [7]. The almost explosive development of modern proteomics technologies over recent years was associated with the completion of the human genome project and the availability of genome sequence databases, the progress in mass spectrometry technologies including the development of "soft" ionization technologies, such as electrospray and matrix-assisted laser desorption ionization, as well as advances in bioinformatics [8]. At present, mass spectrometry in combination with library searches has evolved as the backbone of proteomics and allows for the simultaneous structural identification of multiple proteins in complex mixtures [9].

Clinical proteomics has focused on the discovery of novel drug targets as well as the discovery of diagnostic and prognostic disease biomarkers [9]. Clinical proteomics also aims at providing the clinician with tools to accurately diagnose, monitor, and predict treatment effects for patients, thus, enabling individualized patient management when properly utilized. The key is that such protein-marker based strategies hold the promise of being highly sensitive, specific, and predictive, and, overall may outperform the currently established clinical diagnostic tests.

Why are Molecular Marker Strategies Considered Predictive?

Most kidney injuries leading to end stage renal disease are characterized by silent and progressive courses and nonspecific symptoms that in their early stages, often remain undetected by current clinical diagnostic tools [10]. The quality of diagnostic tools is determined by their sensitivity and specificity. The sensitivity and specificity of chemical and biochemical molecular markers that are traditionally used in clinical diagnostics, preclinical and clinical drug development is sometimes poor. The reasons include, but are not limited to, the fact that often the following assumptions have been made: (1) one marker detects all disease processes/drug effects targeted against a specific organ, and (2) one marker fits all patient populations and age groups. Also, when these more traditional markers were established in the clinic, the mechanisms of diseases or drug effects, in many cases, were not well understood. Molecular markers did not have to undergo the rigorous validation and qualification procedures as required by current regulatory guidelines today and have, historically, been introduced as diagnostic tools into the clinic based on scientific consensus. A good example of this is creatinine concentrations in serum. Although generally considered a marker of glomerular filtration in the kidney, it is now known that creatinine is also actively secreted in the proximal tubules and even reabsorbed by the kidney [11]. Serum creatinine is not specific for the kidney but can also increase in the case of muscle damage and it is gender and age-dependent. Furthermore, its sensitivity is poor and a rather large amount of glomeruli needs to be destroyed for the serum creatinine to increase by 20%, the value that is considered clinically significant [12]. This can result in a critical delay in therapeutic interventions. This becomes problematic when the disease process or drug toxicity primarily targets other parts of the kidney and glomeruli are only damaged at a later stage by secondary processes, such as inflammation. There has never been, and there will never be, a single molecular marker that is able to adequately assess all aspects of the kidney's function and detect all types of kidney injury with adequate sensitivity. Although these shortcomings of serum creatinine as a molecular marker for kidney injury are well documented, it still remains the primary marker for preclinical and clinical drug development.

Poor sensitivity and specificity relate directly with poor predictive value. To better understand how molecular markers can be more predictive, it is important to look at the stages of kidney injury caused by a disease or a drug. This is illustrated in Fig. 4.1. The development of a disease process or drug injury can roughly be divided into three stages: a genetic, biochemical, and symptomatic stage [15].

A genetic predisposition may increase the risk for an individual to develop a disease, modify the efficacy or tolerability of a drug, or influence its tissue distribution and pharmacokinetics; however, in most cases, other factors, such as diseases, drugs, nutritional status, and/or environmental factors, will also be required to trigger a pathologic biochemical process.

During the biochemical stage, changes in gene expression, protein expression, and biochemical profiles occur, but the cells and organs are still able to compensate for this. At this stage, an injury process should be detectable if sufficiently sensitive assays are available. During the biochemical phase, no notable histological damage has occurred yet, and the disease process may be fully reversible if an appropriate therapeutic intervention is available.

In the symptomatic stage, biochemical changes on a cellular, organ, or systemic level can no longer be compensated for. This leads to pathophysiological and histological changes that define the symptoms of the injury

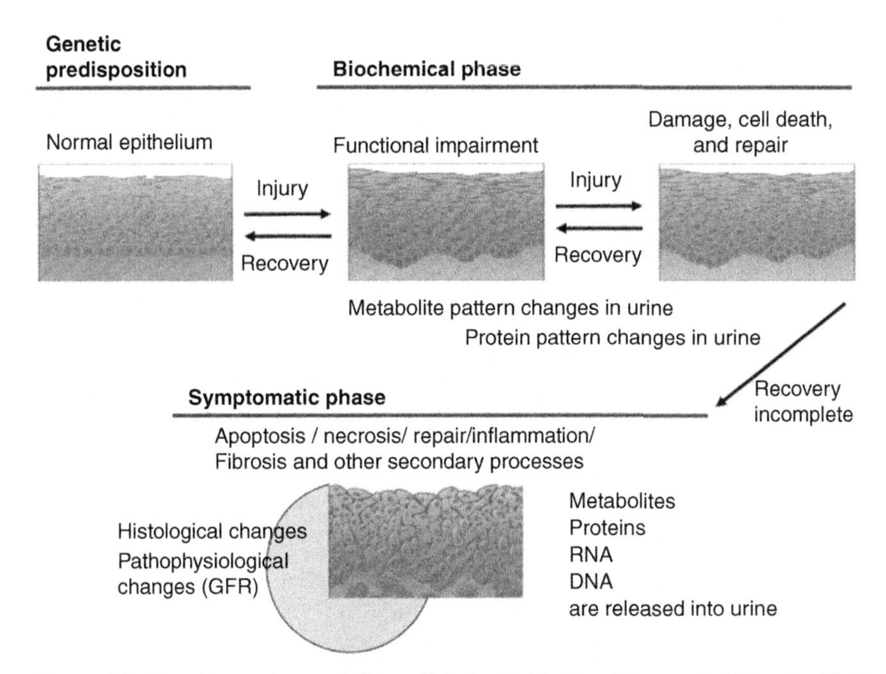

Figure 4.1 ***Time-Dependency of Kidney Tubular Epithelium Injury and Molecular Markers in Urine [13,14].*** Injury will affect cell function before histological and pathophysiological damage can be detected. At an early point in the process, this is reflected in protein and metabolite patterns in urine, as absorption and excretion are altered, repair proteins are formed and cells release proteins into urine. The resulting extent of urine metabolite and protein pattern changes depends on the intensity of the injury and how many cells/tubules are affected. Proteins that have been found to be changed in urine and that may serve as early kidney injury markers are listed in Table 4.5 and are shown in Fig. 4.4. As increasing numbers of cells die by necrosis and/or apoptosis, the biochemical phase of injury will progress toward the symptomatic phase. These cells will release at least some of their contents, such as metabolites, proteins, RNA, and DNA, into the urine. Cell death will also trigger secondary reactions such as inflammation and fibrosis. Once this occurs, a complete recovery may no longer be possible. The injury results in histological changes and kidney function will be reduced. It is not until the symptomatic phase that currently established diagnostic markers, such as, serum creatinine concentrations and blood urea nitrogen will significantly change.

process. Most established outcome metrics used presently during preclinical and clinical drug development detect injury processes in their symptomatic stage. The concept of monitoring biochemical changes and detecting an injury process before detectable histological or pathophysiological damage occurs is attractive. If the cause-effect relationships between protein expression, biochemical changes, the symptoms of a disease, and a drug effect or toxicity are known, then detecting specific changes in protein and cell biochemistry patterns has the potential to predict development of the symptomatic injury.

Technologies, such as genomics/transcriptomics, proteomics, and biochemical profiling (metabolomics) have the potential for the development of molecular marker strategies that allow for monitoring early changes in cell signal transduction, regulation, and biochemistry with high sensitivity and specificity and, therefore, may detect an injury process at a much earlier stage than currently established clinical diagnostic markers.

NONTARGETED AND TARGETED PROTEOMICS

Nontargeted

Nontargeted proteomics approaches try to evaluate a whole proteome. This concept embraces the acknowledgment that the complexity of protein networks and their interactions can only be assessed and fully understood if information covering all aspects of a biological system is available. By definition, targeted approaches that evaluate only one or several specific pathways are biased and may miss critical information. Nontargeted approaches, however, are nonbiased and seek to capture as much information as possible. Nevertheless, given the large numbers and varying abundance of different proteins in biological samples, as of today, there is no single experimental approach that enables the visualization of a complete proteome [2]. The three basic pillars (Fig. 4.2) of mass spectrometry-based nontargeted proteomics are [16]:

- the front end fractionation of complex mixtures,
- mass spectral data acquisition, and
- protein identification and characterization by database searching.

Proteomics inherently is a hypothesis-generating discovery technology. Proteomic studies can be classified as comparative studies that try to establish quantitative or qualitative protein differences between samples and descriptive studies that focus on the identification of proteins [2] (Fig. 4.3). In both cases, study designs can be either pathway-driven (targeted) or nonpathway-driven (nontargeted). Most clinical molecular marker discovery studies have no predetermined hypothesis of which pathways or proteins might be of interest. Nonpathway-driven studies are often conducted with the goal to offer new insight into previously unknown mechanisms. Attempts have been made to utilize the protein patterns to detect a pathobiochemical process without further protein identification and mechanistic qualification. This approach of pattern-based patient classification is also called

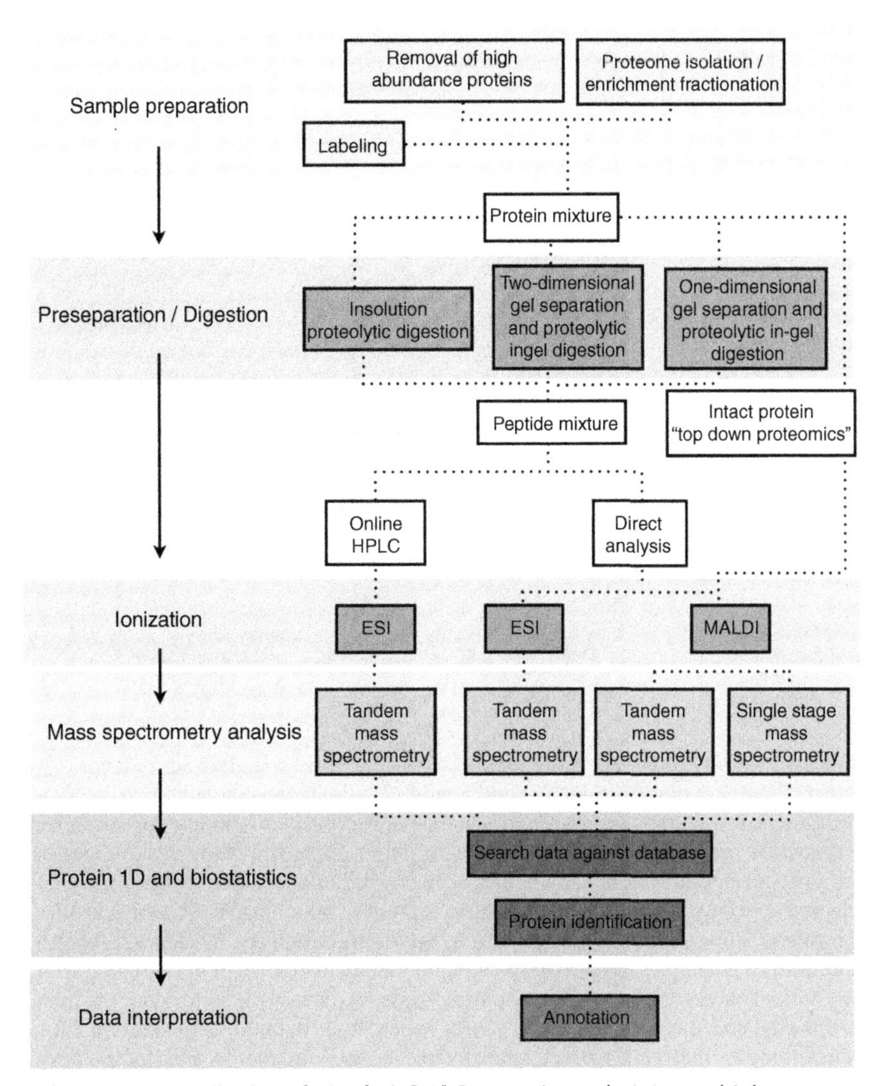

Figure 4.2 ***Proteomics Sample Analysis [17].*** Proteomics analysis is a multiple step procedure that typically involves sample preparation, preseparation and/or digestion, ionization, mass spectrometry analysis, protein identification, biostatistics, and annotation. Proteomics strategies can be divided into "bottom-up" and "top-down" approaches. Bottom-up approaches are most frequently used and involve digestion of the proteins of interest and, after mass spectrometry analysis, identification of proteins using database searching based on the detected peptides. Top-down proteomics does not involve a digestion step and analyzes the intact proteins. As discussed, both strategies have their advantages and limitations.

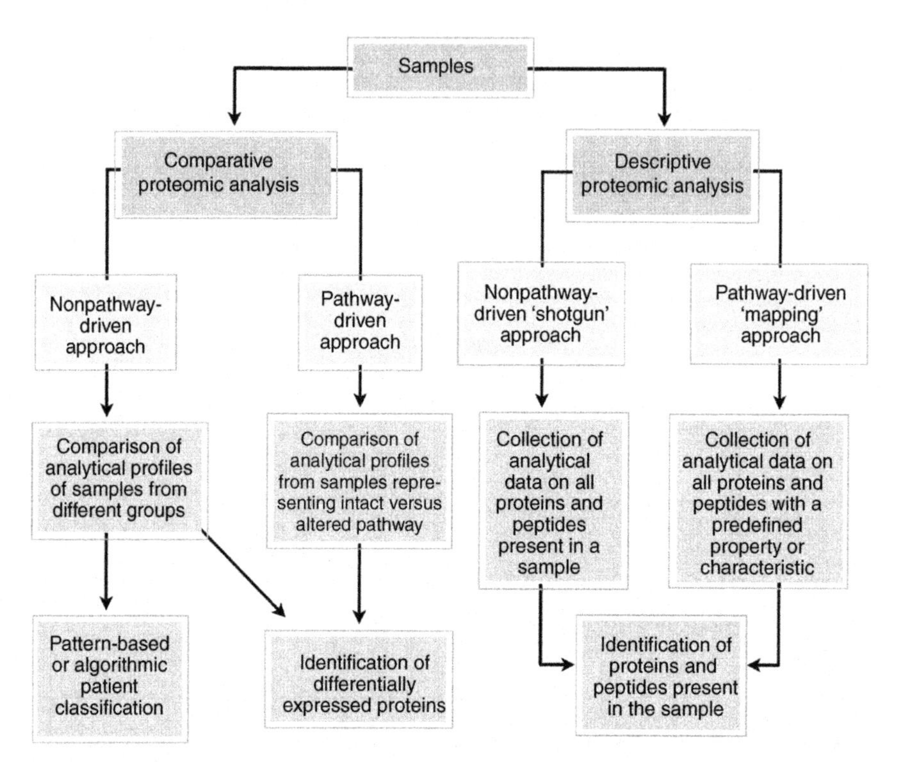

Figure 4.3 ***Main Proteomics Strategies [2].*** The goal of comparative approaches is to detect differences between samples and, therefore, requires semiquantitative comparison. Descriptive studies are usually qualitative and provide information about which proteins are present in a defined sample. In either approach, study designs can be pathway or nonpathway driven. Pathway driven studies are targeted—they focus on selected specific pathways, a protein interaction network or a specific subpopulation of proteins. Some previous knowledge or a hypothesis is required. By contrast, no prior biological knowledge is used in the design of nonpathway-driven or nontargeted studies. Global analysis is undertaken (although steps are usually taken to reduce sample complexity) and the data generated can be regarded as hypothesis-generating. Most clinical protein marker discovery studies have been nontargeted and comparative, and they identify proteins differing between study groups. Often such studies do not produce protein identities, but generate algorithms to classify samples on the basis of protein separation profiles (fingerprinting). The output in descriptive studies is a list of proteins. This list typically represents the catalogue of all proteins detectable with a particular technology [2].

"fingerprinting." Although straight forward, given the biological variability of a proteome and the many potential confounding factors in complex patient populations, fingerprinting has its usually poorly controlled risks. Furthermore, it is difficult to validate and qualify such diagnostic approaches to an extent that they will be acceptable for approval by regulatory agencies

[18]. In contrast, pathway-driven studies seek to achieve more in-depth mechanistic or functional insight [2]. They focus on specific proteins or protein networks and usually use a more targeted strategy.

Targeted

Targeted assays do not seek to capture a whole proteome, but assess a set of known proteins that typically have common pathways, protein network or context, such as inflammation or kidney dysfunction markers. In many cases, immunoassays are used for this purpose and only those proteins can be detected, against which antibodies are included. Thus, the use of targeted assays for research purposes requires preexisting knowledge about a disease process or drug effect or at least a hypothesis. Limitations are the availability of antibodies, their specificity and the sometimes poor batch-to-batch reproducibility of more complex assays.

In clinical proteomics, after the proteins of interest are identified, there may no longer be a need to assess the whole proteome because the desired information can be obtained by measuring a set of well-defined and qualified proteins. Another advantage of targeted assays is that these usually require less sample preparation, are quantitative, can be validated, are relatively high throughput, and can be run using instrumentation that may already be readily available in a clinical laboratory, such as ELISA readers, multiplexing platforms, or mass spectrometry.

An ideal case scenario would be the availability of quantitative targeted protein arrays that contain the whole human proteome and that can be scanned in a high-throughput fashion, similar to those already available for genome array analysis. Unfortunately, the information currently known about the human proteomes is insufficient. There is more than one relevant proteome, and the technology for building such comprehensive protein chips is not yet available.

PROTEINS AND THE KIDNEY

Kidney research has mainly focused on two proteomes: the kidney and urine. Although the kidney extensively communicates with the blood compartment, blood or plasma proteomes have only been of minor interest, simply because proteome changes originating from the kidney are quickly diluted and mixed with protein populations from other organs. This may create opportunities for a more systemic and holistic analysis, but it also complicates the interpretation of such data.

The renal proteome is made up by multiple cell types that comprise the kidney. The kidney can be viewed as an assembly of subproteomes of lesser complexity than the whole, released by or contained in kidney cell compartments such as plasma membranes, nuclei, cytosol, and mitochondria [19]. Proteomic studies have sought to improve our understanding of kidney function, attempted to map proteins in the cortex of the human kidney [20] and compared protein expression in the cortex and medulla of the rat kidney. The function and regulation of specific cell populations in the glomerulus, proximal tubule, thick ascending loop of Henle, and inner medullar collecting duct of the kidney have been studied using cell culture models. They have also been studied after isolation of the cells and tissues of interest using sieving or microdissection techniques. For a comprehensive review, see Janech et al. [7].

Structures in the cortex mainly reabsorb water, electrolytes, glucose, and amino acids and they produce hormones that regulate blood pressure (renin), hematopoiesis (erythropoietin), and calcium homeostasis (1,25 dihydroxy vitamin D3) [21]. The inner medulla is mainly responsible for concentrating urine and is characterized by high osmolarity and relatively low oxygen tension. In contrast to the relatively leaky proximal tubule, the inner medullary collecting duct of the mammalian nephron is characterized by low sodium permeability and by a large transepithelial resistance [22]. A tight epithelial barrier is critical for the control of sodium excretion. To maintain functionality and cellular viability, inner medulla cells have a unique metabolism that ensures the maintenance of intracellular ATP concentrations (high expression levels of the γ-subunit of Na^+/K^+-ATPase), intracellular osmolyte concentrations (NUP88), and tight junction integrity (MUPP1) [22]. As indicated, these distinct functions require the expression of specific sets of proteins. Exact knowledge of these distinct proteomes will not only allow for characterizing the type of injury and yield information regarding the associated mechanisms, but also for locating the injury. As shown in Fig. 4.4, patterns of protein kidney injury markers can be mapped in the nephron.

As urine can harbor proteins from all kidney subproteomes, and the protein composition of urine is perturbed by kidney injury or disease, the urine proteome can subsequently signal the status of kidney health as well as the onset, nature, and location of injury and dysfunction [19]. Even though an intact glomerular membrane will prevent larger proteins from entering the so-called "primitive" urine, the urine of healthy individuals contains a significant amount of peptides and proteins. These proteins originate from

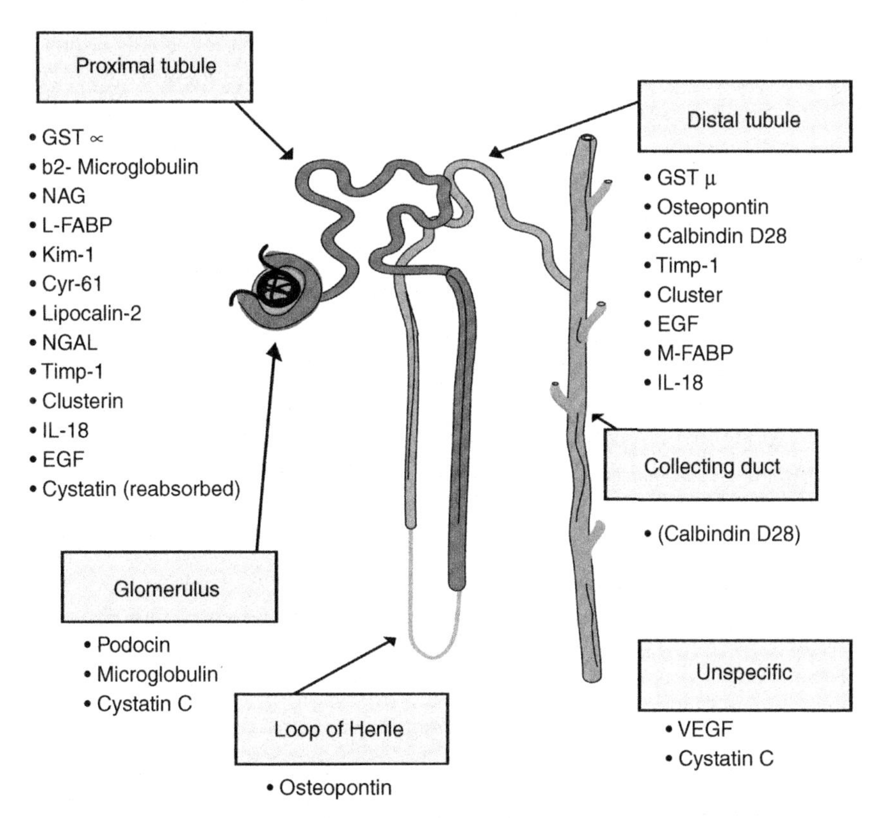

Figure 4.4 ***Protein Markers of Kidney Injury and Their Mapping to the Nephron.*** Potential marker proteins frequently mentioned in the literature are shown. Thus, this list should not be considered complete. The mapping represents the most abundant locations; however, in the case of some proteins, this may be an over-simplification. For more information about these proteins, please see Table 4.5.

three main sources: extra-renal (such as filtration of plasma proteins), the kidney, and the lower urinary tract [23]. It has been estimated that 49% of the proteins in urine are soluble proteins that enter the urine *via* glomerular filtration or tubular secretion, such as Tamm-Horsfall protein, 48% are urinary sediment proteins (mainly due to sloughing of epithelial cells, shedding of microvilli and apoptosis of epithelial cells resulting in cell membrane fragments), and 3% stems from urinary exosomes [24]. However, the handling of proteins by the kidney involving the complex glomerular slit diaphragm and podocytes as well as the role of proximal tubule epithelium in protein secretion, breakdown, and reabsorption is not fully understood yet [25,26].

The human urine proteome probably contains more than 100,000 proteins and peptides of which 5,000 are considered high frequency and have been observed in more than 40% of the individuals examined in different studies [27–29]. However, not all of these have been identified yet. In a 2D electrophoresis study, 1,118 protein spots were reproducibly found in normal urine samples. Two hundred seventy-five of those were characterized as isoforms of 82 proteins [30]. Although there is extensive knowledge regarding the handling of small molecules by the kidney, there is surprisingly little data regarding the handling of proteins. Using proteomics technologies, plasma and urine proteomes, which were considered the input and output proteomes, were studied [31]. After removal of proteins secreted downstream of the kidney, 2611 proteins were found in plasma and 1522 proteins were found in urine. These could be separated into three subproteomes: plasma-only (2280 proteins), plasma and urine (394 proteins), and urine-only (1128 proteins). It seemed reasonable to assume that the plasma-only subproteome was derived mainly from soluble proteins and proteins in solid plasma components that do not pass through the glomerular membrane. The plasma-only proteome also contained proteins that had a molecular weight of <30 kDa, based on their molecular weight, should have been filtered, but were probably retained due their charge, shape, interactions, or associations with other proteins. The plasma-and-urine subproteome probably contained soluble proteins that were filtered from plasma or secreted by the kidney. The urine-only subproteome was most likely constituted by soluble proteins that were released into the urine by epithelial secretion or shedding and/or were from solid-phase components in urine [31] (Table 4.1).

It is well-established that aging induces morphological changes of the kidney and results in reduced kidney function. This includes the glomerular filtration rate that declines 20–25% during the age range of 40–80 years and the ability of the kidney to concentrate urine. In a study based on 324 apparently healthy subjects between the ages of 2 and 73 years of age, the low molecular weight proteome in urine was assessed using capillary electrophoresis-mass spectrometry [34]. Five thousand polypeptide signals could be separated, and 325 of them showed age-dependent differences. Most of these changes were associated with the development of the kidney before puberty. Forty-nine proteins were found to be associated with aging in adults and several of these were associated with proteolytic activity and uromodulin targeting. Interestingly, several subjects did not have urinary protein patterns that matched their age and this may have reflected undetected chronic disease processes [34].

Table 4.1 Sources of urinary proteins

Soluble proteins	
• Glomerular filtration of plasma proteins	• Normally present (<150 mg/day) • Defects in glomerular filter increases high molecular weight protein concentrations in urine, such as albumin • Defects in proximal tubule reabsorption or abnormal production of low molecular weight plasma proteins increase low molecular weight proteins, such as β2-microglobulin
• Epithelial cell secretion of soluble proteins	• Via exocytosis (e.g., epidermal growth factor) or glycosylphosphatidylinositol-anchored protein detachment (e.g., Tamm-Horsfall protein), proteolytic fragments
• Interstitial processes, cell injury, and other cells	• Leakage of proteins during injury, such as inflammation, immune reactions, necrosis, apoptosis, and repair; products of prostate gland
Solid phase components	
• Whole cell shedding of epithelial cells • Plasma membranes and intracellular organs • Exosome secretion • Other cells and bacteria	• Increased cell number during diseases, such as acute tubular necrosis (renal tubular cell shedding) and glomerular diseases (podocyte shedding) • Nonspecific nephrotoxic, necrotic, and apoptotic processes • Normal process [33] • During certain disease processes: red blood cells, white blood cells, tumor cells, bacterial infections

Epithelial cells include all epithelial cells along urinary tract from podocytes to urethral epithelia.
Source: Based on O'Riordan E, Goligorsky MS. Emerging studies of the urinary proteome: the end of the beginning? Curr Opin Hypertens 2005;14: 579–585 [23] and Pisitkun T, Johnstone R, Knepper MA. Discovery of urinary biomarkers. *Mol Cell Proteomics* 2006;5:1760–71 [32].

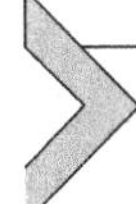

THE PROTEOMICS SAMPLE

Urine

As discussed earlier, urine is an attractive matrix. It is considered a "proximal" matrix, a biofluid that is close to, or in direct contact with, the site of disease [18]. Proximal fluids are local sinks for metabolites, proteins or peptides secreted, shed or leaked from the tissue of interest. The nephron is capable of filtering smaller proteins and reabsorbing proteins. Proteins are distinctly formed and distributed throughout the different parts of the nephron, the reasons of which are defined by differences in function, availability of oxygen, and osmolarity. Thus, changes in protein patterns in the urine allow for localizing the injury in the kidney (Fig. 4.4).

The gold standard for the quantification of proteinuria has been 24-h urine collection, but 24-h urine collection is time-consuming, inconvenient, and dependent on the patient's compliance [24]. Collection of midstream urine is widely considered the standard for urine proteomics analysis [35]. While no difference has been observed between first-void and mid-stream urine in males, there was marked variation in females, most likely due to bacterial contamination of first-void urine [36]. However, this may also depend on which disease is targeted with the analysis. Prostate cancer markers were found in higher abundance in first-void than in mid-stream urine [37], indicating that urine collection protocols affect the results of proteomics studies and need to be assessed during method development [38].

Urine as a source of protein markers has several advantages and disadvantages [28]. Urine can easily be obtained noninvasively in relatively large quantities and there are no limits for how often urine can be collected from the same patient. Typically, urinary peptides and proteins are water soluble. Thus solubilization that can pose major problems in the proteomics analysis of cells and tissues is not a problem with urinary peptides and proteins. Urinary proteins are usually small with molecular weights of less than <30 kDa and can be analyzed with time-of-flight mass spectrometers without enzymatic digestion to peptides (top down proteomics). In most cases, they are also stable since urine stagnates in the bladder at body temperature for hours. Degradation and proteolytic processes are often complete by the time the urine is collected [28]. The addition of protease inhibitors to stabilize urine samples for proteomics analysis is not recommended anymore [38,39]. In contrast, activation of proteases constitutes a major problem in the collection of blood samples for proteome analysis. Nonetheless, urine is considered one of the most difficult proteomic samples to work with due to its highly variable contents, dilution due to fluid intake, and pH, as well as the presence of various proteins in low abundance or modified forms [28].

It has independently been shown that the urine proteomics samples can be stored for up to 6 h at room temperature, up to 3 days at +4°C and several years at −20°C [28,36,39–41], but this may not be the case for individual proteins, peptides [13], and urinary exosomes that have been described to be less stable [33]. Long-term storage at −80°C seems to be a safer approach [33,42]. When frozen samples were thawed, an initial loss of minor protein signals was observed [43]. Hereafter, urine samples for proteomics analysis were generally found to be stable for at least three freeze-thaw cycles, but marked losses of proteins were found if samples were frozen and thawed more often [36,38,43]. Overall, it is recommended to avoid freeze-thaw cycles whenever possible [35].

A challenge is that the dynamic range of protein concentrations in body fluids spans, several orders of magnitude and urine is no exception to this rule [18,28,44,45]. Most analytical approaches assessing the urine proteome include an initial sample preparation step enriching the proteins of interest. A common strategy utilized is the removal of high-abundance proteins that confer little diagnostic information using techniques, such as column purification (size exclusion, ion exchanger, affinity columns), selective surfaces, immunodepletion and equalizer beads [28,41,46]. Immunodepletion has the inherent risk of also loosing proteins of interest by codepletion [28] that may be caused by protein–protein interactions independent of the desired specific antibody interactions. The preparation of urine samples for proteomic analysis has systematically been studied and is described in detail by Kushnir et al. [46], Thongboonkerd et al. [47,48], Pieper et al. [49], and Khan and Packer [50].

Kidney Tissues and Cell Culture

The basic principle of the preparation of a tissue sample is that the heterogeneity should be diminished as much as possible and that the sample should be pure and relevant. The first step in the proteome analysis of tissue samples is homogenization. Homogenization methods used for proteomics purposes can be divided in five major categories: mechanical, ultrasonic, pressure, freeze-thaw, and osmotic or detergent lysis [51]. It is critical to protect the samples from proteolysis during processing [39]. The most common protective measures are protein denaturation and the addition of protease inhibitors. Later steps include the removal of contaminants, such as salts, detergents, abundant proteins, lipids, polysaccharides, nucleic acids, and other contaminants, as well as protein enrichment, using precipitation, centrifugation, prefractionation, electrophoretic, antibody-based procedures, and/or chromatographic techniques [51]. The sample preparation approach also depends on the intended analysis. For comprehensive reviews, please see Bodzon-Kulakowska et al. [51], Ahmed [52], Matt et al. [53], and Hu et al. [54].

Cell cultures are of interest for mechanistic and molecular marker qualification studies. It is assumed that a cell on average expresses 10,000 proteins [51]. If a cell culture contains multiple types of cells, then this number is higher. The preparation of cell cultures is simpler than that of tissue and often involves direct lysis of the cells in the dish after removal of the cell culture medium as a first step. After solubilization, the sample is transferred and sonicated. The following steps may involve those as described for tissue sample preparation earlier [51–53]. Tissue samples, cells, and purified samples should be stored for long term at −80°C.

It should always be kept in mind that the results of proteomics analysis may be influenced by sample preparation (e.g., 2D-gels, enzymatic digestion, and isolation of cell organelles), the selectivity of separation technologies preceding mass spectrometry analysis (e.g., activated surfaces and ion exchangers), and ionization methods.

ANALYTICAL TECHNOLOGIES

To date, more than 228,000 human proteins have been described in the literature [55]. It has been estimated that the number of different components that comprise the human proteome adds up to approximately one million, vastly exceeding the number of different genes in the human genome [56]. To date, 20,687 human protein-coding genes have been identified [57]. As aforementioned, this is because of single nucleotide polymorphisms and posttranslational modifications. Currently, more than 300 different posttranslational modifications have been described. Posttranslational modifications, such as phosphorylation, can be temporary and diseases can influence protein reactions, for example, through radical formation, as well as somatic mutations.

Table 4.2 gives an overview of the major proteomics technologies. Proteomic approaches can broadly be separated into techniques that are based on separation and detection of the intact proteins (top-down proteomics) and techniques that involve digestion of the proteins into peptides and analysis of the resulting peptide patterns (bottom-up proteomics) [16,59,60]. In the traditional bottom up approach, intact proteins are digested into peptides and peptide mixtures are then separated by HPLC and analyzed utilizing mass spectrometry [60] (Fig. 4.2). The peptide patterns are analyzed using database searches and protein hits are identified based on the peptide patterns. Some disadvantages of the bottom-up approach are:

- the ability to quantify proteins is limited, unless labels are used;
- the sometimes difficult identification of proteins based on complex peptide mixtures; and
- posttranslational modifications may get lost or remain undetected.

Table 4.2 Proteomics technologies

Technology	Description	Advantages	Disadvantages
2-Dimensional electrophoresis	Separation by isoelectric point and size. Proteins are stained after separation and compared between gels.	Widely available. Easiest form of "top down" proteomics. Separation of charge reflects posttranslational modifications.	Low-abundance, molecules < 10 kDa as well as large, basic, and hydrophobic proteins, such as membrane proteins are difficult to detect.
DIGE	2DE with fluorescent labeling of proteins before separation in gel. Two proteomes (e.g., treatment and control) as well as an internal standard can be separated on the same gel.	Improved spot alignment, improved reproducibility, better quantification, and spot abundance in comparison to 2D gels.	Low-abundance, large and hydrophobic proteins are difficult to detect, requires three color imaging system and other additional equipment compared to 2D-electrophoresis.
LC-MS	Proteins are digested before separation by HPLC. The HPLC can be one dimensional (also after protein spots are cut from gels) or multidimensional using column switching. HPLC separation can also be done offline.	Sensitive and more likely to see low abundance proteins and other proteins than 2D electrophoresis. Can easily be automated. Allows for protein identification in combination with database search.	Quantification and measurement of posttranslational modifications require additional tools. Not very quantitative. Reassembly of tryptic peptides into molecules can lead to incorrect results.
ICAT/iTRAQ	LC-MS with isotopic labeling of the peptides after digestion. Like DIGE, this allows for simultaneous analysis of several proteomes. The labeled peptides are chemically identical but have predictable mass differences. The abundance of the differently labeled peptides is detected in the mass spectrometer by comparison of the intensity of the same peptides with different labels.	Relative quantification of low-abundance and hydrophobic samples. Since several proteomes are analyzed simultaneously, peak alignment is less of a problem.	The number of direct comparisons is limited. Quantification and measurement of posttranslational modifications will require additional tools.

(*Continued*)

Table 4.2 Proteomics technologies (*cont.*)

Technology	Description	Advantages	Disadvantages
SELDI	Proteins are bound to affinity surfaces on a MALDI chip. Different surfaces are available. Bound proteins are detected in the mass spectrometer (usually time-of-flight).	Samples can be enriched for specific low-abundance proteins. High-throughput platform for protein marker discovery.	Proteins that do not bind to the selected surfaces will not be detected. Thus, this technology is more biased than those discussed earlier. Not intrinsically quantitative. Large amount of variability between laboratories has historically been a problem. One of the reasons is that the manufacturer combined it with a low-resolution TOF instrument and its susceptibility to interferences.
Capillary electrophoresis-MS	Separation of proteins by elution time in capillary electrophoresis and by size in mass spectrometer, highly sensitive, low sample volume.	Reproducible and sensitive. Good technology for protein marker discovery.	Limited to proteins < 20 kDa.
Protein arrays	Proteins, antibodies, or aptamers printed on a microchip or bead in a multiplexed format.	Sensitive and rapid. Allows for semiquantitative comparison among samples.	Antibody specificity is variable and difficult to control. In contrast, aptamers have the advantage that they are synthesized and thus more reproducible. Is bias-based and depends on the proteins, antibodies or aptamers bound. It does not detect proteins that do not bind to the array.

Abbreviations: DIGE, Difference gel electrophoresis; ICAT, isotope-coded affinity tags; iTRAQ, isobaric tag for relative and absolute quantitation; LC-MS, liquid chromatography-mass spectrometry; MALDI, matrix-assisted laser desorption/ionization; SELDI, surface-enhanced laser desorption/ionization; 2D, two-dimensional.

Source: Based on Janech MG, Raymond JR, Arthur JM. Proteomics in renal research. Am J Physiol Renal Physiol 2007;292: F501–F512 [7]; Domon B, Aebersold R. Mass spectrometry in protein analysis. Science 2006; 312: 212–217 [9]; Decramer S, Gonzalez de Peredo A, Breuil B, Mischak H, Monsarrat B, Bascands J., Schanstra JP. Urine in clinical proteomics. Mol Cell Proteomics 2008;7: 1850–1862 [28], de Hoog CL, Mann M. Proteomics. Annu Rev Genomics Hum Genet 2004;5: 267–293 [58].

Top-down proteomics avoids these problems by analyzing the intact protein. It allows for detecting posttranslational modifications and provides a "bird's-eye" semiquantitative view of a proteome [16,60]. The simplest form of top-down proteomics is 2D-gel electrophoresis. In most cases, mass spectrometry-based top down proteomics involves high-resolution mass spectrometers and works best for proteins <100 kDa and less complex proteomes [16,60]. In the past, technical limitations have caused top-down proteomics to stay in the background, however, due to recent advances in separation technology, mass spectrometry instrumentation and bioinformatics tools, top-down proteomics has emerged as complementary and even alternative technology to digestion-based bottom-up approaches [60]. Most top-down proteomics studies have involved denaturization of proteins before mass spectrometry-based analysis. Although this preserves most posttranslational modifications, biologically relevant protein–protein and protein–ligand information are lost [60,61]. Analysis of native proteins after native size-exclusion or ion exchange chromatography seem to be a viable alternative. It has been shown that native protein complexes can be transferred into the gas phase and analyzed by mass spectrometry [60].

The combination of different proteomics methods may result in some overlap but also will give significant additional information.

Electrophoresis

In the early days of proteomics, 2D gel electrophoresis-based approaches were the standard [62]. 2D gel electrophoresis is based on the separation of individual proteins contained in a proteome by isoelectric point (first dimension) and then by molecular weight (second dimension). The intact proteins are visualized and quantified by staining and densitometry. Thus, 2D-gel electrophoresis is a top-down proteomics technique. 2D-gels can resolve 1,500–3,000 proteins. By spreading the pH range across several gels, also known as zoom gels, between 5,000 and 10,000 proteins can be resolved [63]. 2D-gels of different proteomes that run on separate gels are compared and the intensities of stained protein spots are analyzed using statistical procedures. Next, relevant protein spots are cut out, destained, digested and the peptides are eluted from the gel matrix [64]. The result is a peptide mixture from substantially purified proteins, which is further analyzed by peptide fingerprinting using MALDI-TOF or HPLC-ion trap mass spectrometry, then followed by database search. 2D-gels and subsequent mass spectrometry-based identification is best suited for samples of limited complexity where specific proteins that need to be characterized do not require high throughput [9]. Inherent problems include that certain groups of proteins, such as membrane proteins usually are not captured, the reproducibility among separations run on different gels, and achieving the correct alignment of corresponding protein spots.

The latter problem has been improved by the introduction of difference gel electrophoresis (DIGE) [65–67]. Although over recent years, LC-MS has emerged as the most widely used method for proteomics, DIGE is still relevant [67]. DIGE allows for the visualization of hundreds to thousands of protein species, including information about changes in pI and molecular weight due to truncation, degradation, genetic variation, alternate splicing, posttranslational processing, and modifications [67]. In contrast, LC-MS has difficulties to quantify truncated or degraded protein species, as absence of peptide detection does not confirm absence from a sample [67]. For DIGE analysis, two or more proteomes are stained with different fluorescent dyes. The proteomes and internal standards are then pooled and simultaneously separated on the same 2D gel. After separation, the different proteins are visualized separately using respective discrete excitation and emission wavelengths. Fluorescent labeling, inclusion of internal standards and simultaneous separation have not only improved reproducibility, but also the reliability of semiquantitative comparison of proteomes [7,67].

The use of capillary electrophoresis coupled to a mass spectrometer (CE-MS) as an alternative to LC-MS has substantially increased over recent years, mostly due to the availability of novel and more effective mass spectrometry interfaces [64,68]. In comparison to HPLC-MS, CE-MS requires lower samples volumes and usually allows for faster separation. On the other hand, limitations include small sample capacity and difficulties in analyzing of proteins >20 kDa [69]. The most commonly used CE-MS ionization sources are electrospray-ionization and matrix assisted laser desorption/ionization (MALDI) [64,68]. CE-electrospray ionization interfaces include sheathless and sheathflow liquid-junction interfaces [70,71]. The sheath liquid in the latter provides electrical contact between the CE separation and electrospray source, albeit at the expense of diluting the CE effluent [64,71]. In most cases CE-MALDI-MS couplings have been offline, with the CE effluent collected by a fraction collector and then analyzed in a MALDI-MS instrument [64]. CE-MS has been used for both bottom-up and top-down proteomics [64]. Microfluidic chip CE-MS and LC-CE-MS devices for proteomics analysis have been developed [64,72,73].

LC-MS

Mass spectrometry-based protein identification relies on the digestion of protein samples into peptides using a sequence-specific protease, such as trypsin [58]. Trypsin cleaves at the C-termini of ariginine and lysine residues. Based on the occurrence of these two amino acids in proteins, an average of 10 peptides is expected for a stretch of 100 amino acid residues [74]. There are the following advantages of analyzing peptides rather than proteins by LC-MS:

- If proteins have been separated by 2D-gels, peptides are easier to elute from the gels than proteins.
- The molecular weight of proteins alone is often not sufficient information to identify a protein.
- Proteins are heterogeneous and do not necessarily possess a single molecular weight.
- The peptides add structural information.
- Peptides fall into the effective mass range of most mass spectrometers, that is, between 1 and 5 kDa, unless a time-of-flight mass spectrometry detector is used.

Mass spectrometers consist of the following components: an inlet, an ionization source, sections to focus, separate, select and fragment ions, and a detector [75]. The most common inlets and ionization methods used in proteomics are high-performance liquid chromatography (one- or multidimensional) in combination with electrospray ionization or MALDI. Mass spectrometers measure the mass-to-charge ratio of an ion. This is achieved by manipulating ions in an electric and/or magnetic field or by measuring time-of-flight. In contrast to small molecules that in most cases, are singly charged (this means that the mass-to-charge ratio reflects their molecular weight), large molecules are usually multiply charged. The intensity of the signal caused by a specific molecule reflects the abundance of the ion. As aforementioned, a problem with ionization in a liquid phase or matrix such as electrospray ionization and MALDI is that the abundance of ions varies with ionization efficiency, which may depend on other molecules that are in the ionization source at the same time (ion suppression, ion enhancement). Therefore, mass spectrometry cannot be considered a quantitative technology for proteins when complex mixtures are analyzed [75]. In addition to quadrupole and time-of-flight mass spectrometers, ion traps and hybrids, such as quadrupole- time-of-flight (QTOF), quadrupole linear ion traps (QTRAP), ion trap-orbitrap, ion trap-cyclotron resonance Fourier transformation mass spectrometers (FTMS) are used (Table 4.3). Much of the uncertainty of peptide identification is directly related to the accuracy of the mass spectrometer used. The more accurate the mass, the less potential false positive matches in a data base search are possible. High-resolution mass spectrometers, such as FTMS and orbitraps greatly increase the confidence in peptide and, subsequently, protein identification [60,75]. In addition, the better resolution of peptides with similar masses allows for

Table 4.3 Comparison of mass spectrometry detectors used for proteomics

	3D-ion trap	Q-TOF	TOF-TOF	FTMS	Orbitrap	QQQ	QTRAP
Mass accuracy	+	+++	+++	++++	++++	++	++
Resolving power	+	++	+++	++++	++++	+	+
Sensitivity	++	+	+++	++	+++	++++	++++
Dynamic range	+	++	++	++	+++	++++	++++
Ionization source	API (MALDI)	API (MALDI)	MALDI	API	API	API	API
MS/MS capabilities	MS^n	Yes	Yes	Yes	Yes[a]	Yes	MS^n
Protein identification	++	++	+++	++++	++++	+	+++
Quantification	+	+++	++	+++	+++	++++	++++
Throughput	++	++/+++[b]	++++	++	++	++	++
Detection of posttranslational modifications	+/+++[c]	+	+	+	+	+	+++

Abbreviations: API, Atmospheric pressure ionization source such as electrospray ionization and atmospheric pressure chemical ionization; FTMS, Fourier transformation ion cyclotron resonance mass spectrometer; Q-TOF, quadrupole-time-of-flight mass spectrometer; TOF-TOF, tandem time-of-flight mass spectrometer; QQQ, triple stage quadrupole mass spectrometer; QTRAP, triple stage quadrupole/ linear ion trap mass spectrometer; MALDI, matrix assisted laser desorption/ionization.
+, Possible or low; ++, good; +++, high; ++++, excellent.
[a] only in combination with collision cell
[b] with MALDI
[c] in combination with electron transfer dissociation (ETD)
Source: Based on Domon B, Aebersold R. Mass spectrometry in protein analysis. Science 2006;312: 212–217 [9]; Catherman AD, Skinner OS, Kelleher NL. Top Down proteomics: facts and perspectives. Biochem Biophys Res Commun 2014;445: 683–693 [60]

the detection of more signals when compared with lower resolution mass spectrometers [9].

Tandem mass spectrometry is the basis of shotgun proteomics or MudPIT (multidimensional protein identification technology) approaches, a strategy that attempts to analyze the complete proteome of a cell, tissue, or organism in a single experiment [74,76,77]. Although different approaches have been described, they all have the same basic strategy in common [78]: the proteins in the sample are digested usually using trypsin, the resulting peptide mixtures are subjected to one-, two- and three-dimensional fractionation (online or offline), and the peptides from the last separation step are usually separated using reversed-phase chromatography and analyzed by tandem mass spectrometry, in most cases, including a linear ion trap [74]. The MS/MS spectra are assigned to peptide sequences and software tools using search algorithms to assign the detected peptides to proteins [79,80].

For the last two decades, mainly two LC-MS/MS data acquisition strategies have been used, data-dependent acquisition (DDA) for nontargeted/shotgun proteomics and multireaction monitoring (MRM) for the targeted detection and precise and reproducible quantification of proteins [81,82]. In an MS/MS detector, MS1 is usually set to scan the abundance and mass/charge of all ions eluting from the chromatographic column at any given time. In MS2 selected or all ions (precursors) are fragmented and then the abundances and mass/charge-values of the fragments are monitored. DDA and MRM have in common that a single precursor is isolated in MS1, fragmented and analyzed in MS2. While in the case of MRM, the user selects precursor ions of interest, in DDA mode, the precursor ions are chosen by the instrument on the basis of abundance [82]. In MRM, only a specific, user-selected fragment is detected, while DDA scans for all fragment ions in MS2. Most MS/MS instruments can perform a DDA cycle (MS1 scan and 10 MS2 scans) within 2 s. This means that if too many peptide species coelute, then DDA detects only the most abundant peptides and misses all others. This reduces reproducibility and the chance that low-abundance peptides are detected [82]. Moreover, to capture as many peptides as possible, DDA algorithms limit sampling of each peptide species to only once or twice. This precludes reliable quantification, which would require unlimited sampling of the most abundant peptides. Ideally, an acquisition strategy would have the breadth of DDA and the precision of MRM quantification. Thus, over the last years, data independent acquisition (DIA) strategies have been developed. In contrast to DDA and MRM, all precursors within a certain relatively large mass/charge range are isolated, fragmented and analyzed in a single MS2 scan. DIA repeatedly samples the same peptide mixtures within the selected MS1 mass/charge window and thus allows for more precise quantification than DDA [82]. Low abundance peptides that fall into the mass/charge window are not excluded and can be identified as long as their spectra show sufficient signal intensity for proper interpretation. The complexity of the MS2 spectra affect sensitivity and DIA is more vulnerable to interference from other peptides [82]. Moreover, the highly complex MS2 spectra are much more difficult to interpret than in DDA [81,82]. Traditional DDA algorithms for peptide identification cannot be used for DIA, as such algorithms assume that all fragments detected in MS2 originate from a single precursor peptide [82]. Hence, the development of new computational approaches for the analysis of DIA spectra is required. For a detailed review, please see Hu et al. [82]. Although a promising approach, the broader acceptance of DIA will require future improvements in hardware and software, such as instruments with faster duty cycle speed, better sensitivity, and peak resolutions at the MS2 level [82].

While, as aforementioned, 2D gel electrophoresis can separate proteins with different posttranslational modifications, the identification of modified proteins in complex mixtures using LC-MS/MS remains a challenge. Among the known posttranslational modifications, glycolysation and phosphorylation are most important; however, these modifications are labile and can already be lost in the ionization source under normal electrospray ionization conditions [83]. This becomes even more of a problem when collision energy is applied. The result is cleavage of the labile modification and detection of the peptide fragment lacking the modification. There are two scenarios: if a neutral species is lost after cleavage, only the peptide backbone can be detected and if the lost species is charged, two signals can be detected. One signal again corresponds to the peptide backbone and the second signal in a low mass range is the posttranslational modification-specific fragment ion, also referred to as reporter ion. Even in this case, information regarding the location of the modification is lost. Precursor ion scanning and neutral loss scanning in combination with mild fragmentation conditions have been employed to assess posttranslational modifications using LC-MS [83]. Electron capture and electron transfer dissociation are technologies that are complimentary to classical fragmentation, and they tend to result in fragmentation more evenly distributed over the entire peptide backbone. This makes them particularly useful in localizing posttranslational modifications [9,84].

Matrix Assisted Laser Desorption/Ionization Mass Spectrometry (MALDI-MS)

Although LC-electrospray-MS is the most frequently used proteomics bioanalytical technology to date, MALDI remains a valuable technology that, in addition to its ease of use, high sensitivity and high-throughput, is an alternative ionization method that often yields complementary protein hits [9,69]. For MALDI-MS analysis, the protein solution is mixed with a matrix solution that catalyzes the (co)crystallization of matrix molecules and proteins onto a target plate [85,86]. The target plate is then analyzed by MALDI-TOF MS. In MALDI-TOF MS, the proteins are liberated in an ionized form from the target surface by firing a laser pulse at the crystallized proteins. The ionized proteins are accelerated and separated based on mass through the vacuum tube of the time-of-flight mass spectrometer by an electrical field and reach the detector [85].

One of the major problems has been its low reproducibility that, among other reasons, is caused by inhomogenous cocrystallization of analytes with the MALDI matrix [69,86]. Over recent years these variability issues have improved due to the development of novel sample treatment methods, improved MALDI matrices and target plates. Especially promising is the use of nanoparticle-based MALDI matrices (1–100 nm particle diameter) that provide a high surface-to-volume ratio. These have been shown to increase sensitivity and reproducibility [69].

Surface-enhanced laser desorption/ionization (SELDI) is a variation of MALDI and has been widely used in discovery studies to identify new protein molecular markers [28,86]. The basic principle is that the sample is exposed to chips with different active surfaces that enrich certain groups of proteins. These are then eluted onto a MALDI plate and analyzed using a time-of-flight mass spectrometer. SELDI can be automated. Although the system is highly integrated and easy to use, it is prone to producing artifacts [86–89]. The most likely reason is that the time-of-flight mass spectrometer that was included in the SELDI system lacked appropriate resolution [28,86]. Also, due to the selectiveness of the activated surfaces, only a fraction of the proteome is analyzed and potentially critical information may be missed. A significant source of variability and artifacts is that binding of proteins to the active surfaces is not very robust and is easily influenced by even small variations in pH, salt concentrations, and interfering compounds [86–89]. Due to these short-comings, SELDI has lost most of its importance.

The use of MALDI-MS in urine proteomics is discussed in detail by Gopal et al. [69]. Importantly, MALDI-MS-based urine proteomics allows for the identification of bacterial biomarkers in urine samples, enabling rapid and direct identification of bacteria in urine samples, yielding results for the diagnosis of urinary tract infections within 1 h [90,91]. Moreover, MALDI-MS imaging allows for combining proteomics with histology [92]. Combined with mathematical algorithms, MALDI/MS spectra can be processed in order to give semiquantitative mapping information on tissue sections. Mass spectrometry imaging can be carried out using fresh/frozen as well as formalin-fixed paraffin embedded tissues. The MALDI matrix is applied on tissue sections followed by raster-stepped MALDI analysis of the surface. For more detail, please see Longuespée et al. [92].

Labeling Technologies for LC-MS-Based Quantitative Proteomics

Several strategies for LC-MS-based quantitative proteomics have been developed over the last years [93,94]. These involve [94]:

- Stable isotope labels: isotope-coded affinity labeling (ICAT), nonselective isotope-coded protein labeling (ICPL), isotope-differentiating binding energy shift tag (IDBEST), tandem mass tags (TMT), isobaric peptide termini labeling (IPTL) and, most importantly, isobaric tags for absolute and relative quantification (iTRAQ), and stable isotope labeling by amino acids in cell culture (SILAC).
- Inductively coupled plasma mass spectrometry (ICP-MS) in combination with element tags.
- Label-free statistical approaches.
- Absolute quantitation by mass spectrometry (AQUA).

Stable isotope labeling strategies can roughly be categorized into two groups, postharvest and metabolic labeling methods that involve incorporation of isotopically labeled amino acids into metabolically active cells [94]. Isotope-coded affinity tags (ICAT) contain three functional regions—an affinity purification region, a peptide-binding region, and an isotopically distinct linker region. Typically, a biotin tag is used for affinity purification and a thiol-specific binding moiety covalently links the reagent to cysteine in the target peptide. The linker region is isotopically labeled with ^{12}C or ^{13}C [95]. Thus, after reacting with differently labeled tags, the same peptides remain chemically identical but can be distinguished by the mass spectrometer based on their different tag masses. This allows for labeling of two samples. The samples are mixed and analyzed simultaneously in one run, eliminating the problem of peak shifts and alignment that may occur when samples are independently analyzed and improving quantification.

The disadvantage of ICAT is that it is restricted to labeling peptides that contain cysteine and, thus, less peptides for protein identification may be available. Another approach that has similar advantages and disadvantages as ICAT is proteolytic ^{18}O labeling. A protease and $H_2{}^{18}O$ are used to generate labeled peptides [96]. Isobaric tag for relative and absolute quantitation (iTRAQ) labels all peptides and thus increases the confidence in protein identification by labeling a larger number of peptides per protein of interest. Samples are trypsin digested, labeled and then the iTRAQ-labeled peptide samples of 4–8 proteomes are combined, fractionated using a strong cation exchange column and analyzed using nano-LC-MS/MS [94].

While ICAT and iTRAQ can be applied to serum and other biofluids, cell homogenates as well as tissue samples and are "postharvest" labels, stable isotope labeling with amino acids in cell culture (SILAC) has specifically been developed to detect proteome differences in cell cultures. The cells are incubated with isotopically distinct forms of amino acids until the complete proteome contains amino acids with a specific label. Proteomes with different labels can then be mixed and be analyzed simultaneously [94,97,98].

Most of the earlier mentioned quantitative proteomic labeling methods use electrospray ionization or MALDI mass spectrometry [94]. Inductively coupled plasma-mass spectrometry (ICP-MS) is mainly used for detecting and quantifying metals and its value in proteomics has emerged only recently [99]. The use of ICP-MS in proteomic studies is possible whenever there is an ICP-detectable heteroatom present in the proteins. ICP-MS allows for studying metal–biomolecule interactions, but also for absolutely quantifying proteins/peptides that carry heteroatoms. This includes the absolute quantification of some posttranslational modifications in biomolecules that contain biologically important heteroatoms covalently attached to the protein, like phosphorylation, iodination, or metalation can be performed [94,99]. Interest has mostly focused on the use of ICP-MS in phosphoproteomics, partially due to the fact that ion suppression often interferes with the detection of phosphopeptides in "soft ionization" mass spectrometry such as electrospray ionization and MALDI, but not with ICP-MS [99].

Microarrays

Microarrays technology can be used for the targeted quantification of proteins, but can also be used as a nontargeted discovery tool. Most targeted microarrays are based on immobilized antibodies, while nontargeted microarrays are often referred to as protein microarrays. Protein function assays are based on immobilized recombinant proteins, peptides, or libraries hereof as well as antibodies. They are used for screening for novel substrates, for enzyme activities, for protein–protein interactions, for protein–lipid interactions, and for protein–small molecule interactions. Arrays with up to 8000 human proteins are available [100]. In reverse-phase assays cell proteomes are coated on the array after extraction and then the immobilized proteins are screened using antibody-based detection. For more details, please see Korf and Wiemann [56]. Gupta et al. [101] provide a comprehensive overview over available protein arrays, instrumentation, and software.

A more recent and very promising development are aptamer-based proteomics technologies, specifically the SOMAScan assay [102,103]. Aptamers were discovered 25 years ago and are short single-stranded oligonucleotides, which fold into diverse molecular structures that bind with high affinity and specificity to proteins, peptides, and small molecules [102]. SOMAmers are chemically modified aptamers and these chemical modifications greatly expand the repertoire of targets to which SOMAmers can bind compared to unmodified DNA. SOMAmers bind to proteins by folding into diverse and intricate shapes that interact specifically with protein surfaces. SOMAmers are selected in vitro from large libraries of randomized sequences. Once a SOMAmer is selected and its sequence is known, in contrast to antibodies, SOMAmers can be manufactured, thus avoiding the relatively large batch-to-batch variability inherent to antibody-based proteomics assays. Proteins in complex matrices such as plasma are measured with a process that transforms a signature of protein concentrations into a corresponding signature of DNA aptamer concentrations, which is quantified using a DNA microarray [102]. SOMAmer-based proteomic technology for biomarker discovery is capable of simultaneously measuring thousands of proteins from small sample volumes (15 μL of serum or plasma). The current version of the SOMAScan assay measures 1,310 proteins with lower limits of detection (1 pM median), 7 logs of overall dynamic range (~100 fM-1 μM), and 5% median coefficient of variation [102,103]. Once the protein markers of interest are identified, more targeted versions of the SOMAScan assay for diagnostics can easily be developed [104].

Technologies for Targeted Proteomics

After the molecular markers of interest have been identified, the next step is to establish targeted and validated assays that are capable of quantifying these specific compounds with acceptable total imprecision and sensitivity. It has been reported that there are 105 FDA-cleared or approved tests for the quantification of proteins and additional tests for 65 proteins and 32 peptides have been listed in the Directory of Rare Analyses [105,106]. For a complete list, please see Anderson [105]. As aforementioned, the human genome has 20,687 protein-coding genes [57], this means that the assays approved and cleared by the FDA and listed in the Directory of Rare Analyses cover less than 1% of the human proteome. For the targeted analysis and quantification of proteins, in most cases, antibody-based (80% of all approved) assays and enzymatic assays are used; however, mass spectrometry-based assays are gaining importance for the quantification of proteins [105].

Targeted strategies measure well-defined molecular markers to detect pattern changes. To achieve this goal, the analytical strategies have to be at least semiquantitative or quantitative.

Enzyme-linked immunosorbent assays are frequently used for the quantitative measurement of proteins [107,108]. Most of these assays are single-analyte assays. Multiplexing assays are targeted assays that can test multiple analytes in a single test using a single sample. In principle, current protein multiplexing assays are simultaneous ELISA microassays coated adjacently on a surface. This surface can be an array or a bead.

The basic principle, briefly, is that [107,108] microspot arrays can be based on several flat surfaces, such as polylysine and aminopropylsilane, epoxysilane-treated glass, and other surfaces that may be covalently or noncovalently linked. Spot sizes are between 50 and 250 μm; 50 and 300 spots, or even more can be printed in predefined geometric patterns into a 96-well plate. Fluorescence or chemiluminescent labels are used for detection. The spots in each well are resolved by microimaging. These assays can be developed on a microarray surface and microarray analyzers and software can be used for analysis. Among antibody microarrays, analog to sandwich ELISAs, sandwich arrays are the most quantitative; however, they require a second antibody [58,100]. Most commercially multiplexed sandwich microarrays quantify cytokines and chymokines [56,58].

Bead assays are microspheres with a diameter of approximately 6 μm that can be color-coded to assign individual addresses for up to 100 different populations mixed in solution. In a typical set up, each bead population is coated with an analyte binding capture agent that can be either DNA probes or antigen/antibody capture for protein assays. These populations of coupled beads are then mixed into a solution to form an array. Fluorescent labels can be used to detect signals. The analyzer sorts out the populations based on the color code that reads the signal on the bead.

Interestingly, in comparison to ELISAs using the same antibody/antigen combination, microspot assays were found to be more sensitive and have a wider linear range [108]. They also have better sensitivity and specificity than conventional ELISAs in low-analyte samples.

In comparison to microspot assays, the signal intensity of beads is much lower since the beads are dispersed throughout the entire volume of the assay fluid that typically is 50–200 μL for a 96-well format assay. Another reason is that only one half of the bead surface area is excited and the bead does not always flow past the detector with the excited half optimally exposed [108]. Also, as previously indicated, bead-based assays are limited to about 100 analytes that can be measured simultaneously.

In addition to protein microarrays, two frequently used technologies are the Luminex xMAP and the Mesoscale Discovery electro-chemiluminescence detection system. The Luminex xMAP technology is based on polystyrene bead sets encoded with different intensities of red and infrared dyes and coated with a specific-capture antibody against one of the analytes of interest. Interrogation of the beads by two lasers identifies the spectral property of the bead and thus the associated analyte, in addition to the R-phycoerythrin labeled secondary antibody against the specific analyte. The Mesoscale Discovery assay platform uses plates fitted with up to 10 carbon electrodes per well, each electrode being coated with a different capture-antibody. The assay procedure follows that of a sandwich ELISA, with any analytes of interest captured on the electrode being detected with an analyte-specific ruthenium-conjugated secondary antibody. Upon electrochemical stimulation, the ruthenium label emits light at the surface of the electrodes, allowing the concentration of the analyte to be determined relative to the particular electrode [109].

Current protein multiplexing assays vary in three key characteristics [108], the number of analytes that can simultaneously be measured, the type of platform, and qualitative versus quantitative. Due to differences in sensitivity, specificity, robustness, and dynamic ranges, some technologies are more appropriate for quantitative assays than others and quantitation depends on the development and validation of an appropriate calibration and quality control strategy. Protein multiplexing assays are limited by the number of suitable antibodies that are highly specific and bind their cognate antigens with comparable binding constants [28]. Another limitation is the wide concentration differences in cellular proteins that, as already mentioned earlier, cover several orders of magnitude. Therefore, a single multiplex assay can only semiquantitatively compare proteins that are present in a cell or body fluid in the same concentration range.

In the vast majority of protein detection platforms, the binding event of a protein to a specific recognition molecule must be detected with a signal transducer. In ELISAs, protein microarrays and quantum dot detection platforms [110], the readout is based on a fluorescent or colorimetric signal [111]. Inherent autofluorescence or optical absorption of the matrix of many biological samples or reagents may become a limiting factor. Similarly, nanowires [112], microcantilevers [113], carbon nanotubes [114], and electrochemical biosensors [115] rely on charge-based interactions between the protein or tag of interest and the sensor, making each system dependent on conditions of varying pH and ionic strength. Since the matrices of even complex biological samples usually lack a detectable magnetic background signal, a magnetic field–based detection platform for protein detection in clinical samples has been described [111,116], that is, matrix insensitive, yet still capable of rapid, multiplex protein detection with resolution down to attomolar concentrations and a wide linear dynamic range.

Notwithstanding the already discussed problems with specificity and batch-to-batch consistency of antibodies, as of today, most assays for the targeted quantification of proteins have been based on affinity reagents such as antibodies. These assays currently still slightly surpass the sensitivity of LC-MS/MS-based approaches, although lower limits of detection for multireaction mode (MRM)-based assays as low as the low-attomole range having been reported. [117] The advantages of the quantification of proteins using LC-MS/MS over immunological assays are that LC-MS/MS does not require the availability of antibodies, does not depend on the quality of antibodies, and that the quality of the data is easier to control. Interferences can usually be easily detected by inspection of the ion chromatograms, while potential cross-reactivity or other interference with an antibody reaction is much more difficult to detect. Among the LC-MS/MS technologies for the quantification of proteins already discussed earlier, such as DIA, isobaric tags, and parallel reaction monitoring (PRM), LC-MS/MS assays based on MRM are the most frequently used strategy for the quantification of specific proteins of interest [81,117]. Typically, the specific peptides of the candidate protein molecular markers have already been identified [117] and can be found in databases, such as Peptide Atlas [118] and the Global Proteome Machine Database [119]. Purity and enrichment of the sample subjected to LC-MS/MS analysis greatly reduces interferences and increases sensitivity. Sample preparation may be based on subcellular fractionation, protein precipitation, antibody or apatamer-based protein enrichment, removal of high-abundance proteins, size-exclusion chromatography, and/or liquid–solid extraction [81]. A semiautomated assay for online enrichment and MRM LC-MS/MS analysis using specific antibody-based capture of individual tryptic peptides from a digest of whole human plasma has been described, the stable isotope standards and capture by antipeptide antibodies (SISCAPA) method [120]. This method uses a simplified magnetic bead protocol and a novel rotary magnetic bead trap device. Following offline equilibrium binding of peptides by antibodies and subsequent capture of the antibodies on magnetic beads, the bead trap permits washing of the beads and elution of bound peptides inside a 150 μm inner diameter capillary that forms part of a nanoflow LC-MS/MS system [120].

After sample purification and enrichment, proteins are digested and the specific peptides are analyzed using LC-MS/MS [121]. Absolute quantification is possible by utilizing synthetic isotopically labeled versions of the specific peptides [18,117,122]. Heavy isotope-labeled peptides can be used as calibrators and/or as an internal standards. In most cases, the isotope-labeled peptide is added after digestion of the proteins with trypsin, and the digest is then separated by reversed-phase HPLC.

For selection of the best ion transitions for MRM, a number of criteria should be considered [117,123]. The target peptides should not exhibit any enzyme missed cleavage sites, and they should not be susceptible to post-translational modification, unless it is the purpose of the assay to quantify those [117]. They should be of a size that accommodates the mass range of the triple stage quadrupole MS/MS instrument, usually in the range from 7 to 30 amino acids, and they should uniquely identify the protein of interest. Each MRM transition requires an optimized set of mass spectrometry parameters for maximum sensitivity. Not to replicate the substantial effort required for the setup of an MRM assay for protein quantification, several online resources such as assay repositories are available as well as software suites that guide through the process of selecting appropriate MRM ion transitions. For a detailed list, please see Picotti and Aebersol [117]. Most studies have used at least two peptides per protein and up to two different charge states for each of the parent ions. Considering the complexity of tryptic digests of whole proteomes and the likelihood of isobaric peptide sequences, it is impossible to rely on the detection of a single MRM transition. Three to five intense transitions for each endogenous peptide should be measured [117]. MRM measurements are usually carried out using a 1-Da window to select the target precursor and fragment ions.

Hence, monitoring of multiple transitions for a peptide and the application of several predefined constraints such as chromatographic retention times and fragment ion relative intensities reduce the likelihood of false positives [117]. Moreover, long gradients and detection using high-resolution mass spectrometry increases assay specificity [81].

Due to the specificity of MRM, LC-MS/MS assays are capable of simultaneously quantifying multiple proteins [121]. As discussed earlier, quantification of a protein is often based on more than 10 MRM transitions. Due to this, designing and validating hundreds of individual peptide transitions for the quantitative analysis of complex samples is challenging [117]. In silico algorithms for the MRM-based targeted analysis of multiple proteins are available [123]. The simultaneous quantification of up to 45 proteins in plasma samples using MRM LC-MS/MS has been reported [124]. Again, one of the challenges with targeted multianalyte LC-MS/MS assays is the potentially large concentration differences of proteins in the matrix of interest [125,126].

Database Searches, Biostatistics, and Annotation

Accurate, consistent, and transparent data processing and analysis are critical parts of the proteomics workflow in general and for molecular marker discovery in particular [127]. Although sophisticated and powerful data analysis tools are available today, it should not be forgotten that, as discussed in more detail in Section "Validation of Analytical Assays, Quality Control, and Standardization." later, the quality of the results is determined by the quality of the analyzed samples and the quality of their analysis. It is only when sufficient quality can be assured that meaningful results and conclusions can be expected.

Proteomic data analysis typically includes the following steps [127]:

Data processing: This includes signal processing ensuring that high sensitivity, resolution, and mass accuracy are fully retained and exploited during downstream data analysis and conversion of signals into appropriate and preferably standardized data formats.

Peptide identification: Assignment of MS/MS spectra to peptide sequences. As described earlier, "bottom up" proteomics is based on digestion and subsequent detection of the resulting peptides by mass spectrometry methods. The next step is to identify the detected peptides and assign these to protein structures. In general, there are three strategies to achieve this goal. The most common peptide identification method is sequence database searching [128]. This is achieved by using a data base engine such as Sequest, Mascot, Comet, X!Tandem, or Spectrum Mill. In general, the underlying algorithms are matching and scoring the experimental MS/MS spectra with predicted mass-to-charge ratios of fragment ions of peptide sequences derived from protein databases [127]. These predicted spectra are a very rough estimate of the peptide's "true" spectrum. Nevertheless, sequence-based search engines are theoretically able to identify all peptides that may be present in the sample [128]. The second approach is "de novo" sequencing. These algorithms try to derive the peptide's sequence directly from the MS/MS spectrum by taking all possible amino acid combinations into consideration. As a result, the search space of de novo sequencing algorithms is considerably larger compared to sequence database searching. De novo sequencing algorithms often derive multiple sequences per spectrum at equal probability, and are substantially slower than other approaches, which renders de novo sequencing unsuited as a primary analysis tool for the identification of MS/MS spectra [128]. The third approach is to identify MS/MS spectra by spectral library searching [128]. Spectral library search engines use libraries of identified, generally experimental MS/MS spectra to identify observed MS/MS spectra. Since the recorded library spectra contain intensity information as well as peaks from noncanonical fragment ions, higher identification rates are observed, especially in low-resolution datasets. As spectral libraries only contain peptides that are detectable, spectral library search engines can identify more spectra than sequence database search engines while being significantly faster [128]. Griss [128] provides an overview over available spectral library search engines. High quality experimental data allow for more effective searches.

Validation: False discovery rates (false positives) are a major problem in proteomics and can be caused by: (1) the statistical process used to identify significant protein signal differences, and (2) the algorithms used for identifying the structures of such proteins. For example, 2D-gels from treatment and controls or from different treatment groups are usually compared using multiple Student's t-tests with a significance threshold of 0.05%. This means that, theoretically, 5% of protein spots may be falsely identified as different [129]. False discovery rates can be reduced by more robust experimental design, improved quality of samples and analysis, the use of technologies that allow for a direct comparison of proteomes, such as DIGE and labeling [129–131] and the use of appropriate sample sizes [132]. Another major source of errors is protein identification. This is caused by the fact that several peptides may be common to more than one protein. Thus, it is important to assess the validity of the protein assignment and to associate a probability with the identification. Naturally, if more peptide matches for a specific protein can be identified, then there is greater confidence in its correct identification. Statistical postprocessing procedures such as the PeptideProphet, Percolator, and LFDR (local false discovery rate) algorithms are available that estimate the rates of false positive and false negative errors [133]. Alternative approaches such as "reversed database" searches have been explored [134]. Also, databases are still fraught with problems such as redundancies, inconsistencies in nomenclature, fused genes, and inappropriately translated introns [135]. Overall, it has to be noted that a positive "hit" and its associated proposed structure can only be viewed as a hypothesis. Important "hits" should always be confirmed using independent technologies such as Western blot or ELISA.

Quantification: Quantification and protein identification may not be possible in the same experiment. As described, in bottom up proteomics, there are several LC-MS/MS-based quantification strategies, such as DIA, isobaric tags, and PRM, and MRM. Quantification in bottom up proteomics is based on peptides. Especially DIA requires specialized software solutions. Software packages supporting quantitative LC-MS/MS-based proteomics are summarized in Shi et al. [81], Hu et al. [82], and Picotti and Aebersol [117]. While after stable isotope labeling, multiple proteomes are compared in the same analytical run, unlabeled approaches require more rigorous control of the analysis and data collection conditions including control of instrument drifts during multiple analytical runs (mass calibration, elution times) and normalization of ion abundances to compensate for differences in instrument performance (for example, accumulating contamination of the ionization source). The use of isotope-labeled internal standards is an effective strategy to compensate for and to control some of these problems.

Analysis of proteomic data sets. After successful analysis of the proteomes of interest, the next step is to analyze the interaction of proteins, protein complexes, signaling pathways, and networks. The analytical tools available have been categorized in three types [136]:

Basic traditional statistical analysis. This is commonly used as a first pass to identify the "low hanging fruit" in a data set [136]. This involves tests, such as t-test, Wilcoxon test, and/or analysis of variance (ANOVA). Such statistical approaches alone are insufficient to discover all of the biologically relevant information in a proteomics dataset, but usually constitute an important first step.

Machine learning classification. These approaches complement traditional statistical approaches. These allow for consideration of numerous variables simultaneously and remove potential investigator bias. Supervised machine learning involves training a model based on sample sets with known labels, for example, treatment and control. In contrast, unsupervised classification or clustering is based on an unlabeled dataset and seeks to group samples with similar attribute profiles. The most important unsupervised clustering algorithms are principal component analysis, independent component analysis, K-means, and hierarchical clustering. Important supervised classification algorithms are partial least square, random forests, and support vector machine [136]. It is reasonable to expect that the proteins in a proteomics data set are not fully independent of each other in vivo. At the beginning each protein represents a data point or feature. The purpose of machine leaning is to combine features that have certain commonalities into dimensions, basically by combining groups of possibly correlated features (proteins) into a smaller number of uncorrelated dimensions, which then facilitates interpretation and visualization of the datasets.

Assignment of functional and biological information. Putting the results into context with existing knowledge about molecular interaction networks, such as metabolic pathways and signaling pathways is an important last step. Several pathway and network analysis software tools can help to address the challenges of proteomics data interpretation [136]. Nevertheless, although such tools are rapidly increasing in sophistication, a true understanding of the biological and pathological implications and concepts will still require review of the relevant literature [136].

Machine learning, clustering methods and pathway analysis tools as relevant for proteomics have comprehensively been reviewed by Karimpour-Fard et al. [136]. Moreover, the analysis and interpretation of proteomics data sets is greatly facilitated by the availability of reliable and complete protein databases and proteome maps. Important reference databases and large community-based proteomic data sets are PeptideAtlas [118,137], GPMDB [119,138], neXtProt [139], and Human Proteome Map [140]. Information about the urine proteome are publicly available [141–143]. Many of these databases are continuously expanded and the repository of proteomics data is promoted by initiatives, such as the Human Proteome Organization and the Chromosome-Centric Human Proteome Project [144–146].

Normalization of Urine Data

Protein concentrations in urine depend on the dilution of urine samples and, thus, normalization strategies, mostly based on creatinine, have extensively been used to compensate for variation in urine dilution [147]. It has become evident, however, that creatinine concentrations in urine can be affected by tubular excretion, aging, gender, and disease processes. As an alternative strategy, normalization based on urinary "housekeeping" peptides that are ubiquitously present in human urine, and seem to be more robust than creatinine, has been recommended [148,149]. Nevertheless, these are only single studies and there does not seem to be general consensus. Given the fact that this is a critical problem, there has been surprisingly little discussion and more systematic studies assessing this important issue are lacking.

Validation of Analytical Assays, Quality Control, and Standardization

In general, every proteomics assay should be validated as considered fit for purpose. Accordingly, an exploratory discovery assay will require less validation than a quantitative assay used for a clinical trial or as a clinical diagnostic tool. Since nonbiased proteomics approaches are not truly quantitative, validation is limited to assessing sample stability and reproducibility, which should be considered the required minimum for every exploratory proteomics assay. It has been estimated that the technical variation in 2D gel electrophoresis typically is associated with a coefficient of variance of 20–30% [150]. In the case of immuno-based assays, the interaction of the antibody with the analyte will require validation including potential cross-reactivities and interferences with other drugs, metabolites, or matrix components. It has to be noted that disease processes may interfere with immuno-based assays, for example, by changing the hematocrit, increasing endogenous compounds, such as lipids and bilirubin, and by the formation of antibodies, such as rheumatoid factor [151–153]. Such interferences may not be noticed if an assay is exclusively validated based on samples from healthy subjects. A potential problem especially with more exploratory-type, semiquantitative immuno-based multiplexing assays is also the batch-to-batch comparability of the antibodies. A semiquantitative comparison of results generated at different times or in different laboratories may not always be possible. It is important to note that the analysis of macromolecules is inherently more variable. Based on generally accepted guidelines for the HPLC/UV and LC-MS analysis of small molecules, an assay is only acceptable if, except at the lower limit of quantitation, interday precision is ≤15% and interday accuracy is within ±15% of the nominal value. Accordingly, an analytical run of study samples is accepted if at least two-thirds of the quality control samples fall within 15% of their nominal value. For the quantification of macromolecules using immuno-based assays, regulatory agencies have accepted limits of ±25 and ±30% [154].

It must be realized that current regulatory guidelines have been written mostly with the quantification of single drug compounds in mind and may be too rigid for emerging multianalyte technologies [109,155]. In fact, it has been suggested that biomarker methods should not be classified as "good laboratory practice" assays; nor should they be validated by the same guiding principles developed for drug analysis by HPLC/UV or LC-MS [154,156]. The challenge with multianalyte assays is that several compounds are quantified simultaneously and that it is not possible to optimize the assay for each compound to the extent that is possible for analysis of single compounds [157]. On the other hand, there is potential benefit in the additional information conveyed by molecular marker assays. Thus the risk-benefit-ratio must be evaluated for the individual markers included in a multianalyte assay using different criteria than with standard assays designed to measure single drug compounds. This has been recognized by regulatory agencies. The U S FDA guidelines have suggested that "further research is needed to establish the validity of available tests and determine whether improvements in biomarkers predict clinical benefit" [158].

Proteomics studies in similar patient populations with the same diseases have generated different protein fingerprints and have identified different sets of potential protein markers. The reasons may include differences in sample handling, preparation, and analytical technologies for proteome analysis. In fact, most proteomics-based publications cannot be compared, thus greatly reducing their value [4]. The problems with interlaboratory comparability of data was further emphasized by a cross-validation study involving 27 proteomics laboratories conducted by the Human Proteome Organization Sample Working Group [159]. An equimolar test sample containing 20 highly purified recombinant proteins and tryptic peptides of 1250 kDa size was distributed to the test laboratories. Only seven laboratories reported all 20 proteins correctly and only one laboratory reported all of the 1250 kDa tryptic peptides. Missed identifications (false negatives), environmental contaminations, database matching, and curation of protein identification were identified in this study as the major problems with this process [159]. Interestingly, after the problems had been identified, laboratories that missed proteins during the original analysis identified all 20 proteins correctly during a second round, emphasizing the importance of experience and expertise in this complex field [131].

The task of truly understanding proteomes and their association and mechanistic relationships with the different stages of disease development is a monumental task that cannot be carried out by a single laboratory. Therefore, as also discussed earlier, sharing of proteomics data sets and their deposit into databases will be important and greatly facilitate generating a larger understanding of the proteome and to translate this proteomics data into clinical benefits [132]. Today, consortia play a critical role in the development, qualification, and acceptance of a molecular marker. Consortia depend on pooling and comparison of data from different sites [133,134]. The quality of such databases and the validity of the decisions derived from such data greatly depend on the quality of the individual data sets. A critical tool is the cross-validation of the laboratories involved in the measurement of molecular markers to control data quality and to ensure comparability. Since successful cross-validation for single analytes can already be a challenging task, it can be expected that consistent measurement of more complex analyte mixtures in multiple laboratories is even more difficult [131,159].

To assure comparability between laboratories, two strategies can be pursued, harmonization and standardization [155]. Whereas, harmonization is less restricting, standardization requires compliance with strict guidelines and rules. The harmonization or standardization of biomarker assays should focus not only on assay conditions and performance but also on preanalytical issues, such as the sample matrix used and sample handling, as well as on postanalytical issues, such as data analysis and reporting. The outcome of a harmonization approach should be the establishment of common standard operating procedures (SOPs) [155]. Standardization is more rigorous and usually requires use of the same procedures, consumables, calibrators, controls, instrumentation, and data analysis algorithms. Regular cross-validation of the participating laboratories ensures that minimum performance requirements are consistently met [155].

Harmonization and standardization are required to ensure comparability of data. This will also have to include sufficiently stringent quality control to avoid the "dilution" of databases by inferior data sets [4,141]. Since the correct identification of proteins is greatly improved with the quality of the spectral data, such quality criteria should include minimum requirements for mass accuracy and mass spectrometry resolution. Steps toward standardization of sample collection, processing, and proteomics analysis have been taken at consensus conferences and through publication of guidelines [160–163] as well as by the establishment of the Human Kidney and Urine Proteome Project (HKUPP) [164,165]. Proteomics associations, such as the Human Proteome Organization (HUPO) [166] support proteomics efforts through systematic research in sample handling, technologies, procedures, protocols, and defining standards [58,167,168]. Harmonization and standardization efforts also include study design, infrastructure requirements, minimum information about a proteomics experiment [169], minimum reporting requirements, standard data formats, common sets of vocabularies and ontologies, annotations, and validation guidelines [170]. Likewise, standards for the publication of proteomics data, such as the Paris Report have been established [171].

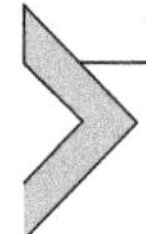

PROTEOMICS IN RENAL RESEARCH AND AS A MARKER FOR KIDNEY FUNCTION, DISEASE, AND INJURY

The use and role of proteomics in the discovery of clinical markers and as a potential clinical diagnostic tool have extensively been reviewed before [2,24,172–176].

Identification of Disease, Pharmacodynamic, and Toxicodynamic Molecular Mechanisms

The ability to characterize subcellular, cellular, and organ proteomes in an unbiased fashion has led to important insights into biological processes and signal transduction pathways [17]. There are two fundamentally distinct concepts in proteomics. First, the concepts as described earlier are most important for molecular marker discovery in nephrology and can be termed expression-based proteomics. Expression-based proteomics seeks to describe the proteome at a given moment and its changes in response to disease and drug exposure. Ultimately, it only results in correlative relationships and its use as a clinical molecular marker will require more in-depth mechanistic qualification. Second, functional proteomics seeks to assess the interactions of proteins and the interactions within and among protein networks. In many cases, functional proteomics studies complement genetics and functional genomics studies that often lead to a gene product with a putative biochemical function, but a poorly characterized biochemical mode of action [17]. Functional proteomics allows for the identification of interacting proteins and for mapping proteins to specific biochemical pathways and protein networks. The methods employed include, but are not limited to, affinity purification, the binary yeast two hybrid approach, phage display technology, protein arrays, tandem affinity purification tags, and computational prediction models, some of which are based on the known three-dimensional structure and binding motifs. For a more detailed overview, please see Köcher and Superti-Furga [17] and Sanderson [177].

Activity-based protein profiling for the functional annotation of enzymes uses site-directed, small molecule-based covalent probes that can be used in native biological systems [178]. These probes are designed to target a subset of the proteome with shared principles of binding and/or reactivity [178]. These probes consist of three components: a binding group that directs the probe toward the target protein, a reactive group (electrophilic or photoreactive) for covalent labeling and an analytical tag. Typically, samples are affinity purified and proteins with the covalent tags are identified by mass spectrometry and database search [178].

Posttranslational modifications determine the functionality of most eukaryote proteins and, thus, are critical for understanding the mechanistic role of proteins and the functionality of protein networks. Posttranslational modifications are covalent processing events that change the properties of a protein by proteolytic cleavage or by the addition of a modifying group to one or more amino acids [179]. Posttranslational modifications may determine a protein's activity state, localization, turnover, and interactions with other proteins. Standard proteomics approaches are usually not suitable for identifying or mapping posttranslational modifications and specific purification procedures, analytical technologies or databases have been developed. For detailed reviews, please see Mann and Jensen [179], Witze et al. [180], Ruttenberg et al. [181], Hoffert and Knepper [182], and Doll and Burlingame [183].

Proteomics has been used to study the role of proteins—protein functions, interactions, and protein networks—in the physiological functions of kidney cells and to study disease mechanisms in the kidney. Representative studies include the:

- role of calmodulin in glucose uptake in human mesangial cells [184];
- role of proteins in the physiology and function of glomerular cells [185];
- role of proteins in function of the renal tubule [186];

- adaptive response of the tubule to acidosis [187];
- adaptation of cells in Henle's loop to osmotic stress [188];
- role of proteins in the physiology of collecting duct cells [189];
- regulation and function of aquaporin-2 in collecting duct cells [190–195];
- adaptation of collecting duct cells to osmotic stress [196,197];
- evaluation of molecular mechanisms underlying renal fibrosis [198,199];
- evaluation of molecular mechanisms of nephropathies [200–203]; and
- cytotoxicity of calcium oxalate monohydrate [204,205].

The use of proteomics for evaluating molecular mechanisms underlying physiological processes in the kidney and their response to disease and xenobiotic challenges provides the basis for protein marker discovery and qualification.

Acute and Chronic Kidney Injury

Today, kidney biopsies are still the gold standard for the diagnosis of chronic and/or acute kidney diseases [206]. It has been discussed that proteomics analysis of urine samples can be considered a noninvasive biopsy. Indeed, data in the literature suggests that, in the future, in many cases, urine may replace kidney biopsies or at least may provide guidance for when a biopsy should be collected. Renal biopsies have their limitations. They require an invasive procedure that usually involves hospitalization, resampling is difficult, depending on a patient's habitus it may be impossible to collect a biopsy and, although they are helpful as a diagnostic tool, biopsies provide little guidance in terms of treatment and prognosis. Urine proteomics may overcome several of these shortcomings. Urine is easily available, is obtained noninvasively and does not require hospitalization, can be frequently sampled, can be collected from any patient unless anuric, can be used to closely guide treatment and can be used to monitor treatment efficacy, tolerability, disease progression, or recovery [206].

Proteomics has been used to establish and qualify animal models, understand renal disease mechanisms, for molecular marker discovery in animal models, and in clinical studies. Clinical discovery studies have mostly focused on the urine proteome. Representative proteomics studies in patients with chronic kidney disease and acute kidney injury are summarized in Table 4.4.

Proteomics, in many cases in combination with transcriptomics studies, have suggested several promising novel molecular markers for the clinical investigation of acute, and possibly chronic, kidney injury [13,238–240].

Table 4.4 Selected clinical proteomics studies for the discovery of renal disease protein markers

Diagnostic target	Study population	Matrix	Analytical technology	Identified correlations	References
ADPKD	34 ADPKD and 10 CKD patients at different CKD stages, 10 healthy controls	Urine, extracellular vesicles	HILIC fractionation, LC-MS/MS	• 1048 proteins were identified in urine, 1245 proteins in extracellular vesicles, 527 of these proteins overlapped. • Periplakin, envoplakin, villin-1, and complements C3 and C9 were selected as candidates for further confirmation by immunoassays and were found to correlate with total kidney volume.	[207]
ADPKD	13 patients with PKD1 mutation, 18 healthy controls	Urine, exosomes	SDS page, LC-MS/MS	• 2008 proteins were identified in the exosomes, of which 9 proteins differed between both groups. • Most notably, transmembrane protein 2 (TMEM-2) was consistently higher in PKD1 patients (average 209%) and polycystin-1 and polycystin-2 were lower in the PKD-1 patients than in controls. • The polycystin-1/TMEM-2 and polycystin-2/TMEM-2 ratios were inversely correlated with height adjusted total kidney volume and may be potential biomarkers.	[208]
ADPKD	16 early stage, 24 late stage, and 6 tolvaptan-treated ADPKD patients, 11 controls	Urine (pooled), extracellular vesicles	iTRAQ label, MuDPIt	• 83 proteins were found to be different in urine extracellular vesicles from ADPKD patients and healthy controls. • Most importantly, compared to healthy controls, polycystin-1, and polycystin-2 expression, together with that of other Ca^{2+}-binding proteins such as annexin A1, annexin A2, protein S100-A9, protein S100-A8, and retinoic acid induced protein 3 were found to be significantly altered in ADPKD patients. • Differences in apolipoprotein A1 and aquaporin-2 between ADPKD patients and healthy controls indicate impaired concentration capabilities of the kidneys in ADPKD patients.	[209]

(*Continued*)

Table 4.4 Selected clinical proteomics studies for the discovery of renal disease protein markers (*cont.*)

Diagnostic target	Study population	Matrix	Analytical technology	Identified correlations	References
ADPKD	41 ADPKD patients, 189 healthy controls (discovery cohort), 251 ADPKD patients and 89 healthy controls (validation cohort)	Urine	CE-MS, LC-MS/MS	• 657 peptides with significantly altered excretion in ADPKD patients in comparison to healthy controls were detected. 209 of these peptides could be sequenced using LC-MS/MS. • A support-vector-machine-based diagnostic biomarker model based on the 142 most consistent peptide markers achieved a diagnostic sensitivity of 84.5% and specificity of 94.2% in the independent validation cohort.	[210]
ADPKD	30 ADPKD patients, 30 healthy controls	Urine (pooled)	iTRAQ label, isoelectric focusing, LC-MS/MS	• 1700 proteins with at least 2 peptides were identified. • 155 proteins were found different between urine from ADPKD patients and controls. • These included proteins of the complement system, apolipoproteins, serpins, growth factors, collagens, and extracellular matrix components.	[211]
Steroid resistant/steroid sensitive nephrotic syndrome; minimal change disease, FSGS	Pediatric and adult patients, 19 subjects in remission, 19 with relapse, 5 with orthostatic proteinuria	Urine	SELDI-TOF	• Five peaks were found that distinguish steroid resistant from steroid sensitive patients (mass/charge 3917, 4155, 6330, 7037, and 11117). • The peak with mass/charge 11117 was identified as β2-microglobulin. • No other proteins were identified.	[212]

FSGS	10 patients ($n = 6$ steroid-sensitive, $n = 4$ steroid-resistant) with biopsy-proven FSGS	Urine	LC-MS/MS	• 21 protein discriminated between urine samples from steroid resistant and sensitive patients. • Apolipoprotein A-1 and matrix-remodeling protein 8 were identified as the proteins with the greatest difference.	[213]
FSGS	11 patients with FSGS	Urine	LC-MS/MS	• 54 proteins correlating with glomerular filtration rates were identified. • Ribonuclease 2 and haptoglobin showed the greatest differences associated with glomerular filtration rates.	[214]
Steroid resistant/steroid sensitive nephrotic syndrome	25 patients with idiopathic nephrotic syndrome, 17 control patients	Urine	SELDI-TOF	• A protein with the mass of 4144 Da was identified as the most important qualifier. • The structure of this protein was not identified. • Nephrotic syndrome patients were distinguished from controls with 92.3% sensitivity and 93.7% specificity. • 100% of the steroid resistant and sensitive patients were classified correctly.	[215]
Steroid resistant/steroid sensitive nephrotic syndrome	19 patients with steroid resistant and 15 patients with steroid sensitive nephrotic syndrome, 10 controls	Urine	SELDI-TOF	• The 13.8 kDa fragment of alpha 1B glycoprotein was found to have high discriminatory power for steroid resistance.	[216]

(Continued)

Table 4.4 Selected clinical proteomics studies for the discovery of renal disease protein markers (*cont.*)

Diagnostic target	Study population	Matrix	Analytical technology	Identified correlations	References
Lupus nephritis	98 children with SLE (78% African–American) and 30 controls with juvenile idiopathic arthritis	Urine	SELDI-TOF	• Transferrin, ceruloplasmin, alpha 1-acid-glycoprotein, lipocalin • Type, prostaglandin-D synthetase, albumin, and albumin-related fragments were identified as the protein signature differentiating both urines from both patient groups. • Confirmation with immunonephelometry or ELISA suggested transferrin, alpha 1-acid-glycoprotein, lipocalin-type prostaglandin-D synthetase as the most promising biomarker candidates.	[217]
Lupus nephritis	Patients with inactive (n = 49) and active (n = 26) Lupus nephritis	Urine	SELDI-TOF	• Proteins with $m/z = 3340$ and $m/z = 3980$ distinguished active from inactive LN with 92% sensitivity and specificity of 92% each. • No attempt was made to further identify these proteins.	[218]
Membranous nephropathy	Five patients with idiopathic membranous nephropathy, five patients with idiopathic FSGS, three healthy controls	Urine, microvesicles	iTRAQ label, strong cation exchange, LC-MS/MS	• 245 proteins were identified in the microvesicles. • Of these 16 proteins were found different among the three groups. • Lysosome membrane protein 2 (LIMP-2) distinguished patients with membranous nephropathy from FSGS patients and controls.	[219]
Fanconi syndrome	7 pediatric patients with cystinosis, 6 patients with isofosfamide-induced Fanconi syndrome, 45 patients with other renal diseases	Urine	CE-MS	• 24 peptides and proteins in the urine samples from Fanconi syndrome patients differed significantly from the controls. • Structure of 9 of these 24 peptides were successfully identified using an iontrap-orbitrap hybrid mass spectrometer. • Patients with Fanconi syndrome were identified with 89% specificity and 82% sensitivity.	[220]

Contrast nephropathy	12 patients that underwent therapeutic cardiac catheterization and required a radiocontrast agent, two urine samples were collected: before and after the procedure, 31 controls	Urine	DIGE, linear trap mass spectrometry	• Compared to preprocedure, 39 protein spots were increased, 17 were decreased. • 21 of these 56 spots could be identified, all of which represented proteins derived from albumin. • Among these proteins known to activate the complement pathway were found.	[221]
Diabetic nephropathy	Four groups: patients with type 2 diabetes and no microalbuminuria (n = 45), type 2 diabetes with micro- and macroalbuminuria (n = 38), proteinuria due to nondiabetics disease (n = 34), healthy controls (n = 45)	Urine	SELDI-TOF, protein arrays	• In contrast to diabetic patients with proteinuria, a highly abundant protein with a mass/charge of 6188 was present in urine of the other groups. • A protein with a mass/charge of 14766 was selectively excreted in diabetic patients with proteinuria. • A protein with a mass/charge of 11774 was selectively excreted in the urine of diabetic and nondiabetic patients with proteinuria. • The peak with a mass/charge of 11774 was identified as β2-microglobulin, the peak with a mass/charge as UbA52, an ubiquitin ribosomal fusion protein, and the peak with mass/charge of 6188 was identified as a processed for of ubiquitin. • UbA52 concentrations in urine were considered the most promising protein marker.	[222]

(Continued)

Table 4.4 Selected clinical proteomics studies for the discovery of renal disease protein markers (*cont.*)

Diagnostic target	Study population	Matrix	Analytical technology	Identified correlations	References
Diabetic nephropathy	Nested case control study including 14 patients with type 2 diabetes and 14 controls (training cohort) and 17 patients with type 2 diabetes and 17 controls (validation cohort)	Urine	SELDI-TOF	• SELDI detected 714 unique urine protein peaks • Of these a 12-peak set correctly predicted diabetic nephropathy with 93% sensitivity and 86% specificity • Proteins were not identified • Urine proteomic profiles identified norm-albuminuric individuals with type 2 diabetes who will subsequently develop diabetic nephropathy	[223]
Diabetic nephropathy	Type 2 diabetic patients without proteinuria ($n = 10$), with microalbuminuria ($n = 13$), with macroalbuminuria ($n = 13$) and controls ($n = 10$)	Urine	DIGE, QTOF	• 195 protein spots unique to the urine of diabetic patients were found, representing 62 unique proteins • These proteins belonged to several functional groups, such as cell development, cell organization, defense response, metabolism, and signal transduction • Seven proteins were found to be progressively upregulated with increasing albuminuria and four proteins exhibited progressive downregulation • The majority of the marker candidates were glycoproteins	[224]
Diabetic nephropathy	44 type I diabetic patients with more of 5 years of diabetes, age matched control group	Urine	CE-MS	• Overall more than 1000 different polypeptides (800 Da–66.5 kDa) were found in urine. • 54 polypeptides were only found in diabetic patients. • Another set of 88 polypeptides were either present or absent patients with beginning albuminuria (albumin-to-creatinine ratio > 35 mg/mmol). • Polypeptides were not further characterized.	[225]

Diabetic nephropathy	15 type I diabetic patients (5 with retinopathy, 5 with retinopathy and nephropathy, and 5 without complications), 5 healthy controls, all 4 groups matched by age and gender	Urine	SDS-page, LC-MS/MS	• Gelsolin and antithrombin III as markers for type I diabetes • Ganglioside GM2 activator and beta hexosaminidase subunit beta as urinary marker for retinopathy • Ephrin type-B receptor 4 and vitamin K-dependent protein Z as marker for type I diabetes with retinopathy and nephropathy	[226]
Diabetic nephropathy	Five patients with CKD stage III–V and five matched healthy subjects for learning phase, three patients with CKD stage III–V and three matched healthy subjects for validation phase	Urine, exosome	2D-gel, LC-MS/MS	• 352 proteins were identified in the exosomes. • Label-free, quantitative LC-MS/MS identified a panel of 3 proteins (AMBP, MLL3 and VDAC1) that distinguished patients with diabetic nephropathy from healthy controls.	[227]
Diabetic nephropathy	737 patients with type 2 diabetes, 89 patients developed microalbuminuria during a mean follow up of 4.1 years	Urine	CE-MS	• A vector-machine driven classifier based on 273 distinct urinary peptides (CKD273) was able to predict based on the urine samples collected at baseline who will develop microalbuminuria during the study period and who will not (89% negative predictive value).	[228]
IgA nephropathy	49 IgA nephropathy and 42 CKD patients for learning data set, 14 IgA nephropathy and 24 CKD patients for validation data set	Urine	SELDI-TOF	• Compared to CKD patients, in IgA nephropathy patients perlecan laminin G-like 3 peptide and Ig kappa light chains were significantly lower and were found inversely correlated with the severity of IgA nephropathy.	[229]

(Continued)

Table 4.4 Selected clinical proteomics studies for the discovery of renal disease protein markers (*cont.*)

Diagnostic target	Study population	Matrix	Analytical technology	Identified correlations	References
IgA nephropathy	13 patients with biopsy-proven IgA nephropathy	Urine	2D-gel, LC-MS/MS	• The following urine proteins correlated with the biopsy score: afamin, leucine-rich alpha-2-glycoprotein, ceruloplasmin, alpha-1-microgolbulin, hemopexin, apolipoprotein A-I, complement C3, vitamin D-binding protein, beta-2-microglobulin, and retinol-binding protein 4. • Pathway analysis suggested impairment of extracellular matrix (ECM)-receptor interaction pathways as well as activation of complement and coagulation pathway in progression of IgA nephropathy.	[230]
IgA nephropathy	30 patients with IgA nephropathy, 30 healthy controls	Urine	iTRAQ label, Isoelectric focusing, LC-MS/MS,	• 1238 proteins identified by at least 2 peptides were detected. • 18 proteins were found to differentiate patients with IgA nephropathy from healthy controls. • These proteins included complement components, coagulation factors, ECM, intracellular, and transmembrane proteins.	[231]
IgA nephropathy	43 patients with IgA nephropathy, 23 with membranous nephropathy, 13 with lupus nephropathy, 15 with focal segmental glomerulosclerosis (FSGS), 14 with diabetic nephropathy, 6 with nonglomerular disease, and 30 healthy controls	Urine	2D-gel, MALDI-TOF/TOF	• Higher concentrations of albumin fragments, alpha-1-antitrypsin and alpha-1-beta glycoprotein, lower concentrations of the laminin G-like 3 (LG3) fragment of endorepellin in the urine of patients with IgA nephropathy. • Significant inverse correlation between LG3 levels and glomerular filtration rate in the 43 patients with IgA nephropathy, which was not observed in 65 patients with other glomerular diseases.	[232]

Acute kidney injury	60 patients undergoing cardiopulmonary bypass	Urine	SELDI-TOF	• 15 patients (25%) developed acute renal failure 2–3 days after surgery. Proteins with protein m/z= 6400, 28500, 43000, and 66000 were significantly different in urine samples collected 2 and 6 h after cardiopulmonary bypass in these patients than in the urine samples collected at the same time from patients without acute renal failure. • The sensitivity and specificity of the 28.5-, 43-, and 66-kDa biomarkers for the prediction of ARF at 2 h following cardiopulmonary bypass was 100%. • No attempt was made to identify the structure of these proteins.	[233]
Acute kidney injury	30 patients in intensive care, 16 of whom developed acute kidney injury (training cohort), 20 patients in intensive care of whom 13 developed acute kidney injury, and 31 allogeneic hematopoietic stem cell transplant patients of whom 18 developed acute kidney injury (validation cohort)	Urine	CE-MS	• In the urine samples from the training cohort, 20 peptides were significantly associated with acute kidney injury. These were identified to be degradation products of six proteins: albumin, alpha-1-antitrypsin, beta-2-microglobulin, fragments of fibrinogen alpha and collagens 1 alpha (I), and 1 alpha (III). • In the validation cohort, it was found that in comparison to more established markers of acute kidney injury such as serum cystatin C and urinary kidney injury molecule-1, interleukin-18, and neutrophil gelatinase associated-lipocalin (Table 4.5), the proteomic marker pattern was found to be of superior prognostic value, detecting acute kidney injury up to 5 days in advance of the rise in serum creatinine.	[234]

(Continued)

Table 4.4 Selected clinical proteomics studies for the discovery of renal disease protein markers (*cont.*)

Diagnostic target	Study population	Matrix	Analytical technology	Identified correlations	References
Acute kidney injury	36 patients undergoing cardiopulmonary bypass, 6 of whom developed acute kidney injury	Urine	DIGE, isoelectric focusing, MALDI-MS/MS	• Inflammation-associated (zinc-alpha-2-glycoprotein, leucine-rich alpha-2-glycoprotein, mannan-binding lectin serine protease 2, basement membrane-specific heparan sulfate proteoglycan, immunoglobulin kappa) and tubular dysfunction-associated (retinol-binding protein, adrenomedullin-binding protein, and uromodulin) proteins were found differentially regulated. • Decreased excretion of zinc-alpha-2-glycoprotein in patients with acute kidney injury was confirmed by Western blot and enzyme-linked immunosorbent assay in an independent cohort of 22 patients with and 46 patients without acute kidney injury.	[235]
Acute kidney injury	12 patients with acute kidney injury and early recovery and 12 matching patients with late/nonrecovery (learning cohort), 14 patients with acute kidney injury and early recovery and 14 matching patients with late/nonrecovery, 12 controls (validation cohort)	Urine	DIGE, isoelectric focusing, MALDI-MS/MS	• Eight prognostic candidates were discovered including alpha 1 microglobulin, alpha-1 antitrypsin, apolipoprotein D, calreticulin, cathepsin D, CD59, insulin-like growth factor-binding protein 7 (IGFBP-7), and neutrophil gelatinase-associated lipocalin (NGAL). • These proteins were quantified using immune-based assays in samples from the validation cohort. IGFBP-7 and NGAL discriminated between early and late/nonrecovery patients and patients with and without acute kidney injury.	[236]

Acute kidney injury	Patients with sepsis of whom 17 patients had or developed acute injury within 5 days and 17 patients with no change in renal function (training cohort), 44 patients with and 17 patients without acute kidney injury (validation cohort)	Urine	MALDI-MS, CE-MS	• In urine samples from the test cohort, 51 peptides were significantly different. Of these 39 were selected via cross-validation. • Based on these 39 peptides, a support vector machine classifier was constructed. • Prognostic performance of this classifier was tested in samples from the validation cohort. It was found that the area under the receiver operating characteristics curve was 0.82 and sensitivity and specificity were 86% and 76%, respectively. • The identified peptides were fragments from the collagen chains alpha-1(I) and alpha-1(II), alpha-1-antitrypsin, beta-2-microglobulin (B2M), and fibrinogen alpha chain.	[237]

Abbreviations: ADPKD, Autosomal dominant polycystic kidney disease; CE, capillary electrophoresis; CKD, chronic kidney disease; DIGE, differential gel electrophoresis; FSGS, focal segmental glomerulosclerosis; HILIC, hydrophilic interaction liquid chromatography; iTRAQ, isobaric tags for relative and absolute quantification; MudPIT, multidimensional protein identification technology; QTOF, quadrupole time-of-flight mass spectrometry; Ref., reference; SELDI, surface-enhanced laser desorption ionization; SLS, systemic lupus erythematosus; TOF, time-of-flight mass spectrometry; 2D, two-dimensional.

Many kidney proteins appearing in urine during injury are either proteins that are usually reabsorbed in the proximal tubules, released by cell damage, leaked into the urine during inflammation or immune reactions, or are repair proteins that are formed and released during the healing process. The most important urinary proteins that have been described as kidney injury markers in the literature are summarized in Table 4.5. Most of these can be measured by ELISA or protein multiplexing assays [13], and for some, such as neutrophil gelatinase-associated lipocalin (NGAL) assays, on analytical platforms established in clinical laboratories are available [311–313]. Although Table 4.5 lists single molecular makers, the rational design of a panel of these markers seems superior to the analysis of them individually. If designed correctly, analysis of such a panel will also result in information regarding the nature and location of kidney injury (Fig. 4.4) and repair processes.

Nephrotoxicity and Drug Development

Presently, regulatory agencies still rely primarily on traditional markers, such as creatinine concentrations in serum, blood urea nitrogen, kidney histology, and estimated glomerular filtration rates to assess a drug candidate's nephrotoxic potential during preclinical and clinical drug development. The problem is that these are very insensitive markers with rather poor predictive value. As aforementioned, an increase in serum creatinine considered clinically relevant will require significant kidney damage [239,314]. Serum creatinine is not a specific marker that depends on many other factors and will be delayed in rising if glomerular filtration is not the primary target of a disease or toxicity process. Likewise, changes in histology and blood urea nitrogen are rather late markers that require significant kidney damage before they can be appreciated. There is consensus that there is an urgent need in drug development for better molecular markers that have better sensitivity and specificity [315]. There is evidence that metabolomics and proteomics-based kidney injury markers are more sensitive, specific, and predictive than the currently established markers [316–320].

The use of proteomics in toxicology has also been referred to as "toxicoproteomics" [317]. Toxicoproteomics is a promising concept for two reasons:

While the drug target and therefore the pharmacodynamic mechanisms are usually already known early in the drug development process, toxicity is usually detected during the later stages of preclinical development

Table 4.5 Important protein markers of kidney dysfunction [13,14,238–240].

Protein	Description	PSTC	References
Calbindin	• Calbindin D is a vitamin D-dependent calcium-binding protein of 28 kDa that is found predominantly in the epithelial cells of the distal tubules of the kidney. • Nephrotoxic drugs and diseases involving the distal tubule have been shown to change calbindin concentrations in urine.	No	[242–245]
Clusterin	• A glycoprotein first isolated in Sertoli cells • Is present in most tissues • Is synthesized after tubular injury and protects the tubule • Urine concentrations correlate with tubular damage	Yes	[246–248]
Cystatin C	• 13 kDa extracellular inhibitor of cysteine proteases. Serum concentrations are independent of gender, muscle mass, and age. • Is freely filtered, reabsorbed and catabolized by the proximal tubule; there is no active excretion. • Urinary cystatin C concentrations are elevated in patients with tubular injury.	Yes	[249,250]
Cystein-Rich Protein (Cyr61)	• Is a heparin binding protein that is secreted and associated with cell surfaces and extracellular matrix. • Was found to be secreted in the straight proximal tubulus only a few hours after injury. • It must be considered a limitation that urinary concentrations were found to decrease over time although kidney injury was progressing.	No	[251,252]
Epidermal growth factor (EGF)	• EGF is a 53-amino-acid peptide that is produced by the ascending portion of Henle's loop and by the distal convoluted tubule. • It seems to modulate tissue response to injury in kidneys with tubulo-interstitial damage.	No	[253–256]
α-glutathione-S-transferase (α-GST)	• Cytosolic enzyme in the proximal tubule. • The appearance of α-GST is due to leakage of cytosolic content into the urine, dying cells or due to shedding of viable or apoptotic cells into the urine.	No	[257–260]

(Continued)

Table 4.5 Important protein markers of kidney dysfunction [13,14,238–240]. (*cont.*)

Protein	Description	PSTC	References
π-glutathione-S-transferase (π-GST)	• Cytosolic enzyme in the distal tubule and collection duct. • Is released into the urine likely *via* the same mechanisms as α-GST. • Has been used together with α-GST to differentiate between proximal and distal tubule damage.	No	[261,262]
Interleukin 18	• IL-18 is a proinflammatory cytokine and its 24 kDa precursor is cleaved in the proximal tubule. • Urinary concentrations predict delayed transplant kidney function and acute kidney injury and correlated with its severity. • Seems most sensitive to ischemic injury and seems less (or not) affected by nephrotoxins, chronic kidney disease, and urinary tract infections. • Promotes tubule cell apoptosis and necrosis.	No	[239,240,263–265]
Kidney injury molecule-1 (KIM-1)	• A type 1 transmembrane protein not detected in normal kidney tissue. • Is expressed at very high levels in case of dedifferentiated proximal tubule cells, after ischemic or toxic injury and in case of renal cell carcinoma. • A soluble form of cleaved KIM-1 can then be detected in urine. • Promotes epithelial regeneration, regulates apoptosis	Yes	[240,266–268]
Liver-type fatty acid binding protein (L-FABP)	• L-FABP is a 14-kDa protein that is normally expressed in the kidney proximal convoluted and straight tubules. • Antioxidant, suppresses tubule-interstitial damage. • Increased urinary L-FABP concentrations were found in patients with acute kidney injury, nondiabetic chronic kidney disease, early diabetic nephropathy, idiopathic focal glomerulosclerosis, and polycystic kidney disease. • A challenge is that due to its size L-FABP can be filtered, but is mainly taken up by the proximal tubule; there is some evidence that plasma concentration may not affect urine concentration.	No	[240,269–272]

Microalbumin	• Established molecular marker defined as urinary albumin concentrations between 30–300 mg/L. • Although originally believed only to be a measure of intraglomerular pressure and/or structural changes of the glomerular basement membrane, there is evidence that glomerular membranes normally leak albumin and that albumin is retrieved by the proximal tubule and thus may also be a marker of proximal tubule function.	Yes	[273–275]
β2- microglobulin	• It is the 11.8 kDa light chain of the MHC I molecule expressed on the surfaces of nucleated cells. • Its monomeric form is filtered and reabsorbed in the proximal tubule. • Has been shown to be an early marker of tubular dysfunction.	Yes	[276–278]
N-acetyl-β-glucosaminidase (NAG)	• NAG (> 130 kDa) is a proximal tubule lysosomal enzyme. • Sensitivity, subtle alterations in the epithelial cells in the brush border of the proximal result in shedding of the enzyme into urine. • Increased NAG concentrations in urine have been found after exposure to nephrotoxic drugs, in patients with delayed renal allograft function, with acute kidney injury, with chronic glomerular disease, with diabetic nephropathy and following cardio-pulmonary bypass.	No	[279–282]
Neutrophil gelatinase-associated lipocalin (NGAL)	• NGAL is a lysosomal enzyme that regulates iron trafficking, promotes tubule cell survival, triggers nephrogenesis by stimulating the conversion of mesenchymal cells into kidney epithelium and, in the kidney, is mainly located in the proximal tubule. • Its size is about 25kD and it is protease resistant; it is filtered by the kidney and its plasma/urine concentration relationship will require further clarification. • There is evidence that NGAL may be useful as a sensitive and predictive marker of ischemia/reperfusion, acute kidney injury, nephrotoxicity, and chronic kidney disease.	No	[240,283–285]

(*Continued*)

Table 4.5 Important protein markers of kidney dysfunction [13,14,238–240]. (*cont.*)

Protein	Description	PSTC	References
Osteopontin	• Is synthesized at highest levels in bone and epithelial tissues (44 kDa bone phosphoprotein). • Is found at relatively high concentrations in urine and is believed to act as an inhibitor of mineral precipitation and stone formation. • In human and rodent kidneys, expression is limited to the thick ascending loop of Henle and distal convoluted tubules. • Was found upregulated in rodent models after kidney injury, such as ischemia/reperfusion and drug nephrotoxicity.	No	[286–289]
Retinol binding protein	• A 21 kDa protein that is synthesized in the kidney and is involved in vitamin A transport. • It is freely filtrated and reabsorbed in the proximal tubule. • Plasma and urine concentrations may be associated and vitamin A deficiency may cause false negatives.	No	[277,290]
Podocin	• Podocin is a stomatin family member and is an important component of the glomerular slit-diaphragm complex which colocalizes and interacts with nephrin and CD2AP in the lipid rafts of the podocyte foot process cell membrane. • Damage to the podocyte releases podocin into the urine. • Its mRNA in urine has also been shown to be a molecular marker of kidney dysfunction.	No	[291–294]
Tissue inhibitor of metalloproteinase 1 (TIMP-1)	• TIMP-1 (28.5 kDa) is an inhibitor of matrix metallo-proteinases and is expressed in the proximal tubule. • TIMP-1 mRNA and protein is upregulated in different models of renal disease and human sclerotic glomeruli.	No	[295–298]

Trefoil factor 3	• The trefoil factor family (TFF) peptides are important proteins involved in the regeneration and repair of the urinary tract. • TFF peptides are secretory products of various mucine-producing epithelial cells and promote restitution and regeneration processes of mucous epithelia via induction of cell migration, resistance to proapoptotic stimuli, and angiogenesis. • TFF3 is the most abundant TFF in the urinary tract followed by TFF1 • TFF1 and TFF3 seem to be differently regulated and show potential to predict various chronic kidney disease stages in patients.	Yes	[299–305]
Vascular endothelial growth factor (VEGF)	• VEGF is an important stimulator of angiogenesis; circulating and urinary VEGF levels have been suggested as clinically useful predictors of tumor behavior. • VEGF is also a mediator during inflammation. • Urinary VEGF seems to be of advantage over plasma since venipuncture activates platelets and may release cytokines, including VEGF, artificially elevating measured VEGF levels. • VEGF concentrations in urine were found to be associated with alloimmune processes against kidney transplants. • VEGF may be involved in remodeling after injury leading to increased urinary concentrations.	No	[306–310]

Abbreviations: PSTC, Predictive Safety Testing Consortium, see also European Medicines Agency, Committee for Medicinal Agency, Committee for Medicinal Products for Human Use [241].

or, in many cases, during clinical development. The challenge is that the unknown toxicodynamic mechanism (mechanistic toxicology) remains to be identified [318]. Proteomics is a powerful strategy to achieve this goal. Proteomics can be used for marker discovery or as a molecular marker itself that can be used to support drug development.

As of today, proteomics has been used to evaluate the toxicity of the following xenobiotics: acyclovir [321], 4-aminophenol [317], cisplatin [317,322,323], cyclosporine [324,325], dichlorovinyl-L-cysteine [326,327], gentamycin [317,322,323,328], puromycin [317], and uranium [329]. For a comprehensive review of toxicoproteomics, please see Merrick and Witzmann [316].

The US FDA and European Medicines Agency (EMEA) approved a set of seven urinary proteins as biomarkers of nephrotoxicity that were submitted by the Predictive Safety Testing Consortium (PSTC) in collaboration with multiple pharmaceutical companies to the Voluntary Exploratory Data Submission (VXDS) committee of the US FDA [241,319]. These biomarkers are for regulatory use in certain preclinical settings [330] and are discussed in Table 4.5. These markers are urinary total protein, albumin, β2-microglobulin, cystatin C, kidney injury molecule-1 (KIM-1), clusterin, and trefoil factor-3. Data indicating that these markers add information to serum creatinine and blood urea nitrogen and that, as indicated by receiver operating characteristics, six of the seven outperformed one or both of the established clinical markers was submitted. The submission was supported by data of up to 14-day GLP (good laboratory practice) toxicology studies in rats, validation reports of the analytical assays and a review of the scientific literature; however, several limitations were acknowledged in the EMEA/FDA document [241]. These include a lack of data demonstrating that these molecular markers can be used for monitoring the evolution of kidney changes over time, can be used for monitoring the reversibility of the injury and kidney recovery, or can be transferred to other species. Moreover, their general use for monitoring of nephrotoxicity in a clinical setting was not recommended at this stage [241]. Incremental qualification potential was acknowledged and will require the submission of additional data. It is interesting to note that several kidney function markers listed in Table 4.5 were not included and also that the choice of molecular markers is essentially focused on the proximal tubule and does not allow for mapping of the damage to a specific location in the kidney.

Despite the limitations, approval of these kidney injury protein markers has been considered a “door opening safety biomarker success story” [330] and is a good example for how molecular marker tools for drug development can be expected to be developed, reviewed, and approved in the near future.

Kidney Transplantation

Over the past 30 years, 1-year outcomes after kidney transplantation have markedly improved. Although there has also been progress in terms of long-term kidney graft survival over the last decades, this improvement has mostly been driven by better short-term graft survival, while long-term attrition is only slowly improving [331]. The pathogenesis of these graft losses is multifactorial [332]. The early detection of injury to the transplant kidney is critical to minimize permanent injury and to maintain long-term function [332–334]. The current strategy of treating most transplant recipients still leads in a relatively high proportion of patients to either underimmunosuppression or overimmunosuppression [334]. Among many additional factors that can damage a transplant kidney, allograft immune reactions, infections, such as BK and CMV virus, and recurrent or de novo glomerulopathy are most important [335]. These contribute to chronically progressive scarring processes and ultimately allograft dysfunction. Currently, serum creatinine is still the gold standard biomarker for monitoring kidney transplant patients [336]. Unfortunately, the rise of creatinine in serum is a late event and occurs when kidney function is already severely and often irreversibly impaired [336]. As previously discussed, serum creatinine also lacks specificity and a subsequent kidney biopsy must be procured. But even kidney biopsies are not necessarily conclusive [336]. Biopsies face many dilemmas: they sample only a fraction of the kidney but injury processes are often patchy, different injury processes may present the similar histological changes, and histology analysis and grading does not use objective metrics [337]. Urine represents an average of the processes occurring at a given moment in the kidney and proteomics metrics is objective and nonbiased. Therefore, the development of new molecular marker strategies for the specific and early detection of antiallograft immune processes and immunosuppressant toxicity has generated substantial interest in the field of transplantation [332,334]. Three types of biomarkers are discussed: (1) those associated with the risk of rejection (alloreactivity/tolerance), (2) those reflecting individual response to immune-suppressants, and (3) those associated with graft dysfunction and injury [334]. The proteomes of interest are plasma and serum, mostly used for the detection and monitoring of alloimmune processes, kidney biopsies, to provide better discrimination between underlying injury processes and to complement histology, and urine, for the monitoring of kidney injury caused by immune processes, disease processes, and drug toxicities.

There are two approaches to monitor alloimmune reactions against a renal allograft: one is to monitor aspects of the immune system, which includes alloimmune recognition and activation pathways as well as the effector pathways of inflammation. For a comprehensive list of individual markers, please see Brunet et al. [334] and Gwinner [338]. The other approach is to monitor kidney injury markers. Rejection is a complex, heterogeneous, and variable process. Often insufficient sensitivity and specificity in the use of single rejection markers has led to the concept of combining markers. This is not limited to the mathematical fact that this will increase the overall sensitivity and specificity. From a biological perspective, appropriately designed sets of markers can also capture variation in the rejection process better [339].

The effect of kidney injury on urine, kidney biopsies, and plasma proteomes has been studied in animal models [332,339] and in multiple clinical trials using nontargeted proteomics. Representative clinical proteomics studies are summarized in Table 4.6. For comprehensive reviews and overview tables, please also see Sidgel et al. [365,366], Kienzel-Wagner et al. [367,368], and Kim et al. [369]. In addition, targeted approaches based on one or several of the protein kidney injury markers described earlier (Table 4.5) have shown promising results [370,371].

Most proteomics studies in kidney transplant patients, such as those listed in Table 4.6, have focused on the urine proteome. Such studies have typically used mass spectrometry-based assays. Many of the earlier studies used surface-enhanced laser desorption/ionization time-of-flight mass spectrometry (SELDI-TOF) [366]. As aforementioned [Section "Matrix Assisted Laser Desorption/Ionization Mass Spectrometry (MALDI-MS)], one of the problems with this technology is that the time-of-flight mass spectrometer often does not have sufficient resolution to allow for identification of protein and peptide marker candidates [366]. Such studies have typically characterized marker proteins based only on their mass/charge. A more recent example for such a study is Tetaz et al. [357]. However, verification of marker protein identity is desirable. An example for a proteomics study that identified specific proteins, in this case a urinary protein biomarker pattern including β2-microglobulin, neutrophil gelatinase-associated lipocalin (NGAL), kidney injury molecule 1 (KIM-1), and clusterin to detect and diagnose chronic allograft nephropathy, using high-resolution mass

Table 4.6 Identification of proteomics-based biomarkers for diagnosis after kidney transplantation in clinical studies

Diagnostic target	Study population	Matrix	Analytical technology	Identified correlations	References
Acute allograft rejection	17 patients with and 15 patients without rejection	Urine	SELDI-TOF	• Polypeptides with 6.5, 6.6, 6.7, 7.1, and 13.4 kDa were identified that allowed for classification; • those peptides/proteins were not further identified; • sensitivity 83%, specificity 100%.	[340]
Acute allograft rejection	19 patients with different grades of rejection (Banff 1997 Ia to IIb), 10 patients with urinary tract infection, 29 patients without rejection, 66 nontransplant subjects	Urine	CE-TOF	• 17 Urinary polypeptides discriminated between renal transplant patients and nontransplant patients • 10 between urinary tract infection and samples without infection or rejection (control) • 16 between renal allograft patients with and without rejection • 10 between acute allograft rejection and urinary tract infection • one protein differentiating between healthy subjects and renal allograft patients, a fragment of collagen alpha 5(IV) protein, was identified	[341]
Acute allograft rejection	23 patients with and 22 patients without rejection, 20 healthy subjects	Urine	SELDI-TOF	• 7 polypeptides with 2.0, 2.8, 4.8, 5.9, 7.0, 19.0, and 25.7 kDa were identified that allowed for classification. • Acute rejection could be distinguished from stable renal allograft patients with sensitivity of 90.5–91.3% and a specificity of 77.2–83.3%. • A protein with the 78.5 kDa was found that distinguished between renal allograft patients and healthy subjects. Sensitivity and specificity were 100%.	[342]

(*Continued*)

Table 4.6 Identification of proteomics-based biomarkers for diagnosis after kidney transplantation in clinical studies (*cont.*)

Diagnostic target	Study population	Matrix	Analytical technology	Identified correlations	References
Acute allograft rejection	18 patients with and 22 patients without rejection, 5 patients with tubular necrosis, 5 patients with glomerulopathy, 5 non-transplant patients with urinary tract infections, 28 healthy subjects	Urine	SELDI-TOF	• Patients with rejection showed prominent peak clusters in regions of $m/z = 5{,}270–5{,}550$, $7{,}050–7{,}360$, and $10{,}530–11{,}100$. • In urine from normal subjects, those clusters were missing. • 82% from the stable transplant group and 6% from the acute rejection group did not show those clusters. • the peptides/proteins in the clusters were further structurally identified in Schaub et al. [344] and were found to be mostly associated with β2-microglobulin.	[343]
Acute allograft rejection	34 samples were collected from 32 renal transplant patients, 17 of these samples were from 15 patients with acute rejection	Urine	SELDI-TOF, protein chip arrays	• 45 protein peaks of interest were identified. • 16 of these peaks showed promise as candidate molecular markers to detect acute rejection. • 13 of these proteins (3.4, 4.1, 6.5, 6.6, 6.7, 7.0, 7.1 7.3, 7.5, 7.8, 8.0, 10.8, and 13.4 kDa) were present in the majority of urine samples during rejection, but absent in nonrejection samples. • Three proteins (9.0, 9.7, and 9.8 kDa) were present in nonrejection urine samples, but were absent in samples collected during rejection. • Urine samples collected during rejection could be distinguished from those without rejection with a sensitivity of 91.3%.	[345]

Acute allograft rejection	10 patients with biopsy-proven acute rejection, 10 stable renal transplant patients, 10 patients with nonspecific proteinuria, 10 age-matched healthy controls	Urine	MudPIT	• 1446 proteins were identified in urine, of which 67 proteins were only identified in the urine of healthy controls. • Differences characteristic for urine from patients with acute rejection included proteins associated with MHC antigens, the complement cascade and extracellular matrix. • A subset of proteins, uromodulin, SERPINF1 and CD44, was further cross-validated using ELISA and was found different in urine from patients with acute rejection than in the urine samples from the other patient groups.	[346]
Acute allograft rejection	20 patients with biopsy-proven acute rejection, 20 stable renal transplant patients, 10 patients with BK virus infection	Urine	LC-fractionation, MALDI-TOF, and MS/MS	• A total of 20,937 unique peptide peaks with distinct m/z and HPLC fractions were resolved in the 900- to 4000-Da range. • A 40-peptide panel was found to discriminate urine samples from patients with acute rejection from urine samples from the other patient groups. • A 6-gene biomarker panel (collagen1A2, collagen 3A1, uromodulin, metalloproteinase-7, serpin-peptidase inhibitor SERPING1, tissue inhibitor of metalloproteinase 1) classified AR with high specificity and sensitivity (area under ROC curve = 0.98).	[347]

(Continued)

Table 4.6 Identification of proteomics-based biomarkers for diagnosis after kidney transplantation in clinical studies (*cont.*)

Diagnostic target	Study population	Matrix	Analytical technology	Identified correlations	References
Acute allograft rejection	11 patients with biopsy-proven acute rejection, 21 patients without biopsy-proven acute rejection	Plasma	iTRAQ-MALDI-TOF/TOF	• Three proteins, titin, kininogen-1, and lipopolysaccharide-binding protein, showed the best discrimination between the patient cohorts. • Longitudinal monitoring over the first 3 months posttransplant based on ratios of these three proteins showed clear discrimination between the two patient cohorts at time of rejection. • The score then declined to baseline following treatment and resolution of the rejection episode and remained comparable between cases and controls throughout the period of quiescent follow-up. • Results were validated using ELISA and initial cross-validation estimated a sensitivity of 80% and specificity of 90% for classification of biopsy-confirmed acute rejection based on a four-protein ELISA protein classifier.	[348]

Acute allograft rejection	Samples from patients with: acute rejection ($n = 30$), stable graft function ($n = 30$), chronic allograft injury ($n = 30$), BK virus nephropathy ($n = 18$) (test set), acute rejection ($n = 44$), stable graft function ($n = 44$), chronic allograft injury ($n = 18$), BK virus nephropathy ($n = 20$), calcineurin inhibitor nephrotoxicity ($n = 8$), healthy controls ($n = 10$) (validation set)	Urine	iTRAQ label, LC-MS/MS	• 389 proteins displayed differential abundances across urine specimens of the injury types. • SUMO2 (small ubiquitin-related modifier 2) was a protein marker for graft injury irrespective of causation. • 69 urine proteins had differences in abundance in urine samples form patients with acute rejection compared to samples from patients with stable grafts. • 9 urine proteins were highly specific for acute rejection because of their significant differences ($p < 0.01$; fold increase > 1.5) from all other patient cohorts. • Increased levels of three of these proteins, fibrinogen beta ($p = 0.04$), fibrinogen gamma ($p = 0.03$), and HLA DRB1 ($p = 0.003$) were validated by ELISA in urine samples from patients with acute rejection using an independent sample set. • These fibrinogen proteins further segregated acute rejection from BK virus nephritis.	[349]
Acute allograft rejection	8 patients, 5 patients with acute rejection	Plasma	iTRAQ label, 2D-LC-MS/MS	• Among the 179 proteins identified using iTRAQ labeling, 66 proteins were at least twofold different between patients with or without acute rejection. • These proteins were associated with inflammation and complement activation in acute rejection. • Transcription factors that discriminated between the two patient cohorts included nuclear factor-κB, signal transducer and activator of transcription 1, signal transducer and activator of transcription 3.	[350]

(Continued)

Table 4.6 Identification of proteomics-based biomarkers for diagnosis after kidney transplantation in clinical studies (*cont.*)

Diagnostic target	Study population	Matrix	Analytical technology	Identified correlations	References
Allograft rejection	Patients with biopsy proven interstitial fibrosis and tubular atrophy, stages 0 ($n = 8$), I ($n = 8$), and II/III ($n = 8$)	Urine	2D-DIGE, MALDI-TOF, CE-linear ion trap	• 62% of the urinary proteins detected were identified using mass spectrometry and data base searches. • 44% were secreted, 17% membrane, 13% plasma, 10% cytoplasmic, 3% cytoskeletal, and 1% mitochondrial proteins. • 19 proteins with differential concentrations depending on the stage of renal graft injury were found. • Among those were β2 microglobulin, MASP-2, α-1-B-glycoprotein, leucine-rich α-2-glycoprotein-1, α-1-antitrypsin, immunoglobulin lambda light chain, transferrin, and Zn-α-2-glycoprotein.	[351]
Allograft rejection	11 stable kidney transplant patients, 10 patients with biopsy-proven acute rejection, 11 patients with biopsy-proven chronic graft injury, 8 healthy controls	Urine	Antibody microarray	• ANXA11, integrin α3, integrin β3, and TNF-α, were identified as markers differentiating between acute and chronic rejection. • These 4 proteins were qualified using reverse capture protein microarrays.	[352]
Allograft rejection	16 patients with biopsy-proven subclinical rejection, 23 patients without rejection (training cohort), 18 patients with subclinical rejection, 10 patients with clinical acute rejection, and 36 patients without rejection (validation	Urine	CE-MS	• Of the 5010 polypeptides detected in urine, 13 with most discriminatory power were selected. • Sequence information of these polypeptides identified altered collagen a (I) and a (III) chain fragments in rejection samples. • These fragments suggested an involvement of matrix metalloproteinase-8 (MMP-8), which was confirmed by immunostaining of biopsies.	[353]

Allograft rejection	Patients with biopsy-confirmed acute rejection ($n = 12$), chronic rejection ($n = 12$), stable graft function ($n = 12$), and healthy individuals ($n = 13$)	Serum	MALDI-TOF	• 18 differential peptide peaks were selected as potential molecular markers for acute allograft rejection. • 6 differential peptide peaks were selected as potential molecular markers for chronic rejection. • The peptides were only identified by their m/z and not further characterized. • A classifier algorithm recognized 82.6% of acute rejections and 99.0% of chronic rejection episodes correctly.	[354]
Chronic allograft injury	34 renal transplant patients with histologically proven chronic allograft nephropathy and 36 patients with normal renal transplant function	Urine	SELDI-TOF	• An 11.7 kDa protein identified as β2 microglobulin was the primary protein distinguishing urine samples from patients with chronic allograft nephropathy from the controls.	[355]
Chronic allograft injury	39 patients with chronic allograft nephropathy and 32 control individuals	Urine	LC-MS/MS	• 6000 polypeptide ions in undigested urine samples could be identified. • Unsupervised hierarchical clustering differentiated between the groups when including all the identified peptides. • Specifically, peptides derived from uromodulin and kininogen were found to be significantly more abundant in urine from controls than in urine from patients with chronic allograft nephropathy and correctly identified the two groups. • Sensitivity and specificity were found to be 90%.	[356]

(Continued)

Table 4.6 Identification of proteomics-based biomarkers for diagnosis after kidney transplantation in clinical studies (*cont.*)

Diagnostic target	Study population	Matrix	Analytical technology	Identified correlations	References
Chronic allograft injury	Urine samples from patients without (n = 14), with possible (n = 12) and with confirmed chronic allograft injury (n = 15)	Urine	SELDI-TOF	• The biomarker demonstrating the highest diagnostic performance was a protein of 8860 Da that predicted chronic allograft injury with a sensitivity of 93% and a specificity of 65%. • No attempt was made to identify the structure of this protein marker.	[357]
Chronic allograft injury	34 renal patients with histologically proven chronic allograft injury and 36 renal patients with normal renal function	Urine	LC-MS/MS	• β2 microglobulin was found as the protein with the highest discriminatory power to distinguish between the two patient cohorts. • Other important proteins that discriminated between the two cohorts were neutrophil gelatinase-associated lipocalin, clusterin, and kidney injury biomarker 1. • Significantly higher urinary concentrations of these proteins were found in patients with chronic allograft injury compared to those with normal kidney function.	[358]
Chronic allograft injury	32 patients with chronic allograft dysfunction, 14 with interstitial fibrosis, 18 with chronic active antibody mediated rejection, 18 controls: eight stable renal transplant patients, and 10 healthy individual	Urine	MALDI-TOF	• 14 proteins ions were identified that discriminated between the samples from patients with interstitial fibrosis and chronic rejection. • 100% of both patient groups were identified correctly. • These proteins were characterized only by their mass/charge and were not further identified.	[359]

Chronic allograft injury	Two cohorts with a total of 77 renal allograft patients with mild or moderate/severe chronic allograft nephropathy as confirmed by biopsy	Density purified blood cells	MudPIT in combination with linear ion trap	• The study used a proteogenomic approach. • 302 proteins unique to mild and 509 proteins unique to moderate/severe chronic allograft nephropathy were detected and identified using database searching.	[360]
Stable kidney transplant patients	Serial urine samples from healthy individuals, kidney donors before and after surgery ($n = 20$), recipients immediately after surgery, kidney transplant patients 1 months to 4 years after transplantation ($n = 16$)	Urine	MALDI-TOF, iTRAQ	• Several protein peaks were detected that were associated with transplantation (mass/charge: 3370, 3441, 3385, 4303, 10350, 11732). • The protein with a mass/charge of 11732 was β2 microglobulin, none of the other proteins was identified. • Although there were differences in the urinary protein patterns among individuals, longitudinal comparison of protein patterns in the same individuals over time suggested that individual urine protein patterns are remarkably stable. • It was concluded that the longitudinal follow up of urinary protein patterns in individual patients may be a sensitive biomarker.	[361]
BK-virus nephropathy	21 patients with BK-virus nephropaty, 28 patients with acute rejection, and 29 patients with stable graft function.	Urine	SELDI-TOF	• Peaks that corresponded to $m/z = 5872, 11311, 11929, 12727$, and 13349 were significantly higher in the urine of patients with BK virus nephropathy than in urine from patients with acute rejection. • No attempt was made to structurally identify these proteins.	[362]

(*Continued*)

Table 4.6 Identification of proteomics-based biomarkers for diagnosis after kidney transplantation in clinical studies (*cont.*)

Diagnostic target	Study population	Matrix	Analytical technology	Identified correlations	References
Differential diagnosis of kidney dysfunction after renal allograft	Biobanked samples from patients with: acute rejection ($n = 70$), stable graft function ($n = 70$), chronic allograft injury ($n = 70$), BK virus nephropathy ($n = 35$) (test set), acute rejection ($n = 42$), stable graft function ($n = 47$), chronic allograft injury ($n = 46$), BK virus nephropathy ($n = 16$) (validation set)	Urine	iTRAQ label, LC-MS/MS	• From a group of over 900 proteins identified in transplant injury, a set of 131 peptides were assessed by selected reaction monitoring for their significance in accurately segregating organ injury causation and pathology. • A set of 35 proteins were identified for their ability to segregate the three major transplant injury clinical groups, comprising the final panel of 11 urinary peptides for acute rejection (93% area under the curve [AUC]), 12 urinary peptides for chronic allograft nephropathy (99% AUC), and 12 urinary peptides for BK virus nephritis (83% AUC).	[363]
Differential diagnosis of kidney dysfunction after renal allograft	2 sets of Banff'97 graded biopsies. Set 1:4 Banff 0, 4 Banff 1 and 5 Banff 2,3; set 2: 4 Banff 0, 5 Banff 1, 10 Banff 2,3	Renal biopsies	MudPIT in combination with linear ion trap	• Genome-wide expression analysis was conducted in parallel to proteomics. • Proteins were mapped to multiple pathways including immune response, inflammatory cell activation and apoptosis as observed during chronic rejection. • The extent of changes increased with the severity of renal allograft injury.	[364]

Abbreviations: CE, Capillary electrophoresis; DIGE, differential gel electrophoresis; iTRAQ, isobaric tags for relative and absolute quantification; MudPIT, multidimensional protein identification technology; Ref., reference; SELDI, surface-enhanced laser desorption ionization; TOF, time-of-flight mass spectrometry; 2D, two-dimensional.

spectrometry in combination with immunological assays is Cassidy et al. [358].

A nontargeted urine proteomics test based on two-dimensional capillary electrophoresis/mass spectrometry (CE-MS) is commercially offered for the clinical diagnosis of chronic kidney diseases [372]. This technology has also successfully been used in clinical trials in transplant patients [341,353]. The data from one of these studies indicated that urine samples from patients with subclinical acute T-cell-mediated tubule-interstitial rejection could be distinguished from nonrejection controls mainly based on altered collagen α(I) and α(III) chain fragments suggesting an involvement of matrix metalloproteinase-8 (MMP-8) [353]. Nevertheless, this CE-MS proteomics assay has mainly been studied in chronic kidney disease patients and its value specifically for transplant patients still remains to be evaluated.

While most of these protein markers discovered in the studies listed in Table 4.6 have not yet progressed beyond the discovery/early proof-of-concept stage, several promising candidates have emerged [334,373]. Among these the most advanced marker candidates discovered in proteomics studies in kidney transplant patients are the CXC-receptor 3 chemokines CXCL-9 and CXCL-10 in the urine of kidney transplant patients. CXCL-9 and CXCL-10 are secreted by leukocytes in the kidney graft and are inflammation markers [374]. As such they have been shown to be early markers of kidney graft dysfunction including, but not limited to, kidney graft inflammation associated with subclinical tubulitis [375], subclinical rejection [376], and subclinical BK virus infections [377]. The sensitivity and specificity of CXCL-9 and CXCL-10 exceeded that of creatinine concentrations in serum [375]. CXCL-9 was clinically qualified/validated in a multicenter observational study in 280 adult and pediatric kidney transplant patients [378]. The results showed that low urinary CXCL-9 protein concentrations collected 6 months after transplantation from stable allograft recipients classified individuals least likely to develop acute rejection or a reduction in estimated glomerular filtration rate between 6 and 24 months [378]. In a prospective study of nonsensitized stable living donor kidney transplant patients randomized to stay on or to be withdrawn from the immunosuppressant drug tacrolimus, high urinary CXCL-9 levels predated clinical detection of acute rejection by a median of 15 days [379]. It was also found that the combination of urinary CXCL-10 levels normalized to urine creatinine with donor-specific antibody monitoring significantly improved the noninvasive diagnosis of antibody-mediated rejection and may allow for the stratification of patients at high risk for graft loss [380].

Proteomics studies also resulted in other candidate protein markers that have shown promise in clinical validation trials, such as urinary CCL-2 (predictor of interstitial fibrosis and atrophy and kidney graft loss) [381], urinary β2-microglobulin (chronic allograft nephropathy) [355], and serum aminoacylase-1 (long-term outcome in patients with delayed kidney graft function) [382].

Cancer

Renal cell carcinoma is the most common cancer of the kidney. The main histological subtypes are clear cell, papillary and chromophobe renal cell carcinoma [383,384]. Renal cell carcinomas are often diagnosed at a later stage when approximately 40% of the patients already have local or advanced metastasis. The prognosis of patients with metastatic disease is poor with a 5-year survival of less than 10%. Renal tumors are a challenge for the pathologist since most common benign and malignant renal tumors cannot easily be distinguished [383,384].

There are many opportunities where proteomics can contribute to the diagnosis, treatment and monitoring of renal cancer patients. Today, histological diagnosis, staging, detection of relapse, and monitoring of therapeutic response require either invasive procedures or the use of radiology and cross-sectional imaging [383,385]. Ideally, a comprehensive set of protein tumor markers would have the following characteristics [386]:

- is secreted or shed by the malignant cells;
- can be detected in an easily available body fluid;
- is detected as soon as the tumor becomes active;
- is detected by a simple, robust, and sensitive assay;
- can diagnose a tumor with high specificity during the early stages;
- detects the reoccurrence of a tumor;
- establishes and monitors therapeutic success;
- correlates with the clinical stage of a tumor;
- is objectively measured and independent of the experience level of the examiner; and
- predicts clinical outcome.

Proteomics in combination with tumor cell lines and animal models have been used to better understand tumor biology and treatment response. Clinical proteomics, with the goal of discovering potential protein markers, have been based on biopsy samples as well as on urine, serum, and plasma. To date, however, most proteomics studies in renal cancer have focused on tissue samples and clear cell carcinomas. Table 4.7 shows an overview

Table 4.7 Identification of proteomics-based biomarkers for diagnosis of renal cell carcinoma.

Study population	Matrix	Analytical technology	Identified correlations	References
29 patients (clear cell renal cell carcinoma tissue and adjacent normal kidney tissue sample)	Tissue	2-nitrobenzenesulfenyl labeling, MALDI-TOF	• 34 proteins were found at significantly higher concentrations and 58 proteins had significantly lower concentrations in the tumor than in the normal kidney tissue. • Among these proteins galectin-1 and carnosinase CNDP-2 were considered the most promising potential markers to differentiate between clear cell renal cell carcinoma and normal tissue.	[387]
6 patients with primary tumor and healthy kidney control tissue from the same kidney, 6 patients with metastasis	Tissue	iTRAQ label, LC-MS/MS	• 1256 nonredundant proteins were identified and 456 of these were quantified. • 29 proteins were differentially expressed (12 overexpressed and 17 underexpressed) in metastatic and primary renal cell carcinoma. • Profilin-1 (Pfn1), 14-3-3 zeta/delta (14-3-3ζ), and galectin-1 (Gal-1) were further verified in two independent sets of tissues using Western blot and immunohistochemical analysis. • 14-3-3ζ and Gal-1 also showed higher expression in tumors with poor prognosis than in those with good prognosis.	[388]
199 patients with clear cell renal cell carcinoma, 30 matching normal controls	Tissue, serum, urine	iTRAQ label, LC-MS/MS	• 55 proteins discriminated clear cell renal cell carcinoma from normal kidney tissue. • 54 were previously reported to play a role in carcinogenesis. • 39 are secreted proteins. • Dysregulation of alpha-enolase (ENO1), L-lactate dehydrogenase A chain (LDHA), heat shock protein beta-1 (HSPB1/Hsp27), 10 kDa heat shock protein, and mitochondrial (HSPE1) was confirmed in two independent sets of patients by western blot and immunohistochemistry. • Hsp27 was found elevated in urine and serum and high serum Hsp27 was associated with high grade (Grade 3-4) tumors. Hsp27 was considered the most promising potential marker for clear cell renal cell carcinoma in this study.	[389]

Table 4.7 Identification of proteomics-based biomarkers for diagnosis of renal cell carcinoma. (*cont.*)

Study population	Matrix	Analytical technology	Identified correlations	References
8 patients with renal cell carcinoma tissue and adjacent autologous normal renal tissue	Tissue	Label-free, quantitative LC-MS/MS	• 1761 proteins were identified and quantified. • 596 proteins were identified as differentially expressed between cancer and noncancer tissues. • Of these, two proteins, adipose differentiation-related protein and coronin 1A, were further validated by immunohistochemistry and were considered the most promising potential markers for renal cell carcinoma.	[390]
98 patients with clear cell renal cell carcinoma and 100 healthy controls	Serum	SELDI-TOF	• Proteins with $m/z = 5911, 7987, 8948, 9304$, and 15953 discriminated serum from patients with clear cell renal cell carcinoma and controls. • These five proteins were identified as Bcl-2 family apoptosis regulatory proteins, WAP four-disulfide core protein, Krueppel-like factor 8, monocyte chemotactic protein-1, serum amyloid β -protein-4. • The sensitivity and specificity in predicting clear-cell renal cell carcinoma was 88.8% (79/89) and 91.0% (91/100), respectively.	[391]
114 patients, 102 with malignant renal cell carcinoma, 12 with benign tumor	Serum	Magnetic bead enrichment, MALDI-TOF, LC-MS/MS	• 200 unique endogenous peptides, originating from 32 proteins, were identified • Among these, serum deprivation-response protein (SDPR) and zyxin were higher, while seroglycin and thymosin beta-4-like protein 3 were lower in serum from patients with renal cell carcinoma in comparison to serum from the controls.	[392]
30 patients with renal cell carcinoma, 30 healthy controls	Serum	LC-MS/MS	• 19 candidate peptides are highly correlated with the occurrence of RCC cancer. • These peptides had 100% sensitivity and 93.3% specificity for discriminating serum samples from patients with renal cell carcinoma from the controls. • Among these, 2 peptides with the sequence IYQLNSKLV and AGISMRSGDSPQD were reported for the first time in the detection of cancer.	[393]

118 patients with renal cell carcinoma, 137 healthy controls, 35 patients with other renal tumors (16 benign renal masses, and 19 with other malignant renal tumors)	Urine	Magnetic bead enrichment, MALDI-TOF, LC-MS/MS	• A cluster of 12 signals could differentiate urine from patients with renal cell carcinoma from urine from the other patient cohorts with a sensitivity of 76% and a specificity of 87%. • Among these peptides were fragments of proteins involved in tumor pathogenesis and progression such as meprin 1α, probable G-protein coupled receptor 162, osteopontin, phosphorylase β kinase regulatory subunit alpha, and transmembrane protein 1.	[394]
20 patients with clear cell renal cell carcinoma tissue and tissue from a healthy region of the kidney	Tissue	MALDI imaging, LC-MS/MS	• 108 proteins were different in tumor tissue and normal tissue, 56 of these proteins were identified using LC-MS/MS. • Based on these results, a refined panel of 26 proteins and 39 lipid species was identified that could distinguish tumor from nontumor tissues.	[395]

Abbreviations: iTRAQ, Isobaric tags for relative and absolute quantification; SELDI, surface-enhanced laser desorption ionization; Ref., reference; TOF, time-of-flight mass spectrometry.

of representative clinical proteomics biomarker discovery studies in patients with renal cell carcinoma. For comprehensive reviews, please also see Raimondo et al. [396] and Di Meo et al. [397].

Relatively few studies have used biological fluids, such as urine, albeit results have been encouraging. The combination of serology with proteomics technologies represents a powerful tool to identify protein markers of renal cell carcinoma [385,398,399]. A good example is a study described by Sakissan et al. [399]. The immunogenic protein expression profile of the human renal cell carcinoma cell line CAL54 was assessed using 2D-gel electrophoresis combined with immunoblotting using sera from healthy individuals and patients with renal cell carcinoma. Prometalloproteinase 7 was identified as a potential marker. An immunoassay was developed, and the sera of 30 healthy individuals, 40 control patients, and 30 clear cell carcinoma patients were analyzed. Sensitivity of 93% and a specificity of 75% were found [399].

The urine proteome is also of interest for the discovery of molecular markers for urothelial, ovarian, and prostate cancer [37,40,149,397,400–402].

The Effects of Extra-Renal Proteome Changes on the Urine Proteome

Since some proteins can cross over from blood into urine, the urine proteome is also affected by extra-renal changes. This can provide diagnostic opportunities such as the measurement ofVEGF in urine as a tumor marker [307] and the detection of potential protein markers of coronary artery disease in urine [403]. But this also must be considered as a confounding factor, such as changes of the urine proteome caused by cigarette smoking [404], age [34], gender, and even circadian changes, just to name a few examples. Although the urine proteome can be changed by extrarenal factors, this, surprisingly, has not been taken into account in many clinical proteomics studies even if they were based on complex patient populations. This emphasizes the requirement for appropriate qualification studies before a protein marker can be used as a clinical diagnostic tool.

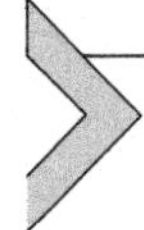

PROTEOMICS AS CLINICAL DIAGNOSTIC TOOL IN NEPHROLOGY

During the last decade, proteomics research and clinical proteomics have been expected to lead to new disease and diagnostic markers that translate into new and improved clinical tests [405] (Table 4.8); however,

Table 4.8 Purposes of in vitro diagnostic markers [405].

Marker Type	
Acute marker	• used when an acute disease event occurs • helps in the process of differential diagnosis • assists in decision making regarding best treatment option
Screening marker	• identifies the diseased, preferably in a still asymptomatic stage, within a population • assists decision for initiation of treatment; early treatment usually correlates with high probability of treatment success • are generally applied in population subgroups with increased risk and disease frequency
Primary risk assessment marker	• assesses the risk of a healthy individual to suffer from a disease, drug effect, or environmental challenge in the future
Secondary risk assessment marker	• used to determine how a disease may develop • used to determine the risk of a patient to suffer recurrent disease or secondary complications
Disease staging/classification marker	• diagnoses and classifies different disease stages
Treatment response stratification marker	• predicts the probability to respond to a drug • predicts tolerability of a drug treatment • assists in the selection of the most effective and/or drug regimen with the best risk/benefit ratio
Treatment or therapeutic monitoring marker	• monitors the long-term efficacy of a drug treatment • monitors for the development of potential chronic toxicodynamic effects • may guide dosing or long-term adjustment/individualization of drug effects
Compliance marker	• provides information on treatment compliance

many published clinical studies have led to confusion and constructive discussions about the suitability of technologies, such as protein arrays and SELDI [366], in clinical practice instead. It is interesting to note that most molecular marker discovery studies are descriptive and have substantial design flaws, such as being underpowered and/or not appropriately taking potential confounding factors into account. Many publications, especially those based on SELDI-TOF analysis, provide only mass/charge values of peaks of interest without any further protein identification or long lists of protein hits as generated by database searches without further confirmation or qualification. Only rarely have further steps been taken to further develop such markers into clinical diagnostics. Meanwhile, there are regulatory guidelines for how to develop molecular markers into markers that can be used to guide drug development and what is required to develop a marker into a clinical diagnostic tool [18,406–410]. Regulatory guidelines clearly distinguish between validation and qualification. As aforementioned, while validation confirms the validity of the analytical assay and that the key performance parameters are within acceptable ranges, qualification proves that a marker, or set of markers, is linked to the disease process or drug effect that it is intended to monitor. Qualification is the key to a molecular marker development and can be a complex and daunting task that may require substantial resources [410]. Like in the case of the promising field of metabolomics, clinical proteomics research has to shift focus from pure discovery to mechanistic and clinical qualification of the discovered marker to establish their clinical indication, collection schedule, limitations, robustness, sensitivity, specificity, and potential predictive value using appropriately validated analytical assays to ultimately meet regulatory acceptance criteria [18,334,409].

The idea of building expert systems based on nonbiased proteome analyses that will generate a holistic view of a patient's plasma and urine proteome, and their combination with other omics technologies such as genomics, transcriptomics, and metabolomics that will allow for a systems biology-based approach to medicine is attractive. It is not unrealistic to expect that this is where the future of medicine lies, especially since this will open the doors for predictive, preventive, and individualized medicine; however, the current utilization of truly nonbiased proteomics as a clinical tool is unrealistic. Even though substantial progress has been made over the last several years, we do not have any technologies yet that will enable a nonbiased proteome analysis. While the human proteome has more than 200,000 proteins that can be in different states of folding, posttranslational modification, interaction, and allosteric regulation, current proteomics technologies can only capture a few thousand simultaneously at best. Current proteomics approaches have also suffered from the wide dynamic range that is required to measure the whole proteome with regulatory proteins often hidden by highly abundant proteins [17]. As aforementioned, the typical high-abundance proteins are removed to unmask those in lower abundance that are considered of more significance for detecting pathobiochemical processes. This means that the methods themselves introduce significant bias into proteome analysis. True nontargeted proteomics is still a complex multistep process and, besides more targeted approaches, no clinically feasible high-throughput technology is available [8,411]. The other problem is that no computational approaches are presently available that convert the highly complex data into clinically useful specific and robust information. One of the reasons is that there is still a significant lack of understanding in the biological meaning of specific changes in protein patterns. This is due to our current lack of understanding of the intra- and interindividual variability of pathways, plasma, and urine proteomes [8] as well as where the limits of normal and pathological states are. In addition, valid interpretation will require detailed knowledge of protein interaction patterns [3]. The changes that are caused by a disease or drug are often of high complexity and will not only affect the primarily targeted pathways, but due to compensatory regulation and cross talk at the cellular, organ and systemic level, as well as secondary processes such as inflammation and oxidative stress, may affect a multitude of different pathways [26]. The current knowledge of pathways, protein interactions, and networks is still incomplete and proteomics itself has proven to be a valuable tool to expand knowledge in this area.

At the moment, a more realistic approach than nontargeted proteomics is the targeted analysis of known arrays of protein markers that are well-qualified [409]. Analyses of such "combinatorial" markers can be realized using antibody, SOMAmer [412], and mass spectrometry-based protein multiplexing platforms as described earlier. In general, specific combinatorial protein patterns will confer more information than the measurement of a single protein and, thus, can be expected to have better specificity and sensitivity. It can be anticipated that such combinatorial markers are comprised of 5–20 proteins [409].

As described for metabolite (metabolomics) molecular markers, protein marker discovery, qualification and determination of sensitivity and specificity needs to take time-dependency of the proteome changes into account. While the end-stage of an injury is usually static or only slowly changing, during the early stages of injury development, the proteome changes can be relatively rapid and extensive. It is important to understand the dynamics of the mechanisms associated since this will determine sample collection strategies. In addition, it may be necessary to develop several sets of molecular markers that are specific for certain stages of a disease process.

After discovery, a critical part of the development of protein markers into clinical diagnostic tools is mechanistic and clinical qualification as required for regulatory approval [18,410]. Clinical qualification is based on the determination of specificity and sensitivity in clinical trials, usually using receiver operating characteristic (ROC) curves [413]. These assess the performance of the molecular markers compared with a current clinical outcomes gold standard. For the development of molecular markers, this means that it will be assessed to which extent a certain molecular marker pattern will be successful in predicting the development of a certain symptomatic disease end-stage, such as kidney dysfunction. The problem is that these end-stage injuries may alternately be caused by distinct underlying biochemical mechanisms that ultimately cause the same symptoms. Several of these distinct and alternate biochemical processes may not even be fully understood yet and may require a more detailed classification of the symptomatic disease process. Alternatively, during later stages, symptomatic injuries caused by different drug toxicities and diseases often involve the same pathobiochemical and pathological mechanisms, such as mitochondrial dysfunction, the formation of oxygen radicals, necrosis, apoptosis, inflammation, and other immune reactions. The further a pathological process progresses, the more difficult it may be to find specific molecular marker changes. One of the problems with the gold standard outcome being less specific than the molecular marker is that there is no 1:1 relationship between a molecular marker and the predicted clinical outcome. Several molecular marker patterns that are caused by distinct biochemical disease processes that ultimately lead to the same symptoms may be valid predictors of a single clinical outcome. Such a scenario will lead to good specificity—a specific marker pattern will be able to reliably predict a certain clinical outcome; however, sensitivity might be poor since the same outcomes caused by other distinct biochemical processes may be missed. Following current practices and regulatory guidance, this may lead to the rejection of a valid, highly specific molecular marker while, ironically, a less predictive and specific molecular marker that is a surrogate for later and more common disease processes may be acceptable.

Also, there is only poor consensus in terms of definition of the "end-stage disease" endpoint. For example, there are more than 30 different definitions of acute renal failure, or, now, acute kidney injury, in the published literature [13,414]. Therefore, it will be difficult to establish sensitivity and specificity for a candidate protein marker if the gold standard itself is potentially inconsistent.

Proteomics plays an important role in molecular marker discovery and qualification during their development into potential clinical diagnostic tools.

As evidenced by the regulatory approval of a protein kidney injury marker panel for preclinical rat drug toxicity studies, proteins in urine as diagnostic markers are starting to have an impact [14,241]. It is reasonable to expect that proteins and the analysis of protein panels, especially in urine, will play an increasingly important role as clinical diagnostic tools in nephrology in the near future.

Even if proteomics and systems biology-based expert systems may still be unavailable for years in the future, it is likely that nephrology will be among the first to benefit from progress in proteomics. This is due to the fact that urine, a proximal matrix that has a selective proteome, which is in direct communication with, and, to a large extent, reflective of biochemical processes in the kidney, is noninvasively and readily available.

Although several promising protein biomarkers have been discovered using proteomics-based discover strategies, only very few are about to slowly find their way into clinical practice. The bottleneck is clinical implementation [334,410]. In addition to the regulatory approval process and the required large clinical validation studies [410], there are two other important hurdles: sample handling and analysis and clinical acceptance [334].

First, recommendations for guidelines how to appropriately collect and handle proteomics samples have been developed [415]. Nevertheless, these are mostly intended to guide the collection of research samples. Complex sampling procedures are more difficult to establish in clinical routine environment. Moreover, diagnostic proteomics, especially when based on LC-MS/MS bioanalytical technologies, will mostly be laboratory-developed tests. However, as already discussed, most guidelines for bioanalytical assay validation have been written having drug compounds in mind. The validation and quality control of multianalyte assays for the qualitative and quantitative analysis of endogenous compounds is challenging, especially in the case of nontargeted assays. First regulatory guidance addressing this issue are emerging [416]. Biomarker assays used for medical decisions should be fully validated [155,334]. Especially for more complex proteomics assays, laboratory-to-laboratory variability will be an important issue since small changes in the sample processing procedure (e.g., removal of high-abundance proteins for proteomics analysis), instrumentation, analytical method, and data processing algorithms can lead to differences among laboratories. Another, in comparison trivial yet unresolved problem, is the normalization of marker levels in urine samples to compensate for differences in urine concentration in individual samples. As discussed earlier, the commonly used normalization strategy based on creatinine urine concentrations may become misleading in certain cases, as kidney injury may also affect creatinine concentrations in urine [366]. So far no consensus has been reached how to effectively address this problem.

In most cases "predictive" means that a biomarker is more sensitive and specific than established clinical markers, such as creatinine concentrations in serum or kidney biopsies. Such predictive biomarkers show a clinically relevant change preceding the increase of established, less sensitive clinical markers. One of the thresholds for their clinical use is that many clinicians do often feel uncomfortable to base decisions on such new, predictive biomarkers and are unsure about the potential therapeutic consequences of the results [410]. It is reasonable to expect that this will be the case the more complex and the more specific, sensitive, and predictive a new biomarker is so that training and education will become an important component of successful clinical protein biomarker implementation.

REFERENCES

[1] Knepper MA. Proteomics and the kidney. J Am Soc Nephrol 2002;13:1398–408.
[2] Welberry Smith MP, Banks RE, Wood LS, Lewington AJP, Selby PJ. Application of roteomic analysis to the study of renal diseases. Nat Rev Nephrol 2009;5:701–12.
[3] Bilello JA. The agony and ecstasy of "omic" technologies in drug development. Curr Mol Med 2005;5:39–52.
[4] Fliser D, Novak J, Thongboonkerd, et al. Advances in urinary proteome analysis and biomarker discovery. J Am Soc Nephrol 2007;18:1057–71.
[5] Peng J, Gygi SP. Proteomics: The move to mixtures. J Mass Spectrom 2001;36:1083–91.
[6] O'Farrell PH. High resolution two dimensional electrophoresis of proteins. J Biol Chem 1975;250:4007–21.
[7] Janech MG, Raymond JR, Arthur JM. Proteomics in renal research. Am J Physiol Renal Physiol 2007;292:F501–12.
[8] Beretta L. Proteomics from the clinical perspective: many hopes and much debate. Nat Methods 2007;10:785–6.
[9] Domon B, Aebersold R. Mass spectrometry in protein analysis. Science 2006;312:212–7.
[10] Stojnev S, Pejcic M, Dolicanin Z, et al. Challenges of genomics and proteomics in nephrology. Ren Failure 2009;31:765–72.
[11] Musso CG, Michelángelo H, Vilas M, Reynaldi J, Martinez B, Algranati L, Macías Núñez JF. Creatinine reabsorption by the aged kidney. Int Urol Nephrol 2009;41:727–31.
[12] Berl T. American Society of Nephrology Renal Research Report. J Am Soc Nephrol 2005;16:1886–903.
[13] Vaidya VS, Ferguson MA, Bonventre JV. Biomarkers of acute kidney injury. Annu Rev Pharmacol Toxicol 2008;48:463–8.
[14] Müller PY, Dieterle F. Tissue-specific, non-invasive toxicity biomarkers: translation from preclinical safety assessment to clinical safety monitoring. Expert Opin Drug Metab Toxicol 2009;5:1023–38.
[15] Christians U, Klawitter J, Bendrick-Peart J, Schöning W, Schmitz V. Toxicodynamic therapeutic drug monitoring of immunosuppressants: promises, reality and challenges. Ther Drug Monit 2008;30:151–8.
[16] Siuti N, Kelleher NL. Decoding protein modifications using top-down mass spectrometry. Nat Methods 2007;10:817–21.

[17] Köcher T, Superti-Furga G. Mass spectrometry-based functional proteomics: from molecular machines to protein networks. Nat Methods 2007;4:807–15.
[18] Rifai N, Gillette MA, Carr SA. Protein biomarker discovery and validation: the long and uncertain path to clinical utility. Nat Biotechnol 2006;24:971–83.
[19] O'Riordan E, Gross SS, Goligorsky MS. Technology insight: renal proteomics at the crossraods between promise and problems. Nat Clin Pract Nephrol 2006;2: 445–57.
[20] Magni F, Sarto C, Valsecchi C, et al. Expanding the proteome two-dimensional gel electrophoresis reference map of human renal cortex by peptide mass fingerprinting. Proteomics 2005;5:816–25.
[21] Thongboonkerd V, Malasit P. Renal and urinary proteomics: current applications and challenges. Proteomics 2005;5:1033–42.
[22] Berl T. How do kidney cells adapt to survive in hypertonic inner medulla? Trans Am Clin Climatol Assoc 2009;120:389–401.
[23] O'Riordan E, Goligorsky MS. Emerging studies of the urinary proteome: the end of the beginning? Curr Opin Hypertens 2005;14:579–85.
[24] Barratt J, Topham P. Urine proteomics: the present and future of measuring urinary protein components in disease. Can Med Assoc J 2007;177:361–8.
[25] Birn H, Christensen EI. Renal albumin absorption in physiology and pathology. Kidney Int 2006;69:440–9.
[26] Goligorsky MS, Addabbo F, O'Riordan E. Diagnostic potential of urine proteome: a broken mirror of renal diseases. J Am Soc Nephrol 2007;18:2233–9.
[27] Adachi J, Kumar C, Zhang Y, Olsen JV, Mann M. The human urinary proteome contains more than 1500 proteins including a large portion of membrane proteins. Genome Biol 2006;6:R80.
[28] Decramer S, Gonzalez de Peredo A, Breuil B, Mischak H, Monsarrat B, Bascands J, Schanstra JP. Urine in clinical proteomics. Mol Cell Proteomics 2008;7:1850–62.
[29] Coon JJ, Zurbig P, Dakana M, et al. CE-MS analysis of the human urinary proteome for biomarker discovery and disease diagnostics. Proteomics Clin Appl 2008;2:964–73.
[30] Candiano G, Santucci L, Petretto A, Bruschi M, Dimuccio V, Urbani A, Bagnasco S, Ghiggeri GM. 2D-electrophoresis and the urine proteome map: where do we stand? J Proteomics 2010;73:829–44.
[31] Jia L, Zhang L, Shao C, Song E, Sun W, Li M, Gao Y. An attempt to understand kidney's protein handling function by comparing plasma and urine proteomes. PLoS One 2009;4:e5146.
[32] Pisitkun T, Johnstone R, Knepper MA. Discovery of urinary biomarkers. Mol Cell Proteomics 2006;5:1760–71.
[33] Zhou H, Yuen PS, Pisitkun T, et al. Collection, storage, preservation, and normalization of human urinary exosomes for biomarker discovery. Kidney Int 2006;69:1471–6.
[34] Zürbig P, Decramer S, Dakna M, et al. The human urinary proteome reveals high similarity between kidney aging and chronic kidney disease. Proteomics 2009;9:2108–17.
[35] Wu J, Chen YD, Gu W. Urinary proteomics as a novel tool for biomarker discovery in kidney diseases. J Zhejiang Univ Sci B 2010;11:227–37.
[36] Schaub S, Wilkins J, Weiler T, Sangster K, Rush D, Nickerson P. Urine protein profiling with surface-enhanced laser-desorption / ionization time-of-flight mass spectrometry. Kidney Int 2004;65:323–32.
[37] Theodorescu D, Schiffer E, Bauer HW, et al. Discovery and validation of urinary biomarkers for prostate cancer. Proteomics Clin Appl 2008;2:556–70.
[38] Thongboonkerd V. Practical points in urinary proteomics. J Proteome Res 2007;6: 3881–90.
[39] Havanapan P, Thongboonkerd V. Are protease inhibitors required for gel-based proteomics of the kidney and urine? J Proteome Res 2009;8:3109–17.

[40] Theodorescu D, Wittke S, Ross MM, et al. Discovery and validation of new protein biomarkers for urothelial cancer: a prospective analysis. Lancet Oncol 2006;7:230–40.
[41] Weissinger EM, Schiffer E, Herenstein B, et al. Proteomic patterns predict acute graft-versus host-disease after allogenic hematopoetic stem cell transplantation. Blood 2007;109:5511–9.
[42] Zerefos PG, Vlahou A. Urine sample preparation and protein profiling by two-dimensional electrophoresis and matrix-assisted laser desorption ionization time of flight mass spectrometry. Methods Mol Biol 2008;428:141–57.
[43] Fiedler GM, Baumann S, Leichtle A, et al. Standardized peptidome profiling of human urine by magnetic bead separation and matrix assisted laser desorption/ionization time-of-flight mass spectrometry. Clin Chem 2007;53:421–8.
[44] Lescuyer P, Hochstrasser D, Rabilloud T. How shall we use the proteomics toolbox for biomarker discovery? J Proteome Res 2007;6:3371–6.
[45] Anderson NL, Anderson NG. The human plasma proteome: history, character, and diagnostic prospects. Mol Cell Proteomics 2002;1:845–67.
[46] Kushnir MM, Mrozinski P, Rockwood AL, Crockett DK. A depletion strategy for improved detection of human proteins from urine. J Biomol Tech 2009;20:101–8.
[47] Thongboonkerd V, Mungdee S, Chiangjong W. Should urine pH be adjusted prior to gel-based proteome analysis? J Proteome Res 2009;8:3206–11.
[48] Thongboonkerd V, Chutipongtanate S, Kanlaya R. Systematic evaluation of sample preparation methods for gel-based human urinary proteomics: quantity, quality, and variability. J Proteome Res 2006;5:183–91.
[49] Pieper R. Preparation of urine samples for proteomic analysis. Methods Mol Biol 2008;425:89–99.
[50] Khan A, Packer NH. Simple urinary sample preparation for proteomic analysis. J Proteome Res 2006;5:2824–38.
[51] Bodzon-Kulakowska A, Bierczynska-Krzysik A, Dylag T, et al. Methods for samples preparation in proteomic research. J Chromatogr B 2007;849:1–31.
[52] Ahmed FE. Sample preparation and fractionation for proteome analysis and cancer biomarker discovery by mass spectrometry. J Sep Sci 2009;32:771–98.
[53] Matt P, Fu Z, Fu Q, Van Eyck JE. Biomarker discovery: proteome fractionation and separation in biological samples. Physiol Genomics 2008;33:12–7.
[54] Hu S, Loo JA, Wong DT. Human body fluid proteome analysis. Proteomics 2006;6:6326–53.
[55] Mathivanan S, Ahmed M, Ahn NG, et al. Human Proteinpedia enables sharing of human protein data. Nat Biotechnol 2008;26:164–7.
[56] Korf U, Wiemann S. Protein microarrays as a discovery tool for studying protein-protein interactions. Expert Rev Proteomics 2005;2:13–26.
[57] The ENCODE Project Consortium. An integrated encyclopedia of DNA elements in the human genome. Nature 2012;489:57–74.
[58] de Hoog CL, Mann M. Proteomics. Annu Rev Genomics Hum Genet 2004;5:267–93.
[59] Thongboonkerd V. Proteomics. Forum Nutr 2007;60:80–90.
[60] Catherman AD, Skinner OS, Kelleher NL. Top Down proteomics: facts and perspectives. Biochem Biophys Res Commun 2014;445:683–93.
[61] Catherman AD, Durbin KR, Ahlf DR, Early BP, Fellers RT, Tran JC, Thomas PM, Kelleher NL. Large-scale top-down proteomics of the human proteome: membrane proteins, mitochondria, and senescence. Mol Cell Proteomics 2013;12:3465–73.
[62] Anderson L. Six decades searching for meaning in the proteome. J Proteomics 2014;107:24–30.
[63] Kolch W, Mischak H, Pitt AR. The molecular make-up of a tumor: proteomics in cancer research. Clin Sci (Lond) 2005;108:369–83.

[64] Štěpánová S, Kašička V. Recent developments and applications of capillary and microchip electrophoresis in proteomic and peptidomic analyses. J Sep Sci 2016;39:198–211.
[65] Lilley KS, Friedman DB. All about DIGE: quantification technology for differential display 2D-gel proteomics. Expert Rev Proteomics 2004;1:401–9.
[66] Friedman DB, Lilley KS. Optimizing the difference gel electrophoresis (DIGE) technology. Meth Mol Biol 2008;428:93–124.
[67] Arentz G, Weiland F, Oehler MK, Hoffmann P. State of the art of 2D DIGE. Proteomics Clin Appl 2015;9:277–88.
[68] Robledo VR, Smyth WF. Review of the CE-MS platform as a powerful alternative to conventional couplings in bio-omics and target-based applications. Electrophoresis 2014;35:2292–308.
[69] Gopal J, Muthu M, Chun SC, Wu HF. State-of-the-art nanoplatform-integrated MALDI-MS impacting resolutions in urinary proteomics. Proteomics Clin Appl 2015;9: 469–81.
[70] Krenkova J, Foret F. On-line CE/ESI/MS interfacing: recent developments and applications in proteomics. Proteomics 2012;12:2978–90.
[71] Bonvin G, Schappler J, Rudaz S. Capillary electrophoresis-electrospray ionization-mass spectrometry interfaces: fundamental concepts and technical developments. J Chromatogr A 2012;1267:17–31.
[72] Mellors JS, Black WA, Chambers AG, Starkey JA, Lacher NA, Ramsey JM. Hybrid capillary/microfluidic system for comprehensive online liquid chromatography-capillary electrophoresis-electrospray ionization-mass spectrometry. Anal Chem 2013;85:4100–6.
[73] Chambers AG, Mellors JS, Henley WH, Ramsey JM. Monolithic integration of two-dimensional liquid chromatography-capillary electrophoresis and electrospray ionization on a microfluidic device. Anal Chem 2011;83:842–9.
[74] Picotti P, Aebersold R, Domon B. The implications of proteolytic background in shotgun proteomics. Mol Cell Proteomics 2007;6:1589–98.
[75] Cravat BF, Simon GM, Yates JR. The biological impact of mass-spectrometry-based proteomics. Nature 2007;450:991–1000.
[76] Washburn MP, Wolters D, Yates JR III. Large-scale analysis of the yeast proteome by multidimensional protein identification technology. Nat Biotechnol 2001;19:242–7.
[77] Wolters DA, Washburn MP, Yates JR III. An automated multi-dimensional protein identification technology for shotgun proteomics. Anal Chem 2001;73:5683–90.
[78] Aebersol R, Mann M. Mass spectrometry-based proteomics. Nature 2003;422:198–207.
[79] Sadygov RG, Cociorva D, Yates JR III. Large-scale database searching using tandem mass-spectra: looking up the answer in the back of the book. Nat Methods 2004;1: 195–202.
[80] Nesvizhskii AI. Protein identification by tandem mass spectrometry and sequence database searching. Methods Mol Biol 2006;367:87–120.
[81] Shi T, Song E, Nie S, Rodland KD, Liu T, Qian WJ, Smith RD. Advances in targeted proteomics and applications to biomedical research. Proteomics 2016;16:2160–82.
[82] Hu A, Noble WS, Wolf-Yadlin A. Technical advances in proteomics: new developments in data-independent acquisition. F1000Res 2016; 5 (F1000 Faculty Rev): 419.
[83] Carapito C, Klemm C, Aebersold R, Domon B. Systematic LC-MS analysis of labile post-translational modifications in complex mixtures. J Proteome Res 2009;8:2608–14.
[84] Witze ES, Old WM, Resing K, Ahn NG. Mapping protein post-translational modifications with mass spectrometry. Nat Mehods 2007;10:798–806.
[85] Karas M, Bachman D, Bahr U, Hillenkamp F. Matrix assisted ultraviolet laser desorption of non-volatile compounds. Int J Mass Spectrom Ion Process 1987;78:53–68.
[86] Albrethsen J. Reproducibility in protein profiling by MALDI-TOF mass spectrometry. Clin Chem 2007;53:852–8.

[87] Poon TC. Opportunities and limitations of SELDI-TOF-MS in biomedical research: practical advices. Expert Rev Proteomics 2007;4:51–65.
[88] Check E. Proteomics and cancer: running before we can walk? Nature 2004;429:496–7.
[89] Kiehntopf M, Siegmund R, Deufel T. Use of SELDI-TOF mass spectrometry for identification of new biomarkers: potential and limitations. Clin Chem Lab Med 2007;45:1435–49.
[90] Ferreira L, Sanchez-Juanes F, Gonzalez-Avila M, Cembrero-Fucinos D, et al. Direct identification of urinary tract pathogens from urine samples by matrix-assisted laser desorption ionization-time of flight mass spectrometry. J Clin Microbiol 2010;48: 2110–5.
[91] Nomura F. Proteome-based bacterial identification using matrix-assisted laser desorption ionization-time of flight mass spectrometry (MALDI-TOF MS): a revolutionary shift in clinical diagnostic microbiology. Biochim Biophys Acta 2015;1854:528–37.
[92] Longuespée R, Casadonte R, Kriegsmann M, Pottier C, Picard de Muller G, Delvenne P, Kriegsmann J, De Pauw E. MALDI mass spectrometry imaging: A cutting-edge tool for fundamental and clinical histopathology. Proteomics Clin Appl 2016;10:701–19.
[93] Gygi SP, Rist B, Gerber SA, Turecek F, Gelb MH, Aebersold R. Quantitative analysis of complex protein mixtures using isotope-coded affinity tags. Nat Biotechnol 1999;17:994–9.
[94] Chahrour O, Cobice D, Malone J. Stable isotope labelling methods in mass spectrometry-based quantitative proteomics. J Pharm Biomed Anal 2015;113:2–20.
[95] Haqqani AS, Kelly JF, Stanimirovic DB. Quantitative protein profiling by mass spectrometry using isotope-coded affinity tags. Methods Mol Biol 2008;439:225–40.
[96] Alex A, Gucek M, Li X. Applications of proteomics in the study of inflammatory bowel diseases: current status and future directions with available technologies. Inflamm Bowel Dis 2009;15:616–29.
[97] Gruhler S, Kratchmarova I. Stable isotope labeling by amino acids in cell culture. Methods Mol Biol 2008;424:101–11.
[98] Ong SE, Mann M. A practical recipe for stable isotope labeling by amino acids in cell culture (SILAC). Nat Protoc 2006;1:2650–60.
[99] Maes E, Tirez K, Baggerman G, Valkenborg D, Schoofs L, Encinar JR, Mertens I. The use of elemental mass spectrometry in phosphoproteomic applications. Mass Spectrom Rev 2016;35:350–60.
[100] Dieterle F, Marrer E. New technologies around biomarkers and their interplay with drug development. Anal Bioanal Chem 2008;390:141–54.
[101] Gupta S, Manubhai KP, Kulkarni V, Srivastava S. An overview of innovations and industrial solutions in Protein Microarray Technology. Proteomics 2016;16:1297–308.
[102] Gold L, Ayers D, Bertino J, et al. Aptamer-based multiplexed proteomic technology for biomarker discovery. PLoS One 2010;5:e15004.
[103] Gold L, Walker JJ, Wilcox SK, Williams S. Advances in human proteomics at high scale with the SOMAscan proteomics platform. N Biotechnol 2012;29:543–9.
[104] Kraemer S, Vaught JD, Bock C, Gold L, Katilius E, Keeney TR, Kim N, Saccomano NA, Wilcox SK, Zichi D, Sanders GM. From SOMAmer-based biomarker discovery to diagnostic and clinical applications: a SOMAmer-based, streamlined multiplex proteomic assay. PLoS One 2011;6:e26332.
[105] Anderson NL. The clinical plasma proteome: A survey of clinical assays for proteins in plasma and serum. Clin Chem 2010;56:177–85.
[106] Young DS, Hicks JM. DORA2005-2007: directory of rare analysis. Washington, DC: American Association for Clinical Chemistry; 2007.
[107] Yu X, Petritis B, LaBaer J. Advancing translational research with next-generation protein microarrays. Proteomics 2016;16:1238–50.

[108] Ling MM, Ricks C, Lea P. Multiplexing molecular diagnostics and immunoassays using emerging microarray technologies. Expert Rev Mol Diagn 2007;7:87–98.
[109] Chowdhury F, Williams A, Johnson P. Validation and comparison of two multiplex technologies, Luminex and Mesoscale Discovery, for human cytokine profiling. J Immunol Methods 2009;340:55–64.
[110] Shingyoji M, Gerion D, Pinkel D, Gray JW, Chen F. Quantum dots-based reverse phase protein microarray. Talanta 2005;67:472–8.
[111] Gaster RS, Hall DA, Nielsen CH, et al. Matrix-insensitive protein assays push the limits of biosensors in medicine. Nat Med 2009;15:1327–32.
[112] Zheng G, Patolsky F, Cui Y, Wang WU, Lieber CM. Multiplexed electrical detection of cancer markers with nanowire sensor arrays. Nat Biotechnol 2005;10:1294–301.
[113] Ji HF, Gao H, Buchapudi KR, Yang X, Xu X, Schulte MK. Microcantilever biosensors based on conformational change of proteins. Analyst 2008;133:434–43.
[114] Ghosh S, Sood AK, Kumar N. Carbon nanotube flow sensors. Science 2003;299:1042–4.
[115] Drummond TG, Hill MG, Barton JK. Electrochemical DNA sensors. Nat Biotechnol 2003;21:1192–9.
[116] Osterfeld SJ, Yu H, Gaster RS, et al. Multiplex protein assays based on real-time magnetic nanotag sensing. Proc Natl Acad Sci USA 2008;105:20637–40.
[117] Picotti P, Aebersold R. Selected reaction monitoring-based proteomics: workflows, potential, pitfalls and future directions. Nat Methods 2012;9:555–66.
[118] Available from: http://www.peptideatlas.org/
[119] Available from: http://gpmdb.thegpm.org/
[120] Anderson NL, Jackson A, Smith D, Hardie D, Borchers C, Pearson TW. SISCAPA peptide enrichment on magnetic beads using an in-line bead trap device. Mol Cell Proteomics 2009;8:995–1005.
[121] Lee JW, Figeys D, Vasilescu J. Biomarker assay translation from discovery to clinical studies in cancer drug development: quantification of emerging protein biomarkers. Adv Cancer Res 2007;96:269–98.
[122] Barr JR, Maggio VL, Patterson DG Jr, et al. Isotope dilution--mass spectrometric quantification of specific proteins: model application with apolipoprotein A-I. Clin Chem 1996;42:1676–82.
[123] Kitteringham NR, Jenkins RE, Lane CS, Elliott VL, Park BK. Multiple reaction monitoring for quantitative biomarker analysis in proteomics and metabolomics. J Chromatogr B Analyt Technol Biomed Life Sci 2009;877:1229–39.
[124] Kuzyk MA, Smith D, Yang J, Cross TJ, Jackson AM, Hardie DB, Anderson NL, Borchers CH. Multiple reaction monitoring-based, multiplexed, absolute quantitation of 45 proteins in human plasma. Mol Cell Proteomics 2009;8:1860–77.
[125] Hortin GL, Sviridov D. The dynamic range problem in the analysis of the plasma proteome. J Proteomics 2010;73:629–36.
[126] Hortin GL, Sviridov D, Anderson NL. High-abundance polypeptides of the human plasma proteome comprising the top 4 logs of polypeptide abundance. Clin Chem 2008;54:1608–16.
[127] Domon B, Aebersold R. Challenges and opportunities in proteomics data analysis. Mol Cell Proteomics 2006;5:1921–6.
[128] Griss J. Spectral library searching in proteomics. Proteomics 2016;16:729–40.
[129] Fuxius S, Eravci M, Broedel O, et al. Technical strategies to reduce the amount of "false significant" results in quantitative proteomics. Proteomics 2008;8:1780–4.
[130] Karp NA, Lilley KS. Design and analysis issues in quantitative proteomics studies. Pract Proteomics 2007;1:42–50.
[131] Mann M. Comparative analysis to guide quality improvements in proteomics. Nat Methods 2009;6:717–9.

[132] Horgan GW. Sample size and replication in 2D gel electrophoresis studies. J Proteome Res 2007;6:2884–7.
[133] Tu C, Sheng Q, Li J, Ma D, Shen X, Wang X, Shyr Y, Yi Z, Qu J. Optimization of search engines and postprocessing approaches to maximize peptide and protein identification for high-resolution mass data. J Proteome Res 2015;14:4662–73.
[134] Elias JE, Haas W, Faherty BL, Gygi SP. Comparative evaluation of mass spectrometry platforms used in large scale proteomics investigations. Nat Methods 2005;2:667–75.
[135] Bell AW, Nilsson T, Kearney RE, Bergeron JJM. The protein microscope: incorporating mass spectrometry into cell biology. Nat Methods 2007;10:783–4.
[136] Karimpour-Fard A, Epperson LE, Hunter LE. A survey of computational tools for downstream analysis of proteomic and other omic datasets. Hum Genomics 2015;9:28.
[137] Farrah T, Deutsch EW, Omenn GS, Sun Z, Watts JD, Yamamoto T, Shteynberg D, Harris MM, Moritz RL. State of the human proteome in 2013 as viewed through PeptideAtlas: comparing the kidney, urine, and plasma proteomes for the biology- and disease-driven Human Proteome Project. J Proteome Res 2014;13:60–75.
[138] Craig R, Cortens JP, Beavis RC. Open source system for analyzing, validating, and storing protein identification data. J Proteome Res 2004;3:1234–42.
[139] Gaudet P, Argoud-Puy G, Cusin I, Duek P, Evalet O, Gateau A, Gleizes A, Pereira M, Zahn-Zabal M, Zwahlen C, Bairoch A, Lane L. neXtProt: organizing protein knowledge in the context of human proteome projects. J Proteome Res 2013;12:293–8.
[140] Kim MS, Pinto SM, Getnet D, et al. A draft map of the human proteome. Nature 2014;509(7502):575–81.
[141] Zhang Y, Zhang Y, Adachi J, et al. MAPU : Max-Planck unified database of organellar, cellular, tissue and body fluid proteomes. Nucl Acid Res 2006;35(Database issue):D771–9.
[142] Li SJ, Peng M, Li H, et al. Sys-BodyFluid: a systematical database for human body fluid proteome research. Nucleic Acid Res 2009;37(Database issue):D907–12.
[143] Kalantari S, Jafari A, Moradpoor R, Ghasemi E, Khalkhal E. Human urine proteomics: analytical techniques and clinical applications in renal diseases. Int J Proteomics 2015;2015:782798.
[144] Legrain P, Aebersold R, Archakov A, et al. The human proteome project: current state and future direction. Mol Cell Proteomics 2011;10. M111.009993.
[145] Paik YK, Jeong SK, Omenn GS, et al. The Chromosome-Centric Human Proteome Project for cataloging proteins encoded in the genome. Nat Biotechnol 2012;30:221–3.
[146] Marko-Varga G, Omenn GS, Paik YK, Hancock WS. A first step toward completion of a genome-wide characterization of the human proteome. J Proteome Res 2013;12:1–5.
[147] Vestergaard P, Leverett R. Constancy of urinary creatinine excretion. J Lab Clin 1958;51:211–8.
[148] Jantos-Siwy J, Schiffer E, Brand K, et al. Quantitative urinary proteome analysis for biomarker evaluation in chronic kidney disease. J Proteome Res 2009;8:268–81.
[149] Theodorescu D, Fliser D, Wittke S, et al. Pilot study of capillary electrophoresis coupled to mass spectrometry as a tool to detect potential prostate cancer biomarkers in urine. Electrophoresis 2005;26:2797–808.
[150] Molloy MP, Brzezinski EE, Hang J, et al. Overcoming technical variation and biological variation in quantitative proteomics. Proteomics 2003;3:1912–9.
[151] Martín BB, Marquet P, Ferrer JM, et al. Rheumatoid factor interference in a tacrolimus immunoassay. Ther Drug Monit 2009;31:743–5.
[152] Cavalier E, Carlisi A, Chapelle JP, Delanaye P. False positive PTH results: an easy strategy to test and detect analytical interferences in routine practice. Clin Chim Acta 2008;387:150–2.

[153] Berth M, Bosmans E, Everaert J, Dierick J, Schiettecatte J, Anckaert E, Delanghe J. Rheumatoid factor interference in the determination of carbohydrate antigen 19-9 (CA 19-9). Clin Chem Lab Med 2006;44:1137–9.
[154] Cummings J, Ward TH, Greystoke A, Ranson M, Dive C. Biomarker method validation in anticancer drug development. Br J Pharmacol 2008;153:646–56.
[155] Shipkova M, López OM, Picard N, Noceti O, Sommerer C, Christians U, Wieland E. Analytical aspects of the implementation of biomarkers in clinical transplantation. Ther Drug Monit 2016;38(Suppl. 1):S80–92.
[156] Lee JW, Weiner RS, Sailstad JM, et al. Method validation and measurement of biomarkers in non-clinical and clinical samples in drug development: a conference report. Pharm Res 2005;22:499–511.
[157] Christians U, Klepacki J, Shokati T, Klawitter J, Klawitter J. Mass spectrometry-based multiplexing for the analysis of biomarkers in drug development and clinical diagnostics- how much is too much? Microchem J 2012;105:32–8.
[158] U.S. Department of Health and Human Services, Food and Drug Administration, Center for Drug Evaluation and Research (2007) Guidance for Industry: Clinical Trial Endpoints for the Approval of Cancer Drugs and Biologics. Version May 2007. Available from: http://www.fda.gov/downloads/Drugs/GuidanceComplianceRegulatoryInformation/Guidances/ucm071590.pdf
[159] Bell AW, Deutsch EW, Au CE, et al. A HUPO test sample study reveals common problems in mass spectrometry-based proteomics. Nat Methods 2009;6:423–30.
[160] Martens L, Hermjakob H. Proteomics data validation: why all must provide data. Mol Biosyst 2007;3:518–22.
[161] Wilkins MR, Appel RD, Van Eyk JE, et al. Guidelines for the next 10 years of proteomics. Proteomics 2006;6:4–8.
[162] Mischak H, Apweiler R, Banks RE, et al. Clinical proteomics: a need to define the field and to begin to set adequate standards. Proteomics 2007;1:148–56.
[163] Gibson F, Anderson L, Babnigg G, et al. Guidelines for reporting the use of gel electrophoresis in proteomics. Nat Biotechnol 2008;26:863–4.
[164] Human Kidney and Urine Proteome Project (HKUPP). Available from: http://hkupp.kir.jp/
[165] Yamamoto T, Langham RG, Ronco R, Knepper MA, Thongboonkerd V. Towards standard protocols and guidelines for urine proteomics: a report on the Human Kidney and Urine Proteome Project (HKUPP) symposium and workshop. Proteomics 2008;8:2156–9.
[166] Human Proteome Organization (HUPO). Available from: www.hupo.org
[167] Omenn GS, States DJ, Adamski J, et al. Overview of the HUPO Plasma Proteome Project: results from the pilot phase with 35 collaborating laboratories and multiple analytical groups, generating a core data set of 3020 proteins and a publicly-available database. Proteomics 2005;5:3226–45.
[168] States DJ, Ommen GS, Blackwell TW, Fermin D, Eng J, Speicher DW, Hanash SM. Challenges in deriving high-confidence protein identifications from data gathered by HUPO plasma proteome collaboration studies. Nat Biotech 2006;24:333–8.
[169] Taylor CF, Paton NW, Lilley KS, et al. The minimum information about a proteomics experiment (MIAPE). Nat Biotechnol 2007;25:887–93.
[170] Rodriguez H, Snyder M, Uhlén M, et al. Recommendations from the 2008 international summit on proteomics data release and sharing policy. The Amsterdam principles. J Proteome Res 2009;8:3689–92.
[171] Paris Report. Available from: http://www.mcponline.org/site/misc/ParisReport_ Final.xhtml
[172] Caubet C, Lacroix C, Decramer S, et al. Advances in urine proteome analysis and biomarker discovery in renal disease. Pediatr Nephrol 2010;25:27–35.

[173] Niwa T. Biomarker discovery for kidney diseases by mass spectrometry. J Chromatogr B 2008;870:148–53.
[174] Cayer DM, Nazor KL, Schork NJ. Mission critical: the need for proteomics in the era of next-generation sequencing and precision medicine. Hum Mol Genet 2016;25:R182–9.
[175] Sabbagh B, Mindt S, Neumaier M, Findeisen P. Clinical applications of MS-based protein quantification. Proteomics Clin Appl 2016;10:323–45.
[176] Jain KK. Role of proteomics in the development of personalized medicine. Adv Protein Chem Struct Biol 2016;102:41–52.
[177] Sanderson CM. The cartographer's toolbox: building bigger and better human protein interaction networks. Brief Funct Genomic Proteomic 2009;8:1–11.
[178] Barglow KT, Cravatt BF. Activity-based protein profiling for the functional annotation of enzymes. Nat Methods 2007;10:822–7.
[179] Mann M, Jensen ON. Proteomic analysis of post-translational modifications. Nat Biotechnol 2003;21:255–61.
[180] Witze ES, Old WM, Resing KA, Ahn NG. Mapping protein post-translational modifications with mass spectrometry. Nat Methods 2007;10:798–806.
[181] Ruttenberg BE, Pisitikun T, Knepper MA, Hoffert JD. PhosphoScore: an open-source phosphorylation site assignment tool for MSn data. J Proteome Res 2008;7:3054–9.
[182] Hoffert JD, Knepper MA. Taking aim at shotgun proteomics. Anal Biochem 2008;375:1–20.
[183] Doll S, Burlingame AL. Mass spectrometry-based detection and assignment of protein posttranslational modifications. ACS Chem Biol 2015;10:63–71.
[184] Ramachandra Rao SP, Wassell R, Shaw MA, Sharma K. Profiling of human mesangial cell subproteomes reveals a role for calmodulin in glucose uptake. Am J Physiol Renal Physiol 2007;292:F1182–9.
[185] Miyamoto M, Yoshida Y, Taguchi I, et al. In-depth proteomic profiling of the normal human kidney glomerulus using two-dimensional protein prefractionation in combination with liquid chromatography-tandem mass spectrometry. J Proteome Res 2007;6:3680–90.
[186] Brooks H, Sorensen AM, Terris J, et al. Profiling if renal tubule Na^+ transporter abundances in NHE3 and NCC null mice using targeted proteomics. J Physiol 2001;530:359–66.
[187] Curthoys NP, Taylor L, Hoffert JD, Knepper MA. Proteomic analysis of the adaptive response of rat renal proximal tubules to metabolic acidosis. Am J Physiol Renal Physiol 2007;292:F140–7.
[188] Dihazin H, Asif AR, Agarwal NK, Doncheva Y, Müller GA. Proteomic analysis of cellular response to osmotic stress in thick ascending limb of Henle's loop (TALH) cells. Mol Cell Proteomics 2005;4:1445–58.
[189] Yu MJ, Pisitkun T, Wang G, Shen RF, Knepper MA. LC-MS/MS analysis of apical and basolateral plasma membranes of rat renal collecting duct cells. Mol Cell Proteomics 2006;5:2131–45.
[190] Hoffert JD, Chou CL, Knepper MA. Aquaporin-2 in the "omics" era. J Biol Chem 2009;284:14683–7.
[191] Sachs AN, Pisitkun T, Hoffert JD, Yu MJ, Knepper MA. LC-MS/MS analysis of differential centrifugation fractions from native inner medullary collecting duct of rat. Am J Physiol Renal Physiol 2008;295:F1799–806.
[192] Yu MJ, Pisitkun T, Wang G, et al. Large-scale LC-MS/MS analysis of detergent-resistant membrane proteins from rat renal collecting duct. Am J Physiol Cell Physiol 2008;295:661–78.
[193] Pisitkun T, Bieniek J, Tchapyjnikov D, et al. High-throughput identification of IMCD proteins using LC-MS/MS. Physiol Genomics 2006;25:263–76.

[194] Bansal AD, Hoffert JD, Pisitkun T, et al. Phosphoproteomic profiling reveals vasopressin-regulated phosphorylation sites in collecting duct. J Am Soc Nephrol 2010;21:303–15.
[195] Hoffert JD, Wang G, Pisitkun T, Shen RF, Knepper MA. Am automated platform for analysis of phosphoproteomic datasets: application to kidney collecting duct phosphoproteins. J Proteome Res 2007;6:3501–8.
[196] Valkova N, Kültz D. Constitutive and inducible stress proteins dominate the proteome of the murine inner medullary collecting duct-3 (mIMDC3) cell line. Biochim Biophys Acta 2006;1764:1007–20.
[197] Klawitter J, Rivard CJ, Brown LM, et al. A metabonomic and proteomic analysis of changes in IMCD3 cells chronically adapted to hypertonicity. Nephron Physiol 2008;109:1–10.
[198] Kypreou KP, Kavvadas P, Karamessinis P, et al. Altered expression of calreticulin during the development of fibrosis. Proteomics 2008;8:2407–19.
[199] Chen YX, Li Y, Wang WM, et al. Phosphoproteomic study of human tubular epithelial cell in response to transforming growth factor beta 1-induced epithelial-to-mesenchymal transition. Am J Nephrol 2010;31:24–35.
[200] Feng D, Imasawa T, Nagano T, et al. Citrullination preferentially proceeds in glomerular Bowman's capsule and increases in obstructive nephropathy. Kidney Int 2005;68:84–95.
[201] Tilton RG, Haidacher SJ, LeJeune WS, et al. Diabetes-induced changes in the renal cortical proteome assessed with two-dimensional gel electrophoresis and mass spectrometry. Proteomics 2007;7:1729–42.
[202] Barati MT, Merchant ML, Klain AB, Jevans AW, McLeish KR, Klein JB. Proteomic analysis defines altered cellular redox pathways and advanced glycation end-product metabolism in glomeruli db/db diabetic mice. Am J Physiol Renal Physiol 2007;293:F1157–65.
[203] Thongboonkerd V, Chutipongtanate S, Kanlaya R, et al. Proteomic identification of alterations in metabolic enzymes and signaling proteins in hypokalemic nephropathy. Proteomics 2006;6:2273–85.
[204] Thongboonkerd V, Smangoen T, Sinchaikul S, Chen ST. Proteomic analysis of calcium oxalate monohydrate crystal-induced cytotoxicity in distal renal tubular cells. J Proteome Res 2008;7:4689–700.
[205] Chen S, Gao X, Sun Y, Xu C, Wang L, Zhou T. Analysis of HK-2 cells exposed to oxalate and calcium oxalate crystals: proteomic insights into the molecular mechanisms of renal injury and stone formation. Urol Res 2010;38:7–15.
[206] Bramham K, Mistry HD, Poston L, Chappell LC, Thompson AJ. The non-invasive biopsy will urinary proteomics make the renal tissue biopsy redundant. QJ Med 2009;102:523–38.
[207] Salih M, Demmers JA, Bezstarosti K, Leonhard WN, Losekoot M, van Kooten C, Gansevoort RT, Peters DJ, Zietse R, Hoorn EJ. DIPAK Consortium. Proteomics of urinary vesicles links plakins and complement to polycystic kidney disease. J Am Soc Nephrol 2016;27(10):3079–92.
[208] Hogan MC, Bakeberg JL, Gainullin VG, Irazabal MV, Harmon AJ, Lieske JC, Charlesworth MC, Johnson KL, Madden BJ, Zenka RM, McCormick DJ, Sundsbak JL, Heyer CM, Torres VE, Harris PC, Ward CJ. Identification of biomarkers for PKD1 using urinary exosomes. J Am Soc Nephrol 2015;26:1661–70.
[209] Pocsfalvi G, Raj DA, Fiume I, Vilasi A, Trepiccione F, Capasso G. Urinary extracellular vesicles as reservoirs of altered proteins during the pathogenesis of polycystic kidney disease. Proteomics Clin Appl 2015;9:552–67.
[210] Kistler AD, Serra AL, Siwy J, Poster D, Krauer F, Torres VE, Mrug M, Grantham JJ, Bae KT, Bost JE, Mullen W, Wüthrich RP, Mischak H, Chapman AB. Urinary proteomic

biomarkers for diagnosis and risk stratification of autosomal dominant polycystic kidney disease: a multicentric study. PLoS One 2013;8:e53016.
[211] Bakun M, Niemczyk M, Domanski D, Jazwiec R, Perzanowska A, Niemczyk S, Kistowski M, Fabijanska A, Borowiec A, Paczek L, Dadlez M. Urine proteome of autosomal dominant polycystic kidney disease patients. Clin Proteomics 2012;9:13.
[212] Khurana M, Traum AZ, Aivado M, et al. Urine proteomic profiling of pediatric nephrotic syndrome. Pediatr Nephrol 2006;21:1257–65.
[213] Kalantari S, Nafar M, Rutishauser D, Samavat S, Rezaei-Tavirani M, Yang H, Zubarev RA. Predictive urinary biomarkers for steroid-resistant and steroid-sensitive focal segmental glomerulosclerosis using high resolution mass spectrometry and multivariate statistical analysis. BMC Nephrol 2014;15:141.
[214] Kalantari S, Nafar M, Samavat S, Rezaei-Tavirani M, Rutishauser D, Zubarev R. Urinary prognostic biomarkers in patients with focal segmental glomerulosclerosis. Nephrourol Mon 2014;6:e16806.
[215] Woroniecki RP, Orlova TN, Mendelev N, et al. Urinary proteome of steroid-sensitive and steroid-resistant idiopathic nephrotic syndrome of childhood. Am J Nephrol 2006;26:258–67.
[216] Piyaphanee N, Ma Q, Kremen O, Czech K, Greis K, Mitsnefes M, Devarajan P, Bennett MR. Discovery and initial validation of α 1-B glycoprotein fragmentation as a differential urinary biomarker in pediatric steroid-resistant nephrotic syndrome. Proteomics Clin Appl 2011;5:334–42.
[217] Suzuki M, Wiers K, Brooks EB, Greis KD, Haines K, Klein-Gitelman MS, Olson J, Onel K, O'Neil KM, Silverman ED, Tucker L, Ying J, Devarajan P, Brunner HI. Initial validation of a novel protein biomarker panel for active pediatric lupus nephritis. Pediatr Res 2009;65:530–6.
[218] Mosley K, Tam FW, Edwards RJ, Crozier J, Pusey CD, Lightstone L. Urinary proteomic profiles distinguish between active and inactive lupus nephritis. Rheumatology 2006;45:1497–504.
[219] Rood IM, Merchant ML, Wilkey DW, Zhang T, Zabrouskov V, van der Vlag J, Dijkman HB, Willemsen BK, Wetzels JF, Klein JB, Deegens JK. Increased expression of lysosome membrane protein 2 in glomeruli of patients with idiopathic membranous nephropathy. Proteomics 2015;15:3722–30.
[220] Drube J, Schiffer E, Mischak H, et al. Urinary proteome pattern in children with renal Fanconi syndrome. Nephrol Dial Transplant 2009;24:2161–9.
[221] Wang L, Ni Z, Xie Z, et al. Analysis of the urine proteome of human contrast-induced kidney injury using two-dimensional fluorescence differential gel electrophoresis/matrix-assisted laser desorption time-of-flight mass spectrometry/liquid chromatography mass spectrometry. Am J Nephrol 2010;31:45–52.
[222] Dihazi H, Müller GA, Lindner S, et al. Characterization of diabetic nephropathy by urinary proteomic analysis: identification of a processed ubiquitin form as a differentially excreted protein in diabetic nephropathy patients. Clin Chem 2007;53:1636–45.
[223] Otu HH, Can H, Spentzios D, et al. Prediction of diabetic nephropathy using urine proteomic profiling 10 years prior to development of nephropathy. Diabetes Care 2007;30:638–43.
[224] Rao P, Lu X, Standley M, et al. Proteomic identification of urinary biomarkers of diabetic nephropathy. Diabetes Care 2007;30:629–37.
[225] Meier M, Kaiser T, Herrmann A, et al. Identification of urinary protein pattern in type 1 diabetic adolescents with early diabetic nephropathy by a novel combined proteome analysis. J Diabetes Complications 2005;19:223–32.
[226] Caseiro A, Barros A, Ferreira R, Padrão A, Aroso M, Quintaneiro C, Pereira A, Marinheiro R, Vitorino R, Amado F. Pursuing type 1 diabetes mellitus and related complications through urinary proteomics. Transl Res 2014;163:188–99.

[227] Zubiri I, Posada-Ayala M, Sanz-Maroto A, Calvo E, Martin-Lorenzo M, Gonzalez-Calero L, de la Cuesta F, Lopez JA, Fernandez-Fernandez B, Ortiz A, Vivanco F, Alvarez-Llamas G. Diabetic nephropathy induces changes in the proteome of human urinary exosomes as revealed by label-free comparative analysis. J Proteomics 2014;96:92–102.
[228] Lindhardt M, Persson F, Zürbig P, Stalmach A, Mischak H, de Zeeuw D, Lambers Heerspink H, Klein R, Orchard T, Porta M, Fuller J, Bilous R, Chaturvedi N, Parving HH, Rossing P. Urinary proteomics predict onset of microalbuminuria in normoalbuminuric type 2 diabetic patients, a sub-study of the DIRECT-Protect 2 study. Nephrol Dial Transplant 2016;. (Epub ahead of print).
[229] Rocchetti MT, Papale M, d'Apollo AM, Suriano IV, Di Palma AM, Vocino G, Montemurno E, Varraso L, Grandaliano G, Di Paolo S, Gesualdo L. Association of urinary laminin G-like 3 and free K light chains with disease activity and histological injury in IgA nephropathy. Clin J Am Soc Nephrol 2013;8:1115–25.
[230] Kalantari S, Rutishauser D, Samavat S, Nafar M, Mahmudieh L, Rezaei-Tavirani M, Zubarev RA. Urinary prognostic biomarkers and classification of IgA nephropathy by high resolution mass spectrometry coupled with liquid chromatography. PLoS One 2013;8:e80830.
[231] Mucha K, Bakun M, Jaźwiec R, Dadlez M, Florczak M, Bajor M, Gala K, Pączek L. Complement components, proteolysis-related, and cell communication-related proteins detected in urine proteomics are associated with IgA nephropathy. Pol Arch Med Wewn 2014;124:380–6.
[232] Surin B, Sachon E, Rougier JP, Steverlynck C, Garreau C, Lelongt B, Ronco P, Piedagnel R. LG3 fragment of endorepellin is a possible biomarker of severity in IgA nephropathy. Proteomics 2013;13:142–52.
[233] Nguyen MT, Ross GF, Dent CL, Devarajan P. Early prediction of acute renal injury using urinary proteomics. Am J Nephrol 2005;25:318–26.
[234] Metzger J, Kirsch T, Schiffer E, Ulger P, Mentes E, Brand K, Weissinger EM, Haubitz M, Mischak H, Herget-Rosenthal S. Urinary excretion of twenty peptides forms an early and accurate diagnostic pattern of acute kidney injury. Kidney Int 2010;78:1252–62.
[235] Aregger F, Pilop C, Uehlinger DE, Brunisholz R, Carrel TP, Frey FJ, Frey BM. Urinary proteomics before and after extracorporeal circulation in patients with and without acute kidney injury. J Thorac Cardiovasc Surg 2010;139:692–700.
[236] Aregger F, Uehlinger DE, Witowski J, Brunisholz RA, Hunziker P, Frey FJ, Jörres A. Identification of IGFBP-7 by urinary proteomics as a novel prognostic marker in early acute kidney injury. Kidney Int 2014;85:909–19.
[237] Carrick E, Vanmassenhove J, Glorieux G, Metzger J, Dakna M, Pejchinovski M, Jankowski V, Mansoorian B, Husi H, Mullen W, Mischak H, Vanholder R, Van Biesen W. Development of a MALDI MS-based platform for early detection of acute kidney injury. Proteomics Clin Appl 2016;10:732–42.
[238] Devarajan P, Williams LM. Proteomics for biomarker discovery in acute kidney injury. Semin Nephrol 2007;6:637–51.
[239] Devarajan P. Emerging urinary biomarkers in the diagnosis of acute kidney injury. Expert. Opin Med Diagn 2008;2:387–98.
[240] Devarajan P, Murray P. Biomarkers in acute kidney injury: are we ready for prime time? Nephron Clin Pract 2014;127:176–9.
[241] European Medicines Agency, Committee for Medicinal Products for Human Use. Final conclusions on the pilot joint EMEA/FDA VXDS experience on qualification of nephrotoxicity biomarkers, 2009. Available from: http://www.ema.europa.eu/docs/en_GB/document_library/Regulatory_and_procedural_guideline/2009/10/WC500004205.pdf

[242] Sooy K, Kohut J, Christakos S. The role of calbindin and 1,25 dihydroxy vitamin D3 in the kidney. Curr Opin Nephrol Hyperten 2000;9:341–7.
[243] Roth J, Brown D, Norman AW, Orci L. Localization of the vitamin D-dependent calcium-binding protein in mammalian kidney. Am J Physiol 1982;12:F243–52.
[244] Betton GR, Kenne K, Somers R, Marr A. Protein biomarkers of nephrotoxicity; a review of findings with cyclosporin A, a signal transduction kinase inhibitor and N-phenylanthranilic acid. Cancer Biomark 2005;1:59–67.
[245] Takashi Y, Zhu K, Miyake K, Kato K. Urinary 28-kD calbindin-D as a new marker for damage to distal renal tubules caused by cisplatin-based chemotherapy. Urol Int 1996;56:174–9.
[246] Trougakos IP, Gonos ES. Regulation of clusterin/apolipoprotein J, a functional homologue to the small heat shock proteins, by oxidative stress in ageing and age-related diseases. Free Radic Res 2006;40:1324–34.
[247] Girton RA, Sundin DP, Rosenberg ME. Clusterin protects renal tubular epithelial cells from gentamicin-mediated cytotoxicity. Am J Physiol Renal Physiol 2002;282: F703–9.
[248] Silkensen JR, Skubitz KM, Skubitz AP, et al. Clusterin promotes the aggregation and adhesion of renal porcine epithelial cells. J Clin Invest 1995;96:2646–53.
[249] Shlipak MG, Sarnak MJ, Katz R, Cystatin C, et al. and the risk of death and cardiovascular events among elderly persons. N Engl J Med 2005;352:2049–60.
[250] Conti M, Moutereau S, Zater M, et al. Urinary cystatin C as a specific marker of tubular dysfunction. Clin Chem Lab Med 2006;44:288–91.
[251] Yang GP, Lau LF. Cyr61, product of a growth factor-inducible immediate early gene, is associated with the extracellular matrix and the cell surface. Cell Growth Differ 1991;2:351–7.
[252] Muramatsu Y, Tsujie M, Kohda Y, et al. Early detection of cysteine rich protein 61 (CYR61, CCN1) in urine following renal ischemic reperfusion injury. Kidney Int 2002;62:1601–10.
[253] Di Paolo S, Gesualdo L, Stallone G, Ranieri E, Schena FP. Renal expression and urinary concentration of EGF and IL-6 in acutely dysfunctioning kidney transplanted patients. Nephrol Dial Transplant 1997;12:2687–93.
[254] Gesualdo L, Di Paolo S, Calabró A, et al. Expression of epidermal growth factor and its receptor in normal and diseased human kidney: an immunohistochemical and in situ hybridization study. Kidney Int 1996;49:656–65.
[255] Grandaliano G, Gesualdo L, Bartoli F, et al. MCP-1 and EGF renal expression and urine excretion in human congenital obstructive nephropathy. Kidney Int 2000;58: 182–92.
[256] Stangou M, Alexopoulos E, Papagianni A, et al. Urinary levels of epidermal growth factor, interleukin-6 and monocyte chemoattractant protein-1 may act as predictor markers of renal function outcome in immunoglobulin A nephropathy. Nephrology 2009;14:613–20.
[257] Sundberg A, Appelkvist EL, Dallner G, Nilsson R. Glutathione transferases in the urine: sensitive methods for detection of kidney damage induced by nephrotoxic agents in humans. Environ Health Perspect 1994;102(Suppl. 3):293–6.
[258] Sundberg AG, Nilsson R, Appelkvist EL, Dallner G. Immuno-histochemical localization of alpha and pi class glutathione transferases in normal human tissues. Pharmacol. Toxicol 1993;72:321–31.
[259] Prozialeck WC, Edwards JR, Vaidya VS, Bonventre JV. Preclinical evaluation of novel urinary biomarkers of cadmium nephrotoxicity. Toxicol Appl Pharmacol 2009;238:301–5.
[260] Prozialeck WC, Edwards JR, Lamar PC, Liu J, Vaidya VS, Bonventre JV. Expression of kidney injury molecule-1 (Kim-1) in relation to necrosis and apoptosis during

the early stages of Cd-induced proximal tubule injury. Toxicol Appl Pharmacol 2009;238:306–14.
[261] Harrison DJ, Kharbanda R, Cunningham DS, McLellan LI, Hayes JD. Distribution of glutathione S-transferase isoenzymes in human kidney: basis for possible markers of renal injury. J Clin Pathol 1989;42:624–8.
[262] Eijkenboom JJ, van Eijk LT, Pickkers P, Peters WH, Wetzels JF, van der Hoeven HG. Small increases in the urinary excretion of glutathione S-transferase A1 and P1 after cardiac surgery are not associated with clinically relevant renal injury. Intensive Care Med 2005;31:664–7.
[263] Parikh CR, Abraham E, Ancukiewicz M, Edelstein CL. Urine IL-18 is an early diagnostic marker for acute kidney injury and predicts mortality in the intensive care unit. J Am Soc Nephrol 2005;16:3046–52.
[264] Parikh CR, Jani A, Mishra J, et al. Urine NGAL and IL-18 are predictive biomarkers for delayed graft function following kidney transplantation. Am J Transplant 2006;6:1639–45.
[265] Washburn KK, Zappitelli M, Arikan AA, et al. Urinary Interleukin-18 is an acute kidney injury biomarker in critically ill children. Nephrol Dial Transplant 2008;23: 566–72.
[266] Ichimura T, Asseldonk EJ, Humphreys BD, Gunaratnam L, Duffield JS, Bonventre JV. Kidney injury molecule-1 is a phosphatidylserine receptor that confers a phagocytic phenotype on epithelial cells. J Clin Invest 2008;118:1657–68.
[267] Vaidya VS, Ramirez V, Ichimura T, Bobadilla NA, Bonventre JV. Urinary kidney injury molecule-1: a sensitive quantitative biomarker for early detection of kidney tubular injury. Am J Physiol Renal Physiol 2006;290:F517–29.
[268] Han WK, Bailly V, Abichandani R, Thadhani R, Bonventre JV. Kidney Injury Molecule-1 (KIM-1): a novel biomarker for human renal proximal tubule injury. Kidney Int 2002;62:237–44.
[269] Maatman RG, van de Westerlo EM, van Kuppevelt TH, Veerkamp JH. Molecular identification of the liver- and the heart-type fatty acid-binding proteins in human and rat kidney. Use of the reverse transcriptase polymerase chain reaction. Biochem J 1992;288:285–90.
[270] Kamijo A, Sugaya T, Hikawa A, et al. Clinical evaluation of urinary excretion of liver-type fatty acid binding protein as a marker for monitoring chronic kidney disease: a multi-center trial. J Lab Clin Med 2005;145:125–33.
[271] Oyama Y, Takeda T, Hama H, Tanuma A, et al. Evidence for megalin-mediated proximal tubular uptake of L-FABP, a carrier of potentially nephrotoxic molecules. Lab Invest 2005;85:522–31.
[272] Kamijo A, Sugaya T, Hikawa A, et al. Urinary liver-type fatty acid binding protein as a useful biomarker in chronic kidney disease. Mol Cell Biochem 2006;284: 175–82.
[273] Russo LM, Sandoval RM, McKee M, et al. The normal kidney filters nephrotic levels of albumin retrieved by proximal tubule cells: retrieval is disrupted in nephrotic states. Kidney Int 2007;71:504–13.
[274] Russo LM, Sandoval RM, Brown D, Molitoris BA, Comper WD. Controversies in nephrology: response to 'renal albumin handling, facts, and artifacts'. Kidney Int 2007;72:1195–7.
[275] Russo LM, Sandoval RM, Campos SB, Molitoris BA, Comper WD, Brown D. Impaired tubular uptake explains albuminuria in early diabetic nephropathy. J Am Soc Nephrol 2009;20:489–94.
[276] Davey PG, Cowley DM, Geddes AM, Terry J. Clinical evaluation of beta 2-microglobulin, muramidase, and alanine aminopeptidase as markers of gentamicin nephrotoxicity. Contrib Nephrol 1984;42:100–6.

[277] Tolkoff-Rubin NE, Rubin RH, Bonventre JV. Non-invasive renal diagnostic studies. Clin Lab Med 1988;8:507–26.
[278] Miyata T, Jadoul M, Kurokawa K, Van Ypersele de Strihou C. Beta-2 microglobulin in renal disease. J Am Soc Nephrol 1998;9:1723–35.
[279] Palmieri L, Ronca G, Cioni L, Puccini R. Enzymuria as a marker of renal injury and disease: studies of N-acetyl-beta-glucosaminidase, alanine aminopeptidase and lysozyme in patients with renal disease. Contrib Nephrol 1984;42:123–9.
[280] Diener U, Knoll E, Ratge D, Langer B, Wisser H. Urinary excretion of alanine-aminopeptidase and N-acetyl-beta-D-glucosaminidase during sequential combination chemotherapy. J Clin Chem Clin Biochem 1982;20:615–9.
[281] Liangos O, Perianayagam MC, Vaidya VS, et al. Urinary N-acetyl-beta-(D)-glucosaminidase activity and kidney injury molecule-1 level are associated with adverse outcomes in acute renal failure. J Am Soc Nephrol 2007;18:904–12.
[282] Bazzi C, Petrini C, Rizza V, Arrigo G, Napodano P, Paparella M, D' Amico G. Urinary N-acetyl-beta-glucosaminidase excretion is a marker of tubular cell dysfunction and a predictor of outcome in primary glomerulonephritis. Nephrol Dial Transplant 2002;17:1890–6.
[283] Haase M, Bellomo R, Devarajan P, Schlattmann P, Haase-Fielitz A. NGAL Meta-analysis Investigator Group. Accuracy of neutrophil gelatinase-associated lipocalin (NGAL) in diagnosis and prognosis in acute kidney injury: a systematic review and meta-analysis. Am J Kidney Dis 2009;54:1012–24.
[284] Mishra J, Dent C, Tarabishi R, et al. Neutrophil gelatinase-associated lipocalin (NGAL) as a biomarker for acute renal injury after cardiac surgery. Lancet 2005;365(9466): 1231–8.
[285] Mishra J, Ma Q, Prada A, et al. Identification of neutrophil gelatinase-associated lipocalin as a novel early urinary biomarker for ischemic renal injury. J Am Soc Nephrol 2003;14:2534–43.
[286] Asplin JR, Arsenault D, Parks JH, Coe FL, Hoyer JR. Contribution of human uropontin to inhibition of calcium oxalate crystallization. Kidney Int 1998;53:194–9.
[287] Xie Y, Sakatsume M, Nishi S, Narita I, Arakawa M, Gejyo F. Expression, roles, receptors, and regulation of osteopontin in the kidney. Kidney Int 2001;60:1645–57.
[288] Hudkins KL, Giachelli CM, Cui Y, Couser WG, Johnson RJ, Alpers CE. Osteopontin expression in fetal and mature human kidney. J Am Soc Nephrol 1999;10:444–57.
[289] Thomas SE, Lombardi D, Giachelli C, Bohle A, Johnson RJ. Osteopontin expression, tubulo interstitial disease, and essential hypertension. Am J Hypertens 1998;11:954–61.
[290] Bernard AM, Vyskocil AA, Mahieu P, Lauwerys RR. Assessment of urinary retinol-binding protein as an index of proximal tubular injury. Clin Chem 1987;33:775–9.
[291] Sato Y, Wharram BL, Lee SK, et al. Urine podocyte mRNAs mark progression of renal disease. J Am Soc Nephrol 2009;20:1041–52.
[292] Wang G, Lai FM, Kwan BC, Lai KB, Chow KM, Li PK, Szeto CC. Podocyte loss in human hypertensive nephrosclerosis. Am J Hypertens 2009;22:300–6.
[293] Zhou H, Cheruvanky A, Hu X, et al. Urinary exosomal transcription factors, a new class of biomarkers for renal disease. Kidney Int 2008;74:613–21.
[294] Nakatsue T, Koike H, Han GD, et al. Nephrin and podocin dissociate at the onset of proteinuria in experimental membranous nephropathy. Kidney Int 2005;67:2239–53.
[295] Frederick J, Woessner JF. Matrix metalloproteinases and their inhibitors in connective tissue remodeling. FASEB J 1991;5:2145–54.
[296] Sharma K, Mauer SM, Kim Y, Michael AF. Altered expression of matrix metalloproteinase-2, TIMP, and TIMP-2 in obstructive nephropathy. J Lab Clin Med 1995;125:754–76.
[297] Sanders JS, Huitema MG, Hanemaaije R, van Goor H, Kallenberg CG, Stegeman CA. Urinary matrix metalloproteinases reflect renal damage in anti-neutrophil cytoplasm autoantibody-associated vasculitis. Am J Physiol Renal Physiol 2007;293:F1927–34.

[298] Hörstrup JH, Gehrmann M, Schneider B. Elevation of serum and urine levels of TIMP-1 and tenascin in patients with renal disease. Nephrol Dial Transplant 2002;17: 1005–13.
[299] Mashimo H, Wu DC, Podolsky DK, Fishman MC. Impaired defense of intestinal mucosa in mice lacking intestinal trefoil factor. Science 1996;274:262–5.
[300] Mashimo H, Podolsky DK, Fishman MC. Structure and expression of murine intestinal trefoil factor: high evolutionary conservation and postnatal expression. Biochem Biophys Res Commun 1995;210:31–7.
[301] Debata PR, Panda H, Supakar PC. Altered expression of trefoil factor 3 and cathepsin L gene in rat kidney during aging. Biogerontology 2007;8:25–30.
[302] Kjellev S. The trefoil factor family - small peptides with multiple functionalities. Cell Mol Life Sci 2009;66:1350–69.
[303] Taupin D, Podolsky DK. Trefoil factors: initiators of mucosal healing. Nat Rev Mol Cell Biol 2003;4:721–32.
[304] Rinnert M, Hinz M, Buhtz P, Reiher F, Lessel W, Hoffmann W. Synthesis and localization of trefoil factor family (TFF) peptides in the human urinary tract and TFF2 excretion into the urine. Cell Tissue Res 2010;339:639–47.
[305] Lebherz-Eichinger D, Tudor B, Ankersmit HJ, Reiter T, Haas M, Roth-Walter F, Krenn CG, Roth GA. Trefoil factor 1 excretion is increased in early stages of chronic kidney disease. PLoS One 2015;10:e0138312.
[306] Gunsilius E, Petzer A, Stockhammer G, Nussbaumer W, Schumacher P, Clausen J, Gastl G. Thrombocytes are the major source for soluble vascular endothelial growth factor in peripheral blood. Oncology 2000;58:169–74.
[307] Chan LW, Moses MA, Goley E, et al. Urinary VEGF and MMP levels as predictive markers of 1-year progression-free survival in cancer patients treated with radiation therapy: a longitudinal study of protein kinetics throughout tumor progression and therapy. J Clin Oncol 2004;22:499–506.
[308] Neufeld G, Cohen T, Gengrinovitch S, Poltorak Z. Vascular endothelial growth factor (VEGF) and its receptors. FASEB J 1999;13:9–22.
[309] Peng W, Chen J, Jiang Y, Shou Z, Chen Y, Wang H. Acute renal allograft rejection is associated with increased levels of vascular endothelial growth factor in the urine. Nephrology 2008;13:73–9.
[310] Kitamoto Y, Tokunaga H, Miyamoto K, Tomita K. VEGF is an essential molecule for glomerular endothelial cells and its excretion in urine might be a unique marker of glomerular injury. Rinsho Byori 2000;48:485–90.
[311] Krzeminska E, Wyczalkowska-Tomasik A, Korytowska N, Paczek L. Comparison of two methods for determination of NGAL levels in urine: ELISA and CMIA. J Clin Lab Anal 2016;. doi: 10.1002/jcla.21962.
[312] Cruz DN, Virzì GM, Brocca A, Ronco C, Giavarina D. A comparison of three commercial platforms for urinary NGAL in critically ill adults. Clin Chem Lab Med 2016;54:353–62.
[313] Makris K, Stefani D, Makri E, Panagou I, Lagiou M, Sarli A, Lelekis M, Kroupis C. Evaluation of a particle enhanced turbidimetric assay for the measurement of neutrophil gelatinase-associated lipocalin in plasma and urine on Architect-8000: analytical performance and establishment of reference values. Clin Biochem 2015;48:1291–7.
[314] Gobe GC, Coombes JS, Fassett RG, Endre ZH. Biomarkers of drug-induced acute kidney injury in the adult. Expert Opin Drug Metab Toxicol 2015;11:1683–94.
[315] U.S. Department of Health and Human Services, Food and Drug Administration (2004) Challenge and opportunity on the critical path to new medical products. Available from: http://www.fda.gov/downloads/ScienceResearch/SpecialTopics/CriticalPathInitiative/CriticalPathOpportunitiesReports/UCM113411.pdf

[316] Gibbs A. Comparison of the specificity and sensitivity of traditional methods for assessment of nephrotoxicity in the rat with metabonomic and proteomic methodologies. J Appl Toxicol 2005;25:277–95.
[317] Merrick BA, Witzmann FA. The role of toxicoproteomics in assessing organ specific toxicity. EXS 2009;99:367–400.
[318] Kennedy S. The role of proteomics in toxicology: identification of biomarkers of toxicity by protein expression analysis. Biomarkers 2002;7:269–90.
[319] Hamdam J, Sethu S, Smith T, et al. Safety pharmacology--current and emerging concepts. Toxicol Appl Pharmacol 2013;273:229–41.
[320] Wishart DS. Emerging applications of metabolomics in drug discovery and precision medicine. Nat Rev Drug Discov 2016;15:473–84.
[321] Lu H, Han YJ, Xu JD, Xing WM, Chen J. Proteomic characterization of acyclovir-induced nephrotoxicity in a mouse model. PLoS One 2014;9(7):e103185.
[322] Rouse R, Siwy J, Mullen W, Mischak H, Metzger J, Hanig J. Proteomic candidate biomarkers of drug-induced nephrotoxicity in the rat. PLoS One 2012;7:e34606.
[323] Ferreira L, Quiros Y, Sancho-Martínez SM, García-Sánchez O, Raposo C, López-Novoa JM, González-Buitrago JM, López-Hernández FJ. Urinary levels of regenerating islet-derived protein III β and gelsolin differentiate gentamicin from cisplatin-induced acute kidney injury in rats. Kidney Int 2011;79:518–28.
[324] Puigmulé M, López-Hellin J, Suñé G, Tornavaca O, Camaño S, Tejedor A, Meseguer A. Differential proteomics analysis of cyclosporine A-induced toxicity in renal proximal tubule cells. Nephrol Dial Transplant 2009;24:2672–86.
[325] Klawitter J, Klawitter J, Kushner E, et al. Association of immunosuppressant-induced protein changes in the rat kidney with changes in urine metabolite patterns: A proteo-metabonomic study. J Proteome Res 2010;9:865–75.
[326] De Graauw M, Le Dévédec S, Tijdens I, Smeets MB, Deelder AM, van de Water B. Proteomic analysis of alternative protein tyrosine phosphorylation in 1,2-dichlorovinyl-cystein-induced cytotoxicity in primary cultured rat renal proximal tubular cells. J Pharmacol Exp Ther 2007;322:89–100.
[327] Korrapati MC, Chilakapati J, Witzmann FA, Rao C, Lock EA, Mehendale HM. Proteomics of S-(1,2-dichlorovinyl)-l-cysteine-induced acute renal failure and autoprotection in mice. Am J Physiol Renal Physiol 2007;293:F994–F1006.
[328] Com E, Boitier E, Marchandeau JP, Brandenburg A, Schroeder S, Hoffmann D, Mally A, Gautier JC. Integrated transcriptomic and proteomic evaluation of gentamicin nephrotoxicity in rats. Toxicol Appl Pharmacol 2012;258:124–33.
[329] Malard V, Gaillard JC, Bérenguer F, Sage N, Quéméneur E. Urine proteomic profiling of uranium nephrotoxicity. Biochim Biophys Acta 2009;1794:882–91.
[330] Marrer E, Dieterle F. Impact of biomarker development on drug safety assessment. Toxicol Appl Pharmacol 2010;243:167–79.
[331] Lamb KE, Lodhi S, Meier-Kriesche HU. Long-term renal allograft survival in the United States: a critical reappraisal. Am J Transplant 2011;11:450–62.
[332] Bohra R, Klepacki J, Klawitter J, Klawitter J, Thurman JM, Christians U. Proteomics and metabolomics in renal transplantation-quo vadis? Transpl Int 2013;26:225–41.
[333] Nickerson P. Prost-transplant monitoring of renal allografts: are we there yet? Curr Opin Immunol 2009;21:563–8.
[334] Brunet M, Shipkova M, van Gelder T, et al. Barcelona consensus on biomarker-based immunosuppressive drugs management in solid organ transplantation. Ther Drug Monit 2016;38(Suppl. 1):S1–S20.
[335] El-Zoghby ZM, Stegall MD, Lager DJ, et al. Identifying specific causes of kidney allograft loss. Am J Transplant 2009;9:527–35.
[336] Nankivell BJ, Kuypers DR. Diagnosis and prevention of chronic kidney allograft loss. Lancet 2011;378(9800):1428–37.

[337] Sidgel TK, Sarwal MM. The proteogenomic path towards biomarker discovery. Pediatr Transplantation 2008;12:737–47.
[338] Gwinner W. Renal transplant rejection markers. World J Urol 2007;25:445–55.
[339] Voshol H, Brendlen N, Müller D, et al. Evaluation of biomarker discovery approaches to detect protein biomarkers of acute renal allograft rejection. J Proteome Res 2005;4:1192–9.
[340] Clarke W, Silverman BC, Zhang Z, et al. Characterization of renal allograft reception by urinary proteomic analysis. Ann Surg 2003;237:660–5.
[341] Wittke S, Haubitz M, Walden M, et al. Detection of acute tubulointerstitial rejection by proteomic analysis of urinary samples in renal transplant recipients. Am J Transplant 2005;5:2479–88.
[342] O'Riordan E, Orlova TN, Mei JJ, et al. Bioinformatic analysis of the urine proteome of acute allograft rejection. J Am Soc Nephrol 2004;15:3240–8.
[343] Schaub S, Rush D, Wilkins J, et al. Proteomic-based detection of urine proteins associated with acute renal allograft rejection. J Am Soc Nephrol 2004;15:219–27.
[344] Schaub S, Wilkins JA, Antonovici M, et al. Proteomic-based identification of cleaved urinary β2-microglobulin as a potential marker for acute injury in renal allografts. Am J Transplant 2005;5:729–38.
[345] Clarke W. Proteomic research in renal transplantation. Ther Drug Monit 2006;28:19–22.
[346] Sigdel TK, Kaushal A, Gritsenko M, Norbeck AD, Qian WJ, Xiao W, Camp DG 2nd, Smith RD, Sarwal MM. Shotgun proteomics identifies proteins specific for acute renal transplant rejection. Proteomics Clin Appl 2010;4:32–47.
[347] Ling XB, Sigdel TK, Lau K, Ying L, Lau I, Schilling J, Sarwal MM. Integrative urinary peptidomics in renal transplantation identifies biomarkers for acute rejection. J Am Soc Nephrol 2010;21:646–53.
[348] Freue GV, Sasaki M, Meredith A, Günther OP, Bergman A, Takhar M, Mui A, Balshaw RF, Ng RT, Opushneva N, Hollander Z, Li G, Borchers CH, Wilson-McManus J, McManus BM, Keown PA, McMaster WR. Genome Canada Biomarkers in Transplantation Group. Proteomic signatures in plasma during early acute renal allograft rejection. Mol Cell Proteomics 2010;9:1954–67.
[349] Wu D, Zhu D, Xu M, Rong R, Tang Q, Wang X, Zhu T. Analysis of transcriptional factors and regulation networks in patients with acute renal allograft rejection. J Proteome Res 2011;10:175–81.
[350] Sigdel TK, Salomonis N, Nicora CD, Ryu S, He J, Dinh V, Orton DJ, Moore RJ, Hsieh SC, Dai H, Thien-Vu M, Xiao W, Smith RD, Qian WJ, Camp DG 2nd, Sarwal MM. The identification of novel potential injury mechanisms and candidate biomarkers in renal allograft rejection by quantitative proteomics. Mol Cell Proteomics 2014;13:621–31.
[351] Bañón-Maneus E, Diekmann F, Carrascal M, et al. Two-dimensional difference gel electrophoresis urinary proteomic profile in the search of nonimmune chronic allograft dysfunction biomarkers. Transplantation 2010;89:548–58.
[352] Srivastava M, Eidelman O, Torosyan Y, Jozwik C, Mannon RB, Pollard HB. Elevated expression levels of ANXA11, integrins β3 and α3, and TNF-α contribute to a candidate proteomic signature in urine for kidney allograft rejection. Proteomics Clin Appl 2011;5:311–21.
[353] Metzger J, Chatzikyrkou C, Broecker V, Schiffer E, Jaensch L, Iphoefer A, Mengel M, Mullen W, Mischak H, Haller H, Gwinner W. Diagnosis of subclinical and clinical acute T-cell-mediated rejection in renal transplant patients by urinary proteome analysis. Proteomics Clin Appl 2011;5:322–33.
[354] Sui W, Huang L, Dai Y, Chen J, Yan Q, Huang H. Proteomic profiling of renal allograft rejection in serum using magnetic bead-based fractionation and MALDI-TOF MS. Clin Exp Med 2010;10:259–68.

[355] Johnston O, Cassidy H, O'Connell S, O'Riordan A, Gallagher W, Maguire PB, Wynne K, Cagney G, Ryan MP, Conlon PJ, McMorrow T. Identification of β2-microglobulin as a urinary biomarker for chronic allograft nephropathy using proteomic methods. Proteomics Clin Appl 2011;5:422–31.
[356] Quintana LF, Campistol JM, Alcolea MP, Bañon-Maneus E, Sol-González A, Cutillas PR. Application of label-free quantitative peptidomics for the identification of urinary biomarkers of kidney chronic allograft dysfunction. Mol Cell Proteomics 2009;8:1658–73.
[357] Tetaz R, Trocmé C, Roustit M, Pinel N, Bayle F, Toussaint B, Zaoui P. Predictive diagnostic of chronic allograft dysfunction using urinary proteomics analysis. Ann Transplant 2012;17:52–60.
[358] Cassidy H, Slyne J, O'Kelly P, Traynor C, Conlon PJ, Johnston O, Slattery C, Ryan MP, McMorrow T. Urinary biomarkers of chronic allograft nephropathy. Proteomics Clin Appl 2015;9:574–85.
[359] Quintana LF, Solé-Gonzalez A, Kalko S, et al. Urine proteomics to detect biomarkers for chronic allograft dysfunction. J Am Soc Nephrol 2009;20:428–35.
[360] Kurian SM, Heilman R, Mondala TS, et al. Biomarkers for early and late stage chronic allograft nephropathy by proteogenomic profiling of peripheral blood. PLoS One 2009;4:e6212.
[361] Akkina SK, Zhang Y, Nelsestuen GL, Oetting WS, Ibrahim HN. Temporal stability of the urine proteome after kidney transplant: more sensitive than protein composition? J Proteome Res 2009;8:94–103.
[362] Jahnukainen T, Malehorn D, Sun M, Lyons-Weiler J, Bigbee W, Gupta G, Shapiro R, Randhawa PS, Pelikan R, Hauskrecht M, Vats A. Proteomic analysis of urine in kidney transplant patients with BK virus nephropathy. J Am Soc Nephrol 2006;17:3248–56.
[363] Sigdel TK, Gao Y, He J, Wang A, Nicora CD, Fillmore TL, Shi T, Webb-Robertson BJ, Smith RD, Qian WJ, Salvatierra O, Camp DG 2nd, Sarwal MM. Mining the human urine proteome for monitoring renal transplant injury. Kidney Int 2016;89:1244–52.
[364] Nakorchevsky A, Hewel JA, Kurian SM, et al. Molecular mechanisms of chronic kidney transplant rejection via large-scale proteogenomic analysis of tissue biopsies. J Am Soc Nephrol 2010;21:362–73.
[365] Sigdel TK, Gao X, Sarwal MM. Protein and peptide biomarkers in organ transplantation. Biomark Med 2012;6:259–71.
[366] Sigdel TK, Lee S, Sarwal MM. Profiling the proteome in renal transplantation. Proteomics Clin Appl 2011;5:269–80.
[367] Kienzl-Wagner K, Pratschke J, Brandacher G. Biomarker discovery in transplantation--proteomic adventure or mission impossible? Clin Biochem 2013;46:497–505.
[368] Kienzl-Wagner K, Pratschke J, Brandacher G. Proteomics—a blessing or a curse? Application of proteomics technology to transplant medicine. Transplantation 2011;92: 499–509.
[369] Kim SC, Page EK, Knechtle SJ. Urine proteomics in kidney transplantation. Transplant Rev 2014;28:15–20.
[370] Nashan B, Abbud-Filho M, Citterio F. Prediction, prevention and management of delayed graft function: where are we now? Clin Transplant 2016;30(10):1198–208.
[371] Ramirez-Sandoval JC, Herrington W, Morales-Buenrostro LE. Neutrophil gelatinase-associated lipocalin in kidney transplantation: a review. Transplant Rev 2015;29:139–44.
[372] Mischak H, Vlahou A, Ioannidis JP. Technical aspects and inter-laboratory variability in native peptide profiling: the CE-MS experience. Clin Biochem 2013;46:432–43.

[373] Lo DJ, Kaplan B, Kirk AD. Biomarkers for kidney transplant rejection. Nat Rev Nephrol 2014;10:215–25.
[374] Lo DJ, Weaver TA, Kleiner DE, et al. Chemokines and their receptors in human renal allotransplantation. Transplantation 2011;91:70–7.
[375] Schaub S, Nickerson P, Rush D, et al. Urinary CXCL9 and CXCL10 levels correlate with the extent of subclinical tubulitis. Am J Transplant 2009;9:1347–53.
[376] Blydt-Hansen TD, Gibson IW, Gao A, et al. Elevated urinary CXCL10-to-creatinine ratio is associated with subclinical and clinical rejection in pediatric renal transplantation. Transplantation 2015;99:797–804.
[377] Jackson JA, Kim EJ, Begley B, et al. Urinary chemokines CXCL9 and CXCL10 are noninvasive markers of renal allograft rejection and BK viral infection. Am J Transplant 2011;11:2228–34.
[378] Hricik DE, Nickerson P, Formica RN, et al. Multicenter validation of urinary CXCL9 as a risk-stratifying biomarker for kidney transplant injury. Am J Transplant 2013;13:2634–44.
[379] Hricik DE, Formica RN, Nickerson P, et al. Clinical Trials in Organ Transplantation-09 Consortium. Adverse outcomes of tacrolimus withdrawal in immune-quiescent kidney transplant recipients. J Am Soc Nephrol 2015;26:3114–22.
[380] Rabant M, Amrouche L, Lebreton X, et al. Urinary C-X-C motif chemokine 10 independently improves the noninvasive diagnosis of antibody-mediated kidney allograft rejection. J Am Soc Nephrol 2015;26:2840–51.
[381] Ho J, Wiebe C, Gibson IW, et al. Elevated urinary CCL2: Cr at 6 months is associated with renal allograft interstitial fibrosis and inflammation at 24 months. Transplantation 2014;98:39–46.
[382] Welberry Smith MP, Zougman A, Cairns DA, et al. Serum aminoacylase-1 is a novel biomarker with potential prognostic utility for long-term outcome in patients with delayed graft function following renal transplantation. Kidney Int 2013;84:1214–25.
[383] Banks RE, Craven RA, Harnden P, Madaan S, Joyce A, Selby PJ. Key clinical issues in renal cancer: a challenge for proteomics. Word J Urol 2007;25:537–56.
[384] Mancini V, Battaglia M, Ditonno P, et al. Current insights in renal cell cancer pathology. Urol Oncol 2008;26:225–38.
[385] Seliger B, Dressler SP, Lichtenfels R, Kellner R. Candidate biomarkers in renal cell carcinoma. Proteomics 2007;7:4601–12.
[386] Kashyap MK, Kumar A, Emelianenko N, et al. Biochemical and molecular markers in renal cell carcinoma: an update and future prospects. Biomarkers 2005;10:258–94.
[387] Okamura N, Masuda T, Gotoh A, et al. Quantitative proteomic analysis to discover potential diagnostic markers and therapeutic targets in human renal cell carcinoma. Proteomics 2008;8:3194–203.
[388] Masui O, White NM, DeSouza LV, Krakovska O, Matta A, Metias S, Khalil B, Romaschin AD, Honey RJ, Stewart R, Pace K, Bjarnason GA, Siu KW, Yousef GM. Quantitative proteomic analysis in metastatic renal cell carcinoma reveals a unique set of proteins with potential prognostic significance. Mol Cell Proteomics 2013;12:132–44.
[389] White NM, Masui O, Desouza LV, Krakovska O, Metias S, Romaschin AD, Honey RJ, Stewart R, Pace K, Lee J, Jewett MA, Bjarnason GA, Siu KW, Yousef GM. Quantitative proteomic analysis reveals potential diagnostic markers and pathways involved in pathogenesis of renal cell carcinoma. Oncotarget 2014;30(5):506–18.
[390] Atrih A, Mudaliar MA, Zakikhani P, Lamont DJ, Huang JT, Bray SE, Barton G, Fleming S, Nabi G. Quantitative proteomics in resected renal cancer tissue for biomarker discovery and profiling. Br J Cancer 2014;110:1622–33.

[391] Nuerrula Y, Rexiati M, Liu Q, Wang YJ. Differential expression and clinical significance of serum protein among patients with clear-cell renal cell carcinoma. Cancer Biomark 2015;15:485–91.
[392] Gianazza E, Chinello C, Mainini V, Cazzaniga M, Squeo V, Albo G, Signorini S, Di Pierro SS, Ferrero S, Nicolardi S, van der Burgt YE, Deelder AM, Magni F. Alterations of the serum peptidome in renal cell carcinoma discriminating benign and malignant kidney tumors. J Proteomics 2012;76:125–40.
[393] Huang Z, Zhang S, Hang W, Chen Y, Zheng J, Li W, Xing J, Zhang J, Zhu E, Yan X. Liquid chromatography-mass spectrometry based serum peptidomic approach for renal clear cell carcinoma diagnosis. J Pharm Biomed Anal 2014;100:175–83.
[394] Chinello C, Cazzaniga M, De Sio G, Smith AJ, Gianazza E, Grasso A, Rocco F, Signorini S, Grasso M, Bosari S, Zoppis I, Dakna M, van der Burgt YE, Mauri G, Magni F. Urinary signatures of renal cell carcinoma investigated by peptidomic approaches. PLoS One 2014;9:e106684.
[395] Jones EE, Powers TW, Neely BA, Cazares LH, Troyer DA, Parker AS, Drake RR. MALDI imaging mass spectrometry profiling of proteins and lipids in clear cell renal cell carcinoma. Proteomics 2014;14:924–35.
[396] Raimondo F, Corbetta S, Chinello C, Pitto M, Magni F. The urinary proteome and peptidome of renal cell carcinoma patients: a comparison of different techniques. Expert Rev Proteomics 2014;11:503–14.
[397] Di Meo A, Pasic MD, Yousef GM. Proteomics and peptidomics: moving toward precision medicine in urological malignancies. Oncotarget 2016;. doi: 10.18632/oncotarget.8931.
[398] Seliger B, Lichtenfels R, Kellner R. Detection of renal cell carcinoma-associated markers via proteome and other 'ome'-based analyses. Brief Funct Genomic Proteomic 2003;2:194–212.
[399] Sakissan G, Fergelot P, Lamy PJ, et al. Identification of Pro-MMP-7 as a serum marker for renal cell carcinoma by use of proteome analysis. Clin Chem 2008;54:574–81.
[400] Lin YC, Tsui KH, Yu CC, Yeh CW, Chang PL, Yung BYM. Searching cell-secreted proteomes for potential urinary bladder tumor markers. Proteomics 2006;6:4381–9.
[401] Rehman I, Azzouzi AR, Catto JWF, Allen S, Cross SS, Feeley K, Meuth M, Hamdy FC. Proteomic analysis of voided urine after prostatic massage from patients with prostate cancer: a pilot study. Urology 2004;64:1238–43.
[402] Chambers AF, Vanderhyden BC. Ovarian cancer biomarkers in urine. Clin Canc Res 2006;12:323–7.
[403] Zimmerli LU, Schiffer E, Zürbig P, et al. Urinary proteome biomarkers in coronary artery disease. Mol Cell Proteomics 2008;7:290–8.
[404] Airoldi L, Magagnotti C, Iannuzzi AR, et al. Effects of cigarette smoking on the human urinary proteome. Biochem Biophys Res Commun 2009;381:397–402.
[405] Vitzthum F, Behrens F, Anderson NL, Saw JH. Proteomics: from basic research to diagnostic application. A review of requirements and needs. J Proteome Res 2005;4:1086–97.
[406] Wagner JA, Williams SA, Webster CJ. Biomarkers and surrogate end points for fit-for-purpose development and regulatory evaluation of new drugs. Clin Pharmacol Ther 2007;81:104–7.
[407] Goodsaid FM, Frueh FW, Mattes W. Strategic paths for biomarker qualification. Toxicology 2008;245:219–23.
[408] Manolis E, Vamvakas S, Isaac M. New pathway for qualification of novel methodologies in the European Medicines Agency. Proteomics Clin Appl 2011;5:248–55.
[409] Molitoris BA, Melnikov VY, Okusa MD, Himmelfarb J. Technology insight: biomarker development in acute kidney injury- what can we anticipate? Nat Clin Practice Nephrol 2008;4:154–65.

[410] Mischak H, Ioannidis JP, Argiles A, et al. Implementation of proteomic biomarkers: making it work. Eur J Clin Invest 2012;42:1027–36.
[411] Anderson NL. The roles of multiple proteomics platforms in a pipeline of new diagnostics. Mol Cell Proteomics 2005;4:1441–4.
[412] Mehan MR, Ostroff R, Wilcox SK, Steele F, Schneider D, Jarvis TC, Baird GS, Gold L, Janjic N. Highly multiplexed proteomic platform for biomarker discovery, diagnostics, and therapeutics. Adv Exp Med Biol 2013;735:283–300.
[413] Fawcett T. Introduction to ROC analysis. Pattern Recognit Lett 2006;27:861–74.
[414] Kellum JA, Levin N, Bouman C, Lameire N. Developing a consensus classification system for acute renal failure. Curr Opin Crit Care 2002;8:509–14.
[415] Mullen W, Delles C, Mischak H. EuroKUP COST action. Urinary proteomics in the assessment of chronic kidney disease. Curr Opin Nephrol Hypertens 2011;20:654–61.
[416] U.S. Department of Health and Human Services, Food and Drug Administration. Bioanalytical Method Validation. Draft Guidance, 2013. Available from: http://www.fda.gov/downloads/drugs/guidancecomplianceregulatoryinformation/guidances/ucm368107.pdf

CHAPTER FIVE

Cystatin C as a Multifaceted Biomarker in Kidney Disease and Its Role in Defining "Shrunken Pore Syndrome"

A. Grubb, MD, PhD
Department of Clinical Chemistry and Pharmacology, University Hospital, Lund University, Lund, Sweden

Contents

Cystatin C, a basic nonglycosylated 13.3 kDa protein comprising a single polypeptide chain of 120 amino acid residues, is produced at a stable rate from a housekeeping gene in all nucleated cells and eliminated from the circulation by free filtration through the glomerular membranes as described in the first edition of "Biomarkers of Kidney Disease" [1]. This discourse will report some of the progress in the use of cystatin C as a multifaceted marker of kidney disease, which has occurred since then, including its role in identifying the novel syndrome called "Shrunken Pore Syndrome."

Biomarkers of Kidney Disease. http://dx.doi.org/10.1016/B978-0-12-803014-1.00005-4

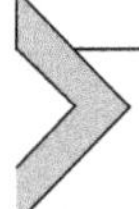

FACTORS INFLUENCING THE DIAGNOSTIC PERFORMANCE OF CYSTATIN C- OR CREATININE-BASED GFR-ESTIMATING EQUATIONS AND CAUSING THE PLETHORA OF EQUATIONS: THE CONCEPTS OF "INTERNAL" OR "EXTERNAL" VALIDATION

More than 100 different creatinine- and cystatin C-based estimating equations for glomerular filtration rate (GFR) have been described [2]. This has caused confusion and debate about which equation(s) is (are) to be preferred. In order to arrive at the best decision concerning selection of an estimating equation, the reasons for the generation of a multitude of cystatin C- and creatinine-based GFR-estimating equations must be known and they are as follows.

A. Not using (or lack of) international reference materials ("calibrators") for cystatin C or creatinine.

B. Use of varying, noncommutable assays of cystatin C or creatinine.

C. Use of different methods to measure GFR.

D. Use of different statistical approaches to generate the equations.

E. Use of different populations to generate the equations. The populations may vary in parameters, such as, disease panorama, proportion between healthy and sick persons, age distribution, GFR distribution, ethnicity, etc.

Of these reasons, those in A have contributed heavily in producing differing GFR-estimating equations. Standard reference materials for creatinine were released in 2006 [3] and for cystatin C (ERM-DA471/IFCC) in 2010 [4–6]. Therefore, it was only recently possible for diagnostic companies to offer assay methods for cystatin C based on such reference materials [2]. This has resulted in decreased variation among the results for cystatin C obtained with different assays at different laboratories as reported by external quality assessment organizations, such as Equalis [2,7]. However, the problems mentioned in A will persist for a few years until the ongoing replacement of old reference materials with new international reference material will finish.

The use of noncommutable assays for cystatin C and creatinine is now diminishing inter alia due to the availability of international reference materials for both cystatin C and creatinine. A further contributing factor is the availability of an internationally recognized reference method, isotope dilution mass spectrometry (IDMS) for the determination of plasma creatinine levels [8]. No such method is yet available for the concentration of any protein in any biologic fluid, although some efforts are being made concerning the cystatin C concentration in biologic fluids [9,10].

As all GFR-estimating equations are generated using measured GFR values, it is obvious that if different methods to measure GFR produce different results, the corresponding generated GFR-estimating equations will also differ. Although renal clearance of inulin is generally accepted as the best way to measure GFR, it has rarely been used in efforts to generate GFR-estimating equations, as it requires expensive reagents and is based on a technically complicated and slow procedure [11,12]. Instead, several faster and less expensive methods to measure GFR have been used, for example, renal clearance of ^{51}Cr–EDTA, ^{125}I–iothalamate, ^{99}Tc–DTPA or iohexol; or plasma clearance of iohexol, iothalamate, ^{51}Cr–EDTA, or ^{99}Tc–DTPA; and endogenous clearance of creatinine [11,12]. They produce different results and it is surprising that, only recently, a careful analysis has been done to evaluate which of these compounds produce results that agree with those obtained by the renal clearance of inulin [11–13]. The results of some of the methods studied differed very significantly from those produced by the renal clearance of inulin [11–13] and would therefore result in widely differing GFR-estimating equations, even if all other factors mentioned in A, B, D and E were identical in the development of the equations. In particular, endogenous creatinine clearance was found to strongly deviate from the results obtained by renal inulin clearance [11,12]. Although this has been known since 1935 [14,15], endogenous creatinine clearance continues to be used both to measure GFR and to produce GFR-estimating equations based on cystatin C or creatinine. Even those methods that were judged to be useful to replace renal inulin clearance in clinical practice [11–13] displayed small, but statistically significant, differences in measured GFR [11–13]. Also these small differences produce different GFR-estimating equations, although these differences probably would be clinically insignificant in most situations.

Different statistical approaches to generate the GFR-estimating equations may be used, for example, to increase the diagnostic performance in certain GFR intervals at the expense of the diagnostic performance in other GFR intervals. One example is the use of the creatinine-based MDRD equation for a population dominated by persons with a decrease in GFR, whereas the creatinine-based CKD–EPI equation was developed to produce a better diagnostic performance in a population of persons with both normal- and abnormal GFR [16,17]. However, although the CKD–EPI equation displayed a significantly improved diagnostic performance in patients with normal GFR, it was at the expense of a poorer performance in patients with a decrease in GFR, at least in some populations [18].

Different characteristics between populations concerning, for example, disease panorama, proportion between healthy and sick persons, age distribution, GFR distribution, ethnicity, etc., will result in the generation of different GFR-estimating equations even if all the other factors mentioned in A, B, C and D are equal. This cannot be considered problematic as it only reflects the varying and complex biomedical properties of such populations. As a matter of fact, if you know the characteristics of the population for which you are going to select the best GFR-estimating equation, it can be an advantage to be able to choose from a set of equations, which are specialized for different types of populations.

To characterize a GFR-estimating equation, the concepts of internal or external validation are often used. "Internal" validation means that the diagnostic performance of the equation is tested in a population, which is a part of, or very similar, to the population used to generate the equation; whereas "external" validation means that the diagnostic performance has been tested in a population different from the one that was used to generate the equation. It is assumed, that the results of an "external" validation are more reliable than the results of an "internal" validation in assessing the diagnostic performance of an equation in other external populations. This is, however, an oversimplification, as all the factors in A, B, C, D and E must be considered, when you will try to anticipate the diagnostic performance of an estimating equation in a specific context. For example, if an estimating equation based on endogenous creatinine clearance to measure GFR of the individuals in a population, is "externally" validated in another population, also using endogenous creatinine clearance to measure GFR, the "external" validation may produce a good result. But the equation would, nevertheless, most probably display a bad diagnostic performance in another population for which renal clearance of inulin, rather than endogenous creatinine clearance, was used to measure GFR.

OPTIMIZING THE USE OF CYSTATIN C- AND CREATININE-BASED GFR-ESTIMATING EQUATIONS

The first reports on cystatin C as a marker of GFR [19–22] used a slow assay, and it was not until a rapid and automated assay was developed [23] that its diagnostic performance as a marker of GFR could be carefully studied in large patient cohorts. Further development of several automated assays and the production of an international reference preparations have allowed a rapid growth of knowledge concerning the use of cystatin C as a marker of GFR and the search string "cystatin C AND (renal OR glomerular)" resulted in 2995 items on July 18, 2016 at www.ncbi.nlm.nih.gov/pubmed.

A similar development has occurred for creatinine so that old, biased assays based on the Jaffe reaction, to a large part have been replaced by enzyme-based specific assays and international reference preparations have become available [3,24]. These advances have allowed the development of cystatin C- or creatinine-based GFR-estimating equations, which produce estimated GFR values, designated as $eGFR_{cystatin\ C}$ or $eGFR_{creatinine}$, 80–85% of which are between ±30% of GFR measured by invasive gold-standard methods [25–30]. Some of these equations use assays based on international reference preparations of cystatin C and creatinine and populations of many thousands individuals to generate and test the equations [2,31]. However, as shown in several investigations, the highest percentages of estimated GFR values between ±30% of measured GFR values are obtained using GFR-estimating equations based on both cystatin C and creatinine [25–32]. Such equations might produce estimated GFR values, 90–91% of which are between ±30% of GFR measured by gold-standard methods [25,29]. The imprecision of all gold-standard procedures to measure GFR means that even if a gold-standard procedure is repeated within a short interval on patients with stable kidney function, less than 100% of the second determination will be within ±30% of the first. It should also be noted that in evaluations of GFR values produced by GFR-estimating equations, it is axiomatically assumed that the imprecision of the gold-standard procedure used to measure GFR is 0%. This means that the calculated percentage of estimated GFR values between ±30% of the measured GFR values, obtained by any GFR-estimating equation, generally is lower than the true one, as the imprecision of the gold-standard procedure usually increases the number of estimated GFR values outside the ±30% interval.

A GFR-estimating equation producing GFR values, 90–91% of which are within ±30% of GFR measured by gold-standard methods, is therefore close to what is theoretically possible. It has been demonstrated that the arithmetic mean of a cystatin C- and a creatinine-based equation, $eGFR_{mean}$, displays a diagnostic performance at least as good as that displayed by more complex equations [29,33]. This observation has been used to further improve the diagnostic performance of cystatin C- and creatinine-based estimating equations [34,35]. Although GFR-estimating equations using both cystatin C and creatinine clearly seem to have a better diagnostic performance than equations based on only one of these GFR markers, such combined equations do not perform optimally in a number of clinical situations, for example, if the patient has an abnormally low muscle mass or is treated with a high dose of glucocorticoids [34]. A strategy for GFR estimation based on the automatic use of cystatain C- and creatinine-based equations will, in these cases, have a worse diagnostic performance than a strategy that only uses the cystatin C- or creatinine-based GFR-estimating equation not influenced by the specific patient characteristics [34]. Such a strategy requires that GFR is estimated by both a cystatin C- and creatinine-based equation, producing $eGFR_{cystatin\ C}$ or $eGFR_{creatinine}$, and that the results are compared. If the two equations produce similar estimates, their average is a very reliable estimate of GFR. If the estimates do not agree and a specific factor known to disturb either the cystatin C- or creatinine-based estimate is present, only the estimate produced by the equation not disturbed by this factor, is used [34]. This has been shown to further increase the estimates, which are within ±30% of GFR measured by gold-standard methods [35].

As a matter of fact, during the 22 years since 1994, when we introduced cystatin C-based estimations of GFR in parallel with creatinine-based estimations, we have had about 20 cases for which the GFR estimates based on cystatin C or creatinine agreed, but disagreed with GFR measured by our invasive gold-standard procedure (plasma clearance of iohexol). In all cases in which relevant information was available, it turned out that the error had to do with some technical problems in the execution of the gold-standard procedure. We therefore consider that, in practice, agreeing cystatin C- and creatinine-based estimates of GFR are at least as reliable as GFR measured by invasive gold-standard procedures. This strategy is described at the multilingual site www.egfr.se, which can also be implemented to calculate absolute GFR from relative GFR, which might be required in, for example, for dosing of medicines cleared by the kidneys. The site uses a creatinine-based GFR-estimating equation, the LMrev equation [18,36], which, in contrast to most or all other creatinine-based equations, works for both children and adults [37]. The cystatin C-based equation, the CAPA equation [2], was developed using a large international population of children and adults with known GFR.

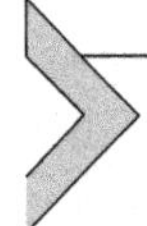

CYSTATIN C AND CREATININE ($eGFR_{CYSTATIN\ C}$ AND $eGFR_{CREATININE}$) AS MARKERS OF END-STAGE RENAL DISEASE (ESRD), HOSPITALIZATION, CARDIOVASCULAR DISEASE, AND DEATH

A decrease in GFR signals increased risks for the development of ESRD, cardiovascular disease, hospitalization, and death and GFR estimations based on cystatin C ($eGFR_{cystatin\ C}$) are consistently superior to GFR estimations based on creatinine ($eGFR_{creatinine}$) to predict these conditions [38–40]. The cause for the superiority of cystatin C as a risk marker is unknown, but observational studies have suggested that inflammation, old age, male gender, greater weight, and cigarette smoking increase the cystatin C level, thereby augmenting the potential of cystatin C as a risk marker [41]. However, statistical correlations in observational studies do not prove causal connections. As a matter of fact, a study of elective surgery of patients displayed a postoperative sharp rise in inflammation of the patients, with large increases in the levels of CRP and other inflammatory markers, but with no increase in the level of cystatin C, thus allowing rejection of the hypothesis that inflammation causes a raised cystatin C level [42]. The statistical correlations between inflammation, old age, male gender, greater weight, and cigarette smoking and cystatin C might be due to that all these factors promote the development of atherosclerosis, also in the renal arteries, thus producing a decrease in GFR and an increase in cystatin C [42]. Another hypothesis for the superiority of cystatin C as a risk marker is that an increase in cystatin C, with a molecular mass of 13.3 kDa, signals a shrinking of the pores in the glomerular membranes earlier than an increase in creatinine with a molecular mass of 113 Da [43,44]. The use of $eGFR_{mean}$ in clinical practice, as one of the best ways to estimate GFR, has allowed the recent identification of a new syndrome called "Shrunken Pore Syndrome," which might be connected to the superiority of cystatin C to predict ESRD, cardiovascular disease, hospitalization, and death. This is described in the next section.

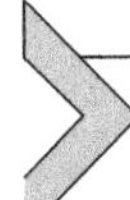

IDENTIFICATION OF "SHRUNKEN PORE SYNDROME": ITS INFLUENCE ON MORTALITY

The use of $eGFR_{mean}$ and the simultaneous comparison of $eGFR_{cystatin\ C}$ and $eGFR_{creatinine}$, as the best way to estimate GFR in clinical practice [34,35], identifies a significant number of patients with clear differences between $eGFR_{cystatin\ C}$ and $eGFR_{creatinine}$ [33]. Some of these differences can be explained by factors, such as, muscle wasting or treatment with large doses of glucocorticoids, known to invalidate the GFR estimations based on creatinine or cystatin C [34]. However, most of the patients showing such differences between $eGFR_{cystatin\ C}$ and $eGFR_{creatinine}$, do not display such known factors and their $eGFR_{mean}$ is, despite the differences between $eGFR_{cystatin\ C}$ and $eGFR_{creatinine}$, still the best way to estimate GFR (Figure 5.1) [33]. A large part of the patients displaying these differences has a pattern of $eGFR_{cystatin\ C}$ and $eGFR_{creatinine}$ in which $eGFR_{cystatin\ C}$ is less or equal to 60% of $eGFR_{creatinine}$

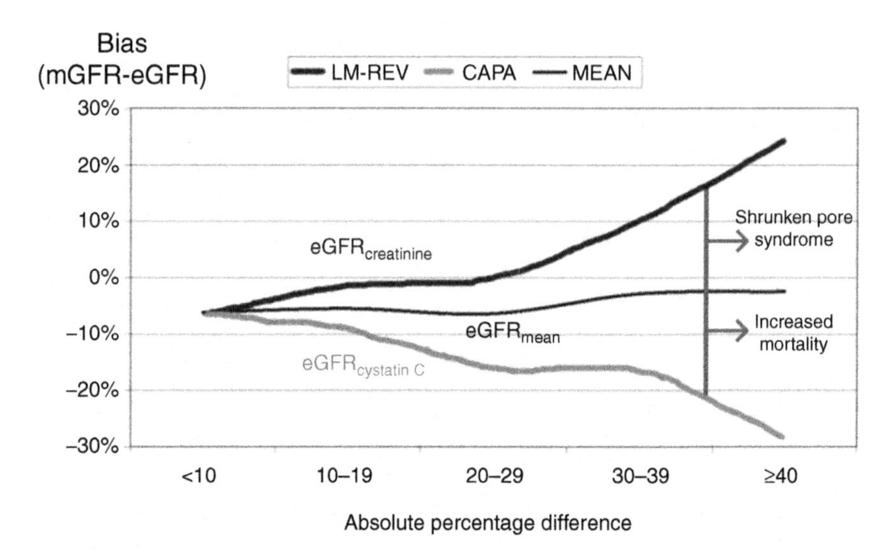

Figure 5.1 ***Estimations of GFR (eGFR) by Cystatin C- ($eGFR_{cystatin\ C}$) and Creatinine ($eGFR_{creatinine}$)-Based Estimation Equations and by the Mean of $eGFR_{cystatin\ C}$ and $eGFR_{creatinine}$ ($eGFR_{mean}$).*** The CAPA equation [2] and the LMrev equation [36] were used to estimate GFR in a patient population of 1112 individuals with GFR measured by plasma clearance of iohexol [33]. The bias between measured GFR (mGFR) and estimated GFR, when eGFR is estimated as $eGFR_{mean}$, is small, even when there are big differences between $eGFR_{cystatin\ C}$ and $eGFR_{creatinine}$. When $eGFR_{cystatin\ C} \leq 60\%\ eGFR_{creatinine}$ the "Shrunken Pore Syndrome" [44] is diagnosed and these patients display a high increase in mortality [59].

($eGFR_{cystatin\ C} \leq 60\%\ eGFR_{creatinine}$) [44]. These patients display higher cystatin C–creatinine concentration ratios than patients with similar $eGFR_{mean}$, but with agreeing values of $eGFR_{cystatin\ C}$ and $eGFR_{creatinine}$ ($eGFR_{cystatin\ C} \approx eGFR_{creatinine}$) [44]. When the concentrations of other low-molecular mass proteins, for example, β_2-microglobulin, β-trace protein, and retinol-binding protein, are measured in patients with $eGFR_{cystatin\ C} \leq 60\%\ eGFR_{creatinine}$, it can be observed that the concentration ratios of these proteins to creatinine are also higher than in patients in whom $eGFR_{cystatin\ C} \approx eGFR_{creatinine}$ (Figure 5.2) [44]. The genes for these proteins are located at different chromosomes [45–48], have different regulation elements, and the synthesis of these proteins is not generally influenced by the factors that affect the production of cystatin C, in the same way [49–53]. This indicates that the production of these proteins is not coregulated and thus cannot explain the concordant increases of their plasma levels. But this concurrent increase can be explained if the proteins have a common clearance mechanism.

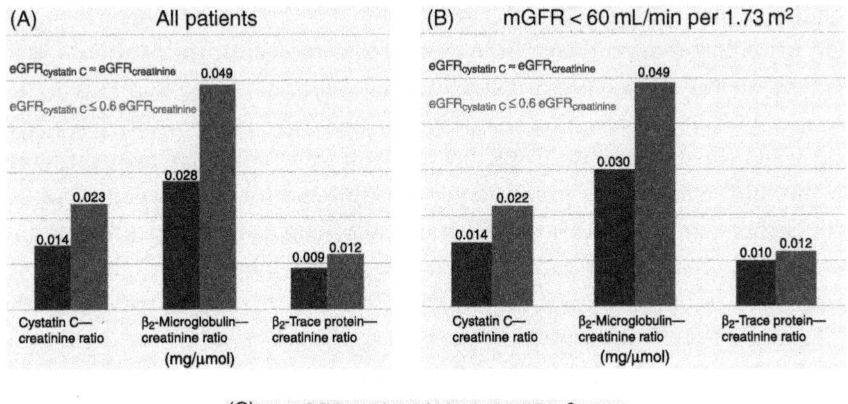

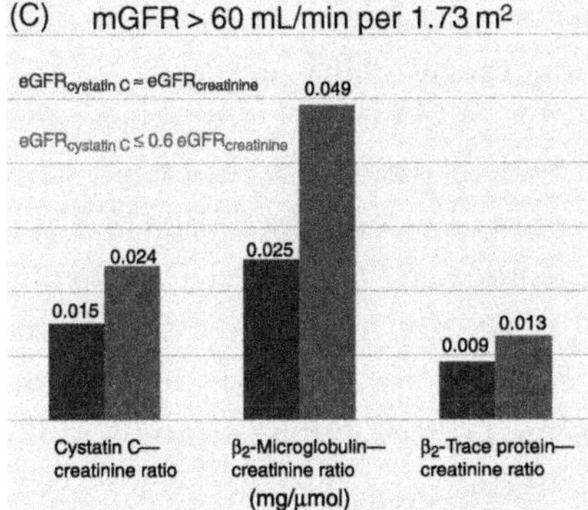

Figure 5.2 ***The Ratios (mg/μmol) Between Cystatin C and Creatinine; β_2-Microglobulin and Creatinine; and β-Trace Protein and Creatinine, When $eGFR_{cystatin\ C} \approx eGFR_{creatinine}$ and When $eGFR_{cystatin\ C} \leq 60\%\ eGFR_{creatinine}$.*** All differences between patients with $eGFR_{cystatin\ C} \approx eGFR_{creatinine}$ and with $eGFR_{cystatin\ C} \leq 60\%\ eGFR_{creatinine}$ are statistically significant. The same pattern is present, whether (A) all patients, (B) patients with GFR < 60 mL/min per 1.73 m^2, or (C) patients with GFR > 60 mL/min per 1.73 m^2 are studied.

As proteins below ≈20 kDa in molecular mass (<22 Å in Stokes–Einstein radius) are mainly excreted via glomerular transport [54], a reduction in their GFR would result in a simultaneous increase of their plasma levels. The simplest pathophysiologic way of interpreting this is that the normally high-sieving coefficients of these proteins drop significantly. According to the two-pore model of glomerular permeability, this can easily be explained by a reduction in the radii of the small pores of the glomerular filtration barrier. The explanation that creatinine and other small molecules do not simultaneously increase in concentration would then be, that their sieving coefficients are still close to unity (i.e., one) despite the shrunken pores.

Therefore, it seems that the observation of $eGFR_{cystatin\ C} \leq 60\%\ eGFR_{creatinine}$ and a simultaneous increase of the level of low-molecular mass proteins in a patient, identify a new syndrome, tentatively called as "Shrunken Pore Syndrome" [44]. It is interesting that a similar mechanism previously has been suggested for the increase in plasma levels of low-molecular mass proteins in the third trimester of pregnancy [43,55,56] and for the development of still higher concentrations of low-molecular mass proteins in preeclampsia [57,58]. This might indicate that the (patho-)physiologic changes in pregnancy and preeclampsia are similar to those occurring in patients with "Shrunken Pore Syndrome."

As "Shrunken Pore Syndrome" was identified recently [44], few studies of its clinical consequences have been performed. However, one recent investigation shows, that the mortality in patients undergoing elective coronary artery bypass grafting and suffering from the syndrome preoperatively, is much higher than that in patients without the syndrome, whether the preoperative GFR is normal or decreased (Figs. 5.3 and 5.4) [59]. Another recent study also indicates that "Shrunken Pore Syndrome" in a population of healthy seniors predicts increased risks for mortality and overall morbidity [60]. More studies are required, and are underway, to elucidate the full clinical consequences of the syndrome and if cut-offs other than $eGFR_{cystatin\ C} \leq 60\%\ eGFR_{creatinine}$ can be used to identify patients with increased mortality more efficiently (Figs. 5.1, 5.3, and 5.4) [59]. It can be expected that the "Shrunken Pore Syndrome," in addition to signaling a high mortality in different patient cohorts, also generally will indicate higher risks for the development of ESRD, cardiovascular disease, and for hospitalization.

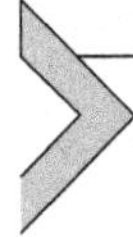

CYSTATIN C AS AN INDICTOR OF THE CIRCADIAN RHYTHM OF GFR

In healthy individuals and in most patients with renal disease, GFR displays a circadian rhythm such that GFR during the day is 20–40% higher than during the night [61,62]. This is mirrored by the diurnal variation of the cystatin C level, which is higher during the night in individuals with a normal circadian rhythm, thus mirroring the lower GFR [63]. In contrast, the creatinine level does not mirror the diurnal variation of GFR, as the tubular secretion of creatinine varies inversely with the GFR during a 24-h period [62,63].

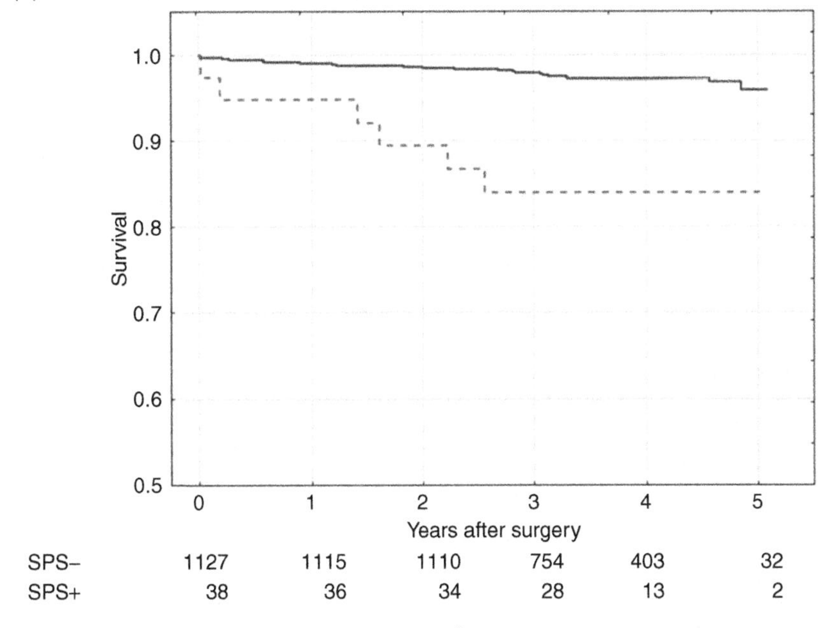

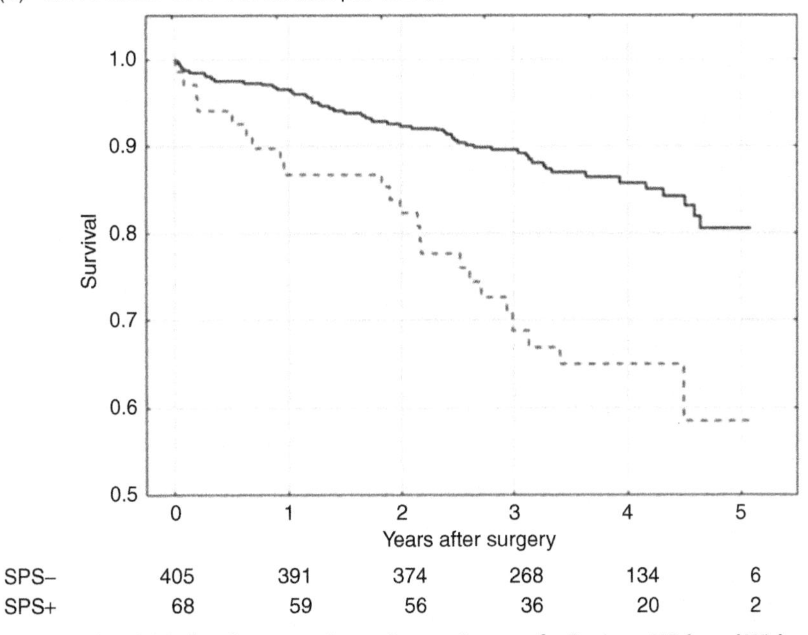

Figure 5.3 ***Survival after Coronary Artery Bypass Surgery for Patients With and Without Shrunken Pore Syndrome (SPS).*** $eGFR_{cystatin\ C}$ was estimated using the CAPA equation and $eGFR_{creatinine}$ using the LMrev equation. The cut-off level for SPS was $eGFR_{cystatin\ C} \leq 70\%$ of $eGFR_{creatinine}$. (A) Patients with GFR > 60 mL/min per 1.73 m^2 with Shrunken Poor Syndrome [*SPS, broken line* (red broken line in web version)] and without [*solid line* (blue solid line in web version)]. (B) Patients with GFR < 60 mL/min per 1.73 m^2 with Shrunken Poor Syndrome [*SPS, broken line* (red broken line in web version)] and without [*solid line* (blue solid line in web version)]. For both levels of GFR: $p < 0.001$ with log-rank test.

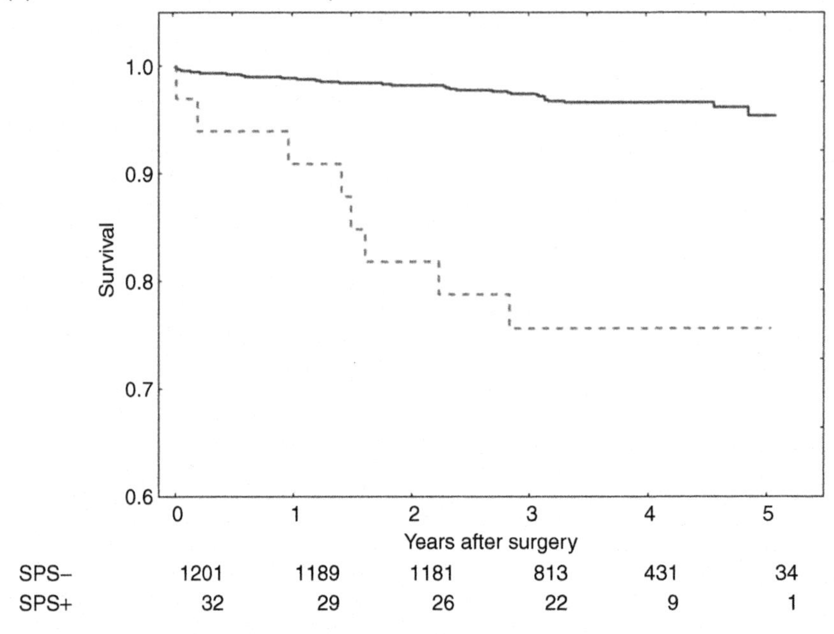

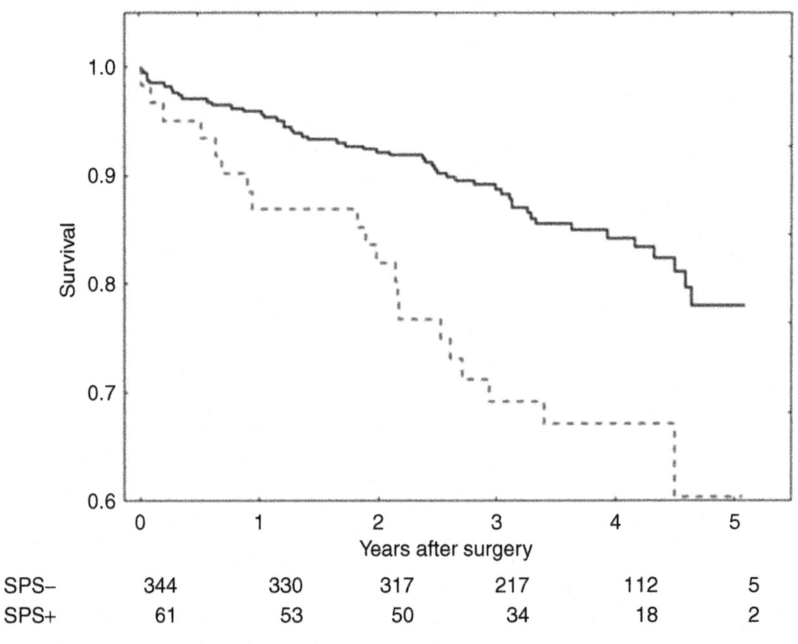

Figure 5.4 ***Survival After Coronary Artery Bypass Surgery for Patients With and Without Shrunken Pore Syndrome (SPS).*** $eGFR_{cystatin\ C}$ was estimated using the $CKD–EPI_{cystatin\ C}$ equation and $eGFR_{creatinine}$ using the $CKD–EPI_{creatinine}$ equation. The cut-off level for SPS was $eGFR_{cystatin\ C} \leq 60\%$ of $eGFR_{creatinine}$. (A) Patients with GFR > 60 mL/min per 1.73 m^2 with Shrunken Poor Syndrome [*SPS, broken line* (red broken line in web version)] and without [*solid line* (blue solid line in web version)]. (B) Patients with GFR < 60 mL/min per 1.73 m^2 with Shrunken Poor Syndrome [*SPS, broken line* (red broken line in web version)] and without [*solid line* (blue solid line in web version)]. For both levels of GFR: $p < 0.001$ with log-rank test.

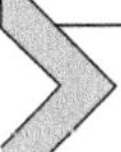

CYSTATIN C AS AN INDICATOR OF "RENAL RESERVE"

The increase in GFR measured by, for example, renal inulin clearance, occurring after consumption of a large amount of boiled meat (1.5–2 g/kg) or intravenous infusion of amino acids is called "renal reserve" [64–66]. A loss of renal reserve is considered to predispose patients to chronic kidney disease but the technical difficulties in measuring "renal reserve" has made accumulation of reliable clinical data difficult. Estimations based on the level of plasma creatinine cannot be used, as the creatine/creatinine absorbed from meat will conceal the decrease in creatinine caused by an increase in GFR. However, a recent report by Fuhrman, Maier, and Schwartz [67] suggests that measuring cystatin C before and after a protein load demonstrates an increase in GFR, the "renal reserve," as shown by a decrease in the cystatin C level. The small amount of cystatin C in meat is probably not absorbed and even if small amounts were absorbed, the antibodies used in the assays for cystatin C only recognize human cystatin C.

REFERENCES

[1] Grubb A. Cystatin C as a biomarker in kidney disease. In: Edelstein CL, editor. Biomarkers in kidney disease. Amsterdam; Boston: Academic Press; Elsevier; 2011. p. 291–312.

[2] Grubb A, Horio M, Hansson LO, et al. Generation of a new cystatin C-based estimating equation for glomerular filtration rate using seven assays standardized to the international calibrator. Clin Chem 2014;60:974–86.

[3] Myers GL, MillerWG, Coresh J, et al. Recommendations for improving serum creatinine measurement: a report from the Laboratory Working Group of the National Kidney Disease Education Program. Clin Chem 2006;52:5–18.

[4] Blirup-Jensen S, Grubb A, Lindström V, et al. Standardization of cystatin C. Development of primary and secondary reference preparations. Scand J Clin Lab Invest 2008;241:67–70.

[5] Grubb A, Blirup-Jensen S, Lindström V, et al. First certified reference material for cystatin C in human serum ERM-DA471/IFCC. Clin Chem Lab Med 2010;48:1619–21.

[6] Grubb A, Blirup-Jensen S, Lindström V, Schmidt C, Althaus H, Zegers I. First certified reference material for cystatin C in human serum ERM-DA471/IFCC. Clin Chem Lab Med 2010;48(11):1619–21.

[7] Available from: www.eqaulis.se

[8] Stokes P, O'Connor G. Development of a liquid chromatography-mass spectrometry method for the high-accuracy determination of creatinine in serum. J Chromatogr B 2003;794:125–36.

[9] Tyrefors N, Michelsen P, Grubb A. Two new types of assays to determine protein concentrations in biological fluids using mass spectrometry of intact proteins. Cystatin C in spinal fluid as an example. Scand J Clin Lab Invest 2014;74:546–54.

[10] Gonzalez-Antuna A, Rodrigues-Gonzalez P, Ohlendorf R, et al. Determination of cystastin C in human serum by isotope dilution mass spectrometry using mass overlapping peptides. J Proteomics 2015;112:141–55.

[11] Swedish Council on Health Technology Assessment. Methods to estimate and measure renal function (glomerular filtration rate). A systematic review (English Abstract). Report 214. 2013. Available from: http://www.sbu.se/en/Published/Yellow/Methods-to-Estimate-and-Measure-Renal-Function-Glomerular-Filtration-Rate
[12] Soveri I, Berg U, Björk J, et al. Measuring GFR: a systematic review. Am J Kidney Dis 2014;64:411–24.
[13] Delanaye P. How measuring glomerular filtration rate? In: Sahay M, editor. Comparison of reference methods, basic nephrology and acute kidney injury. InTech; Available from: http://www.intechopen.com/books/basic-nephrology-and-acute-kidney-injury/how-measuring-glomerular-filtration-rate-comparison-of-reference-methods
[14] Shannon JA, Smith HW. The excretion of inulin, xylose and urea by normal and phlorizinized man. J Clin Invest 1935;14:393–401.
[15] Shannon JA. The renal excretion of creatinine in man. J Clin Invest 1935;14:403–10.
[16] Levey AS, Bosch JP, Lewis JB, et al. A more accurate method to estimate glomerular filtration rate from serum creatinine: a new prediction equation. Modification of Diet in Renal Disease Study Group. Ann Intern Med 1999;130:461–70.
[17] Levey AS, Stevens LA, Schmid CH, et al. CKD-EPI. A new equation to estimate glomerular filtration rate. Ann Intern Med 2009;150:604–12.
[18] Nyman U, Grubb A, Larsson A, et al. The revised Lund–Malmö GFR estimating equation outperforms MDRD and CKD-EPI across GFR, age and BMI intervals in a large Swedish population. Clin Chem Lab Med 2014;52:815–24.
[19] Löfberg H, Grubb A. Quantitation of γ-trace in human biological fluids: indications for production in the central nervous system. Scand J Clin Lab Invest 1979;39:619–26.
[20] Grubb A, Simonsen O, Sturfelt G, et al. Serum concentration of cystatin C, factor D and beta-2-microglobulin as a measure of glomerular filtration rate. Acta Med Scand 1985;218:499–503.
[21] Simonsen O, Grubb A, Thysell H. The blood serum concentration of cystatin C (gamma-trace) as a measure of the glomerular filtration rate. Scand J Clin Lab Invest 1985;45:97–101.
[22] Grubb A, Löfberg H. Human γ-trace. Structure, function and clinical use of concentration measurements. Scand J Clin Lab Invest 1985;45(Suppl. 177):7–13.
[23] Kyhse-Andersen J, Schmidt C, Nordin G, et al. Serum cystatin C, determined by a rapid, automated particle-enhanced turbidimetric method, is a better marker than serum creatinine for glomerular filtration rate. Clin Chem 1994;40:1921–6.
[24] Panteghini M. IFCC Scientific Division. Enzymatic assays for creatinine: time for action. Clin Chem Lab Med 2008;46:567–72.
[25] Stevens LA, Coresh J, Schmid CH, et al. Estimating GFR using cystatin C alone and in combination with serum creatinine: a pooled analysis of 3,418 individuals with CKD. Am J Kidney Dis 2008;51:395–406.
[26] Bouvet Y, Bouissou F, Coulais Y, et al. GFR is better estimated by considering both serum cystatin C and creatinine levels. Pediatr Nephrol 2006;21:1299–306.
[27] Ma YC, Zuo L, Chen JH, et al. Improved GFR estimation by combined creatinine and cystatin C measurements. Kidney Int 2007;72:1535–42.
[28] Tidman M, Sjöström P, Jones I. A Comparison of GFR estimating formulae based upon s-cystatin C and s-creatinine and a combination of the two. Nephrol Dial Transplant 2008;23:154–60.
[29] Nyman U, Grubb A, Sterner G, et al. Different equations to combine creatinine and cystatin C to predict GFR. Arithmetic mean of existing equations performs as well as complex equations. Scand J Clin Lab Invest 2009;69:619–27.
[30] Schwartz GJ, Munoz A, Schneider MF, et al. New equations to estimate GFR in children with CKD. J Am Soc Nephrol 2009;20:629–37.
[31] Inker LA, Schmid CH, Tighiouart H, et al. Estimating glomerular filtration rate from serum creatinine and cystatin C. N Engl J Med 2012;367:20–9.

[32] Rule AD, Bergstralh EJ, Slezak JM, et al. Glomerular filtration rate estimated by cystatin C among different clinical presentations. Kidney Int 2006;69:399–405.
[33] Björk J, Grubb A, Larsson A, et al. Accuracy of GFR estimating equations combining standardized cystatin C and creatinine assays: A cross-sectional study in Sweden. Clin Chem Lab Med 2015;53:403–14.
[34] Grubb A. Non-invasive estimation of glomerular filtration rate (GFR). The Lund model: simultaneous use of cystatin C- and creatinine-based GFR-prediction equations, clinical data and an internal quality check. Scand J Clin Lab Invest 2010;70:65–70.
[35] Grubb A, Nyman U, Björk J. Improved estimation of glomerular filtration rate (GFR) by comparison of $eGFR_{cystatin\ C}$ and $eGFR_{creatinine}$. Scand J Clin Lab Invest 2012;72:73–7.
[36] Björk J, Grubb A, Sterner G, Nyman U. Revised equations for estimating glomerular filtration rate based on the Lund-Malmö study cohort. Scand J Clin Lab Invest 2011;71:232–9.
[37] Nyman U, Björk J, Lindström V, Grubb A. The Lund-Malmö creatinine-based glomerular filtration rate prediction equation for adults also performs well in children. Scand J Clin Lab Invest 2008;68:568–76.
[38] Shlipak MG, Sarnak MJ, Katz R, et al. Cystatin C and the risk of death and cardiovascular events among elderly persons. N Engl J Med 2005;352:2049–60.
[39] Peralta CA, Shlipak MG, Judd S, et al. Detection of chronic kidney disease with creatinine, cystatin C, and urine albumin-to-creatinine ratio and association with progression to end-stage renal disease and mortality. JAMA 2011;305:1545–52.
[40] Shlipak MG, Matsushita K, Ärnlöv J, et al. Cystatin C versus creatinine in determining risk based on kidney function. N Engl J Med 2013;369:932–43.
[41] Knight EL, Verhave JC, Spiegelman D, et al. Factors influencing cystatin C levels other than renal function and the impact on renal function measurement. Kidney Int 2004;65:1416–21.
[42] Grubb A, Björk J, Nyman U, et al. Cystatin C, a marker for successful aging and glomerular filtration rate, is not influenced by inflammation. Scand J Clin Lab Invest 2011;71:145–9.
[43] Grubb A, Lindström V, Kristensen K, et al. Filtration quality: a new measure of renal disease. Clin Chem Lab Med 2007;45(Suppl.):S273–4.
[44] Grubb A, Lindström V, Jonsson M, et al. Reduction in glomerular pore size is not restricted to pregnant women. Evidence for a new syndrome: "Shrunken pore syndrome". Scand J Clin Lab Invest 2015;75:333–40.
[45] Abrahamson M, Islam MQ, Szpirer J, et al. The human cystatin C gene (CST3), mutated in hereditary cystatin C amyloid angiopathy, is located on chromosome 20. Hum Genet 1989;82:223–6.
[46] Rocchi M, Covone A, Romeo G, et al. Regional mapping of RBP4 to 10q23-q24 and RBP1 to 3q21-q22 in man. Somat Cell Molec Genet 1989;15:185–90.
[47] Goodfellow PN, Jones EA, Van Heyningen V, et al. The beta-2-microglobulin gene is on chromosome 15 and not in the HL-A region. Nature 1975;254:267–9.
[48] White DM, Mikol DD, Espinosa R, et al. Structure and chromosomal localization of the human gene for a brain form of prostaglandin D-2 synthase. J Biol Chem 1992;267:23202–8.
[49] Abbink FC, Laarman CA, Braam KI, et al. Beta-trace protein is not superior to cystatin C for the estimation of GFR in patients receiving corticosteroids. Clin Biochem 2008;41:299–305.
[50] Gobin SJ, Biesta P, Van den Elsen PJ. Regulation of human beta 2-microglobulin transactivation in hematopoietic cells. Blood 2003;101:3058–64.
[51] Hokland M, Larsen B, Heron I, Plesner T. Corticosteroids decrease the expression of beta 2-microglobulin and histocompatibility antigens on human peripheral blood lymphocytes in vitro. Clin Exp Immunol 1981;44:239–46.

[52] Kotnik P, Fischer-Posovszky P, Wabitsch M. RBP4: a controversial adipokine. Eur J Endocrinol 2011;165:703–11.
[53] Juraschek SP, Coresh J, Inker LA, et al. Comparison of serum concentrations of β-trace protein, β2-microglobulin, cystatin C, and creatinine in the US population. Clin J Am Soc Nephrol 2013;8:584–92.
[54] Lund U, Rippe A, Venturoli D, et al. Glomerular filtration rate dependence of sieving of albumin and some neutral proteins in rat kidneys. Am J Physiol 2003;284:F1226–34.
[55] Strevens H, Wide-Swensson D, Torffvit O, Grubb A. Serum cystatin C for assessment of glomerular filtration rate in pregnant and non-pregnant women. Indications of altered filtration process in pregnancy. Scand J Clin Lab Invest 2002;62:141–7.
[56] Kristensen K, Lindström V, Schmidt C, et al. Temporal changes of the plasma levels of cystatin C, beta-trace protein, b_2-microglobulin, urate and creatinine during pregnancy indicate continuous alterations in the renal filtration process. Scand J Clin Lab Invest 2007;67:612–8.
[57] Strevens H, Wide-Swensson D, Grubb A. Serum cystatin C is a better marker for pre-eclampsia than serum creatinine or serum urate. Scand J Clin Lab Invest 2001;61:575–80.
[58] Kristensen K, Wide-Swensson D, Schmidt C, et al. Cystatin C, beta-2-microglobulin and beta-trace protein in pre-eclampsia. Acta Obstet Gynecol Scand 2007;86:921–6.
[59] Dardashti A, Nozohoor S, Grubb A, Bjursten H. Shrunken Pore Syndrome is associated with a sharp rise in mortality in patients undergoing elective coronary artery bypass grafting. Scand J Clin Lab Invest 2016;76:74–81.
[60] Purde MT, Nock S, Risch L, et al. The cystatin C/creatinine ratio, a marker of glomerular filtration quality: associated factors, reference intervals, and prediction of morbidity and mortality in healthy seniors. Transl Res 2016;169:80–90.
[61] Koopman MG, Koomen GC, Krediet RT, et al. Circadian rhythm of glomerular filtration rate in normal individuals. Clin Sci 1989;77:105–11.
[62] van Acker BA, Koomen GC, Koopman MG, et al. Discrepancy between circadian rhythms of inulin and creatinine clearance. J Lab Clin Med 1992;120:400–10.
[63] Larsson A, Akerstedt T, Hansson LO, Axelsson J. Circadian variability of cystatin C, creatinine, and glomerular filtration rate (GFR) in healthy men during normal sleep and after an acute shift of sleep. Chronobiol Int 2008;25:1047–61.
[64] Bosch JP, Saccaggi A, Lauer A, et al. Renal functional reserve in humans. Effect of protein intake on glomerular filtration rate. Am J Med 1983;75:943–50.
[65] Englund MS, Berg UB, Arfwidson K. Renal functional reserve in transplanted and native single kidneys of children and adults. Pediatr Nephrol 1997;11:312–7.
[66] ter Wee PM, Geerlings W, Rosman JB, et al. Testing renal reserve filtration capacity with an amino acid solution. Nephron 1985;41:193–9.
[67] Fuhrman DY, Maier PS, Schwartz GJ. Rapid assessment of renal reserve in young adults by cystatin C. Scand J Clin Lab Invest 2013;73:265–8.

CHAPTER SIX

Biomarkers in Acute Kidney Injury

C.L. Edelstein, MD, PhD
Division of Renal Diseases and Hypertension, University of Colorado Denver, Aurora, CO, United States

Contents

Biomarkers of Kidney Disease. http://dx.doi.org/10.1016/B978-0-12-803014-1.00006-6

INTRODUCTION

Blood urea nitrogen (BUN) and serum creatinine (SCr) are not very sensitive and specific markers of kidney function in acute kidney injury (AKI) as they are influenced by many renal and nonrenal factors independent of kidney function. A biomarker that is released into the blood or urine by the injured kidney and is analogous to the troponin release by injured myocardial cells after myocardial ischemia or infarction, may be a more sensitive and specific early marker of AKI than BUN and SCr. Numerous biomarkers of AKI have been detected in the urine or serum of patients with AKI. Urine interleukin-18 (IL-18), urine neutrophil gelatinase-associated lipocalin (NGAL), urine kidney injury molecule-1 (KIM-1), and urine liver-type fatty acid-binding protein (L-FABP) are biomarkers of renal tubular injury in early AKI and identify patients with AKI before the rise in BUN or SCr. Urine IL-18, urine NGAL, and urine KIM-1 are biomarkers of poor prognosis in patients with AKI. A combination of urinary tissue inhibitor of metalloproteinases-2 (TIMP2) and insulin-like growth factor-binding protein 7 (IGFBP7), which is known as NephroCheck is the first FDA-approved biomarker of AKI. Urinary

[TIMP2] × [IGFBP7] of greater than 0.3 is an early biomarker of AKI and predicts outcomes. In the future, the combined use of functional and damage markers may advance the field of biomarkers of AKI. More studies are required to definitively demonstrate the association between early kidney damage biomarkers and clinical outcomes and to determine whether randomization to a treatment for AKI based on high structural/damage biomarker levels results in an improvement in kidney function and improves clinical outcomes.

DEFINITION AND CLASSIFICATION OF AKI

AKI is common and the absolute incidence of AKI has increased during the last decade [1,2]. Between 5 and 20% of critically ill patients in the intensive care unit (ICU) have an episode of AKI. Up to 4.9 % of critically ill patients in the ICU will require renal replacement therapy (RRT). AKI requiring RRT in the ICU has a high mortality of over 50%. The most common causes of AKI in the ICU are septic shock, ischemia, and nephrotoxins.

AKI has been defined conceptually as a rapid decline in the glomerular filtration rate (GFR) that occurs over hours or days [3,4]. In the RIFLE criteria for AKI, AKI is defined as a 50% increase in SCr from the baseline [4,5] (Fig. 6.1). The RIFLE classification of AKI divides AKI into the following stages (1) risk, (2) injury, (3) failure, (4) loss of function, and (5) end-stage kidney disease (ESKD) [3–5] (Fig. 6.1). The term AKI replaces the term acute renal failure (ARF) and ARF is restricted to patients that have AKI and need RRT. The RIFLE criteria have been validated in multiple studies [3–5].

The acute kidney injury network (AKIN) has also developed a classification of AKI [4,5] (Table 6.1). The AKIN group attempted to increase the sensitivity of the RIFLE criteria by recommending that a smaller change in SCr (0.3 mg/dL) be used as a threshold to define the presence of AKI and identify patients with Stage 1 AKI (analogous to RIFLE—Risk) [4,5]. In the AKIN classification of AKI, a time of 48 h over which AKI occurs (compared to 1–7 days for the RIFLE criteria) was proposed. In addition, patients receiving RRT were classified as Stage 3 AKI (RIFLE—Failure). The AKIN criteria differ from the RIFLE criteria in several ways: (1) The RIFLE criteria are defined as changes within 7 days, whereas the AKIN criteria use changes over 48 h, (2) The AKIN classification includes less

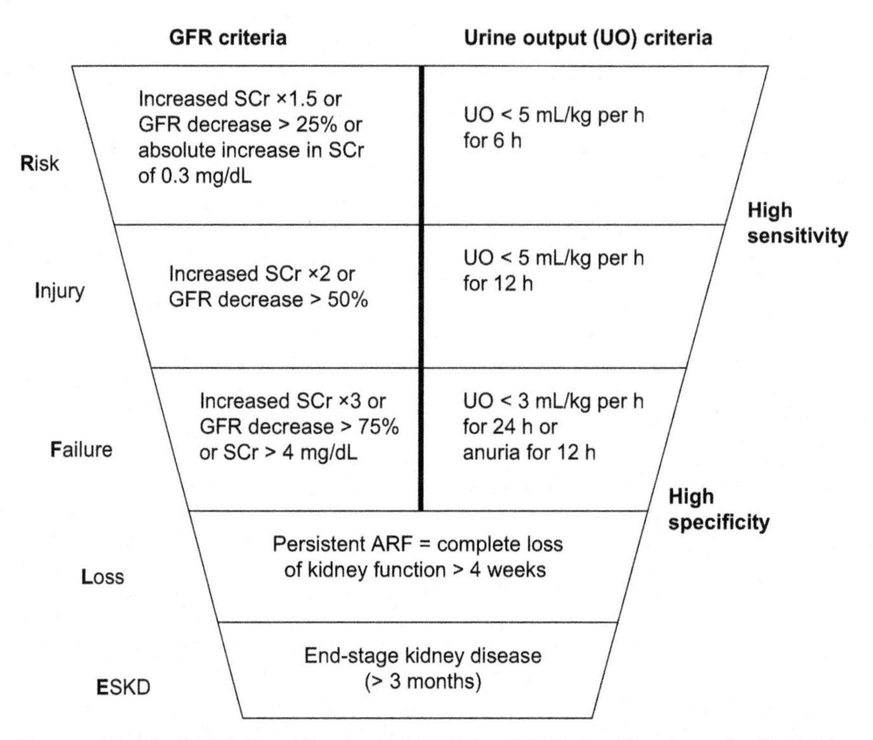

Figure 6.1 ***The RIFLE Classification of AKI.*** The RIFLE classification of AKI divides AKI into the five stages depending on kidney function as determined by serum creatinine (SCr) and urine output. The RIFLE criteria have been validated in multiple studies. Most biomarker studies in AKI have used the RIFLE or AKIN classification of AKI.

severe injury in the criteria, (3) AKIN avoids using GFR as a marker in AKI, as there is no dependable way to measure GFR in AKI and equations to measure GFR in AKI are not reliable if the SCr change is not in a steady state, and (4) AKIN notes that the volume status should be optimized and urinary tract obstructions be excluded when using oliguria as diagnostic criteria.

The Kidney Disease Improving Global Outcomes (KDIGO) classification of AKI builds on the RIFLE and AKIN classifications. The KDIGO classification has both the increase in SCr (0.3 mg/dL) over 48 h and the 1.5–1.9-fold increase in SCr known or presumed to have occurred over 1–7 days (Table 6.1).

Most of the studies referenced in this chapter have used the RIFLE, AKIN, or KDIGO definitions of AKI.

Table 6.1 The AKIN and Kidney Disease Improving Global Outcomes (KDIGO) classifications of AKI

Stage	Kidney function	UO
Stage 1	Increase in SCr ≥ 0.3mg/dL (within 48 h; AKIN and KDIGO) or increase to ≥ 150–199% (1.5–1.9-fold) from baseline (within 1–7 days; KDIGO)	<0.5 mL/kg per h for ≥6 h
Stage 2	Increase in SCr to 200–299% (>2–2.9 fold) from baseline	<0.5 mL/kg per h for ≥12 h
Stage 3	Increase in SCr to ≥ 300% (≥3-fold) from baseline or SCr ≥ 4 mg/dL with an acute rise of at least 0.5 mg/dL or initiation of RRT	<0.3 mL/kg per h for ≥ 24 h or anuria ≥ 12 h

A change in SCr of 0.3 mg/dL is used to define the presence of AKI and identify patients with Stage 1 AKI (analogous to RIFLE—Risk). In the AKIN classification of AKI, a time of 48 h over which AKI occurs (compared to 1–7 days for the RIFLE criteria) was proposed. Patients receiving renal replacement therapy (RRT) were classified as Stage 3 AKI (RIFLE—Failure). Most biomarker studies in AKI have used the RIFLE or AKIN classification of AKI.

SERUM CREATININE IN AKI

SCr and BUN have typically been used to diagnose AKI. Creatinine is a small molecule of 113 Da. It is distributed in the body throughout the body water. It is generated in muscle from the nonenzymatic conversion of creatine and phosphocreatine. The reasons why SCr is not sensitive or specific for the diagnosis of AKI will be outlined further.

Interference with the creatinine assay may give false SCr values [6]. The classic method of creatinine measurement is the Jaffe reaction that uses a colorimetric method, which detects creatinine, as well as, noncreatinine chromogens in the serum. In the Jaffe reaction, creatinine acts directly with the picrate ion under alkaline conditions to form a red–orange complex. Up to 20% of the color reaction may be due to substances other than creatinine, for example, glucose, uric acid, ketones, cephalosporines, furosemide, hemoglobin, paraproteins, paraquat, and diquat [6]. Plasma ketosis and cephalosporines may result in an increase in SCr due to interference with the picric acid assay for creatinine. Very-high bilirubin levels can cause false low SCr. The kinetic alkaline picrate method is the most widely used method for creatinine determination in clinical laboratories in the United States. This method reduces interference from noncreatinine chromogens.

SCr may change due to nonrenal factors independent of kidney function, for example, age, gender, race, muscle mass, nutritional status, total parentral nutrition, and infection [6,7]. Vigorous prolonged exercise may result in increased SCr due to an increase in muscle-creatinine generation. Ingestion of creatine supplements may increase SCr. Ingestion of cooked meat may increase SCr as during cooking creatine in meat is converted to creatinine, which is absorbed by the gastrointestinal tract. Restriction of dietary protein may result in a decrease in SCr.

SCr may change due to renal factors that are independent of kidney function. For example, several medications, such as, trimethoprim, cimetidine, and salicylates alter the tubular secretion of creatinine leading to changes in SCr independent of GFR [6,7]. In addition, SCr is not sensitive to the loss of kidney reserve as evidenced by the small change in SCr after the loss or donation of one kidney with a normal remaining kidney [8]. Alterations in SCr may lag several days behind actual changes in GFR [7,9].

BUN is also suboptimal for the diagnosis of AKI and is also dependent on nonrenal factors independent of kidney function, for example, protein intake, catabolic state, upper gastrointestinal bleeding, volume status, and therapy with high-dose steroids [6,10–12]. Thus alterations in SCr and BUN in AKI are not particularly sensitive or specific for small changes in GFR.

A biomarker that is released into the blood or urine by the injured kidney and is analogous to the troponin release by injured myocardial cells after myocardial ischemia or infarction, may be a more sensitive and specific marker of AKI than BUN and SCr. In addition, earlier detection of AKI with a kidney-specific biomarker may result in an earlier nephrology consultation, more optimal dosing of antibiotics, avoidance of nephrotoxic agents, and even earlier specific therapies for the repair of the damaged kidney. Earlier diagnosis of AKI may identify patients with mild AKI that have increases in SCr in the normal range that may not be recognized by clinicians, for example, an increase in serum creatnine from 0.4 to 0.8 mg/dL. An ideal biomarker of AKI would allow the early detection of kidney injury before an increase in SCr and/or BUN. It would differentiate AKI from acute glomerulonephritis or acute interstitial nephritis, would predict the need for dialysis, mortality, and long-term kidney outcome and would be useful to monitor the effects of an intervention or treatment.

Major interventional trials in AKI, for example, anaratide [13,14] and fenoldopam [15], have failed in humans. A possible reason for their failure in AKI is the late intervention in the course of AKI due to the dependence on SCr and BUN for diagnosis.

In this chapter, the use of biomarkers to assist the differential diagnosis of AKI, to predict short- and long-term outcomes, and to allow risk stratification in AKI will be discussed. Biomarkers of AKI in a transplanted kidney will be discussed separately in Chapter 8.

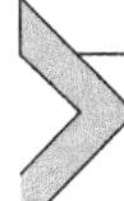

BIOLOGY OF BIOMARKERS

IL-18

Caspase-1 (previously known as interleukin-1β-converting enzyme or ICE) activates the proinflammatory cytokines IL-1β and IL-18 [16]. Caspase-1-deficient mice developed less ischemic AKI as determined by SCr and acute tubular necrosis (ATN) scores than wild-type mice [17]. As IL-1β does not play an injurious role in ischemic AKI in mice [18], a lack of the active form of IL-18 was investigated as the mechanism of the protection against AKI in caspase-1-deficient mice. In an electrochemiluminescence assay of whole kidney, IL-18 increased more than 100% in wild-type AKI as compared to sham-operated controls. On immunoblot analysis, there was a conversion of the precursor to the active form of IL-18 in AKI wild-type mice, but not in the caspase-1-deficient AKI mice and sham-operated controls. To determine whether IL-18 plays an injurious role in ischemic AKI, wild-type mice were injected with IL-18-neutralizing antiserum prior to the ischemic insult. IL-18-neutralizing antiserum-treated mice were protected against AKI to a similar degree as caspase-1-deficient mice. The study concluded that IL-18 is a mediator of ischemic AKI in mice [17,19]. The use of IL-18-binding protein (IL-18BP) in two subsequent studies, involving transgenic mice or IL-18-deficient mice, has confirmed that IL-18 is a mediator of ischemic AKI in mice [20,21].

Immunohistochemistry of mouse kidneys demonstrated an increase in IL-18 protein in injured tubular epithelial cells in AKI kidneys compared to normal controls. In a separate study using freshly isolated proximal tubules from mice, it was determined that hypoxic proximal tubules had high levels of IL-18 [22]. On the basis of the demonstration of IL-18 in injured proximal tubules, IL-18 was measured in the urine. Urine IL-18 was increased in mice with ischemic AKI compared to sham-operated mice [17]. Thus the hypothesis was developed that IL-18 could be released from the injured tubular epithelial cells into the urine and serve as a urinary biomarker of AKI in humans.

IL-18 is a proinflammatory cytokine that plays a role in both the innate- and acquired immune responses [23,24]. IL-18 plays an important role in host defenses against tumors and infections. Activated macrophages express high levels of IL-18. A wide variety of cells express IL-18, including, mononuclear cells, keratinocytes, osteoblasts, intestine, renal epithelial cells, and dendritic cells. Neutralization of IL-18 has potential therapeutic effects. Blockade of IL-18 using neutralizing antibodies [25], exogenous IL-18BP [26], or caspase-1 inhibition [27] protects mice from liver necrosis. Exogenously administered recombinant human IL-18BP is therapeutically effective in a mouse model of collagen-induced arthritis [28]. IL-18BP reduces ischemic dysfunction in a suprafused human atrial myocardium model [29]. Strategies to block IL-18 using IL-18BP are underway in clinical trials of rheumatoid arthritis [30].

In summary, the proinflammatory cytokine IL-18 is both a mediator and a biomarker of ischemic AKI. IL-18 is a mediator of ischemic AKI in mice as evidenced by the studies that IL-18 expression increases in the kidney in AKI and inhibition of IL-18 is protective against AKI in mice [17,19–21].

NGAL

NGAL is a 21-kDa protein of the lipocalin superfamily. NGAL is a critical component of innate immunity to bacterial infection and is expressed by immune cells, hepatocytes, and renal tubular cells in various disease states [31]. NGAL is a small-secreted polypeptide that is protease resistant and thus may be easily detected in the urine.

NGAL protein increases massively in the renal tubules and in the first UO after ischemic AKI in rats and mice [32]. The appearance of NGAL in the urine preceded the appearance of other urinary markers, such as, the tubular proteins *N*-acetyl-β-D-glucosaminidase (NAG) and β2-microglobulin (β2M). Studies in cultured human proximal tubule cells subjected to in vitro hypoxic injury confirmed the origin of NGAL from tubule cells. NGAL was also detected in the urine of mice in the early stage of cisplatin-induced nephrotoxicity [33]. These animal studies demonstrated that NGAL may represent an early, sensitive, noninvasive urinary biomarker for ischemic- and nephrotoxic kidney injury.

An NGAL reporter mouse that accurately recapitulates the endogenous message and demonstrates injury in vivo in real time was developed [34]. In the kidney, NGAL imaging showed a sensitive, rapid, dose-dependent, reversible, and organ- and cell-specific relationship with tubular stress, which correlated with the level of urinary NGAL. Unexpectedly, tubular cells of the distal nephron were the source of urinary NGAL.

KIM-1

KIM-1 is a putative epithelial cell-adhesion molecule containing a novel immunoglobulin domain. KIM-1 mRNA and protein are expressed at a low level in normal kidney but are increased dramatically in postischemic kidney [35]. KIM-1 has recently been identified as the first nonmyeloid-phosphatidylderine receptor that confers a phagocytic phenotype on injured epithelial cells both in vivo and in vitro [36].

Urinary KIM-1 is a noninvasive, rapid, sensitive, and reproducible biomarker for the early detection of both cisplatin-induced AKI and ischemic AKI in rats [37]. In this study, a sandwich-KIM-1 ELISA test was used. At 1 day after cisplatin administration, there was a three- to fivefold increase in the urinary KIM-1 compared to plasma creatinine, BUN, urinary NAG, glycosuria, and proteinuria that were not increased in the urine. At 24 h of postischemic reperfusion, after 10 min of bilateral renal pedicle clamping, urine KIM-1 levels were 10-fold higher than control rats and plasma creatinine, BUN, glycosuria, proteinuria, and urinary NAG levels were not yet increased.

KIM-1 is also a tissue and urinary biomarker for nephrotoxicant-induced kidney injury. Tissue and urinary expression were measured with three different nephrotoxins in the rat: S-(1,1,2,2-tetrafluoroethyl)-L-cysteine, folic acid (FA), and cisplatin. Marked increases in KIM-1 expression localized to proximal tubule epithelial cells were detected. In addition, KIM-1 protein was detected in urine of nephrotoxin-treated rats [38]. KIM-1 is a sensitive and tissue-specific biomarker of early AKI compared to BUN, SCr, and NAG in rats injected with gentamicin, mercury, or chromium [39]. A rapid dipstick test for the detection of urinary KIM-1 (rat) or KIM-1 (human) has been developed [40]. On dipstick, the urinary KIM-1-band intensity significantly correlated with levels of KIM-1 in a dose-and time-dependent manner as measured by histopathologic damage and immunohistochemical assessment of renal KIM-1. KIM-1 was detected in rats with cadmium, gentamicin, or ischemic AKI. In humans, the urinary KIM-1-band intensity was significantly greater in patients with AKI compared to healthy volunteers. KIM-1 was measured in rats with adriamycin-induced nephropathy before and after angiotensin-converting enzyme (ACE) inhibition [41]. Renal and urinary KIM-1 correlated with proteinuria and interstitial damage. Reduction of proteinuria correlated with a decrease in renal and urinary KIM-1. KIM-1 has been accepted by the Food and Drug Administration and European Medicines Agency as a highly sensitive and specific urinary biomarker to monitor drug-induced kidney injury in preclinical studies and on a case-by-case basis in clinical trials.

Tubular Enzymes

The apical membrane of proximal tubular epithelial cells contains numerous microvilli that form the brush border. The brush border contains enzymes that carry out the specialized functions of the proximal tubule. Intracellular enzymes can be released into the urine due to injury either by exocytosis or leakage. The detection of proteins, especially enzymes, released from damaged proximal and/or distal tubular cells has been used as a biomarker of AKI for many years. Examples of tubular enzymes are glutathione-S-transferase (GST) isomers that are cytoplasmic enzymes found in proximal and distal tubular cells, NAG that is a lysosomal enzyme found mostly in proximal tubules, and alkaline phosphatase (AP) and γ-glutamyl transpeptidase (γGT).

Cystatin C

Butler et al., in 1961, studied the urine proteins of 223 individuals by starch-gel electrophoresis and found a new urine protein fraction in the post-γ-globulin fraction [42]. This protein was named cystatin C. Cystatin C is a 13 kDa protein produced by all nucleated cells. It is a polypeptide chain with 120 amino acid residues. It is freely filtered by the glomerulus, completely reabsorbed by the proximal tubules, and is not secreted by the renal tubules [43]. Thus some of the limitations of SCr, for example, effect of muscle mass, diet, gender, and tubular secretion may not be a problem with cystatin C. Cystatin C is best measured by an immunonephelometric assay.

In studies using Cr–EDTA clearance as the reference standard, the blood concentration of cystatin C, was identified as a measure of GFR [44]. Cystatin C is a better marker of GFR than SCr as demonstrated in the following studies: serum cystatin C- and cystatin C-based formulae were as good in estimating GFR as the Modification of Diet in Renal Disease (MDRD) formula [45]. Uzan et al. studied the diagnostic significance of cystatin C using noncreatinine measures of GFR. Serum (99m)Tc–DTPA clearance was compared with serum cystatin C, creatinine, β2M levels, and creatinine clearance in a group of patients with GFRs of 10–60 mL/min per 1.73 m^2 and healthy controls. Reference clearance, determined by serum (99m)Tc–DTPA, was best correlated with creatinine clearance ($r = 0.957$) and cystatin C ($r = 0.828$), compared to β2M ($r = 0.767$) and creatinine ($r = 0.682$). As these patients had a GFR < 60, it was concluded that the serum cystatin C level can be used as a marker of GFR in patients with kidney failure [46]. Artunc et al. compared SCr, serum cystatin C, and the clearance of the iodinated contrast dye iopromide (reference standard) in 127 patients undergoing cardiac catheterization. Serum cystatin C showed a higher nonparametric correlation ($r = 0.805$) to the iopromide clearance compared to SCr ($r = 0.652$) and compared to GFR estimated by the Cockcroft–Gault formula ($r = 0.690$). A serum cystatin C value of greater than 1.3 mg/L demonstrated 88% sensitivity and 96% specificity for the detection of kidney failure (an iopromide clearance of less than 80 mL/min per m^2) [47]. At a multinational meeting held in 2002 in Germany [48], it was decided that (1) cystatin C is at least equal, if not superior, to SCr as a marker of GFR; (2) the independence from height, gender, age, and muscle mass is advantageous; and (3) select patient groups, such as, children, the elderly, and patients with reduced muscle mass may benefit from its use as a marker of GFR.

Cystatin C is discussed in more detail in Chapter 5.

L-FABP

FABPs are a family of carrier proteins for fatty acids and other lipophilic substances, such as, eicosanoids and retinoids. FABPs facilitate the transfer of fatty acids between extra- and intracellular membranes. Some FABPs transport lipophilic molecules from the outer cell membrane to certain intracellular receptors, such as PPAR. L-FABP binds fatty acids and transports fatty acids to the mitochondria or peroxisomes, where the fatty acids are metabolized via β-oxidation and provide energy for tubular epithelial cells. Besides kidney, FABPs have been identified in liver, intestine, muscle, heart, adipocytes, skin, ileum, brain, myelin, and testis.

L-FABP, was correlated with the degree of tubulointerstitial damage in a model of FA-induced nephropathy in mice [49]. The protein expression levels of human L-FABP in both the kidney and urine significantly correlated with the degree of tubulointerstitial damage, the infiltration of macrophages, and the deposition of type I collagen.

Urinary L-FABP was measured in mice with ischemic AKI and cisplatin-induced AKI [50]. In both ischemic AKI and cisplatin-induced AKI, urinary L-FABP was increased in the urine before the increase in BUN. Renal histology scores worsened with longer ischemic time or with the increased dose of cisplatin. In both AKI models, urinary L-FABP showed a better correlation with histology injury scores and GFR, as measured by fluorescein isothiocyanate-labeled inulin injection, than with BUN and urinary NAG.

L-FABP was evaluated as a biomarker of renal ischemia in both human kidney transplant patients and a mouse model of AKI [51]. In 12 living, related kidney-transplant patients, intravital video analysis of peritubular capillary blood flow was performed immediately after reperfusion of the transplanted organs. A significant direct correlation was found between urinary L-FABP level and both peritubular capillary blood flow and the ischemic time of the transplanted kidney, as well as, hospital stay. Human L-FABP transgenic mice demonstrated lower BUN levels and less histologic injury than injured wild-type mice. In addition, human-L-FABP-transgenic mice subjected to AKI demonstrated the transition of L-FABP from the cytoplasm of proximal tubular cells to the tubular lumen on immunohistochemistry. These data show that increased urinary L-FABP after ischemic reperfusion injury is a biomarker of AKI.

Netrin

Netrins are laminin-like molecules with a distinctive domain organization. Netrins belong to the laminin-related family of axon-guidance molecules. Netrin-1, -3, and -4 are encoded by distinct genes. Mouse netrin-1 shares 52% amino acid identity with mouse netrin-3. Netrins act via two receptors, "deleted in colon cancer" (DCC) and UNC5. Netrins play a role in axonal guidance including development of mammary gland, lung, pancreas, and blood vessels; inhibition of leukocyte migration; and chemoattraction of endothelial cells. Netrin-1 is a potent inhibitor of leukocyte chemoattraction. The kidney has high levels of netrin expression.

The role of netrin in ischemic AKI in mice was determined [52]. In ischemic AKI, netrin-1 and -4 mRNA expression was downregulated while expression of netrin-3 was upregulated. Netrin-1-protein levels were increased between 3 and 24 h of reperfusion. Immunolocalization showed that netrin-1 increased in tubular epithelial cells early in AKI. Administration of recombinant netrin-1 significantly improved kidney function and histology, suggesting that the downregulation of netrin-1 in vascular endothelial cells may promote endothelial cell activation and infiltration of leukocytes into the kidney resulting in tubular injury. In another study, it was demonstrated that netrin-1 overexpression protects against ischemic AKI in mice by inhibition of apoptosis [53].

As netrin-1 expression is increased early in the tubules during ischemic AKI, netrin-1 was investigated as an early biomarker of AKI [54]. Urinary netrin-1 excretion was determined in ischemic-, cisplatin-, FA-, and endotoxin-induced AKI in mice. Urinary netrin-1 levels increased markedly within 3 h of ischemia reperfusion, reached a peak level at 6 h, and returned near to baseline by 72 h. SCr significantly increased only after 24 h of reperfusion.

Exosomes

Urinary exosomes can be released from every segment of the nephron, including podocytes. Exosomes are 50–90 nm vesicles. An exosome is created inside the cell when a segment of the cell membrane invaginates and is endocytosed. The internalized segment is broken into smaller vesicles that can be expelled from the cell. The released vesicles are called exosomes. Exosomes consist of a lipid raft and are secreted by cells under normal and pathologic conditions under the control of RNA called "exosomal shuttle RNA." The detection of urinary exosomal transcription factors may provide an understanding of the cellular regulatory pathways, in addition to, being biomarkers of disease.

Exosomes were isolated by differential centrifugation in rats and humans with AKI [55]. The exosomes were found to contain activating transcription factor 3 (ATF3) detected by western blot. ATF3 was found in the concentrated exosomal fraction but not in whole urine. ATF3 was present in urine exosomes in rat models of AKI before the increase in SCr. ATF3 was found in exosomes isolated from patients with AKI but not from patients with chronic kidney disease (CKD) or controls. Exosomal Fetuin-A was increased in rats with cisplatin-induced AKI compared to control rats [56]. By immunoelectron microscopy and elution studies, Fetuin-A was localized inside urinary exosomes. Urinary exosomal Fetuin-A was increased in three ICU patients with AKI compared to patients without AKI. The study concluded that proteomic analysis of urinary exosomes can provide candidate biomarkers for the diagnosis of AKI [56]. Na/H exchanger isoform 3 in urinary exosomal vesicles is a marker of AKI in humans [57]. The discovery of urinary exosomal vesicles has opened a new field of biomarker research [58].

TIMP2 and IGFBP7

TIMP2 is a human gene and a member of a gene family that encodes proteins that are natural inhibitors of the matrix metalloproteinases. Metalloproteinases are peptidases that play a role in degradation of the extracellular matrix. Also, TIMP2 can suppress proliferation of endothelial cells in response to angiogenic factors TIMP2 also inhibits protease activity in tissues undergoing remodeling of the extracellular matrix.

IGFBP7 regulates the availability of insulin-like growth factors in tissues and modulates IGF binding to its receptors. IGFBP7 stimulates cell adhesion and cancer growth.

TIMP2 and IGFBP7 are also markers of cell-cycle arrest. Renal tubular cells enter a period of G1 cell-cycle arrest after ischemia or sepsis. TIMP2 and IGFBP7 are involved in G1 cell-cycle arrest during the early phase of cell injury. The G1 cell-cycle arrest may prevent the division of cells with DNA damage until the DNA damage is repaired [59]. The role of the cell cycle in AKI is now an area of interest. Quiescent cells are normally in the $G_{0.}$ phase. For the process of cell division and repair, the cell enters the cell cycle in a specific time schedule. Arrest of a cell in the G_1- or G_2 phase, favors a hypertrophic and fibrotic phenotype. Exit of the cell in the late G_1 phase leads to apoptosis. Cyclins and cyclin-dependent kinases control each phase of the cell cycle. There is a protective response, during which cells enter cell-cycle arrest thus avoiding cell division during stress and injury. Both TIMP2 and IGFBP7 are involved in the G_1 cell-cycle arrest phase that occurs during the very early phases of cellular stress. Detection of cell-cycle arrest may serve as a biomarker of impending tubular damage in AKI. It is also believed that manipulation of the cell cycle may have therapeutic potential [60]. In summary, the following model is proposed for the involvement of TIMP2 and IGFBP7 in AKI [60]. TIMP2 and IGFBP7 are expressed in the tubular cells in response to DNA damage and other forms of injury. TIMP2 and IGFBP7 block the effect of the cyclin-dependent protein-kinase complexes on cell-cycle promotion, which results in G1 cell-cycle arrest for short periods of time to prevent damaged cells from dividing.

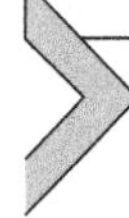

BIOMARKERS FOR THE DIFFERENTIAL DIAGNOSIS OF AKI (TABLE 6.2)

Urine IL-18

Urine IL-18 was measured in 72 patients and was significantly increased in patients with AKI versus normal controls, prerenal azotemia, urinary tract infection, CKD, and nephrotic syndrome [61]. This study in humans demonstrated the association of urine IL-18 with established tubular injury and formed the basis for examining urine IL-18 in more detail.

Urine NGAL

A multicenter study of serum NGAL was performed in 143 critically ill children with systemic inflammatory response syndrome (SIRS) or septic

Table 6.2 Biomarkers for the differential diagnosis of AKI

Situation	Biomarker	Number	Study	References
AKI	Urine IL-18	50	Increased in ATN versus normal, prerenal azotemia, urinary tract infection, CKD, and nephrotic syndrome	[61]
Children in ICU	Serum NGAL	143	Increased in sepsis and septic shock	[62]
Adults in ICU	Serum NGAL	151	Increased in sepsis without AKI	[63]
Adults in ICU	Serum NGAL	65	Increased in SIRS, sepsis, and septic shock	[64]
Adults and children	Serum- and urine NGAL		Increased in ADPKD, diarrhea-associated HUS, HIV infection, HIV nephropathy, HIV-associated HUS, and active lupus nephritis	[65–68]
Adults with liver failure	Urine NGAL and IL-18	55	The biomarker with best accuracy for ATN diagnosis was NGAL followed by IL-18	[69]
ED	Urine NGAL	635	Increased in AKI compared to patients with prerenal azotemia, CKD, or normal kidney function	[70]
Adults	Urine KIM-1	109	Increased in glomerulonephritis, acute rejection, chronic allograft nephropathy, SLE, diabetic nephropathy, and hypertension	[71]
Adults with ATN	Urine KIM-1	40	Higher in patients with ischemic AKI compared to other renal diseases, urine brush-border enzymes did not correlate with AKI	[72]
Adults	Urine brush-border enzymes		Increased in both acute- and chronic kidney conditions	[44,72]
Sepsis patients in the ICU	Urine brush-border enzymes	440	Urinary α-GST and π-GST are elevated early in sepsis syndrome, not predictive of AKI	[73]
Adults in ICU	Urine L-FABP	80	Increased in septic shock	[74]
Liver cirrhosis	Serum cystatin C	36	Accurate GFR marker, better marker of renal function than SCr	[75]
Liver cirrhosis	Serum cystatin C	97	Better marker of renal function than SCr	[76]

ADPKD, autosomal dominant polycystic kidney disease; ATN, acute tubular necrosis; CKD, chronic kidney disease; ED, emergency department; HUS, hemolytic uremic syndrome; ICU, intensive care unit; L-FABP, liver-type fatty acid-binding protein; SCr, serum creatinine; SLE, systemic lupus erythematosus.

shock during the first 24 h of admission to the pediatric ICU [62]. There was a significant difference in serum NGAL between healthy children, critically ill children with SIRS, and critically ill children with septic shock. Serum NGAL was significantly increased in critically ill children with AKI compared with those without AKI. The study concludes that serum NGAL is a highly sensitive but nonspecific predictor of AKI in critically ill children with septic shock. However, other studies have demonstrated that serum NGAL should be used with caution as a biomarker of AKI in patients with sepsis. In 151 adults in the ICU with or without sepsis, it was found that serum NGAL is increased in patients with sepsis in the absence of AKI and should be used with caution as a marker of AKI in septic ICU patients [63]. In another study of serum NGAL, the influence of sepsis on NGAL was determined. Sixty-five patients admitted to a general ICU with normal SCr were studied [64]. Data from 27 patients with SIRS, severe sepsis, or septic shock without AKI, and 18 patients with septic shock and concomitant AKI were analyzed. Serum NGAL was raised in patients with SIRS, severe sepsis, and septic shock. The study also concluded that plasma NGAL should be used with caution as a marker of AKI in ICU patients with septic shock.

NGAL is also increased in other conditions besides ischemic AKI. Serum- and urine-NGAL levels were increased in 26 patients with autosomal dominant polycystic kidney disease (ADPKD) and a significant correlation was found between urine- and plasma-NGAL levels and residual renal function [65]. In a study of 34 children with diarrhea-associated hemolytic uremic syndrome (HUS), the majority of patients with HUS had renal tubular epithelial injury as evidenced by elevated urinary NGAL, which was associated with higher BUN, SCr concentrations, and a more frequent need for dialysis [66]. In a study of HIV-infected children, elevated levels of NGAL were found in HIV-associated nephropathy and HIV-associated HUS [67]. In a study of 85 patients, urinary NGAL, but not plasma NGAL, was found to be a biomarker of activity in lupus nephritis [68].

The sensitivity and specificity of a single urine NGAL measurement for diagnosing AKI was determined in 635 patients in the emergency department (ED) [70]. Patients with AKI had a significantly higher urine-NGAL level than patients with prerenal azotemia, CKD, or normal kidney function. At a NGAL-cutoff value of 130 μg/g creatinine, the sensitivity and specificity of NGAL for detecting acute injury were 0.900 (95% CI, 0.73–0.98) and 0.995 (CI, 0.990–1.00), respectively, and these values were superior to those for NAG, α1-microglobulin (α1M), α1-acid glycoprotein, fractional excretion of sodium (FENa), and SCr.

A biomarker that differentiates the different causes of AKI in liver-transplant patients would be useful to clinicians in managing AKI in liver-transplant patients. Fifty-five cirrhotic patients with acute decompensation were studied: 39 had AKI, 16 had acute decompensation without AKI, and 34 had acute or chronic liver failure [69]. Causes of AKI were prerenal, type-1 hepatorenal syndrome and ATN. A panel of 12 urinary biomarkers was assessed using a multiplex assay. The biomarker with best accuracy for ATN diagnosis was NGAL. Other attractive biomarkers for ATN diagnosis were IL-18, albumin, trefoil-factor-3, and π-GST. Biomarkers with less accuracy for ATN [area under the receiver operating characteristic (ROC) curve (AUROCC) < 0.8] were cystatin-C and KIM-1.

KIM-1

KIM-1 is also a biomarker of AKI in humans. Urine samples were collected from 32 patients with various acute and CKDs, as well as, from 8 normal controls. There was extensive expression of KIM-1 in proximal tubule cells in kidney biopsies from all six patients with biopsy-confirmed ATN. Urinary KIM-1 levels were significantly higher in patients with ischemic ATN compared to patients with other forms of ARF or CKD [72]. KIM-1 is also a sensitive biomarker of tubular injury in other renal diseases besides AKI. Renal KIM-1 and urine KIM-1 expression was significantly increased in human kidney tissue in patients with focal glomerulosclerosis IgA nephropathy, membranoproliferative glomerulonephritis, membranous glomerulonephritis, acute rejection, chronic allograft nephropathy, systemic lupus erythematosus (SLE), diabetic nephropathy, hypertension, and Wegener's granulomatosis compared to normal kidney tissue [71]. KIM-1 was primarily expressed at the luminal side of dedifferentiated proximal tubules, in areas with fibrosis and in areas of inflammation in macrophages. Renal KIM-1 positively correlated with renal damage, negatively with renal function, but not with proteinuria. Urinary KIM-1 was increased in the same group of patients and correlated positively with tissue KIM-1 and macrophages, negatively with renal function, but not with proteinuria. This study demonstrates that KIM-1 is upregulated in renal disease and is associated with renal fibrosis and inflammation and that urinary KIM-1 can be used as a noninvasive biomarker of tubular injury in multiple renal diseases.

Brush-Border Enzymes

Nearly 30 years ago, tubular enzymes in the urine were measured as a biomarker of AKI [77]. In acute tubular disorders, for example, renal failure from acute pancreatitis, the concentrations of α1M were high in patients with acute tubular injury compared to normal controls. Tubular enzymuria may be very sensitive to tubular injury from multiple causes. Dipeptidyl aminopeptidase was increased in the urine in patients with tubulointerstitial nephritis and chronic glomerulonephritis [44]. Of the five brush-border enzymes investigated, AP was the most sensitive to detect contrast nephropathy [78]. Urinary α- and π-GST did not show good discrimination for early detection of acute kidney injury following cardiopulmonary bypass [79]. Concentrations of urinary brush-border enzymes, such as, γ-glutamyl transferase (GGT) and AP, were increased in both acute and chronic kidney conditions [72]. In summary, measurement of tubular enzymuria is inexpensive and easy to measure. However, tubular enzymuria may be increased in multiple causes of tubular injury, including, ATN, acute rejection, and acute tubulointerstitial nephritis.

Serum Cystatin C

The following studies have determined the use of cystatin C as a marker of low GFR in patients with AKI. Changes in cystatin C occur sooner after changes in kidney function than SCr. Herget-Rosenthal studied patients after uninephrectomy for living kidney donation. Serum cystatin C increased 1 day after uninephrectomy compared to SCr that increased 2 days after uninephrectomy [8]. SCr concentration and calculated creatinine clearance are thought to be of limited value as GFR markers in patients with decompensated liver cirrhosis. Thirty-six patients with decompensated liver cirrhosis and 56 noncirrhotic controls were studied. Inulin clearance, serum cystatin C, and creatinine clearances were studied. Plasma-cystatin C concentration was found to be an accurate GFR marker in cirrhotic patients. Plasma-creatinine concentration and calculated creatinine clearance were of no practical value, as their reference values varied with the severity of the liver disease [75].

L-FABP

The presence of sepsis may increase some biomarkers in the absence of AKI. Urinary- and serum L-FABP were measured in 80 critically ill patients [74]. Urinary L-FABP levels in patients with septic shock were significantly higher than those in patients with severe sepsis without shock, patients with ARF, and healthy subjects ($p < 0.001$). Serum-L-FABP levels showed no significant differences between patients with septic shock, patients with severe sepsis, patients with ARF, and healthy subjects.

In 121 patients with newly diagnosed biopsy-proven chronic glomerulonephritis followed for 5 years, urinary L-FABP was increased in patients with proteinuria and predicts progression of renal function [80].

Urinary L-FABP is increased in AKI, as well as, other kidney diseases. The cut offs of urinary L-FABP that detect AKI versus other kidney diseases need to be determined.

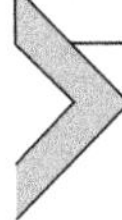

BIOMARKERS FOR THE EARLY DIAGNOSIS OF AKI (TABLE 6.3)

Urine IL-18

Studies in humans demonstrated that urine IL-18 is an early predictive biomarker of AKI [81]. The ARDS network had collected urine samples in patients after the initiation of mechanical ventilation. These urine samples were obtained and a nested case–control study within the ARDS network trial was performed to determine whether urinary IL-18 is an early diagnostic biomarker for AKI in critically ill patients in the ICU [82]. On multivariable analysis, urine IL-18 values predicted development of AKI (defined as a 50% increase in SCr) 24 and 48 h later. On diagnostic performance testing, urine IL-18 demonstrated an AUROCC of 73% to predict AKI in the next 24 h. The presence of sepsis in both control and AKI patients did not have a significant effect on urinary IL-18. On multivariable analysis, the urine IL-18 value on the day of initiation of mechanical ventilation for ARDS was a strong predictor of mortality [82].

Subsequently it was determined whether the finding that urine IL-18 is an early biomarker of AKI in critically ill adults could be reproduced in children. One hundred thirty-seven children with an average age of 6.5 years (53% male) were studied. The peak levels of IL-18 correlated with the severity of AKI by the pediatric RIFLE (pRIFLE) classification. In nonseptic AKI patients, urinary IL-18 was higher than controls at 2 days prior to a significant rise in SCr. Urinary IL-18 was associated with increased mortality. Urinary IL-18 was also increased in patients with sepsis. In conclusion, urinary IL-18 rises prior to SCr in nonseptic critically ill children, predicts severity of AKI, and is an independent predictor of mortality [88].

Table 6.3 Biomarkers for the early diagnosis of AKI

Study	Biomarker	Number	Result	AUROCC	References
Adults post-CPB, TRIBE-AKI	Urine IL-18 and NGAL, plasma NGAL	1219	Increased within 6 h after surgery	0.75–0.76	[83]
Children post cardiac surgery, TRIBE-AKI	Urine IL-18 and NGAL, plasma NGAL	311	Increased within 6 h after surgery	0.71–0.72	[84]
ARDS patients in ICU	Urine IL-18	138	Predicted development of AKI 24 and 48 h before AKI, strong predictor of mortality	0.73	[82]
Children post-CPB	Urine IL-18	55	Increased at 4–6 h, peaked at 12 h, and remained elevated up to 48 h	0.73	[85]
Adults post-CPB	Urine IL-18 and NGAL	33	Increased at 2–4 h postoperatively		[86]
Adults post-CPB	Urine IL-18	100	Did not predict AKI on arrival in ICU	0.53	[87]
Children in the ICU	Urine IL-18	137	Rises prior to SCr, predicts severity of AKI, independent predictor of mortality	0.31–0.77	[88]
Contrast nephropathy	Urine IL-18	51	Not increased at 24 and 72 h after cardiac catheterization		[89]
Children post-CPB	Urine NGAL	71	Independent predictor of AKI at 2 h after CPB	0.998	[90]
Children post-CPB	Plasma NGAL	71	Increase at 2 h correlated with change in SCr, duration of AKI, length of hospital stay, and mortality	0.96	[91]
Adults post-CPB	Urine NGAL	81	Increased at 3 and 18 h in patients who later developed AKI	0.8	[92]
Adults post-CPB	Plasma- and urine NGAL	50	Predicts AKI at 2 h post-CPB	0.8–0.96	[93]

Adults post-CPB	Urine NGAL	426	Peaked immediately and remained higher 3, 18, and 24 h after surgery	0.506–0.611	[94]
Adults post-CPB	Urine NGAL	90	Not a good predictor of AKI at 3 h postsurgery	0.59, 0.65	[95]
Contrast nephropathy	Urine- and serum NGAL	35	Increased at 2 and 4 h after contrast		[96]
Children in the ICU	Urine NGAL	140	Increases 2 days before a 50% or greater rise in SCr	0.78	[97]
Adults in ICU	Serum NGAL	88	Predicted development of AKI	0.956	[98]
Adults in the ICU	Plasma- and urine NGAL	194	Predicts AKI up to 72 h post-ICU admission	0.77–0.87	[99]
Adults post-CPB	Urine KIM-1	90	Predicts AKI immediately and 3 h post-CPB	0.68, 0.65	[95]
Children/adults post-CPB, TRIBE-AKI	Urine KIM-1	299/1203	Increased at 6–12 h, peaked at 2 days in adults, and 1 day in children		[100]
Miscellaneous metaanalysis	Urine KIM-1	2979	Early detection of AKI	0.86	[101]
Uninephrectomy for living kidney donation	Serum cystatin C	10	Increased after 1 day versus SCr that increased after 2 days		[8]
Hospitalized patients	Serum cystatin C	85	Increased 1 day earlier than SCr	0.82–0.97	[102]
Adults post-CPB	Urine cystatin C	72	At 6 h predicted AKI or RRT, plasma cystatin C did not predict AKI at 6 h	0.734	[103]
Critically ill children	Serum cystatin C	25	Better than SCr to identify a creatinine clearance of under 80 mL/min	0.792–0.851	[104]
Adults post-CPB, TRIBE-AKI	Urine cystatin C	1203	Not significantly associated with the development of AKI		[105]

(Continued)

Table 6.3 Biomarkers for the early diagnosis of AKI (*cont.*)

Study	Biomarker	Number	Result	AUROCC	References
Children post-CPB, TRIBE-AKI	Serum cystatin C	288	Predicts severe AKI following pediatric cardiac surgery		[106]
Children/adults post-CPB	Urine L-FABP	299/1203	Peaked within 6 h in both age groups		[100]
Children after CPB	Urine L-FABP	40	Increased at 4 and 12 h after CPB in patients that developed AKI		[107]
Adults in the ICU	GGT, AP, NAG, and α-GST	26	Predict AKI on admission to the ICU		[108]
Adults with AKI	Urine cystatin C and α1M	73	Predict requirement for dialysis	0.86–0.92	[109]
Children post-CPB, TRIBE-AKI	Urine albumin/creatinine ratios	294	Predicted development of AKI		[110]

AP, alkaline phosphatase; AUROCC, area under ROC curve; CPB, cardiopulmonary bypass; GGT, γ-glutamyl transferase; GST, α-glutathione-S-transferase; L-FABP, liver-type fatty acid-binding protein; α1M, α1-microglobulin; NAG, *N*-acetyl-β-D-glucosaminidase; NEP, neutral endopeptidase; RBP, retinol-binding protein; TRIBE-AKI, Translational Research Investigating Biomarkers Endpoints in Early Acute Kidney Injury.

AKI, as defined by a 50% increase in SCr, occurs in about 25% of patients after CPB. In this high-risk group, it was tested whether urine IL-18 is a predictive biomarker for AKI in children following CPB. Urine IL-18 increased at 4–6 h after CPB, peaked at over 25-fold at 12 h, and remained markedly elevated up to 48 h after CPB. In contrast, using SCr, AKI was detected only 48–72 h after CPB. The results indicate that IL-18 is an early, predictive biomarker of AKI after CPB [85]. On multivariate analysis, urine IL-18 was independently associated with the number of days in AKI among cases suggesting that it may be a marker of AKI severity [85].

In another study it was determined that urine IL-18 and NGAL predict AKI after cardiac surgery in adults [86]. Thirty-three patients undergoing CPB were classified as AKI (50% increase in SCr within 48 h after CPB) and no AKI. Urine NGAL and IL-18 were increased the AKI group at 2–4 h postoperatively. The concentrations of IL-18 and NGAL at 2 h postoperatively correlated with increased SCr at 12 h postoperatively.

However, in another prospective observational cohort study in adults, it was determined that urine IL-18 does not predict AKI after cardiac surgery. One hundred patients undergoing CPB at a single center were studied. Twenty patients developed AKI. On arrival in the ICU and at 24 h postoperatively, urine IL-18 was not different in patients who developed AKI compared to non-AKI patients [87].

The Translational Research Investigating Biomarkers Endpoints in Early Acute Kidney Injury (TRIBE-AKI) Clinical Consortium was established to accelerate the development of biomarkers. The consortium is a National Institutes of Health-funded multidisciplinary group and includes investigators from nine major academic centers who have expertise in preclinical, translational, epidemiologic, and health-services research. In the TRIBE-AKI study, urine IL-18 and NGAL were studied as early biomarkers of AKI in a prospective multicenter observational cohort study of 1219 patients receiving cardiac surgery. In addition, the hypothesis was tested that urine IL-18 and NGAL levels will be better markers of postoperative AKI than SCr and predict the severity of AKI and short-term mortality [83]. It was demonstrated that urine IL-18, urine NGAL, and plasma NGAL associate with subsequent AKI and poor outcomes. Urine IL-18 and urine- and plasma-NGAL levels peaked within 6 h after surgery. After multivariable adjustment, the highest quintiles of urine IL-18 and plasma NGAL associated with 6.8-fold and 5-fold higher odds of AKI, respectively, compared with the lowest quintiles. Elevated urine IL-18 and urine- and plasma-NGAL levels associated with longer length of hospital stay, longer ICU stay, and higher risk for dialysis or death. Urine IL-18 and plasma NGAL significantly improved the AUROCC to 0.76 and 0.75, respectively. Urine IL-18 and plasma NGAL significantly improved risk prediction over the clinical models alone.

Urine IL-18 and urine NGAL also predict AKI and poor outcomes after pediatric cardiac surgery [84]. The TRIBE-AKI Consortium conducted a prospective, multicenter cohort study in 311 children undergoing surgery for congenital cardiac lesions. Urine IL-18 and urine- and plasma-NGAL levels peaked within 6 h after surgery. Severe AKI, defined by dialysis or doubling in SCr during hospital stay, occurred in 53 patients at a median of 2 days after surgery. The first postoperative urine IL-18 and urine-NGAL levels strongly associated with severe AKI. After multivariable adjustment, the highest quintiles of urine IL-18 and urine NGAL associated with 6.9- and 4.1-fold higher odds of AKI, respectively, compared with the lowest quintiles. Elevated urine IL-18 and urine-NGAL levels were associated with longer hospital stay, longer ICU stay, and duration of mechanical ventilation. The AUROCC of urine IL-18 and urine NGAL for diagnosis of severe AKI was 0.72 and 0.71, respectively. The addition of these urine biomarkers improved risk prediction over clinical models alone.

In a nested case–control study of 15 patients with contrast-induced nephropathy (CIN) and 36-matched controls, urinary IL-18 was measured before, as well as, 24 and 72 h after cardiac catheterization. No statistically significant differences in urine IL-18 were detected between cases and controls or between the patient samples obtained before and after the cardiac catheterization [89].

In summary, the majority of published studies demonstrate that urine IL-18 is an early biomarker of AKI in humans and that urine IL-18 increases before SCr in critically ill adults and children in the ICU, in adults and children after CPB, and in adults after contrast administration.

Urine NGAL

Urinary NGAL is also an early biomarker of AKI in adult postcardiac surgery [92]. In 81 cardiac surgery patients, urine samples were collected immediately, preoperatively, and at various time intervals after surgery. Mean urinary NGAL concentrations in patients who developed AKI were significantly higher early after surgery and remained significantly higher at 3 and 18 h after cardiac surgery compared with patients who did not develop AKI.

There are other studies confirming the value of NGAL as an early biomarker of AKI in cardiac surgery patients. Serum NGAL and cystatin C were measured in 100 adult patients after cardiac surgery [111]. On arrival in the ICU, serum NGAL and cystatin C were independent predictors of AKI and were superior to BUN and SCr for the prediction of AKI. In the same group of patients, it was also determined that NGAL and cystatin C correlated with and were independent predictors of duration and severity of AKI and duration of ICU stay after cardiac surgery [112]. The combination of NGAL and cystatin C did not add to the predictive value. Also in the same group of patients, the predictive value of NGAL increased with the grade of AKI [113]. For example, plasma NGAL was higher for more severe AKI (greater than 50% increase in SCr) compared to less severe AKI (greater than 25% increase in SCr). NGAL also increased with increasing RIFLE classes of AKI. In 50 adult patients undergoing CPB, urinary- and serum NGAL were predictive biomarkers of AKI as early as 2 h postoperation [93].

NGAL may also represent an early sensitive biomarker of AKI after contrast administration for coronary angiography [96]. There was a significant rise in serum NGAL 2 and 4 h after contrast administration, and a rise in urinary NGAL 4 and 12 h after contrast.

Urine NGAL is an early biomarker of AKI in critically ill children aged between 1 month and 21 years who were on mechanical ventilation [97]. In 140 patients, mean- and peak urine-NGAL concentrations increased with worsening pediatric RIFLE maximum status. Urine NGAL concentrations rose in AKI, 2 days before and after a 50% or greater rise in SCr. Urine NGAL was a good diagnostic marker for AKI development with an AUROCC of 0.78. Urine NGAL was a marker of persistence of AKI for 48 h or longer with an AUROCC of 0.79. Urine NGAL was not a good marker for AKI severity when it was recorded after a rise in SCr had occurred (AUROCC of 0.63). In 88 critically ill adults, serum NGAL had an AUROCC of 0.96, sensitivity of 85%, specificity of 97% to predict the development of AKI [98]. Median urinary π-GST was higher in critically ill patients compared to normal controls. However, the an AUROCC for urinary π-GST indicated that it was not a good predictor of AKI.

NGAL is an early predictive biomarker of CIN in children [114]. Ninety-one children (0–18 years) with congenital heart disease undergoing elective cardiac catheterization and angiography with contrast administration were studied. CIN, defined as a 50% increase in SCr from baseline, was found in 11 subjects (12%). A significant elevation of NGAL concentrations in urine and plasma was noted within 2 h after cardiac catheterization. In contrast, detection of CIN by an increase in SCr was only possible 6–24 h after cardiac catheterization. By multivariate analysis, the 2-h NGAL concentrations in the urine and plasma, but not patient demographics or contrast volume, were found

The diagnosis of AKI is problematic in premature infants. Urinary NGAL was measured in 20 premature infants [115]. Neonates born at an earlier gestational age (GA) or low birth-weight infants had higher urine-NGAL concentrations. The study concludes that the use of NGAL as a biomarker of AKI in premature infants merits further investigation.

Urinary NGAL was evaluated as an early biomarker for prediction of AKI in preterm infants. Urine NGAL was measured in 50 preterm infants, GA between 28 and 34 weeks [116]. The median urine-NGAL levels were significantly higher in the preterm infants with AKI compared to controls on postnatal days 1 and 7.

In a prospective observational study, 194 consecutive adult admissions to the ICU, with absence of CKD, renal transplant or AKI as defined by RIFLE criteria were studied [99]. The admission plasma- and urine NGAL were significantly higher in the patients who developed AKI compared to the non-AKI patients. Hospital mortality was higher in the AKI group (17% vs. 4%). Plasma NGAL performed fairly on admission (AUROCC of 0.77) and at 24 and 48 h (AUROCC of 0.88 and 0.87) following ICU admission. Urine NGAL had a fair predictive value on admission (AUROCC of 0.79), at 24 h (AUROCC of 0.78), and was good at 48 h (AUROCC of 0.82). Thus, in critically ill patients without preexisting kidney disease, both plasma- and urine NGAL measured at admission can predict AKI (defined by RIFLE criteria) occurrence up to 72 h post-ICU admission.

The current status of NGAL as an early biomarker of AKI was reviewed in a metaanalysis. Fifty-eight manuscripts reporting on NGAL as a biomarker of AKI in more than 16,500 patients were analyzed [117]. Following cardiac surgery, NGAL measurement in over 7000 patients was predictive of AKI and its severity, with an overall AUROCC of 0.82–0.83. Similar results were obtained in over 8500 critically ill patients. In over 1000 patients undergoing kidney transplantation, NGAL measurements predicted delayed graft function with an overall AUC of 0.87. In all three settings, NGAL significantly improved the prediction of AKI risk over the clinical model alone.

In summary, urine NGAL is an early biomarker of AKI in children and adults post-CPB, after contrast administration, in critically ill ICU patients, in patients presenting to the ED, and in trauma patients (Table 6.3).

KIM-1

KIM-1 was measured in 90 patients undergoing cardiac surgery [95]. Thirty-six patients developed AKI within 72 h after surgery. The AUROCC to predict AKI immediately and 3 h postoperatively was 0.68 and 0.65 for KIM-1, 0.61 and 0.63 for NAG, and 0.59 and 0.65 for NGAL. Combining the three biomarkers, KIM-1, NAG, and NGAL, increased the sensitivity for early detection of AKI to 0.75 and 0.78.

In a metaanalysis, urinary KIM-1 was analyzed in 2979 patients from 11 eligible studies [101]. The estimated sensitivity of urinary KIM-1 for the diagnosis of AKI was 74% and specificity was 86% and the AUROCC was 0.86. Subgroup analysis suggested that population settings and detection time were the key factors affecting the efficiency of KIM-1 for AKI diagnosis.

Brush-Border Enzymes

Neutral endopeptidase (NEP) and retinol-binding protein (RBP) were increased in the urine of patients after open-heart surgery independent of AKI [118]. It has also been demonstrated that hemodialysis exacerbates tubular enzymuria in patients with AKI [119]. AP, GGT, leucine aminopeptidase (LAP), and dipeptidyl peptidase IV (DPP) were measured in kidney-transplant patients with normal graft function, ATN, acute rejection, and healthy controls. Enzymuria was increased with both acute rejection and ATN. Successful treatment of rejection resulted in a decrease in the enzymuria [120]. In a prospective pilot study of 26 consecutive critically ill adult patients admitted to the ICU, urinary levels of GGT, AP, NAG, and α- and π-GST but not lactate dehydrogenase (LDH), were higher in the AKI group on admission and were useful in predicting AKI [108].

Urinary α- and π-GST were measured during the 48 h after ICU admission in 40 consecutive patients who were admitted with a diagnosis of sepsis [73]. AKI was diagnosed according to the AKIN criteria. Nineteen patients developed AKI, all within 24 h of ICU admission. Urinary α-GST level was not increased in patients who developed AKI versus non-AKI patients. Median urinary π-GST level was significantly higher in those who developed Stage 1 AKI, and in those who developed Stage 3 AKI compared to the non-AKI group. Median urinary π-GST level at ICU admission was higher in all groups than in healthy control subjects. The AUROCC for urinary π-GST level indicated that it was not a good predictor of AKI. The conclusion of this study was that urinary π-GST is elevated early in all patients with sepsis syndrome, but is not predictive of AKI as defined by AKIN.

Cystatin C

In patients with AKI, serum cystatin C rises prior to SCr. In 85 patients at high risk to develop AKI, it was determined whether serum cystatin C detected AKI earlier than SCr. AKI was defined according to the RIFLE classification. Serum cystatin C increased by more than 50% at 0.6 days earlier than the increase in SCr. Serum cystatin C also demonstrated a high-diagnostic value to detect AKI as indicated by AUROCC on the 2 days before the R or "risk of renal dysfunction" criteria was fulfilled by creatinine. This study concluded that serum cystatin C is useful for the detection of AKI and may detect AKI 1–2 days earlier than creatinine [102].

Serum cystatin C and β2M were measured in 25 children in the ICU in a prospective, observational study [104]. The ability of serum cystatin C and β2M to identify a creatinine clearance rate and a Schwartz–creatinine clearance rate under 80 mL/min per 1.73 m^2 was better than that of creatinine (AUROCC: 0.851 and 0.792 for cystatin C, 0.802 and 0.799 for β2M, and 0.633 and 0.625 for creatinine). This study concluded that serum cystatin C and β2M were better than SCr, to detect AKI in critically ill children.

A study evaluated the use of urine cystatin C for the early diagnosis of AKI. Plasma and urine were prospectively collected from 72 adults undergoing elective cardiac surgery [103]. AKI was defined as a 25% or greater increase in plasma creatinine or RRT within the first 72 h following surgery. Plasma cystatin C and NGAL did not predict the development of AKI within the first 6 h following surgery. However, both urinary cystatin C and NGAL were increased in the 34 patients who later developed AKI, compared to patients with no AKI. The urinary cystatin C at 6 h after ICU admission was the most useful for the prediction of AKI.

In a study by the TRIBE-AKI Consortium of 1203 adults and 299 children undergoing cardiac surgery, it was found that urinary cystatin C was not significantly associated with the development of AKI after cardiac surgery [105]. In the same group of patients, the performance of KIM-1 and L-FABP, alone or in combination with other injury biomarkers during the perioperative period was evaluated [100]. AKI was defined as doubling of SCr or need for acute dialysis. KIM-1 peaked 2 days after surgery in adults and 1 day after surgery in children, whereas L-FABP peaked within 6 h after surgery in both age groups. In multivariable analyses, the highest quintile of the first postoperative KIM-1 level was associated with AKI compared with the lowest quintile in adults, whereas the first postoperative L-FABP was not associated with AKI. The study concluded that postoperative elevations of KIM-1 associate with AKI and adverse outcomes in adults but were not independent of other AKI biomarkers, urinary IL-18, and NGAL. A panel of multiple biomarkers provided moderate discrimination for AKI.

However, in contrast to adults, early postoperative serum cystatin C predicts severe AKI following pediatric cardiac surgery. In a prospective study of 288 children (50% under 2 years of age) undergoing cardiac surgery, it was determined whether measurement of pre- and postoperative serum cystatin C in AKI over that obtained by SCr [106]. Preoperative cystatin C was not associated with AKI. The highest quintile of postoperative cystatin C, measured within 6 h postoperatively, predicted Stage 1 and 2 AKI. Postoperative serum cystatin C independently predicted longer duration of ventilation and ICU length of stay. Thus, postoperative serum cystatin C may be used in the future to risk-stratify patients for AKI treatment trials.

In another study in critically ill patients, SCr, serum cystatin C, and 24-h creatinine clearance were determined. Serum cystatin C correlated better with GFR than did creatinine and was diagnostically superior to creatinine [121]. During continuous veno-venous hemofiltration (CVVH), the quantity of cystatin C removed is less than 30% of its production and no rapid changes in its serum concentration are observed [122]. This study suggests that CVVH is unlikely to significantly influence serum concentrations of cystatin C and that cystatin C can be used to monitor residual kidney function during CVVH.

The reason why cystatin C rises before SCr, is not clear. A possible explanation is that cystatin C represents the ideal endogenous marker of GFR: it is produced by all nucleated cells at a constant rate, is not affected by changes in body mass, nutrition or gender, and is not degraded or secreted by the renal tubules. In contrast, SCr is affected by many nonrenal factors that affect generation of creatinine and tubular secretion.

There are limitations to the use of cystatin C as a marker of GFR. Abnormalities of thyroid function [123] and glucocorticoid therapy [124,125] may affect cystatin C independently of kidney function. Levels of C-reactive protein (CRP) may increase cystatin C levels and it has been suggested that cystatin C is a marker of inflammation [126].

Urine L-FABP

Urine L-FABP was measured in 40 pediatric patients prior to and following CBP [107]. Enzyme-linked immunosorbent assay (ELISA) analysis demonstrated increased L-FABP levels of about 94- and 45-fold at 4 and 12 h, respectively, following surgery in the 21 patients who developed AKI. Western blot analysis, confirmed the presence of L-FABP in the urine. Both bypass time and urinary L-FABP were significant independent risk factors for AKI. This study demonstrates that urinary L-FABP represents a sensitive and predictive early biomarker of AKI after cardiac surgery.

In the TRIBE-AKI study, urine L-FABP peaked within 6 h after surgery in both adults and children [100].

In summary, urinary L-FABP is increased in rodents with AKI before the increase in SCr. Urinary L-FABP is also increased in humans with AKI. Urinary L-FABP, but not serum L-FABP, is increased in patients with severe sepsis. In view of studies that L-FABP is also a biomarker of progression of CKD, larger and multicenter studies of L-FABP as an early biomarker of AKI in patients are warranted.

Other Biomarkers

In 294 children undergoing cardiac surgery that were studied by the TRIBE-AKI Consortium, it was determined whether preoperative and postoperative urine albumin/creatinine ratios (ACRs) predict postoperative AKI as defined by AKIN criteria [110]. Preoperative ACR did not predict AKI in younger or older children. In children aged <2 years, first postoperative ACR $\geq$ 908 mg/g (103 mg/mmol) predicted Stage 2 AKI development. In children aged $\geq$2 years, postoperative ACR $\geq$ 169 mg/g (19.1 mg/mmol) predicted Stage 1 AKI. Postoperative ACR is a rapidly available and easy-to-perform biomarker of AKI that performs similarly as other AKI biomarkers. Its use is enhanced in children aged $\geq$2 years and in combination with serum cystatin C.

Renal near-infrared spectroscopy was used to evaluate regional kidney oximetry in a noninvasive continuous real-time fashion and reflected tissue perfusion. Fifty-nine infants undergoing CBP were studied. Renal near-infrared spectroscopy was continuously measured intraoperatively and for at least 24 h postoperatively. Renal oximetry intraoperatively and in the first 12, 24, and 48 h postoperatively was significantly lower in AKI patients compared with controls ($p < 0.05$). The study concluded that prolonged low-renal oximetry values during cardiac surgery correlate with the development of AKI [127].

In mice with FA-induced AKI, plasma FGF23 levels increased significantly from baseline already after 1 h of AKI, with an 18-fold increase at 24 h [128]. FGF23 levels were also increased in mice with AKI induced with osteocyte-specific parathyroid hormone receptor ablation, the global deletion of parathyroid hormone, or the vitamin D receptor, indicating that the increase in FGF23 was independent of parathyroid hormone- and vitamin D signaling. FGF23 levels increased to a similar extent in wild-type mice on a normal phosphate diet, indicating that the marked FGF23 elevation is independent of dietary phosphate. Bone production of FGF23 was significantly increased in AKI. The half-life of intravenously administered recombinant FGF23 was only modestly increased. On the basis of animal studies, a human study was performed. FGF23 levels rise in humans with AKI following cardiac surgery [128]. FGF23 was measured before and after cardiac surgery in 14 patients, 4 of whom developed AKI, which was defined as a 50% increase in SCr within 5 days postoperatively. FGF23 levels rose significantly during the postoperative period in both AKI cases and controls, but were significantly higher in the AKI cases beginning at 24 h postoperatively and at all subsequent time-points. Although SCr also increased significantly by 24 h, the observed 1.4-fold change was of substantially smaller magnitude than the simultaneous 15.9-fold increase in FGF23.

Netrin-1 was also measured in patients with AKI. Urinary netrin-1 was increased in 13 patients with AKI and no urinary netrin-1 was found in 6 healthy volunteer-urine samples [54]. Urine netrin-1 was measured in 10 healthy controls, 22 recipients of a renal allograft, 11 patients with ischemic AKI, 13 with AKI associated with sepsis, 9 with radiocontrast-induced AKI, and 8 with drug-induced AKI [129]. Urinary netrin-1 levels normalized for urinary creatinine were significantly higher in all subject groups, the highest values were in patients with sepsis and in transplant patients immediately postoperatively. In 150 septic patients in the ICU, SCr was increased after 24 h of ICU admission while netrin-1 levels increased significantly as early as 1 h, peaked at 3–6 h and remained elevated up to 48 h of ICU admission in septic AKI patients [130]. KIM-1 increased significantly by 6 h, peaked at 24 h, and remained significantly elevated until 48 h of ICU admission. Urinary netrin-1 is an early predictive biomarker of AKI after cardiac surgery. In patients undergoing CPB, netrin-1 was measured in 26 patients who developed AKI (defined as a 50% or greater increase in SCr after CPB) and 34 controls (patients who did not develop AKI after CPB) [131]. SCr increased at 48 h after CPB. In contrast, urine netrin-1 increased at 2 h after CPB, peaked at 6 h and remained elevated up to 48 h after CPB. The AUROCC for diagnosis of AKI at 2, 6, and 12 h after CPB was 0.74, 0.86, and 0.89, respectively. The 6-h urine netrin-1 measurement strongly correlated with duration and severity of AKI, as well as, length of hospital stay. Netrin-1 predicts the development of AKI in liver-transplant patients [132]. In 63 patients after liver transplant, AKI was detected at 48 h after liver transplantation using SCr. Urine netrin-1 and NGAL were increased significantly and peaked at 2 h after liver transplantation but were no longer significantly elevated at 6 h after transplantation. The AUROCC for urine netrin-1 for diagnosis of AKI at 2, 6, and 24 h after liver transplantation was 0.66, 0.57, and 0.59, respectively.

The insertion/deletion (I/D) genetic polymorphism of ACE and the I allele, which is associated with lower ACE activity, was measured in 181 consecutive patients admitted to the ICU for an expected stay >48 h [133]. II, ID, and DD genotype frequencies were 25, 48, and 27%. AKI was defined in terms of the RIFLE classification. II and ID genotypes were associated with lower baseline circulating rates of ACE. There was a significantly greater II genotype proportion in AKI patients (42%) compared to patients without AKI. After an adjustment for the identified prognostic factors, II genotype was independently associated with increased risk of AKI and death among patients with AKI. This study suggests that genetic factors may affect the susceptibility to and prognosis of AKI.

SCr, -cystatin C, -IL-6, and -IL-8 and urine IL-18, -NGAL, -IL-6, and -IL-8 were measured before and within 24 h after liver transplantation in 40 patients [134]. AKI was defined as a ≥50% sustained increase in creatinine above preoperative values occurring within 24 h of transplantation and persisting for at least 24 h. The AUROCCs for biomarkers measured 24 h after surgery were 0.749 for urine IL-18, 0.833 for urine NGAL, 0.745 for urine IL-6, 0.682 for serum IL-6, 0.773 for urine IL-8, and 0.742 for serum IL-8. Postoperative cystatin C was not significantly different between AKI and no AKI groups. Thus, serum IL-8 and urine IL-18, NGAL, IL-6, and IL-8 are elevated in AKI within the first 24 h following liver transplantation. Whether these biomarkers rise within 6 h after AKI in liver-transplant patients remains to be studied.

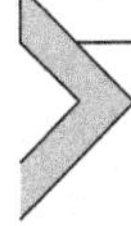

BIOMARKERS THAT PREDICT SHORT-TERM OUTCOMES (TABLE 6.4)

IL-18 and NGAL

It was determined that urinary IL-18 and NGAL are early predictive biomarkers of CIN after coronary angiography and whether IL-18 and NGAL predict later cardiac events [135]. CIN was diagnosed in 13 of 150

Table 6.4 Biomarkers that predict short-term outcomes

Situation	Number	Biomarker	Short-term outcome	References
Adults post-CPB	100	Plasma NGAL and cystatin C	Duration and severity of AKI and duration of ICU stay	[112,113] [111]
Children post-CPB	71	Plasma NGAL at 2 h	Change in creatinine, duration of AKI, length of hospital stay, and mortality	[91]
ED	635	Urinary NGAL	Clinical outcomes, including nephrology consultation, dialysis, and admission to the ICU	[136]
Nonoliguric AKI	73	Urine cystatin C	Patients requiring RRT	[109]
Contrast nephropathy	150	Urine IL-18	Independent predictor of later major cardiac events	[135]
Adults in ICU	138	Urine IL-18	Strong predictor of mortality	[82]
Children in ICU	137	Urine IL-18	Predicts severity of AKI and independent predictor of mortality	[88]
Adults post-CPB, TRIBE-AKI	1219	Urine IL-18, urine- and plasma NGAL	Longer length of hospital stay, longer ICU stay, and higher risk for dialysis or death	[83]
Children post-CPB, TRIBE-AKI	311	Urine IL-18 and NGAL	Longer hospital stay, longer ICU stay, and duration of mechanical ventilation	[84]
Adults in ICU	1439	Urine IL-18	Poor-to-moderate ability to predict AKI, RRT or 90-day mortality	[137]
Children post-CPB	288	Serum cystatin C	Predicts longer duration of ventilation and ICU length of stay	[106]
Adults post-CPB, TRIBE-AKI	1150	Serum cystatin C	Identifies adverse outcomes	[138]

(Continued)

Table 6.4 Biomarkers that predict short-term outcomes (*cont.*)

Situation	Number	Biomarker	Short-term outcome	References
Cirrhosis	188	Urine NGAL, IL-18, KIM-1, L-FABP, and albuminuria	AKI progression and death	[139]
Adults post-CPB	97	Urine angiotensinogen/creatinine ratio	Worsening of AKI, AKIN stage 3, need for RRT or death	[140]
Hospitalized patients	201	Urinary NAG and KIM-1	Dialysis requirement or hospital death	[141]
Adults on RRT in ICU	817	Plasma IL-8, IL-18, and TNFR	Independently associated with slower renal recovery	[142]
Adults on RRT in ICU	817	Plasma IL-6, IL-8, IL-10 and IL-18; MIF; TNFR-I; and DR-5	Associated with mortality	[142]
Critically ill adults	152	L-FABP	Predicts doubling of SCr, dialysis or death within 7 days	[143]
Adults post-CPB	95	IL-18, L-FABP, NGAL, and KIM-1	Worsening AKI or death, AKIN stage 3 or death	[144]

ATN, acute tubular necrosis; CPB, cardiopulmonary bypass; L-FABP, liver-type fatty acid-binding protein; NAG, *N*-acetyl-β-D-glucosaminidase; RRT, renal replacement therapy; SCr, serum creatinine.

patients. At 24 h after the procedure, urinary IL-18 and NGAL levels were significantly increased in the CIN group compared to non-CIN patients. The time of AKI as predicted by urine IL-18 was 24 h earlier than the time of AKI as determined by the rise in SCr. IL-18, but not SCr, was also found to be an independent predictor of later major cardiac events up to 17 months after CPB. The AUROCC for the early diagnosis of AKI was 74.9% for urine IL-18 and 73.4% for urinary NGAL.

NGAL is the most extensively studied biomarker in AKI. Urinary- and serum NGAL were demonstrated to be sensitive, specific, and highly predictive early biomarkers of AKI in children after cardiac surgery [90]. Seventy-one children undergoing cardiopulmonary bypass (CPB) were studied. Serial urine and blood samples were analyzed by western blots and ELISA for NGAL expression. Diagnosis of AKI, defined as a 50% increase in SCr from baseline, developed 1–3 days after CPB. In contrast, urinary NGAL rose significantly at 2 h after CPB. By multivariate analysis, the urinary NGAL at 2 h after CPB was a powerful independent predictor of AKI. In addition, 2-h postoperative plasma-NGAL levels strongly correlated with change in creatinine, duration of AKI, length of hospital stay, and mortality after CPB [91].

The sensitivity and specificity of a single urine NGAL measurement for diagnosing AKI was determined in 635 patients in the ED [70]. In multiple logistic regression, urinary NGAL was highly predictive of clinical outcomes, including nephrology consultation, dialysis, and admission to the ICU. In 31 multiple trauma patients, urinary NGAL concentration on admission was significantly higher in patients who subsequently developed AKI [145].

AKI is a common and severe complication in patients with cirrhosis. Using kidney biomarkers to predict the patients that are at highest risk for AKI progression may allow expedited management. One hundred eighty-eight patients with cirrhosis were studied [139]. Out of the 188, 44 patients (23%) developed AKI progression and 39 patients (21%) experienced both AKI progression and death. NGAL, IL-18, KIM-1, L-FABP, and albuminuria were significantly higher in patients with AKI progression and death. FENa was not associated with worsening of AKI. The study concluded that multiple biomarkers of kidney injury, but not FENa, are independently associated with the progression of AKI and mortality in patients with cirrhosis.

Urine IL-18 was measured in 1439 critically ill patients at ICU admission and 24 h [137]. The highest urine IL-18 during the first 24 h in the

ICU associated with the development of AKI with an AUROCC of 0.586 and with the development of Stage 3 AKI with AUROCC of 0.667. IL-18 predicted the initiation of RRT with an AUROCC of 0.655 and 90-day mortality with an AUROCC of 0.536. The study concluded that IL-18 had poor-to-moderate ability to predict AKI, RRT, or 90-day mortality in this large cohort of critically ill patients.

The concentration of 32 candidate biomarkers in the urine of 95 patients with AKIN Stage 1 after cardiac surgery, was measured [144]. IL-18 was the best predictor of worsening AKI or death and AKIN Stage 3 or death (AUROCC of 0.74 and 0.89). L-FABP (AUROCC of 0.67 and 0.85), NGAL (AUROCC of 0.72 and 0.83), and KIM-1 (AUROCC of 0.73 and 0.81) were also good predictors. The combination of IL-18 and KIM-1 had a very good predictive value (AUROCC of 0.93) to predict AKIN Stage 3 or death.

Brush-Border Enzymes

In 73 consecutive patients with nonoliguric AKI, urinary excretion of α1- and β2M, cystatin C, RBP, α-GST, γ-GT, LDH, and NAG was measured early in the course of the AKI [109]. Urinary excretion of cystatin C and α1M had the highest diagnostic accuracies as indicated by the largest AUROCCs in identifying patients requiring dialysis. This study concluded that in nonoliguric AKI, increased urinary excretion of cystatin C and α1M may predict an unfavorable outcome, as indicated by the requirement for dialysis [109].

The relationship between urinary NAG and KIM-1 level and adverse clinical outcomes was determined prospectively in 201 hospitalized patients with AKI. Patients with the highest levels in urinary NAG and KIM-1 had the higher odds for dialysis requirement or hospital death. This study demonstrates that urinary biomarkers of AKI, such as, NAG and KIM-1 can predict adverse clinical outcomes in patients with AKI [141].

Cystatin C

SCr and serum cystatin C were measured at the preoperative visit and daily on postoperative 1–5 days in 1150 high-risk adult cardiac surgery patients in the TRIBE-AKI Consortium [138]. SCr detected more cases of AKI than cystatin C. Clinical outcomes generally were not statistically different for AKI cases detected by creatinine- or cystatin C level. However, for each AKI threshold, patients with AKI confirmed by both markers had a significantly higher risk of the combined mortality/dialysis outcome compared with patients with AKI detected by creatinine level alone. In conclusion, it was found that cystatin C level was less sensitive for AKI detection than creatinine level. However, confirmation of SCr with serum cystatin C level appeared to identify a subset of patients with AKI with a substantially higher risk of adverse outcomes.

L-FABP

In 152 critically ill patients, the ability of L-FABP to predict injury progression, dialysis, or death within 7 days in critically ill adults with early AKI was determined [143]. Urine L-FABP has an AUROCC of 0.79, which improved to 0.82, when added to the clinical model to predict doubling of SCr, dialysis, or death within 7 days. Urine NGAL, IL-18, and KIM-1 had AUROCC of 0.65, 0.64, and 0.62, respectively and did not significantly improve discrimination of the clinical model.

Other Biomarkers

A biomarker that detects a lack of renal recovery or mortality in patients on RRT was investigated in a multicenter, prospective, cohort study of 817 critically ill patients receiving RRT [142] The association between day-1 plasma inflammatory apoptosis and growth factor biomarkers and renal recovery and mortality at day 60 was determined. Increased concentrations of plasma IL-8, IL-18, and TNFR-I were independently associated with slower renal recovery. Higher concentrations of IL-6, IL-8, IL-10, IL-18, MIF, TNFR-I, and DR-5 were associated with mortality. Elevated plasma concentrations of inflammatory and apoptosis biomarkers are associated with RRT dependence and death.

Liquid chromatography/tandem mass spectrometry was used to identify 30 potential prognostic urinary biomarkers of severe AKI in a group of patients that developed AKI after cardiac surgery [140]. Urinary angiotensinogen measured an average of 1 day postoperatively had the best discriminative characteristics. The urine angiotensinogen/creatinine ratio predicted worsening of AKI, AKIN Stage 3, need for RRT, discharge >7 days from sample collection, and composite outcomes of AKIN Stage 2 or -3, AKIN Stage 3 or death, and RRT or death. Thus, elevated urinary angiotensinogen is a prognostic biomarker and is associated with adverse outcomes in patients with AKI.

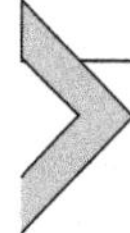

BIOMARKERS FOR RISK STRATIFICATION OF PATIENTS WITH EXISTING AKI (TABLE 6.5)

The ability to predict whether AKI will progress may improve management, guide patient and family counseling, and identify patients for enrollment into therapeutic trials in AKI.

Results from the TRIBE-AKI studies, the largest AKI biomarker studies performed to date, have significantly advanced the field of biomarkers in AKI by determining traditional and novel biomarkers that contribute to improvement in risk prediction [151]. Biomarkers measured on the day of AKI diagnosis improve risk stratification and identify patients at higher risk for progression of AKI and worse patient outcomes [146]. Samples from the TRIBE-AKI study in adults were analyzed. It was determined whether AKI biomarkers (IL-18, urine- and plasma NGAL, and urinary albumin to creatinine ratio) measured at the time of the first clinical diagnosis of early AKI after cardiac surgery could determine the severity of the AKI. Biomarker measurement was on the day of AKI diagnosis in 380 patients who

Table 6.5 Biomarkers for risk stratification

Situation	Number	Biomarker	Risk	References
Adults post-CPB, TRIBE-AKI	380	Urine IL-18, urine- and plasma NGAL, and urinary albumin to creatinine ratio	Worsening AKIN stage	[146]
Adults prior to CPB, TRIBE-AKI	1139	Preoperative BNP	Mild and severe AKI	[147]
Adults prior to CPB, TRIBE-AKI	1219	Perioperative plasma H-FABP	Risk stratification of AKI and mortality	[148]
Children prior to CPB, TRIBE-AKI	106	Preoperative CK-MB and H-FABP	Risk of postoperative AKI	[149]
ED	1635	Urine NGAL, KIM-1, L-FABP, IL-18, and cystatin C	Severity and duration of AKI and dialysis initiation and death	[150]

H-FABP, heart-type fatty acid-binding protein; L-FABP, liver-type fatty acid-binding protein; TRIBE-AKI, Translational Research Investigating Biomarkers Endpoints in Early Acute Kidney Injury.

developed at least AKIN Stage 1 AKI. The primary endpoint (progression of AKI defined by worsening AKIN stage) occurred in 45 (11.8%) patients. Using multivariable logistic regression, and after adjustment for clinical predictors, the highest quintiles of the three biomarkers remained associated with AKI progression compared with biomarker values in the lowest two quintiles. IL-18 (odds ratio = 3.0, 95% CI = 1.3–7.3), urinary albumin to creatinine ratio (odds ratio = 3.4, 95% CI = 1.3–9.1), and plasma NGAL (odds ratio = 7.7, 95% CI = 2.6–22.5). Each biomarker improved risk classification compared with the clinical model alone, with plasma NGAL performing the best (category-free net reclassification improvement of 0.69, $p < 0.0001$).

It was determined whether preoperative brain natriuretic peptide (BNP) levels predict postoperative AKI among patients undergoing cardiac surgery [147]. Preoperative BNP was measured in 1139 patients from the TRIBE-AKI study undergoing cardiac surgery. Preoperative BNP was a strong and independent predictor of mild and severe AKI. Compared with the lowest BNP quintile, the highest quintile had a significantly higher risk of at least mild AKI and severe AKI. Compared with clinical parameters alone, BNP modestly improved risk prediction of AKI cases into lower and higher risk. Preoperative BNP may improve preoperative risk stratification and discrimination among surgical candidates.

Heart-type fatty acid binding protein (H-FABP) is a myocardial protein that detects cardiac injury. It was determined whether plasma H-FABP was associated with AKI in the TRIBE-AKI cohort of 1219 patients at high risk for AKI who underwent cardiac surgery [148]. In analyses adjusted for known AKI risk factors, the first postoperative log (H-FABP) was significantly associated with severe AKI, while preoperative log (H-FABP) was not associated with any AKI. The study concluded that perioperative plasma-H-FABP levels may be used for risk stratification of AKI and long-term mortality after discharge in patients undergoing cardiac surgery.

In 106 children undergoing cardiac surgery as part of the TRIBE-AKI study the association of cardiac biomarkers [N-type pro-B-type natriuretic peptide, creatine kinase-MB (CK-MB), H-FABP, and troponins I and T] with the development of postoperative AKI was investigated [149]. Patients who developed AKI had higher median levels of pre- and postoperative cardiac biomarkers compared with patients without AKI preoperative CK-MB and H-FABP were associated with increased risk of postoperative AKI and provide good discrimination of patients who develop AKI. These biomarkers may be useful for risk stratifying patients undergoing cardiac surgery.

BIOMARKERS OF AKI AND LONG-TERM OUTCOMES

AKI biomarkers have prognostic value for in-hospital outcomes. The association of AKI biomarkers with long-term mortality is less well known. Long-term mortality was determined in 1199 adults who underwent cardiac surgery and were enrolled in the TRIBE-AKI studies [152]. NGAL, IL-18, KIM-1, L-FABP, and albumin were measured on postoperative days 1–3. During a median follow-up of 3 years, 139 participants died. In patients with clinical AKI, the highest tertiles of peak urinary NGAL, IL-18, KIM-1, L-FABP, and albumin were independently associated with a 2.0- to 3.2-fold increased risk for mortality compared with the lowest tertiles. In patients without clinical AKI, the highest tertiles of peak IL-18 and KIM-1 were independently associated with long-term mortality Thus, urinary IL-18, KIM-1, and NGAL, in the immediate postoperative period give prognostic information about 3-year mortality in patients with or without clinical AKI.

BIOMARKERS OF SUBCLINICAL AKI (TABLE 6.6)

It is biologically possible that biomarkers of tubular injury can increase in the urine in patients that do not have a rise in SCr. This condition has been called "subclinical" AKI because it is below detection levels using SCr [153]. There is evidence that patients with subclinical AKI have worse clinical outcomes [153]. Also patients with increased creatinine and increased markers of tubular injury have a worse prognosis [153]. It is suggested that tubular injury biomarkers could be a useful tool for risk prediction and monitoring the course of renal function in critically ill patients. Improved risk assessment could result in an earlier nephrology consult or earlier initiation of future therapies [153].

Some biomarkers of AKI are increased in prerenal tubular injury. Prerenal AKI is a physiologic response to underperfusion before a sustained increase in SCr. The diagnosis of prerenal AKI is made retrospectively after a transient rise in plasma creatinine. Cystatin C, NGAL, γGT, IL-18, and KIM-1 were measured at 0, 12, and 24 h following ICU admission. Five hundred twenty-nine patients were stratified into groups having no AKI, AKI with recovery by 24 h, recovery by 48 h, or the composite of AKI greater than 48 h, or dialysis [154]. Prerenal AKI was defined in 61 patients as AKI with recovery within 48 h and a fractional sodium excretion $< 1\%$. Biomarker concentrations significantly and progressively increased with the

Table 6.6 Biomarkers of subclinical AKI

Situation	Number	Biomarker	Study	References
Adults in ICU	529	KIM-1, cystatin C, and IL-18	Increased in prerenal AKI compared with no AKI	[154]
Children post-CPB	287	Serum cystatin C defined AKI	Strongly associated with urine IL-18 and KIM-1	[155]
ED	1635	Urine NGAL, KIM-1, L-FABP, IL-18, and cystatin C	Identified patients with low SCr who were at risk of adverse events	[150]
AAA surgery		Urinary L-FABP and H-FABP	Indicates renal proximal and distal tubule injury in the absence of an increase in SCr	[156]
Adults in ICU	380	Urine NGAL and L-FABP	Associated with the development of AKI and the need for dialysis in patients with eGFR > 60 mL/min	[157]
Adults post-CPB	34	Urine NEP and RBP	Increased independent of increase in SCr	[118]
Contrast nephropathy	127	Serum cystatin C	Detects AKI (an iopromide clearance of less than 80 mL/min per m^2)	[47]

AAA, abdominal aortic aneurysm; GFR, glomerular filtration rate; H-FABP, heart-type fatty acid-binding protein; L-FABP, liver-type fatty acid-binding protein; β2M, β2-microglobulin; NEP, neutral endopeptidase; RBP, retinol-binding protein; SCr, serum creatinine.

duration of AKI even in the patients with prerenal AKI. The median concentration of KIM-1, cystatin C, and IL-18 were significantly greater in prerenal AKI compared with no AKI, while NGAL and γGT concentrations were not significant. The results suggest that prerenal AKI represents a milder form of "subclinical" injury.

It was determined whether defining clinical AKI by increases in cystatin C versus SCr alters associations with biomarkers and clinical outcomes [155]. Two hundred eighty-seven pediatric patients without preoperative AKI or ESKD who were undergoing cardiac surgery were studied. The

SCr-defined versus cystatin C-defined AKI incidence differed substantially (43.6% vs. 20.6%). Percentage agreement was 71% ($\kappa = 0.38$); Stage 2 or worse AKI percentage agreement was 95%. IL-18 and KIM-1 discriminated for cystatin C-defined AKI better than for SCr-defined AKI. For IL-18 and KIM-1, the AURROCs were 0.74 and 0.65, respectively, for cystatin C-defined AKI, and 0.66 and 0.58, respectively, for SCr-defined AKI. Fifth (vs. first) quintile concentrations of both biomarkers were more strongly associated with cystatin C-defined AKI. The cystatin C definitions and SCr definitions were similarly associated with clinical outcomes of resource use. Compared with the SCr-based definition, the cystatin C-based definition is more strongly associated with urine IL-18 and KIM-1 in children undergoing cardiac surgery.

In a multicenter-prospective cohort study, 5 urinary biomarkers, NGAL, KIM-1, L-FABP, IL-18, and cystatin C were measured in 1635 unselected ED patients at the time of hospital admission [150]. It was determined whether the biomarkers diagnosed intrinsic AKI and predicted adverse outcomes during hospitalization. All biomarkers were elevated in intrinsic AKI. Urinary NGAL and KIM-1 predicted a composite outcome of dialysis initiation or death during hospitalization, and both improved the net risk classification compared with conventional assessments. These biomarkers also identified a substantial subpopulation with low-SCr at hospital admission, but who were at risk of adverse events. In conclusion, urinary biomarkers enable prospective diagnostic and prognostic stratification in the ED.

Urinary L-FABP and H-FABP were measured before, during, and within 3 days after abdominal aortic aneurysm (AAA) surgery in 22 patients [156]. There was an abrupt and significant elevation of both urine FABPs normalized to urinary creatinine at 2 h after aortic clamp release. The significant rise of both urinary L-FABP and H-FABP after AAA surgery indicates renal proximal and distal tubule injury in this population in the absence of an increase in SCr.

The ability of urine NGAL, L-FABP, and cystatin C to predict AKI development, death, and dialysis in a nested case–control study of 380 critically ill adults with an eGFR over 60 mL/min per 1.73 m^2 was determined [157]. One hundred thirty AKI cases were identified by biomarker measurement and were compared with 250 controls without AKI. Urine NGAL and L-FABP were both independently associated with the development of AKI and the need for dialysis in a critically ill population without prevalent kidney injury and may add incremental information to current risk prediction tools. However, both markers had only modest utility for discriminating incident injury from noninjury using current conventional definitions.

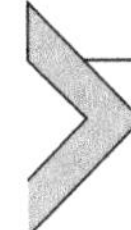

THE EFFECT OF INTERVENTIONS ON BIOMARKERS OF AKI (TABLE 6.7)

The next step in biomarker measurement in cardiac surgery patients has been to determine whether interventions prevent AKI and also whether interventions lower plasma or urine biomarkers. In this regard, in a prospective study of 60 patients undergoing either standard CPB or miniaturized CPB, kidney function was better protected during miniaturized CPB as determined by urinary NGAL [158]. However, in 60 patients undergoing either off-pump or on-pump CABG surgery, urine NGAL was not

Table 6.7 Effect of interventions on AKI biomarkers

Intervention	Number	Biomarker	Result	References
Standard versus miniaturized CPB	60	Urine NGAL	Lower during miniaturized CPB	[158]
Off-pump versus on-pump CPB	60	Urine NGAL and SCr	Not different between the groups	[159]
Cold blood or cold crystalloid in aortic surgery	172	Urinary RBP, α1M, NAG, and AP	Not different between the groups	[160]
Continuing versus stopping preoperative statins	625	Urine IL-18, NGAL, and KIM-1 and plasma NGAL	Lower risk of elevation of biomarkers	[161]
EPO administration prior to CPB	98	Cystatin C and NGAL	Not different between the groups	[162]
EPO administration prior to CPB	40	Urine NGAL	Not different between the groups	[163]
Sodium bicarbonate versus sodium chloride prior to CPB	100	SCr, plasma urea, and urinary NGAL	Lower in patients receiving sodium bicarbonate	[164]

AP, intestinal alkaline phosphatase; CPB, cardiopulmonary bypass; EPO, erythopoetin; NAG, *N*-acetyl-β-D-glucosaminidase; RBP, retinol-binding protein.

different between the groups [159]. SCr and the incidence of AKI were also not different between the groups.

Five urinary biomarkers, RBP, α1M, microalbumin, NAG, and intestinal AP were measured in 172 patients randomized to receive cold blood or cold crystalloid for renal perfusion during thoracoabdominal aortic aneurysm repair [160]. Twenty-seven patients in the cold blood group and 21 patients in the cold crystalloid group developed AKI ($p = 0.4$). Changes in renal biomarkers were similar in the groups.

The effect of preoperative statins on AKI postcardiac surgery is controversial [165]. In a multicenter prospective cohort study of 625 adult patients undergoing elective cardiac surgery, patients were grouped according to whether statins were continued or held in the 24 h before operation [161]. Continuing, as opposed to stopping, a statin before operation was not associated with a lower risk of AKI, as defined by a doubling of SCr or dialysis. However, continuing a statin was associated with a lower risk of elevation of urine IL-18, urine NGAL, urine KIM-1, and plasma NGAL. The study concluded that statins may prevent kidney injury after cardiac surgery, as evidenced by lower levels of AKI biomarkers.

In a single-site, randomized, case–control, and double-blind study, the effect of preemptive erythopoetin (EPO) administration on the incidence of postoperative AKI in patients with risk factors for AKI undergoing complex valvular heart surgery was determined [162]. Ninety-eight patients with preoperative risk factors for AKI were studied. The patients were randomly allocated to either the EPO group ($n = 49$) or the control group ($n = 49$). AKI was defined as an increase in SCr > 0.3 mg/dL or > 50% from baseline. The incidence of postoperative AKI was 32.7% versus 34.7%, ($p = 0.831$) for EPO versus no EPO and biomarkers of renal injury including cystatin C and NGAL showed no significant differences between EPO versus no EPO groups. The postoperative increase in IL-6 and myeloperoxidase was similar between the groups.

In a double-blind randomized control study, 80 patients admitted to the ICU postcardiac surgery were randomized by computer to receive intravenously isotonic saline ($n = 40$) versus α-EPO ($n = 40$) [163]. The primary outcome was the change in urinary NGAL concentration from baseline and 48 h after EPO injection. EPO treatment did not significantly modify the difference in urine NGAL between 48 h and randomization compared to placebo and the incidence of AKI was similar between groups.

In a study of 100 cardiac surgical patients at increased risk of postoperative AKI, patients were randomized to either 24 h of intravenous infusion of sodium bicarbonate (4 mmol/kg) or sodium chloride (4 mmol/kg) [164]. The primary outcome was AKI as defined by a postoperative increase in plasma creatinine concentration > 25% of baseline within the first five postoperative days. Secondary outcomes included urinary NGAL. The increase in plasma creatinine, plasma urea, urinary NGAL, and urinary NGAL/creatinine ratio was less in patients receiving sodium bicarbonate.

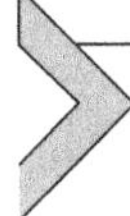

BIOMARKERS OF AKI IN THE ICU (TABLE 6.8)

NGAL

NGAL is the most widely studied biomarker of AKI in the ICU.

The ability of plasma NGAL at ICU admission to predict AKI in adult ICU patients was determined in 88 consecutive patients [170]. The sensitivity and specificity of plasma NGAL to predict AKI was 82 and 97%, respectively (AUROCC = 0.92). In patients without AKI at admission and who developed or did not develop AKI, the AUROCC was 0.956. The study concluded that in adult-ICU patients, plasma NGAL at ICU admission was an early biomarker of AKI. Plasma NGAL increased 48 h before RIFLE criteria [170].

In a prospective cohort study, the predictive value of urine- and plasma NGAL and plasma cystatin C to differentiate between sustained, transient, and absent AKI was determined in 700 ICU patients [169]. Urine NGAL was the only biomarker that significantly differentiated between sustained and transient AKI on ICU admission.

Evaluation of serum NGAL as an outcome-specific biomarker at the initiation of RRT, was performed in ICU patients [168]. There was a significant difference in serum NGAL between healthy subjects, critically ill patients with SIRS and critically ill patients with sepsis. Multiple linear regressions showed that serum-NGAL levels were independently related to the severity of AKI and the extent of systemic inflammation-NGAL levels were higher in nonsurvivors compared to survivors. Serum NGAL was a strong independent predictor for 28-day survival.

The diagnostic accuracy of plasma NGAL for early detection of AKI and need for RRT was determined in an adult ICU [167]. Plasma NGAL was a good diagnostic marker for AKI that developed within the next 48 h (AUROCC of 0.78) and for RRT therapy (AUROCC of 0.82). Peak plasma-NGAL concentrations increased with worsening AKI severity.

In 632 consecutive patients in a prospective cohort study, the ability of plasma- and urine NGAL compared to estimated GFR (eGFR) to predict

Table 6.8 Biomarkers of AKI in the ICU

Situation	Number	Biomarker	Study	AUROCC	References
Adults at ICU admission	632	Plasma- and urine NGAL	Predicts development of severe AKI (similarly to SCr-derived eGFR)	0.77–0.88	[166]
Adults	301	Plasma NGAL	Marker for AKI development and RRT use	0.78–0.82	[167]
Adults	109	Serum NGAL	Independent predictor of 28-day survival	ND	[168]
Adults	700	Urine NGAL	Predicts sustained AKI	ND	[169]
Adults	88	Urine NGAL	More useful than serum NGAL in predicting AKI	0.86	[64]
Adults at ICU admission	65	Plasma NGAL	Early biomarker of AKI	0.96	[170]
Adults	88	Plasma NGAL	Increased 48 h before RIFLE criteria	0.92	[170]
Children	140	Urine NGAL	Increased 2 days before 50% rise in SCr	0.78	[97]
Children	168	Serum NGAL	Increased in AKI versus no AKI		[62]
Adults	88	Serum NGAL	Predicts the development of AKI	0.96	[98]
Children with sepsis	143	Serum NGAL	Increases in sepsis		[62]
ED	635	Urine NGAL	Increased in AKI versus other renal diseases, predicts clinical outcomes	0.948	[150]
Critically ill, trauma	31	Urine NGAL	Predicts AKI	0.98	[145]
Adults on RRT	101	Urine IL-18	Correlates with hospital mortality	ND	[171]
ARDS adults	138	Urine IL-18	Predicts development of AKI 24 and 48 h later, predicts mortality	0.73	[82]
Children	137	Urine IL-18	Correlated with the severity of AKI, increases 2 days prior to SCr	ND	[88]
Infants postoperative	47	Urine IL-18, NGAL and cystatin C	Differentiates patients with good versus poor outcomes	0.62	[172]
Adults	451	Urine IL-18	Did not predict AKI development, but did predict poor clinical outcomes	0.67	[173]

Adults on RRT	101	Urine IL-18	Correlates with hospital mortality	ND	[171]
Adults	4512	Urine IL-18	Predicts AKI	0.66	[174]
Adults at ICU admission	150	Urine KIM-1	Increases by 6 h, peaks at 24 h, and remains elevated at 48 h	ND	[175]
Infants postoperative	49	Urine KIM-1	Does not differentiate good versus poor outcomes	ND	[172]
Adults	700	Urine KIM-1	Increases at the time of AKI	0.73	[176]
Adults	38	Urinary α-GST	Increased at ICU admission in all RIFLE groups	ND	[73]
ED versus ICU	77	Plasma NGAL	Measurement in ED not better than ICU for AKI diagnosis, 30-day mortality and dialysis	ND	[177]
Adults	26	GGT, AP, NAG, and GST	Predicts AKI	0.845–0.950	[108]
Sepsis adults	40	Urinary α-GST and π-GST	Increased in sepsis, not predictive of AKI	ND	[73]
Noncardiac children	160	NGAL	Diagnosis cystatin C-defined AKI	0.69	[178]
Adults	151	Serum- and urine cystatin C	Poor biomarkers for AKI and RRT	0.66	[179]
Adults	327	Plasma cystatin C	Not different between septic and nonseptic patients, predicts a composite outcome	0.78–0.80	[180]
Adults	47	Serum cystatin C	Outperforms SCr for the detection of an impaired GFR	0.94	[181]
Adults	50	Serum cystatin C	Correlates better with GFR (creatinine clearance) than SCr	0.927	[121]
Adults	422	Serum cystatin C	Rapidly detects AKI and predicts sustained AKI	0.80	[182]
Children	25	Serum cystatin C	Better than SCr to identify a creatinine clearance < 80 mL/min	0.792–0.851	[104]
Adults	80	Urinary L-FABP	Higher in septic shock versus sepsis without shock, AKI and controls	ND	[74]

(Continued)

Table 6.8 Biomarkers of AKI in the ICU (*cont.*)

Situation	Number	Biomarker	Study	AUROCC	References
Adults	145	Urinary NGAL and L-FABP	Higher in AKI, predicts mortality	0.73–0.78	[183]
Adults	337	Urinary L-FABP and NAG	Predicts mortality better than SCr	0.75	[184]
Adults	sCD163	80	Potential value of urine sCD163 levels for identifying sepsis and AKI	0.83	[185]
Adults	sTREM-1	104	Early diagnosis of sepsis, predicts AKI in sepsis patients	ND	[186]
Adults	RBP4	123	Decreased in critically ill ICU versus normal controls	ND	[187]
Adults on RRT	Ang-2	117	Higher in AKI patients with RIFLE category—Injury or Failure, predictor of mortality	ND	[188]
Adults	Resistin	230	Serum resistin was elevated in sepsis, associated with renal failure and unfavorable outcomes	ND	[189]

Ang-2, angiopoietin-2; AOPPs, advanced oxidation-protein products; AP, alkaline phosphatase; ATF3, activating transcription factor 3; ATN, acute tubular necrosis; BNP, brain natriuretic peptide; CKD, chronic kidney disease; CPB, cardiopulmonary bypass; DPP, dipeptidyl peptidase IV; GFR, glomerular filtration rate; GGT, γ-glutamyl transferase; γGT, γ-glutamyl transpeptidase; GST, glutathione-S-transferase; IGFBP7, urine insulin-like growth factor-binding protein 7; LAP, leucine aminopeptidase; L-FABP, liver-type fatty acid-binding protein; β2M, β2-microglobulin; NAG, *N*-acetyl-β-D-glucosaminidase; ND, not determined; NEP, neutral endopeptidase; PKD, polycystic kidney disease; RBP, retinol-binding protein; RBP4, retinol-binding protein 4; sCD163, soluble hemoglobin scavenging receptor; SCr, serum creatinine; TIMP2, tissue inhibitor of metalloproteinases-2; TREM-1, triggering receptor expressed on myeloid cells.

severe AKI was determined on ICU admission [166]. Plasma- and urine NGAL values at ICU admission were significantly related to AKI severity as determined by RIFLE criteria. NGAL measured at ICU admission predicted the development of severe AKI similarly to SCr-derived eGFR. It was also found that NGAL in combination with eGFR alone or with other clinical parameters has increased predictive ability.

It was determined whether plasma NGAL predicts adverse outcome in a heterogeneous multicenter group of 369 critically ill adult patients with acute respiratory failure [190]. The AUROCCs of baseline NGAL were as follows: 0.73 for RRT, 0.63 for hospital, and 0.58 for 90-day mortality. It was concluded that baseline plasma NGAL gives no additional value in prediction of hospital and 90-day mortality compared with RIFLE or SAPS II, and has only moderate predictive power regarding RRT in critically ill adult patients with acute respiratory failure.

A systematic review of the literature was performed to determine the value of plasma- and urinary NGAL to predict AKI in ICU patients [191]. The primary outcome measure was an AUROCC for NGAL to predict study outcomes. Eleven studies of plasma- or urine NGAL with a total of 2875 patients were included. The AUROCC for the prediction of AKI ranged from 0.54 to 0.98. Five studies reported the AUROCC for use of RRT ranging from 0.73 to 0.89, and four studies reported the AUROCC for mortality ranging from 0.58 to 0.83. There were no differences in the predictive values of urinary- and plasma NGAL. The study concluded that differences in study design and results made it difficult to evaluate the value of NGAL to predict AKI in the ICU and that NGAL seems to have reasonable value in predicting use of RRT but not mortality.

In summary, in the ICU, there is variation among studies and conditions like sepsis, COPD, cardiac dysfunction, age (NGAL seems superior in children), sex, and baseline renal function that may affect the sensitivity and specificity of NGAL as a biomarker of AKI in the ICU [192]. NGAL may be increased in sepsis without AKI [193]. Plasma NGAL is a marker of severity in patients with severe acute pancreatitis without AKI. Plasma NGAL may predict development of multiple organ failure and fatal outcome. Plasma NGAL may be increased in patients with COPD without AKI and is related to COPD severity. Plasma NGAL should be used with caution in critically ill patients due to the confounding factors discussed earlier [193]. In addition, the plasma- and urinary NGAL assays detected different molecular forms of NGAL and the different molecular forms have different predictive values as a biomarker of AKI [194]. Thus, while urine

NGAL is an early biomarker of AKI and a predictor of outcomes in critically ill children and adults in the ICU, further work is needed to describe the natural history of AKI, NGAL physiology in the ICU, and cutoff values of NGAL for AKI detection and prognostication [195].

IL-18

The ability of urine IL-18, measured within 24 h of ICU admission, to predict AKI, death, and receipt of acute dialysis was prospectively investigated in 451 ICU patients [173]. The AUROCC for urine IL-18 to predict subsequent AKI within 24 h was 0.62 and improved modestly to 0.67 in patients whose enrollment eGFR was greater than 75 mL/min. Urine IL-18 was an independent predictor of a composite outcome of death or acute dialysis within 28 days. Thus, urine IL-18 does not reliably predict AKI development but does predict poor clinical outcomes in a broadly selected, critically ill adult population.

A strong correlation between serum IL-18 and the hospital mortality of ICU patients with dialysis-dependent AKI was found [171] in serum samples collected from 101 critically ill patients at the initiation of RRT in the ICU. Serum IL-18- and cystatin C concentrations and Acute Physiology and Chronic Health Evaluation III (APACHE III) scores determined on the first day of RRT were independent predictors of hospital mortality.

Urine IL-18 as a biomarker of AKI in various clinical settings was analyzed in a metaanalysis of prospective studies [174]. Subgroup analysis showed the AUROCC of urinary IL-18 to predict AKI was 0.66 in ICU or coronary care-unit patients.

In a large multicenter study, urine IL-18 was measured in 1439 critically ill patients at ICU admission and at 24 h [196]. The highest urine IL-18 during the first 24 h in the ICU associated with the development of AKI with an AUROCC of 0.59 and with the development of Stage 3 AKI with an AUROCC of 0.67. IL-18 predicted the initiation of RRT with an AUROCC of 0.66 and 90-day mortality with an AUROCC of 0.54 Thus, in this study, IL-18 had poor-to-moderate ability to predict AKI, RRT, or 90-day mortality in this large cohort of critically ill patients.

To better understand the diagnostic and predictive performance of urinary biomarkers of kidney injury, biomarkers were stratified by baseline renal function and time after renal insult [197]. Five hundred twenty-nine patients in two general ICUs were studied. On ICU entry, no biomarker (GGT, AP, NGAL, cystatin C, KIM-1, and IL-18), had an AUROCC above 0.7 in the diagnosis or prediction of AKI. NGAL, cystatin

C, and IL-18 predicted dialysis with an AUROCC > 0.7. All, except KIM-1, predicted death at 7 days (AUROCC between 0.61 and 0.69). Performance was improved by stratification for eGFR or time or both. With eGFR < 60 mL/min, cystatin C and KIM-1 had an AUROCC of 0.69 and 0.73, respectively, within 6 h of injury. Between 12 and 36 h, the AUROCC was 0.88 for cystatin C, 0.85 for NGAL, and 0.94 for IL-18. The study concluded that the duration of injury and baseline renal function should be considered in evaluating biomarker performance to diagnose AKI in the ICU [197].

KIM-1

Urinary KIM-1 and netrin-1 were studied as early biomarkers of septic AKI in the ICU [175]. SCr started to increase at 24 h after of ICU admission. KIM-1 increased significantly by 6 h, peaked at 24 h, and remained significantly elevated until 48 h of ICU admission. Netrin-1 levels increased significantly at 1 h after ICU admission, peaked at 3–6 h, and remained elevated up to 48 h of ICU admission in septic AKI patients. Urinary KIM-1 levels in AKI patients at 24 h and 48 h were higher in nonsurvivors compared to survivors [175].

The influence of timing on the predictive values of KIM-1, brush-border enzymes, and NGAL measured before the rise of SCr in critically ill, nonseptic patients, was determined [176]. Both NGAL and KIM-1 concentrations gradually increased until AKI diagnosis. The brush-border enzymes, π- and α-GST, peaked at 24 h before AKI and declined rapidly afterward. NGAL predicted AKI at 24 h before AKI with an AUROCC of 0.79. KIM-1 was a good discriminator at the time of AKI only (AUROCC of 0.73) [176].

The predictive value of AKI biomarkers may depend on the time interval following tubular injury. In 700 adult critically ill patients, biomarkers were measured at 4 time-points prior to the rise in SCr ($T = 0$, −16, −20, and −24 h) [198]. Patients with sepsis and or AKI at ICU entry were excluded. Both NGAL ($p = 0.001$ at $T = -24$ versus non-AKI patients) and KIM-1 ($p < 0.0001$ at $T = 0$ versus non-AKI patients) concentrations gradually increased until AKI diagnosis, whereas π- and α-GST peaked at $T = -24$ before AKI and showed a rapid decline afterward. Thus, KIM-1 was a good discriminator at $T = 0$ only (AUROCC of 0.73). The study concluded that the time relationship between the biomarker measurements and the injurious event influences the individual test results.

In summary, while KIM-1 is a very promising biomarker of AKI in non-ICU patients, KIM-1 remains to be proven as an early biomarker of AKI and prognosis in the ICU in large study populations.

Brush-Border Enzymes

Biomarker performance on arrival in the ED was compared with subsequent performance in the ICU [177]. Urinary- and plasma NGAL, urinary cystatin C, urinary AP, urinary GGT and GST, and albumin were measured on ED presentation, and at 0, 4, 8, and 16 h, and 2, 4, and 7 days in the ICU in patients. It was determined that early measurement of biomarkers of AKI in the ED has utility, but is not better than measurement later in the ICU.

Urinary levels of π- and α-GST—markers of proximal and distal renal tubule damage, respectively—for the early diagnosis of AKI in the ICU were investigated in 38 patients [73]. Urinary α-GST level was not increased in patients who developed AKI versus non-AKI patients and was elevated early in all patients with sepsis syndrome.

Cystatin C

In 442 patients, cystatin C and creatinine were measured on admission to the ICU and then daily for 7 days [182]. Cystatin C predicted sustained AKI with an AUROCC of 0.80.

The effect of sepsis on levels of plasma cystatin C in AKI and non-AKI patients was determined [180]. Three hundred twenty-seven ICU patients were classified on the basis of the presence or absence of sepsis and AKI. The change in cystatin C or creatinine did not differ significantly between septic and nonseptic patients without or without AKI.

In another study, in 151 ICU patients, serum- and urine cystatin C were found to be poor biomarkers for AKI and RRT in the ICU [179]. The AUROCC to predict AKI for serum cystatin C 2 days before AKI was 0.72 and at 1 day before AKI was 0.62. Serum- and urine cystatin C on day 0 were poor predictors for the need for RRT (under the ROC curve ≤ 0.66) [179].

There are limitations to the use of cystatin C as a marker of GFR. Abnormalities of thyroid function [123] and glucocorticoid therapy [124,125] may affect cystatin C independently of kidney function. Levels of CRP may increase cystatin C levels and it has been suggested that cystatin C is a marker of inflammation [126].

L-FABP

Urine L-FABP and NGAL were measured in 145 patients at the time of admission to the medical and surgical ICU [183]. AKI patients had a significantly higher level of urinary NGAL and L-FABP and also higher mortality than non-AKI patients. The AUROCC was 0.773 for NGAL and 0.780 for L-FABP for the diagnosis of AKI. In multivariate Cox analysis, urinary L-FABP was an independent predictor for 90-day mortality. Urinary L-FABP is promising both for the diagnosis of AKI and for the prediction of prognosis in heterogeneous ICU patients [183]. Studies of L-FABP as a biomarker of AKI in the ICU are summarized in Table 6.8.

CD163

CD163 is a scavenger receptor for haptoglobin–hemoglobin complexes that is mostly expressed by monocytes and macrophages and is shedded [as soluble CD163 (sCD163)] by inflammatory stimuli. The diagnostic value of urine sCD163 for the identification of sepsis, severity of sepsis, AKI, and prognosis was determined in 60 ICU patients and 20 controls [185]. There were higher levels of urine sCD163 on the day of ICU admission in the sepsis group compared with the SIRS group. Urine-sCD163 levels at AKI diagnosis were significantly higher than the sepsis patients at ICU admission. The study shows the potential value of urine-sCD163 levels for identifying sepsis and AKI [185].

TREM-1

Triggering receptor expressed on myeloid cells-1 (TREM-1) is an amplifier of the innate immune response. Its soluble form (sTREM-1) acts as a decoy for the natural TREM-1 ligand and dampens the activation of TREM-1. sTREM-1 has received attention as a biomarker of sepsis in the ICU [199]. Urine sTREM-1 was evaluated for early sepsis identification, severity and prognosis assessment, and for the diagnosis of AKI in 104 patients in the ICU [186]. On the day of admission to the ICU, and compared with the SIRS group, the sepsis group exhibited higher levels of urine sTREM-1 and APACHE II scores. Urine sTREM-1 was higher in the severe sepsis group compared to the sepsis group. Urine sTREM-1 was increased in nonsurvivors. For 17 patients with AKI, urine-sTREM-1, SCr, and BUN levels at 48 h before AKI diagnosis were higher than in non-AKI subjects. Urine sTREM-1 was more sensitive than WBC and serum CRP for the early diagnosis of sepsis and for prognosis. Thus, urine sTREM-1 may also be an early marker of AKI in sepsis patients [186].

ANG-2

Endothelial activation is an early event in the pathogenesis of microcirculatory dysfunction, capillary leakage, and multiorgan dysfunction syndrome. Angiopoietin-2 (Ang-2), is a circulating antagonistic ligand of the endothelial-specific Tie2 receptor and has been identified as an important gatekeeper of endothelial activation. It was determined whether Ang-2 was as an outcome-specific biomarker in 117 critically ill patients requiring RRT in the ICU [188]. It was determined that circulating Ang-2 is a strong and independent predictor of mortality in dialysis-dependent ICU patients [188].

HSP-72

Urinary heat shock protein (HSP-72) was measured in 56 critically ill patients from 3 days before and until 2 days after the AKI diagnosis and in control patients without AKI [200]. Urinary HSP-72 levels rose at 3 days before the AKI diagnosis in critically ill patients compared to KIM-1, IL-18, and NGAL that were not increased. HSP-72 remained elevated during the AKI diagnosis. The sensitivity, specificity, and accuracy in the validation test for HSP-72 were 100, 83.3, and 90.9%, respectively.

Urinary GAGs

The endothelial glycocalyx plays a role in lung and kidney homeostasis [201,202]. The glycocalyx (or the endothelial surface layer in vivo) is a layer of glycosaminoglycans (GAGs) and associated proteoglycans lining the vascular lumen. The glycocalyx maintains the endothelial barrier and regulates leukocyte–endothelial adhesion. In a mouse model of sepsis, there is induction of renal heparanase, a heparan sulfate (HS)-specific endoglucuronidase [203]. HS is the most abundant glycocalyx GAG. The early pathogenic induction of glomerular heparanase was detected in mouse urine [203]. A high-sensitivity, mass spectrometry-based approach to detect and characterize fragmented GAGs within biologic samples from critically ill patients was developed [204]. In further studies, it was hypothesized that septic shock in humans would be associated with the pathologic degradation of the glomerular endothelial glycocalyx and excretion of GAG fragments into the urine and that urinary GAGs would be associated with the development of AKI and mortality. To test the hypothesis urinary GAGs were measured in two separate cohorts of critically ill trauma patients and ARDS patients [205]. It was found that septic shock was associated with an early increase in urinary GAGs and predictive of ongoing/progressive renal dysfunction in the ensuing 72 h. The ability of urinary GAGs to predict renal dysfunction persisted even after controlling for severity of illness. Urine HS and HA were associated with septic shock mortality and HS remained predictive of mortality after controlling for severity of illness. Urinary GAG fragmentation was present in ARDS patients with normal renal function at baseline who later developed AKI. These mass spectrometry-based measures could be largely replicated using an inexpensive, rapidly performed colorimetric assay of sulfated glycosaminoglycans. The study concluded that urinary GAGs are promising biomarkers with both diagnostic and prognostic implications in critical illness [205].

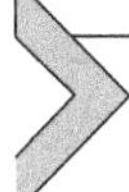

URINARY STABILITY STUDIES FOR BIOMARKERS OF AKI

Variations in storage and processing have the potential to influence biomarker measurements and may confound comparisons between different studies. The effect of centrifugation and storage at 4°C for 48 h before freezing at −80°C, centrifugation and storage at 25°C for 48 h before freezing at −80°C, and uncentrifuged samples immediately frozen at −80°C on urine concentrations of NGAL, IL-18, KIM-1, L-FABP, and cystatin C by established ELISA techniques, was determined [206]. Neither storing samples at 4°C for 48 h nor centrifugation had a significant effect on measured levels. Storing samples for NGAL, cystatin C, L-FABP, and KIM-1 at 25°C for 48 h, did not have a significant effect on measured levels. However, concordance correlation coefficients for IL-18 between samples stored at 25°C for 48 h and the reference standard was 0.81 (95% CI, 0.66–0.96). The study concluded that all candidate markers tested using the specified assays showed high stability with both short-term storage at 4°C and without centrifugation prior to freezing. For optimal accuracy, it was recommended that urine for IL-18 measurement should not be stored at 25°C before long-term storage or analysis.

The stability of biomarkers of AKI in children was determined [207]. Urine IL-18, NGAL, and KIM-1 were stable in urine stored at 4°C for 24 h, but showed significant degradation (5.6–10.1% from baseline) when stored at 25°C. All three biomarkers showed only a small although significant decrease in concentration (0.77–2.9% from baseline) after three freeze–thaw cycles. Similarly, all three biomarkers displayed only a small but significant decrease in concentration (0.84–3.2%) after storage for 5 years.

COMBINATIONS OF AKI BIOMARKERS

The classical biomarker paradigm is that one test detects one disease, for example, troponin for acute myocardial infarction, prostate-specific antigen for prostate cancer. However, AKI is a complex disease with multiple causes and it is possible that one biomarker will not be sufficient to make an early diagnosis. It has also been proposed that a panel of GFR markers be used to facilitate the detection of reduced GFR at various stages and in different populations [208].

Both urinary NGAL and IL-18 were measured in children that developed AKI after CPB [85]. NGAL increased 25-fold within 2 h and declined within 6 h after surgery. In contrast, urine IL-18 increased at 4–6 h after CPB, peaked at over 25-fold at 12 h, and remained markedly elevated up to 48 h after CPB. Also, on multivariate analysis, both IL-18 and NGAL were independently associated with number of days in AKI among cases. These results indicate that NGAL and IL-18 are increased in tandem after CPB. The combination of these two biomarkers may allow for the reliable early diagnosis and prognosis of AKI at all times after CPB, much before the rise in SCr [85]. A panel of biomarkers of AKI may improve the early diagnosis of AKI in different populations of patients with AKI.

One hundred nine patients undergoing open-heart surgeries were studied. Twenty-six patients developed AKI, defined as an increase in SCr of $\geq$ 0.3 mg/dL or $\geq$150% of baseline creatinine [209]. SCr, urinary L-FABP, and NGAL were tested preoperatively, at 0 and 2 h postoperation. The levels of urinary L-FABP and NGAL were significantly higher in AKI patients than non-AKI patients at 0 and 2 h postoperative. The AUROCC for L-FABP was 0.844 at 0 h and 0.832 at 2 h, 0.866 for NGAL at 0 h, and 0.871 at 2 h to predict AKI occurrence. A combination of L-FABP and NGAL analyzed at the same time point as mentioned earlier, had an AUROCC of 0.911–0.927 to predict AKI and an AUROCC of 0.81–0.87 to predict AKI Stage 2–3. Thus, the combination of the two biomarkers enhanced the accuracy of the early detection of postoperative AKI after cardiac surgery before a rise in SCr.

Urine NGAL at 2, 6, and 12 h postoperative and preoperative BNP was studied 135 children undergoing to cardiac surgery for congenital heart disease [210]. Primary endpoints were development of AKI (defined as a

1.5-fold increase in SCr) and intubation time. AKI occurred in 39% of patients. Urine NGAL values at 2 h had a good diagnostic accuracy for early diagnosis of AKI with an AUROCC of 0.85. Urine NGAL combined with preoperative BNP values were significantly associated with adverse outcome (longer intubation time and mortality).

Urinary levels of matrix metalloproteinase-9, NAG, and KIM-1 were examined in 44 patients with various acute and CKDs and 30 normal subjects in a cross-sectional study [211]. In addition, a case–control study of children undergoing CPB surgery was performed. AKI was defined as a greater than 50% increase in the SCr within the first 48 h after surgery. In the cross-sectional study, combining all three biomarkers achieved a perfect score, as determined by AUROCC, for diagnosing AKI. In the case–control study, KIM-1 was better than NAG at all time points for early diagnosis of AKI after CPB, but combining both was no better than KIM-1 alone. Urinary matrix metalloproteinase-9 was not a sensitive marker in the case–control study.

The diagnostic performance of 9 urinary biomarkers of AKI was evaluated in 204 patients with or without AKI: healthy volunteers, patients undergoing cardiac catheterization, and patients admitted to the ICU [212]. The biomarkers studied were: KIM-1, NGAL, IL-18, hepatocyte growth factor, cystatin C, NAG, VEGF, chemokine interferon-inducible protein 10, and total protein. Using a logic regression model, the area under the curve (0.94) was greater for the combination of biomarkers than for the individual biomarkers. Age-adjusted levels of urinary KIM-1, NAG, hepatocyte growth factor, VEGF, and total protein were significantly higher in patients who died or required RRT compared to those who survived or did not need RRT.

Urinary KIM-1, NGAL, and NAG were measured at five time points for the first 24 h after surgery in 90 adults undergoing cardiac surgery [213]. Thirty-six patients developed AKI as defined by an increase of SCr of 0.3 within 72 h after surgery. The AUROCCs to predict AKI immediately and 3 h postoperatively were 0.68 and 0.65 for KIM-1, 0.61 and 0.63 for NAG, and 0.59 and 0.65 for NGAL. Combining the three biomarkers improved the AUROCCs to 0.75 and 0.78. This study demonstrates that a combination of biomarkers may be better than individual biomarkers for the early detection of AKI before a rise in SCr.

In a systematic review of biomarkers of AKI, Parikh et al. determined methodologic quality of biomarker studies reported on MEDLINE and EMBASE databases between 2000 and 2006 [214]. In total, 31 studies evaluated

21 novel urine- and serum biomarkers of AKI. Urine IL-18, KIM-1, and NAG performed best in some studies for the diagnosis of established AKI. Serum cystatin C, urine NGAL, IL-18, and brush-border enzymes (GST) performed best for the early diagnosis of AKI. KIM-1 and IL-18 performed best for the prediction of mortality risk in patients with AKI.

The value of combining a functional damage biomarker (plasma cystatin C) with a tubular damage biomarker (urine NGAL) was determined. The goal was to develop a composite biomarker for prediction of discrete characteristics of AKI. Three hundred forty-five children undergoing CPB were analyzed. Severe AKI was defined as KDIGO Stages 2–3 within 7 days of CPB. The composite of plasma cystatin C and urinary NGAL had a greater likelihood than the change in SCr for severe AKI and persistent AKI. Biomarker composites had a greater probability for specific outcomes than the change in SCr.

In summary, more than one biomarker may be necessary to obtain sufficient sensitivity and specificity for AKI screening. In combination, a panel of injury biomarkers and functional biomarkers like cystatin C may result in a greater potential for the earlier diagnosis of AKI. In the future, clinicians that are aware of the limitations of different biomarkers in different diseases and at different time points after the AKI insult, may request a specific test or a panel of tests at specific time points.

TIMP2 AND IGFBP7 (TABLE 6.9)

Twelve patients with early recovery and 12 matching patients with late/nonrecovery were selected and their proteome analyzed by gel electrophoresis and mass spectrometry [215]. Eight prognostic candidates including α1M, α1-antitrypsin, apolipoprotein D, calreticulin, cathepsin D, CD59, IGFBP-7, and NGAL were identified. IGFBP7 predicted mortality (AUROCC of 0.68), recovery (AUROCC of 0.74;), and severity of AKI (AUROCC of 0.77), and was associated with the duration of AKI. IGFBP-7 was a more accurate predictor of renal outcome than NGAL. It was concluded that IGFBP-7 is a novel prognostic urinary marker that warrants further investigation.

Next, it was demonstrated that urinary TIMP2 is a biomarker of AKI in the ICU. Ninety-eight adult critically ill patients admitted to the ICU were prospectively studied. Urinary TIMP-2 was measured on ICU admission [220]. Urinary TIMP-2 was able to distinguish severe AKI from non-severe AKI with an AUROCC of 0.80. Urinary TIMP-2, NAG, and plasma

Table 6.9 Urinary TIMP-2 and IGFBP7

Situation	Number	Result	AUROCC	References
Adults in ICU, Sapphire study	728	Predicts AKI (KDIGO Stage 2–3) within 12 h	0.80	[216]
Adults with late AKI	12	IGFBP7 predicted mortality, recovery, severity of AKI, and duration of AKI, more accurate than NGAL	0.68–0.77	[215]
Adults in ICU, Topaz study	420	Urinary [TIMP-2] × [IGFBP7] > 0.3 had 7 times the risk for AKI		[217]
High risk surgical patients	107	[TIMP-2] × [IGFBP7] > 0.3 predicts risk of any AKI, early use of RRT, 28-day mortality 0.77	0.77–0.85	[218]
Children post-CPB	51	Diagnoses AKI at 4 h	0.85	[219]
High risk adults post-CPB	50	Diagnoses AKI at 4 h	0.84	[59]
Adults in ICU	98	Distinguishes severe AKI from nonsevere AKI, predicts in-hospital mortality	0.80	[220]
Adults in ICU, Opal study	154	[TIMP-2] × [IGFBP7] > 0.3 identifies patients at high risk for AKI		[221]
Adults in ICU	692	Identifies AKI patients at increased risk for mortality or receipt of RRT over the next 9 months	0.7	[222]

NGAL, were significantly higher in nonsurvivors than in survivors. Plasma IL-6 and EPO were not higher in nonsurvivors than in survivors. Urinary TIMP-2 was able to predict in-hospital mortality significantly better than SCr. Urinary TIMP-2 detected severe AKI with performance equivalent to plasma NGAL and urinary NAG, with an AUROCC value higher than 0.80.

IGFBP7 and TIMP2 were subsequently shown to be biomarkers of risk stratification in AKI in three studies: Sapphire, Topaz, and Opal studies. Three hundred forty candidate biomarkers were measured in critically ill ICU patients with sepsis or one or more risk factors for AKI, for example, hypotension, sepsis, and major trauma. In a discovery study, the biomarkers were ranked by the ability to predict RIFLE—Injury and Failure within 12–36 h. The two best biomarkers that were discovered were the cell-cycle arrest proteins, urinary IGF-BP7 and TIMP-2, both inducers of G1 cell-cycle arrest, a key mechanism implicated in AKI [216]. In the Sapphire validation study, in 728 critically ill patients, the primary endpoint was moderate to severe AKI (KDIGO Stage 2–3) within 12 h of sample collection [216]. IGFBP7 and TIMP-2 demonstrated an AUC of 0.80 (0.76 and 0.79 alone) for the primary endpoint. Urine concentrations of IGFBP7 and TIMP-2 were significantly superior to all previously described markers of AKI, for example, IL-18, NGAL and KIM-1, none of which achieved an AUC >0.72.

In the Topaz study, in 420 ICU patients, a predefined cutoff value of IGFBP7 and TIMP-2, was prospectively validated for risk assessment in AKI diagnosed by a clinical adjudication committee [217]. Critically ill patients with urinary [TIMP-2] × [IGFBP7] greater than 0.3 had 7 times the risk for AKI compared with critically ill patients with a test result below 0.3. Urinary [TIMP-2] × [IGFBP7] greater than 0.3 identified patients at risk for imminent AKI.

In a subsequent study, it was determined whether IGFBP7 and TIMP-2 predict AKI in high-risk surgical patients. A predefined cutoff value of [TIMP-2] × [IGFBP7] > 0.3 was used for assessing diagnostic accuracy. One hundred seven patients were included in the study, of whom 45 (42%) developed AKI. The AUROCC for the risk of any AKI was 0.85, for early use of RRT 0.83 and for 28-day mortality 0.77. Thus, urinary [TIMP-2] × [IGFBP7] test detects patients with risk of AKI after major noncardiac surgery [218].

TIMP2 and IGFBP7 were confirmed as biomarkers of AKI postcardiac surgery in children. Serial urine samples were analyzed for [TIMP-2] × [IGFBP7] in 51 children undergoing cardiac surgery with CPB [219]. The primary outcome measure was AKI defined by the pRIFLE criteria within 72 h after surgery. Urinary [TIMP-2] × [IGFBP7], 4 h following surgery demonstrated an AUROCC of 0.85 for the diagnosis of AKI. It was concluded that urinary [TIMP-2] × [IGFBP7] represent sensitive, specific, and highly predictive early biomarkers for AKI after surgery for congenital heart disease.

TIMP2 and IGFBP7 are biomarkers of AKI postcardiac surgery in adults. Serial urine samples were analyzed for [TIMP-2] × [IGFBP7] concentrations in 50 patients at high risk for AKI undergoing CPB. Twenty-six patients (52%) developed AKI. The diagnosis of AKI based on SCr and/or oliguria did not occur until 1–3 days after surgery. In contrast, the urine concentration of [TIMP-2] × [IGFBP7] rose from a mean of 0.49 at baseline to 1.51 at 4 h after surgery in patients who developed AKI. The maximum urinary [TIMP-2] × [IGFBP7] concentration achieved in the first 24 h following surgery had an AUROCC of 0.84. The sensitivity was 0.92 and specificity was 0.81 for a cutoff value of 0.50. The decline in urinary [TIMP-2] × [IGFBP7] values was the strongest predictor for renal recovery. In summary, urinary [TIMP-2] × [IGFBP7] serves as a sensitive and specific biomarker to predict AKI early after cardiac surgery and to predict renal recovery [59].

The development and diagnostic accuracy of two clinical cutoffs of TIMP2 and IGFBP2 was determined. Cutoffs for the sensitivity and specificity for prediction of KDIGO Stages 2–3 AKI within 12 h were determined using data from the multicenter Sapphire-cohort study. The cutoffs were then verified in a new study (Opal) enrolling 154 critically ill adults from 6 sites in the United States. One hundred subjects (14%) in Sapphire and 27 (18%) in Opal met the primary endpoint. The results of the Opal study replicated those of the Sapphire study. The relative risk in both studies for subjects testing at ≤0.3 versus >0.3 were 4.7 and 4.4. The relative risk for ≤0.3 versus >2 was 12 and 18. For the 0.3 cutoff, the sensitivity was 89% in both studies, and the specificity was 50 and 53%. For the cutoff of 2.0, the sensitivity was 42% and 44%, and specificity 95% and 90%. The study concluded that urinary [TIMP-2] × [IGFBP7] values of 0.3 or greater identify patients at high risk and those >2 at highest risk for AKI [221].

TIMP2 and IGFBP7 have now been validated for long term-outcomes in AKI. [TIMP-2] × [IGFBP7] levels are associated with adverse long-term outcomes in patients with AKI [222]. The 9-month incidence of a composite endpoint of all-cause mortality or the need for RRT in a secondary analysis of a prospective observational international study of critically ill adults was determined. Baseline [TIMP-2] × [IGFBP7] values were available for 692 subjects, of whom 382 (55.2%) subjects developed Stage 1 AKI within 72 h of enrollment and 217 (31.4%) subjects met the composite endpoint. Univariate analysis showed that [TIMP-2] × [IGFBP7] > 2 was associated with increased risk of the composite endpoint (hazard ratio, 2.11). In a multivariate analysis adjusted for the clinical model, [TIMP-2] × [IGFBP7] levels > 0.3 were associated with death or RRT only in subjects who developed AKI. In conclusion, [TIMP-2] × [IGFBP7] measured early in the setting of critical illness may identify patients with AKI at increased risk for mortality or receipt of RRT over the next 9 months.

The measurement of [TIMP-2] × [IGFBP7] has been marketed as NephroCheck and has recently been FDA approved for the detection of AKI in ICU patients. An FDA news release in September 2014, said the following: "Traditional laboratory tests catch AKI after the fact, and the results often do not come in until after the condition has progressed to a moderate or severe stage. In contrast, NephroCheck works as an early-warning system, identifying in the urine the presence of two proteins IGFBP7 and TIMP-2 that predict risk of developing AKI within the next 12 h, NephroCheck identified 92% of patients with AKI. However, NephroCheck reported false positives in roughly one in two patients without AKI in both tests."

A recent review nicely summarizes the FDA approval of NephroCheck [223]. The FDA approval of NephroCheck overcomes many of the laboratory standardization issues that have confronted other biomarkers of AKI [223]. The FDA approval of TIMP-2 and IGFBP7 and the point-of-care device to measure these biomarkers may accelerate NephroCheck use globally, which is a step forward in the biomarker field. Robust thresholds for AKI diagnosis and intervention have long been awaited. Whether the threshold used in Nephrocheck, which is based on a creatinine-dependent definition of AKI, will also identify "biomarker-positive, creatinine-negative" individuals at high risk of dialysis and death, remains uncertain. At present, there is not much information about the performance of these or other cell-cycle arrest markers in patients with impaired baseline renal function and in other situations that cause AKI, for example, contrast, sepsis, and nephrotoxins.

CONCLUSIONS

Biomarkers for AKI have a role to play both at the bedside in taking care of patients and in the design and conduct of clinical trials [224]. Biomarkers of AKI can be utilized for the initial diagnosis and staging, differential diagnosis, and prognosis. The FDA approval of the biomarker combination [TIMP2] × [IGFBP7] marketed as NephroCheck and the point-of-care apparatus to measure [TIMP2] × [IGFBP7] at the bedside has greatly advanced the field of biomarkers in AKI. The appropriate thresholds for [TIMP2] × [IGFBP7] for diagnosis and determination of severity are being established. However, the appropriate thresholds (cutoffs) for diagnosis or categories of severity have not been determined for damage biomarkers, such as, IL-18, NGAL, and KIM-1 [225]. Thresholds for diagnosis and determination of severity can vary with cause and context of AKI.

As measured by the AUROCC, biomarker performance varies considerably especially in heterogeneous populations. It is believed that even a modestly performing biomarker with an AUC less than 0.70 can assist in clinical decision making, simply by selecting an appropriate cutoff [225]. The question of how to optimally utilize biomarkers of AKI remains unresolved [226]. Biomarker performance is influenced by patient case mix, comorbid illness, the cause of the AKI, timing of measurement, the specific biomarker being measured, and the selected thresholds for diagnosis [226].

In an ADQI Consensus Conference, the use of biomarkers in AKI was discussed [227]. It was mentioned that several biomarkers are now available for assessing changes in kidney function (e.g., serum cystatin C) and detecting kidney damage (e.g., urinary NGAL, KIM-1, or IL-18). It was believed that a combination of kidney functional and damage markers simultaneously provides a simple method to stratify patients with AKI [227]. Utilization of a combination of functional and damage markers to evaluate patients with AKI will provide an improved understanding of the mechanisms and pathophysiology of AKI and the determination of prognosis [227]. Patients can be stratified as (1) no functional changes or damage, (2) loss of function with no damage, (3) damage without loss of function, and (4) damage with loss of function. Functional and damage markers can be used to assist in decisions related to triage of patients with AKI and identifying patients with who are at risk for progression. Set cutoffs for various biomarkers and their bedside utility are needed [228]. Combined use of functional and damage markers may advance the field of biomarkers of AKI in the future.

Are biomarkers ready for prime time [229]? It is believed that further studies are required to definitively demonstrate the association between early kidney damage biomarkers and clinical outcomes and to determine whether randomization to a treatment for AKI based on high structural/damage biomarker levels results in an improvement in kidney function and improves clinical outcomes and mortality [229]. Large multicenter interventional clinical trials to prevent AKI using biomarkers are needed [230]. Further research is needed before biomarkers can be routinely used in clinical practice [231].

REFERENCES

[1] Waikar SS, Curhan GC, Wald R, et al. Declining mortality in patients with acute renal failure, 1988 to 2002. J Am Soc Nephrol 2006;17(4):1143–50.
[2] Lameire N, Van Biesen W, Vanholder R. The changing epidemiology of acute renal failure. Nat Clin Pract Nephr 2006;2:364–77.

[3] Lassnigg A, Schmidlin D, Mouhieddine M, et al. Minimal changes of serum creatinine predict prognosis in patients after cardiothoracic surgery: a prospective cohort study. J Am Soc Nephrol 2004;15:1597–605.
[4] Van Biesen W, Vanholder R, Lameire N. Defining acute renal failure: RIFLE and beyond. Clin J Am Soc Nephrol 2006;1(6):1314–9.
[5] Bellomo R, Kellum JA, Ronco C. Defining and classifying acute renal failure: from advocacy to consensus and validation of the RIFLE criteria. Intensive Care Med 2007;33(3):409–13.
[6] Stevens LA, Lafayette RA, Perrone RD, Levey AS. Laboratory evaluation of kidney function. In: Schrier RW, editor. Diseases of the kidney and urinary tract. 8th ed. Philadelphia: Lippincott, Williams and Wilkins; 2007. p. 299–336.
[7] Star RA. Treatment of acute renal failure. Kidney Int 1998;54:1817–31.
[8] Herget-Rosenthal S, Pietruck F, Volbracht L, et al. Serum cystatin C—a superior marker of rapidly reduced glomerular filtration after uninephrectomy in kidney donors compared to creatinine. Clin Nephrol 2005;64(1):41–6.
[9] Moran SM, Myers BD. Course of acute renal failure studied by a model of creatinine kinetics. Kidney Int 1985;27:928–37.
[10] Waikar SS, Bonventre JV. Can we rely on blood urea nitrogen as a biomarker to determine when to initiate dialysis? Clin J Am Soc Nephrol 2006;1(5):903–4.
[11] Walser M. Determinants of ureagenesis, with particular reference to renal failure. Kidney Int 1980;17:709–21.
[12] Luke RG. Uremia and the BUN. New Engl J Med 1981;305:1213–5.
[13] Allgren RL, Marbury TC, Rahman SN, et al. Anaritide in acute tubular necrosis. N Engl J Med 1997;336:828–34.
[14] Lewis J, Salem MM, Chertow GM, et al. Atrial natriuretic factor in oliguric acute renal failure. Anaritide Acute Renal Failure Study Group. Am J Kidney Dis 2000;36: 767–74.
[15] Kellum JA. Prophylactic fenoldopam for renal protection? No, thank you, not for me—not yet at least. Critical Care Med 2005;33(11):2681–3.
[16] Dinarello CA. Biologic basis for interleukin-1 in disease. Blood 1996;87:2095–147.
[17] Melnikov VY, Ecder T, Fantuzzi G, et al. Impaired IL-18 processing protects caspase-1-deficient mice from ischemic acute renal failure. J Clin Invest 2001;107: 1145–52.
[18] Haq M, Norman J, Saba SR, et al. Role of IL-1 in renal ischemic reperfusion injury. J Am Soc Nephrol 1998;9:614–9.
[19] Melnikov VY, Faubel SG, Siegmund B, et al. Neutrophil-independent mechanisms of caspase-1- and IL-18-mediated ischemic acute tubular necrosis in mice. J Clin Invest 2002;110:1083–91.
[20] He Z, Altmann C, Hoke TS, et al. Interleukin-18 (IL-18) binding protein transgenic mice are protected against ischemic AKI. Am J Physiol Renal Physiol 2008;295: F1414–21.
[21] Wu H, Craft ML, Wang P, et al. IL-18 contributes to renal damage after ischemia-reperfusion. J Am Soc Nephrol 2008;19:2331–41.
[22] Edelstein CL, Hoke TS, Somerset H, et al. Proximal tubules from caspase-1 deficient mice are protected against hypoxia-induced membrane injury. Nephrol Dial Transplant 2007;22:1052–61.
[23] Dinarello CA, Fantuzzi G. Interleukin-18 and host defense against infection. J Infect Dis 2003;187(Suppl. 2):S370–84.
[24] Boraschi D, Dinarello CA. IL-18 in autoimmunity: review. Eu Cytokine Netw 2006;17:224–52.
[25] Faggioni R, Jones-Carson J, Reed DA, et al. Leptin-deficient (ob/ob) mice are protected from T cell-mediated hepatotoxicity: role of tumor necrosis factor alpha and IL-18. Proc Natl Acad Sci USA 2000;97:2367–72.

[26] Faggioni R, Cattley RC, Guo J, et al. IL-18-binding protein protects against lipopolysaccharide- induced lethality and prevents the development of Fas/Fas ligand-mediated models of liver disease in mice. J Immunol 2001;167:5913–20.
[27] Fiorucci S, Santucci L, Antonelli E, et al. NO-aspirin protects from T cell-mediated liver injury by inhibiting caspase-dependent processing ofTh1-like cytokines. Gastroenterology 2000;118:404–21.
[28] Dinarello CA. Interleukin-18 and the treatment of rheumatoid arthritis. Rheum Dis Clin North Am 2004;30:417–34.
[29] Dinarello CA. Novel targets for interleukin 18 binding protein. Ann Rheum Dis 2001;60(Suppl. 3):iii18–24.
[30] de Vries B, Matthijsen RA, van Bijnen AA, et al. Lysophosphatidic acid prevents renal ischemia-reperfusion injury by inhibition of apoptosis and complement activation. Am J Pathol 2003;163:47–56.
[31] Schmidt-Ott KM, Mori K, Li JY, et al. Dual action of neutrophil gelatinase-associated lipocalin. J Am Soc Nephrol 2007;18:407–13.
[32] Mishra J, Ma Q, Prada A, et al. Identification of neutrophil gelatinase-associated lipocalin as a novel early urinary biomarker for ischemic renal injury. J Am Soc Nephrol 2003;14:2534–43.
[33] Mishra J, Mori K, Ma Q, et al. Neutrophil gelatinase-associated lipocalin: a novel early urinary biomarker for cisplatin nephrotoxicity. Am J Nephrol 2004;24(3):307–15.
[34] Paragas N, Qiu A, Zhang Q, et al. The Ngal reporter mouse detects the response of the kidney to injury in real time. Nat Med 2011;17:216–22.
[35] Ichimura T, Bonventre JV, Bailly V, et al. Kidney injury molecule-1 (KIM-1), a putative epithelial cell adhesion molecule containing a novel immunoglobulin domain, is up-regulated in renal cells after injury. J Biol Chem 1998;273:4135–42.
[36] Ichimura T, Asseldonk EJ, Humphreys BD, et al. Kidney injury molecule-1 is a phosphatidylserine receptor that confers a phagocytic phenotype on epithelial cells. J Clin Invest 2008;118(5):1657–68.
[37] Vaidya VS, Ramirez V, Ichimura T, et al. Urinary kidney injury molecule-1: a sensitive quantitative biomarker for early detection of kidney tubular injury. Am J Physiol Renal Physiol 2006;290:F517–29.
[38] Ichimura T, Hung CC, Yang SA, et al. Kidney injury molecule-1: a tissue and urinary biomarker for nephrotoxicant-induced renal injury. Am J Physiol Renal Physiol 2004;286(3):F552–63.
[39] Zhou Y, Vaidya VS, Brown RP, et al. Comparison of kidney injury molecule-1 and other nephrotoxicity biomarkers in urine and kidney following acute exposure to gentamicin, mercury, and chromium. Toxicol Sci 2008;101(1):159–70.
[40] Vaidya VS, Ford GM, Waikar SS, et al. A rapid urine test for early detection of kidney injury. Kidney Int 2009;76(1):108–14.
[41] Kramer AB, van Timmeren MM, Schuurs TA, et al. Reduction of proteinuria in adriamycin-induced nephropathy is associated with reduction of renal kidney injury molecule (Kim-1) over time. Am J Physiol Renal Physiol 2009;296(5):F1136–45.
[42] Butler FA, Flynn FV. The occurrence of post-gamma protein in urine: a new protein abnormality. J Clin Pathol 1961;14:172–8.
[43] Westhuyzen J. Cystatin C: a promising marker and predictor of impaired renal function. Ann Clin Lab Sci 2006;36(4):387–94.
[44] Simonsen O, Grubb A, Thysell H. The blood serum concentration of cystatin C (gamma-trace) as a measure of the glomerular filtration rate. Scand J Clin Lab Inv 1985;45(2):97–101.
[45] Grubb A, Nyman U, Bjork J, et al. Simple cystatin C-based prediction equations for glomerular filtration rate compared with the modification of diet in renal disease prediction equation for adults and the Schwartz and the Counahan-Barratt prediction equations for children. Clin Chem 2005;51:1420–31.

[46] Uzun H, Ozmen KM, Ataman R, et al. Serum cystatin C level as a potentially good marker for impaired kidney function. Clin Biochem 2005;38(9):792–8.
[47] Artunc FH, Fischer IU, Risler T, Erley CM. Improved estimation of GFR by serum cystatin C in patients undergoing cardiac catheterization. Int J Cardiol 2005;102(2):173–8.
[48] Filler G, Bokenkamp A, Hofmann W, et al. Cystatin C: as a marker of GFR—history, indications, and future research. Clin Biochem 2005;38(1):1–8.
[49] Yokoyama T, Kamijo-Ikemori A, Sugaya T, et al. Urinary excretion of liver type fatty acid binding protein accurately reflects the degree of tubulointerstitial damage. Am J Pathol 2009;174(6):2096–106.
[50] Negishi K, Noiri E, Doi K, et al. Monitoring of urinary L-type fatty acid-binding protein predicts histological severity of acute kidney injury. Am J Pathol 2009;174(4):1154–9.
[51] Yamamoto T, Noiri E, Ono Y, et al. Renal L-type fatty acid-binding protein in acute ischemic injury. J Am Soc Nephrol 2007;18(11):2894–902.
[52] Wang W, Brian RW, Ramesh G. Netrin-1 and kidney injury. I. Netrin-1 protects against ischemia-reperfusion injury of the kidney. Am J Physiol Renal Physiol 2008;294(4):F739–47.
[53] Wang W, Reeves WB, Pays L, et al. Netrin-1 overexpression protects kidney from ischemia reperfusion injury by suppressing apoptosis. Am J Pathol 2009;175(3):1010–8.
[54] Brian RW, Kwon O, Ramesh G. Netrin-1 and kidney injury. II. Netrin-1 is an early biomarker of acute kidney injury. Am J Physiol Renal Physiol 2008;294(4):F731–8.
[55] Zhou H, Cheruvanky A, Hu X, et al. Urinary exosomal transcription factors, a new class of biomarkers for renal disease. Kidney Int 2008;74(5):613–21.
[56] Zhou H, Pisitkun T, Aponte A, et al. Exosomal Fetuin-A identified by proteomics: a novel urinary biomarker for detecting acute kidney injury. Kidney Int 2006;70(10):1847–57.
[57] du Cheyron D, Daubin C, Poggioli J, et al. Urinary measurement of Na^+/H^+ exchanger isoform 3 (NHE3) protein as new marker of tubule injury in critically ill patients with ARF. Am J Kidney Dis 2003;42:497–506.
[58] Salih M, Zietse R, Hoorn EJ. Urinary extracellular vesicles and the kidney: biomarkers and beyond. Am J Physiol Renal Physiol 2014;306:F1251–9.
[59] Meersch M, Schmidt C, Van AH, et al. Urinary TIMP-2 and IGFBP7 as early biomarkers of acute kidney injury and renal recovery following cardiac surgery. PLoS ONE 2014;9:e93460.
[60] Kellum JA, Chawla LS. Cell-cycle arrest and acute kidney injury: the light and the dark sides. Nephrol Dial Transplant 2015;31(1):16–22.
[61] Parikh CR, Jani A, Melnikov VY, et al. Urinary interleukin-18 is a marker of human acute tubular necrosis. Am J Kidney Dis 2004;43:405–14.
[62] Wheeler DS, Devarajan P, Ma Q, et al. Serum neutrophil gelatinase-associated lipocalin (NGAL) as a marker of acute kidney injury in critically ill children with septic shock. Critical Care Med 2008;36:1297–303.
[63] Aydogdu M, Gursel G, Sancak B, et al. The use of plasma and urine neutrophil gelatinase associated lipocalin (NGAL) and Cystatin C in early diagnosis of septic acute kidney injury in critically ill patients. Dis Markers 2013;34:237–46.
[64] Martensson J, Bell M, Oldner A, et al. Neutrophil gelatinase-associated lipocalin in adult septic patients with and without acute kidney injury. Intensive Care Med 2010;36:1333–40.
[65] Bolignano D, Coppolino G, Campo S, et al. Neutrophil gelatinase-associated lipocalin in patients with autosomal-dominant polycystic kidney disease. Am J Nephrol 2007;27:373–8.

[66] Trachtman H, Christen E, Cnaan A, et al. Urinary neutrophil gelatinase-associated lipocalcin in D + HUS: a novel marker of renal injury. Pediatr Nephrol 2006;21:989–94.
[67] Soler-Garcia AA, Johnson D, Hathout Y, Ray PE. Iron-related proteins: candidate urine biomarkers in childhood HIV-associated renal diseases. Clin J Am Soc Nephrol 2009;4(4):763–71.
[68] Suzuki M, Wiers KM, Klein-Gitelman MS, et al. Neutrophil gelatinase- associated lipocalin as a biomarker of disease activity in lupus nephritis. Pediatr Nephrol 2008;23:403–12.
[69] Ariza X, Sola E, Elia C, et al. Analysis of a urinary biomarker panel for clinical outcomes assessment in cirrhosis. PLoS ONE 2015;10:e0128145.
[70] Nickolas TL, O'Rourke MJ, Yang J, et al. Sensitivity and specificity of a single emergency department measurement of urinary neutrophil gelatinase-associated lipocalin for diagnosing acute kidney injury. Ann Int Med 2008;148:810–9.
[71] van Timmeren MM, van den Heuvel MC, Bailly V, et al. Tubular kidney injury molecule-1 (KIM-1) in human renal disease. J Pathol 2007;212:209–17.
[72] Han WK, Bailly V, Abichandani R, et al. Kidney Injury Molecule-1 (KIM-1): a novel biomarker for human renal proximal tubule injury. Kidney Int 2002;62:237–44.
[73] Walshe CM, Odejayi F, Ng S, Marsh B. Urinary glutathione S-transferase as an early marker for renal dysfunction in patients admitted to intensive care with sepsis. Crit Care Resusc 2009;11:204–9.
[74] Nakamura T, Sugaya T, Koide H. Urinary liver-type fatty acid-binding protein in septic shock: effect of polymyxin B-immobilized fiber hemoperfusion. Shock 2009;31(5):454–9.
[75] Orlando R, Mussap M, Plebani M, et al. Diagnostic value of plasma cystatin C as a glomerular filtration marker in decompensated liver cirrhosis. Clin Chem 2002;48(6 Pt 1):850–8.
[76] Gerbes AL, Gulberg V, Bilzer M, Vogeser M. Evaluation of serum cystatin C concentration as a marker of renal function in patients with cirrhosis of the liver. Gut 2002;50:106–10.
[77] Yu H, Yanagisawa Y, Forbes MA, et al. Alpha-1-microglobulin: an indicator protein for renal tubular function. J Clin Pathol 1983;36(3):253–9.
[78] Hartmann HG, Braedel HE, Jutzler GA. Detection of renal tubular lesions after abdominal aortography and selective renal arteriography by quantitative measurements of brush-border enzymes in the urine. Nephron 1985;39(2):95–101.
[79] Susantitaphong P, Perianayagam MC, Tighiouart H, et al. Urinary alpha- and piglutathione s-transferases for early detection of acute kidney injury following cardiopulmonary bypass. Biomarkers 2013;18:331–7.
[80] Mou S, Wang Q, Li J, et al. Urinary excretion of liver-type fatty acid-binding protein as a marker of progressive kidney function deterioration in patients with chronic glomerulonephritis. Clin Chim Acta 2012;413:187–91.
[81] Mehta RL. Urine IL-18 levels as a predictor of acute kidney injury in intensive care patients. Nat Clin Pract Nephrol 2006;2(5):252–3.
[82] Parikh CR, Abraham E, Ancukiewicz M, Edelstein CL. Urine IL-18 is an early diagnostic marker for acute kidney injury and predicts mortality in the ICU. J Am Soc Nephrol 2005;16:3046–52.
[83] Parikh CR, Devarajan P, Zappitelli M, et al. Postoperative biomarkers predict acute kidney injury and poor outcomes after adult cardiac surgery. J Am Soc Nephrol 2011;22(9):1748–57.
[84] Parikh CR, Devarajan P, Zappitelli M, et al. Postoperative biomarkers predict acute kidney injury and poor outcomes after pediatric cardiac surgery. J Am Soc Nephrol 2011;22(9):1737–47.

[85] Parikh CR, Mishra J, Thiessen-Philbrook H, et al. Urinary IL-18 is an early predictive biomarker of acute kidney injury after cardiac surgery. Kidney Int 2006;70:199–203.
[86] Xin C, Yulong X, Yu C, et al. Urine neutrophil gelatinase-associated lipocalin and interleukin-18 predict acute kidney injury after cardiac surgery. Renal Failure 2008;30(9):904–13.
[87] Haase M, Bellomo R, Story D, et al. Urinary interleukin-18 does not predict acute kidney injury after adult cardiac surgery: a prospective observational cohort study. Crit Care 2008;12(4):R96.
[88] Washburn KK, Zapitelli M, Arikan AA, et al. Urinary interleukin-18 as an acute kidney injury biomarker in critically ill children. Nephrol Dial Transplant 2008;23: 566–72.
[89] Bulent Gul C, Gullulu M, Oral B, et al. Urinary IL-18: a marker of contrast-induced nephropathy following percutaneous coronary intervention. Clin Biochem 2008;41:544–7.
[90] Mishra J, Dent C, Tarabishi R, et al. Neutrophil gelatinase-associated lipocalin (NGAL) as a biomarker for acute renal injury after cardiac surgery. Lancet 2005;365:1231–8.
[91] Dent C, Dastrala S, Bennet M, et al. Plasma NGAL predicts AKI, morbidity and mortality after pediatric cardiac surgery: a prospective uncontrolled cohort study. Crit Care 2007;11:R127–32.
[92] Wagener G, Jan M, Kim M, et al. Association between increases in urinary neutrophil gelatinase-associated lipocalin and acute renal dysfunction after adult cardiac surgery. Anesthesiology 2006;105(3):485–91.
[93] Tuladhar SM, Puntmann VO, Soni M, et al. Rapid detection of acute kidney injury by plasma and urinary neutrophil gelatinase-associated lipocalin after cardiopulmonary bypass. J Cardiovasc Pharmacol 2009;53(3):261–6.
[94] Wagener G, Gubitosa G, Wang S, et al. Urinary neutrophil gelatinase-associated lipocalin in acute kidney injury after cardiac surgery. Am J Kidney Dis 2008;52:425–33.
[95] Han WK, Wagener G, Zhu Y, et al. Urinary biomarkers in the early detection of acute kidney injury after cardiac surgery. Clin J Am Soc Nephrol 2009;4(5):873–82.
[96] Bachorzewska-Gajewska H, Malyszko J, Sitniewska E, et al. Neutrophil-gelatinase-associated lipocalin and renal function after percutaneous coronary interventions. Am J Nephrol 2006;26(3):287–92.
[97] Zapitelli M, Washburn KK, Arikan AA, et al. Urine neutrophil gelatinase-associated lipocalin is an early marker of acute kidney injury in critically ill children: a prospective cohort study. Crit Care 2007;11:R84.
[98] Constantin JM, Futier E, Perbet S, et al. Plasma NGAL is an early marker of acute kidney injury in adult critically ill patients: a prospective study. J Crit Care 2010;25(1):176.
[99] Matsa R, Ashley E, Sharma V, et al. Plasma and urine neutrophil gelatinase-associated lipocalin in the diagnosis of new onset acute kidney injury in critically ill patients. Crit Care 2014;18:R137.
[100] Parikh CR, Thiessen-Philbrook H, Garg AX, et al. Performance of kidney injury molecule-1 and liver fatty acid-binding protein and combined biomarkers of AKI after cardiac surgery. Clin J Am Soc Nephrol 2013;8:1079–88.
[101] Shao X, Tian L, Xu W, et al. Diagnostic value of urinary kidney injury molecule 1 for acute kidney injury: a meta-analysis. PLoS ONE 2014;9:e84131.
[102] Herget-Rosenthal S, Marggraf G, Husing J, et al. Early detection of acute renal failure by serum cystatin C. Kidney Int 2004;66:1115–22.
[103] Koyner JL, Bennet MR, Worcester EM, et al. Urinary cystatin c as an early biomarker of acute kidney injury following adult cardiothoracic surgery. Kidney Int 2008;74(8):1059–69.
[104] Herrero-Morin JD, Malaga S, Fernandez N, et al. Cystatin C and beta2-microglobulin: markers of glomerular filtration in critically ill children. Crit Care 2007;11:R59.

[105] Koyner JL, Garg AX, Shlipak MG, et al. Urinary cystatin C and acute kidney injury after cardiac surgery. Am J Kidney Dis 2013;61:730–8.

[106] Zappitelli M, Krawczeski CD, Devarajan P, et al. Early postoperative serum cystatin C predicts severe acute kidney injury following pediatric cardiac surgery. Kidney Int 2011;80:655–62.

[107] Portilla D, Dent C, Sugaya T, et al. Liver fatty acid binding protein as a biomarker of acute kidney injury after cardiac surgery. Kidney Int 2008;73:465–72.

[108] Westhuyzen J, Endre ZH, Reece G, et al. Measurement of tubular enzymuria facilitates early detection of acute renal impairment in the intensive care unit. Nephrol Dial Transplant 2003;18:543–51.

[109] Herget-Rosenthal S, Poppen D, Husing J, et al. Prognostic value of tubular proteinuria and enzymuria in nonoliguric acute tubular necrosis. Clin Chem 2004;50(3):552–8.

[110] Zappitelli M, Coca SG, Garg AX, et al. The association of albumin/creatinine ratio with postoperative AKI in children undergoing cardiac surgery. Clin J Am Soc Nephrol 2012;7:1761–9.

[111] Haase-Fielitz A, Bellomo R, Devarajan P, et al. Novel and conventional serum biomarkers predicting acute kidney injury in adult cardiac surgery—a prospective cohort study. Crit Care Med 2009;37(2):553–60.

[112] Haase M, Bellomo R, Devarajan P, et al. Novel biomarkers early predict the severity of acute kidney injury after cardiac surgery in adults. Ann Thorac Surg 2009;88(1): 124–30.

[113] Haase-Fielitz A, Bellomo R, Devarajan P, et al. The predictive performance of plasma neutrophil gelatinase-associated lipocalin (NGAL) increases with grade of acute kidney injury. Nephrol Dial Transplant 2009;24(11):3349–54.

[114] Hirsch R, Dent C, Pfriem H, et al. NGAL is an early predictive biomarker of contrast-induced nephropathy in children. Pediatr Nephrol 2007;22:2089–95.

[115] Lavery AP, Meinzen-Derr JK, Anderson E, et al. Urinary NGAL in premature infants. Pediatr Res 2008;64(4):423–8.

[116] Tabel Y, Elmas A, Ipek S, et al. Urinary neutrophil gelatinase-associated lipocalin as an early biomarker for prediction of acute kidney injury in preterm infants. Am J Perinatol 2014;31:167–74.

[117] Haase-Fielitz A, Haase M, Devarajan P. Neutrophil gelatinase-associated lipocalin as a biomarker of acute kidney injury: a critical evaluation of current status. Ann Clin Biochem 2014;51:335–51.

[118] Blaikley J, Sutton P, Walter M, et al. Tubular proteinuria and enzymuria following open heart surgery. Intensive Care Med 2003;29(8):1364–7.

[119] Fink JC, Cooper MA, Zager RA. Hemodialysis exacerbates enzymuria in patients with acute renal failure: brief report. Renal Failure 1996;18(6):947–50.

[120] Sarvary E, Borka P, Sulyok B, et al. Diagnostic value of urinary enzyme determination in renal transplantation. Transplant Int 1996;9(Suppl. 1):S68–72.

[121] Villa P, Jimenez M, Soriano MC, et al. Serum cystatin C concentration as a marker of acute renal dysfunction in critically ill patients. Crit Care 2005;9(2):R139–43.

[122] Baas MC, Bouman CS, Hoek FJ, et al. Cystatin C in critically ill patients treated with continuous venovenous hemofiltration. Hemodial Int 2006;10(Suppl. 2):S33–7.

[123] Manetti L, Pardini E, Genovesi M, et al. Thyroid function differently affects serum cystatin C and creatinine concentrations. J Endocrinol Invest 2005;28:346–9.

[124] Risch L, Herklotz R, Blumberg A, Huber AR. Effects of glucocorticoid immunosuppression on serum cystatin C concentrations in renal transplant patients. Clin Chem 2001;47:2055–9.

[125] Risch L, Huber AR. Glucocorticoids and increased serum cystatin C concentrations. Clin Chim Acta 2002;320:133–4.

[126] Knight EL, Verhave JC, Spiegelman D, et al. Factors influencing serum cystatin C levels other than renal function and the impact on renal function measurement. Kidney Int 2004;65:1416–21.
[127] Ruf B, Bonelli V, Balling G, et al. Intraoperative renal near-infrared spectroscopy indicates developing acute kidney injury in infants undergoing cardiac surgery with cardiopulmonary bypass: a case-control study. Crit Care 2015;19:27.
[128] Christov M, Waikar SS, Pereira RC, et al. Plasma FGF23 levels increase rapidly after acute kidney injury. Kidney Int 2013;84:776–85.
[129] Ramesh G, Kwon O, Ahn K. Netrin-1: a novel universal biomarker of human kidney injury. Transplant Proc 2010;42:1519–22.
[130] Tu Y, Wang H, Sun R, et al. Urinary netrin-1 and KIM-1 as early biomarkers for septic acute kidney injury. Renal Failure 2014;36:1559–63.
[131] Ramesh G, Krawczeski CD, Woo JG, et al. Urinary netrin-1 is an early predictive biomarker of acute kidney injury after cardiac surgery. Clin J Am Soc Nephrol 2010;5:395–401.
[132] Lewandowska L, Matuszkiewicz-Rowinska J, Jayakumar C, et al. Netrin-1 and semaphorin 3A predict the development of acute kidney injury in liver transplant patients. PLoS ONE 2014;9:e107898.
[133] du CD, Fradin S, Ramakers M, et al. Angiotensin converting enzyme insertion/deletion genetic polymorphism: its impact on renal function in critically ill patients. Crit Care Med 2008;36(12):3178–83.
[134] Sirota JC, Walcher A, Faubel S, et al. Urine IL-18, NGAL, IL-8 and serum IL-8 are biomarkers of acute kidney injury following liver transplantation. BMC Nephrol 2013;14:17.
[135] Ling W, Zhaohui N, Ben H, et al. Urinary IL-18 and NGAL as early predictive biomarkers in contrast-induced nephropathy after coronary angiography. Nephron 2008;108:c176–81.
[136] Awad AS, Okusa MD. Distant organ injury following acute kidney injury. Am J Physiol Renal Physiol 2007;293(1):F28–9.
[137] Nisula S, Yang R, Poukkanen M, et al. Predictive value of urine interleukin-18 in the evolution and outcome of acute kidney injury in critically ill adult patients. Br J Anaesth 2015;114:460–8.
[138] Spahillari A, Parikh CR, Sint K, et al. Serum cystatin C- versus creatinine-based definitions of acute kidney injury following cardiac surgery: a prospective cohort study. Am J Kidney Dis 2012;60:922–9.
[139] Belcher JM, Garcia-Tsao G, Sanyal AJ, et al. Urinary biomarkers and progression of AKI in patients with cirrhosis. Clin J Am Soc Nephrol 2014;9:1857–67.
[140] Alge JL, Karakala N, Neely BA, et al. Urinary angiotensinogen and risk of severe AKI. Clin J Am Soc Nephrol 2013;8:184–93.
[141] Liangos O, Perianayagam MC, Vaidya VS, et al. Urinary N-Acetyl-beta-(D)-glucosaminidase activity and kidney injury molecule-1 level are associated with adverse outcomes in acute renal failure. J Am Soc Nephrol 2007;18:904–12.
[142] Murugan R, Wen X, Shah N, et al. Plasma inflammatory and apoptosis markers are associated with dialysis dependence and death among critically ill patients receiving renal replacement therapy. Nephrol Dial Transplant 2014;29:1854–64.
[143] Parr SK, Clark AJ, Bian A, et al. Urinary L-FABP predicts poor outcomes in critically ill patients with early acute kidney injury. Kidney Int 2015;87:640–8.
[144] Arthur JM, Hill EG, Alge JL, et al. Evaluation of 32 urine biomarkers to predict the progression of acute kidney injury after cardiac surgery. Kidney Int 2014;85:431–8.
[145] Makris K, Markou N, Evodia E, et al. Urinary neutrophil gelatinase-associated lipocalin (NGAL) as an early marker of acute kidney injury in critically ill multiple trauma patients. Clin Chem Lab Med 2009;47(1):79–82.

[146] Koyner JL, Garg AX, Coca SG, et al. Biomarkers predict progression of acute kidney injury after cardiac surgery. J Am Soc Nephrol 2012;23:905–14.
[147] Patel UD, Garg AX, Krumholz HM, et al. Preoperative serum brain natriuretic peptide and risk of acute kidney injury after cardiac surgery. Circulation 2012;125:1347–55.
[148] Schaub JA, Garg AX, Coca SG, et al. Perioperative heart-type fatty acid binding protein is associated with acute kidney injury after cardiac surgery. Kidney Int 2015;88(3):576–83.
[149] Bucholz EM, Whitlock RP, Zappitelli M, et al. Cardiac biomarkers and acute kidney injury after cardiac surgery. Pediatrics 2015;135:e945–56.
[150] Nickolas TL, Schmidt-Ott KM, Canetta P, et al. Diagnostic and prognostic stratification in the emergency department using urinary biomarkers of nephron damage: a multicenter prospective cohort study. J Am Coll Cardiol 2012;59(3):246–55.
[151] Hsu RK, Hsu CY. We can diagnose AKI "early". Clin J Am Soc Nephrol 2012;7:1741–2.
[152] Coca SG, Garg AX, Thiessen-Philbrook H, et al. Urinary biomarkers of AKI and mortality 3 years after cardiac surgery. J Am Soc Nephrol 2014;25:1063–71.
[153] Haase M, Kellum JA, Ronco C. Subclinical AKI—an emerging syndrome with important consequences. Nat Rev Nephrol 2012;8:735–9.
[154] Nejat M, Pickering JW, Devarajan P, et al. Some biomarkers of acute kidney injury are increased in pre-renal acute injury. Kidney Int 2012;81:1254–62.
[155] Zappitelli M, Greenberg JH, Coca SG, et al. Association of definition of acute kidney injury by cystatin C rise with biomarkers and clinical outcomes in children undergoing cardiac surgery. JAMA Pediatr 2015;169:583–91.
[156] Kokot M, Biolik G, Ziaja D, et al. Assessment of subclinical acute kidney injury after abdominal aortic aneurysm surgery using novel markers: L-FABP and H-FABP. Nefrologia 2014;34:628–36.
[157] Siew ED, Ware LB, Bian A, et al. Distinct injury markers for the early detection and prognosis of incident acute kidney injury in critically ill adults with preserved kidney function. Kidney Int 2013;84:786–94.
[158] Capuano F, Goracci M, Luciani R, et al. Neutrophil gelatinase-associated lipocalin levels after use of mini-cardiopulmonary bypass system. Interact Cardiovasc Thorac Surg 2009;9(5):797–801.
[159] Wagener G, Gubitosa G, Wang S, et al. A comparison of urinary neutrophil gelatinase-associated lipocalin in patients undergoing on- versus off-pump coronary artery bypass graft surgery. J Cardiothorac Vasc Anesth 2009;23(2):195–9.
[160] Lemaire SA, Jones MM, Conklin LD, et al. Randomized comparison of cold blood and cold crystalloid renal perfusion for renal protection during thoracoabdominal aortic aneurysm repair. J Vasc Surg 1919;49(1):11–9.
[161] Molnar AO, Parikh CR, Coca SG, et al. Association between preoperative statin use and acute kidney injury biomarkers in cardiac surgical procedures. Ann Thorac Surg 2014;97:2081–7.
[162] Kim JH, Shim JK, Song JW, et al. Effect of erythropoietin on the incidence of acute kidney injury following complex valvular heart surgery: a double blind, randomized clinical trial of efficacy and safety. Crit Care 2013;17:R254.
[163] de SS, Ponte B, Weiss L, et al. Epoetin administrated after cardiac surgery: effects on renal function and inflammation in a randomized controlled study. BMC Nephrol 2012;13:132.
[164] Haase M, Haase-Fielitz A, Bellomo R, et al. Sodium bicarbonate to prevent increases in serum creatinine after cardiac surgery: a pilot double-blind, randomized controlled trial. Crit Care Med 2009;37:39–47.
[165] Keys DO, Edelstein CL. High-potency statins are associated with increased hospitalisations with acute kidney injury. Evid Based Med 2014;19:28.

[166] de Geus HR, Bakker J, Lesaffre EM, le Noble JL. Neutrophil gelatinase-associated lipocalin at ICU admission predicts for acute kidney injury in adult patients. Am J Respir Crit Care Med 2011;183:907–14.
[167] Cruz DN, de CM, Garzotto F, et al. Plasma neutrophil gelatinase-associated lipocalin is an early biomarker for acute kidney injury in an adult ICU population. Intensive Care Med 2010;36:444–51.
[168] Kumpers P, Hafer C, Lukasz A, et al. Serum neutrophil gelatinase-associated lipocalin at inception of renal replacement therapy predicts survival in critically ill patients with acute kidney injury. Crit Care 2010;14:R9.
[169] de Geus HR, Woo JG, Wang Y, et al. Urinary neutrophil gelatinase-associated lipocalin measured on admission to the intensive care unit accurately discriminates between sustained and transient acute kidney injury in adult critically ill patients. Nephron Extra 2011;1:9–23.
[170] Constantin JM, Futier E, Perbet S, et al. Plasma neutrophil gelatinase-associated lipocalin is an early marker of acute kidney injury in adult critically ill patients: a prospective study. J Crit Care 2010;25:176.
[171] Lin CY, Chang CH, Fan PC, et al. Serum interleukin-18 at commencement of renal replacement therapy predicts short-term prognosis in critically ill patients with acute kidney injury. PLoS ONE 2013;8:e66028.
[172] Hazle MA, Gajarski RJ, Aiyagari R, et al. Urinary biomarkers and renal near-infrared spectroscopy predict intensive care unit outcomes after cardiac surgery in infants younger than 6 months of age. J Thorac Cardiovasc Surg 2013;146:861–7.
[173] Siew ED, Ikizler TA, Gebretsadik T, et al. Elevated urinary IL-18 levels at the time of ICU admission predict adverse clinical outcomes. Clin J Am Soc Nephrol 2010;5:1497–505.
[174] Liu Y, Guo W, Zhang J, et al. Urinary interleukin 18 for detection of acute kidney injury: a meta-analysis. Am J Kidney Dis 2013;62:1058–67.
[175] Tu Y, Wang H, Sun R, et al. Urinary netrin-1 and KIM-1 as early biomarkers for septic acute kidney injury. Renal Failure 2014;36:1559–63.
[176] de Geus HR, Fortrie G, Betjes MG, et al. Time of injury affects urinary biomarker predictive values for acute kidney injury in critically ill, non-septic patients. BMC Nephrol 2013;14:273.
[177] Ralib A, Pickering JW, Shaw GM, et al. The clinical utility window for acute kidney injury biomarkers in the critically ill. Crit Care 2014;18:601.
[178] Lagos-Arevalo P, Palijan A, Vertullo L, et al. Cystatin C in acute kidney injury diagnosis: early biomarker or alternative to serum creatinine? Pediatr Nephrol 2015;30(4):665–76.
[179] Royakkers AA, Korevaar JC, van Suijlen JD, et al. Serum and urine cystatin C are poor biomarkers for acute kidney injury and renal replacement therapy. Intensive Care Med 2011;37:493–501.
[180] Martensson J, Martling CR, Oldner A, Bell M. Impact of sepsis on levels of plasma cystatin C in AKI and non-AKI patients. Nephrol Dial Transplant 2012;27:576–81.
[181] Delanaye P, Cavalier E, Morel J, et al. Detection of decreased glomerular filtration rate in intensive care units: serum cystatin C versus serum creatinine. BMC Nephrol 2014;15:9.
[182] Nejat M, Pickering JW, Walker RJ, Endre ZH. Rapid detection of acute kidney injury by plasma cystatin C in the intensive care unit. Nephrol Dial Transplant 2010;25:3283–9.
[183] Cho E, Yang HN, Jo SK, et al. The role of urinary liver-type fatty acid-binding protein in critically ill patients. J Korean Med Sci 2013;28:100–5.
[184] Hiruma T, Asada T, Yamamoto M, et al. Mortality prediction by acute kidney injury biomarkers in comparison with serum creatinine. Biomarkers 2014;19:646–51.

[185] Su L, Feng L, Liu C, et al. Diagnostic value of urine sCD163 levels for sepsis and relevant acute kidney injury: a prospective study. BMC Nephrol 2012;13:123.
[186] Su LX, Feng L, Zhang J, et al. Diagnostic value of urine sTREM-1 for sepsis and relevant acute kidney injuries: a prospective study. Crit Care 2011;15:R250.
[187] Koch A, Weiskirchen R, Sanson E, et al. Circulating retinol binding protein 4 in critically ill patients before specific treatment: prognostic impact and correlation with organ function, metabolism and inflammation. Crit Care 2010;14:R179.
[188] Kumpers P, Hafer C, David S, et al. Angiopoietin-2 in patients requiring renal replacement therapy in the ICU: relation to acute kidney injury, multiple organ dysfunction syndrome and outcome. Intensive Care Med 2010;36:462–70.
[189] Koch A, Gressner OA, Sanson E, et al. Serum resistin levels in critically ill patients are associated with inflammation, organ dysfunction and metabolism and may predict survival of non-septic patients. Crit Care 2009;13:R95.
[190] Linko R, Pettila V, Kuitunen A, et al. Plasma neutrophil gelatinase-associated lipocalin and adverse outcome in critically ill patients with ventilatory support. Acta Anaesthesiol Scand 2013;57:855–62.
[191] Hjortrup PB, Haase N, Wetterslev M, Perner A. Clinical review: predictive value of neutrophil gelatinase-associated lipocalin for acute kidney injury in intensive care patients. Crit Care 2013;17:211.
[192] Legrand M, Darmon M, Joannidis M. NGAL and AKI: the end of a myth? Intensive Care Med 2013;39:1861–3.
[193] Darmon M, Gonzalez F, Vincent F. Limits of neutrophil gelatinase-associated lipocalin at intensive care unit admission for prediction of acute kidney injury. Am J Respir Crit Care Med 2011;184:142–3.
[194] Glassford NJ, Schneider AG, Xu S, et al. The nature and discriminatory value of urinary neutrophil gelatinase-associated lipocalin in critically ill patients at risk of acute kidney injury. Intensive Care Med 2013;39:1714–24.
[195] Glassford NJ, Eastwood GM, Young H, et al. Rationalizing the use of NGAL in the intensive care unit. Am J Respir Crit Care Med 2011;184:142.
[196] Nisula S, Yang R, Poukkanen M, et al. Predictive value of urine interleukin-18 in the evolution and outcome of acute kidney injury in critically ill adult patients. Br J Anaesth 2015;114:460–8.
[197] Endre ZH, Pickering JW, Walker RJ, et al. Improved performance of urinary biomarkers of acute kidney injury in the critically ill by stratification for injury duration and baseline renal function. Kidney Int 2011;79(10):1119–30.
[198] de Geus HR, Fortrie G, Betjes MG, et al. Time of injury affects urinary biomarker predictive values for acute kidney injury in critically ill, non-septic patients. BMC Nephrol 2013;14:273.
[199] Wu Y, Wang F, Fan X, et al. Accuracy of plasma sTREM-1 for sepsis diagnosis in systemic inflammatory patients: a systematic review and meta-analysis. Crit Care 2012;16:R229.
[200] Morales-Buenrostro LE, Salas-Nolasco OI, Barrera-Chimal J, et al. Hsp72 is a novel biomarker to predict acute kidney injury in critically ill patients. PLoS ONE 2014;9:e109407.
[201] Schmidt EP, Yang Y, Janssen WJ, et al. The pulmonary endothelial glycocalyx regulates neutrophil adhesion and lung injury during experimental sepsis. Nat Med 2012;18:1217–23.
[202] Schmidt EP, Li G, Li L, et al. The circulating glycosaminoglycan signature of respiratory failure in critically ill adults. J Biol Chem 2014;289:8194–202.
[203] Lygizos MI, Yang Y, Altmann CJ, et al. Heparanase mediates renal dysfunction during early sepsis in mice. Physiol Rep 2013;1:e00153.

[204] Sun X, Li L, Overdier KH, et al. Analysis of total human urinary glycosaminoglycan disaccharides by liquid chromatography-tandem mass spectrometry. Anal Chem 2015;87:6220–7.
[205] Schmidt EP, Overdier KH, Sun X, et al. Urinary glycosaminoglycans predict outcomes in septic shock and ARDS. Am J Respir Crit Care Med 2016.
[206] Parikh CR, Butrymowicz I, Yu A, et al. Urine stability studies for novel biomarkers of acute kidney injury. Am J Kidney Dis 2014;63:567–72.
[207] Schuh MP, Nehus E, Ma Q, et al. Long-term stability of urinary biomarkers of acute kidney injury in children. Am J Kidney Dis 2015;67(1):56–61.
[208] Herget-Rosenthal S, Bokenkamp A, Hofmann W. How to estimate GFR-serum creatinine, serum cystatin C or equations? Clin Biochem 2007;40(3–4):153–61.
[209] Liu S, Che M, Xue S, et al. Urinary L-FABP and its combination with urinary NGAL in early diagnosis of acute kidney injury after cardiac surgery in adult patients. Biomarkers 2013;18:95–101.
[210] Cantinotti M, Storti S, Lorenzoni V, et al. The combined use of neutrophil gelatinase-associated lipocalin and brain natriuretic peptide improves risk stratification in pediatric cardiac surgery. Clin Chem Lab Med 2012;50:2009–17.
[211] Han WK, Waikar SS, Johnson A, et al. Urinary biomarkers in the early diagnosis of acute kidney injury. Kidney Int 2008;73(7):863–9.
[212] Vaidya VS, Waikar SS, Ferguson MA, et al. Urinary biomarkers for sensitive and specific detection of acute kidney injury in humans. Clin Transl Sci 2008;1:200–8.
[213] Kirchner GI, Meier-Wiedenbach I, Manns MP. Clinical pharmacokinetics of everolimus. Clin Pharmacokinet 2004;43(2):83–95.
[214] Coca SG, Yalavarthy R, Concato J, Parikh CR. Biomarkers for the diagnosis and risk stratification of acute kidney injury: a systematic review. Kidney Int 2008;73(9): 1008–16.
[215] Aregger F, Uehlinger DE, Witowski J, et al. Identification of IGFBP-7 by urinary proteomics as a novel prognostic marker in early acute kidney injury. Kidney Int 2014;85:909–19.
[216] Kashani K, Al-Khafaji A, Ardiles T, et al. Discovery and validation of cell cycle arrest biomarkers in human acute kidney injury. Crit Care 2013;17:R25.
[217] Bihorac A, Chawla LS, Shaw AD, et al. Validation of cell-cycle arrest biomarkers for acute kidney injury using clinical adjudication. Am J Respir Crit Care Med 2014;189:932–9.
[218] Gocze I, Koch M, Renner P, et al. Urinary biomarkers TIMP-2 and IGFBP7 early predict acute kidney injury after major surgery. PLoS ONE 2015;10:e0120863.
[219] Meersch M, Schmidt C, Van AH, et al. Validation of cell-cycle arrest biomarkers for acute kidney injury after pediatric cardiac surgery. PLoS ONE 2014;9:e110865.
[220] Yamashita T, Doi K, Hamasaki Y, et al. Evaluation of urinary tissue inhibitor of metalloproteinase-2 in acute kidney injury: a prospective observational study. Crit Care 2014;18:716.
[221] Hoste EA, McCullough PA, Kashani K, et al. Derivation and validation of cutoffs for clinical use of cell cycle arrest biomarkers. Nephrol Dial Transplant 2014;29:2054–61.
[222] Koyner JL, Shaw AD, Chawla LS, et al. Tissue inhibitor metalloproteinase-2 (TIMP-2) IGF-binding protein-7 (IGFBP7) levels are associated with adverse long-term outcomes in patients with AKI. J Am Soc Nephrol 2015;26(7):1747–54.
[223] Endre ZH, Pickering JW. Acute kidney injury: cell cycle arrest biomarkers win race for AKI diagnosis. Nat Rev Nephrol 2014;10:683–5.
[224] Kellum JA, Devarajan P. What can we expect from biomarkers for acute kidney injury? Biomark Med 2014;8:1239–45.
[225] Endre ZH. Novel biomarkers of acute kidney injury: time for implementation? Biomark Med 2014;8:1185–8.

[226] Bagshaw SM, Zappitelli M, Chawla LS. Novel biomarkers of AKI: the challenges of progress 'amid the noise and the haste'. Nephrol Dial Transplant 2013;28:235–8.

[227] Murray PT, Mehta RL, Shaw A, et al. Potential use of biomarkers in acute kidney injury: report and summary of recommendations from the 10th Acute Dialysis Quality Initiative consensus conference. Kidney Int 2014;85:513–21.

[228] Cruz DN, Bagshaw SM, Maisel A, et al. Use of biomarkers to assess prognosis and guide management of patients with acute kidney injury. Contrib Nephrol 2013;182:45–64.

[229] Devarajan P, Murray P. Biomarkers in acute kidney injury: are we ready for prime time? Nephron Clin Pract 2014;127:176–9.

[230] Koyner JL, Parikh CR. Clinical utility of biomarkers of AKI in cardiac surgery and critical illness. Clin J Am Soc Nephrol 2013;8:1034–42.

[231] Vanmassenhove J, Vanholder R, Nagler E, Van BW. Urinary and serum biomarkers for the diagnosis of acute kidney injury: an in-depth review of the literature. Nephrol Dial Transplant 2013;28:254–73.

CHAPTER SEVEN

Biomarkers of Extra-Renal Complications of AKI

S. Faubel, MD
Medicine, Division of Renal Diseases and Hypertension, University of Colorado Denver, Veteran Affairs Medical Center, Denver, CO, United States

Contents

Hospital-acquired acute kidney injury (AKI) is a common complication that is associated with significant morbidity and mortality [1]. Hospital stay is longer in patients with AKI [2] and patients are more likely to be discharged to short- or long-term care facilities [2,3]. A wealth of epidemiologic data has accumulated that AKI is independently associated with increased mortality. Increased in-hospital mortality occurs in patients with both mild [4–6] or severe (requiring renal replacement therapy) AKI [7,8]. Long-term mortality is also increased in patients with AKI [9,10]. For example, patients with normal renal function who developed AKI after cardiothoracic surgery had increased risk of death at 10 years when controlling for other variables; increased long-term mortality risk occurred even in

Biomarkers of Kidney Disease. http://dx.doi.org/10.1016/B978-0-12-803014-1.00007-8

patients who had complete recovery of kidney function [10]. In critically ill patients requiring RRT who became RRT-independent, mortality after a median follow up of 3 years was 57% versus 34% in well-matched critically ill patients not requiring RRT [9].

Although AKI is clearly associated with increased risk of in-hospital and long-term mortality, the mechanisms by which AKI contributes to death are unclear. Notably, however, clinical and experimental data indicate that AKI contributes to distant organ injury. Thus, the high mortality of AKI may be due to deleterious systemic effects of AKI that may have both short- and long-term negative consequences. In experimental models of AKI, AKI has been demonstrated to adversely affect the lung [11,12], heart [13], liver [14,15], brain [16], and intestine [17]. In patients with AKI, numerous short- and long-term "extra-renal" complications have been observed to occur. For example, short-term (i.e., in-hospital) "extra-renal" complications include sepsis [6,18], heart failure [19], immunoparalysis [20], and respiratory failure requiring mechanical ventilation [7,21]. Long-term (i.e., after hospital discharge) "extra-renal" complications include bone fractures [22], upper gastrointestinal bleeding [23], malignancy [24], severe sepsis [25], active tuberculosis [26], and major cardiovascular events [9,27] including stroke [28], myocardial infarction [9,27], and congestive heart failure [9,27]. Thus, an important approach to reducing the significant short- and long-term mortality of AKI will be to identify and target its systemic complications [13,29,30]. These complications may be considered "nontraditional" complications that are in contrast to the traditional complications of AKI such as acidosis, electrolyte abnormalities, and fluid overload [31]. Although the traditional complications have been recognized for over half a century and can generally be corrected with RRT, these nontraditional complications have been only recently examined and will likely require therapies beyond RRT to correct [31].

The focus of biomarker development in AKI has been to detect AKI early in its course with the goal of initiating therapy to improve kidney function; however, early identification of AKI may also facilitate treatments to target extra-renal complications. Since AKI is diagnosed so late in its course, the inflammatory and other systemic consequences of AKI may be greatly underappreciated. Thus, the development of biomarkers that diagnose AKI earlier in its course will not only assist in the treatment of AKI, but may also be used to better establish the course of extra-renal complications and the role of kidney failure in the development of other organ dysfunctions. Furthermore, an additional area of potential development is the identification of biomarkers that might predict specific extra-renal complications. Although numerous extra-renal complications may occur in patients with AKI, the best studied complications with associated biomarkers are from the effects of AKI on inflammation and lung injury. Therefore, in this section, the inflammatory and pulmonary complications of AKI as well as their potential biomarkers will be discussed. Data from animal as well as clinical studies will be reviewed.

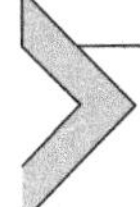

AKI AND INFLAMMATION

Proinflammatory Cytokines Mediate Organ Dysfunction

An exuberant inflammatory response is one mechanism behind the development of organ failure in patients with multiple organ dysfunction syndrome (MODS). Proinflammatory cytokines such as TNF and IL-1β initiate the cascade of events resulting in the systemic inflammatory response syndrome (SIRS) that can lead to MODS. In animals, injection of either TNF or IL-1β results in a shock-like state characterized by fever, hypotension, cardiac dysfunction, and lung injury with pulmonary edema and inflammation [32]. Increased production of proinflammatory cytokines occurs after both infectious and noninfectious assaults such as sepsis, pancreatitis, and trauma.

Proinflammatory Cytokines are Increased in the Serum in Animal Models of AKI

The effect of AKI on the production and elimination of proinflammatory cytokines may be a key mechanism by which patients with AKI have increased distant organ dysfunction and increased mortality. In animals with AKI, TNF [13,33], IL-1β [13,33,34], IL-6 [34–36], CXCL1 (known as IL-8 in humans) [34,36], and GCSF [37] increase in the serum after AKI. Since cytokine production also increases in the kidney [35,38–40], renal cytokine production may contribute to renal injury and cause the increase in serum cytokines. Circulating cytokines may then contribute to extra-renal organ injury (discussed in the subsequent sections). Although the kidney is likely an important source of cytokine production in AKI, data indicate that extra-renal cytokine production also occurs. For example, serum cytokines such as IL-6 are increased after bilateral nephrectomy [34,36]; As both kidneys are removed in this model of acute renal failure, the kidney cannot be the source of increased serum cytokines in this model. Indeed, data indicate that the cytokine production is increased in both the liver and spleen after bilateral nephrectomy [41,42]; furthermore, macrophage depletion significantly reduced serum cytokines after bilateral nephrectomy [41] indicating that macrophages are a specific cell type responsible for cytokine production in AKI. The increase in serum cytokines after bilateral nephrectomy is notable as it demonstrates that the systemic milieu of acute renal failure itself results in a proinflammatory state [43–45].

Clearance of Proinflammatory Cytokines is Impaired in Acute Kidney Injury

The metabolism and clearance of IL-1β, IL-6, IL-10, GCSF, and TNF have been examined in animals and data suggest that the kidney plays a role in the elimination of these cytokines [46–51]. In patients with chronic renal failure not on dialysis, serum IL-6 and TNF are increased [52], suggesting that impaired kidney function results in increased cytokine levels. Additionally, increasing levels of serum IL-6 are significantly correlated with decreasing levels of glomerular filtration rate [53], further suggesting that impaired kidney function may affect cytokine clearance. Animal studies have specifically demonstrated that the kidneys are important for cytokine clearance, as serum levels of IL-6, IL-1β, and IL-8 were higher in mice with bilateral nephrectomy administered these cytokines intravenously versus control mice given these cytokines intravenously [41]. Furthermore, data indicate that IL-6 is normally filtered and then resorbed and metabolized by the proximal tubule [54]. Notably, mice with prerenal azotemia and normally functioning proximal tubules did not have increased levels of serum IL-6 after IV administration of IL-6 indicating that the ability to clear IL-6 was intact; these data suggest that normally functioning proximal tubules are necessary for elimination of circulating IL-6 [54].

Excess Production and Impaired Clearance of Proinflammatory Cytokines May Occur in AKI

Although many insults affect cytokine production, AKI may be a unique scenario where *both* production and clearance of cytokines are affected. As numerous insults that lead to cytokine production may occur in patients already with AKI (e.g., hemorrhage, infection), impaired elimination and accumulation of cytokines would have significant clinical consequences. The excess cytokine burden due to increased production and impaired cytokine clearance may explain the development of distant organ dysfunction in patients with AKI.

SERUM CYTOKINES ARE INCREASED IN PATIENTS WITH AKI

Serum IL-6, IL-8, and IL-10 may be particularly relevant biomarkers during AKI that are associated with adverse outcomes and thus have the potential to be biomarkers of extra-renal complications of AKI. It should be noted that these and other biomarkers discussed herein are increased in insults other than AKI; however—as discussed previously—AKI may directly affect the production and clearance of these biomarkers and they have been demonstrated to be specifically helpful in identification of AKI and stratifying outcome as discussed in the subsequent section.

Serum IL-6, IL-8, and IL-10 are Increased in Patients With Established AKI and Predict Mortality

One of the first studies to examine serum cytokine levels and outcomes in patients with AKI was an analysis of a subset of patients in the Program to Improve Care in Acute Renal Disease (PICARD) study. PICARD was a prospective multicenter cohort study designed to examine the natural history and outcomes of critically ill ICU patients with established AKI. In order to examine the effect of AKI on inflammation, serum IL-1β, TNF, IL-6, IL-8, C-reactive protein, and IL-10 were determined in a subset of 98 patients from the PICARD study at the time of enrollment and then weekly for the duration of hospital stay. Patients were enrolled into the PICARD at the time of nephrology consultation, indicating that the patients studied were those with established AKI. For patients with a baseline serum creatinine less than 1.5 mg/dL, AKI was defined as an increase in serum creatinine of at least 0.5 mg/dL within 48 h; for those with a baseline creatinine of greater than 1.5, AKI was defined as an increase in serum creatinine of at least 1.0 mg/dL within 48 h. IL-1β, TNF-α, IL-6, and IL-8 are proinflammatory cytokines, CRP is an acute phase reactant that is typically increased in inflammatory conditions, and IL-10 is an antiinflammatory cytokine. Compared to healthy controls ($n = 48$), patients with AKI had significantly elevated levels of all serum markers as determined at baseline (i.e., at the time of entry into the study). Compared to stable end stage renal disease patients, patients with AKI had significantly increased IL-6, IL-10, and CRP (IL-1β, TNF, and IL-8 were not determined in the ESRD patients).

To determine if increases in cytokines or CRP might portend worse outcomes, a multivariate analysis of cytokines for predictors of in-hospital mortality and adjusted for demographics and sepsis status was performed. After adjustment, increased serum levels of IL-6, IL-8, and IL-10 at baseline were significantly correlated with increased in-hospital morality in patients with AKI. Specifically, increasing quartiles of cytokine values were associated with increasing risk of mortality. For example, the IL-6 values (pg/mL) of 65.4 (quartile 1), 110.8 (quartile 2), 227.4 (quartile 3), and 641.7 (quartile 4) were associated with an increased odds ratio of death of 1.0, 1.3, 1.8, and 3.0, respectively. The fact that increases in proinflammatory cytokines (IL-6 and IL-8) and an antiinflammatory cytokine (IL-10) predicted increased mortality, suggests that the immune response in patients with AKI is significantly dysregulated. Given that increased proinflammatory cytokines are associated with organ dysfunction and that increased antiinflammatory cytokines might be associated with increased risk of infection, it is plausible that the increase in these cytokines are not just biomarkers of poor outcome, but may play a role in mediating extra-renal complications (e.g., lung injury, cardiac dysfunction, infections). In the PICARD study, it is important to note that when cytokine values were further adjusted for severity of illness (APACHE III scores), only IL-6 remained an independent predictor of mortality. Thus, IL-6 may have particular clinical relevance regarding outcomes in patients with AKI.

Another study has also examined the relationship between increases in plasma IL-6, IL-8, and IL-10 on outcomes in critically ill ICU patients with AKI [55]. In this study, HLA-DR expression and plasma IL-6, IL-8, and IL-10 were determined in 103 consecutive critically ill ICU patients with the SIRS, with and without AKI [55]. HLA-DR and plasma cytokines were determined prospectively on the day of admission and 2 days after. Patients with AKI had significantly lower HLA-DR expression and higher plasma levels of IL-6, IL-8, and IL-10 than patients without AKI. Thus, even in patients with SIRS, the cytokines IL-8, IL-6, and IL-10 emerged as notably elevated in patients with AKI. On day 2, serum levels of IL-6 and IL-10 demonstrated moderate predictive power to predict survival (AUCs of 0.703 and 0.749, respectively).

The Biological Markers for Recovery of Kidney (BioMaRK) study was a multicenter, prospective, nested observational cohort study which was an ancillary study to the VA/NIH Acute Renal Failure Trial Network (ATN) clinical trial [56]. In the BioMaRK cohort of 817 critically ill patients, several plasma biomarkers were collected within 1 day of RRT initiation and were analyzed relative to kidney function recovery and mortality. IL-1, IL-6, IL-8, IL-10, IL-18, macrophage inhibitory factor (MIF), TNF-α, TNFR-I and TNFR-II, death receptor 5 (DR-5), and GCS-F were determined; in their analysis, increased levels of plasma IL-8 and TNFR-I were associated with delayed kidney function recovery; increased levels of IL-6, IL-8, IL-10, IL-18, MIF, TNFR-I, and DR-5 were associated with mortality [57]. Thus, these results are in line with the PICARD data that also demonstrated increased mortality with increased plasma IL-6, IL-8, and IL-10.

In a subsequent BioMaRK analysis, the same plasma biomarkers were analyzed on day 8 and associations between biomarkers and kidney function recovery and mortality were assessed [58]. Increased levels of IL-6, IL-8, IL-18, IL-10, TNFR-I, and TNFR-II on day 8 were associated with higher rates of RRT dependence and increased concentrations of IL-6, IL-8, IL-10, IL-18, TNFR-I, and DR-5 were associated with increased mortality.

Serum IL-6 is an Early Biomarker of Acute Kidney Injury

In the studies discussed previously, serum IL-6 and other inflammatory markers were found to be increased in patients with *established* AKI. A number of subsequent studies have found that serum IL-6 and other proinflammatory factors are also increased *early* after AKI and may be used to identify patients with AKI prior to a rise in serum creatinine.

In an analysis of patients from the Prospective Recombinant Human Activated Protein C Worldwide Evaluation in Severe Sepsis (PROWESS) data set, predictors of AKI in critically ill patients with severe sepsis were examined. PROWESS was a prospective randomized controlled study of the use of drotrecogin α to treat severe sepsis. For their analysis, the characteristics of the 547 patients who developed AKI in the placebo arm of the study were examined (there were 840 total patients in the placebo arm of this study). AKI was defined as an increase in serum creatinine of 25% or 0.3 mg/dL during the first week. Data analysis included biochemical, clinical, and demographic data, platelet count, protein C concentration, APACHE II scores, and plasma IL-6 concentration. Interestingly, increasing quartiles of plasma IL-6 were significantly correlated with the development of AKI as judged by an increase in serum creatinine.

IL-6 and other inflammatory markers as predictors of the development of AKI were also studied in an analysis of 879 patients involved in the low-tidal volume versus high-tidal volume mechanical ventilation study database of the first National Heart, Lung, and blood Institute Acute Respiratory Distress Syndrome Clinical Network (ARDS-net) trial. In this trial, 209 (24%) patients developed AKI as defined by an increase in serum creatinine of at least 50% from baseline. Baseline values of IL-6, IL-8, IL-10, von

Willebrand factor, TNF-α, type I and II soluble TNF receptors (sTNF-I and –II), protein C, plasminogen activator inhibitor-1 (PAII), surfactant protein-A, surfactant protein-D, and intercellular adhesion molecule-1 were correlated with the development of AKI. After adjustments for demographics, interventions, and severity of illness, increased levels of IL-6, sTNFR-I, sTNFR-II, and PAI-1 levels were independently associated with the development of AKI.

Plasma IL-6 and IL-10 as predictors of AKI were examined in the TRIBE-AKI Consortium adult cohort [59]. In this study, 960 patients after cardiovascular surgery were studied and biomarkers of AKI were examined. In their analysis, elevated levels of either IL-6 or IL-10 not only predicted AKI, but also predicted increased long-term mortality. Serum IL-6 and IL-10 were also examined in the TRIBE-AKI cohort of children with AKI; in this analysis of 106 children (24 with AKI), neither plasma IL-6 nor IL-10 (measured within 6 h postcardiopulmonary bypass (CPB) or on day 3) were predictive of AKI [60]. This is in contrast to another study of children receiving CPB in which plasma IL-6 levels were greater in patients with AKI than controls and increased by 2 h after the initiation of CPB [61]. Animal studies indicate that the peak levels of plasma IL-6 are reached rapidly after AKI (by 2 h) [34]—thus, a potential reason for the discrepant result in these two trials is the late measurement of plasma IL-6 in the former study.

In summary, analysis of the large cohorts in the PROWESS study, ARDS-net trial, and TRIBE-AKI Consortium demonstrated that increased serum IL-6 was independently associated with the development of AKI in patients with sepsis, ARDS, and postcardiovascular surgery, respectively.

PULMONARY COMPLICATIONS OF AKI

Pulmonary complications are the most common and well-recognized extra-renal complication of AKI [62]. In fact, respiratory failure requiring mechanical ventilation occurs twice as often in patients with AKI than similarly ill patients without AKI [7,21]. The requirement for mechanical ventilation is even higher for AKI patients who require renal replacement therapy (74% vs. 30%) [7]. The development of respiratory failure in patients with AKI is a particularly ominous occurrence and is associated with a marked increased mortality [7,21,63–70]. The need for mechanical ventilation is an independent predictor of mortality in patients with AKI, even when adjusted for severity of illness [7,21,63–70]. In one study, the mortality rate for AKI with mechanical ventilation was 81% versus 29%

for those not requiring mechanical ventilation [63]; and respiratory failure was associated with the worst prognosis of all associated organ failures with an odds ratio of death of 10.3 for associated respiratory failure versus 1.7 for associated nonrespiratory organ failure [63]. The important effect of respiratory failure on mortality in patients with AKI is also highlighted by an analysis of factors affecting mortality derived from the large Veterans Affairs/National Institutes of Health (VA/NIH) ATN; hypoxia without mechanical ventilation or the requirement for mechanical ventilation were the factors with the highest point levels compared to the 19 other variables used in the mortality prediction score (the higher the score, the higher the overall risk of mortality) [71]. Specifically, patients not on mechanical ventilation requiring fractional inspired oxygen (FiO_2) of ≥60% are assigned 14 points whereas patients requiring mechanical ventilation are assigned 12 points. All other factors that contribute to the mortality score carry a weight of five points or fewer. In another mortality prediction score derived from the ATN trial data, the use of just four clinical factors was useful to predict mortality—these factors included mechanical ventilation requirement, increasing age, increasing levels of bilirubin, and lower mean arterial pressure [72]. Thus, respiratory complications are key markers of increased mortality in patients with AKI and emphasize the important and potentially unique relationship between AKI and respiratory failure. Notably, adding serum IL-8 (but not other biomarkers) levels increased the power of the 4-variable model to predict mortality.

For those who do survive, the development of respiratory failure is also associated with increased morbidity. Patients with AKI and respiratory failure also have an increased likelihood of being discharged to an extended care facility [3].

Data suggest that AKI can both cause and exacerbate pulmonary function that may ultimately lead to respiratory failure requiring mechanical ventilation. For example, a recent study demonstrated that patients with AKI require longer mechanical ventilation and have an impaired ability to wean from mechanical ventilation [73]. In this observational, retrospective analysis, the outcomes of critically ill cancer patients with respiratory failure were compared between those with and without AKI (defined as an increase in serum creatinine to at least 1.5). The median duration of mechanical ventilation was 10 days in AKI patients versus 7 in patients without AKI; the duration of weaning was 41 days in patients with AKI versus 21 days in patients without AKI. Furthermore, the presence of AKI at the initiation of mechanical ventilation is an independent predictor of prolonged mechanical ventilation (defined as greater than 21 days) with an odds ratio of 5.630 (CI: 1.378–22.994) [74].

Pulmonary complications due to renal failure have been recognized for over 100 years [75]; however, the pathogenesis of AKI-associated respiratory failure remains to be clearly explained. Fluid retention and overload leading to hydrostatic (cardiogenic) pulmonary edema is well known to cause respiratory failure in patients with both AKI and end stage kidney disease [30]. This form of pulmonary edema is typically characterized by signs of fluid overload, including lower extremity edema, increased pulmonary capillary occlusion (wedge) pressure, and increased central venous pressure (CVP). The presence of this form of respiratory compromise is typically confirmed by resolution of symptoms with fluid removal via diuretics or ultrafiltration by dialysis. Although fluid overload is one mechanism of pulmonary edema that occurs in patients with AKI, data suggest that other mechanisms of lung injury may also occur. Pulmonary edema and shortness of breath in the presence of normal or low pulmonary capillary wedge pressure have been demonstrated [76,77], suggesting that AKI may also cause noncardiogenic pulmonary edema.

Noncardiogenic pulmonary edema is the hallmark of acute lung injury (ALI). Although caused by a wide variety of insults (e.g., trauma, sepsis, pneumonia), central to the pathogenesis of ALI is an exuberant proinflammatory response that results in the upregulation of adhesion molecules and chemokines in the lung which facilitates neutrophil infiltration. Lung neutrophil infiltration and activation directly injures the capillary endothelial barrier leading to the influx of proteinaceous edema fluid in the interstitial and alveolar space. In patients with AKI and pulmonary edema, lung neutrophil infiltration has been documented in autopsy studies suggesting that neutrophil mediated capillary injury may occur in AKI [78,79]. Thus, clinical data suggest that AKI causes inflammation, endothelial damage, and noncardiogenic pulmonary edema.

Lung Inflammation in Experimental AKI

Animal data support the notion that AKI may cause noncardiogenic pulmonary edema via neutrophil infiltration. Lung injury has been examined after ischemic AKI [11,12,34,36,80–84] as well as bilateral nephrectomy [11,34,36,82,84,85] and is characterized by pulmonary edema and neutrophil infiltration. Ischemic AKI is a common cause of AKI in hospitalized patients, and as such, it is a clinically relevant model. It is well known, however, that ischemia reperfusion injury of other organs (e.g., hind limb, gut, liver) is also associated with lung injury. Bilateral nephrectomy is a useful model to study the systemic effects of acute renal failure because renal failure occurs in the absence of renal ischemia. Remarkably, lung injury is similar after ischemic AKI and bilateral nephrectomy and is characterized by neutrophil infiltration and pulmonary edema within 4 h [34,36]. Unilateral renal ischemia, a model of renal ischemia without renal failure (serum creatinine and BUN are normal), is not associated with lung injury [34]. Thus, renal ischemia in the absence of renal failure is insufficient to cause lung injury.

Further supporting the role of inflammatory mediators in the pathogenesis of AKI-mediated lung injury, is the demonstration that cytokines (e.g., TNF-α [80]), adhesion molecules (e.g., ICAM-1 [80]), chemokines (e.g., KC [34,36], MIP-2 [34,36], CINC2 [82], CXCR2 [82]), heat shock proteins (e.g., HSP70 [82], HSP47 [82]), and NFκB [80] are all increased in the lung after AKI. Genomic responses in the lung after ischemic AKI or bilateral nephrectomy are associated with increased inflammatory genes [11]. In addition, multiple antiinflammatory agents reduce lung injury in animal models of AKI and include: CNI-1493, an inhibitor of the p38 MAP kinase pathway [12]; α-MSH [80]; IL-6 inhibition (via anti-IL-6 antibodies or genetic deficiency) [86]; IL-10 [34]; IL-8 inhibition (via anti-IL-8 antibodies or genetic deficiency of the IL-8 receptor) [87]; dexmedetomidine (an antiinflammatory anesthetic agent) [88]; anti-HMGB1 antibodies [89]; inhibition of TLR4 (via mutation inhibiting TLR4 function) [89]; inhibition of NF-kB [90]; and entanercept (inhibitor of TNF) [90].

Role of IL-6 and CXCL1 (IL-8) in AKI-Mediated Lung Injury

To determine if IL-6 mediates lung injury after AKI, IL-6 deficient mice and IL-6 antibody treated mice have been studied. Both IL-6 deficient mice and IL-6 antibody treated mice had improved lung injury after ischemic AKI and bilateral nephrectomy. The improvement in lung injury with IL-6 inhibition was associated with reduced lung neutrophil accumulation, reduced pulmonary edema, and reduced lung CXCL1. CXCL1 is the murine analog of human IL-8. CXCL1 and IL-8 are neutrophil chemokines. Alveolar macrophage production of IL-8 is thought to be a key mediator of ALI in patients [91–93]. Subsequent studies in mice demonstrated that circulating (i.e., serum) IL-6 mediates lung injury in AKI [87]. Specifically, intravenous injection of IL-6 to IL-6 deficient mice with AKI restored lung inflammation [87] yet intrapulmonary administration of IL-6 (via intratracheal instillation) did not affect lung inflammation [94]. Studies indicate that IL-6 mediates lung inflammation in AKI via upregulation of CXCL1 production in the lung endothelium via classic IL-6 signaling. In classic IL-6 signaling, IL-6 binds to the soluble IL-6R and signals via gp130 (which is ubiquitously expressed on all cells). In vitro, addition of IL-6 with soluble IL-6 receptor to endothelial cells increased CXCL1 levels in the media [87]. Thus, in AKI, circulating IL-6 likely binds to the circulating soluble IL-6 receptor and signals via gp130 on lung endothelial cells via gp130 to stimulate CXCL1 production [87]. The role of CXCL1 in AKI-mediated lung injury has been specifically studied; mice treated with CXCL1 antibodies or deficient in CXCR2 (a receptor for CXCL1) are protected against AKI mediated lung injury [87].

Serum IL-6 and IL-8 Increase 2 h After Cardiopulmonary Bypass-Associated AKI and Predict Prolonged Mechanical Ventilation

To determine whether serum cytokines might be early biomarkers or AKI and predict the adverse outcome of prolonged mechanical ventilation, a case control study of serum cytokines in pediatric patients undergoing cardiac surgery was performed [61]. Levels of serum interleukin (IL)-1α, IL-5, IL-6, IL-8, IL-10, IL-17, IL-18, interferon (IFN)-γ, TNF, granulocyte colony-stimulating factor (G-CSF), and granulocyte-macrophage colony stimulating factor (GM-CSF) were determined in 18 cases (with AKI) and 21 controls (without AKI) at 2, 12, and 24 h following CPB. AKI was defined as a 50% increase in serum creatinine within 3 days of CPB. Serum IL-6 levels at 2 and 12 h and serum IL-8 levels at 2, 12, and 24 h were significantly associated with the development of AKI. Of note, none of the other cytokines was significantly changed in cases versus controls at these time points.

To determine if the increases in serum IL-6 and IL-8 might be associated with complications in patient with AKI, the relationship of levels of these cytokines with prolonged mechanical ventilation (greater than 24 h) then was determined. In patients with AKI, serum IL-6 levels were significantly associated with prolonged mechanical ventilation with an area under the receiver operator curve (ROC) of 0.95. IL-8 levels at 2 h predicted prolonged mechanical ventilation in all patients. Although several previous studies had determined that certain serum cytokines were increased in patients with AKI, this is the first study to document that serum cytokines increase very early (within 2 h of AKI) and predict an adverse outcome (prolonged mechanical ventilation). It is remarkable that the pattern of cytokine increase and decline noted in patients with AKI is similar to the rise and fall of serum cytokines in animal models of AKI [34]. Specifically, serum IL-6 and CXCL1 (the murine analog of human IL-8) are also increased by 2 h after AKI, where the levels are the highest, and then begin to decline at 12 and 24 h.

Another study has examined whether increases in plasma IL-8, specifically, might be a biomarker of AKI [95]. In this study, plasma IL-8 was determined before and at 2, 24, and 48 h in 143 adult patients following CPB. AKI was defined by two criteria: (1) increase in serum creatinine by at least 0.3 mg/dL or 50% [AKI network (AKIN) stage 1] or (2) increase in serum creatinine by at least 50% alone. Increased serum IL-8 at 2 h predicted the identification of AKI by both criteria.

SUMMARY

In both patient and animal models, early AKI is associated with a proinflammatory burst that is characterized by the early increase in serum IL-6 and KC/IL-8. A proinflammatory burst such as this is the common link and accepted mechanism by which disparate inciting events (e.g., hemorrhage, trauma, and pancreatitis) mediate respiratory complications and ALI in other settings [96,97]. Other proinflammatory cytokines have not been shown to be increased early in patients with AKI. Thus, the increase in serum IL-6 and IL-8 in patients with AKI who develop prolonged mechanical ventilation is particularly relevant as these cytokines may be both biomarkers of AKI and prolonged mechanical ventilation as well as therapeutic targets of pulmonary complications of AKI.

REFERENCES

[1] Uchino S, Bellomo R, Goldsmith D, Bates S, Ronco C. An assessment of the RIFLE criteria for acute renal failure in hospitalized patients. Crit Care Med 2006;34(7):1913–7.
[2] Chertow GM, Burdick E, Honour M, Bonventre JV, Bates DW. Acute kidney injury, mortality, length of stay, and costs in hospitalized patients. J Am Soc Nephrol 2005;16(11):3365–70.
[3] Liangos O, Wald R, O'Bell JW, Price L, Pereira BJ, Jaber BL. Epidemiology and outcomes of acute renal failure in hospitalized patients: a national survey. Clin J Am Soc Nephrol 2006;1(1):43–51.
[4] Bates DW, Su L, Yu DT, et al. Mortality and costs of acute renal failure associated with amphotericin B therapy. Clin Infect Dis 2001;32(5):686–93.
[5] Coca SG, Peixoto AJ, Garg AX, Krumholz HM, Parikh CR. The prognostic importance of a small acute decrement in kidney function in hospitalized patients: a systematic review and meta-analysis. Am J Kidney Dis 2007;50(5):712–20.
[6] Levy EM, Viscoli CM, Horwitz RI. The effect of acute renal failure on mortality. A cohort analysis. JAMA 1996;275(19):1489–94.

[7] Metnitz PG, Krenn CG, Steltzer H, et al. Effect of acute renal failure requiring renal replacement therapy on outcome in critically ill patients. Crit Care Med 2002;30(9):2051–8.
[8] du Cheyron D, Bouchet B, Parienti JJ, Ramakers M, Charbonneau P. The attributable mortality of acute renal failure in critically ill patients with liver cirrhosis. Intensive Care Med 2005;31(12):1693–9.
[9] Wu VC, Wu CH, Huang TM, et al. Long-term risk of coronary events after AKI. J Am Soc Nephrol 2014;25(3):595–605.
[10] Hobson CE, Yavas S, Segal MS, et al. Acute kidney injury is associated with increased long-term mortality after cardiothoracic surgery. Circulation 2009;119(18):2444–53.
[11] Hassoun HT, Grigoryev DN, Lie ML, et al. Ischemic acute kidney injury induces a distant organ functional and genomic response distinguishable from bilateral nephrectomy. Am J Physiol Renal Physiol 2007;293(1):F30–40.
[12] Kramer AA, Postler G, Salhab KF, Mendez C, Carey LC, Rabb H. Renal ischemia/reperfusion leads to macrophage-mediated increase in pulmonary vascular permeability. Kidney Int 1999;55(6):2362–7.
[13] Kelly KJ. Distant effects of experimental renal ischemia/reperfusion injury. J Am Soc Nephrol 2003;14(6):1549–58.
[14] Kim M, Park SW, D'Agati VD, Lee HT. Isoflurane activates intestinal sphingosine kinase to protect against bilateral nephrectomy-induced liver and intestine dysfunction. Am J Physiol Renal Physiol 2011;300(1):F167–76.
[15] Yildirim A, Gumus M, Dalga S, Sahin YN, Akcay F. Dehydroepiandrosterone improves hepatic antioxidant systems after renal ischemia-reperfusion injury in rabbits. Ann Clin Lab Sci 2003;33(4):459–64.
[16] Liu M, Liang Y, Chigurupati S, et al. Acute kidney injury leads to inflammation and functional changes in the brain. J Am Soc Nephrol 2008;19(7):1360–70.
[17] Park SW, Kim M, Kim JY, et al. Paneth cell-mediated multiorgan dysfunction after acute kidney injury. J Immunol 2012;189(11):5421–33.
[18] Mehta RL, Bouchard J, Soroko SB, et al. Sepsis as a cause and consequence of acute kidney injury: Program to Improve Care in Acute Renal Disease. Intensive Care Med 2011;37(2):241–8.
[19] Ronco C, House AA, Haapio M. Cardiorenal syndrome: refining the definition of a complex symbiosis gone wrong. Intensive Care Med 2008;34(5):957–62.
[20] Himmelfarb J, Le P, Klenzak J, Freedman S, McMenamin ME, Ikizler TA. Impaired monocyte cytokine production in critically ill patients with acute renal failure. Kidney Int 2004;66(6):2354–60.
[21] Waikar SS, Liu KD, Chertow GM. The incidence and prognostic significance of acute kidney injury. Curr Opin Nephrol Hypertens 2007;16(3):227–36.
[22] Wang WJ, Chao CT, Huang YC, et al. The impact of acute kidney injury with temporary dialysis on the risk of fracture. J. Bone Miner. Res. 2014;29(3):676–84.
[23] Wu PC, Wu CJ, Lin CJ, Wu VC. Long-term risk of upper gastrointestinal hemorrhage after advanced AKI. Clin J Am Soc Nephrol.
[24] Chao CT, Wang CY, Lai CF, et al. Dialysis-requiring acute kidney injury increases risk of long-term malignancy: a population-based study. J Cancer Res Clin Oncol 2014;140(4):613–21.
[25] Lai TS, Wang CY, Pan SC, et al. Risk of developing severe sepsis after acute kidney injury: a population-based cohort study. Crit Care 2013;17(5):R231.
[26] Wu VC, Wang CY, Shiao CC, et al. Increased risk of active tuberculosis following acute kidney injury: a nationwide, population-based study. PLoS One 2013;8(7):e69556.
[27] Chawla LS, Amdur RL, Shaw AD, Faselis C, Palant CE, Kimmel PL. Association between AKI and long-term renal and cardiovascular outcomes in United States veterans. Clin J Am Soc Nephrol 2014;9(3):448–56.

[28] Wu VC, Wu PC, Wu CH, et al. The impact of acute kidney injury on the long-term risk of stroke. J Am Heart Assoc 2014;3(4):1–11.
[29] Awad AS, Okusa MD. Distant organ injury following acute kidney injury. Am J Physiol Renal Physiol 2007;293(1):F28–9.
[30] Van Biesen W, Lameire N, Vanholder R, Mehta R. Relation between acute kidney injury and multiple-organ failure: the chicken and the egg question. Crit Care Med 2007;35(1):316–7.
[31] Faubel S, Edelstein CL. Mechanisms and mediators of lung injury after acute kidney injury. Nat Rev Nephrol.
[32] Okusawa S, Gelfand JA, Ikejima T, Connolly RJ, Dinarello CA. Interleukin 1 induces a shock-like state in rabbits. Synergism with tumor necrosis factor and the effect of cyclooxygenase inhibition. J Clin Invest 1988;81(4):1162–72.
[33] Kelly KJ, Williams WW Jr, Colvin RB, et al. Intercellular adhesion molecule-1-deficient mice are protected against ischemic renal injury. J Clin Invest 1996;97(4):1056–63.
[34] Hoke TS, Douglas IS, Klein CL, et al. Acute renal failure after bilateral nephrectomy is associated with cytokine-mediated pulmonary injury. J Am Soc Nephrol 2007;18(1):155–64.
[35] Kielar ML, John R, Bennett M, et al. Maladaptive role of IL-6 in ischemic acute renal failure. J Am Soc Nephrol 2005;16(11):3315–25.
[36] Klein CL, Hoke TS, Fang WF, Altmann CJ, Douglas IS, Faubel S. Interleukin-6 mediates lung injury following ischemic acute kidney injury or bilateral nephrectomy. Kidney Int.
[37] Zhang Y, Woodward VK, Shelton JM, et al. Ischemia-reperfusion induces G-CSF gene expression by renal medullary thick ascending limb cells in vivo and in vitro. Am J Physiol Renal Physiol 2004;286(6):F1193–201.
[38] Molls RR, Savransky V, Liu M, et al. Keratinocyte-derived chemokine is an early biomarker of ischemic acute kidney injury. Am J Physiol Renal Physiol 2006;290(5):F1187–93.
[39] Miura M, Fu X, Zhang QW, Remick DG, Fairchild RL. Neutralization of Gro alpha and macrophage inflammatory protein-2 attenuates renal ischemia/reperfusion injury. Am J Pathol 2001;159(6):2137–45.
[40] Donnahoo KK, Meng X, Ayala A, Cain MP, Harken AH, Meldrum DR. Early kidney TNF-alpha expression mediates neutrophil infiltration and injury after renal ischemia-reperfusion. Am J Physiol 1999;277(3 Pt 2):R922–9.
[41] Andres-Hernando A, Dursun B, Altmann C, et al. Cytokine production increases and cytokine clearance decreases in mice with bilateral nephrectomy. Nephrol Dial Transplant 2012;27(12):4339–47.
[42] Andres-Hernando A, Altmann C, Ahuja N, et al. Splenectomy exacerbates lung injury after ischemic acute kidney injury in mice. Am J Physiol Renal Physiol 2011;301(4):F907–16.
[43] Brunet P, Capo C, Dellacasagrande J, Thirion X, Mege JL, Berland Y. IL-10 synthesis and secretion by peripheral blood mononuclear cells in haemodialysis patients. Nephrol Dial Transplant 1998;13(7):1745–51.
[44] Morita Y, Yamamura M, Kashihara N, Makino H. Increased production of interleukin-10 and inflammatory cytokines in blood monocytes of hemodialysis patients. Res Commun Mol Pathol Pharmacol 1997;98(1):19–33.
[45] Horl WH. Hemodialysis membranes: interleukins, biocompatibility, and middle molecules. J Am Soc Nephrol 2002;13(Suppl. 1):S62–71.
[46] Bemelmans MH, van Tits LJ, Buurman WA. Tumor necrosis factor: function, release and clearance. Crit Rev Immunol 1996;16(1):1–11.
[47] Bocci V. Interleukins. Clinical pharmacokinetics and practical implications. Clin Pharmacokinet 1991;21(4):274–84.

[48] Hepburn TW, Hart TK, Horton VL, et al. Pharmacokinetics and tissue distribution of SB-251353, a novel human CXC chemokine, after intravenous administration to mice. J Pharmacol Exp Ther 2001;298(3):886–93.
[49] Poole S, Bird TA, Selkirk S, et al. Fate of injected interleukin 1 in rats: sequestration and degradation in the kidney. Cytokine 1990;2(6):416–22.
[50] Rachmawati H, Beljaars L, Reker-Smit C, et al. Pharmacokinetic and biodistribution profile of recombinant human interleukin-10 following intravenous administration in rats with extensive liver fibrosis. Pharm Res 2004;21(11):2072–8.
[51] Tanaka H, Tokiwa T. Influence of renal and hepatic failure on the pharmacokinetics of recombinant human granulocyte colony-stimulating factor (KRN8601) in the rat. Cancer Res 1990;50(20):6615–9.
[52] Descamps-Latscha B, Herbelin A, Nguyen AT, et al. Balance between IL-1 beta, TNF-alpha, and their specific inhibitors in chronic renal failure and maintenance dialysis. Relationships with activation markers of T cells, B cells, and monocytes. J Immunol 1995;154(2):882–92.
[53] Pecoits-Filho R, Heimburger O, Barany P, et al. Associations between circulating inflammatory markers and residual renal function in CRF patients. Am J Kidney Dis 2003;41(6):1212–8.
[54] Dennen P, Altmann C, Kaufman J, et al. Urine interleukin-6 is an early biomarker of acute kidney injury in children undergoing cardiac surgery. Crit Care 2010;14(5):R181.
[55] Ahlstrom A, Hynninen M, Tallgren M, et al. Predictive value of interleukins 6, 8 and 10, and low HLA-DR expression in acute renal failure. Clin Nephrol 2004;61(2):103–10.
[56] Palevsky PM, Zhang JH, O'Connor TZ, et al. Intensity of renal support in critically ill patients with acute kidney injury. N Engl J Med 2008;359(1):7–20.
[57] Murugan R, Wen X, Shah N, et al. Plasma inflammatory and apoptosis markers are associated with dialysis dependence and death among critically ill patients receiving renal replacement therapy. Nephrol Dial Transplant 2014;29(10):1854–64.
[58] Murugan R, Wen X, Keener C, et al. Associations between intensity of RRT, inflammatory mediators, and outcomes. Clin J Am Soc Nephrol 2015;10(6):926–33.
[59] Zhang WR, Garg AX, Coca SG, et al. Plasma IL-6 and IL-10 concentrations predict AKI and long-term mortality in adults after cardiac surgery. J Am Soc Nephrol.
[60] Greenberg JH, Whitlock R, Zhang WR, et al. Interleukin-6 and interleukin-10 as acute kidney injury biomarkers in pediatric cardiac surgery. Pediatr Nephrol 2015;30(9):1519–27.
[61] Liu KD, Altmann C, Smits G, et al. Serum interleukin-6 and interleukin-8 are early biomarkers of acute kidney injury and predict prolonged mechanical ventilation in children undergoing cardiac surgery: a case-control study. Crit Care 2009;13(4):R104.
[62] Faubel S. Pulmonary complications after acute kidney injury. Adv Chronic Kidney Dis 2008;15(3):284–96.
[63] Chertow GM, Christiansen CL, Cleary PD, Munro C, Lazarus JM. Prognostic stratification in critically ill patients with acute renal failure requiring dialysis. Arch Intern Med 1995;155(14):1505–11.
[64] Mehta RL, Pascual MT, Gruta CG, Zhuang S, Chertow GM. Refining predictive models in critically ill patients with acute renal failure. J Am Soc Nephrol 2002;13(5):1350–7.
[65] Neveu H, Kleinknecht D, Brivet F, Loirat P, Landais P. Prognostic factors in acute renal failure due to sepsis. Results of a prospective multicentre study. The French Study Group on Acute Renal Failure. Nephrol Dial Transplant 1996;11(2):293–9.
[66] Uchino S, Kellum JA, Bellomo R, et al. Acute renal failure in critically ill patients: a multinational, multicenter study. JAMA 2005;294(7):813–8.
[67] Lins RL, Elseviers MM, Daelemans R, et al. Re-evaluation and modification of the Stuivenberg Hospital Acute Renal Failure (SHARF) scoring system for the prognosis of acute renal failure: an independent multicentre, prospective study. Nephrol Dial Transplant 2004;19(9):2282–8.

[68] Chertow GM, Lazarus JM, Paganini EP, Allgren RL, Lafayette RA, Sayegh MH. Predictors of mortality and the provision of dialysis in patients with acute tubular necrosis. The Auriculin Anaritide Acute Renal Failure Study Group. J Am Soc Nephrol 1998;9(4):692–8.
[69] Paganini EP, Halstenberg WK, Goormastic M. Risk modeling in acute renal failure requiring dialysis: the introduction of a new model. Clin Nephrol 1996;46(3):206–11.
[70] Liano F, Gallego A, Pascual J, et al. Prognosis of acute tubular necrosis: an extended prospectively contrasted study. Nephron 1993;63(1):21–31.
[71] Demirjian S, Chertow GM, Zhang JH, et al. Model to predict mortality in critically ill adults with acute kidney injury. Clin J Am Soc Nephrol 2011;6(9):2114–20.
[72] Pike F, Murugan R, Keener C, et al. Biomarker enhanced risk prediction for adverse outcomes in critically ill patients receiving RRT. Clin J Am Soc Nephrol 2015;10(8):1332–9.
[73] Vieira JM Jr, Castro I, Curvello-Neto A, et al. Effect of acute kidney injury on weaning from mechanical ventilation in critically ill patients. Crit Care Med 2007;35(1):184–91.
[74] Pan SW, Kao HK, Lien TC, Chen YW, Kou YR, Wang JH. Acute kidney injury on ventilator initiation day independently predicts prolonged mechanical ventilation in intensive care unit patients. J Crit Care 2011;26(6):586–92.
[75] Schulz JB, Bremen D, Reed JC, et al. Cooperative interception of neuronal apoptosis by BCL-2 and BAG-1 expression: prevention of caspase activation and reduced production of reactive oxygen species. J Neurochem 1997;69(5):2075–86.
[76] Rackow EC, Fein IA, Sprung C, Grodman RS. Uremic pulmonary edema. Am J Med 1978;64(6):1084–8.
[77] Gibson DG. Haemodynamic factors in the development of acute pulmonary oedema in renal failure. Lancet 1966;2(7475):1217–20.
[78] Zettergren L. Uremic lung; report of four cases reaching autopsy. Acta Soc Med Ups 1955;60(3–4):161–71.
[79] Hopps HC, Wissler RW. Uremic pneumonitis. Am J Pathol 1955;31(2):261–73.
[80] Deng J, Hu X, Yuen PS, Star RA. Alpha-melanocyte-stimulating hormone inhibits lung injury after renal ischemia/reperfusion. Am J Respir Crit Care Med 2004;169(6):749–56.
[81] Grigoryev DN, Liu M, Hassoun HT, Cheadle C, Barnes KC, Rabb H. The local and systemic inflammatory transcriptome after acute kidney injury. J Am Soc Nephrol 2008;19(3):547–58.
[82] Kim do J, Park SH, Sheen MR, et al. Comparison of experimental lung injury from acute renal failure with injury due to sepsis. Respiration 2006;73(6):815–24.
[83] Nath KA, Grande JP, Croatt AJ, et al. Transgenic sickle mice are markedly sensitive to renal ischemia-reperfusion injury. Am J Pathol 2005;166(4):963–72.
[84] Rabb H, Wang Z, Nemoto T, Hotchkiss J, Yokota N, Soleimani M. Acute renal failure leads to dysregulation of lung salt and water channels. Kidney Int 2003;63(2):600–6.
[85] Heidland A, Heine H, Heidbreder E, et al. Uremic pneumonitis. Evidence for participation of proteolytic enzymes. Contrib Nephrol 1984;41:352–66.
[86] Klein CL, Hoke TS, Fang WF, Altmann CJ, Douglas IS, Faubel S. Interleukin-6 mediates lung injury following ischemic acute kidney injury or bilateral nephrectomy. Kidney Int 2008;74(7):901–9.
[87] Ahuja N, Andres-Hernando A, Altmann C, et al. Circulating IL-6 mediates lung injury via CXCL1 production after acute kidney injury in mice. Am J Physiol Renal Physiol 2012;303(6):F864–72.
[88] Gu J, Chen J, Xia P, Tao G, Zhao H, Ma D. Dexmedetomidine attenuates remote lung injury induced by renal ischemia-reperfusion in mice. Acta Anaesthesiol Scand 2011;55(10):1272–8.
[89] Doi K, Ishizu T, Tsukamoto-Sumida M, et al. The high-mobility group protein B1-toll-like receptor 4 pathway contributes to the acute lung injury induced by bilateral nephrectomy. Kidney Int 2014;86(2):316–26.

[90] White LE, Santora RJ, Cui Y, Moore FA, Hassoun HT. TNFR1-dependent pulmonary apoptosis during ischemic acute kidney injury. Am J Physiol Lung Cell Mol Physiol 2012;303(5). L449-59.
[91] Meduri GU, Kohler G, Headley S, Tolley E, Stentz F, Postlethwaite A. Inflammatory cytokines in the BAL of patients with ARDS. Persistent elevation over time predicts poor outcome. Chest 1995;108(5):1303–14.
[92] Meduri GU, Headley S, Kohler G, et al. Persistent elevation of inflammatory cytokines predicts a poor outcome in ARDS. Plasma IL-1 beta and IL-6 levels are consistent and efficient predictors of outcome over time. Chest 1995;107(4):1062–73.
[93] Zemans RL, Matthay MA. Bench-to-bedside review: the role of the alveolar epithelium in the resolution of pulmonary edema in acute lung injury. Crit Care 2004;8(6):469–77.
[94] Bhargava R, Janssen W, Altmann C, et al. Intratracheal IL-6 protects against lung inflammation in direct, but not indirect, causes of acute lung injury in mice. PLoS One 2013;8(5):e61405.
[95] Liangos O, Kolyada A, Tighiouart H, Perianayagam MC, Wald R, Jaber BL. Interleukin-8 and acute kidney injury following cardiopulmonary bypass: a prospective cohort study. Nephron Clin Pract 2009;113(3):c148–54.
[96] Ware LB, Matthay MA. The acute respiratory distress syndrome. N Engl J Med 2000;342(18):1334–49.
[97] Bellingan GJ. The pulmonary physician in critical care ★ 6: The pathogenesis of ALI/ARDS. Thorax 2002;57(6):540–6.

CHAPTER EIGHT

Biomarkers in Kidney Transplantation

S. Jain, PhD* and A. Jani, MD*,**
*University of Colorado, Aurora, CO, United States **Denver Veteran Affairs Medical Center, Denver, CO, United States

Contents

Biomarkers of Kidney Disease. http://dx.doi.org/10.1016/B978-0-12-803014-1.00008-X

BIOMARKERS: AN OVERVIEW

The use of biomarkers in the diagnosis and prognostication of specific diseases is an area of intense research. Simple methods to noninvasively diagnose and monitor diseases hold obvious clinical appeal. In transplantation, the ability to noninvasively diagnose and monitor the various causes of allograft dysfunction might result in improved graft and patient outcomes.

A number of authors have attempted to define the characteristics of a biomarker. Parikh and Deverajan [1] suggested the following characteristics were desirable for biomarkers of acute kidney injury: (1) the biomarker should be easily detectable, using simple bedside or standard clinical laboratory techniques, in readily available clinical samples (such as urine and blood), (2) the biomarker would easy to measure the rapidly and reliably, (3) the biomarker should possess a high sensitivity to detect the relevant disease early, (4) the biomarker would have a range of cutoff that would allow for risk stratification, (5) and finally the biomarker would demonstrate strong performance on statistical analysis.

In discussing potential biomarkers of brain injury, Bakay and Ward [2] suggested a biomarker should: (1) be specific and sensitive for the brain, (2) appear in serum rapidly, and (3) demonstrate a reliable temporal relationship with injury. These cogent definitions could be more broadly applied to all biomarkers.

Sandler et al. [3] suggested that biomarkers should also allow a clinician to determine the prognosis of a particular disease and enable physicians to plan diagnostic and treatment interventions.

Kidney transplantation (KTx) is distinct from acute kidney injury (AKI) of native kidneys and from brain injury in that the clinician is able to obtain tissue more easily. Many investigators have therefore suggested the use of specific biomarkers obtained from the allograft (as opposed to serum or urine) to identify disease states early, prognosticate outcome, and modify treatment.

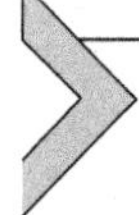

BIOMARKERS OF AKI POSTTRANSPLANTATION

Pretransplant Biomarkers

Tissue Markers

Oberbauer et al. [4] assessed whether apoptosis of tubular epithelial cells in donor kidney biopsies was associated with early renal-allograft function. Donor biopsies of patients with biopsy-proven acute tubular damage after engraftment but no signs of rejection (*n* = 23) had significantly greater apoptotic tubular epithelial cells when compared to patients with immediate transplant function (*n* = 44) or early rejection (*n* = 22). A significantly greater percentage of apoptotic cells were found in the distal versus the proximal tubule in all groups. The authors suggested that the number of apoptotic renal tubular epithelial cells in donor biopsies prior to engraftment was predictive of subsequent acute tubular injury in the early postoperative course of patients undergoing KTx.

Schwarz et al. [5] examined the contribution of adhesion molecule expression in donor kidney biopsies to early allograft dysfunction. Biopsies were obtained from living— (*n* = 20) and deceased—(*n* = 53) donor kidneys before engraftment and examined for the expression of intercellular adhesion molecule 1 (ICAM-1), vascular cell adhesion molecule 1 (VCAM-1), and endothelial leukocyte adhesion molecule (E-selectin). Living donor biopsies uniformly demonstrated significantly lower expression of ICAM-1 and VCAM-1 versus deceased-donor biopsies. There was no difference in tubular epithelial cell expression of adhesion molecules between transplants with primary function versus allografts with early rejection in either living- or deceased-donor kidneys. Significantly less expression of tubular epithelial cell ICAM-1 was seen in deceased-donor kidney that subsequently had prompt function (38 ± 29%) versus delayed graft function (DGF) in cadaveric kidneys (65 ± 24, $P < 0.05$). The authors suggested that tubular epithelial cell adhesion molecule expression was not a predictor of acute rejection but could predict posttransplant AKI due to ischemia.

Nguyen et al. evaluated whether pretransplantation recipient Treg suppressive function and frequency could predict DGF and slow graft function (SGF) in deceased-donor kidney transplant recipients with AKI (*n* = 37; DGF, *n* = 10; SGF, *n* = 27) versus immediate graft function (IGF) (*n* = 16). In pretransplant samples, Treg suppressive function, but not frequency, was decreased in recipients with AKI ($P < 0.01$). Treg suppressive function could discriminate DGF from IGF (odds ratio, 0.77; $P < 0.01$), accurately predicted AKI in receiver operating characteristic (ROC) curve (area under the curve, 0.82; $P < 0.01$), and predicted 14-day estimated glomerular filtration rate in linear regression ($P < 0.01$) in univariate and multivariate analyses that accounted for the effects of cold ischemic time and donor age in multinomial logistic regression. This study suggests that recipient Treg suppressive function is a potential independent pretransplant predictor of SGF and DGF [6].

Earlier studies of biomarkers of acute rejection (AR) were often small and single center. San Segundo et al. conducted a multicenter study, in which different Treg subpopulations were retrospectively correlated with AR in kidney transplant recipients during the first year after transplantation. AR occurred in 8% of transplant recipients ($n = 75$) and was associated with an increase in the activated Treg (aTreg) subpopulation (CD4CD25CD62LCD45RO). Furthermore, an activated Treg percentage greater than 1.46% prior to transplantation was associated with an increased risk of AR. Inclusion of thymoglobulin induction and recipient age in a multivariate logistic regression model was associated with prediction of AR in 92.4% of recipients. Thus, identification and evaluation of CD4CD25CD62LCD45RO-activated Treg cells may serve as a pretransplantation biomarker to predict AR in kidney transplant patients [7].

Plasma Markers

Morgan et al. [8] examined whether progressively rising serum creatinine in expanded criteria donors predicted outcome. Allografts from donors with peak serum creatinine levels of >2.0 mg/dL were divided into two groups defined by the terminal donor serum creatinine: group 1 had decreasing creatinine ($n = 27$) with a terminal creatinine ≤0.2 mg/dL than the peak serum creatinine, and group 2 had increasing serum creatinine ($n = 24$) with a terminal creatinine equal to the peak creatinine. Donor peak serum creatinine was not significantly different between the two groups (group 1 = 3.1 ± 1.3, group 2 = 3.2 ± 1.3; $P = 0.6521$. As expected, the mean terminal creatinine was significantly higher in group 2 (3.2 ±1.3 mg/dL) versus group 1(1.9 ± 0.9 mg/dL; $P < 0.0001$). Surprisingly, the outcomes were not statistically different between recipients of allografts from either group. DGF occurred more frequently in the group with falling donor serum creatinine (group1 = 32%, group 2 = 24%; $P = 0.7881$) although the difference did not achieve statistical significance. Mean recipient serum creatinine (group 1 = 1.6 ± 0.6, group 2 = 1.6 ± 0.4; $P = 0.3533$) and allograft survival were also not significantly different at follow-up (group 1 = 89%, group 2 = 92%; $P =$ NS). The findings of proportionately more donor hypertension (group 1, 30%, group 2, 13%; $P = 0.1331$) and more chronic lesions in the biopsies of donors with a rising serum creatinine (group 1, 41%; group 2, 0%; $P = 0.0023$) may explain the poorer outcomes in the allografts from group 1.

Sadeghi et al. [9] examined the association between pretransplant plasma levels of the antiinflammatory molecule, soluble glycoprotein 130 (sgp130), and posttransplant acute tubular necrosis (ATN) in 105 first-time, deceased-donor kidney transplant recipients. ATN was diagnosed in 29% (30/105) of patients and acute rejection was diagnosed in 18/130 patients. Pretransplant sgp130 plasma levels were significantly reduced in patients who went on to have ATN as compared with patients who had IGF ($P = 0.004$) or acute rejection ($P = 0.009$). The odds ratio of ATN was 4.3 on multivariable logistic regression analysis with a pretransplant sgp130 of ≤ 250 pg/mL.

Fonseca et al. examined the longitudinal changes of the renally cleared adipokines, leptin, and adiponectin in the first week post-KTx. Forty consecutive adult patients with end-stage renal disease who were undergoing KTx were studied. Leptin and adiponectin were measured in blood samples that were collected before (day 0) and after KTx (days 1, 2, 4, and 7). At posttransplant day 1, leptinemia and adiponectinemia declined 43 and 47%, respectively. At all times studied after KTx, the median leptin levels, but not adiponectin levels, were significantly higher in patients developing DGF ($n = 18$). Shortly after KTx (day 1), leptin values were significantly higher in DGF recipients in contrast to patients with promptly functioning kidneys, approximately 2 times higher when controlling for gender and BMI. The leptin reduction rate between pretransplant and 1 day after KTx moderately predicted DGF (AUC = 0.73). On day 1, serum leptin predicted DGF (AUC-ROC = 0.76) with a performance slightly better than serum creatinine (AUC-ROC = 0.72), even after correcting for BMI (AUC-ROC = 0.73). Separating this analysis by gender showed that the performance of leptin in predicting DGF for male gender (AUC-ROC = 0.86) improved. Kidney graft function is an independent determinant of leptin levels, but not of adiponectin. Leptin levels at day 1 slightly outperformed serum creatinine in predicting the occurrence of DGF, and more accurately in male gender. No significant association was detected with acute rejection [10].

Genetic Markers

Naesens et al. used microarrays to examine whole genome expression profiles in 53 human renal-allograft protocol biopsies obtained both at implantation and after transplantation. Gene expression profiles of living-donor kidneys and pristine deceased-donor kidneys with normal histology were significantly different prior to reperfusion. Deceased-donor kidneys and posttransplantation biopsies from well-functioning, nonrejecting kidneys, regardless of donor source, demonstrated a significant increase in renal complement expression. Complement gene expression at the time of implantation was associated with both early and late graft function. The authors therefore suggested that posttransplant graft function might be improved by therapies that modulated complement [11].

Scian et al. used microarray profiling to examine the preimplantation transcriptome of 112 kidney transplant recipient biopsies from 100 deceased-donor kidneys. Subjects were divided into either an estimated glomerular filtration rate (eGFR)-high group (n = 74; eGFR > 45 mL/min/1.73 m^2), or an eGFR-low group (n = 35; eGFR ≤45 mL/min/1.73 m^2) based on eGFR at 1 month posttransplant. Three genes—CCL5, CXCR4, and ITGB2—were found to predict posttransplant outcomes, suggesting that their presence in pretransplant molecular gene expression profiles might allow prognostication of future graft function [12].

Posttransplant Biomarkers

Tissue Markers

Kidney injury molecule-1 (KIM-1) is a transmembrane type 1 epithelial cell protein that belongs to the immunoglobulin gene superfamily. The extracellular component contains a novel six-cysteine immunoglobulin-like domain and a mucin domain. Normal rat kidneys express low levels of KIM-1 mRNA and protein. In contrast, postischemic rat kidneys express significantly increased levels of KIM-1 mRNA and protein in regenerating proximal tubule epithelial cells at 48 h [13]. The extracellular component of KIM-1can be cleaved by metalloproteinases, resulting in its appearance in urine [14].

Zhang et al. [15] investigated the expression of KIM-1as a biomarker for diagnosing early tubular injury in randomly selected kidney transplant renal biopsies by immunohistochemistry. Expression of KIM-1 was compared with morphological findings of tubular injury and acute cellular rejection (ACR). The authors also determined whether KIM-1 staining intensity correlated with renal function. Three groups were examined: group 1 (n = 25)—a control group of protocol renal-transplant biopsies without any obvious injury obtained within the first year posttransplant; group 2 (n = 25)—biopsies demonstrating obvious tubular injury without ACR; and group 3 (n = 12)—biopsies demonstrating Banff criteria IA and IB mild ACR. KIM-1 was absent in 72% of the protocol biopsies (group 1). In the remaining 28% of biopsies, KIM-1 expression was focal, low-grade, and localized to proximal tubules. No morphologic difference was detected between KIM-1-positive and -negative cases, leading the authors to suggest that KIM-1 expression was more sensitive than routine histology examination for detection of low-grade proximal tubule injury; group 2 biopsies demonstrated obvious tubular injury and KIM-1 expression was seen in

100% of cases. Expression localized to the plasmalemmal surface of proximal luminal epithelium but extended to the lateral cellular membranes if epithelial junctions were disrupted. The basal aspect of the epithelium was negative for KIM-1staining even when there was marked tubular injury morphologically. In the acute rejection group (group 3), KIM-1 expression was seen in 11/12 biopsies (92%). KIM-1 expression was not seen on infiltrating inflammatory cells, including lymphocytes, monocytes, and plasma cells in all groups. The highest intensity for KIM-1 staining was seen in the tubular injury group followed by the acute rejection group, and lowest in the protocol biopsies. In the protocol biopsy group, greater levels of KIM-1 staining portended better recovery of function over 18 months. KIM-1 immunoreactivity correlated with BUN and creatinine when all three groups were combined as well as in each individual group.

Obeidat et al. compared the ability of traditional clinical and histopathology scores to transcriptome biomarkers of kidney injury, to predict early posttransplant renal function measured by 99mTc-mercaptoacetyltriglycine (MAG3) or 99mTc-diethylenetriaminepentaacetic acid (DTPA) scans (n = 143 consecutive kidney transplant recipients). Grafts with severe posttransplant dysfunction were more likely to be from deceased donors ($P < 0.001$), have a higher HLA antigen mismatch ($P < 0.001$), and to be transplanted into older recipients ($P = 0.040$). Severe dysfunction was characterized by a lower urine output during the first 8 h posttransplant ($P < 0.001$), a higher postop day 7 serum creatinine ($P < 0.001$), and a higher incidence of DGF ($P < 0.001$). Clinical- and pathology-based scores did not differentiate grafts with normal or mild–moderate dysfunction from severe dysfunction. However, the overall transcriptome ($P < 0.001$) and transcripts of preselected AKI genes were significantly different between normal function, mild–moderate dysfunction, and severe dysfunction. Kidney injury molecule-1 (KIM-1; $P = 0.001$) and neutrophil gelatinase-associated lipocalin (NGAL; $P = 0.002$) were the most highly expressed genes, whereas genes associated with glutathione metabolism (GSTA1, 3 and 4) were most downregulated in grafts with subsequent severe dysfunction. The authors suggested that renal nuclear medicine scans were more objective measures of posttransplant function, and that analysis of the transcriptome and specific AKI genes better correlated with posttransplant function than traditional clinical- or histopathology-based scores [16].

Plasma Markers

Boom et al. [17] investigated whether serum calcium levels were a risk factor for the development of DGF in a cohort of 585 cadaveric transplants.

Serum calcium metabolism and the presence of nephrocalcinosis, ATN, or acute rejection in biopsies obtained the first posttransplant week were related to the occurrence of DGF. The incidence of DGF was 31%. Serum calcium levels correlated independently with DGF [odds ratio = 1.14 (95% confidence interval = 1.04–1.26) per 0.1 mmol/L]. The use of calcium channel blockers prior to transplantation protected against DGF [OR 0.5 (95% CI 0.29– 0.87)]. Nephrocalcinosis was found in 17% (12/71) of biopsies but was not associated with DGF or serum calcium levels.

Cystatin C (CysC) is a low molecular mass protein that is an alternative measure of graft function in adult transplant patients [18]. The ability of cystatin C to identify graft function and recovery from DGF has been evaluated in a number of studies. Thervet et al. [19] determined that cystatin C identified renal recovery earlier than sCr in patients with DGF. Le Bricon et al. evaluated cystatin C as a marker of allograft function during the first posttransplant month in 30 recipients and 56 healthy controls. Plasma cystatin C was found to correlate with both the reciprocal of the creatinine clearance estimated by the Cockcroft–Gault formula ($r = 0.882$; $P < 0.001$), and with plasma creatinine ($r = 0.741$; $P < 0.0001$). Cystatin C was also found to be more sensitive than serum creatinine for detecting decreased GFR and for predicting DGF [18]. Hall et al. confirmed these findings in 78 deceased-donor kidney transplant recipients, and demonstrated that cystatin C was superior to serum creatinine as a predictor of poor early graft function and the need for renal replacement therapy within the first week of KTx. Cystatin C was also shown to be a good prognostic marker of graft function at 3 months posttransplant [20].

Liu et al. compared the ability of cystatin C to diagnose acute rejection with serum creatinine, blood urea nitrogen, β(2) microglobulin and uric acid in 76 kidney transplant recipients, including 43 without and 33 with acute rejection. In recipients without rejection, cystatin C was found to decrease a greater amount than the other variables on postop day 1. Cystatin C was also found to increase earlier than the other variables in recipients with rejection, and to correlate better with ^{99m}Tc-DTPA GFR [21].

Welberry Smith et al. identified aminoacylase-1 (ACY-1) as a potential outcome biomarker during mass spectrometry analysis of serum from recipients with DGF. Subsequent longitudinal analysis of samples from 55 recipients patients confirmed that ACY-1 level on posttransplant day 1 or 2 was a moderate predictor of DGF that complemented the strongest predictor, cystatin C. The complementary relationship between ACY-1 and cystatin C was further validated in 194 patients, with ROC curves (95% CI) for day 1 of 0.74 (0.67–0.85; $n = 138$) for ACY-1, 0.9 (0.84–0.95) for cystatin C, and 0.93 (0.88–0.97) for both ACY-1 and cystatin C combined. A highly significant association was found between posttransplant day 1 or 3 serum ACY-1 and dialysis-free survival of recipients with DGF. Serum ACY-1 levels were able to discriminate between delayed, slow, and IGF [22].

Urine Markers

Mishra et al. [23] hypothesized that the expression of NGAL could serve as an early biomarker of AKI following transplantation. NGAL expression was assessed in specimens obtained from 13 deceased- and 12 living-donor kidneys approximately 1 h after engraftment. Staining intensity of NGAL was correlated with the need for dialysis, peak serum creatinine posttransplant, and cold ischemia time. NGAL expression was significantly increased in deceased-donor biopsies compared with living-donor kidneys (2.3 ± 0.8 vs. 0.8 ± 0.7, respectively, $P < 0.001$). NGAL staining intensity correlated with cold ischemia time ($R = 0.87$, $P < 0.001$) and with peak posttransplant serum creatinine that occurred days later ($R = 0.86$, $P < 0.001$). The most intense staining for NGAL was seen in four patients who developed DGF and required dialysis in the first posttransplant week. The authors concluded that NGAL staining intensity in early protocol biopsies was a novel predictive biomarker of early AKI in renal allografts.

Parikh et al. [24] examined whether urinary IL-18 might serve as a biomarker of ATN in a study of 72 subjects, including healthy controls, patients with different forms of AKI, and patients with other renal diseases. Patients with ATN had significantly greater median urinary IL-18 concentrations (644 pg/mg creatinine; $P < 0.0001$) compared with all other subjects, including healthy controls (16 pg/mg creatinine), patients with prerenal azotemia (63 pg/mg creatinine), patients with urinary tract infection (63 pg/mg creatinine), patients with chronic renal insufficiency (12 pg/mg creatinine), and patients with nephrotic syndrome (34 pg/mg creatinine). Median urinary IL-18 concentrations measured in the first 24 h after KTx were significantly greater in recipients of deceased-donor kidneys with DGF versus recipients with prompt graft function. Recipients of deceased-donor kidneys with DGF had a median urinary IL-18 of 924 pg/mg creatinine versus 171 pg/mg creatinine in patients who received a deceased-donor kidney with prompt graft function and 73 pg/mg creatinine in patients who received a living-donor kidney with prompt graft function ($P < 0.002$). Lower urinary IL-18 levels were associated with a steeper decline in serum creatinine concentrations on postoperative days 0–4 following KTx ($P = 0.009$).

In a follow-up study, the same authors [25] assessed whether urine NGAL and IL-18 were predictive biomarkers for DGF (defined as dialysis requirement within the first posttransplant week). Urinary NGAL and IL-18 from recipients of living donor kidneys ($n = 23$), deceased-donor kidneys with prompt graft function ($n = 20$), and deceased-donor kidneys with DGF ($n = 10$) were assessed on postop day 0. Peak postoperative serum creatinine requiring dialysis was found to occur 2–4 days after transplant in recipients with DGF. Urine NGAL and IL-18 values were significantly elevated levels in the DGF group ($P < 0.0001$). ROC analysis for the prediction of DGF based on urinary NGAL or IL-18 on postoperative day 0 showed an area under the curve of 0.9. Both urine NGAL and IL-18 on day 0 predicted the postoperative trend in serum creatinine by multivariate analysis, after adjusting for effects of age, gender, race, urine output, and cold ischemia time ($P < 0.01$).

Hall et al. [26] performed a prospective, multicenter, observational cohort study of deceased-donor kidney transplant recipients to assess IL-18, urinary NGAL, and KIM-1 as biomarkers for predicting graft recovery and the need for dialysis within 1 week of transplant. Serial urine samples were collected on the first three posttransplant days and analyzed for the putative biomarkers. Graft recovery was defined as DGF, SGF, or IGF. Of the 91 recipients studied, 34 had DGF, 33 had SGF, and 24 had IGF. Median levels of urine NGAL and IL-18 showed significant separation at all time points in all the three groups. Median urine KIM-1 levels on the other hand were not statistically different between groups. ROC curve analysis suggested that the urine NGAL or IL-18 measured on the first postoperative day were moderately accurate when used to predict dialysis within 1 week. Multivariate analysis confirmed that elevated levels of urine IL-18 or NGAL predicted the need for dialysis even after adjustment for serum creatinine, cold ischemia time, urine output, and recipient and donor age. Furthermore, NGAL and IL-18 quantiles also predicted graft recovery up to 3 months after transplantation.

Hall et al. also studied urine CysC in the prospective multicenter study mentioned previously, that included 91 deceased-donor kidney transplants. Serial urine samples were collected for 2 days following transplant and on the first postoperative day urine CysC was a predictor of DGF and of 3-month allograft function [27].

Schaub et al. [28] investigated whether noninvasive screening of urinary biomarkers of tubular injury correlated with subclinical tubulitis found in protocol biopsies. Recipients were divided into four groups: (1) recipients with stable graft function and normal tubular histology ($n = 24$); (2) recipients with stable graft function and subclinical tubulitis on protocol biopsy ($n = 38$); (3) recipients with clinical tubulitis Ia/Ib ($n = 18$); and (4) recipients with other clinical tubular pathologies ($n = 20$). Urine was examined for intact/cleaved beta2-microglobulin (i/cβ2m), retinol-binding protein (RBP), NGAL, and alpha1-microglobulin (α1m). Tubular proteinuria was found in 38% (RBP)—79% (α1m) of group 1. Group 2 had slightly higher but nonsignificant levels of i/cβ2m ($P = 0.11$), RBP ($P = 0.17$), α1m ($P = 0.09$), and NGAL ($P = 0.06$) than group 1with substantial overlap. Groups 3 and 4 had significantly greater levels of RBP, NGAL, and α1m than stable transplants with normal tubular histology or stable transplants with subclinical tubulitis ($P < 0.002$). The authors concluded that none of the biomarkers allowed for clear differentiation between stable transplants with normal tubular histology and stable transplants with subclinical tubulitis.

Kwon et al. [29] found that urinary actin, interleukin-6 (IL-6), and interleukin-8 (IL-8) were associated with sustained ischemic AKI in renal allografts. Urine specimens were collected in the first posttransplant week from 30 recipients of deceased-donor (including 9 with "sustained ARF" and 21 patients deemed "recovery"), and 10 recipients of living-donor kidneys. Urine was analyzed for actin, gamma-glutamyl transpeptidase (GGTP), lactate dehydrogenase (LDH), IL-6, tumor necrosis factor-alpha (TNF-α), and IL-8. Posttransplant day 0 urinary actin, GGTP, IL-6, and IL-8 were elevated in recipients who subsequently had sustained ARF versus recipients who subsequently recovered, although these did not reach statistical significance. In contrast, recipients with recovering function had increased urinary TNF-α and LDH compared to recipients with sustained ARF. Receiver operating characteristic curve analysis demonstrated that elevated urinary actin, IL-6, and IL-8 on day 0 were predictors of sustained ARF. Using a cutoff value for actin of 24.8 μg/g urine creatinine, the sensitivity and specificity were 0.67 and 0.86, respectively. The AUC was 0.75, whereas the AUC for predicting recovery was 0.25. Using a cutoff value for IL-6 of 60.2 ng/g urine creatinine, the sensitivity and specificity were 0.83 for both parameters, and the AUC was 0.91. Using a cutoff value for IL-8 of 78.3 ng/g urine creatinine the sensitivity and specificity were 1.00 and 0.61, respectively, while the AUC was 0.82. The authors concluded that increased urinary actin, IL-6, and IL-8 on postoperative day 0 could be biomarkers for the prediction of sustained ischemic AKI posttransplant. The study also reported the urinary levels of the putative biomarkers in patients excluded from the analysis due to minimal urine flow. Unfortunately the results were generally reported per mililiter of urine output and could therefore not be compared directly with the cutoff values. Nevertheless, such patients are often not reported in biomarker studies, which are often conducted on selected patients. This obviously indicates a limitation of all urinary biomarker studies.

miRNA Biomarkers

Yadav et al. examined the predictive value of serial urinary KIM-1 normalized to urinary creatinine (uKIM-1, pg/mg) levels at 0, 6, 12, 18, 24, and 48 h of posttransplant. Normalized uKIM-1 and AUC-ROC of uKIM-1 progressively increased for up to 48 h in both DGF and IGF. Urinary KIM-1values were significantly greater at 6, 12, 18, 24, and 48 h posttransplant in recipients with DGF versus IGF except at 30 min posttransplant. Despite a progressive increase in uKIM-1 values in both DGF and IGF, values overlapped between two groups during the first 12 h posttransplant. The earliest nonoverlapping values of uKIM-1 first occurred at 18 h and the earliest predictive AUC-ROC of uKIM-1 in DGF without overlap with IGF also occurred at 18 h posttransplant with a specificity of 100% and sensitivity of 89.9% (cutoff value of normalized KIM-1 = 923.43 pg/mg) [30].

Pianta et al. demonstrated that the relative performance of biomarkers for prediction of graft function is time-dependent. The utility of clusterin expression for prediction of DGF was analyzed in a cohort study of 81 renal-transplant recipients, and compared with urinary IL-18, NGAL, KIM-1, serum creatinine, and clinical variables. Anuria was highly specific but had a low sensitivity for diagnosing DGF 4 h after implantation of the graft. At 4 h, postreperfusion, ROC analysis indicated that urinary clusterin, IL-18, kidney injury molecule-1, and NGAL concentration were all predictive of DGF. Following adjustment for preoperative clinical variables and anuria, clusterin and IL-18 independently improved the clinical model for prediction of DGF. In contrast KIM-1 only modestly improved the prediction of DGF, whereas NGAL, serum creatinine, and the creatinine reduction ratio failed to improve the clinical model. The creatinine reduction ratio was found to independently predict DGF by 12 h postreperfusion [31].

Wilflingseder et al. prospectively compared biopsies taken at zero hour and again within 12 days of transplant in patients with AKI to ten matched allografts without pathology. Genome wide microRNA and mRNA profiling and analysis, followed by validation in independent expression profiles, was performed on 42 AKI and 21 protocol biopsies. Follow-up biopsies of AKI allografts compared to time-matched protocol biopsies revealed a molecular AKI signature that comprised 20 mRNAs and 2 miRNAs (miR-182-5p and miR-21-3p). The secretory leukocyte peptidase inhibitor (SLPI) was found in biopsies from allografts with AKI, and SLPI protein was detected in both plasma and urine. MiR-182-5p strongly correlated with global gene expression changes during AKI suggesting it may serve as a molecular regulator of posttransplant AKI, and serve as a biomarker [32].

Genetic Biomarkers of DGF

The molecule, Cluster of Differentiation 28 or CD28, is a costimulatory molecule found in T cells. Interaction with B7-1 and B7-2 on antigen-presenting cells results in T-cell activation, whereas interaction with cytotoxic T-lymphocyte antigen 4 (CTLA4) causes T-cell inhibition [33]. The genes for CD28, cytotoxic T-lymphocyte associated antigen 4 (CTLA4), programmed cell death 1 and inducible costimulator (ICOS) are all found on chromosome 2q [34].

Haimila et al. [34] examined the association of genetic variations in inducible costimulator genes with kidney transplant outcomes. Single nucleotide polymorphisms (SNPs) of the genes for CD28, CTLA4, ICOS, and PPCD1 were investigated in 678 deceased-donor recipients and correlated with kidney transplant outcome. DGF was defined as a serum creatinine of > 500 μmol/L (>5.65 mg/dL) in the first posttransplant week, the need for more than one dialysis session, or the presence of oliguria (defined as <1 L/day for more than 2 days). The occurrence of DGF was associated with two SNPs on the ICOS gene, rs10183087 and rs4404254 (odds ratio = 5.8; $P = 0.020$ and odds ratio = 5.8; $P = 0.019$, respectively). However, the associations reported were no longer significant after a Bonferroni correction for multiple tests. ICOS expression has been found to be decreased in rs4404254 TT homozygotes compared with CT or CC genotypes [35]. The authors suggested therefore that rs4404254 TT homozygotes perhaps expressed lower levels of ICOS and thus endured more DGF. The ICOS SNP rs10932037 was associated with decreased graft survival ($P = 0.026$). None of the SNPs examined were associated with acute rejection.

Israni et al. [36] performed a cross-sectional study of DGF in 965 recipients of deceased-donor kidneys from 512 donors. DGF was defined as the need for dialysis therapy in the first week after transplantation. SNPs in the donor genes for TNF-α, transforming growth factor beta1 (TGFB1), interleukin 10 (IL10), p53 (TP53), and heme oxygenase 1 (HMOX1) were correlated with the occurrence of DGF, as well as secondary outcomes including acute rejection and estimated glomerular filtration rate. DGF was significantly associated with the G allele of TNF SNP rs3093662 (odds ratio = 1.85 compared with A allele; 95% CI = 1.16–2.94; $P = 0.009$; $n = 965$) after adjustment for cold ischemia time, recipient race, extended criteria donor, donor cause of death, donor race, donor age, and source of DGF information. This association, however, became nonsignificant after adjustment for multiple comparisons. The authors suggested that the study had inadequate sample size for the study of infrequent genotypes and multiple comparisons.

Oxidative Stress Markers

Oxidative stress is an important feature of ischemia–reperfusion [37,38]. Kidney transplants are subjected not only to periods of ischemia and reperfusion, but also to allogenicity that can act as an adjuvant for oxidative damage. An imbalance in the production of ROS and the ability of the graft to scavenge ROS may trigger a robust inflammatory response within the transplanted organ and cause cell death, tissue injury, and graft dysfunction [39,40]. Several investigators have therefore investigated whether detection of ROS could be an early biomarker of graft injury. Waller et al. found that plasma carbonyl and 8-isporostane (products of protein and lipid damage by free radicals, respectively) were reliable biomarkers of reperfusion injury [41].

BIOMARKERS OF ACUTE REJECTION

A number of biomarkers derived from a variety of sources have been used both in the donor and in the recipient to determine the possible risk of acute rejection.

Genetic Biomarkers of Acute Rejection

Dmitrienko et al. [42] used a case–control design to examine the polymorphic frequencies of the T-cell signaling genes CD45, CD40L, and CTLA4, and the cytokine genes TNF-α, IFN-γ, IL-10, and TGF-β in 100 deceased- and living-donor recipients of first kidney transplants. Fifty recipients with biopsy-proven acute rejection (BPAR) were compared with 50 recipients who did not have AR. Fifty normal subjects were included as an indicator of local polymorphic gene frequency. Multivariate analysis showed no significant association between BPAR and single nucleotide polymorphisms in CTLA4, TGF-β, IL-10, or TNF-α genes or dinucleotide repeat polymorphisms in IFN-γ and CD40L genes. Allele TGFb-25pro was significantly associated with increased graft failure ($P = 0.0007$) whereas CD40L-147 was associated with reduced graft failure ($P = 0.004$). No subject had a CD45G (guanosine instead of cytosine) allele detected, likely due to the inclusion of only Caucasian patients. Thus immune response gene polymorphisms examined in this study showed no significant association with BPAR in subjects receiving triple immunosuppression.

The Fc gamma receptor IIA (FcγRIIA) is a member of the Fc receptor family. Unlike the Fc gamma receptors FcγRI and FcγRIIIa which are common to both mice and humans, the FcγRIIA is unique to humans [43]. These receptors activate cells via src family kinases and are thought to play a central role in leukocyte activation and cytotoxicity, and the initiation of the complement cascade [44].

Yuan et al. [44] examined whether FcγRIIA genotypes were associated with renal-allograft rejection. The distribution of the genotypes in the study patient group differed from the control groups. The study included 53 recipients who had suffered graft loss within 1 year of transplant (including 42 recipients who had lost their graft within 3 months) due to histologically confirmed acute rejection and 46 renal-allograft recipients with well-functioning grafts for at least 1 year. A group of 58 normal, random blood donors were also included in the analysis. Homozygosity for FcγRIIA-R/R131 was significantly more frequent in recipients with acute rejection than in nonrejection with well-functioning grafts and blood donors ($P < 0.05$). Renal-allograft recipients with well-functioning grafts followed the predicted distribution of FcγRIIA genotypes and allele frequencies when compared with normal, random blood donors ($P = 0.989$). Recipients with acute rejection were found to have a distinct distribution of FcγRIIA genotypes: the distribution of FcγRIIA-R/R131, FcγRIIA-R/H131, and FcγRIIA-H/H131 was 45, 42, and 13% in patients with acute rejection versus 20, 52, and 28%, respectively, in the 46 recipients with well-functioning grafts ($P < 0.05$), and 21, 52, and 27%, respectively, in the normal blood donors ($P < 0.05$). The frequency of the R/R131 genotype was significantly greater in recipients with graft loss compared to both control groups (45% vs. 20 and 21%, respectively, $P < 0.05$). The frequencies of FcγRIIA-R131 and FcγRIIA-H131 were 0.66 and 0.34, respectively, in patients with acute rejection, and were significantly different from recipients with well-functioning grafts ($P < 0.05$). The authors suggested that FcγRIIA polymorphisms could be useful markers for potential risk of rejection.

Slavcheva et al. [33] retrospectively examined the association between acute rejection and two polymorphisms in the CTLA4 gene, the dinucleotide (AT)n repeat polymorphism in exon 3 and the single nucleotide polymorphism A/G at position 49 in exon 1. The study included 207 liver- and 167 renal-transplant recipients. Both populations had a higher-than-expected rate of acute rejection (53.7 and 34% for liver and renal grafts, respectively). The authors acknowledged this and suggested it was due to the use of azathioprine-based triple-drug therapy. With respect to the (AT)n repeat polymorphism, there was an increased incidence of acute rejection in association with alleles 3 and 4 in both liver and kidney ($P = 0.002$ and 0.05, respectively). Allele 1 was less frequently observed in African American recipients versus Caucasian liver and kidney transplant recipients [frequency of 33.8 and 69%, respectively ($P < 0.0001$)]. Patients with allele 1 had a tendency toward a lower rate of acute rejection (42% vs. 57.8%, $P = 0.058$) suggesting that allele 1 was potentially protective. The A/G single nucleotide polymorphism was not associated with acute rejection in the patients studied.

Gao et al. investigated CTLA4 SNPs in 167 Chinese deceased-donor renal-transplantation recipients. The association between five CTLA4 SNPs (rs733618T/C, rs4553808A/G, rs5742909C/T, rs231775G/A, rs3087243G/A) and early acute rejection (AR) was determined during a 6-month period. AR was associated with a higher frequency of the rs733618TT genotype and T allele ($P = 0.000$ and $P = 0.002$, respectively), and the frequency of haplotype TACGG was significantly higher in recipients experiencing AR than those without ($P = 0.018$). CTLA4 SNPs may therefore represent genetic biomarkers of either increased risk of AR in Chinese renal transplantation, or protection from AR in the case of haplotype CACAG [45].

TIM-3 is a molecule that is selectively expressed on IFN-γ-producing CD4+ T helper 1 and CD8+ T cytotoxic 1 T cells, which play important roles in the development of acute allograft rejection. Luo et al. examined sequential changes in TIM-3 gene expression in peripheral blood lymphocytes (PBL) in three groups of renal-transplant recipients: an AR group ($n = 24$), a "no AR" group ($n = 20$), and a "stable" group ($n = 18$). PBL TIM-3 mRNA was significantly increased during AR compared with TIM-3 mRNA from recipients in the "no AR" and "stable" groups. ROC curve analysis demonstrated a specificity of 100% and a sensitivity of 87.5% for PBL TIM-3 expression in the diagnosis of rejection. PBL TIM-3 mRNA expression correlated with serum creatinine and was reduced by antirejection therapy [46].

Misra et al. correlated the HLA-G 14-bp insertion/deletion (I/D) polymorphism in patients with acute allograft rejection (AR). Risk of AR was increased almost fourfold (OR = 3.62, 95% CI = 1.61–8.14, P = 0.0039) for the 14-bp I/I (insertion-II) genotype. Three genotypes—the 14-bp I/I, 14-bp I/D, and 14-bp D/D showed significantly higher levels of soluble HLA-G in patients without AR versus those with AR [47].

The same group examined six additional CTLA4 SNPs (CTLA4+49 A > G (rs231775), −318 C > T (rs5742909), −658 C > T (rs11571317), −1147 C > T (rs16840252), −1661 A > G (rs4553808), +6230 A > G (rs3087243), and microsatellite (AT)n repeat polymorphism in the setting of ESRD, acute allograft rejection (AR), and DGF cases. The mutant genotype GG of CTLA4+49A > G, +6230 A > G, longer alleles of (AT)n repeats polymorphisms, and the haplotype + 49G: + 6230G and GCTTGG demonstrated risk association for ESRD, DGF, and AR cases. A meta-analysis also revealed risk associations for AR cases with the GG genotype of CTLA4+49A > G SNP, whereas CTLA4 −318 C > T polymorphism was not correlated against TT genotype in AR cases. The authors suggested that CTLA4 variants may confer increased susceptibility to ESRD, AR, and DGF [48].

CD28 (Cluster of Differentiation 28) is a protein expressed on T cells that is involved in the generation of costimulatory signals necessary for T-cell activation and survival. Pawlik et al. examined the association of the CD 28 polymorphism—IVS3 + 17T/C (rs3116496:T/C)—in 270 renal transplant recipients with DGF, AR, and chronic allograft nephropathy. Acute rejection was found in 21.7% of TT genotype carriers, 33.3% of CT carriers, and 60% of CC homozygotes. The risk of acute rejection was significantly higher in carriers of the C allele (CT or CC genotype) versus TT homozygotes (CC + CT vs. TT: OR = 1.93, 95% CI = 1.10–3.39, P = 0.026). In contrast, CD28 gene polymorphisms were not associated with DGF or chronic allograft nephropathy. The IVS3 + 17T/C polymorphism in the CD28 gene was therefore associated with acute kidney allograft rejection, but not with DGF or chronic allograft nephropathy (CAN) [49].

Using QPCR assays and preamplification protocols to overcome the low RNA yield from urine, Lee et al. identified a three-gene signature of 18S rRNA-normalized measures of CD3ε mRNA, IP-10 mRNA, and 18S rRNA that was diagnostic and predictive of ACR in the kidney allograft [50].

Han et al. performed a meta-analysis of studies examining CD28 (IVS3 + 17T/C variant), CTLA4, CD86, and PDCD1 gene polymorphisms with allograft rejection susceptibility. The CD28 IVS3 + 17T/C variant was significantly associated with acute allograft rejection susceptibility (CC +CT/TT OR, 1.45; 95% CI, 1.08–1.94; P = 0.01). In contrast, the CD86+ 1057G/A variant was associated with nonallograft rejection cases and a reduced risk of rejection (AA +AG/GG OR, 0.35; 95% CI, 0.14–0.85; P = 0.02) [51].

Chemokines and Acute Rejection

Chemokines are small proteins characterized by four conserved cysteine residues [52]. They are cytokines that activate G protein–coupled receptors and cause cells to migrate along a concentration gradient. Chemokines therefore allow homing of cells to a specific tissue or tissue compartment where chemokine production is maximal. For example, chemokines direct T or B cells to antigen presented by antigen-presenting cells in the lymphatic system [53]. Other chemokines have proinflammatory properties and are produced by cells during infection or proinflammatory stimuli. These "inflammatory' chemokines then direct leukocytes to areas of tissue damage or inflammation, and may cause the WBCs to become activated [52]. Chemokines have been implicated in the development of allograft rejection [54].

Abdi et al. [54] examined the association of human chemokine receptor genetic variants and outcome in 163 recipients of deceased- and living-donor kidney transplants [55]. The percentage of recipients who had a rejection episode was more than twofold lower in individuals possessing a CCR2–V64I allele compared to recipients who lacked this allele [19% vs. 44%; odds ratio (OR), 0.30; 95% CI, 0.12–0.78; P = 0.014]. In addition, fewer recipients homozygous for the CCR5-59029-A allele experienced at least one episode of rejection versus those possessing a CCR5-59029-G allele (23% for 59029-A/A vs. 44% for 59029-A/G or 59029-G/G; OR, 0.37; 95% CI, 0.16–0.85; P = 0.016). Recipients with a CCR5-59029-G allele also had a significantly higher number of rejection episodes compared to recipients with the CCR5-59029-A allele. These differences persisted after correction for other known risk factors. The authors acknowledged that association of the CCR5-59029-A allele with a lower risk of acute rejection is "counterintuitive" since CCR5-59029-A/A homozygotes demonstrate higher CD4+ T-cell CCR5 cell surface expression [56]. The association of the CCR2–V64I allele with less rejection was more biologically plausible. MCP-1, the ligand of CCR2, is an important monocyte chemoattractant. Both CCR2 and MCP-1 are upregulated in renal-transplant rejection [57]. Additionally, CCR2 knockout mice had a doubling of allograft survival in a fully mismatched MHC murine cardiac transplant model [58]. Abdi et al. postulated that the polymorphism observed in their study could reduce graft allogenicity, possibly by reducing monocyte migration into the graft [55].

Two families of chemokine receptors have been identified: the CC receptors 1–10 (CCR1– CCR10) which bind CC chemokines, and the CXC receptors 1–5 (CXCR1–CXCR5) which bind CXC chemokines [59]. Two important ligands of chemokine receptor CXCR3 are IFN-inducible protein 10 (IP-10 or CXCL10) and monokine induced by IFN-γ (MIG or CXCL9), which are both upregulated in rejecting murine heart allografts. Mice deficient in CXCR3 as well as in IP-10 have prolonged delay in the development of cardiac allograft rejection [59,60]

Hauser et al. [61] examined whether the monokine induced by IFN-γ (MIG, otherwise known as CXCL9) and IFN-γ-inducible protein 10 (IP-10, otherwise known as CXCL10) were early markers of AR. Urine was prospectively collected from 69 de novo renal-transplant recipients for a median of 29 days. Acute rejection was diagnosed clinically in 15 of 69 recipients and confirmed by biopsy in 14. Urine MIG was significantly elevated in 14 of 15 AR patients with a median of 2809 pg/mL (quartiles 25 and 75% = 870 and 13,000; $n = 15$) versus both nonrejecting allograft recipients (median, 25 and 75%: 96, 1.0, and 161, $n = 54$) and healthy controls (median, 25 and 75%: 144, 19, and 208, $n = 13$) ($P < 0.0001$). Urinary MIG predicted AR with a sensitivity of 93% and a specificity of 89%. In patients with acute rejection, urinary MIG was elevated (greater than the cutoff level) 5 days prior to biopsy on average ($P < 0.0001$), and corresponded well with increased urinary IP-10. The authors also suggested that urinary MIG and IP-10 indicated therapeutic success ($P < 0.0001$ and $P < 0.05$, respectively), while neither granzyme B nor serum creatinine were as useful as indicators of adequate antirejection therapy. MIG was also dissociated from infections or other causes of graft dysfunction.

Lazzeri et al. [62] examined the expression and distribution of CXCL10 in tissue specimens obtained from 22 patients with acute rejection or CAN. The authors also retrospectively assayed pretransplant sera from 316 deceased-donor kidney recipients for serum CXCL10 levels. Widespread CXCL10 expression was seen in biopsy specimens obtained from patients with CAN, both in infiltrating inflammatory cells, and also in vessels, tubules, and glomeruli. Recipients with very low pretransplant serum CXCL10 levels (<65 pg/mL; $n = 80$) had significantly better 5-year graft survival rate than recipients with very high (>157 pg/mL; $n = 78$) or intermediate serum CXCL10 levels >97 and <157 pg/mL ($n = 80$) ($P = 0.0002$

and $P = 0.03$, respectively). In addition, pretransplant serum CXCL10 levels > the 75th percentile (>157 pg/mL) were associated with significantly greater acute rejection versus serum CXCL10 levels < the 75th percentile (34.8% vs. 21.4%; $P = 0.01$). More severe and steroid resistant acute rejection (n =14) was associated with significantly higher pretransplant median serum CXCL10 levels compared to 60 recipients with less severe rejection episodes (216.1 vs. 112.4 pg/mL; $P = 0.04$). Multivariate analysis showed that CXCL10 (RR 2.8) and DGF (RR 3.7) had the highest predictive power of graft loss.

A common histopathological feature of acute renal-transplant rejection is mononuclear cell infiltration [63]. Monocyte chemotactic peptide-1 (MCP-1) is a chemotactic and activating chemokine specific for monocytes, that is encoded by the early response gene [64,65]. Grandaliano et al. [57] examined monocyte infiltration, MCP-1 gene and protein expression, and urine MCP-1 levels in kidney transplant recipients biopsied for acute graft dysfunction. Tissues from 13 patients with AR were compared to 7 with acute tubular injury, as well as normal kidney tissue. MCP-1 gene expression was undetectable in normal human kidneys, but increased significantly in tubular injury and AR. MCP-1 in situ hybridization demonstrated MCP-1 interstitial infiltrating mononuclear cells and proximal tubular cells. MCP-1 expression was greater in tissue demonstrating tubular injury than normal tissue, but significantly less than in AR. There was good correlation between expression of the chemokine and the number of infiltrating monocytes ($r = 0.87$, $P < 0.05$). Urinary MCP-1 was measured by ELISA in 8 normal subjects (36 ± 16 pg/mg urine creatinine), 13 clinically stable recipients (33 ± 9 pg/mg, ns vs. normals), 12 recipients with BPAR (250 ± 46 pg/mg, $P < 0.01$ vs. normals), and 5 transplant recipients with acute tubular injury (97 ± 33 pg/mg, $P < 0.05$ vs. normals and patients with BPAR). Successful treatment of BPAR led to a significant decrease in urinary MCP-1 levels.

Peng et al. [66] examined whether the urinary excretion of several chemokines, including fractalkine, chemokine monokine induced by interferon-gamma, interferon-gamma-inducible protein 10 (IP-10), macrophage inflammatory protein-3 alpha, granzyme B, and perforin could predict the occurrence of acute rejection. Urine was collected every 2 weeks during the first 2 months posttransplant, on the day of biopsy, at the end of antirejection therapy (average of 4 days; range = 1–15 days) from 215 allograft recipients and 80 healthy control subjects. Sixty-seven patients developed acute rejection. Areas under the ROC curve to distinguish acute rejection from patients without rejection for fractalkine, Mig, IP-10, MIP-3a, granzyme B, and perforin were 0.834, 0.901, 0.810, 0.734, 0.765, and 0.779, respectively. A cut point for fractalkine of 102.88 ng/mmol creatinine yielded a sensitivity and specificity of 82.1 and 76.5%, respectively ($P < 0.001$). The best set of markers to distinguish acute rejection from the absence of acute rejection was the combination of fractalkine, IP-10, and granzyme B (sensitivity and specificity of 83.6 and 95.0%, respectively). Of all the markers, only changes in urinary fractalkine distinguished recipients with acute rejection from those with acute tubular necrosis. The area under the ROC curve for fractalkine was 0.734 (95% CI: 0.604–0.865), whereas the area under the ROC for the other chemokines and cytotoxic effector molecules was not significant. In addition, among all markers, the area under the ROC curve for fractalkine could best differentiate steroid-resistant (n = 39) from steroid-sensitive acute rejection (n = 28). When a cut point for fractalkine of 233.76 ng/mmol creatinine was used to diagnose steroid-resistant rejection, the specificity and sensitivity were 75.0 and 74.4%, respectively ($P< 0.001$).

Hricik et al. examined urinary mRNAs and protein biomarkers to diagnose biopsy-proven acute rejection (AR), and to stratify patients into groups at risk for developing AR or progressive graft dysfunction, in a multicenter observational study of 280 adult and pediatric primary kidney transplant recipients. Urinary CXCL9 mRNA [OR 2.77, positive predictive value (PPV) 61.5%, NPV 83%] and CXCL9 protein (OR 3.40, PPV 67.6%, NPV 92%) were the most robust in diagnosing AR. Low urinary CXCL9 protein at 6-months posttransplant identified recipients least likely to develop future AR or a decrease in estimated glomerular filtration rate 6–24 months posttransplant (92.5–99.3% NPV). Thus, urinary CXCL9 could be a useful biomarker in the diagnosis of acute rejection, and low urinary CXCL9 at 6 months posttransplant could identify low risk, stable patients [67].

Blydt-Hansen et al. examined urinary CXCL10 in 51 pediatric kidney transplant recipients who had undergone surveillance or indication biopsies. Median urinary CXCL10-to-creatinine (Cr) ratio (ng/mmol) was significantly elevated in subclinical T-cell-mediated rejection (TCMR) [4.4 (2.6, 25.4), $P < 0.001$, n = 17]; clinical TCMR [24.3 (11.2, 44.8), $P < 0.001$, n = 9]; and antibody-mediated rejection (AMR) [6.0 (3.3,13.7), $P = 0.002$, n = 9] versus noninflamed histology [including normal, interstitial fibrosis, and tubular atrophy; 1.4 (0.4,4.2), n = 52], and versus borderline tubulitis [3.3, (1.3,4.9), n = 36]. Increased urinary CXCL10:Cr was independently associated with histological t ($P < 0.001$) and g scores ($P = 0.006$) by multivariate analysis. Area under ROC analysis was 0.81 ($P = 0.045$) for subclinical AR, and 0.88 ($P = 0.019$) for clinical TCMR resulting in a corresponding sensitivity–specificity of 0.59–0.67 and 0.77–0.60 for subclinical and clinical TCMR at cutoffs of 4.82 and 4.72 ng/mmol, respectively. Urinary CXCL10:Cr may serve as a noninvasive biomarker of subclinical and clinical TCMR in pediatric renal-transplant recipients [68].

Lee et al. examined five-gene biomarkers (DUSP1, NKTR, MAPK9, PSEN1, and PBEF1) in blood samples from 143 Korean patients taken at the time of graft biopsy (39 biopsy-proven AR, 84 stable patients, and 20 other graft injuries) at an average of 9 months posttransplantation. Significantly decreased MAPK9 and significantly increased PSEN1 were found in ACR versus controls and recipients with other graft injury. Multivariate logistic regression analysis demonstrated excellent diagnostic accuracy for discrimination between ACR and OGI, especially using the complete five-gene set combined with clinical variables [69].

Roedder et al. examined gene expression data in 558 blood samples from 436 renal-transplant recipients enrolled in the multicenter, international, Assessment of Acute Rejection in Renal Transplantation (AART) study. A 17-gene set—the Kidney Solid Organ Response Test (kSORT) (AUC = 0.94; 95% CI 0.91–0.98)—was identified for AR diagnosis and validated in 124 independent samples (AUC = 0.95; 95% CI 0.88–1.0). kSORT (performed in 191 serial samples) was also able to predict the occurrence of AR up to 3 months prior to graft biopsy. kSORT was able to detect AR in the blood of recipients, independent of time posttransplantation, sample source, and age [AUC = 0.93 (95% CI 0.86–0.99)] [70].

Toll-Like Receptors and Acute Rejection

Toll-like receptors (TLRs) are innate immune system receptors encoded by the germline, that are considered important in host defense [71]. They are expressed on many cell types including antigen presenting cells (APCs), epithelial, and endothelial cells. Activation of TLRs contributes to the upregulation of selectins and chemokines on endothelial cells [72]. TLRs on the surface of APCs are important in the priming of naïve T cells [71].

Hwang et al. examined the impact of TLR4/CD14 and TLR3 polymorphisms on acute rejection in 216 donor–recipient pairs undergoing living donor KTx. TLR4 genotype rs10759932 was associated with higher rejection-free survival rates (log-rank test, $P = 0.0053$) and no episode of acute rejection occurred when the genotype was present in either the donor or the recipient. Single-nucleotide polymorphisms of TLR3 or CD14 were not associated with acute rejection.

Srivastava et al. examined the association of TLR2, TLR3, and TLR9 polymorphisms in donor–recipient pairs and during acute rejection in Indian renal–transplant recipients. The variant allele frequency of TLR2 (−196 to −174 del) was significantly different between donors and recipients (7.5% vs. 5.0%; $P = 0.049$; OR = 3.9; 95% CI = 1.01–15.32). In contrast, TLR3 and TLR9 polymorphisms were not significantly associated with allograft rejection. A low prevalence of AA genotype of TLR9 + 2848 G > A was seen in patients with AR versus patients without, suggesting the polymorphism was associated with protection from AR (OR = 0.30, 95% CI = 0.12–0.88, $P = 0.028$) [73].

In contrast, TLR9 polymorphisms were associated with acute kidney transplant rejection in Korean patients. Two recipient SNPs of TLR9 gene (rs187084-1486; rs352140, G2848A) were associated with acute rejection and a lower eGFR after a year of renal transplantation. The C allele of rs187084-1486 and the A allele of rs352140 G2848A were associated with a protective genotype for acute rejection (OR 0.6, 95% CI 0.40–0.92; $P = 0.018$, OR 0.64, 95% CI 0.42–0.98; $P = 0.04$, respectively) [74].

Gene Transcripts and Acute Rejection

The granules of cytotoxic T lymphocytes contain perforin, a pore-forming protein which, upon release, forms pores in target-cell membranes [75], and granzyme B that is a serine peptidase [76] that causes apoptotic cell death via activation of caspase 3 [77].

Dugre et al. [78] investigated whether cytokine gene transcripts and the mRNA expression of cytotoxic molecules from mitogen-induced PBMCs of renal-transplant recipients could predict acute rejection prior to biopsy. PBMCs were collected twice weekly during the peri-transplant period and weekly thereafter for 3 consecutive months. Interleukins-4, -5, and -6, IFN-γ, perforin, and granzyme B mRNA levels were significantly associated with acute rejection. Upregulation of ≥2 of these cytokines in a given patient identified 75% of rejecting recipients compared with 15% of nonrejecting patients. A limitation of the study was the exclusion of 40/61 enrolled patients for infections, acute tubular necrosis, CsA nephrotoxicity, and "uncertain rejection episodes." In addition some AR episodes were diagnosed clinically, without biopsy confirmation.

Li et al. [79] also examined gene transcript analysis of perforin and granzyme B as a noninvasive diagnostic test for acute rejection, but focused instead on urinary cells. The granules of cytotoxic T lymphocytes contain perforin, a pore-forming protein which, upon release, forms pores in target-cell membranes [75], and granzyme B that is a serine peptidase [76] that causes apoptotic cell death via activation of caspase 3 [77]. Urine specimens ($n = 24$) were collected from 22 renal-allograft recipients with biopsy-proven acute rejection, as well as 127 samples from 63 recipients without evidence of acute rejection. Log-transformed mean (±SE) levels of perforin- and granzyme B mRNA, but not levels of constitutively expressed cyclophilin B mRNA, were greater in the urinary cells from the patients with acute rejection versus recipients without acute rejection (perforin, 1.4 ± 0.3 vs. -0.6 ± 0.2 fg/μg of total RNA; $P < 0.001$; and granzyme B, 1.2 ± 0.3 vs. -0.9 ± 0.2 fg/μg of total RNA; $P < 0.001$). ROC analysis showed that acute rejection could be predicted with a sensitivity and specificity of 83% (for both parameters; using a cutoff value of 0.9 fg of perforin mRNA/μg of total RNA), and with a sensitivity and specificity of 79 and 77%, respectively (using a cutoff value of 0.4 fg of granzyme B mRNA/μg of total RNA). The authors subsequently analyzed sequential urine samples were from 37 allograft recipients during the first 9 days posttransplantation. The levels of perforin- and granzyme B mRNA, but not that of cyclophilin B, were significantly higher in patients who developed acute rejection ($n = 8$) within the first 10 days posttransplant versus 29 recipients in whom acute rejection did not develop (granzyme B, $P = 0.02$ on days 4–6 and $P = 0.009$ on days 7–9; perforin, $P = 0.003$ on days 4–6 and $P = 0.01$ on days 7–9). The authors suggested that mRNA encoding cytotoxic proteins in urinary cells could represent a noninvasive method of diagnosing acute renal-allograft rejection.

Graziotto et al. [80] analyzed the expression of perforin (P), granzyme B (GB), and Fas Ligand (FL) in 68 renal biopsies and 64 samples of PBL in three groups of patients: (1) prereperfusion biopsies reperfusion and PBL from recipient prior to transplantation, (2) biopsies and PBLs collected 5–10 days posttransplant for graft dysfunction, and (3) protocol biopsies and PBLs obtained at 2 months posttransplant in patients with stable renal function. Perforin and granzyme B expression increased significantly over the first 2 months posttransplant in nonrejecting grafts (perforin, $P < 0.05$; granzyme B, $P < 0.01$) perforin overexpression in prereperfusion biopsies was associated with biopsy-proven acute and subclinical rejection in the subsequent 2 months ($chi^2 = 3.93$; $P < 0.05$). No significant increase in CTL transcription was found in PBL samples taken during episodes of AR. The authors noted considerable variability in each sample and suggested that the use of biomarkers may be hindered by time-related variability in their expression, and the need for a sizable quantity of renal tissue to ensure an adequate sensitivity.

In a similar study, Simon et al. [81] examined whether peripheral blood gene expression of perforin and granzyme B transcripts could predict renal-allograft rejection. Peripheral blood was collected twice weekly during the first month posttransplantation. Gene expression was measured using real-time polymerase chain reaction (PCR) in 364 samples from 67 patients. Clinical rejection was either biopsy-proven or based on clinical response to antirejection therapy in one patient.

Recipients with acute rejection ($n = 17$) had increased levels of granzyme B and perforin transcripts on days 5–7, 8–10, 11–13, 17–19, 20–22, and 26–29, versus patients without rejection ($n = 50$, $P < 0.05$ in all cases). The diagnosis of acute rejection, using gene expression criteria, determined by ROC curve analysis, could be made 2–30 days before the diagnosis by standard criteria (median 11 days). The best diagnostic result was achieved with samples taken on postop days 8–10. These samples yielded a sensitivity and specificity of 82% of 90%, respectively, for perforin, and a sensitivity and specificity of 72 and 87%, respectively for granzyme B. Both perforin ($P < 0.01$) and granzyme B gene expression ($P < 0.05$) decreased after initiation of antirejection therapy.

A unique population of suppressor T cells, also known as regulatory T cells or Tregs, has been implicated in AR [82]. Tregs are CD4+CD25+ and are distinguished from other cells by constitutive expression of the fork-head-winged helix transcription factor FOXP3 [83]. Veronese et al. [82] used immunohistochemistry to examine 80 human donor and recipient kidney biopsies for Treg transcription factor FOXP3, as well as CD4 or CD8. FOXP3(+) cells were found in the interstitium of biopsies with types I and II ACR. Ninety-six percent of the FOXP3(+) cells were CD4(+) while a minority expressed CD8. The FOXP3(+) CD4(+) cells were localized primarily to the tubules.

Muthukumar et al. [84] examined urinary cells for FOXP3, CD25, CD3ε, perforin, and 18S ribosomal messenger RNA in 36 subjects with acute rejection (AR), 18 subjects with CAN, and 29 subjects with normal biopsy results (NL). Seventy-five urine samples were collected prior to biopsy (presumably on the day of biopsy—the timing of collection was not further defined) and eight samples were collected after the biopsy. Subjects with acute rejection demonstrated greater urinary log-transformed mean FOXP3/18S ribosomal mRNA copies than patients subjects with CAN or normal histology (AR 3.8 ± 0.5; CAN 1.3 ± 0.7; NL 1.6 ± 0.4, $P < 0.001$). ROC analysis demonstrated that reversal of acute rejection could be predicted with 90% sensitivity and 73% specificity using a cutoff for FOXP3 mRNA of 3.46 ($P = 0.001$). In addition, the 18S-normalized, log-transformed mRNA levels of CD25 (6.9 ± 0.4, 4.0 ± 0.5, and 2.8 ± 0.6, respectively; $P < 0.001$), CD3ε (8.2 ± 0.4, 4.3 ± 0.5, and 1.6 ± 0.5; $P < 0.001$), and perforin (7.6 ± 0.4, 4.5 ± 0.4, and 2.8 ± 0.4; $P < 0.001$) were also greater in subjects with AR compared to subjects with CAN or NL histology. However, CD25, CDε, and perforin did not predict reversal of AR or graft failure.

Metzincins (METS), including matrix metalloproteases (MMP) are involved with extracellular matrix remodeling. Rödder et al. studied the expression of METS, and metzincins and related genes (MARGS) in renal-allograft biopsies with normal histology ($n = 20$), borderline changes ($n = 4$), acute rejection (AR) ($n = 10$), and AR + interstitial fibrosis/tubular atrophy (IF/TA) ($n = 7$). METS and MARGS differentiated AR from borderline, AR + IF/TA and normal histology and their expression changes correlated with Banff t- and i-scores. Thirteen MARGS were significantly enriched in biopsies with AR in all data sets comprising MMP7, -9, TIMP1, -2, thrombospondin2 (THBS2), and fibrillin1. MMP7, -9, and THBS2 were then detected in microdissected glomeruli/tubuli by RT-PCR, confirming the microarray results. Immunohistochemistry also showed increased MMP2, -9, and TIMP1 in AR. TIMP1 and THBS2 were found to be increased in serum from recipients with AR. The authors suggested that METS and MARGS, in particular TIMP1, MMP7/-9 could serve as potential biomarkers for AR [85].

Afaneh et al. compared urinary levels of mRNAs from 21 renal-allograft recipients with biopsy-proven acute rejection with urine from 25 recipients with normal biopsy results and stable graft function (stable). Urinary cell OX40 mRNA ($P < 0.0001$), OX40L mRNA ($P = 0.0004$), and PD-1 mRNA ($P = 0.004$) were significantly greater in the acute rejection group versus stable patients, whereas PD ligand 1 ($P = 0.08$) and PD ligand 2 ($P = 0.20$) were not. ROC curve analysis demonstrated that acute rejection was predicted with a sensitivity of 95% and specificity of 92% using a combination of levels of mRNA for OX40, OX40L, PD-1, and Foxp3 (AUC = 0.98, 95% CI 0.96–1.0, $P < 0.0001$). Furthermore, OX40 ($P = 0.0002$), OX40L ($P = 0.0004$), and Foxp3 ($P = 0.04$) predicted acute rejection reversal, whereas only OX40 mRNA levels ($P = 0.04$) predicted graft loss following acute rejection [86].

CD30 is a transmembrane glycoprotein belonging to the tumor necrosis factor (TNF)/nerve growth factor receptor family [87]. It is 120 kDa in size and is primarily expressed on CD4+ and CD8+ T cells of the Th2-phenotype with little or no expression Th1-type T cells [88]. Activated CD30+ T cells release a soluble form of CD30 (sCD30) into the bloodstream [89].

Cinti et al. [88] retrospectively examined the ability of panel of reactivity antibodies (PRA) and soluble CD30 (sCD30) in stored pretransplant sera to predict the occurrence of AR in the first 6 months following living-donor or deceased-donor KTx. PRA was measured using flow PRA beads and was considered positive when the percentage fluorescence was >5%. BPAR occurred in 58.3% (14/24 patients). sCD30 was found in the pretransplant sera of 37.5% of patients (9/24) and all of these patients subsequently developed AR. PRA was found in the pretransplant sera of 25% patients (6/24), four of whom developed later AR. Both sCD30 and PRA were very specific for AR (sCD30 100%; PRA 80%) whereas sCD30 demonstrated better specificity accuracy (79.1% vs. 50%) and PPV of 100% versus 66.6%. However neither sCD30 nor PRA demonstrated good sensitivity (sCD30 = 64.2%; PRA = 28.5%) nor negative predictive value (NPV) (sCD30 = 66.6%; PRA = 44.4%).

In a separate study, Susal et al. [87] performed a multicenter study involving 29 transplant centers in 15 countries, of whether pretransplant serum sCD30 could predict kidney allograft outcome. Pretransplant sera from 3899 cadaver kidney recipients were examined for serum sCD30 concentration by ELISA and correlated with subsequent allograft survival. Five-year graft survival was significantly lower (64 ± 2%) in 901 recipients with high pretransplant serum sCD30 (≥100 U/mL) than the allograft survival of 75 ± 1% rate in 2998 recipients with low sCD30 (<100 U/mL) ($P < 0.0001$). After the first posttransplant year, recipients with a high pretransplant serum sCD30 had a death-censored half-life of 20.5 years versus 29.4 years for patients with a low sCD30. High pretransplant sCD30 was associated with the need for significantly more rejection treatment (10%) in the second posttransplant year versus patients with a low sCD30 (5% $P = 0.0003$) although this difference did not exist in the first year posttransplant. High pretransplant serum sCD30 appeared to also predict a worse rate of graft loss during the 5-year follow-up period.

Pelzl et al. [90] examined whether soluble CD30 (sCD30) was a useful biomarker of acute rejection in 56 kidney allograft recipients during the early posttransplant period. The recipients were divided into three groups: (1) recipients with primary graft function, an uncomplicated course, and no acute rejection (n = 20), (2) recipients with primary nonfunction due to ATN without evidence of acute rejection (n =11), (3) recipients who experienced an episode of biopsy-proven acute rejection within the first 20 posttransplant days ($n = 25$). Plasma sCD30 levels were measured on postop days 3–5, 7–9, 12–14, and 17–19. ROC analysis revealed that on postop days 3–5, plasma sCD30 allowed recipients who subsequently developed acute allograft rejection ($n = 25$) to be distinguished from recipients with an uncomplicated course ($P < 0.0001$, AUC 0.96, specificity 100%, sensitivity 88%) or those with ATN in the absence of rejection ($P = 0.001$, AUC 0.85, specificity 91%, sensitivity 72%).

Kim et al. [91] retrospectively correlated pretransplant sCD30 levels (high vs. low) with posttransplant graft survival, incidence of acute rejection, and graft function in 120 allograft recipients. During 47.5 ± 11.4 months of follow-up, pretransplant sCD30 was not significantly associated with differences in graft survival rate ($P = 0.5901$). High sCD30 (≥115 U/mL) was associated with a higher incidence of acute rejection (33.9% vs. 22.4% in the low sCD30) but this difference did not reach statistical significance ($P = 0.164$), suggesting the study may have been underpowered. A similar trend was seen in response rate to antirejection therapy. Patients with high sCD30 had an inferior response compared to patients with a low sCD30 (33.3% vs. 7.7%, respectively) but this also failed to reach statistical significance ($P = 0.087$). In contrast, patients with a high sCD30 had significantly elevated serum creatinine 3 years posttransplant versus the low sCD30 group ($P < 0.05$). By multiple regression analysis, pretransplant sCD30 levels, acute rejection episodes, donor age, and kidney weight/recipient body weight ratio, were all independent variables affecting the serum creatinine level 3 years postengraftment.

Kotsch et al. [92] examined whether kinetic real-time RT-PCR–based gene expression profiling of urinary cells from outpatients could predict the occurrence of acute rejection. Urine was collected during the first 3 months posttransplant from 35 kidney transplant recipients, including nine patients who subsequently developed biopsy-proven acute rejection. Granulsin is a substance released from cytotoxic T cells that destroys other cells. Increased granulysin transcription was found in 11 of 14 cases of acute rejection, but was never observed above the CI in any of the 159 urine specimens collected from the nonrejecting group (100% specificity and 80% sensitivity). Granzyme B, perforin, FasL, TNF-α, RANTES, IL-2, IL-10, IFNγ, TGF-β, CD3, and CCR1 all showed less specificity and sensitivity. The authors also suggested that increased urinary granulysin gene expression was predictive of acute rejection occurring more than 4 weeks posttransplant. This was confirmed in only two patients, however, a modification of an RNA extraction protocol was also reported, that permitted a reporting of results within 4–5 h.

CD103 (formerly known as alpha E integrin) is found on the surface of a major subset of CD8+ cytotoxic T lymphocytes (CTL), and functions as a receptor for E-cadherin on epithelial cells [93]. It allows CD103-positive CTLs to bind epithelial cells through E-cadherin [94]. CD103+ T cells have been found exclusively restricted to the tubules during human renal-allograft rejection [95]. CD103+ cells are absent in normal renal tissue [96].

Ding et al. [96] tested the hypothesis that CD103 mRNA would be present in high abundance in urinary cells obtained during an episode of acute rejection. Eighty-nine urine specimens were collected from 79 recipients of renal allografts. Real-time quantitative PCR assay was used to measure CD103 mRNA levels as well as a constitutively expressed 18S ribosomal (r)RNA. CD103 mRNA levels were greater in urinary cells from 30 patients with AR as compared to levels in 12 patients with other findings on allograft biopsy, 12 patients with biopsy-proven CAN, and 25 patients with stable graft function ($P = 0.001$; one-way analysis of variance). In contrast levels of constitutively expressed 18S rRNA did not vary significantly among the four diagnostic categories ($P = 0.44$). However, CD103 mRNA levels predicted AR with a sensitivity of only 59% and a specificity of 75% when a natural log-transformed value of 8.16 CD103 copies/μg was used as the cutoff value ($P = 0.001$). The calculated area under the curve was 0.73 (95% CI, 0.62–0.82) for CD103 mRNA levels and 0.59 for 18S rRNA levels.

Not all biomarkers are able to distinguish between different disease states. Coupes et al. [97] used ELISA to examine whether circulating active TGFβ-1 was detectable in renal-allograft recipients, and whether plasma levels correlated with episodes of AR. Several groups of patients were included in the study: 43 healthy controls, 11 patients with membranous nephropathy (MN) and impaired renal function, 17 transplant recipients with stable renal function, 27 patients with biopsy-proven ACR, 7 patients with biopsy-proven chronic vascular rejection, and 10 patients with biopsy-proven acute tubular necrosis and/or cyclosporine (CsA) toxicity. Plasma samples were collected at the time of biopsy in the latter three groups. Urine TGF β1 was also measured. Plasma TGFβ-1 was not detected in any of the healthy controls or MN patients (detection limit of assay 0.1 ng/mL). In contrast TGFβ-1 was significantly increased in all transplant recipients but could distinguish the different diagnoses. TGF β-1 was found in most of the urine samples including those from healthy controls. The transplant urines had values comparable with normal controls.

Nogare et al. examined whether CTGF, TGF-β, and KIM-1 mRNA in biopsy samples from renal-transplant recipients with graft dysfunction, could be biomarkers of the development and severity of graft fibrosis. Ninety-six kidney transplant recipients who underwent 121 indication graft biopsies were categorized according to the Banff 2007 classification as: ATN (n = 20), acute calcineurin inhibitor nephrotoxicity (CIN; n = 13), acute rejection (AR; n = 58) and interstitial fibrosis and tubular atrophy (IF/TA; n = 30). CTGF and TGF-β mRNA transcripts were significantly greater in biopsies with IF/TA versus all other conditions. KIM-1 mRNA transcripts were higher in biopsies with IF/TA versus CIN. Gene expression of CTGF, TGF-β, and KIM-1 was associated with the severity of fibrosis seen in biopsies [98].

Kurian et al. investigated global gene expression profiles in 148 peripheral blood samples from kidney transplant recipients with excellent function and normal histology (n = 46), AR (n = 63), and no AR (n = 39). Multiple three-way classifier tools identified 200 high-value probe sets with high sensitivity (82–100%), specificity (76–95%), PPV (76–95%), NPV (79–100%), and AUC (84–100% and 0.817–0.968) in a validation cohort [99].

ELISPOT as a Biomarker of Acute Rejection

Gebauer et al. [100] developed an enzyme linked immunoabsorbent spot (ELISPOT) assay for the detection of cytokine secretion by individual, antigen reactive T cells. The assay was specific for antigen-specific IFN-γ-producing T cells expressing a cell surface phenotype of memory T cells (CD45RO+, CD45RA–) [100].

Hricik et al. [101] used the ELISPOT approach to serially measure the frequency of PBLs producing interferon-gamma in response to stimulator cells from donors or third parties in 55 primary kidney transplant recipients. Of this cohort, 37 had donor-stimulated IFN-ELISPOTS measured before transplantation, including 5 recipients who subsequently developed acute rejection. The mean frequency of pretransplant IFN-ELISPOTS was significantly higher in patients who experienced acute rejection (79 ± 69 vs. 30 ± 44 spots per 300,000 cells; P = 0.039 vs. recipients without clinically evident rejection). Pretransplant IFN ELISPOT did not correlate with serum creatinine at 6 −(R = 0.014, P = NS) or 12 months posttransplant (R = 0.114, P = NS).

Platelet Activation and Acute Rejection

Zhang et al. examined the association of pretransplant platelet activation with posttransplant AR and ATN. ELISA was used to determine the expression of the following glycoproteins in the peripheral blood of 203 first kidney transplant recipients of nonheart beating donor kidneys: CD62p (a platelet activation-dependent granule membrane protein); CD63 (a lysosomal enzyme glycoprotein); CD42a (a macula densa granule membrane glycoprotein); and PAC-1 (a fibrinogen receptor monoclonal antibody). Pretransplant expression of CD-63 was 15.45 ± 6.55 in recipients who subsequently developed acute rejection versus 1.74 ± 0.71 and 1.72 ± 1.36 in patients who had subsequent IGF or ATN ($P < 0.01$).

Serum Markers of Inflammation and Acute Rejection

Harris et al. [102] studied whether serial daily measurements of serum C-reactive protein (sCRP) could help differentiate renal dysfunction due to rejection from CsA nephrotoxicity. The total study population of 441 included 187 transplant recipients within 90 days of engraftment, 104 normal controls (healthy blood donors), and 150 patients on renal replacement therapy awaiting transplants [95 on hemodialysis, 55 patients on continuous ambulatory peritoneal dialysis (CAPD)]. Median sCRP concentration in normal controls was 0.5 μg/mL (range, <0.03–10 μg/mL), while HD patients had a median concentration of 3.1 μg/mL (range, <0.03- > 15 μg/mL) for patients on HD and CAPD patients had a median value of 2.9 μg/mL (range, <0.03- > 15 μg/mL). C-reactive protein (CRP) was noted to increase in some patients with inflammatory diseases such as Crohn's disease or CAPD peritonitis. Following transplant sCRP peaked in recipients with excellent primary graft function on postoperative day 2 (median, 29 μg/mL; range, 4 to >200 μg/mL) followed by decline to <20 μg/mL in all patients by day 5 (median, 7 μg/mL; range, 2–19 μg/mL). Median sCRP of recipients with stable graft function (defined as a mean creatinine of 155 μg/mL or 1.7 mg/dL) was 4 μg/mL (range, 1–19 μg/mL). In 30 episodes of steroid-sensitive acute rejection, sCRP was initially significantly increased to a median of 49 μg/mL ($P < 0.001$ vs. uncomplicated controls) but fell rapidly with treatment to a median of 11 μg/mL, with further subsequent decreases. In 19 episodes of steroid-resistant acute rejection, median initial sCRP levels were significantly higher (119 μg/mL, $P < 0.001$ vs. uncomplicated controls) and remained elevated (median = 77 μg/mL) at the end of the treatment. In 24 patients with graft dysfunction attributed to CsA nephrotoxicity there was no increase in sCRP concentrations (median sCRP <5 μg/mL throughout the episodes). Serum CRP levels were not significantly different from uncomplicated controls in six biopsy-proven cases of ATN [median sCRP concentrations for the start, middle, and end of the episode were 7 μg/mL (range, 1–9 μg/mL), 5 μg/mL (range, 1–6 μg/mL), and 2 μg/mL (range, 1–3 μg/mL), respectively]. Other causes of increased sCRP were wound infections, pyelonephritis, and sepsis. It was not clear from the study whether the increases in sCRP preceded the rise in sCr.

Perez et al. [103] examined whether pretransplant serum levels of C-reactive protein would predict the development of acute rejection episodes after kidney transplant. Pretransplant serum CRP was measured in 97 consecutive renal transplant recipients. The mean length of follow-up was 564 days (SD = 274 days) with a range of 6–1059 days. Acute rejection occurred in 39 (40%) recipients, with the majority occurring within the first 100 days posttransplant (median = 85 days). Serum CRP was found to range from 0 to 60 μg/mL, with a median of 9.0 and a mean 14.5 ± 1.6 μg/mL. The lower- and upper quartiles for CRP were <2 μg/mL and >21 μg/mL, respectively. Pretransplant CRP levels were greater in patients who subsequently developed acute rejection versus those who did not (22.2 ± 2.9 vs. 11.7 ± 1.8 μg/mL, respectively, $P = 0.003$). Recipients whose pretransplant CRP was less than the median had a significantly longer time to rejection versus recipients with higher CRP levels ($P = 0.002$). Recipients within the lowest CRP quartile had longer times to rejection versus those in the highest quartile ($P = 0.006$). Similarly, the 3-month incidence of rejection was 13% (3/23) in the lowest CRP quartile group versus 44% (11/25) in the upper quartile group ($P = 0.027$, Fisher exact test). The difference remained significant at 6 months. Of all covariates analyzed by Cox proportional hazards regression multivariate analysis, only pretransplant CRP level was an independent risk factor for rejection ($P = 0.044$).

Myeloid-related protein 8 (MRP8) and MRP14 (MRP14) are S100 family calcium binding proteins abundant in neutrophils and monocytes [104] Upon interaction with activated endothelium these proteins form a heterodimer known as calprotectin (MRP8/14) that becomes associated with endothelium at sites of monocyte and neutrophil adhesion. Calprotectin subsequently increases the endothelial transcription of proinflamatory chemokines and adhesion molecules [105].

Burkhardt et al. [106] used ELISA to measure MRP8/14 serum levels for 28 days in a pilot group of 20 renal-allograft recipients and subsequently in a validation cohort of 36 renal-allograft recipients. Serum MRP8/14,

C-reactive protein and creatinine levels were correlated with biopsy-proven acute rejection. There were seven episodes of acute rejection (five of which were biopsy proven) in the pilot group that occurred a median of 7 days posttransplant (IQR = 21 days). Of the 36 patients in the validation study, 18 experienced at least one acute rejection episode during the first 4 weeks after transplantation. Serum levels of MRP8/14 but not CRP were significantly increased for several days during the first 2 weeks in patients with the acute rejection groups in both studies ($P < 0.005$, on day 6 posttransplant). Using ROC curves, an optimal cutoff of 4.2 μg/mL on posttransplant day 6 for MRP8/14 yielded a specificity of 100% and a sensitivity of 67% for acute rejection in the pilot study. In the validation study, serum MRP8/14 levels were significantly increased on days 2–10 and on days 12–14 in recipients who later developed acute rejection. The best discrimination between the acute rejection and nonrejection groups was found to be on POD 6, as in the pilot study ($P < 0.001$). Plasma CRP did not differ significantly between patients with and without rejection ($P = 0.311$ on POD 6). Serum creatinine was also not able to differentiate significantly between the rejection and nonrejection groups ($P = 0.214$ on day 6 after transplantation). On POD 6 a cutoff of 4.2 μg/mL, a value derived from the pilot study, discriminated between the rejection and no rejection with a specificity of 100% and a sensitivity of 73%. Increased MRP8/14 serum levels preceded acute rejection episodes by a median of 5 days. Serum MRP8/14 was below the cutoff in patients with DGF, urinary tract infections, or cytomegalovirus infections, and these values did not differ significantly from control values.

Roshdy et al. examined the correlation between pre- and posttransplant levels of CRP and early renal-allograft rejection. Pretransplant ($n = 25$, $P = 0.001$) and posttransplant ($n = 33$, $P = 0.001$) CRP levels were significantly higher in acute rejection compared to recipients without rejection, suggesting that CRP may potentially be used as a biomarker of acute rejection [107].

Field et al. analyzed serum biomarkers including NGAL, cystatin C, IP-10, KIM-1, cathepsin L, and VEGF in 94 HLA-incompatible (HLAi) transplant recipients preoperatively, and on posttransplant days 1 and 30. Patients with acute rejection had significantly higher NGAL and IP-10 on day 1 following transplant ($P < 0.005$ and 0.001) and with an AUC of 0.67 and 0.73, respectively. Other biomarkers did not correlate with the development of acute rejection. The authors suggested that IP-10 and NGAL may identify patients at increased risk for rejection who may benefit from increased immunosuppression or investigation [108].

Kishikawa et al. evaluated the rate of microchimerism posttransplantation and observed it over a 12 month follow-up period. Eighty percent of recipients were found to have microchimerism within 2 days of transplantation. In 33% of the patients, microchimerism became negative within 3 months of transplantation and one patient in this cohort experienced acute rejection. Microchimerism persisted for up to 12 months in another 33% of patients. Protocol renal graft biopsy specimens obtained 3 months after transplantation revealed neither ACR nor acute AMR in five patients who demonstrated microchimerism at 3 months [109].

Tissue Biomarkers of AR

Koo et al. [110] compared the expression of adhesion molecules and HLA class II antigens in pretransplant biopsies from deceased- and living-donor kidneys ($n = 65$ and 29, respectively). High levels of intertubular capillary E-selectin expression (grade 2) were detected in 35 out of 65 (54%) deceased-donor donor kidneys compared with no expression in any of the living-donor kidneys ($P < 0.00001$). Expression of HLA-DR antigens, ICAM-1, and VCAM-1 was found in proximal tubules of deceased-donor kidneys, whereas living-donor kidneys had markedly reduced expression. Increased expression of tubular antigens was seen before transplantation in all 11 cadaver renal allografts with biopsy-proven acute rejection within 7 days of engraftment. Tubular antigens were absent in 15 out of 54 (28%) donor kidneys with no rejection in the first 7 days ($P < 0.05$). There was no significant association between tubular antigen expression and 3- and 6-month serum creatinine levels, DGF, and the number of rejection episodes.

Benson et al. [111] performed a prospective immunohistochemical analysis of the correlation between the inflammatory markers in the pretransplant biopsies and subsequent acute rejection in the recipient. Pretransplant biopsies were taken in 77 adult renal-transplant recipients, of whom 29 (38%) rejected. The biopsies were examined for P-selectin (PS), E-selectin (ES), platelets, leukocyte common antigen, macrophages, T cells, and neutrophils. Significantly more recipients rejected if the donor biopsy was positive for PS (63 vs. 24%, $P = 0.0007$), contained ≥5 or more leukocytes/glomerulus (48 vs. 21%, $P = 0.03$), contained >9.3 leukocytes/hpf (46.5 vs. 10.5%, $P = 0.006$), or were both PS positive and contained >9.3 leukocytes/hpf (61.9 vs. 0.0%, $P = 0.0001$). The PS was found to be primarily of platelet origin and most of the leukocytes were macrophages. The authors suggested that immunohistochemical changes present prior to transplantation, could identify those recipients with a greater risk of acute rejection, and allow for tailored immunosuppression.

TNFα is a cytokine synthesized by a number of cell types including monocytes-macrophages [112] and T lymphocytes [113]. TNF-α expression is increased during human acute allograft rejection [114] and it is thought to be involved in the induction of adhesion molecules on graft endothelium and the recruitment of cells into the allograft [115]. TNF-α binds receptors known as TNFRI and TNFRII [116] while naturally occurring soluble TNFRI (sTNFRI) and TNFRII (sTNFRII) released after proteolytic cleavage of cell-surface TNF receptors may regulate its bioactivity [117].

Oliveira et al. [112] studied the expression of IL-1β, soluble IL-1 receptor II (IL-1RII), TNF-α, sTNFRI and sTNFRII, and leukemia inhibitory factor (LIF) in fine-needle aspiration biopsy (FNAB) culture supernatants following KTx. FNABs were performed on 66 kidney transplant recipients on days 7 and 14 after transplant, and again whenever there was acute rejection. The cohort was divided into four groups: group 1— stable patients studied on day POD 7 (n = 30); group 2—patients studied on day 7 after transplantation, and 8 ± 4.5 days before acute rejection diagnosis (n = 12); group 3—patients studied on the first day of acute rejection diagnosis (n = 17); and group 4—stable patients studied on POD 14 (n = 32). Recipients of groups 1 and 4 did not experience any acute rejection during the first 6 months, and every patient in group 2 was studied again on the first day of acute rejection. Serum levels for sIL-1RII, sTNFRI, and sTNFRII were also measured. sTNFRI was found to be significantly higher in group 2 versus 1 (P = 0.002). When the acute rejection groups (groups 2 and 3) were combined, sTNFRI was found to be significantly higher than in the groups representing stable patients (groups 1 and 4) ($P < 0.0001$). A similar pattern was seen with sTNFRII, which was significantly greater in group 2 versus group 1 (P = 0.02), and significantly lower in all stable patients (P = 0.0001). For sTNFRI, a cutoff value for acute rejection of >480 pg/mL resulted in a sensitivity of 89.6%, specificity of 78.3%, PPV of 76.4%, and NPV of 90.6%. For sTNFRII, a cutoff value of > 700 pg/mL resulted in a sensitivity of 91.6%, specificity of 80.6%, PPV of 64.7%, and NPV of 96.1%. IL-1beta and sIL-1RII did not differ significantly among the groups.

Pentraxin-3 (PTX3) is an acute phase reactant produced by cells at sites of local inflammation. Imai et al. examined intrarenal PTX3 expression by immunohistochemistry in graft biopsies with acute rejection (n = 10), protocol biopsies without rejection (n = 37), and peri-operative donor biopsies from the same transplant patients ($n = 94$). PTX3 was primarily found in the interstitium and was maintained at a similar low level pre- and postreperfusion and protocol biopsies without rejection. In contrast, PTX3 expression area was significantly higher in biopsies demonstrating acute rejection ($P < 0.0001$). Acute rejection therapy significantly reduced PTX3 expression area ($P < 0.0001$). PTX3 expression area also positively correlated with allograft dysfunction and the severity of acute rejection [118].

miRNAs as Noninvasive Biomarkers of Acute Rejection

Anglicheau et al. profiled 7 of 33 renal-allograft biopsies (12 AR and 21 normal) using microfluidic cards containing 365 mature human miRNAs (training set). A subset of differentially expressed miRNAs was quantified in the 26 remaining allograft biopsies (validation set). A strong association between intragraft expression of miRNAs and messenger RNAs (mRNAs) was found. Furthermore, AR and renal-allograft function could be predicted with a high level of precision using intragraft levels of miRNAs. miRNAs (miR-142-5p, -155, and -223) were found to be overexpressed in AR biopsies, and also highly expressed in PBMCs. Stimulation with the mitogen phyto-hemagglutinin increased the abundance of miR-155 and decreased miR-223 and let-7c. The authors proposed that miRNA expression patterns could be used as biomarkers of human renal-allograft status [119].

The study by Anglicheau et al. was validated in a separate study by Liu and Xu, who found that miR223 increased in PBMCs of patients with AR versus controls from healthy volunteers. In PBMCs, miR-223 was increased twofold following acute rejection, and 3.76-fold after PHA treatment. Use of miR-223 to predict AR had a specificity of 90% and sensitivity of 92% [120].

Sui et al. used a combined proteomics- and genomics-based approach to examine the transcription factor, microRNA, and long noncoding RNA expression in biopsies of patients with AR and controls. Ninety-nine transcription factors were found in acute rejection biopsies versus normal biopsies. Patients with acute rejection demonstrated expression of 12 miRNAs, 32 lncRNAs, and 5 transcription factors (AP-1, AP-4, STATx, c-Myc, and p53). The authors suggested that overpresentation of transcription factor pathways might be implicated in acute rejection and that an examination of transcription factor, microRNA and long noncoding RNA analysis could identify molecular signaling pathways in acute kidney transplant rejection [121]. The same group also identified an acute rejection signature of

20 miRNAs found in three biopsies demonstrating acute rejection, that was not present in three control biopsies. Two miRNAs—miR-320 and miR-324-3p—were confirmed with qRT-PCR [122].

Oghumu et al. studied miRNA profiles of 11 biopsies with acute pyelonephritis (APN) and 5 with acute rejection (AR). The expression of 25 miRNAs was significantly different between the two conditions, and four miRNAs—MiR-99b, miR-23b let-7b-5p, miR-30a, and miR-145—were validated by qPCR. The study thus suggested that differences in intragraft expression of microRNAs may help distinguish acute rejection from pyelonephritis [123].

Maluf et al. studied 1733 mature miRNAs in the urinary cell pellets of patients with normal kidney transplant function and chronic allograft dysfunction (CAD) with interstitial fibrosis (IF) and tubular atrophy (TA). Twenty two miRNAs demonstrated significantly different expression (≥ twofold change) between CAD with IF/TA and normal graft function. These 22 miRNAs were associated with inflammatory disease, and inflammatory response. A subset of five miRNAs (miR-125b, miR-203, miR-142-3p, miR-204, miR-211) was validated with RT-QPCR and found to have significantly different expression between the two conditions. Furthermore, 12 urinary miRNAs from an independent set of 66 patients, obtained at two time points following transplantation demonstrated significant differences in expression before histological evidence of allograft injury occurred. Urine miRNAs may therefore serve as biomarkers for CAD [124].

In an earlier study, miRNA signatures from tissue and urine were identified in patients with either CAD and IF/TA or normal histology. Fifty-six miRNAs were found in patients with CAD-IF/TA. Five miRNAs were validated by RT-qPCR including miR-142-3p, miR-204, miR-107 and miR-211 ($P < 0.001$), and miR-32 ($P < 0.05$). Expression of miR-142-3p ($P < 0.01$), miR-204 ($P < 0.01$), and miR-211 ($P < 0.05$) was also significantly different between urine samples from CAD+ IF/TA and normal histology. These studies suggest that miRNA profiling may be a noninvasive biomarker of CAD and IFTA [125].

Lorenzen et al. profiled urinary miRNAs and found that miR-10a, miR-10b, and miR-210 were strongly deregulated in the urine of kidney transplant recipients with acute rejection compared to stable patients. Using a validation cohort of patients with acute rejection ($n = 62$), control transplant patients without rejection ($n = 19$), and stable transplant recipients with urinary tract infection ($n = 13$), the authors discovered that miR-10b and miR-210 were downregulated whereas miR-10a was upregulated in acute rejection versus controls. However, only miR-210 could differentiate recipients with acute rejection versus stable patients with a urinary tract infection. A low miR-210 level was also associated with a greater decline in GFR at 1 year posttransplant. This study also suggests that selected urinary miRNAs, such as miR-210, may be biomarkers of acute rejection and predict long-term kidney function [126].

Soltaninejad et al. studied the expression levels of miR-142-5p, miR-142-3p, miR-155, and miR-223 in paired biopsy and peripheral blood mononuclear cell (PBMC) samples from kidney transplant recipients with normal allografts ($n = 17$) or grafts undergoing acute T-cell-mediated rejection ($n = 18$). Intragraft expression levels of all the selected miRNAs were significantly greater in rejecting allografts whereas only miR-142-3p and miR-223 were significantly increased in PBMCs from recipients undergoing rejection. ROC analysis demonstrated that miR-142-5p, miR-142-3p, miR-155, and miR-223 in biopsy samples could differentiate rejection from normal controls with a sensitivity ranging from 90% to 100% and a specificity ranging from 80% to 100%. In PBMCs, miR-142-3p and miR-223 differentiate rejection from normal control with a sensitivity of 100% and a specificity of 65 and 76%, respectively [127].

B-Cell Activation and Acute Rejection (Tissue Biomarkers as Predictors of Response to Therapy)

Hippen et al. [128] examined the relationship between the presence of CD20-positive B-lymphocytes in kidney transplants undergoing ACR and graft survival. Biopsies from 27 recipients with biopsy-proven Banff 1-A or 1-B rejection in the first year posttransplant were stained for CD20 and C4d. The staining patterns were correlated with follow-up data of 4 years for each patient studied. Six patients were found to have interstitial CD20-positive B-cell clusters whereas 21 patients were negative for CD20 infiltrates. Patients in the former group had a significantly greater peak serum creatinine at the time of acute rejection suggesting worse impairment of renal function in the CD20-positive group (median 3.1 mg/dL vs. 2.2 mg/dL in the CD20-negative group, $P = 0.047$). Recipients with CD20-positive interstitial infiltrates were significantly more likely to have steroid-resistant acute rejection ($P = 0.015$ vs. CD20-negative recipients) and to experience immunologic (death censored) allograft loss ($P = 0.024$). However, when death with graft function was included as a cause of graft loss, the trend toward poorer outcomes for the CD20-positive group remained but failed to reach statistical significance ($P = 0.153$), suggesting the study may have been underpowered. The authors suggested that identification of B-cell infiltrates could distinguish a unique subset of patients for whom anti-B-cell therapy may be beneficial.

In a similar study, Kayler et al. [129] determined the influence of lymphocyte depleting therapy on B-cell clusters in 120 allograft biopsies obtained during the first episode of ACR in 120 recipients. Lymphoid clusters (LCs) were found in 59% of the biopsies (71/120). CD20-positive B cells were found in all 71 biopsies and accounted for 5–90% of the cluster leukocyte content. LCs were most frequent in patients who had not received lymphoid depletion or had been treated with thymoglobulin (79 and 75%, respectively) compared to 37% in patients treated with Campath (P = 0.0001). Banff 1a/1b ACR was more frequent in the positive- and negative-LCs (96% vs. 80%, respectively; P = 0.0051). However, over a follow-up of 953 ± 430 days, positive- and negative-LCs did not differ significantly with respect to time to ACR, steroid resistance, serum creatinine, and graft loss. In contrast to the study by Hippen, CD20+ LCs did not predict glucocorticoid resistance or worse outcomes. The authors suggested that LCs contain variable and heterogenous collections of B cells, and suggested a small subset of high-risk patients could potentially exist.

Shabir et al. examined the importance of transitional B lymphocytes (CD19 + CD24hiCD38hi) in augmenting transplant tolerance, and protection from late AMR. Seventy-three primary transplant recipients were studied over 48 ± 6 months at a variety of time points including immediately prior to transplantation, and on five subsequent time points during the first posttransplant year. Transitional B-cell frequencies, but not total B cells or "regulatory" T cells, were associated with protection from acute rejection (any Banff grade; HR: 0.60; 95% CI: 0.37–0.95; P = 0.03). Transitional B-cell proportions were not associated with the development of either de novo donor-specific or nondonor-specific antibody, but preservation of transitional B-cell proportions was associated with decreased reduced rejection rates in patients who developed de novo DSA. This prospective study suggests that transitional (regulatory) B-cell frequencies could serve as biomarkers of immunological status including humoral and cellular immunity and nonadherence [130].

Carpio et al. evaluated the association between B-cell expression patterns and graft function and survival in 110 kidney transplant recipients who had undergone for-cause biopsies. The patients were classified as no rejection (n = 40), TCMR (n = 50), and AMR (n = 20). A CD138-positive plasma cell-rich infiltrate predominated in biopsies with AMR and was associated with positive donor-specific antibodies (DSA) (r = 0.32; $P \leq 0.006$) and stronger reactivity against panel antibodies ($r = 0.41$; $P \leq 0.001$). CD138-positive cell infiltrates were also significantly increased in recipients with late versus early rejection. In contrast, TCMR was associated with CD20-positive lymphocytes, increased human leukocyte antigen mismatch, and retransplantation. CD20 or CD138 did not predict worse graft function or survival, and cellular CD20 and CD138 expressions did not correlate. C4d and DSA did associate with poorer graft function and graft survival 4 years posttransplant. This study suggests an association between CD20-positive B-cell infiltrates and TCMR, and CD138-positive plasma cells with AMR. Graft loss, however, was only associated with C4d [131].

Cytokines as Biomarkers of AR

Hu et al. [132] screened the urine of healthy controls and kidney transplant recipients using an antibody array and a multiplex beads assay. The kidney transplant recipients included 84 patients with renal-allograft injury, 29 patients with stable graft function, and 19 healthy individuals. A number of cytokines were elevated in both acute and chronic injury including interferon-γ induced protein of 10 kDa; monokine induced by interferon-γ; MIP, macrophage inflammatory protein and osteoprotogerin. Unfortunately none of the four biomarkers were able to differentiate specific causes of graft dysfunction. The authors pointed out that since both alloimmune and nonalloimmune causes of graft dysfunction increase cytokine levels, their discriminatory power in general may be limited.

Kutukculer et al. [133] examined whether plasma levels of lymphokines IL-2, IL-3, IL-4, IL-6, IL-8, and soluble CD23 could predict acute rejection in16 renal-transplant recipients during the first 14 days after engraftment. Of the 16 patients, 7 had clinical evidence of acute allograft rejection and 5 showed stable graft function. The remaining four patients had primary nonfunction. Plasma levels of IL-2, whenever detected, were predictive of impending graft rejection. Plasma levels of IL-4 and IL-6 were more reliable for diagnosis of rejection, whereas IL-3, IL-8, and soluble CD23 were not diagnostic or predictive of rejection. The authors pointed out that posttransplant infections could affect the diagnostic performance of these biomarkers.

Crispim et al. [134] evaluated tissue levels of the proinflammatory cytokine interleukin-17 (IL-17) by ELISA in 19 recipients of living- and deceased-donor transplants, and healthy controls. Tissue IL-17 was significantly increased in grafts undergoing rejection (125.7 ± 27.06 pg/mL) versus grafts without rejection group (30 ± 13.32 pg/mL) ($P < 0.05$). Biopsies from healthy controls had no IL-17.

Millan et al. examined intracellular expression of IFN-γ, IL-17, IL-2, and IL-17 in a multicenter study of 142 transplant recipients (63 liver/79 kidney during the pretransplantation period, and during posttransplant weeks 1 and 2, and months 1–3). Pre- and posttransplantation intracellular expression of IFN-γ in CD4(+)CD69(+) and in CD8(+)CD69(+) cells and soluble IL17 was associated with liver and kidney recipients at a high risk of acute rejection. Pretransplant IL-2(+) in CD8(+)CD69(+) cells also identified kidney recipients at high risk of acute rejection [135].

Zhang et al. evaluated whether serum concentrations of interferon-gamma inducible protein-10 (IP-10), fractalkine, and their receptors (CXCR3 and CX3CR1) could predict the occurrence of acute rejection in kidney transplant recipients. Serum samples were obtained from 52 patients (biopsy-proven AR, $n = 15$; no AR, $n = 35$; healthy volunteers, $n = 12$) 1 day prior to transplantation and on postop days 1, 3, 5, 7, and 9. Recipients with BPAR had significantly higher serum fractalkine, CXCR1, IP-10, and CXCR3 levels versus those without AR, and healthy controls. Fractalkine and IP-10 had the largest area under the ROC curve of 0.86 (95% CI: 0.77–0.96). Chemokine levels decreased after steroid therapy. The authors suggested that serum fractalkine, IP-10, and their receptors (in particular the fractalkine/IP-10 combination) may be biomarkers of renal-allograft rejection, and a response to therapy [136].

Flow Cytometry and the Diagnosis of Acute Rejection

Roberti et al. [137] examined the ability of urine flow cytometry (UFC) to diagnose acute rejection. UFC was performed in 30 patients (32 events) admitted for evaluation of graft dysfunction (defined as serum creatinine increment ≥0.6 mg/dL above baseline). The UFC analysis was compared with the subsequent discharge diagnosis. Acute rejection was confirmed by biopsy in all cases. The discharge diagnoses were as follows: acute rejection ($n = 15$); CAN ($n = 8$); drug toxicity ($n = 4$); urine leak ($n = 2$); recurrence of primary disease ($n = 1$); lymphocele ($n = 1$); and unknown ($n = 1$). Urine analysis was performed by FACS analysis and 10,000 cells were counted in each sample. The cells were assessed for anti-HLA-DR, anti-CD3, anti-CD14, anti-CD54 (ICAM-1), and anti-CD25 (IL-2 receptor [IL-2R]). Acute rejection was associated with the presence of ≥5% HLA-DR-positive cells and ICAM-1-positive cells in 100 and 53% of samples, respectively ($P < 0.01$ vs. others). A number of markers were highly specific for the diagnosis of acute rejection including ICAM-1 or CD3-positive cells (100% specificity) and IL-2R receptor- or HLA-DR-positive cells (specificity = 88%). CAN was associated with CD14-positive cells ($P = 0.03$ vs. others; specificity = 87.3%). The most accurate finding associated with the diagnosis of acute rejection was the finding of HLA-DR-positive cells with only a 12% rate of false-positive results (sensitivity = 100%, specificity = 88%, PPV = 88%, and NPV = 100%). Samples from patients with drug toxicity, urological problems, or recurrence of primary disease lacked expression of the antigens studied. The authors suggested that UFC of urinary cells could differentiate acute rejection from other causes of acute allograft dysfunction. HLA-DR was found to be the most sensitive, and ICAM-1 the most specific marker for acute rejection.

In a follow-up study the same authors evaluated whether serial UFC correlated with clinical outcome [138]. A variety of cell surface antigens (anti-CD3, anti-CD14, anti-HLA-DR, anti-CD54, and anti-interleukin 2 receptor) were examined by UFC during a 30-day period after the diagnosis and treatment of 24 acute rejection (AR) episodes. The study included 59 urine specimens, from 17 patients meeting the diagnostic criteria for AR. The most common antigen seen during the first 2 days of AR was HLA-DR (91.7% of the samples), followed by CD14 (50%) and CD54 (41.7%). Expression of HLA-DR-, CD14-, and CD54-positive cells after day 4 correlated with the need for antilymphocytic drugs. The most accurate marker was CD54, with sensitivity = 100% and specificity = 90.9% ($P = 0.001$). CD54- and CD14-positive urinary cells persisted in those patients who had permanent graft injury after treatment of AR.

Stubendorff et al. studied urine samples collected on postkidney transplant days 3–10 from 116 kidney recipients, including 58 with biopsy-proven rejection and 58 with stable transplant function. Recipients with acute rejection had significantly higher protein levels of alpha1-microglobulin (A1MG) and haptoglobin (Hp) (A1MG 29.13 vs. 22.06 μg/mL, $P = 0.001$; Hp 628.34 vs. 248.57 ng/mL, $P = 0.003$). Using both proteins in combination, that is, A1MG and Hp resulted in diagnosis of rejection with a specificity of 80% and a sensitivity of 85%. The authors suggested that urinary A1MG and Hp could be biomarkers of rejection in the early posttransplant period [139].

Proteomic-Based Approaches to Finding Biomarkers of Acute Rejection

Schaub et al. [140] examined whether proteins detected in urine using mass spectrometry could serve as biomarkers of acute rejection. Four patient groups were selected based on allograft function, clinical course, and allograft biopsy result: (1) acute clinical rejection group (n = 18); (2) stable transplant group (n = 22); (3) acute tubular necrosis group (n = 5); and (4) recurrent (or de novo) glomerulopathy group (n = 5). Urine was collected on the day of the allograft biopsy, and the median time to biopsy ranged from 0 to 253 weeks posttransplant. A control group of 28 urines from healthy individuals, as well as 5 urines from nontransplanted patients with lower urinary tract infection were also analyzed. The authors also performed sequential urine analysis in patients in the acute clinical rejection and stable transplant function (groups 1 and 2). Ninety-four percent (17/18 patients) with acute rejection episodes were found to have three prominent peak clusters, whereas only 18% (4/22) of patients without clinical and histologic evidence for rejection, and 0 of 28 normal controls ($P < 0.001$) had a similar finding. The presence or absence of these peak clusters correlated with the clinicopathologic course in most patients. Urine protein profiles in recipients with ATN and glomerulopathy groups were distinct from those with the pattern of rejection. In a follow-up study [141] the protein peaks were found to derive from nontryptic cleaved forms of beta2-microglobulin. Cleavage of intact beta2-microglobulin was found to require a urine pH < 6 and the presence of aspartic proteases. Accordingly, patients with acute tubule-interstitial rejection had lower urine pH than, and greater urine aspartic proteases and intact beta2-microglobulin. The authors proposed that these factors resulted in increased amounts of cleaved urinary beta2-microglobulin.

Chen et al. used microarray analysis of acute rejection (AR) in pediatric renal, adult renal, and adult cardiac transplantation to identify 45 genes that were upregulated in all three conditions. Ten proteins were then chosen for serum ELISA assays in 39 renal-transplant patients, and three (PECAM1, CXCL9, and CD44) were found to be significantly higher in AR. PECAM1, CXCL9, and CD44 were also significantly increased during AR in the 63 cardiac transplant recipients studied. Serum PECAM1 displayed the best biomarker profile for renal AR with a sensitivity of 89% and a specificity of 75%. PECAM1 also demonstrated increased immune-histochemical expression during AR in renal, hepatic, and cardiac transplant biopsies. The authors suggested that integration of gene expression microarray analysis from disease samples and publicly available data sets could serve as a powerful and cost-effective strategy for finding new serum protein biomarkers [142].

Srivastava et al. screened urine samples for candidate biomarkers using large-scale antibody microarrays. Candidates initially identified by the antibody microarray platform including ANXA11, Integrin α3, Integrin β3, and TNF-α were subsequently qualified using Reverse Capture Protein Microarrays. The authors suggested that urinary ANXA11, Integrin α3, Integrin β3, and TNF-α could be used as a quantitative urine proteomic signature for diagnosing chronic or acute rejection based on specificity, sensitivity, and ROC analyses [143].

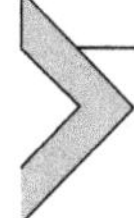

BIOMARKERS OF CHRONIC ALLOGRAFT NEPHROPATHY

Tissue Markers

Kirk et al. [144] hypothesized that clinically stable human kidney transplants were subject to detectable ongoing immune activity which could be correlated with worsening of allograft function. Forty stable renal allografts were biopsied 2–3 years posttransplantation. Biopsies were evaluated by RT-PCR for CD3γ mRNA (a marker of T-cell receptor turnover) as well as genes associated with acute rejection (TNF-α, IFN-γ, IL-1β, IL-2, IL-4, IL-6, and IL-8). Gene expression was then correlated with clinical findings at the time of biopsy and 2 years postbiopsy. Cytokine gene transcription and histological evidence of injury were found in more than two-thirds of clinically stable grafts. Increasing lymphocytic infiltration correlated with the proteinuria ($P = 0.034$) and worsening interstitial fibrosis ($P = 0.005$). The fibrosis demonstrated a significant positive correlation with baseline creatinine ($P = 0.006$) and negatively correlated with the GFR measured on the day of the biopsy ($P = 0.037$). Intragraft CD3γ signal also correlated with increasing proteinuria ($P = 0.043$) implicating increased T-cell activity with deteriorating graft function. On the other hand, CD3γ did not correlate with fibrosis, serum creatinine, or GFR. Both fibrosis ($P = 0.01$) and tubular atrophy ($P = 0.01$) on the original biopsy were correlated with declining renal function at follow-up. CD3γ levels at the time of original biopsy correlated with the highest change in GFR over time ($P = 0.045$). The authors suggested that significant injury and immune activity could be found in patients with clinically stable allografts, and that this injury may be a cause of chronic allograft nephropathy.

Nickel et al. [145] examined whether intragraft expression of perforin, granzyme B, and Fas ligand correlated with long-term clinical outcome following an episode of acute rejection. Gene transcript analysis was performed on 22 human renal biopsies for the expression of perforin, granzyme B, Fas ligand, and Fas. Expression levels were correlated with Banff rejection grades allograft function in the course of acute rejection, and clinical outcome at one year. Fas ligand, but not perforin nor granzyme B was significantly upregulated in therapy-resistant acute rejections ($n = 7$) versus therapy-sensitive acute rejection ($n = 8$). There was no relation between cytotoxic marker expression, Banff rejection grades, or peak serum creatinine.

De Matos et al. assessed whether the morphologic features of biopsies from kidney allograft in recipients with stable renal function could be early markers of allograft dysfunction, after 5 years of follow-up. Forty-nine renal-transplant patients with stable renal function and a median posttransplant time of 105 days were enrolled. Patients were followed thereafter for 64.3 ± 8.5 months. The mean serum Cr at the time of biopsy was 1.44 ± 0.33 mg/dL, whereas the mean sCr after the follow-up period was 1.29 ± 0.27 mg/dL. Nine patients (19%) experienced a reduction of graft function. Eleven biopsies (22%) demonstrated tubulointerstitial alterations whereas 17 biopsies (34%) had evidence of EMT. Fifteen biopsies (32%) demonstrated a high interstitial expression of myofibroblasts and tubular vimentin. By Cox multivariate analysis, high expressions of interstitial myofibroblasts and tubular vimentin were associated with reduced graft function, with risks of 3.3 ($P = 0.033$) and 9.8 ($P = 0.015$), respectively. The authors suggested that fibrogenesis may occur very early after transplantation and represented a risk factor for the long-term deterioration of renal function [146].

Zheng et al. performed a longitudinal study of five biomarkers including: human leukocyte antigen (HLA)-G5 and sCD30 level in sera; intracellular adenosine triphosphate (iATP) release level of CD4(+) T cells; and granzyme B/perforin expression in PBMCs and biopsies, in 84 renal-transplant recipients. Patients with clinical or biopsy-proven rejection had significantly increased iATP release level of CD4(+) T cells, and elevated sCD30 but lowered serum HLA-G5 level versus recipients with stable graft function. Granzyme B and perforin were also increased in PBMCs and graft biopsies of patients with AR. The authors concluded that upregulation of sCD30, iATP, granzyme B, perforin, and downregulation of HLA-G5 might identify recipients at risk of AR. Furthermore, the authors suggested that iATP outperformed the other biomarkers in the prediction of acute rejection [147].

Heng et al. performed a systematic search of 16 studies (680 subjects) involving granzyme B (GZMB) and perforin (PRF) to assess their noninvasive diagnostic performance in the diagnosis of AR. The sensitivity, specificity, positive likelihood ratios, and negative likelihood ratios with 95% CIs for GZMB were as follows: 0.76 (0.71–0.81), 0.86 (0.82–0.89), 4.58 (3.36–6.25), and 0.32 (0.22–0.47), respectively. For PRF the indices were 0.83 (0.78–0.88), 0.86 (0.82–0.89), 4.82 (3.66–6.35), and 0.26 (0.18–0.37), respectively. The combination of GZMB and PRF demonstrated indices of 0.65 (0.53–0.76), 0.96 (0.91–0.98), 12.66 (5.83–27.50), and 0.40 (0.23–0.69), respectively, when both markers were positive. Furthermore, the probability of developing AR in kidney transplant recipients increased from 15% to 73% if both GZMB and PRF were positive, but was reduced to 2% if they were negative. The authors concluded that neither GZMB nor PRF alone were convincing noninvasive diagnostic markers for AR, but that the combination of PRF and GZMB might provide information to direct the performance of an allograft biopsy and possibly result in earlier therapeutic intervention [148].

Djamali et al. measured baseline donor-specific antibody levels as mean fluorescence intensity (MFImax) to define the intensity of immunosuppression. Study includes patients transplanted (n = 146) with a negative flow crossmatch and majority followed for at least 1 year. Mean-calculated panel-reactive antibody and MFImax ranged from 10.3% to 57.2% and 262–1691, respectively, at the time of transplant between low- and high-risk protocols. From transplant to 1 week and 1 year mean MFI-max was increased significantly. Combining both clinical and subclinical rejection, the occurrence of acute rejection was 32% together with 14% cellular, 12% antibody-mediated, and 6% mixed rejection. C4d staining in postreperfusion biopsies and increased specific antibodies at 1-week posttransplant were key predictors of rejection as per regression analyses. Rejection risk increases by 2.8-fold when MFImax goes up by 500. Thus in moderately sensitized patients, an early rise in donor-specific antibodies after transplantation and C4d staining in postreperfusion biopsies are risk factors for rejection [149].

Nogare et al. evaluated mRNA gene expression of kidney injury molecule-1 (KIM-1) in kidney transplant patients with graft dysfunction. Kidney tissue and urinary sediment cells of 77 patients were analyzed which includes cases of acute tubular necrosis (n = 9), acute CIN (n = 10), acute rejection (n = 49), and interstitial fibrosis and tubular atrophy (IF/TA, n = 29). KIM-1 mRNA levels quantified by RT-PCR were higher in the biopsies (26.17; 3.38–294.53) and urinary sediment cells (0.09; 0–5.81) of IF/TA patients compared with all other groups ($P < 0.01$). Thus quantification of KIM-1 mRNA gene can be used as a noninvasive biomarker of IF/TA [150].

Gomez-Alamillo et al. monitored 71 kidney transplant for RNA gene expression of EMT, angiotensinogen, E-cadherin, N-cadherin, transforming growth factor (TGF) beta and bone morphogenetic patients 7 (BMP7). After 1 year, the ratio TGF-beta/BMP7 discriminated patients with a sCr > 1.5 mg/dL ($P = 0.034$). Reversion of EMT at 3 months posttransplantation displayed differences in initial graft progress. At 1 year, the TGF-beta/BMP7 ratio proposed activation of EMT, possible early marker of renal dysfunction. Thus the levels of urinary mRNA of genes that control epithelial–mesenchymal transition (EMT) can detect the development of IF/TA in addition of reflecting early damage [151].

Jia et al. studied 96 kidney transplant recipients including long-term survival group, acute rejection group (AR), and chronic rejection group (CR). Healthy volunteers were used as a control group. Tumor necrosis factor-α–induced protein 8-like 2 (TIPE2) expression was measured in PBMC and kidney biopsy samples. TIPE2 expression was significantly higher in PBMC of CR group than that of the healthy controls ($P < 0.001$); however expression of TIPE2 in AR group was lower than that of healthy controls ($P < 0.05$). The renal expression of TIPE2 in allograft tissues of AR and CR was lower in comparison to normal kidneys though this difference was significant in CR kidneys. Based on the positive correlation between TIPE2 expression in PBMCs and the CR of allo-kidney graft's kidney chronic rejection can be monitored by TIPE2 expression levels in the blood samples. Thus TIPE2 may be used as a diagnosis molecular marker in monitoring kidney chronic rejection clinically [152].

Plasma Markers

Lachmann et al. [153] examined 1014 deceased kidney transplant recipients for HLA antibodies using Luminex Single Antigen beads. Thirty percent of recipients were found to have HLA antibodies, and of these 31% were found to have DSA. DSA-positive recipients had significantly lower graft survival (49% vs. 83% in the HLA antibody negative group; $P \leq 0.0001$). Lower graft survival was also seen in recipients who were DSA-negative but HLA antibody positive (70% vs. 83%; $P = 0.0001$). In a prospective analysis of 195 patients, patients who were repeatedly negative for HLA antibodies had a superior survival probability compared with patients who developed de novo HLA antibodies after the first testing (94% vs. 79%; respectively

$P = 0.05$). The authors concluded that HLA antibodies were detrimental to graft survival, even late in the transplant course.

Lauzurica et al. [154] postulated that cardiovascular disease and CAN are both manifestations of persistent, posttransplant inflammation. The authors studied the role of pregnancy-associated plasma protein A (PAPP-A) in the development of posttransplant cardiovascular events and CAN. PAPP-A is a metalloproteinase linked to zinc that has been used in the diagnosis of fetal Down's Syndrome [155]. PAPP-A has also been found in atheromatous plaques, and circulating levels are increased in acute coronary syndromes [156]. It has also been associated with acute coronary syndrome and atheromatous plaque instability. Lauzurica et al. [154] examined whether serum concentration of pretransplant PAPP-A was associated with posttransplant cardiovascular events and CAN. Pretransplant levels of ultrasensitive CRP, IL-6, TNF-α, and ultrasensitive PAPP-A were measured in 178 renal-transplant recipients. During the follow-up period of 49.3 ± 33.6 months, 19 recipients developed biopsy-proven CAN and 27 recipients had a cardiovascular event. PAPP-A was found to correlate with the other inflammatory markers (PAPP-A vs. CRP, $r = 0.218$; $P = 0.004$; PAPP-A vs. IL-6, $r = 0.235$; $P < 0.001$; PAPP-A vs. TNF-α, $r = 0.372$; $P < 0.001$). Multiple regression analysis showed PAPP-A (RR: 4.27; 95% CI: 1.03–17.60; $P = 0.044$) and TNF-α (RR: 5.6; 95% CI: 1.43–21.83; $P = 0.013$) to be predictors of CAN, whereas PAPP-A (RR : 6.4; 95% CI:1.24–33.11; $P = 0.027$) and CRP (RR: 6.05; 95% CI:1.21–29.74; $P = 0.028$) were predictors of cardiovascular events.

Rotondi et al. [157] examined the role of the chemokine CXCL10/IP-10 in graft failure attributed to both acute and chronic rejection. Pretransplant sera obtained from 316 deceased-donor kidney was retrospectively assayed for serum CXCL10 and CCL22/MDC levels by ELISA. CsA-based immunosuppression was used in 93% of the recipients. The median follow-up time posttransplant (including patients who experienced graft failure) was 39 months. Patients with stable graft function had significantly lower median pretransplant serum CXCL10 levels than recipients who subsequently endured graft failure (93.0 vs. 157.4 pg/mL; $P = 0.0007$). No differences for serum CCL22 levels were observed in the same groups of patients. Patients were grouped based on percentiles of pretransplant serum CXCL10 levels: 0–25th (<64 pg/mL, $n = 80$), 25th–50th (>64 and <97 pg/mL, $n = 78$), 50th–75th (>97 and <157 pg/mL, $n = 78$), and 75th–100th (>157 pg/mL, $n = 80$). Death-censored 5 year-survival rates for grafts in each percentile group were 97.5 93.6, 89.7, 78.7 ($P = 0.0006$).

Pretransplant serum CXCL10 levels did not influence patient survival. The frequency of acute rejection in the first posttransplant month was also increased based on the percentile of pretransplant serum CXCL10 levels in the four groups ($\chi^2 = 11.412$; $P = 0.009$). Patients with pretransplant serum CXCL10 levels > 75th centile (>157 pg/mL) had an increased frequency of acute rejection versus patients with serum CXCL10 levels < the 75th centile (34.8% vs. 21.4%; $P = 0.01$). Multivariate analysis demonstrated that CXCL10 was most predictive of graft loss (RR 2.787) among the variables analyzed.

Grenzi et al. examined 511 kidney transplant recipients, with a graft functioning for at least 2.8 years and a follow-up of 9.3 years. Factors that were independently associated with CAN-graft loss included the presence of class II HLA antibodies, a high sCD30 level (≥34.15 ng/mL), and a serum creatinine of ≥ 1.9 mg/dL (*P* values <0.0001, 0.05, <0.0001, respectively). The hazard ratio of all three factors combined for CAN-graft loss was 20.2. A subsequent analysis of 166 for-cause graft biopsies indicated that serum creatinine and sCD30 levels were independently associated with interstitial lesions. The authors suggested that posttransplant sCD30 levels, particularly when combined with HLA class II antibodies and serum creatinine, could provide valuable prognostic information regarding graft outcome [158].

Genetic Markers

McLaren et al. [159] assessed the frequency of five polymorphisms in ICAM-1, E-selectin, and L-selectin in four groups of patients: renal-allograft recipients with chronic allograft failure ($n = 62$); their matched donors, where available ($n = 33$); kidney allograft recipients with graft survival of greater than 10 years ($n = 110$); and a group of UK controls ($n = 101$). A variant allele in exon 4 of ICAM-1 (R241) was significantly more frequent in recipients with chronic allograft failure versus long-term survivors and UK controls (19.4 vs. 10.0 and 9.4%, respectively, $P = 0.015$ and 0.025). Stratification by time to graft failure demonstrated that more rapid failure was associated with another ICAM-1 variant in the recipient (E469) in exon 6 ($P = 0.033$). No significant association was detected between the selectin polymorphisms studied and chronic allograft failure.

Human beta-defensin-1 (HBD-1) is a 36 amino acid with antimicrobial properties that is found in the loop of Henle, distal tubules, the female genitourinary tract, and plasma [160]. HBD-1 was found to be chemotactic for T cells and dendritic cells through the CCR6 chemokine receptor [161]. Alpha-1-antichymotrypsin (ACT: 4.4 kDa) is a "serpin" or serine protease inhibitor [162] found in liver, kidney [163], and plasma, that may be a potential biomarker of acute liver transplant rejection [164].

Kłoda et al. examined the association of polymorphisms in ICAM1 and VCAM1 genes with renal-allograft function and biopsy-proven CAN. Caucasian renal-transplant recipients (166 men and 104 women) were genotyped for the ICAM1 polymorphism, rs5498 and the VCAM1 gene polymorphisms rs1041163 and rs3170794. Serum creatinine in the first posttransplant month was higher for GG carriers of the rs5498 ICAM1 genotype (AA + AG vs. GG, $P = 0.07$) but did not reach statistical significance. Serum creatinine at posttransplant months 12, 24, and 36 were significantly different among rs5498 ICAM1 genotypes ($P = 0.0046$, $P = 0.016$, and $P = 0.02$), and were significantly greater among GG carriers (AA + AG vs. GG, $P = 0.001$, $P = 0.004$, and $P = 0.006$). The Rs5498 ICAM1 GG genotype and recipient male gender were independently associated with increased serum creatinine [165].

Urine Biomarkers

O'Riordan et al. [166] assessed urinary peptides, human beta-defensin-1 (HBD-1; 4.7 kDa) and alpha-1-antichymotrypsin (ACT: 4.4 kDa), as biomarkers of acute rejection in renal allografts. The paper includes clinical details of 73 patients although the authors indicated that the number of patients included in the different analyses varied due to sample availability. Samples were collected prebiopsy and before treatment was initiated and all cases of acute rejections were confirmed by renal biopsy. The majority (27/34) of acute rejections occurred within 6 months of transplantation. The mean time from urine sampling to biopsy was 1.7 ± 1.6 days. Urine was also collected in patients with clinically stable transplant function, judged by steady serum creatinine during follow-up, as controls. Patients with acute allograft rejection had significantly reduced beta-defensin-1 and increased alpha-1-antichymotrypsin ($P < 0.05$) versus clinically stable transplants. Using both peptides combined, the area under the receiver operator characteristic curve for the diagnosis of acute rejection was 0.912. Urinary beta-defensin-1 levels, quantified by radioimmunoassay, were 176.8 ± 122.3 pg/mL in stable patients versus 83.2 ± 52.2 pg/mL in recipients with acute rejection, for an ROC AUC of 0.749 ($P < 0.01$).

Kurian et al. used proteogenomic approach to examine the blood of kidney transplant recipients with biopsy-proven CAN for biomarkers of mild and moderate/severe CAN. Using DNA microarrays, tandem mass spectroscopy proteomics and bioinformatics, over 2400 genes for mild

CAN, and over 700 for moderate/severe CAN were identified. A consensus analysis revealed 393 (mild) and 63 (moderate/severe) final candidates as CAN markers with a predictive accuracy of 80% (mild) and 92% (moderate/severe). Proteomic profiling also revealed over 500 candidates each, for both mild and moderate/severe CAN, including 302 proteins unique to mild CAN and 509 proteins that were unique to moderate/severe CAN. The authors concluded that the unique signatures of transcript and protein biomarkers for mild and moderate/severe CAN found in the peripheral blood could potentially be used to classify CAN [167].

Ashton-Chess et al. found that TRIB1 mRNA levels in PBMCs can discriminate patients with chronic antibody-mediated rejection (CAMR) from those with other types of late allograft injury with high sensitivity and specificity. Thus TRIB1 can be used as a potential biomarker of chronic antibody-mediated allograft failure [168].

Shi et al. investigated the potential of connective tissue growth factor (CTGF) as a biomarker of chronic renal-allograft injury characterized by tubular atrophy and interstitial fibrosis (TA/IF) using a Fisher to Lewis allogenic rat kidney transplant model. In epithelium CTGF expression was upregulated early and urinary CTGF was significantly elevated from fourth week. SCr is increased at week 12 but was stable before week 8. Urinary CTGF concentration was positively correlated with SCr and degree of interstitial fibrosis. Thus measuring urinary CTGF may offer a potential noninvasive strategy to predict the early onset of chronic renal-allograft injury [169].

Ho et al. studied 111 patients who underwent serial protocol biopsies at 0, 6, and 24 months to identify early urinary predictors for the subsequent development of IFTA. Four urinary proteins [CCL2, CXCL9, CXCL10, and alpha1-microglobulin (alpha1M)] were evaluated using ELISA and immune-nephelometry. CCL2 was associated with late graft outcomes in comparison to other urinary proteins. Hence measuring early urinary CCL2 can independently predict the development of IFTA at 24 months [170].

Satirapoj et al. investigated periostin expression in urine samples from CAN patients ($n = 24$) and renal-transplant patients with normal renal function (transplant controls, $n = 18$). Urine from healthy volunteers with normal renal function was used as a control ($n = 18$). In comparison to transplant and healthy controls urine periostin was significantly higher in CAN ($P < 0.001$) as measured by ELISA. Urine periostin distinguished CAN patients from transplant patients with normal renal function and demonstrated the sensitivity, specificity, and accuracy of 91.7, 77.8, and 85.7%, respectively. Thus urine periostin may be used a potential biomarker for chronic progressive renal injury in transplant recipients [170].

Cassidy et al. used OrbiTrap MS to identify urinary biomarkers that might differentiate transplant recipients with CAN. Thirty-four renal-transplant recipients with biopsy-proven CAN were compared to 36 renal-transplant patients with normal renal function. The primary node of the biomarker pattern was found to be β2 microglobulin. Three other members of this biomarker pattern—namely NGAL, clusterin, and kidney injury biomarker 1—were found in significantly higher urinary concentrations in patients with CAN versus those with normal kidney function. The authors suggested these biomarkers could potentially stratify patients based on transplant function, and might provide a noninvasive method of diagnosing CAN [171].

Oxidative Stress Markers

Reactive oxygen species (ROS) are potential biomarkers of oxidative stress [37].

Simic-Ogrizovic et al. examined plasma and RBCs from hemodialysis patients ($n = 15$, HD group), renal-transplant recipients with stable function ($n = 11$, SF group), renal-transplant recipients with biopsy-proven CAN ($n = 12$, CR group), and healthy controls ($n = 10$, C group) for the expression of malondialdehyde and thiol group levels [markers of oxidative stress (OS)] and glutathione peroxidase and Cu, Zn-superoxide dismutase (markers of antioxidant activity). Recipients with successful renal transplants demonstrated significantly reduced lipid peroxidation. After successful renal transplantation a significant improvement, but not normalization, of antioxidant enzyme activities was accompanied by significantly reduced lipid peroxidation. In contrast, patients with CAN demonstrated increased oxidative stress suggesting that OS may be a relevant pathophysiological factor for the development of CAN [172].

Djamali et al. examined whether EMT and oxidative stress could contribute to kidney tissue fibrosis in the setting of CAN. Using the Fisher 344 → Lewis model of CAN, the authors demonstrated that at 6-month post-transplant, kidney allografts displayed significantly increased tubular atrophy, interstitial fibrosis, and vascular wall thickening versus syngeneic transplants. Allograft recipients also had significantly increased serum creatinine (4.7 ± 1.3 vs. 0.59 ± 0.08 mg/dL, $P = 0.03$) and proteinuria (380 ± 102 vs.

30.2 ± 8 mg/dL, $P = 0.04$) versus syngeneic grafts. Increased alpha-smooth muscle actin (α -SMA) mRNA and protein levels, and reduced E-cadherin mRNA and protein immunoreactivity were found suggesting the presence of CAN-associated EMT. Of note, allograft α-SMA levels increased as early as 1–2 weeks posttransplant. Furthermore, tubular eNOS, iNOS, and interstitial collagen I and III levels were significantly increased in CAN-associated EMT. The authors concluded that CAN-associated EMT may be a link between oxidative stress and allograft fibrosis [173].

miRNAs as Noninvasive Biomarkers of CAN

Danger et al. examined whether microRNA expression patterns were associated with CAMR. Expression profiling of miRNAs in PBMCs of kidney transplant recipients with CAMR or stable graft function revealed 10 miRNAs that were associated with CAMR. Of the selected miRNAs, miR-142-5p was increased in PBMC and biopsies from patients with CAMR, and also in a rodent model of CAMR. ROC curve analysis revealed miR-142-5p as a potential biomarker with an AUC = 0.74; $P = 0.0056$ for patients with CAMR. Furthermore, expression of miR-142-5p was decreased in PHA-activated blood cells and was not modulated in PBMC from recipients with acute. Thus, miR-142-5p may be a potential biomarker in CAMR [174].

Rascio et al. compared the molecular signature of PBMCs and CD4(+) T lymphocytes isolated from kidney transplant recipients with CAMR to recipients with normal graft function and histology. Twenty-nine patients with biopsy-proven CAMR were compared with 29 stable transplant recipients (controls) and 8 transplant recipients with clinical and histological evidence of interstitial fibrosis/tubular atrophy. In PBMCs, 45 genes were differentially expressed between the groups. The genes were primarily upregulated in CAMR and involved type I interferon signaling. Sixteen microRNAs were downregulated in CAMR versus controls, of which four were predicted modulators of six mRNAs identified in the transcriptional analysis. Subsequent transcriptomic analysis of profiles in CD4(+) T lymphocytes in an independent group of patients provided evidence that activation of type I interferon signaling was a specific hallmark of CAMR. Furthermore, recipients with CAMR demonstrated decreased circulating BDCA2(+) dendritic cells, the natural type I interferon-producing cells. The authors thus concluded that type I interferon signaling may be a molecular signature of CAMR [175].

BIOMARKERS OF POLYOMA VIRUS INFECTION

Polyoma viruses are members of the Papovaviridae virus family, and are named for their ability to induce a variety of tumors in newborn mice [176]. The human polyomaviruses BK virus and JC viruses were named with the initials of the patients from whom they were first isolated [177,178]. They are nonenveloped viruses with a circular, double-stranded–DNA genome of 5300 bp and a diameter of 45 nm [179]. JC and BK polyomaviruses share 70% sequence homology with each other and with simian virus 40 (SV40) [179]. The viruses are widespread in immunocompetent hosts in both the United States and Europe, with reported seroprevalence rates of 60–80% [180,181]. Complications of polyoma virus infection typically occur in immunocompromised hosts. BK virus is more commonly associated with the urogenital tract and can cause hemorrhagic cystitis [182], renal-allograft dysfunction, and graft loss [183]. JC virus has been associated with neurologic complications including progressive multifocal leukoencephalopathy [178] but can cause renal-allograft dysfunction. Drachenberg et al. [184] prospectively evaluated polyoma virus infection in a cohort of 103 renal-allograft recipients. Evidence of BKV, JCV, or BKV + JCV shedding was found in 56.3, 27.2, and 16.5% respectively. BK viruria was strongly associated with polyoma virus nephropathy (48%, $P = 0.01$) and graft loss ($P = 0.03$) whereas JCV viruria tended to be asymptomatic ($P = 0.002$). The overall incidence of BKV polyoma virus nephropathy was 5.5% compared with an incidence of 0.9% for JCV polyoma virus nephropathy. Both viruses responded to reduction in immunosuppression.

Polyoma virus nephropathy is therefore an important cause of graft dysfunction [185] and is best diagnosed by biopsy. For prospective monitoring of transplant recipients, however, allograft biopsy would be impractical. Investigators have therefore assessed biomarkers of impending polyoma infection and nephropathy to guide management on when to perform a kidney biopsy.

Nickeleit et al. [185] characterized typical changes caused by polyoma virus in five cases seen within an 8-month period. PCR evidence of BK virus but not JC virus was found in urine samples from all 5 patients. Urinary decoy cells were also found in patients with persistent polyoma virus disease. Decoy cells were characterized by ground-glass type intranuclear inclusions that were positive for polyoma virus by immunohisotchemistry and electron microscopy. The specificity of decoy cell excretion was examined in urine collected from 483 renal-allograft recipients, including five patients with polyoma virus disease. Abundant urinary decoy cells were found in 28 recipients (6%), whereas scant urinary decoy cells were found in a further 72 (15%) allograft recipients. Of the 28 patients with abundant urinary decoy cells, 5 recipients had polyoma virus disease (18%), whereas the remaining 23 (82%) had no cytopathic evidence of polyoma in allograft biopsies by light microscopy or IHC.

In a follow-up study, Nickeleit et al. [179] retrospectively investigated whether BK virus DNA could be found in the plasma of renal-allograft recipients with BK virus nephropathy using PCR. PCR for BK virus was performed on plasma samples from: 9 renal-allograft recipients with BK virus nephropathy; 41 recipients without nephropathy (16 of whom had urinary decoy cells), and urine; and 17 subjects with human immunodeficiency virus type 1 (HIV-1) infection who had not undergone transplantation. The latter served as immunocompromised controls. BK virus DNA was found in the plasma of all nine patients with BK virus nephropathy (diagnosed histologically), in 2 of the 41 renal-allograft recipients without nephropathy and in none of the subjects with HIV-1 infection. Three of the six patients with nephropathy were followed during their posttransplant course. BK virus DNA was initially undetectable but was subsequently found 16–33 weeks prior to the biopsy diagnosis of BK nephropathy.

Batal et al. [186] examined the consequences of increased immunosuppression in 32 allograft recipients with BK viruria, a biopsy diagnosis of ACR, and negative in situ hybridization for viral DNA ($n = 50$). Type IA rejection was seen 24 recipients, type IB in 24, and type IIA in 2 recipients. The presence of high urine viral load (>1.0 E + 05 copies/mL) was associated with development of viremia after antirejection treatment [5/9 (56%) vs. 0/24 (0%) in patients with low urine viral load, $P < 0.001$].

Urinary BKV replication, detected as either decoy cells or DNA PCR antedates BKV viremia by a median of 4 weeks, and biopsy-proven BKV nephropathy by a median of 12 weeks [187].

Drachenberg et al. [184] investigated the frequency and clinical correlation of BKV and JCV replication in a cohort of 103 kidney transplant recipients with urinary decoy cells. Evidence of BKV or JCV DNA by real-time PCR was found in 56.3 and 27.2% of subjects, respectively. A minority of subjects (16.5%) had BKV and JCV coinfection. Subjects with persistent urinary decoy cells (>2 months) or an increase in serum creatinine of greater than 20% underwent allograft biopsy. Subjects with urinary BKV alone had a significantly higher serum creatinine at the time of the biopsy ($P = 0.002$) and at the end of follow-up ($P = 0.05$). BKV viruria was significantly more likely to be associated with viremia at the time of the biopsy compared with pure JCV shedding (93.1% vs. 14.3%, respectively; $P \leq 0.0001$). The absolute level of blood viral copies was less in patients shedding JCV versus BKV (mean of 2.0E + 03 JC copies/mL vs. mean of 2.3E + 06 BKV copies/mL). JCV viremia was also shorter lived and never persisted beyond 1 month. PVAN was more likely if BKV viremia levels were ≥10E4 copies/mL ($P \leq 0.0001$), whereas biopsies were more likely to have normal parenchyma if BKV viremia was <10E4/mL (81% vs. 20% in patients with viremia of ≥10E4, $P < 0.0001$). Polyoma virus-associated nephropathy was more common with BKV (5.5%) versus JCV (0.9%).

Hirsch et al. [188] prospectively examined whether BKV replication was associated with nephropathy in a prospective, single-center study involving 78 renal-transplant recipients. Urine was collected at routine monthly outpatient visits for the first six months posttransplant, and whenever patients required hospitalization, required an allograft biopsy or experienced graft dysfunction. Nested PCR assay was used to measure plasma BKV DNA whenever urinary decoy cells were found. BKV DNA was also measured at 3, 6, and 12 months after transplantation. Twenty-three recipients had urinary decoy-cell shedding at a median of 16 weeks postengraftment (range, 2–69 weeks). BKV viremia was found in 10 patients at a median of 23 weeks (range, 4–73 weeks), and BKV nephropathy was diagnosed in 5 recipients at a median of 28 weeks (range, 8–86 weeks). By Kaplan–Meier analysis, the probability of decoy-cell shedding was 30% (95% CI, 20–40%), the probability of BKV viremia was 13% (95% CI, 5–21%) and the probability of BKV nephropathy was found to be 8% (95% CI 1–15%). The sensitivity and specificity of decoy-cell shedding for the diagnosis of BKV nephropathy was 100 and 71%, respectively, while the PPV was 29%, and the NPV was 100%. The sensitivity and specificity of BKV viremia was 100 and 88%, respectively, with a PPV of 50% and an NPV of 100%. Subjects with biopsy-proven BKV nephropathy had a significantly higher mean plasma viral load compared with subjects without histologic evidence of nephropathy (28,000 copies/mL vs. 2000 copies/mL; $P < 0.001$). On serial testing, BK viral load increased to ≥7700 copies/mL in all subjects who developed BKV nephropathy.

Viscount et al. [189] examined whether detection of urinary BK virus by PCR and urine cytology could identify patients with PVAN. Biopsy-confirmed BK PVAN was diagnosed in four out of 114 patients (3.5%). Using a cutoff value of >1.6E + 04 copies/mL, BKV viremia had a sensitivity and specificity of 100 and 96%, respectively, and a positive- and NPV of 50 and 100%, respectively. A BKV viruria cutoff of >2.5E + 07 copies/mL had a sensitivity and specificity of 100 and 92%, respectively, and a positive- and NPV of 31 and 100%. Urinary decoy cells performed less well. Sensitivity and specificity were 25 and 84%, respectively, and a positive- and NPV of 5 and 97%, respectively, for the diagnosis of concurrent PVAN.

The preceding studies therefore suggest that the absence of decoy-cell shedding or viremia reliably excludes the diagnosis of PVAN.

Singh et al. [190] hypothesized that urinary Haufen was a biomarker of BKV nephropathy. The authors discovered the presence of urinary cast-like polyomavirus aggregates by electron microscopy, that they named "Haufen" after the German word for "cluster or stack". Urine samples from control patients (n = 194 samples from 139 patients) and patients with BK polyomavirus nephropathy (n = 143 samples from 21 patients) were examined for the presence of Haufen, and correlated with histology, BK viruria, and BK viremia. Urinary Haufen correlated with biopsy-proven BK nephropathy, with a concordance rate of 99%. All urinary samples from controls were Haufen-negative, despite the presence of viremia (in 8%) or viruria (in 41%) of control samples. Fifty-four percent (77/143 urine samples) from all 21 patients with BK PVAN contained Haufen. The detection of Haufen had a specificity and sensitivity of 99 and 100%, respectively, and an NPV and PPV of 100 and 97%, respectively.

Jackson et al. examined whether a urine-based chemokine assay could noninvasively differentiate recipients with rejection from other common clinical diagnoses. Urine was collected from 110 adults and 46 children who were: healthy volunteers, stable renal-transplant recipients; recipients with clinical or subclinical acute rejection (AR); BK infection (BKI); calcineurin inhibitor (CNI) toxicity, or interstitial fibrosis (IFTA). Urine CXCL9 and CXCL10 were found to be markedly increased in adults and children with either AR or BKI (P = 0.0002), but not in stable allograft recipients or recipients with CNI toxicity or IFTA. However, neither chemokine could distinguish between AR and BKI. The authors concluded that urinary chemokines may be useful biomarkers to distinguish patients with active clinical issues from those with a benign clinical course [191].

Konietzny et al. examined whether mass spectrometry (MS) of urinary BKV-derived proteins could be used as a noninvasive surveillance method for the detection and monitoring of BKV infection. MS detected peptides derived from Viral Protein 1 (VP1) differentiated between subtypes I and IV. Furthermore, higher urinary decoy cell numbers were associated with VP1 subtype Ib-2 suggesting that viral subtypes may be associated with more severe BKVAN. The authors suggested that MS examination of urinary BKV–derived proteins could potentially distinguish different viral subtypes, and also between BKVAN and acute rejection [192].

Sigdel et al. examined whether noninvasive detection of urine protein biomarkers from 262 renal-allograft recipients with biopsy-confirmed allograft injury could be used to inform immunosuppression titration. Urine samples were randomly split into a training set (n = 108 patients) and an independent validation set (n = 154 patients) with the following biopsy-proven diagnoses: acute rejection (AR) (n = 74), stable graft (STA) (n = 74), chronic allograft injury (CAI) (n = 58), BK virus nephritis (BKVN) (n = 38), nephrotic syndrome (NS) (n = 8), and healthy, normal control (HC) (n = 10). Differential abundances of 389 proteins across urine specimens of the injury types ($P < 0.05$) were found, and SUMO2 (small ubiquitin-related modifier 2) was identified as a "hub" protein for graft injury irrespective of causation. AR was differentiated from stable graft function by 69 urine proteins ($P < 0.01$), of which 12 proteins were upregulated in AR with a mean fold increase of 2.8. Nine urinary proteins (HLA class II protein HLA-DRB1, KRT14, HIST1H4B, FGG, ACTB, FGB, FGA, KRT7, DPP4) were specific for AR versus all other transplant categories ($P < 0.01$; fold increase >1.5). Increased levels of three of the urinary proteins were validated by ELISA in AR using an independent sample set [fibrinogen beta (FGB; $P = 0.04$), fibrinogen gamma (FGG; $P = 0.03$), and HLA DRB1 ($P = 0.003$)]. The fibrinogen proteins also distinguished AR from BKV nephritis (FGB $P = 0.03$, FGG $P = 0.02$). The authors stated that these urinary proteins could serve as biomarkers for the noninvasive diagnosis of acute renal-allograft rejection [193].

Current guidelines suggest that screening for polyomavirus replication may allow PVAN to be detected earlier and graft loss prevented. Recommendations include screening patients at least every 3 months for the first 2 years and annually thereafter until the fifth posttransplant year. However, urine screening is complicated as variations in micturition intervals and urine content may result in interassay variations. The use of urine supernatants, cell pellets, or re-suspended urine may also cause variations in polyoma viral load measurements. Furthermore, PCRs may be inhibited in urinary specimens [187].

SUMMARY

The preceding sections highlight a number of excellent studies that have attempted to advance the use of biomarkers in KTx. In general, most of the studies employed biomarkers that satisfied the characteristics suggested by Parikh [1], Bakay [2], and Sandler [3] discussed in the introduction. Since the field is relatively new, there are a number of limitations in these analyses. In general, the studies reported tended to be small, and included selected patient populations with significant heterogeneity. Another problem is subjectivity in the clinical diagnoses studied. For example, the definition of DGF can vary, and studies of acute rejection may include patients diagnosed clinically, without histological confirmation. A number of studies discussed in the chapter suggest the possibility that genetic influences and differences in the quality of the donor tissue may affect outcome. The latter are rarely, if ever, included in the multivariate studies of posttransplant biomarkers. Therefore, the reader is always left to wonder whether a putative posttransplant biomarker is truly associated with outcomes. There is also the difficult issue of accounting for variations in transplant immunosuppression. It is rare for a population of patients to be on exactly the same immunosuppression, and even rarer for immunosuppressive levels to be the same. Whether variations in the overall level of immunosuppression could account for the reported differences seen with putative biomarkers is unknown. The field of transplantation has an excellent track record of conducting large, well-powered clinical trials with fairly homogenous patient populations. It may be advantageous for future clinical studies to include specimen collection for the express purpose of further defining the role of biomarkers in transplantation. Despite the limitations and the need for refinement, it is clear that studies of biomarkers hold promise. Whether biomarkers ever evolve to the point that they can replace traditional diagnostic methods remains to be seen. The ultimate role of biomarkers may primarily be an adjunct in guiding which diagnostic procedure is best (Tables 8.1–8.4).

Table 8.1 Biomarkers of AKI posttransplantation

Biomarker	*N*	Result	Time of collection	Source	References
Apoptosis of RTEs in donor kidney biopsies	89	Donor biopsies of allografts with subsequent ATN had significantly greater apoptotic RTEs versus immediate transplant function or early AR. A significantly greater percentage of apoptotic cells were found in the distal versus the proximal tubule in all groups.	Donor biopsy	Tissue	[4]
ICAM-1, VCAM-1, and E-selectin	73	LRT donor biopsies had significantly lower expression of ICAM-1 and VCAM-1 versus CRTs. Less expression of RTE cell ICAM-1 in CRTs with subsequent prompt function (38 ± 29%) versus DGF (65 ± 24, $P < 0.05$).	Donor biopsy	Tissue	[5]
Donor serum creatinine	51	DGF more frequent in the group with falling donor sCr. Mean recipient serum creatinine and allograft survival not different in donors with falling versus rising sCr. More donor hypertension and more chronic lesions in the biopsies of donors with a rising serum creatinine.	Expanded criteria donor sCr.	Serum	[8]
Kidney injury molecule—1(KIM-1)	62	KIM-1 absent in 72% of the protocol biopsies. KIM-1 seen in 100% of cases with obvious tubular injury. KIM-1 seen in 92% of biopsies with AR. Highest intensity for KIM-1 seen in tubular injury group followed by AR, and lowest in protocol biopsies. KIM-1 correlated with BUN and sCr	Allograft biopsy	Tissue	[15]
Soluble glycoprotein 130 (sgp130)	105	Pretransplant sgp130 plasma levels significantly reduced in patients who went on to have ATN versus those who had IGF ($P = 0.004$) or acute rejection ($P = 0.009$). Odds ratio of ATN was 4.3 on multivariable logistic regression analysis with a pretransplant sgp130 of ≤250 pg/mL.	Pretransplant plasma	Plasma	[9]

Serum calcium	585	Serum calcium levels correlated independently with DGF (OR = 1.14, 95% CI = 1.04–1.26 per 0.1 mmol/L). The use of calcium channel blockers prior to transplantation protected against DGF (OR 0.5 [95% CI 0.29– 0.87]). Nephrocalcinosis was found in 17% (12/71) of biopsies but was not associated with DGF or serum calcium levels.	Posttransplant serum	Serum	[17]
NGAL	25	NGAL expression significantly increased in CRT biopsies versus LRTs (2.3 ± 0.8 vs. 0.8 ± 0.7, respectively, $P < 0.001$). NGAL staining intensity correlated with cold ischemia time ($R = 0.87$, $P < 0.001$) and with peak posttransplant sCr. ($R = 0.86$, $P < 0.001$). Most intense staining for NGAL seen in four patients with DGF	Allograft biopsy	Tissue	[23]
Urinary IL-18 (uIL-18)	72	ATN had greater uIL-18 (644 pg/mg Cr.; $P < 0.0001$) versus all other conditions including: controls (16 pg/mg Cr.); prerenal azotemia (63 pg/mg Cr.),UTI (63 pg/mg Cr.), CRI (12 pg/mg Cr.), and nephrotic syndrome (34 pg/mg Cr.). CRTs with DGF had a median uIL-18 of 924 pg/mg Cr. versus 171 pg/mg Cr. in CRTs with PGF & 73 pg/mg Cr. in LRTs with PGF ($P < 0.002$). Lower uIL-18 associated with steeper decline in sCr POD 0–4 ($P = 0.009$).	Allograft urine	Urine	[24]
Urinary IL-18 (uIL-18) and NGAL	53	Urine NGAL and IL-18 significantly elevated levels in DGF ($P < 0.0001$). ROC analysis for the prediction of DGF-based on urinary NGAL or IL-18 on POD 0 showed an AUC of 0.9. Urine NGAL and IL-18 on POD 0 predicted the postop trend in sCr by multivariate analysis ($P < 0.01$).	Allograft urine	Urine	[25]

(*Continued*)

Table 8.1 Biomarkers of AKI posttransplantation (*cont.*)

Biomarker	*N*	Result	Time of collection	Source	References
Urine IL-18, NGAL, and KIM-1.	91	Median levels of urine NGAL and IL-18 had significant separation at all time points in DGF, SGF, and hall. Median urine KIM-1 was not statistically different between groups. ROC curve analysis suggested urine NGAL or IL-18 on POD1 were moderately accurate when used to predict dialysis within 1 week. Multivariate analysis confirmed elevated levels of uIL-18 or NGAL predicted the need for dialysis. Urine NGAL and IL-18 quantiles predicted graft recovery up to 3 mos. posttxp.	Allograft urine	Urine	[26]
Beta2-microglobulin (i/cβ2m), RBP, NGAL, and alpha1-microglobulin (α1m)	100	None of the biomarkers allowed for clear differentiation between stable transplants with normal tubular histology and stable transplants with subclinical tubulitis.	Allograft urine	Urine	[28]
Urinary actin, GGTP, LDH, IL-6, TNF-α, and IL-8	40	ROC curve analysis showed elevated urinary actin, IL-6, and IL-8 on POD 0 were predictors of sustained ARF. Sensitivity and specificity for actin = 0.67 and 0.86, respectively (using a cutoff value of 24.8 μg/g urine cr.). AUC = 0.75. Sensitivity and specificity for IL-6 = 0.83 and 0.86 for both (using a cutoff value of 60.2 ng/g urine cr.). AUC =0.91. Sensitivity and specificity for IL-8 = 1.00 and 0.61, respectively (using a cutoff value of 78.3 ng/g urine cr.). AUC = 0.82.	Allograft urine	Urine	[29]

SNPs of genes for CD28, CTLA4, ICOS, and PPCD1	678	DGF associated with two SNPs on the ICOS gene, rs10183087 and rs4404254 (odds ratio = 5.8; $P = 0.020$ and odds ratio = 5.8; $P = 0.019$, respectively). ICOS SNP rs10932037 associated with decreased graft survival ($P = 0.026$). None of the SNPs were associated with acute rejection.	Deceased-donor recipients	SNPs (genes)	[34]
SNPs in donor genes for TNFα, TGFβ1, IL10, p53 TP53, and heme oxygenase 1 (HMOX1)	965	DGF significantly associated with the G allele of TNFα SNP rs3093662 (OR = 1.85 vs. A allele; 95% CI = 1.16–2.94; $P = 0.009$; $n = 965$)	Donors	SNPs (genes)	[36]

IGF, immediate graft function.

Table 8.2 Genetic biomarkers of acute rejection

Allele/SNP	*N*	Result	References
T-cell signaling genes: CD45, CD40L, and CTLA4 Cytokine genes: TNF-α, IFN-γ, IL-10, and TGF-β	150	No significant association between BPAR and SNPs in CTLA4, TGF-β, IL-10 or TNF-α genes or DNPs in IFN-γ, and CD40L genes. TGFb-25pro significantly associated with increased graft failure ($P = 0.0007$). CD40L-147 associated with reduced graft failure ($P = 0.004$).	[42]
FcγRIIA genotypes	157	Homozygosity for FcγRIIA-R/R131 significantly more frequent in AR versus no-AR and blood donors ($P < 0.05$). AR associated with a distinct distribution of FcγRIIA genotypes: the distribution of FcγRIIA-R/R131, FcγRIIA-R/H131, and FcγRIIA-H/H131 was 45, 42, and 13% in AR versus 20, 52, and 28%, respectively, in well-functioning grafts ($P < 0.05$), and 52, and 27%, respectively, in normal blood donors ($P < 0.05$). Frequency of the R/R131 genotype significantly greater in recipients with graft loss versus both control groups (45% vs. 20 and 21%, respectively, $P < 0.05$).	[44]
CTLA4 gene polymorphisms: dinucleotide (AT)n repeat in exon 3 single nucleotide polymorphism A/G at position 49 in exon 1	374	(AT)n repeat polymorphism: increased incidence of AR with alleles 3 and 4 in both liver and kidney ($P = 0.002$ and 0.05, respectively). A/G single nucleotide polymorphism was not associated acute rejection.	[33]
Chemokines: CCR2–V64I and CCR5-59029-A	163	Less acute rejection in human renal transplantation with these alleles.	[55]
Toll-like receptors: TLR4/CD14 and TLR3	216	Higher rejection-free survival rates associated with TLR4 genotype rs10759932 in human allografts. SNPs of TLR3 or CD14 not associated with AR.	[194]

Table 8.3 Biomarkers of acute rejection

	N	Result	Time of collection	Source	References
CXCL9 and CXCL10	69	Urine MIG significantly ↑in 14/15 AR patients: median = 2809 pg/mL) versus 96 in no-AR and 144 in healthy controls ($P < 0.0001$). Sensitivity and specificity of urinary MIG for AR = 93 and 89%. In AR, urinary MIG ↑ (> than cutoff level) 5 days prior to biopsy on average ($P < 0.0001$).	Urine collected for a median 29 days posttransplant	Urine	[61]
Pretransplant serum CXCL10 (CXCL10)	316	CXCL10 > 75th percentile (>157 pg/mL) associated with significantly greater AR versus CXCL10 < 75th percentile (34.8% vs. 21.4%; $P = 0.01$). More severe and steroid resistant AR ($n = 14$) with significantly higher CXCL10 levels versus 60 recipients with less severe AR (216.1 vs. 112.4 pg/mL; $P = 0.04$). multivariate analysis showed CXCL10 (RR 2.8) and DGF (RR 3.7) had highest predictive power of graft loss.	Pretransplant sera	Sera tissue	[62]
Fractalkine, chemokine monokine induced by IFN-γ, IFN-γ-inducible protein 10 (IP-10), macrophage inflammatory protein-3α, granzyme B, and perforin	215	AUC of AR versus no-AR for fractalkine, Mig, IP-10, MIP-3a, granzyme B, and perforin = 0.834, 0.901, 0.810, 0.734, 0.765, and 0.779, respectively. Sensitivity and specificity for fractalkine to diagnose AR = 82.1 and 76.5%, respectively (using cutoff value =102.88 ng/mmol cr.; $P < 0.001$). Fractalkine, IP-10, and granzyme B together were best able to distinguish AR from no-AR (sensitivity and specificity = 83.6 and 95.0%, respectively). Only changes in urinary fractalkine distinguished recipients with AR from ATN.	Protocol urines every 2 weeks × first 2 months, on biopsy day, and at end of anti-AR therapy.	Posttxp urine	[66]

(Continued)

Table 8.3 Biomarkers of acute rejection (*cont.*)

	N	Result	Time of collection	Source	References
sCD30	3899	Five-year graft survival lower (64 ± 2%) with high serum sCD30 (≥100 U/mL) versus graft survival of 75 ± 1% with low sCD30 (<100 U/mL) ($P < 0.0001$). Less need for antirejection therapy in year 2 with low sCD30. This difference was not significant in the first posttransplant year.	Pretransplant	Sera	[87]
sCD30	120	During 47.5 months of follow-up, pretransplant sCD30 was not associated with: differences in graft survival rate ($P = 0.5901$) Higher incidence of AR (33.9% vs. 22.4% in the low sCD30; $P = 0.164$). High sCD30 was associated with significantly elevated serum creatinine 3 years posttransplant versus low sCD30 group ($P < 0.05$).	Pretransplant	Sera	[91]
sCD30	56	ROC analysis on postop days 3–5 showed sCD30 identified recipients who subsequently developed AR ($n = 25$) versus no AR ($P < 0.0001$, AUC 0.96, specificity 100%, sensitivity 88%) or those with ATN in the absence of rejection ($P = 0.001$, AUC 0.85, specificity 91%, sensitivity 72%).	Within the first 20 days posttransplant	Plasma	[90]
TGFβ-1	115	Plasma TGFβ-1 greater in allograft recipients versus normal controls; did not distinguish AR from chronic vascular rejection or ATN. Urine TGFβ-1 was similar in normals and allograft recipients.	At time of biopsy	Plasma urine	[97]

CD103/18S ribosomal (r)RNA	49	CD103 mRNA increased in AR versus other findings on allograft biopsy, CAN, and stable graft function. Using a cutoff value of 8.16 CD103 copies/μg, sensitivity = 59% and specificity = 75% in predicting AR. 18S rRNA did not vary significantly among the four groups.	Within 24 h of biopsy	Urine	[96]
IL-4, -5, and -6, IFN-γ, perforin, & and granzyme B mRNA	61	IL-4, -5, and- 6, IFN-γ, perforin, and granzyme B mRNA were significantly associated with AR. Patients with infections, ATN, CsA nephrotoxicity, and "uncertain rejection episodes" were excluded. Not all AR was confirmed by biopsy.	For 3 months post-transplant	Sera	[78]
Perforin, granzyme B, and cyclophilin B mRNA	122	Mean levels of perforin- and granzyme B mRNA, but not levels of constitutively expressed cyclophliin B mRNA, were greater in the urinary cells from patients with AR versus no AR (perforin, 1.4 vs. −0.6 fg/μg of total RNA; $P < 0.001$; and granzyme B, 1.2 vs. −0.9 fg/μg of total RNA; $P < 0.001$). ROC analysis: AR predicted with a sensitivity and specificity of 83% (for both parameters; using a cutoff value of 0.9 fg perforin mRNA/μg of total RNA), and a sensitivity and specificity of 79 and 77%, respectively (using a cutoff value of 0.4 fg of granzyme B mRNA/μg of total RNA). Levels of perforin- and granzyme B mRNA, but not that of cyclophilin B, were significantly higher in patients who developed AR ($n = 8$) within the first 10 days posttransplant versus 29 recipients in whom AR did not develop (granzyme B, $P = 0.02$ on days 4–6 and $P = 0.009$ on days 7–9; perforin, $P = 0.003$ on days 4–6, and $P = 0.01$ on days 7–9).	First 10 days post-transplant	Urine	[79]

(*Continued*)

Table 8.3 Biomarkers of acute rejection (*cont.*)

	N	Result	Time of collection	Source	References
Perforin and granzyme B	67	Recipients with AR ($n = 17$) had increased levels of granzyme B and perforin transcripts on days 5–7, 8–10, 11–13, 17–19, 20–22, and 26–29, versus patients without AR ($n = 50$, $P < 0.05$ in all cases). Diagnosis of AR, using gene expression criteria determined by ROC curve analysis, could be made 2–30 days before the diagnosis by standard criteria (median 11 days). Best diagnostic result achieved with samples taken on postop days 8–10. These samples yielded a sensitivity and specificity of 82 of 90%, respectively for perforin, and a sensitivity and specificity of 72 and 87%, respectively, for granzyme B. Both perforin ($P < 0.01$) and granzyme B gene expression ($P < 0.05$) decreased after initiation of antirejection therapy.	First month posttransplantation	Blood	[81]
MCP-1	20 (tissue studies), 38 (urine studies)	Urine and tissue MCP-1 significantly higher than that seen with ATN or in normal tissue.	1–60 months. Post-transplant	Tissue urine	[57]

PS, ES, platelets, leukocyte common antigen, macrophages, T cells, and neutrophils.	77	Significantly more recipients rejected if the donor biopsy was positive for PS (63 vs. 24%, $P = 0.0007$), contained ≥5 leukocytes/glomerulus (48 vs. 21%, $P = 0.03$), contained >9.3 leukocytes/hpf (46.5 vs. 10.5%, $P = 0.006$) or were both PS-positive and contained >9.3 leukocytes/hpf (61.9 vs. 0.0%, $P = 0.0001$).	Pretransplant	Tissue	[111]
Serum CRP	441	Median sCRP concentrations: Normal controls = 0.5 μg/mL (range, <0.03–10 μg/mL) HD = 3.1 μg/mL (range, <0.03– > 15 μg/mL) CAPD =2.9 μg/mL (range, <0.03– > 15 μg/mL). Posttransplant: POD2 = 29 μg/mL (range, 4– >200 μg/mL) POD5 = < 20 μg/mL (median, 7 μg/mL; range, 2–19 μg/mL). Stable graft function = 4 μg/mL (range, 1–19 μg/mL). Steroid-sensitive AR = 49 μg/mL l ($P < 0.001$ vs. uncomplicated controls) and 11 μg/mL with treatment. Steroid-resistant AR = 119 μg/mL and remained elevated (median = 77 μg/mL) at end of treatment. CsA nephrotoxicity = < 5 μg/mL throughout the episodes). Biopsy-proven cases of ATN = 7 μg/mL (range, 1–9 μg/mL), 5 μg/mL (range, 1–6 μg/mL), and 2 μg/mL (range, 1–3 μg/mL) at the start, middle, and end of the episode, respectively.	Daily	Serum	[102]

(Continued)

Table 8.3 Biomarkers of acute rejection (*cont.*)

	N	Result	Time of collection	Source	References
CRP	97	Pretransplant CRP levels greater in patients who subsequently developed AR versus those who did not (22.2 ± 2.9 vs. 11.7 ± 1.8 μg/mL, respectively, $P = 0.003$). Recipients within the lowest CRP quartile had longer times to rejection versus those in the highest quartile ($P = 0.006$). 3-month incidence of AR = 13% (3/23) in the lowest CRP quartile group versus 44% (11/25) in the upper quartile group ($P = 0.027$, Fisher exact test). By Cox proportional hazards regression multivariate analysis, only pretransplant CRP level was an independent risk factor for rejection ($P = 0.044$).	Pretransplant	Serum	[103]
CD20	27	AR with CD20-positive interstitial infiltrates was significantly more likely to be steroid resistant ($P = 0.015$ vs. CD20- negative recipients) and to experience immunologic (death censored) allograft loss ($P = 0.024$).	First year posttransplant	Tissue	[128]
CD20+ LC	120	LC found in 59% of biopsies (71/120), all of which were CD20 positive. LC most frequent in patients who had not received lymphoid depletion or were treated with thymoglobulin (79 and 75%, respectively) versus 37% in patients treated with Campath ($P = 0.0001$). Banff 1a/1b ACR more frequent in LC+ versus LC− group (96% vs. 80%, respectively; $P = 0.0051$). No difference in LC-positive and -negative with respect to time to ACR, sCr, steroid resistance, and graft loss.	Biopsy at mean of 8–10 posttransplant months	Tissue	[91]

ELISPOT	55	37/55 donor-stimulated IFN ELISPOTS measured before transplantation, including five who subsequently developed AR. Mean frequency of pretransplant IFN ELISPOTS was significantly > in patients who experienced AR (79 ± 69 vs. 30 ± 44 spots per 300 000 cells; $P = 0.039$ vs. recipients without). Pretransplant IFN ELISPOT did not correlate with sCr. at 6 -($R = 0.014$, P = NS) or 12 months posttransplant ($R = 0.114$, P = NS).	First year posttransplant	Serum	[101]
E-selectin HLA class II antigens in	94	High levels of intertubular capillary E-selectin expression (grade 2) detected in 54% of deceased-donor donor kidneys versus no expression in any living-donor kidney ($P < 0.00001$). Increased expression of tubular antigens seen prior to transplantation in all 11 cadaver renal allografts with biopsy-proven acute rejection. Tubular antigens were absent in 15 out of 54 (28%) donor kidneys with no rejection in the first 7 days ($P < 0.05$). No significant association between tubular antigen expression and 3- and 6-month sCr. levels, DGF, and the number of rejection episodes.	Pretransplant	Tissue	[110]

(Continued)

Table 8.3 Biomarkers of acute rejection (*cont.*)

	N	Result	Time of collection	Source	References
Anti-HLA-DR, anti-CD3, anti-CD14, anti-CD54 (ICAM-1) and anti-CD25 (IL-2R).	30	AR was associated with the presence of ≥5% HLA-DR-positive cells and ICAM-1 positive cells in 100 and 53% of samples, respectively ($P < 0.01$ vs. others). Highly specific for the diagnosis of AR: ICAM-1 or CD3-positive cells (100% specificity) and IL-2R receptor- or HLA-DR-positive cells (specificity = 88%). CAN associated with CD14-positive cells ($P = 0.03$ vs. others; specificity =87.3%). Most accurate finding associated with the diagnosis of AR was the finding of HLA-DR-positive cells: sensitivity = 100%, specificity = 88%, PPV = 88% and NPV = 100%.	On admission for graft dysfunction 10 days–3.5 years posttxp (median of 28 days)	Urine cells	[137]

Table 8.4 Biomarkers of chronic rejection

CD3γ TNF-α, IFN-γ, IL-1β, IL-2, IL-4, IL-6, and IL-8		Cytokine gene transcription and histological evidence of injury found in > two-thirds of clinically stable grafts. Increasing lymphocytic infiltration correlated with the proteinuria ($P = 0.034$) and worsening interstitial fibrosis ($P = 0.005$). Fibrosis showed significant positive correlation with baseline creatinine ($P = 0.006$) and negative correlation with the GFR measured on the day of the biopsy ($P = 0.037$). Intragraft CD3γ signal also correlated with increasing proteinuria ($P = 0.043$). Both fibrosis ($P = 0.01$) and tubular atrophy ($P = 0.01$) on the original biopsy were correlated with declining renal function at follow-up. CD3γ levels at the time of original biopsy correlated with the highest change in GFR over time ($P = 0.045$).	At biopsy 2–3 years post-transplantation	Tissue	[144]
HLA antibodies, DSA	1014 including 195 examined prospectively	30% of recipients found to have HLA antibodies and of these 31% had DSA. DSA-positive recipients had significantly lower graft survival (49% vs. 83% in the HLA antibody negative group; $P \leq 0.0001$). Lower graft survival in DSA-negative but HLA antibody positive patients (70% vs. 83%; $P = 0.0001$). Prospective analysis of 195 patients: patients repeatedly negative for HLA antibodies had a superior survival probability compared with patients who developed de novo HLA antibodies after the first testing (94% vs. 79%; respectively $P = 0.05$).	Median of 5 years post-transplant.	Sera	[153]

(Continued)

Table 8.4 Biomarkers of chronic rejection (*cont.*)

PAPP-A CRP IL-6, TNF-α)	178	PAPP-A found to correlate with the other inflammatory markers (PAPP-A vs. CRP, $r = 0.218$; $P = 0.004$; PAPP-A vs. IL-6, $r = 0.235$; $P < 0.001$; PAPP-A vs. TNF-α, $r = 0.372$; $P < 0.001$). Multiple regression analysis showed PAPP-A (RR: 4.27; 95% CI: 1.03–17.60; $P = 0.044$) and TNF-α (RR: 5.6; 95% CI: 1.43–21.83; $P = 0.013$) to be predictors of CAN PAPP-A (RR: 6.4; 95% CI:1.24–33.11; $P = 0.027$) and CRP (RR: 6.05; 95% CI:1.21–29.74; $P = 0.028$) were predictors of cardiovascular events.	Pretransplant	Sera	[154]
IL-1β soluble IL-1 receptor II (IL-1RII) TNF-α, sTNFRI, sTNFRII, LIF	91	sTNFRI significantly higher in patients with AR versus stable patients ($P = 0.002$). sTNFRII significantly greater in patients with AR versus stable patients ($P = 0.02$). For sTNFRI, a cutoff value for acute rejection of >480 pg/mL resulted in a sensitivity of 89.6%, specificity of 78.3%, PPV of 76.4%, and NPV of 90.6%. For sTNFRII, a cutoff value of > 700 pg/mL resulted in a sensitivity of 91.6%, specificity 80.6%, PPV of 64.7%, and NPV of 96.1%. IL-1beta and sIL-1RII did not differ significantly among the groups	POD 7 and 14	FNAB	[112]
ICAM-1, E-selectin, and L-selectin	306	A variant allele in exon 4 of ICAM-1 (R241) was significantly more frequent in recipients with chronic allograft failure versus long-term survivors and UK controls (19.4 vs. 10.0 and 9.4%, respectively, $P = 0.015$ and 0.025). Stratification by time to graft failure demonstrated that more rapid failure was associated with another ICAM-1 variant in the recipient (E469) in exon 6 ($P = 0.033$). No significant association was detected between the selectin polymorphisms studied and chronic allograft failure.			[159]

HBD-1 and ACT	73	Patients with acute allograft rejection had significantly reduced HBD-1 and increased ACT ($P < 0.05$) versus clinically stable transplants. Using both peptides combined: ROC AUC for the diagnosis of acute rejection = 0.912. Urinary HBD-1 was 176.8 ± 122.3 pg/mL in stable patients versus 83.2 ± 52.2 pg/mL in acute rejection. ROC AUC = 0.749 ($P < 0.01$).	Prebiopsy. Mostly in first 6 months of transplant.	Urine	[166]
CXCL10/IP-10	316	Patients with stable graft function had significantly lower median pretransplant serum CXCL10 levels than recipients who subsequently endured graft failure (93.0 vs. 157.4 pg/mL; $P = 0.0007$). The frequency of acute rejection in the first posttransplant month was also increased based on the percentile of pretransplant serum CXCL10 levels in the four groups ($\chi^2 = 11.412$; $P = 0.009$). Patients with pretransplant serum CXCL10 levels > 75th centile (>157 pg/mL) had increased frequency of acute rejection versus patients with serum CXCL10 levels < the 75th centile (34.8% vs. 21.4%; $P = 0.01$). Multivariate analysis demonstrated that CXCL10 was most predictive graft loss (RR 2.787) among the variables analyzed.	Pretransplant sera: median posttransplant followop of 39 months.	Sera	[157]
Intact/cleaved beta2-microglobulin (i/cβ2m), RBP, NGAL, and alpha1-microglobulin (α1m)	100	Evidence of tubular proteinuria in 38% (RBP)–79% (α1m) of group 1 (recipients with stable graft function and normal tubular histology). Group 2 (recipients with stable graft function and subclinical tubulitis on protocol biopsy) had slightly higher but nonsignificant levels of i/cβ2m ($P = 0.11$), RBP ($P = 0.17$), α1m ($P = 0.09$), and NGAL ($P = 0.06$) group 1with substantial overlap. Groups 3 (recipients with clinical tubulitis Ia/Ib) and 4 (recipients with other clinical tubular pathologies) had significantly increased levels of RBP, NGAL, and α1m than stable transplants with normal tubular histology or stable transplants with subclinical tubulitis ($P < 0.002$).	variable	Tissue and urine	[28]

REFERENCES

[1] Parikh CR, Devarajan P. New biomarkers of acute kidney injury. Crit Care Med 2008;36(4 Suppl):S159–65.
[2] Bakay RA, Ward AA Jr. Enzymatic changes in serum and cerebrospinal fluid in neurological injury. J Neurosurg 1983;58(1):27–37.
[3] Sandler SJ, Figaji AA, Adelson PD. Clinical applications of biomarkers in pediatric traumatic brain injury. Childs Nerv Syst 2009;26(2):205–13.
[4] Oberbauer R, Rohrmoser M, Regele H, Muhlbacher F, Mayer G. Apoptosis of tubular epithelial cells in donor kidney biopsies predicts early renal allograft function. J Am Soc Nephrol 1999;10(9):2006–13.
[5] Schwarz C, Regele H, Steininger R, Hansmann C, Mayer G, Oberbauer R. The contribution of adhesion molecule expression in donor kidney biopsies to early allograft dysfunction. Transplantation 2001;71(11):1666–70.
[6] Nguyen MT, Fryml E, Sahakian SK, et al. Pretransplantation recipient regulatory T cell suppressive function predicts delayed and slow graft function after kidney transplantation. Transplantation 2014;98(7):745–53.
[7] San Segundo D, Millan O, Munoz-Cacho P, et al. High proportion of pretransplantation activated regulatory T cells (CD4+CD25highCD62L+CD45RO+) predicts acute rejection in kidney transplantation: results of a multicenter study. Transplantation 2014;98(11):1213–8.
[8] Morgan C, Martin A, Shapiro R, Randhawa PS, Kayler LK. Outcomes after transplantation of deceased-donor kidneys with rising serum creatinine. Am J Transplant 2007;7(5):1288–92.
[9] Sadeghi M, Daniel V, Lahdou I, et al. Association of pretransplant soluble glycoprotein 130 (sgp130) plasma levels and posttransplant acute tubular necrosis in renal transplant recipients. Transplantation 2009;88(2):266–71.
[10] Fonseca I, Oliveira JC, Santos J, et al. Leptin and adiponectin during the first week after kidney transplantation: biomarkers of graft dysfunction? Metabolism 2015;64(2):202–7.
[11] Naesens M, Li L, Ying L, et al. Expression of complement components differs between kidney allografts from living and deceased donors. J Am Soc Nephrol 2009;20(8):1839–51.
[12] Scian MJ, Maluf DG, Archer KJ, et al. Identification of biomarkers to assess organ quality and predict posttransplantation outcomes. Transplantation 2012;94(8):851–8.
[13] Ichimura T, Bonventre JV, Bailly V, et al. Kidney injury molecule-1 (KIM-1), a putative epithelial cell adhesion molecule containing a novel immunoglobulin domain, is up-regulated in renal cells after injury. J Biol Chem 1998;273(7):4135–42.
[14] Abulezz S. KIM-1 expression in kidney allograft biopsies: improving the gold standard. Kidney Int 2008;73(5):522–3.
[15] Zhang PL, Rothblum LI, Han WK, Blasick TM, Potdar S, Bonventre JV. Kidney injury molecule-1 expression in transplant biopsies is a sensitive measure of cell injury. Kidney Int 2008;73(5):608–14.
[16] Obeidat MA, Luyckx VA, Grebe SO, et al. Post-transplant nuclear renal scans correlate with renal injury biomarkers and early allograft outcomes. Nephrol Dial Transplant 2011;26(9):3038–45.
[17] Boom H, Mallat MJ, de Fijter JW, Paul LC, Bruijn JA, van Es LA. Calcium levels as a risk factor for delayed graft function. Transplantation 2004;77(6):868–73.
[18] Le Bricon T, Thervet E, Benlakehal M, Bousquet B, Legendre C, Erlich D. Changes in plasma cystatin C after renal transplantation and acute rejection in adults. Clin Chem 1999;45(12):2243–9.
[19] Thervet E, Le Bricon T, Hugot V, et al. Early diagnosis of renal function recovery by cystatin C in renal allograft recipients. Transplant Proc 2000;32(8):2779.

[20] Hall IE, Doshi MD, Poggio ED, Parikh CR. A comparison of alternative serum biomarkers with creatinine for predicting allograft function after kidney transplantation. Transplantation 2011;91(1):48–56.
[21] Liu J. Evaluation of serum cystatin C for diagnosis of acute rejection after renal transplantation. Transplant Proc 2012;44(5):1250–3.
[22] Welberry Smith MP, Zougman A, Cairns DA, et al. Serum aminoacylase-1 is a novel biomarker with potential prognostic utility for long-term outcome in patients with delayed graft function following renal transplantation. Kidney Int 2013;84(6):1214–25.
[23] Mishra J, Ma Q, Kelly C, et al. Kidney NGAL is a novel early marker of acute injury following transplantation. Pediatr Nephrol 2006;21(6):856–63.
[24] Parikh CR, Jani A, Melnikov VY, Faubel S, Edelstein CL. Urinary interleukin-18 is a marker of human acute tubular necrosis. Am J Kidney Dis 2004;43(3):405–14.
[25] Parikh CR, Jani A, Mishra J, Urine NGAL. et al. and IL-18 are predictive biomarkers for delayed graft function following kidney transplantation. Am J Transplant 2006;6(7):1639–45.
[26] Hall IE, Yarlagadda SG, Coca SG, et al. IL-18 and urinary NGAL predict dialysis and graft recovery after kidney transplantation. J Am Soc Nephrol 2009;21(1):189–97.
[27] Hall IE, Koyner JL, Doshi MD, Marcus RJ, Parikh CR. Urine cystatin C as a biomarker of proximal tubular function immediately after kidney transplantation. Am J Nephrol 2011;33(5):407–13.
[28] Schaub S, Mayr M, Honger G, et al. Detection of subclinical tubular injury after renal transplantation: comparison of urine protein analysis with allograft histopathology. Transplantation 2007;84(1):104–12.
[29] Kwon O, Molitoris BA, Pescovitz M, Kelly KJ. Urinary actin, interleukin-6, and interleukin-8 may predict sustained ARF after ischemic injury in renal allografts. Am J Kidney Dis 2003;41(5):1074–87.
[30] Yadav B, Prasad N, Agrawal V, et al. Urinary kidney injury molecule-1 can predict delayed graft function in living donor renal allograft recipients. Nephrology (Carlton) 2015;20(11):801–6.
[31] Pianta TJ, Peake PW, Pickering JW, Kelleher M, Buckley NA, Endre ZH. Clusterin in kidney transplantation: novel biomarkers versus serum creatinine for early prediction of delayed graft function. Transplantation 2015;99(1):171–9.
[32] Wilflingseder J, Sunzenauer J, Toronyi E, et al. Molecular pathogenesis of post-transplant acute kidney injury: assessment of whole-genome mRNA and miRNA profiles. PLoS One 2014;9(8):e104164.
[33] Slavcheva E, Albanis E, Jiao Q, et al. Cytotoxic T-lymphocyte antigen 4 gene polymorphisms and susceptibility to acute allograft rejection. Transplantation 2001;72(5):935–40.
[34] Haimila K, Turpeinen H, Alakulppi NS, Kyllonen LE, Salmela KT, Partanen J. Association of genetic variation in inducible costimulator gene with outcome of kidney transplantation. Transplantation 2009;87(3):393–6.
[35] Castelli L, Comi C, Chiocchetti A, et al. ICOS gene haplotypes correlate with IL10 secretion and multiple sclerosis evolution. J Neuroimmunol 2007;186(1–2):193–8.
[36] Israni AK, Li N, Cizman BB, et al. Association of donor inflammation- and apoptosis-related genotypes and delayed allograft function after kidney transplantation. Am J Kidney Dis 2008;52(2):331–9.
[37] Li C, Jackson RM. Reactive species mechanisms of cellular hypoxia-reoxygenation injury. Am J Physiol 2002;282(2):C227–41.
[38] Dolegowska B, Blogowski W, Domanski L. Association between the perioperative antioxidative ability of platelets and early post-transplant function of kidney allografts: a pilot study. PLoS One 2012;7(1):e29779.
[39] Ardalan MR, Estakhri R, Hajipour B, et al. Erythropoietin ameliorates oxidative stress and tissue injury following renal ischemia/reperfusion in rat kidney and lung. Med Princ Pract 2013;22(1):70–4.

[40] Hariharan N, Zhai P, Sadoshima J. Oxidative stress stimulates autophagic flux during ischemia/reperfusion. Antioxid Redox Signal 2011;14(11):2179–90.
[41] Waller HL, Harper SJ, Hosgood SA, et al. Biomarkers of oxidative damage to predict ischaemia-reperfusion injury in an isolated organ perfusion model of the transplanted kidney. Free Radic Res 2006;40(11):1218–25.
[42] Dmitrienko S, Hoar DI, Balshaw R, Keown PA. Immune response gene polymorphisms in renal transplant recipients. Transplantation 2005;80(12):1773–82.
[43] Hogarth PM. Fc receptors are major mediators of antibody based inflammation in autoimmunity. Curr Opin Immunol 2002;14(6):798–802.
[44] Yuan FF, Watson N, Sullivan JS, et al. Association of Fc gamma receptor IIA polymorphisms with acute renal-allograft rejection. Transplantation 2004;78(5):766–9.
[45] Gao JW, Guo YF, Fan Y, et al. Polymorphisms in cytotoxic T lymphocyte associated antigen-4 influence the rate of acute rejection after renal transplantation in 167 Chinese recipients. Transpl Immunol 2012;26(4):207–11.
[46] Luo Y, Shi B, Qian Y, Bai H, Chang J. Sequential monitoring of TIM-3 gene expression in peripheral blood for diagnostic and prognostic evaluation of acute rejection in renal graft recipients. Transplant Proc 2011;43(10):3669–74.
[47] Misra MK, Prakash S, Kapoor R, Pandey SK, Sharma RK, Agrawal S. Association of HLA-G promoter and 14-bp insertion–deletion variants with acute allograft rejection and end-stage renal disease. Tissue Antigens 2013;82(5):317–26.
[48] Misra MK, Kapoor R, Pandey SK, Sharma RK, Agrawal S. Association of CTLA-4 gene polymorphism with end-stage renal disease and renal allograft outcome. J Interferon Cytokine Res 2014;34(3):148–61.
[49] Pawlik A, Dabrowska-Zamojcin E, Dziedziejko V, Safranow K, Domanski L. Association between IVS3+17T/C CD28 gene polymorphism and the acute kidney allograft rejection. Transpl Immunol 2014;30(2–3):84–7.
[50] Lee JR, Muthukumar T, Dadhania D, et al. Urinary cell mRNA profiles predictive of human kidney allograft status. Immunol Rev 2014;258(1):218–40.
[51] Han FF, Fan H, Wang ZH, et al. Association between co-stimulatory molecule gene polymorphism and acute rejection of allograft. Transpl Immunol 2014;31(2):81–6.
[52] Fernandez EJ, Lolis E. Structure, function, and inhibition of chemokines. Annu Rev Pharmacol Toxicol 2002;42:469–99.
[53] Rossi D, Zlotnik A. The biology of chemokines and their receptors. Annu Rev Immunol 2000;18:217–42.
[54] Hancock WW, Lu B, Gao W, et al. Requirement of the chemokine receptor CXCR3 for acute allograft rejection. J Exp Med 2000;192(10):1515–20.
[55] Abdi R, Tran TB, Sahagun-Ruiz A, et al. Chemokine receptor polymorphism and risk of acute rejection in human renal transplantation. J Am Soc Nephrol 2002;13(3):754–8.
[56] McDermott DH, Zimmerman PA, Guignard F, Kleeberger CA, Leitman SF, Murphy PM. CCR5 promoter polymorphism and HIV-1 disease progression. Multicenter AIDS Cohort Study (MACS). Lancet 1998;352(9131):866–70.
[57] Grandaliano G, Gesualdo L, Ranieri E, Monno R, Stallone G, Schena FP. Monocyte chemotactic peptide-1 expression and monocyte infiltration in acute renal transplant rejection. Transplantation 1997;63(3):414–20.
[58] Hancock WW, Gao W, Faia KL, Csizmadia V. Chemokines and their receptors in allograft rejection. Curr Opin Immunol 2000;12(5):511–6.
[59] Hancock WW, Wang L, Ye Q, Han R, Lee I. Chemokines and their receptors as markers of allograft rejection and targets for immunosuppression. Curr Opin Immunol 2003;15(5):479–86.
[60] Hancock WW, Gao W, Csizmadia V, Faia KL, Shemmeri N, Luster AD. Donor-derived IP-10 initiates development of acute allograft rejection. J Exp Med 2001;193(8):975–80.
[61] Hauser IA, Spiegler S, Kiss E, et al. Prediction of acute renal allograft rejection by urinary monokine induced by IFN-gamma (MIG). J Am Soc Nephrol 2005;16(6):1849–58.

[62] Lazzeri E, Rotondi M, Mazzinghi B, et al. High CXCL10 expression in rejected kidneys and predictive role of pretransplant serum CXCL10 for acute rejection and chronic allograft nephropathy. Transplantation 2005;79(9):1215–20.
[63] Platt JL, LeBien TW, Michael AF. Interstitial mononuclear cell populations in renal graft rejection. Identification by monoclonal antibodies in tissue sections. J Exp Med 1982;155(1):17–30.
[64] Rollins BJ, Walz A, Baggiolini M. Recombinant human MCP-1/JE induces chemotaxis, calcium flux, and the respiratory burst in human monocytes. Blood 1991;78(4):1112–6.
[65] Rollins BJ, Stier P, Ernst T, Wong GG. The human homolog of the JE gene encodes a monocyte secretory protein. Mol Cell Biol 1989;9(11):4687–95.
[66] Peng W, Chen J, Jiang Y, et al. Urinary fractalkine is a marker of acute rejection. Kidney Int 2008;74(11):1454–60.
[67] Hricik DE, Nickerson P, Formica RN, et al. Multicenter validation of urinary CXCL9 as a risk-stratifying biomarker for kidney transplant injury. Am J Transplant 2013;13(10):2634–44.
[68] Blydt-Hansen TD, Gibson IW, Gao A, Dufault B, Ho J. Elevated urinary CXCL10-to-creatinine ratio is associated with subclinical and clinical rejection in pediatric renal transplantation. Transplantation 2015;99(4):797–804.
[69] Lee A, Jeong JC, Choi YW, et al. Validation study of peripheral blood diagnostic test for acute rejection in kidney transplantation. Transplantation 2014;98(7):760–5.
[70] Roedder S, Sigdel T, Salomonis N, et al. The kSORT assay to detect renal transplant patients at high risk for acute rejection: results of the multicenter AART study. PLoS Med 2014;11(11):e1001759.
[71] Obhrai J, Goldstein DR. The role of toll-like receptors in solid organ transplantation. Transplantation 2006;81(4):497–502.
[72] Iwasaki A, Medzhitov R. Toll-like receptor control of the adaptive immune responses. Nat Immunol 2004;5(10):987–95.
[73] Srivastava P, Singh A, Kesarwani P, Jaiswal PK, Singh V, Mittal RD. Association studies of toll-like receptor gene polymorphisms with allograft survival in renal transplant recipients of North India. Clin Transplant 2012;26(4):581–8.
[74] Kim TH, Jeong KH, Kim SK, et al. TLR9 gene polymorphism (rs187084, rs352140): association with acute rejection and estimated glomerular filtration rate in renal transplant recipients. Int J Immunogenet 2013;40(6):502–8.
[75] Liu CC, Walsh CM, Young JD. Perforin: structure and function. Immunol Today 1995;16(4):194–201.
[76] Smyth MJ, Trapani JA. Granzymes: exogenous proteinases that induce target cell apoptosis. Immunol Today 1995;16(4):202–6.
[77] Atkinson EA, Barry M, Darmon AJ, et al. Cytotoxic T lymphocyte-assisted suicide. Caspase 3 activation is primarily the result of the direct action of granzyme B. J Biol Chem 1998;273(33):21261–6.
[78] Dugre FJ, Gaudreau S, Belles-Isles M, Houde I, Roy R. Cytokine and cytotoxic molecule gene expression determined in peripheral blood mononuclear cells in the diagnosis of acute renal rejection. Transplantation 2000;70(7):1074–80.
[79] Li B, Hartono C, Ding R, et al. Noninvasive diagnosis of renal-allograft rejection by measurement of messenger RNA for perforin and granzyme B in urine. N Engl J Med 2001;344(13):947–54.
[80] Graziotto R, Del Prete D, Rigotti P, Perforin, Granzyme B. et al. and fas ligand for molecular diagnosis of acute renal-allograft rejection: analyses on serial biopsies suggest methodological issues. Transplantation 2006;81(8):1125–32.
[81] Simon T, Opelz G, Wiesel M, Ott RC, Susal C. Serial peripheral blood perforin and granzyme B gene expression measurements for prediction of acute rejection in kidney graft recipients. Am J Transplant 2003;3(9):1121–7.

[82] Veronese F, Rotman S, Smith RN, et al. Pathological and clinical correlates of FOXP3+ cells in renal allografts during acute rejection. Am J Transplant 2007;7(4):914–22.
[83] Fontenot JD, Gavin MA, Rudensky AY. Foxp3 programs the development and function of CD4+CD25+ regulatory T cells. Nat Immunol 2003;4(4):330–6.
[84] Muthukumar T, Dadhania D, Ding R, et al. Messenger RNA for FOXP3 in the urine of renal-allograft recipients. N Engl J Med 2005;353(22):2342–51.
[85] Rodder S, Scherer A, Korner M, et al. Meta-analyses qualify metzincins and related genes as acute rejection markers in renal transplant patients. Am J Transplant 2010;10(2):286–97.
[86] Afaneh C, Muthukumar T, Lubetzky M, et al. Urinary cell levels of mRNA for OX40, OX40L, PD-1, PD-L1, or PD-L2 and acute rejection of human renal allografts. Transplantation 2010;90(12):1381–7.
[87] Susal C, Pelzl S, Dohler B, Opelz G. Identification of highly responsive kidney transplant recipients using pretransplant soluble CD30. J Am Soc Nephrol 2002;13(6):1650–6.
[88] Cinti P, Pretagostini R, Arpino A, et al. Evaluation of pretransplant immunologic status in kidney-transplant recipients by panel reactive antibody and soluble CD30 determinations. Transplantation 2005;79(9):1154–6.
[89] Romagnani S, Del Prete G, Maggi E, Chilosi M, Caligaris-Cappio F, Pizzolo G. CD30 and type 2 T helper (Th2) responses. J Leukoc Biol 1995;57(5):726–30.
[90] Pelzl S, Opelz G, Daniel V, Wiesel M, Susal C. Evaluation of posttransplantation soluble CD30 for diagnosis of acute renal allograft rejection. Transplantation 2003;75(3):421–3.
[91] Kim MS, Kim HJ, Kim SI, et al. Pretransplant soluble CD30 level has limited effect on acute rejection, but affects graft function in living donor kidney transplantation. Transplantation 2006;82(12):1602–5.
[92] Kotsch K, Mashreghi MF, Bold G, et al. Enhanced granulysin mRNA expression in urinary sediment in early and delayed acute renal allograft rejection. Transplantation 2004;77(12):1866–75.
[93] Hadley GA, Bartlett ST, Via CS, Rostapshova EA, Moainie S. The epithelial cell-specific integrin, CD103 (alpha E integrin), defines a novel subset of alloreactive CD8+ CTL. J Immunol 1997;159(8):3748–56.
[94] Cepek KL, Parker CM, Madara JL, Brenner MB. Integrin alpha E beta 7 mediates adhesion of T lymphocytes to epithelial cells. J Immunol 1993;150(8 Pt 1):3459–70.
[95] Robertson H, Wong WK, Talbot D, Burt AD, Kirby JA. Tubulitis after renal transplantation: demonstration of an association between CD103+ T cells, transforming growth factor beta1 expression and rejection grade. Transplantation 2001;71(2):306–13.
[96] Ding R, Li B, Muthukumar T, et al. CD103 mRNA levels in urinary cells predict acute rejection of renal allografts. Transplantation 2003;75(8):1307–12.
[97] Coupes BM, Newstead CG, Short CD, Brenchley PE. Transforming growth factor beta 1 in renal allograft recipients. Transplantation 1994;57(12):1727–31.
[98] Nogare AL, Dalpiaz T, Pedroso JA, et al. Expression of fibrosis-related genes in human renal allografts with interstitial fibrosis and tubular atrophy. J Nephrol 2013;26(6):1179–87.
[99] Kurian SM, Williams AN, Gelbart T, et al. Molecular classifiers for acute kidney transplant rejection in peripheral blood by whole genome gene expression profiling. Am J Transplant 2014;14(5):1164–72.
[100] Gebauer BS, Hricik DE, Atallah A, et al. Evolution of the enzyme-linked immunosorbent spot assay for post-transplant alloreactivity as a potentially useful immune monitoring tool. Am J Transplant 2002;2(9):857–66.
[101] Hricik DE, Rodriguez V, Riley J, et al. Enzyme linked immunosorbent spot (ELISPOT) assay for interferon-gamma independently predicts renal function in kidney transplant recipients. Am J Transplant 2003;3(7):878–84.
[102] Harris KR, Digard NJ, Lee HA. Serum C-reactive protein. A useful and economical marker of immune activation in renal transplantation. Transplantation 1996;61(11):1593–600.

[103] Perez RV, Brown DJ, Katznelson SA, et al. Pretransplant systemic inflammation and acute rejection after renal transplantation. Transplantation 2000;69(5):869–74.
[104] Hessian PA, Edgeworth J, Hogg N. MRP-8 and MRP-14, two abundant Ca(2 +)-binding proteins of neutrophils and monocytes. J Leukoc Biol 1993;53(2):197–204.
[105] Stroncek DF, Shankar RA, Skubitz KM. The subcellular distribution of myeloid-related protein 8 (MRP8) and MRP14 in human neutrophils. J Transl Med 2005;3:36.
[106] Burkhardt K, Radespiel-Troger M, Rupprecht HD, et al. An increase in myeloid-related protein serum levels precedes acute renal allograft rejection. J Am Soc Nephrol 2001;12(9):1947–57.
[107] Roshdy A, El-Khatib MM, Rizk MN, El-Shehaby AM. CRP and acute renal rejection: a marker to the point. Int Urol Nephrol 2012;44(4):1251–5.
[108] Field M, Lowe D, Cobbold M, et al. The use of NGAL and IP-10 in the prediction of early acute rejection in highly sensitized patients following HLA-incompatible renal transplantation. Transpl Int 2014;27(4):362–70.
[109] Kishikawa H, Kinoshita T, Yonemoto S, et al. Early microchimerism in peripheral blood following kidney transplantation. Transplant Proc 2014;46(2):388–90.
[110] Koo DD, Welsh KI, McLaren AJ, Roake JA, Morris PJ, Fuggle SV. Cadaver versus living donor kidneys: impact of donor factors on antigen induction before transplantation. Kidney Int 1999;56(4):1551–9.
[111] Benson SR, Ready AR, Savage CO. Donor platelet and leukocyte load identify renal allografts at an increased risk of acute rejection. Transplantation 2002;73(1):93–100.
[112] Oliveira JG, Xavier P, Sampaio SM, Mendes AA, Pestana M. sTNFRI and sTNFRII synthesis by fine-needle aspiration biopsy sample cultures is significantly associated with acute rejection in kidney transplantation. Transplantation 2001;71(12):1835–9.
[113] Aggarwal BB, Kohr WJ, Hass PE, et al. Human tumor necrosis factor. Production, purification, and characterization. J Biol Chem 1985;260(4):2345–54.
[114] Strehlau J, Pavlakis M, Lipman M, et al. Quantitative detection of immune activation transcripts as a diagnostic tool in kidney transplantation. Proc Natl Acad Sci USA 1997;94(2):695–700.
[115] Ode-Hakim S, Docke WD, Kern F, Emmrich F, Volk HD, Reinke P. Delayed-type hypersensitivity-like mechanisms dominate late acute rejection episodes in renal allograft recipients. Transplantation 1996;61(8):1233–40.
[116] Kwon B, Youn BS, Kwon BS. Functions of newly identified members of the tumor necrosis factor receptor/ligand superfamilies in lymphocytes. Curr Opin Immunol 1999;11(3):340–5.
[117] Baud L, Fouqueray B, Bellocq A. Switching off renal inflammation by anti-inflammatory mediators: the facts, the promise and the hope. Kidney Int 1998;53(5):1118–26.
[118] Imai N, Nishi S, Yoshita K, et al. Pentraxin-3 expression in acute renal allograft rejection. Clin Transplant 2012;26(Suppl. 24):25–31.
[119] Anglicheau D, Sharma VK, Ding R, et al. MicroRNA expression profiles predictive of human renal allograft status. Proc Natl Acad Sci USA 2009;106(13):5330–5.
[120] Liu XY, Xu J. The role of miR-223 in the acute rejection after kidney transplantation. Xi Bao Yu Fen Zi Mian Yi Xue Za Zhi 2011;27(10):1121–3.
[121] Sui W, Lin H, Peng W, et al. Molecular dysfunctions in acute rejection after renal transplantation revealed by integrated analysis of transcription factor, microRNA and long noncoding RNA. Genomics 2013;102(4):310–22.
[122] Sui W, Dai Y, Huang Y, Lan H, Yan Q, Huang H. Microarray analysis of MicroRNA expression in acute rejection after renal transplantation. Transpl Immunol 2008;19(1):81–5.
[123] Oghumu S, Bracewell A, Nori U, et al. Acute pyelonephritis in renal allografts: a new role for microRNAs? Transplantation 2014;97(5):559–68.
[124] Maluf DG, Dumur CI, Suh JL, et al. The urine microRNA profile may help monitor post-transplant renal graft function. Kidney Int 2014;85(2):439–49.

[125] Scian MJ, Maluf DG, David KG, et al. MicroRNA profiles in allograft tissues and paired urines associate with chronic allograft dysfunction with IF/TA. Am J Transplant 2011;11(10):2110–22.
[126] Lorenzen JM, Volkmann I, Fiedler J, et al. Urinary miR-210 as a mediator of acute T-cell mediated rejection in renal allograft recipients. Am J Transplant 2011;11(10):2221–7.
[127] Soltaninejad E, Nicknam MH, Nafar M, et al. Differential expression of microRNAs in renal transplant patients with acute T-cell mediated rejection. Transpl Immunol 2015;33(1):1–6.
[128] Hippen BE, DeMattos A, Cook WJ, Kew CE 2nd, Gaston RS. Association of CD20+ infiltrates with poorer clinical outcomes in acute cellular rejection of renal allografts. Am J Transplant 2005;5(9):2248–52.
[129] Kayler LK, Lakkis FG, Morgan C, et al. Acute cellular rejection with CD20-positive lymphoid clusters in kidney transplant patients following lymphocyte depletion. Am J Transplant 2007;7(4):949–54.
[130] Shabir S, Girdlestone J, Briggs D, et al. Transitional B lymphocytes are associated with protection from kidney allograft rejection: a prospective study. Am J Transplant 2015;15(5):1384–91.
[131] Carpio VN, Noronha Ide L, Martins HL, et al. Expression patterns of B cells in acute kidney transplant rejection. Exp Clin Transplant 2014;12(5):405–14.
[132] Hu H, Kwun J, Aizenstein BD, Knechtle SJ. Noninvasive detection of acute and chronic injuries in human renal transplant by elevation of multiple cytokines/chemokines in urine. Transplantation 2009;87(12):1814–20.
[133] Kutukculer N, Clark K, Rigg KM, et al. The value of posttransplant monitoring of interleukin (IL)-2, IL-3, IL-4, IL-6, IL-8, and soluble CD23 in the plasma of renal allograft recipients. Transplantation 1995;59(3):333–40.
[134] Crispim JC, Grespan R, Martelli-Palomino G, et al. Interleukin-17 and kidney allograft outcome. Transplant Proc 2009;41(5):1562–4.
[135] Millan O, Rafael-Valdivia L, San Segundo D, et al. Should IFN-gamma, IL-17 and IL-2 be considered predictive biomarkers of acute rejection in liver and kidney transplant? Results of a multicentric study. Clin Immunol 2014;154(2):141–54.
[136] Zhang Q, Liu YF, Su ZX, Shi LP, Chen YH. Serum fractalkine and interferon-gamma inducible protein-10 concentrations are early detection markers for acute renal allograft rejection. Transplant Proc 2014;46(5):1420–5.
[137] Roberti I, Panico M, Reisman L. Urine flow cytometry as a tool to differentiate acute allograft rejection from other causes of acute renal graft dysfunction. Transplantation 1997;64(5):731–4.
[138] Roberti I, Reisman L. Serial evaluation of cell surface markers for immune activation after acute renal allograft rejection by urine flow cytometry—correlation with clinical outcome. Transplantation 2001;71(9):1317–20.
[139] Stubendorff B, Finke S, Walter M, et al. Urine protein profiling identified alpha-1-microglobulin and haptoglobin as biomarkers for early diagnosis of acute allograft rejection following kidney transplantation. World J Urol 2014;32(6):1619–24.
[140] Schaub S, Rush D, Wilkins J, et al. Proteomic-based detection of urine proteins associated with acute renal allograft rejection. J Am Soc Nephrol 2004;15(1):219–27.
[141] Schaub S, Wilkins JA, Antonovici M, et al. Proteomic-based identification of cleaved urinary beta2-microglobulin as a potential marker for acute tubular injury in renal allografts. Am J Transplant 2005;5(4 Pt 1):729–38.
[142] Chen R, Sigdel TK, Li L, et al. Differentially expressed RNA from public microarray data identifies serum protein biomarkers for cross-organ transplant rejection and other conditions. PLoS Comput Biol 2010;6(9).
[143] Srivastava M, Eidelman O, Torosyan Y, Jozwik C, Mannon RB, Pollard HB. Elevated expression levels of ANXA11, integrins beta3 and alpha3, and TNF-alpha contribute to a candidate proteomic signature in urine for kidney allograft rejection. Proteomics Clin Appl 2011;5(5–6):311–21.

[144] Kirk AD, Jacobson LM, Heisey DM, Radke NF, Pirsch JD, Sollinger HW. Clinically stable human renal allografts contain histological and RNA-based findings that correlate with deteriorating graft function. Transplantation 1999;68(10):1578–82.
[145] Nickel P, Lacha J, Ode-Hakim S, et al. Cytotoxic effector molecule gene expression in acute renal allograft rejection: correlation with clinical outcome; histopathology and function of the allograft. Transplantation 2001;72(6):1158–60.
[146] de Matos AC, Camara NO, Tonato EJ, et al. Vimentin expression and myofibroblast infiltration are early markers of renal dysfunction in kidney transplantation: an early stage of chronic allograft dysfunction? Transplant Proc 2010;42(9):3482–8.
[147] Zheng J, Ding X, Tian X, et al. Assessment of different biomarkers provides valuable diagnostic standards in the evaluation of the risk of acute rejection. Acta Biochim Biophys Sin (Shanghai) 2012;44(9):730–6.
[148] Heng B, Li Y, Shi L, et al. A meta-analysis of the significance of granzyme B and perforin in noninvasive diagnosis of acute rejection after kidney transplantation. Transplantation 2015;99(7):1477–86.
[149] Djamali A, Muth BL, Ellis TM, et al. Increased C4d in post-reperfusion biopsies and increased donor specific antibodies at one-week post transplant are risk factors for acute rejection in mild to moderately sensitized kidney transplant recipients. Kidney Int 2013;83(6):1185–92.
[150] Nogare AL, Dalpiaz T, Veronese FJ, Goncalves LF, Manfro RC. Noninvasive analyses of kidney injury molecule-1 messenger RNA in kidney transplant recipients with graft dysfunction. Transplant Proc 2012;44(8):2297–9.
[151] Gomez-Alamillo C, Ramos-Barron MA, Benito-Hernandez A, et al. Relation of urinary gene expression of epithelial-mesenchymal transition markers with initial events and 1-year kidney graft function. Transplant Proc 2012;44(9):2573–6.
[152] Jia L, Gui B, Tian P, et al. TIPE2, a novel biomarker for clinical chronic kidney allograft rejection. Artif Organs 2013;37(2):221–5.
[153] Lachmann N, Terasaki PI, Budde K, et al. Anti-human leukocyte antigen and donor-specific antibodies detected by luminex posttransplant serve as biomarkers for chronic rejection of renal allografts. Transplantation 2009;87(10):1505–13.
[154] Lauzurica R, Pastor C, Bayes B, Hernandez JM, Romero R. Pretransplant pregnancy-associated plasma protein-a as a predictor of chronic allograft nephropathy and posttransplant cardiovascular events. Transplantation 2005;80(10):1441–6.
[155] Wald NJ, Watt HC, Hackshaw AK. Integrated screening for Down's syndrome on the basis of tests performed during the first and second trimesters. N Engl J Med 1999;341(7):461–7.
[156] Bayes-Genis A, Conover CA, Overgaard MT, et al. Pregnancy-associated plasma protein A as a marker of acute coronary syndromes. N Engl J Med 2001;345(14):1022–9.
[157] Rotondi M, Rosati A, Buonamano A, et al. High pretransplant serum levels of CXCL10/IP-10 are related to increased risk of renal allograft failure. Am J Transplant 2004;4(9):1466–74.
[158] Grenzi PC, Campos EF, Tedesco-Silva H, et al. Association of high post-transplant soluble CD30 serum levels with chronic allograft nephropathy. Transpl Immunol 2013;29(1–4):34–8.
[159] McLaren AJ, Marshall SE, Haldar NA, et al. Adhesion molecule polymorphisms in chronic renal allograft failure. Kidney Int 1999;55(5):1977–82.
[160] Valore EV, Park CH, Quayle AJ, Wiles KR, McCray PB Jr, Ganz T. Human beta-defensin-1: an antimicrobial peptide of urogenital tissues. J Clin Invest 1998;101(8):1633–42.
[161] Yang D, Chertov O, Bykovskaia SN, et al. Beta-defensins: linking innate and adaptive immunity through dendritic and T cell CCR6. Science 1999;286(5439):525–8.
[162] Janciauskiene S. Conformational properties of serine proteinase inhibitors (serpins) confer multiple pathophysiological roles. Biochim Biophys Acta 2001;1535(3):221–35.

[163] Conz P, Bevilacqua PA, Ronco C, et al. Alpha-1-antichymotrypsin in renal biopsies. Nephron 1990;56(4):387–90.
[164] Maury CP, Teppo AM, Hockerstedt K. Acute phase proteins and liver allograft rejection. Liver 1988;8(2):75–9.
[165] Kloda K, Domanski L, Pawlik A, Wisniewska M, Safranow K, Ciechanowski K. The impact of ICAM1 and VCAM1 gene polymorphisms on chronic allograft nephropathy and transplanted kidney function. Transplant Proc 2013;45(6):2244–7.
[166] O'Riordan E, Orlova TN, Podust VN, et al. Characterization of urinary peptide biomarkers of acute rejection in renal allografts. Am J Transplant 2007;7(4):930–40.
[167] Kurian SM, Heilman R, Mondala TS, et al. Biomarkers for early and late stage chronic allograft nephropathy by proteogenomic profiling of peripheral blood. PLoS One 2009;4(7):e6212.
[168] Ashton-Chess J, Giral M, Mengel M, et al. Tribbles-1 as a novel biomarker of chronic antibody-mediated rejection. J Am Soc Nephrol 2008;19(6):1116–27.
[169] Shi Y, Tu Z, Bao J, et al. Urinary connective tissue growth factor increases far earlier than histopathological damage and functional deterioration in early chronic renal allograft injury. Scand J Urol Nephrol 2009;43(5):390–9.
[170] Ho J, Rush DN, Gibson IW, et al. Early urinary CCL2 is associated with the later development of interstitial fibrosis and tubular atrophy in renal allografts. Transplantation 2010;90(4):394–400.
[171] Cassidy H, Slyne J, O'Kelly P, et al. Urinary biomarkers of chronic allograft nephropathy. Proteomics Clin Appl 2015;9(5–6):574–85.
[172] Simic-Ogrizovic S, Simic T, Reljic Z, et al. Markers of oxidative stress after renal transplantation. Transpl Int 1998;11(Suppl 1):S125–9.
[173] Djamali A, Reese S, Yracheta J, Oberley T, Hullett D, Becker B. Epithelial-to-mesenchymal transition and oxidative stress in chronic allograft nephropathy. Am J Transplant 2005;5(3):500–9.
[174] Danger R, Paul C, Giral M, et al. Expression of miR-142-5p in peripheral blood mononuclear cells from renal transplant patients with chronic antibody-mediated rejection. PLoS One 2013;8(4):e60702.
[175] Rascio F, Pontrelli P, Accetturo M, et al. A type I interferon signature characterizes chronic antibody-mediated rejection in kidney transplantation. J Pathol 2015;237(1):72–84.
[176] Gross L. Filterable agent A. Recovered from Ak leukemic extracts, causing salivary gland carcinomas in C3H mice. Proc Soc Exp Biol Med 1953;83(2):414–21.
[177] Gardner SD, Field AM, Coleman DV, Hulme B. New human papovavirus (B.K.) isolated from urine after renal transplantation. Lancet 1971;1(7712):1253–7.
[178] Padgett BL, Walker DL, ZuRhein GM, Eckroade RJ, Dessel BH. Cultivation of papova-like virus from human brain with progressive multifocal leucoencephalopathy. Lancet 1971;1(7712):1257–60.
[179] Nickeleit V, Klimkait T, Binet IF, et al. Testing for polyomavirus type BK DNA in plasma to identify renal-allograft recipients with viral nephropathy. N Engl J Med 2000;342(18):1309–15.
[180] Egli A, Infanti L, Dumoulin A, et al. Prevalence of polyomavirus BK and JC infection and replication in 400 healthy blood donors. J Infect Dis 2009;199(6):837–46.
[181] Shah KV, Daniel RW, Warszawski RM. High prevalence of antibodies to BK virus, an SV40-related papovavirus, in residents of Maryland. J Infect Dis 1973;128(6):784–7.
[182] Bogdanovic G, Ljungman P, Wang F, Dalianis T. Presence of human polyomavirus DNA in the peripheral circulation of bone marrow transplant patients with and without hemorrhagic cystitis. Bone Marrow Transplant 1996;17(4):573–6.
[183] Binet I, Nickeleit V, Hirsch HH, et al. Polyomavirus disease under new immunosuppressive drugs: a cause of renal graft dysfunction and graft loss. Transplantation 1999;67(6):918–22.

[184] Drachenberg CB, Hirsch HH, Papadimitriou JC, et al. Polyomavirus BK versus JC replication and nephropathy in renal transplant recipients: a prospective evaluation. Transplantation 2007;84(3):323–30.
[185] Nickeleit V, Hirsch HH, Binet IF, et al. Polyomavirus infection of renal allograft recipients: from latent infection to manifest disease. J Am Soc Nephrol 1999;10(5):1080–9.
[186] Batal I, Franco ZM, Shapiro R, et al. Clinicopathologic analysis of patients with BK viruria and rejection-like graft dysfunction. Hum Pathol 2009;40(9):1312–9.
[187] Hirsch HH, Brennan DC, Drachenberg CB, et al. Polyomavirus-associated nephropathy in renal transplantation: interdisciplinary analyses and recommendations. Transplantation 2005;79(10):1277–86.
[188] Hirsch HH, Knowles W, Dickenmann M, et al. Prospective study of polyomavirus type BK replication and nephropathy in renal-transplant recipients. N Engl J Med 2002;347(7):488–96.
[189] Viscount HB, Eid AJ, Espy MJ, et al. Polyomavirus polymerase chain reaction as a surrogate marker of polyomavirus-associated nephropathy. Transplantation 2007;84(3):340–5.
[190] Singh HK, Andreoni KA, Madden V, et al. Presence of urinary Haufen accurately predicts polyomavirus nephropathy. J Am Soc Nephrol 2009;20(2):416–27.
[191] Jackson JA, Kim EJ, Begley B, et al. Urinary chemokines CXCL9 and CXCL10 are noninvasive markers of renal allograft rejection and BK viral infection. Am J Transplant 2011;11(10):2228–34.
[192] Konietzny R, Fischer R, Ternette N, et al. Detection of BK virus in urine from renal transplant subjects by mass spectrometry. Clin Proteomics 2012;9(1):4.
[193] Sigdel TK, Salomonis N, Nicora CD, et al. The identification of novel potential injury mechanisms and candidate biomarkers in renal allograft rejection by quantitative proteomics. Mol Cell Proteomics 2014;13(2):621–31.
[194] Hwang YH, Ro H, Choi I, et al. Impact of polymorphisms of TLR4/CD14 and TLR3 on acute rejection in kidney transplantation. Transplantation 2009;88(5):699–705.

CHAPTER NINE

Biomarkers of Renal Cancer

N.S. Vasudev, MD, PhD* and R.E. Banks, PhD**
*Medical Oncology, Clinical and Biomedical Proteomics Group, Leeds Institute of Cancer and Pathology, St James's University Hospital, Leeds, United Kingdom
**Biomedical Proteomics, Clinical and Biomedical Proteomics Group, Leeds Institute of Cancer and Pathology, St James's University Hospital, Leeds, United Kingdom

Contents

RENAL CANCER

Renal cancer is the tenth most common cancer in adults, with over one-third of a million new cases and >143,000 deaths worldwide each year (www.globocan.iarc.fr). The incidence of renal cancer has been steadily rising, for example, rates increasing by 168% in females in the United Kingdom between 1975–77 and 2009–11 (www.cancerresearchuk.org/cancer-info/cancerstats/). Peak incidence is in the sixth and seventh decades with a male to female ratio of 3:2. Risk factors for the development of renal cancer include smoking, obesity, and hypertension, as well as end-stage renal failure patients on dialysis who develop acquired renal cystic disease. The clinical investigation and management of patients with renal cancer has undergone major changes in the last 25 years or so following increased understanding of underlying molecular changes, the subsequent appreciation of the heterogeneity of the various subtypes away from purely morphological criteria, and development of targeted therapies [1].

Biomarkers of Kidney Disease. http://dx.doi.org/10.1016/B978-0-12-803014-1.00009-1

The most common type of kidney cancer (~90% of cases) is renal cell carcinoma (RCC), among which the clear cell histological subtype is the most common, accounting for 70–80% of all cases. A further 10–15% are papillary tumors, 4–5% chromophobe tumors, as well as collecting duct (<1%) and the benign oncocytomas (2–5%), each arising from different areas of the kidney and with distinct underlying genetic changes, morphology, and clinical features. In this chapter, the term RCC is used to denote the clear cell phenotype, unless stated otherwise. It is generally accepted that conventional RCC originates from proximal tubules, based on immunohistologic and ultrastructural analysis, as well as more recent DNA methylation and gene expression data [2–5]. The sarcomatoid variant, which can occur with any histologic subtype, is associated with a significantly poorer prognosis. RCCs are graded based on nuclear features, with the most commonly used grading system, the Fuhrman system [6], ranging from 1 (well differentiated) to 4 (poorly differentiated).

Biology

Like several other cancer types, RCC can occur in both sporadic and hereditary forms. The Von Hippel–Lindau (*VHL*) tumor suppressor gene represents an important gene in this regard. Early studies of familial RCC within the rare inherited VHL syndrome localized the genetic defect to the short arm of chromosome 3, with the specific identification of the *VHL* gene at 3p25-26 in 1993 [7]. The *VHL* gene has now also been implicated in most (>80%) of the sporadic clear cell RCC tumors [8–11].

The most well-characterized function of the *VHL* gene product relates to its role in ubiquitination and subsequent proteasomal degradation of the transcription factor hypoxia-inducible factor (HIF) family. Loss of VHL function leads to accumulation of HIF and transcriptional activation of genes including vascular endothelial growth factor (*VEGF*), a feature that has been exploited in RCC therapy, as discussed later. More recently, large-scale RCC genomic sequencing studies have begun to reveal the wider genetic landscape of these cancers. A second major gene implicated in clear cell RCC is the SW1/SNF chromatin remodeling complex gene polybromo1 (*PBRM1*), with mutations found in approximately 40% of cases [12]. Mutations in BRCA-related protein-1 (*BAP1*) occur in 15% of clear cell RCCs and other genes, such as *SETD2* (SET domain-containing protein 2) and *JARID1C* (Jumonji AT-rich interactive domain 1C), are also implicated although with mutations occurring at a much lower frequency [13].

Epigenetic characterization has led to further insights and the molecular complexity of different tumors with different patterns of behavior depending on the combinations of genetic and epigenetic changes is becoming increasingly apparent within tumors of the clear cell phenotype [11]. For the nonclear cell subtypes, distinct genetic and methylation profiles are also starting to emerge [2,14–17], although there appears to be greater heterogeneity and no equivalent driver genes to *VHL* as yet in terms of frequency of involvement.

Genetic subsets of clear cell RCCs are thus beginning to be defined, with divergent biology and clinical outcomes, for example, the association of *BAP1* mutation with poorer survival compared with *PBRM1* with more favorable outcome [18]. It is expected that these novel insights will in turn lead to a new generation of biomarkers and/or novel therapeutic targets, improving survival and personalizing the approach to patient care.

Diagnosis and Treatment

The majority of patients have few, if any, symptoms at diagnosis. The widespread use of imaging techniques such as ultrasound and computerized tomography (CT) has led to a sharp rise in the number of incidental diagnoses although this is not thought to be the sole reason for the rise in incidence. A recent international prospective study of 4288 patients with renal masses in a 2-year period reported an incidental diagnosis rate of 67% [19]. The classic triad of hematuria, flank pain, and abdominal mass are seldom seen. Patients may present with nonspecific symptoms such as weight loss or general malaise, or, more rarely, secondary to paraneoplastic syndromes causing hypercalcemia, polycythemia, or pyrexia.

Once patients present to their physician, diagnosis currently relies on expert radiological review of CT and/or magnetic resonance imaging. For larger tumors this is very reliable and patients usually undergo surgical resection without the need for a preoperative biopsy. Smaller renal masses ($\leq$4 cm), however, are more difficult to accurately diagnose and it is estimated that 20–25% of such lesions are benign [20,21]. This represents a major dilemma to the urological surgeon in terms of balancing the associated morbidity and risks of surgery or ablative procedures, particularly in elderly patients with comorbidities, against the risk of the tumor progressing within the lifetime of the patient.

Surgery remains the mainstay of treatment for patients with RCC. Radical or partial nephrectomy, now often performed laparoscopically, is the gold standard of care in patients with localized and locally advanced tumors, and represents the only means of potential cure. In patients presenting with metastatic disease, removal of the primary tumor should also be considered in carefully selected patients, although this has not been validated in patients treated with more recently introduced therapies. The optimal management of smaller renal masses remains uncertain. Options include surgical excision, radiofrequency ablation, or active surveillance, with each approach carrying its own risks and benefits.

Renal tumors are characteristically resistant to standard chemotherapeutic agents. Advances in our understanding of renal cancer biology have recently led to a revolution in the treatment of this disease. In the past 9 years the Food and Drug Administration (FDA) have approved seven new drugs, namely sorafenib, sunitinib, pazopanib, axitinib, everolimus, temsirolimus, and bevacizumab that are now variably used in the treatment of patients with metastatic RCC (mRCC) and are based on a knowledge of the underlying pathways involved (Fig. 9.1).

Sunitinib (Pfizer) and pazopanib (Novartis) are small molecule tyrosine kinase inhibitors (TKI) that have activity against several receptor kinases including VEGFR-1, 2, and 3; PDGFRβ; c-KIT; and RET. They form the mainstay of first-line treatment for patients with mRCC based on randomized Phase III data [22–24]. In a recent head-to-head study (COMPARZ), the noninferiority of pazopanib to sunitinib was demonstrated, with progression-free survival (PFS) of 8.4 months with pazopanib [95% confidence interval (CI), 8.3–10.9] and 9.5 months with sunitinib (95% CI, 8.3–11.1) [25]. Axitinib (Pfizer) is a second-generation TKI, approved for use in the second-line setting, following initial TKI failure.

Temsirolimus (Wyeth Pharmaceuticals) and everolimus (Novartis) are inhibitors of mammalian target of rapamycin (mTOR) kinase, a component of intracellular signaling pathways. It plays a central role in the phosphoinositide 3-kinase (PI3K)/Akt pathway, which is often aberrantly regulated in cancers, including RCC. Temsirolimus has demonstrated its superiority over interferon-α in a randomized Phase III setting in poor prognosis disease, and is an option in the first-line setting among this group of patients [26]. Everolimus (Novartis), is approved for use in the second-line setting, in patients who have failed on a VEGF-targeted agent [27].

Immunotherapy represents a further treatment modality and has long held promise in RCC. High-dose interleukin-2 is a therapeutic option in carefully selected patients and is associated with durable complete remissions in a small proportion of patients with metastatic disease. More recently,

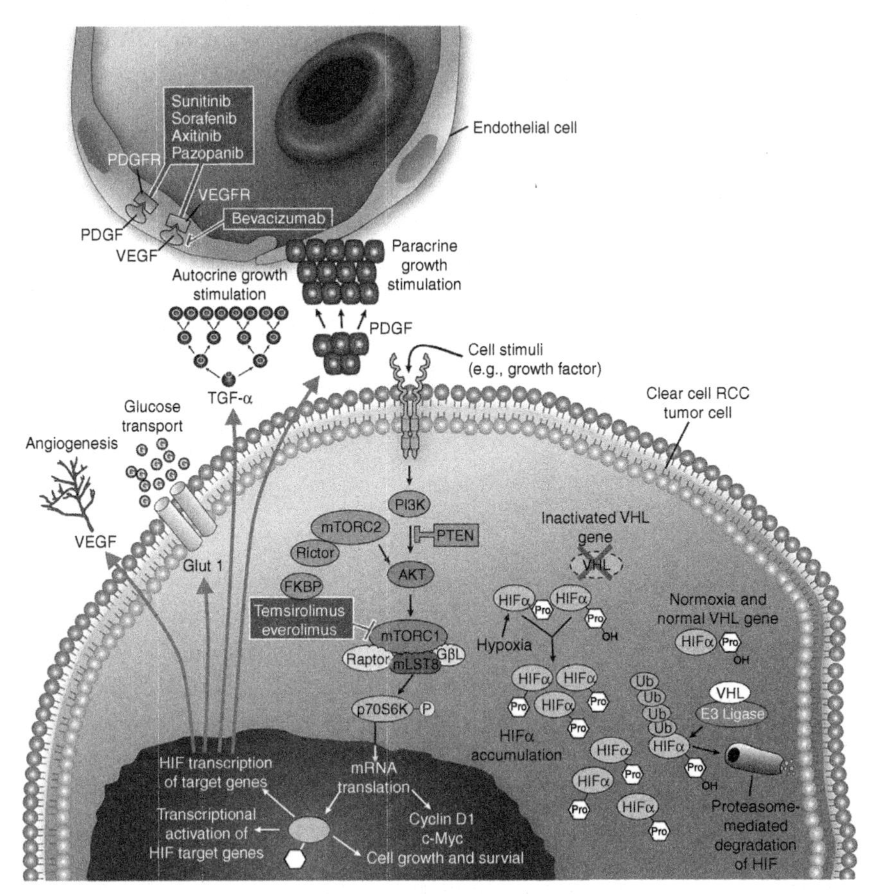

Figure 9.1 ***Schematic illustrating the cellular pathways implicated in conventional (clear cell) renal cell carcinoma (RCC) pathogenesis.*** Also shown are the seven drugs which are approved by the FDA and which were developed specifically to target these pathways [5]. *HIF*, hypoxia-inducible factor; *mTOR*, mammalian target of rapamycin; *mTORC1*, mTOR complex 1; *PDGF*, platelet-derived growth factor; *PTEN*, phosphatase and tensin homolog; *VEGF*, vascular endothelial growth factor; *VEGFR*, vascular endothelial growth factor receptor. *Reprinted from Shuch B, Amin A, Armstrong AJ, et al. Understanding pathologic variants of renal cell carcinoma: distilling therapeutic opportunities from biologic complexity. Eur Urol 2015;67(1):85–97 with permission from ©Elsevier.*

the immune checkpoint inhibitors comprising antibodies to CTLA4, programmed death 1 (PD-1), and programmed death ligand 1 (PD-L1) have shown activity in RCC, either as single agents or in combination. PD-L1 (also known as B7-H1), which is upregulated on many types of tumor cells, interacts with PD-1 on T cells which functions as a negative regulator of T-cell activity and may be important in preventing an effective antitumor

T-cell response. Blocking antibodies to PD-1/PD-L1 are currently in clinical trials. Nivolumab, a PD-1 targeted antibody, has recently shown an improvement in overall survival (OS) in comparison to everolimus (25.0 vs. 19.6 months; $p = 0.002$) in patients who have progressed on previous VEGF TKI therapy [28]. Further clinical trials are ongoing and it is likely these drugs will also join the therapeutic armamentarium.

For patients with localized disease, the standard of care following nephrectomy remains observation. Given their success in the metastatic setting, the use of adjuvant TKI treatment is being explored. The ASSURE trial, examining 1 year of treatment with either sorafenib or sunitinib post-nephrectomy has recently reported, with disappointing results [29]. Results from the European SORCE trial, examining up to 3 years of adjuvant sorafenib, are awaited.

Staging and Prognosis

Cancers of the kidney are characterized by their highly variable natural history and therefore predicting outcomes for individual patients can be difficult. Although 60–70% of patients present with localized disease where surgery or ablation is largely curative, approximately a third of these will relapse after surgery. Detection of relapse is reliant on imaging, with biopsy in some cases, with associated risks from cumulative radiation exposure, expense, and potential morbidity. Many postoperative surveillance imaging regimens have been developed but even with the more recent standardized guidelines, detection of recurrence is often not optimal [30]. This ability to classify patients according to risk is, however, highly desirable. In localized disease, such information could be used to guide the intensity of follow-up and to identify high-risk patients who can be targeted for adjuvant therapy trials. The recent introduction of efficacious but costly treatments also highlights a need to be able to define and target specific patient groups and the concept of genetic profiling (or indeed multi-'omic determinants of personalized medicine) to govern both risk stratification for guiding surveillance protocols and selection of specific targeted treatments is attractive [30].

The staging of renal cancers is, like all solid tumors, based around the TNM system, developed in the 1940s by Pierre Denoix and formalized by the Union for International Cancer Control (UICC) in the 1950s. The work of the UICC and the American Joint Committee on Cancer (AJCC) in this area is now coordinated and revised versions simultaneously published in the *TNM Classification of Malignant Tumours* and the *AJCC Cancer Staging Manual*, respectively, both currently in their seventh editions [31,32].

Table 9.1 AJCC/UICC TNM staging for kidney tumors (7th edition, 2009) [31,32]

Tx	Primary tumor cannot be assessed
T0	No evidence of primary tumor
T1a	Tumor ≤ 4 cm in greatest dimension, limited to the kidney
T1b	Tumor > 4 cm but ≤7 cm in greatest dimension, limited to the kidney
T2	Tumor > 7cm in greatest dimension, limited to the kidney
T2a	Tumor > 7cm but ≤10 cm, limited to the kidney
T2b	Tumor > 10 cm, limited to the kidney
T3	Tumor extends into major veins or perinephric tissues but not beyond Gerota's fascia or into ipsilateral adrenal gland
T3a	Extends into the renal vein or segmental branches or invades perirenal and/or peripelvic fat
T3b	Tumor extends into vena cava below diaphragm
T3c	Tumor extends into vena cava above diaphragm or invades the wall of the vena cava
T4	Tumor invades beyond Gerota's fascia or involves ipsilateral adrenal gland
NX	Regional lymph nodes cannot be assessed
N0	No regional lymph node metastases
N1	Metastasis in any regional lymph nodes
M0	No distant metastasis
M1	Distant metastasis

Stage grouping

Stage I	T1	N0	M0
Stage II	T2	N0	M0
Stage III	T3	N0	M0
	T1,T2,T3	N1	M0
Stage IV	T4	Any N	M0
	Any T	Any N	M1

In this classification, T denotes the extent of the primary tumor, N the extent of regional nodal disease, and M whether metastatic disease is present (Table 9.1).

In addition to tumor stage, tumor size, nodal status, and histological grade are established independent prognostic factors in RCC. Several prognostic models or algorithms have been proposed, variably incorporating these and other factors [33]. Elements such as nuclear grade are, however, subject to intra- and inter-observer variability [34] and additionally the prognostic validity of nuclear grading in histological subtypes other than conventional is questionable [35]. Furthermore, such models typically stratify patients into a limited number of risk groups, based on a "risk score," meaning that estimates of risk can be wide for individual patients. For patients with localized disease, the challenge for clinicians is to identify the 30–40% of patients who will subsequently relapse following radical surgery. A scoring system developed at the Mayo Clinic provides a means for determining prognosis in localized disease, using a model based on 1671 patients, and incorporates pathological T stage, N stage, tumor size, nuclear grade, and histologic tumor necrosis [36]. Patients are divided into three risk groups, with estimated 5-year metastases-free survival rates of 97.1, 73.8, and 31.2% in the low-, intermediate-, and high-risk groups, respectively but a challenge here is the further stratification of patients in the intermediate-risk group. This model was used to stratify intermediate- and high-risk patients in the SORCE trial of adjuvant therapy.

Currently, the model most widely applied to patients with metastatic disease is that proposed by the International Metastatic RCC Database Consortium (IMDC), developed in the era of targeted therapy. The model stratifies patients into three risk groups based on the following prognostic factors: Karnofsky performance status of less than 80%, less than 1 year from diagnosis to treatment, anemia (hemoglobin concentration < lower limit of normal), hypercalcemia (corrected calcium concentration > upper limit of normal), neutrophilia (neutrophil count > upper limit of normal), and thrombocytosis (platelet count > upper limit of normal) [37]. Patients considered to have a favorable prognosis are those with no poor prognostic factors present; intermediate patients have 1–2 factors present; and patients with an unfavorable profile have >2 factors present. The model has been independently validated, demonstrating median OS of 43.2 months in the favorable-risk group, 22.5 months in the intermediate-risk group, and 7.8 months in the poor-risk group [38].

Recent interest has also focused on the development of preoperative nomograms that move away from the traditional reliance on histopathologic criteria. Such information provides early prognostic data to patient and physician, and may guide surgical strategy and consideration of neoadjuvant systemic therapy. The most comprehensive model to date incorporates age, gender, symptoms, tumor size (by CT scan), T stage, and metastasis, to predict RCC-specific mortality with a high degree of accuracy [39]. Ultimately, currently employed prognostic tools are limited by their overreliance on clinicopathologic characteristics, which belie the complexity and varied biology of RCC. Incorporation of robust, well-validated molecular markers into new or existing models is what is now required and should help to further refine prognostication for individual patients.

CANCER BIOMARKERS—GENERAL CONCEPTS

It can be argued that the future of oncological care lies in the delivery of more precise medicine with truly personalized care for patients. Biomarkers are integral to the realization of this vision, with the potential to impact on all aspects of patient care including diagnosis, treatment, and survival. Depending on the intended use, the ideal tumor marker may be a protein (or protein fragment) that can be easily and objectively measured noninvasively in the serum or urine of patients. Alternatively, for differential diagnosis at the histological level, a protein that could be detected by immunohistochemistry (IHC) on tissue sections would be needed. Biomarkers that allow early detection of disease hold tremendous potential, since most solid tumors, including renal cancers, are curable if detected at an early stage. In many cancer types, biomarkers now have established roles in differential diagnosis, tumor subclassification, risk stratification, and disease monitoring. The advent of numerous efficacious but often expensive biological therapies in many cancers means that oncological practice is now heavily influenced by cost-effectiveness. In general, in oncology, therefore, one of the most urgent needs is for the identification of predictive biomarkers, that is, those that are able to identify patients that are likely to respond to a given treatment or not.

The advent of high-throughput–omics (e.g. genomic, transcriptomic, proteomic) technologies has seen a burgeoning of candidate cancer biomarkers in recent years. However, unfortunately this has not been paralleled by an increase in the number of biomarkers approved by the Food and Drug Administration with a steady decline over the last 10 years generally. The reasons for this paradox are multiple and reflect the many obstacles and potential pitfalls associated with successfully taking a marker from the bench to the clinic [40–43]. Some findings from initial discovery fail very early in terms of not being confirmed in subsequent studies and this may arise from poor initial study design, for example, insufficient statistical power, lack of attention to potential confounding factors and inherent bias, or effects of preanalytical factors on the sample quality. Discovery studies must be followed by appropriate prioritization, robust clinical assay development, retrospective validation studies, and finally, prospective validation; and here again issues of study design are paramount, together with the availability of high quality reagents for assays and samples with the appropriate associated clinical data. Furthermore, it is imperative that the marker demonstrates clinical utility and cost-effectiveness, all of which together requires large amounts of time and expenditure, with engagement of multiple stakeholders including academia, health providers, industry, and regulatory authorities [40–42].

Despite these challenges, enthusiasm for biomarker research has not been dampened. To improve the design, conduct, and reporting of such studies, several guidelines have been published, such as reporting recommendations for tumor MARKer prognostic studies (REMARK) [44] and Standards for Reporting Diagnostic Accuracy (STARD) in relation to diagnostic markers [45]. Compliance with these guidelines (now mandated by many journals) should lead to better conducted studies, with earlier identification and prioritization of only the most promising candidate markers for subsequent validation. In parallel, efforts by governments and funding bodies to help facilitate the path to the clinic have increased, with the establishment of large, high-quality research tissue banks and the infrastructure required to help generate evidence on the clinical and cost-effectiveness of a given in vitro test [e.g., the UK National Institute for Health Research funded Diagnostic Evidence Co-operatives (DECs; www.nihr.ac.uk/about/diagnostic-evidence-co-operatives)]. Spanning the whole of the biomarker pipeline, the recently established National Biomarker Development Alliance in the USA [46,47] is a transsector initiative set up to address the issues of a "dysfunctional and disjointed status of biomarker R & D" and aiming to develop widely accepted standards, best practices, and guidelines through an "end-to-end systems approach" [46].

RENAL CANCER BIOMARKERS

The clinical management of patients presenting with RCC presents many challenges that biomarkers are ideally placed to contribute to resolving and several large-scale initiatives are already being employed to systematically address specific aspects [48]. Currently however, no validated biomarkers exist for patients with RCC and their development has been recognized as a challenge and a priority area for research [49,50]. Such biomarkers have the potential to impact all aspects of patient management, from diagnosis, determining prognosis, and detecting recurrence to treatment selection and monitoring. Each of these is considered later, focusing in particular on protein-based biomarkers. This reflects the propensity of such markers in the literature although it is of course recognized that there are transcriptomic profiles such as ClearCode34 [51], which provide prognostic information; increasing numbers of studies highlight miRNA signatures with potential prognostic utility [52] and following the large-scale genomic studies, clinical associations of specific mutations are beginning to be explored.

Diagnostic Markers

In terms of diagnosis, the biomarker need is both at the level of absolute disease detection of RCC through objective noninvasive measurement in the patient's serum or urine, and also for differential diagnosis of the various histological subtypes by the pathologist in some cases. Differential diagnosis histologically is an issue [53,54] as shown by a recent study in which pathologists disagreed on up to 50% of nonclear cell cases [55]. This is particularly the case with eosinophilic renal tumors, characterized by their high content of mitochondria, spanning the full spectrum between benign (oncocytoma) and malignant (conventional or chromophobe variants) and a small percentage of tumors also being unclassifiable. Of course some markers may fulfill both criteria.

Circulating Markers

The relatively low incidence of RCC, even though markedly increasing, means that screening of the general population is unlikely to be feasible and cost-effective. Even if a biomarker for RCC was 100% sensitive and had a specificity of 99.4%, the positive predictive value for men older than 65 years would be only 10% [56]. Thus, screening would need to be targeted at those at high risk of disease, such as those with previous RCC, end-stage renal disease, kidney transplant recipients, familial RCC, or potentially in triaging symptomatic patients. A pan-urological malignancy biomarker panel may be a more viable approach in the screening context.

Efforts to identify circulating diagnostic protein markers for RCC have been few and have been met with limited success or have failed to be confirmed in validation studies. This reflects both the challenge of profiling complex fluids such as plasma and urine, and the heterogeneity and biological complexity of RCC. The overriding challenge in proteomic serum/plasma analysis is the vast dynamic range of protein concentration, starting with albumin at approximately 40 g/L down to cytokines at 1–10 pg/mL. This is a range spanning at least 10 of orders of magnitude and far exceeds the analytical range of any proteomic technology used in biomarker discovery studies. Additionally, just 22 proteins constitute 99% of the entire plasma protein content [57] and prefractionation/enrichment strategies are essential. Urine presents its own analytical challenges in terms of both low protein concentration and high salt content. A recent study demonstrates the promise of proteomic approaches to biomarker discovery, combining analysis of the serum proteome, glycoproteome, and *N*-glycome with comparison of pre- versus postnephrectomy samples in 20 patients with RCC and identifying several differences. The challenge is now the generation of suitable assays for these and validation in larger cohorts [58].

Based on findings from a tissue-based proteomic study [59], the composite 3-marker panel of nicotinamide *N*-methyltransferase, L-plastin, and nonmetastatic cell 1 protein (NM23A) was analyzed in plasma samples from training ($n = 189$) and test groups ($n = 100$) of patients with RCC, benign controls, or healthy volunteers [60]. Although there was marked overlap between the groups for each marker, concentrations in RCC patients were significantly higher ($p < 0.0001$), including in patients with early stage disease, with nicotinamide *N*-methyltransferase showing the highest discrimination. AUC values for the ROC curves were 0.921 and 0.954 for the training and test groups, respectively, for the 3-marker composite score. These findings warrant further independent validation.

Urine is an attractive source of biomarkers for tumors of the urinary tract with the potential for direct shedding of proteins into it and a less complex matrix. Approximately 70% of the urinary proteome is thought to consist of kidney-derived proteins [61]. However there are challenges in terms of normalization of the results with creatinine being most commonly used although not optimal in many patients with varying degrees of renal impairment and being affected by many factors. A promising series of studies have explored the hypothesis that two proteins upregulated in profiling studies of tissue, aquaporin-1 (AQP-1) and perilipin-2 (PLIN-2, also known as adipophilin or ADFP) may be shed into the urine and have potential diagnostically. Initial western blotting studies showed both proteins to be significantly elevated in urine from patients with clear cell or papillary RCC compared with healthy and surgical controls, declining to control levels postnephrectomy [62]. Elevated levels were not seen in patients with nonmalignant conditions of the kidney [63]. Importantly, these results were subsequently confirmed in larger numbers of patients ($n = 61$) with both proteins being significantly correlated with tumor size and reported sensitivities/specificities for diagnosis of 100%/100% and 92%/100% for AQP-1 and PLIN-2, respectively [64]. No overlap with values for patients with other renal masses such as chromophobe RCC or oncocytoma was seen, in line with the proposed proximal tubule origin of these proteins, and similarly concentrations were not elevated in patients with bladder or prostate cancer [65]. Recently, findings have been extended to 720 patients undergoing CT scans for a variety of indications, 80 healthy controls, and 19 patients with confirmed RCC, using an ELISA for AQP-1 [66]. The AUC for the ROC curve was 0.99 or greater for either biomarker alone or in combination and two of three patients in the CT screening cohort with elevated concentrations were subsequently found to have RCC [66]. Elevated concentrations of AQP-1 were also seen in a small number of patients with other malignancies. Clearly these markers show promise and need further prospective independent evaluation with larger numbers of patients and robust immunoassays for both markers. Importantly the stability of AQP-1 in frozen urine samples may be an issue depending on the conditions and would also require further exploration in future studies [67].

Other protein biomarkers of relevance in other diseases have also been investigated. The nuclear matrix protein 22 (NMP22), which forms the basis of the point-of-care NMP22 BladderCheck Test, was found in two initial studies to be significantly elevated in urine samples from patients with RCC compared to those with benign kidney disease or healthy controls [68,69]. However, no confirmation or extension of these findings has been reported since. Kidney injury molecule-1 (KIM-1) which is expressed by dedifferentiated cells of the proximal tubule and is detected in urine, for example, following injury, has been investigated and 91% of 35 clear cell RCC tissues were found to express KIM-1 in the tumor cells or adjacent tubular cells [70]. Prenephrectomy urinary KIM-1 levels were significantly elevated compared with healthy controls or patients with prostate cancer. The elevated urinary KIM-1 concentrations in RCC patients were subsequently confirmed, with neutrophil gelatinase-associated lipocalin shown not to differentiate between RCC patients and healthy controls [71]. Further exploration of urinary KIM-1 in 19 patients with RCC together with IHC demonstrated that 12/19 cases stained positively for KIM-1 and these patients had higher urinary KIM-1 concentrations prenephrectomy which declined postnephrectomy [72]. These findings warrant further study although given the role of KIM-1 as a biomarker of kidney injury, there is likely to be an issue with diagnostic specificity in the context of RCC.

Promising results have been obtained focusing on peptide profiles rather than specific proteins using capillary electrophoresis–mass spectrometric analysis of urine samples. Using a discovery set of 40 samples from patients with RCC and 68 controls a peptide classifier was identified which showed 80% sensitivity and 87% specificity in an independent validation set [73].

Histopathological Diagnosis

The differential diagnosis of the various subtypes of RCC is important as they vary considerably in their clinical behavior, prognosis, and treatment [5]. In terms of the main subtypes, papillary carcinomas have a less aggressive course than clear cell tumors but respond less well to sunitinib. Two main categories are recognized, Type 1 and Type 2, with the latter group of tumors displaying higher nuclear grade, eosinophilic cytoplasm, and a worse outcome than Type 1 tumors. Chromophobe RCCs are thought to arise from the intercalated cell of the collecting duct and also carry a better prognosis than conventional RCC. Many renal epithelial neoplasms can be diagnosed reliably by experienced pathologists on the basis of morphology alone on routine hematoxylin and eosin stained slides. This is particularly the case for the clear cell subtype with a recent central pathology review of 498 cases identified through the SEER program and originally diagnosed as RCC finding a positive predictive value of 95.5% compared with 85.9% for papillary and 65.6% for chromophobe cases [74]. However eosinophilic renal tumors, characterized by their high content of mitochondria, span the full spectrum between benign (oncocytoma) and malignant (clear cell or chromophobe RCC variants) tumors and provide even experienced pathologists with a challenge. Additional differential diagnoses where IHC is considered to be of diagnostic value have been described [75]. Novel subtypes of RCC are continually being defined, both largely on the basis of morphology as illustrated in the recent recommendations of the International Society of Urological Pathology (ISUP) Consensus Conference [75] and with >24 subtypes of renal cancer now defined in the 2013 ISUP Vancouver classification [76], but also increasingly on the basis of genetic changes as described earlier and reviewed recently [77]. Markers would also be useful for diagnosis of metastases in those cases when the primary site is unknown and would be of use in core biopsies where the range of tissue architecture viewed is more limited.

Several markers are already used, particularly for the main recognized subtypes, or are being evaluated [78,79]. For example, to differentiate between clear cell and chromophobe RCC, CD10, a cell surface metalloproteinase localized to the proximal nephron of normal kidney, RCC marker/antigen developed as a monoclonal antibody to gp200 glycoprotein expressed on normal human kidney proximal tubule, and the intermediate filament protein vimentin are expressed in the former whereas chromophobe are negative for these but express E- and kidney-specific cadherins, CK7, CD117, and parvalbumin. Several of these proteins are also expressed on other subtypes in various combinations and with other markers and it is a combination of IHC and morphology that currently enables the diagnosis. Based on the available evidence, it is likely that, at present, a panel of markers may be most informative in any classification and in the future further subdivisions may emerge arising from the integration of molecular genetic findings, for example, proteins such as carbonic anhydrase IX (CAIX) and PAX2 which are upregulated as a consequence of genetic changes in the VHL gene in clear cell RCC.

Prognostic Markers

Many molecules have been proposed as potential prognostic markers of RCC. Few, however, have been taken beyond single studies, often involving relatively small numbers of patient samples, and, as such, none are yet in routine clinical use. The majority of such markers described to date are tissue-based, reflecting both the ready availability of tissue and the challenges of working with serum and urine. In the following section, details of some of the most promising prognostic markers are presented.

Tissue-Based Markers

B7-H Family

The B7-family of T-cell costimulatory and coinhibitory molecules have been identified as promising prognostic RCC biomarkers. Each member has different, although overlapping, functions in controlling the priming, proliferation, and maturation of T cells, which can lead to immune-suppression and evasion of host immune surveillance. We highlighted in the previous edition of this book that a greater understanding of their function may be important in generating a novel approach to therapy. In both RCC and other cancers such as melanoma in particular, this is now the case, with blocking antibodies to PD-1/PD-L1 (B7-H1) now in clinical trials as described earlier.

In terms of prognostic use, immunohistochemical analysis of fresh-frozen tissue sections from 196 RCC patients found that patients with B7-H1-positive tumors were at significantly increased risk of cancer-specific mortality [80]. Extension of the study to paraffin-embedded sections from 306 patients, with a median of 10 years follow-up, confirmed that the 24% of patients positive for B7-H1 were at significantly increased risk of death [hazard ratio (HR) 2.37; $p < 0.001$], with 5-year cancer-specific survival (CSS) rates of 42% versus 83%, in those with and without

B7-H1 expression, respectively. Furthermore, it was found to be an independent predictor of mortality on multivariate analysis [81]. Importantly in a comparison of PD-L1 (B7-H1) expression in matched primary tumor and metastatic tissue from 34 patients with clear cell RCC, the correlation between the sites was relatively poor ($r = 0.24$) suggesting that examination of the primary tumor may not provide the information needed to select the patients most likely to respond to agents targeting the PD-1/PD-L1 axis. In addition the intratumoral heterogeneity seen suggests that a single core biopsy may be insufficient [82]. B7-H4 has also been implicated as a negative regulator of T cell–mediated immunity. Positivity for B7-H4 was correlated with CSS (HR 3.05; $p = 0.002$) in a study of 259 RCC patients. Patients positive for both B7-H1 and B7-H4 were more than 4 times more likely to die from RCC compared to negative or singly positive tumors (HR 4.49; $p < 0.001$) [83]. Importantly, given the increasing diagnosis of small renal masses, B7-H4 expression in 102 clinical stage T1 clear cell RCC cases was found to be significantly associated with prognosis. Of 18 positive cases, 5 (28%) experienced recurrence in the follow-up period compared with 7 (8%) in the 84 cases where expression was absent [84]. B7-H3 has also been shown to have a relatively weak association with CSS, with an HR of 1.38 ($p = 0.029$) in a study involving 743 patients [85]. With the development of immunoassays for the soluble/shed forms of the B7 family, the possibility of use as non-invasive prognostic markers is being explored with promising early results. For example, higher sB7-H1 levels in serum from 172 RCC patients were associated with several features of tumor aggression including tumor size, grade, necrosis, and stage [86] and this represents an area for future exploration.

IMP3

Insulin-like growth factor II mRNA–binding protein 3 (IMP3) is an oncofetal RNA-binding protein, not normally expressed in adult tissues. It is thought to function in the regulation of insulin-like growth factor II production. In an initial study examining IMP3 expression in 371 primary tumors, 5-year survival was significantly longer (82% vs. 27%; $p < 0.0001$) in patients whose tumors did not express the protein compared to those that did. The result was highly statistically significant and IMP3 expression was shown to be a strong, independent predictor of survival on multivariate analysis [87]. These results have since been externally validated in a further cohort of 629 patients with localized renal cancer. A quarter of patients' tumors expressed IMP3 and, again, this was associated with a significantly increased risk of both development of distant metastases and death from RCC. The results were most striking for patients with stage I disease, with positive IMP3 expression associated with a six-fold increased risk of progression to distant metastases [HR 6.46; 95% CI, 3.33–12.53 ($p < 0.001$)] [88]. A prognostic model based on quantitative IMP3 and tumor stage (QITS) has since been proposed [89]. The model, based on a relatively small number of patients ($n = 369$) with localized RCC, combined IMP3 expression (using a computerized image analyzer) and TNM stage in relation to outcome. The model stratifies patients into four groups; patients in quantitative IMP3 and tumor stage group IV ($n = 33$), defined as high-level positivity for IMP3 and any TNM stage or low-level positivity and TNM stages 2 or 3, had 5- and 10-year OS rates of 14 and 4%. The power of the model appears to be in identifying a very high-risk population who would not otherwise be recognized using TNM stage alone. The independent prognostic significance for RFS and OS has more recently been confirmed in a study with 469 cases of localized clear cell RCC in a tissue microarray (TMA) where 16% of cases stained positively for IMP3 (58% of stage III/IV tumors vs. 25% of stage I/II). The 5-year RFS for the 347 patients included in the survival analysis were 87% for patients whose tumors were negative for IMP3 compared with 48% in those with IMP3 positively stained tumors [90]. Whether the addition of IMP3 expression to existing nomograms is beneficial remains to be determined.

In one of few studies to evaluate expression of prognostic markers specifically in chromophobe and papillary RCCs, IMP3 was also found to be an independent prognostic marker in this subset of tumors. The study was large ($n = 334$) considering the rarity of these tumors and suggests that IMP3 positivity is associated with a greater than 10-fold risk of developing distant metastases (HR 13.45; 95% CI, 6.00–133.14; $p < 0.001$) [91].

Carbonic Anhydrase IX

The most studied molecular marker of RCC to date is CAIX. Its expression was originally described in 1986, using a monoclonal antibody (G250), which was noted to bind to RCC but not normal proximal epithelium [92]. It was later shown that the antigen recognized by G250 is CAIX [93] and this is currently the subject of therapeutic targeting and clinical imaging studies. CAIX is a metalloenzyme that contains carbonic anhydrase activity and is capable of catalyzing the reversible hydration of CO_2 to form HCO_3^- and H^+. CAIX is a multidomain transmembrane protein regulated by HIF-1α and thought to play a role in the adaptation of tumors to hypoxic conditions, by regulating intra- and extracellular pH. The ectopic expression of CAIX in many human cancer types (including RCC) but not in corresponding normal tissues has been demonstrated [94]. In renal cancer, immunohistochemical studies have shown that 94–100% of conventional RCC tumors express CAIX at the plasma membrane, with uniformly negative staining in normal kidney tissue [94,95]. Overexpression is associated with tumor aggressiveness and poor outcome across many cancer types, except for RCC, where the opposite appears to be true.

The first large study examined CAIX expression in relation to outcome using a TMA of 321 RCC specimens. A staining percentage of >85% positive cells was used to stratify patients as high [$n = 255$ (79%)] and low [$n = 66$ (21%)] expressers. Among the 149 patients with metastatic disease, low CAIX expression was associated with a significantly worse CSS (median 5.5 vs. 24.8 months; HR 3.10; $p < 0.001$) and was independently prognostic on multivariate analysis, considering T stage, grade, nodal status, and ECOG performance status. Among patients with high-risk localized disease (T stage ≥ 3; grade ≥ 2), low CAIX expression was also associated with poorer outcome but was not independently prognostic [95]. The same authors replicated these results in a study of 224 patients using a TMA. Low CAIX expression was again associated with a significantly worse CSS (22 vs. 67 months) when considering all patients, and was independently prognostic (HR 1.87; $p = 0.006$) [96]. More recently, however, a large study of 730 tumor specimens employing the same antibody (M75) but using whole tissue sections found that although on univariate analysis an association with low CAIX expression (again defined as ≤85%) and poor CSS was demonstrated, this was lost on adjustment for even just one other variable, namely nuclear grade [97]. The reasons for the discrepancy are unclear, but may relate to differences in patient population and the use of TMAs, since expression within larger sections was found to be variable across the tissue. A recent long-term follow-up study of this cohort found the same results in terms of CAIX being significantly prognostic only on univariate analysis [98]. Using samples from 133 patients with mRCC in the TARGET study from both the placebo- and sorafenib-treated arms, CAIX was also not found to be prognostic or predictive of response to treatment in terms of either PFS or tumor shrinkage [99].

Whether CAIX is independently prognostic or not, it is clear that a high CAIX expression is associated with a better outcome in RCC. The reasons for this are uncertain. Patients with tumors expressing low CAIX and absence of VHL mutation have been shown to have a significantly worse outcome compared to those with high expression and VHL mutation ($p = 0.0002$). Since VHL mutation is a hallmark of RCC (and is implicated in the regulation of CAIX expression), it is postulated that this underlying VHL inactivation, rather than the functional consequence of intratumoral hypoxia, leads to high CAIX expression and improved survival in patients with RCC [100].

Ki-67

The nuclear protein Ki-67 which serves as a marker of proliferation has been correlated with survival in many cancers including several studies in RCC. Of the three largest studies to date, the first examined 224 tumors, reporting high Ki-67 expression correlating with increasing tumor stage and grade, with a risk ratio of 2.10 ($p < 0.001$) on multivariate analysis for CSS [96]. Subsequently, in a study of 741 tumors, Ki-67 was independently prognostic for CSS, with an HR of 3.43 (95% CI, 2.64–4.45); $p < 0.001$). The marker retained independent prognostic ability even when controlling for tumor necrosis, which had previously been proposed by others as a surrogate marker for Ki-67 expression [101]. In the most recent study examining Ki-67, together with several markers including CAIX, C-reactive protein (CRP), and HIF-1α and 2α, using a TMA comprising samples from 216 patients with localized RCC of all subtypes, only Ki-67 was significant for disease recurrence in univariate analysis and also remained significant in multivariate analysis with a similar HR to the previous study [102].

Survivin

Survivin is an antiapoptotic protein that belongs to the inhibitor of apoptosis protein (IAP) family and is overexpressed in almost all human malignancies. In RCC, 31% of patients were found to have high survivin expression in a study of 312 patients. On multivariate analysis, expression of the marker remained significantly associated with CSS (HR 2.4; 95% CI, 1.5–3.8; $p < 0.001$) [103]. Similar results were reported in a smaller study of 85 patients [104] and for OS in 75 patients [105] although a slightly larger study with 104 patients found no association with OS [106]. Interestingly, the relationship between survivin and B7-H1 expression has been examined in a study of 298 patients with RCC. Both markers were confirmed to be independently prognostic for CSS; 177 (59.4%) tumors were classed as low or negative and 41 (13.8%) as positive or high for expression of both markers. Five-year CSS rates were 89.3 and 16.2% in these two groups of patients, respectively [107].

Markers in Combination

Ultimately, any single protein marker is unlikely to successfully determine prognosis across all patients. Combining markers to create a molecular signature of disease is likely to be the most promising way forward. Incorporation of such markers into existing prognostic models, which are currently based on standard clinical criteria, is now starting to be explored with promising results.

In a large study of 634 patients with RCC, tumors were examined for expression of B7-H1, survivin, and Ki-67 using whole tissue sections. Each marker was first confirmed to have independent prognostic ability for CSS, even after adjustment for the other two proteins. The panel was then used to create a prognostic algorithm, termed BioScore, based on high or low expression of each protein within each tumor. Patients with a high score were observed to be 5 times more likely to die from their cancer compared to those with a low score (HR 5.03; 95% CI, 3.82–6.61). Most importantly, the BioScore was shown to add prognostic value to three established clinicopathological models, including the SSIGN score. BioScore added little to patients deemed at low risk by the model, but was able to further stratify those deemed at moderate and high risk [108].

In mRCC, a panel of tissue markers has been successfully combined with clinical features to improve standard scoring systems. Using results from 150 patients, a model, consisting of expression of CAIX, PTEN, vimentin, and p53 as well as T stage and performance status, again outperformed the UCLA Integrated Staging System (UISS) (accuracy 0.68 vs. 0.62; $p = 0.0033$) [109]. A later study focusing on localized disease and analyzing a panel of eight proteins in a TMA based on 170 patients found that five proteins, namely Ki-67, p53, endothelial VEGFR-1, epithelial VEGFR-1, and epithelial VEGF-D were independent prognostic indicators of disease-free survival (DFS). The five markers alone predicted DFS with an accuracy of 0.838 (95% CI, 0.813–0.863). This was also more accurate than the UISS, which performed with an accuracy of 0.780 (95% CI, 0.776–0.784). Utilizing a nomogram incorporating the five markers, as well as performance status and T stage, DFS was predicted with an accuracy of 0.904 (95% CI, 0.875–0.932) [110].

One of the caveats of taking forward panels of markers for immunohistochemical analysis is that this often means using TMAs for a variety of reasons including conservation of tissue resources, throughput and efficiencies of staining and review, and costs associated with reagents. Obviously even in single marker studies TMAs may also be used. A systematic analysis recently examined the expression of six of the biomarkers reviewed in this chapter as having potential prognostic use in RCC, namely B7-H1 and H3, survivin, Ki-67, CAIX, and IMP3, using both whole tissue sections and simulated TMA cores from the sections using a masking template [111]. For B7-H3, CAIX, Ki-67, and IMP3, similar results were obtained as for the tissue sections if only 2 or 3 cores were used. However for survivin and B7-H1, even with the maximum of 10 cores, agreement with the staining scores of whole tissue sections was poor, presumably reflecting the focal and rare expression compared with the more diffuse expression patterns seen with the other markers. The number of cores used also impacted on whether the recognized association of these markers with CSS was found if a TMA was used and for B7-H1, even if 10 cores were used, this association was lost. This must be considered when taking forward large biomarker panels but such findings could also account for the variability in conclusions from different studies of single markers even if TMAs were used for some. A further aspect to consider, is the loss of antigenicity of some proteins over time, thereby confounding the interpretation of data generated using archival tissue. Of particular relevance to RCC is the loss of antigenicity of HIFα isoforms in FFPE tissue over time [112].

The prognostic utility of transcript panels for localized clear cell RCC has already been demonstrated with ClearCode34 where a 34 gene classifier assigns cases to good-risk (ccA) or poor-risk (ccB) groups [51] and a 16-gene signature which assigns patients to low-, intermediate-, or high-risk groups [113]. For ClearCode34, significant associations at both univariate and multivariate level were seen for RFS, CSS, and OS and the classifier outperformed several established clinical scoring systems [51]. The 16-gene classifier was also significantly associated with tumor recurrence on multivariable analysis and importantly was able to identify patients with high-risk stage I disease or low-risk stage II/III disease [113]. Interestingly the panels do not include any of the proteins identified here as being the most promising prognostic markers, opening up possibilities for exploration of these gene products at the protein level also.

Circulating Markers

Circulating markers have the advantages of being able to be measured objectively, relatively noninvasively in blood or urine samples and potentially providing prognostic information early, prior to nephrectomy. The number of studies in this area has expanded markedly in the last few years but relatively few examples exist where potential markers have been independently validated in sufficient numbers of patients. Some of the most promising are reviewed in subsequent sections, together with some of the more routinely measured hematological and biochemical parameters.

Routine Hematological and Biochemical Parameters

A number of these have demonstrated an association with outcome in patients presenting with RCC. Such parameters make attractive markers, since they are easily and cheaply measured, as well as being widely available.

Compelling evidence exists demonstrating increased circulating neutrophils or intratumoral neutrophils as an independent poor prognostic factor for patients with mRCC and appearing to be associated with poor outcome following treatment with cytokines or VEGF-targeted therapy [114]. In addition, we described neutrophil/lymphocyte ratio (NLR) as a strong independent predictive factor of CSS (but not OS or DFS) in RCC patients with localized and metastatic disease [115] with the prognostic value of NLR being subsequently confirmed in a number of studies although with variable outcome associations [116–118]. In the largest and most recent of these featuring 678 patients with clear cell nonmetastatic RCC at a single center, NLR was found to be significant on multivariate analysis but only for OS and not CSS or MFS [118] whereas in a study of 192 patients with localized RCC (subtypes not specified), NLR was independently significantly associated with RFS [116]. The reasons for the differences are not clear but may in part reflect the statistical analysis used in terms of whether continuous or cut-offs were applied or the patient populations in terms of stage and subtype mix. A single study has also found NLR to be an independent prognostic factor for DFS for nonclear cell RCC subtypes [117]. Clearly the findings are of potential clinical relevance and should be investigated further.

Several studies have demonstrated that thrombocytosis is a poor prognostic feature in patients with RCC [119]. The underlying biology behind thrombocytosis and RCC survival is uncertain. It is, however, not unique to RCC and applies equally to patients with other tumor types. It is likely, therefore, to reflect a nonspecific, tumor-related, inflammatory response although interleukin-6 production by RCC cells may contribute. In 804 RCC patients prior to nephrectomy, with 126 (15.7%) patients having metastatic disease, a high platelet count (defined as $>450,000\ /\text{mm}^3$) was found to be independently prognostic for poor outcome, in a model containing stage, grade, and performance status, with 5-year survival of 70% versus 38% in patients with a low and high platelet count, respectively [120].

In another large study of 700 patients with mRCC, baseline platelet count (> or <400,000/mm^3) was again reported to be of independent prognostic value (HR 1.65; 95% CI, 1.36–1.99; $p < 0.001$). A quarter of patients were deemed to have thrombocytosis, with a median survival of 8.4 months in this group compared to 14.6 months ($p < 0.001$) [121]. In a very comprehensive study involving 1,828 patients with RCC prior to nephrectomy (27.8% with metastatic disease) with platelet count as a continuous variable or dichotomized around 450,000 mm^{-3}, on multivariate analysis, using most-informative cut-offs, platelet count achieved independent predictor status for CSS. However, adding platelet count to a model composed of age, tumor size, TNM stage, PS, grade, and histologic subtype, increased its accuracy by only 0.3% (from 85.3% to 85.6%). No extra benefit was observed when considering only patients with localized or metastatic disease [122]. In the largest study to date with 3,139 patients, thrombocytosis and continuous platelet number prior to surgery were both independently predictive of CSS in the whole study population and the nonmetastatic group but not in patients with metastatic disease. In agreement with the previous study [122], thrombocytosis, although associated with poor prognosis, did not significantly improve multivariable models of survival [119].

Changes in coagulation in malignancies are widely recognized and recently the prognostic value of plasma fibrinogen in RCC has been described in several studies [123–125]. Significantly higher concentrations were found in patients with metastatic disease [126] and subsequently in a study involving 286 patients with RCC of mixed stages and subtypes, fibrinogen was reported to be significantly independently associated with DFS and OS [125]. This has subsequently been confirmed in a smaller study ($n = 128$) and extended by also showing a significant association with CSS, and with D-dimer also having negative independent prognostic value for OS [123]. Analysis of data from a large cohort ($n = 994$) of patients with localized RCC found plasma fibrinogen to be associated with tumor stage and grade and a significant independent prognostic factor for CSS, MFS, and OS (Fig. 9.2) [124].

Serum LDH, calcium, and hemoglobin have been widely reported to have prognostic utility in patients with metastatic disease and are included within the MSKCC nomogram [127]. More recently, their utility in the era of patients treated with TKIs has also been shown in a study by International Kidney Cancer Working Group involving 3748 patients to develop a prognostic model which was validated against patients treated with TKI therapy [128]. In a large study with 1707 patients with nonmetastatic clear

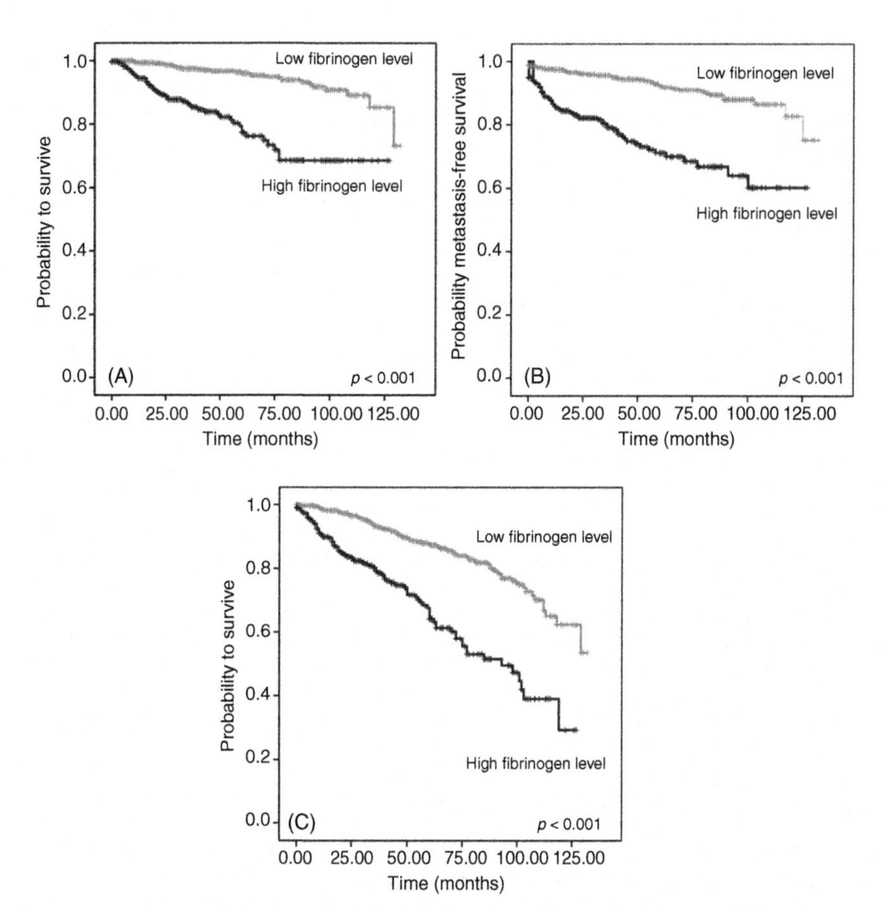

Figure 9.2 Kaplan–Meier curves for (A) cancer-specific survival groups, (B) metastasis-free survival groups, and (C) overall survival (OS), categorized by the preoperative plasma fibrinogen levels [124]. *Adapted from Pichler M, Hutterer GC, Stojakovic T, Mannweiler S, Pummer K, Zigeuner R. High plasma fibrinogen level represents an independent negative prognostic factor regarding cancer-specific, metastasis-free, as well as overall survival in a European cohort of non-metastatic renal cell carcinoma patients. Br J Cancer 2013;109(5):1123–1129 with permission from ©Macmillan Publishers Ltd. on behalf of Cancer Research UK.*

cell RCC, hypercalcemia has also been reported to be independently prognostic for CSS (risk ratio 16.4, $p = 0.002$) together with anemia and ESR [129]. Confirming this and also showing independent prognostic value for OS but not DFS of tissue mRNA levels for parathyroid hormone-like hormone which is recognized as being a major contributing factor to hypercalcemia in RCC, this study also showed a specificity to the clear cell phenotype and not nonclear cell tumors [130].

Other biochemical parameters with evidence of prognostic value include serum sodium, gamma-glutamyltransferase, and total cholesterol. Preoperative sodium was first reported by our own group as being independently prognostic for DFS and OS ($n = 212$) when considered as both a continuous variable and when dichotomized to above and below the median value (139 mmol/L) (HR = 0.44, 95% CI, 0.22–0.88, $p = 0.014$). The majority (92%) of patients had serum sodium values within the laboratory reference range, but patients with values equal to or below the median had significantly poorer survival [131]. Using various cut-offs, the association of hyponatremia with poor outcome has since been confirmed in patients with metastatic disease being treated either with IL-2/IFN-based therapy [132] or targeted therapies [133,134] although it is currently not clear whether it is prognostic or predictive in this context. Recent studies have implicated gamma-glutamyltransferase as having independent prognostic significance in a large study involving 921 patients with RCC of all subtypes and both localized and metastatic disease [135] although in a similar study involving 700 patients with localized disease only, this was found to be significant at the univariate level only [136]. Interestingly, given the association of RCC with obesity and links between body mass index and prognosis, preoperative serum total cholesterol has very recently been reported to be independently prognostic with lower concentrations being associated with poorer survival in a study involving 364 patients with clear cell RCC [137] and 867 patients of all subtypes [138].

CRP, a well-described acute-phase protein, is often raised in patients with cancer and circulating levels have been correlated with outcome. In RCC, raised CRP levels have been correlated with poor survival in a number of studies, in both the early and advanced settings, with some of the major studies reviewed here. It has been shown to be independently prognostic and may be of additional value to currently used nomograms [139]. In a study of 313 patients undergoing nephrectomy for RCC, 66 (21%) of whom had metastatic disease, CRP was independently prognostic for CSS ($p = 0.003$). CRP was treated as three categorical variables, identified using most informative cut-offs, namely ≤ 4.0 mg/L, 4.1–23.0 mg/L, and >23.0 mg/L. Importantly, the addition of CRP to a preexisting prognostic model (UISS) improved its accuracy by 3.8% at 5 years ($p < 0.001$) [140]. A simple scoring system based on CRP and TNM alone, termed TNM-C score, has recently been described based on 249 RCC patients with advanced and localized disease [141]. The investigators dichotomized CRP as $<$ or ≥ 0.5 mg/dL and combined this with TNM to generate four risk groups. The model was externally validated in 290 patients, with a concordance index of 0.865 [141] and subsequently validated in a further 518 patients with clear cell RCC with a c-index of 0.85 [142]. More recently in a cohort of patients with localized clear cell RCC (n = 403), categorical preoperative CRP was independently significantly associated with DFS, and categorically coded preoperative CRP but not continuous CRP increased the prognostic accuracy of the SSIGN score [143]. The ability to increase the predictive accuracy of a score though is likely to be very dependent on the scoring system used however, and in a study with 327 patients and using a base model including TNM stage, grade, and Karnofsky index which already had a high predictive value of 88.1%; addition of CRP to the model did not improve the predictive accuracy [144].

Such studies have measured CRP prior to nephrectomy but then tested the marker against postoperative nomograms and it is relevant to also determine whether CRP adds to published preoperative models [39]. Data from our own group, based on 286 RCC patients (84% clear cell), has demonstrated CRP (dichotomized as ≤ or >15 mg/L) to correspond to 5-year survival rates of 72% (95% CI, 65–78%) and 33% (95% CI, 23–44%), respectively and to be a strong, independent prognostic factor for OS ($p < 0.006$) and CSS ($p < 0.001$) on multivariate analysis, when considering factors incorporated in either pre- or postoperative nomograms and adding significantly to the preoperative score [115]. The largest and most recent study included 1161 patients with both metastatic and localized disease and all subtypes and confirmed CRP as a significant independent prognostic factor for CSS and OS, with stratification to three subgroups on the basis of CRP concentration rather than as a continuous variable increasing the prognostic significance [145].

Although a study of patients with localized disease (n = 263) found CRP preoperatively to be a significant independent prognostic factor for RFS together with nonnormalization of CRP postoperatively; only nonnormalization of CRP postoperatively was a significant prognostic factor for OS on multivariate analysis. In patients whose CRP postoperatively did not normalize, 5-year OS was 30% compared with 96.9% in those who did [146]. Although CRP changes are normally assumed to reflect increased production by the liver as part of an inflammatory response and regulated by cytokines such as interleukin-6 and -10, CRP has also been shown to be produced by renal tumor cells [147]. In the first study examining the prognostic significance of intratumoral CRP staining in 95 patients with localized clear cell RCC, patients identified as being in the high-risk group had a 27-fold increased risk of overall mortality whereas serum CRP was no longer significantly associated with OS when the model was adjusted for CRP staining [148].

Vascular Endothelial Growth Factor

Exploring the possible prognostic utility of the angiogenic factor VEGF is logical given its relationship to VHL and the critical role in angiogenesis which is a determining factor of tumor growth. In an early study measuring serum VEGF in 146 patients with all subtypes of RCC, although increased compared with healthy controls, VEGF was not significantly correlated with stage, grade, or tumor size and was not associated with outcome [149]. However a similar sized study (n = 124) with a similar mix of patients found that higher serum VEGF concentrations were most likely associated with clear cell phenotype and on multivariable analysis, high VEGF was independently significantly associated with CSS and RFS [150].

A significant correlation between plasma VEGF and cytoplasmic VEGF staining has been reported, with cytoplasmic pattern being associated with higher plasma VEGF and high platelet count, compared with membranous staining pattern and lower platelet counts which were associated with lower plasma VEGF. No significant differences were seen in plasma VEGF levels between those who relapsed and those who didn't but VEGF expression in tissue correlated significantly with progression [151]. A later study analyzing plasma samples from 68 patients with clear cell RCC found significantly higher concentrations in patients with N1 or M1 disease and significantly higher concentrations in patients compared with controls. There was no association between VEGF and OS however [152]. Although OS wasn't analyzed, plasma VEGF in 102 patients with clear cell RCC (localized and metastatic), was increased in many patients compared with a healthy control range and was associated with T stage, grade, or ECOG PS but not N or M status. Tissue expression of VEGF was associated not only with T stage and grade but also with metastatic disease. Both tumor expression of VEGF and plasma VEGF were significantly associated with PFS and CSS. In addition although VEGF in tissue and plasma seemed to be lower with VHL genetic changes, patients with wildtype VHL had significantly reduced survival as did patients with altered VHL but high expression of VEGF [153].

Analysis of serum VEGF in 83 patients with nonmetastatic clear cell RCC found serum VEGF to be a significant independent predictor of recurrence (p = 0.013) [154]. The RFS was significantly lower in the cases which stained positively for VEGF or had higher serum concentrations although there was no significant association between VEGF staining and serum VEGF.

The largest study focusing on a subset of 712 RCC patients with metastatic disease in the TARGET study compared sorafenib with placebo in the second-line setting and in an extension of an earlier study exploring solely VEGF [155], found higher baseline plasma concentrations of VEGF ($n = 712$), CAIX ($n = 128$), and TIMP-1 ($n = 123$) in the placebo group to be associated with higher MSKCC score. VEGF and VHL mutation incidence also correlated with ECOG performance status. On univariate analysis utilizing only results from the patients in the placebo arm, VEGF ($n = 348$), CAIX ($n = 66$), and TIMP-1 ($n = 63$) were prognostic for OS [156]. The earlier study had reported baseline plasma VEGF to be independently prognostic for both OS and PFS in placebo-treated patients ($p = 0.042$ and 0.023, respectively) [155] but when only the subset of 59 patients with data available for all 5 plasma biomarkers were examined together with MSKCC score and ECOG PS, only TIMP-1 retained independent prognostic value ($p = 0.002$) [156].

It can be seen from the previous studies that the issue of whether to measure plasma or serum is very much undecided and this highlights an important aspect that is often overlooked. Given that VEGF is contained within platelets and this is released upon clotting, very different concentrations of VEGF are detected in the different matrices [157–160] and neutrophils may also contribute [161]. There is also an issue with sample processing where plasma can become contaminated with platelets and hence have falsely elevated VEGF concentrations and also where delays in processing anticoagulated samples can result in platelet release of VEGF and artefactual increases in plasma VEGF [157,160]. The impact of such effects will be highly variable and depend very much on platelet number and VEGF content of the platelets and it has been proposed that the majority of cancer patients do not have elevated levels of "free" VEGF circulating [160]. This is also supported by the finding of excellent correlations between platelet number and serum VEGF concentrations in patients with advanced cancer [162]. Clearly also given the prognostic value of platelet number/thrombocytosis, measurement of VEGF released from platelets may indirectly become "prognostic" as a reflection of platelet number. An additional issue is that VEGF has been reported to be relatively unstable once frozen and ideally should be measured immediately without freezing, which clearly is impractical, and certainly not subjected to freeze–thaw cycles [163]. This analysis was using serum samples and needs to be investigated further and confirmed as clearly this could have a major impact on the results of research studies and their interpretation.

Carbonic Anhydrase IX

CAIX is a hypoxia-inducible protein upregulated as a consequence of VHL inactivation in RCC and with inconsistent prognostic value in tissue samples as discussed earlier. Known to be shed from the cell surface, it has been detected in conditioned medium from RCC cell lines and primary cultures, and in the serum/urine of RCC patients. An initial small study using serum from patients with localized RCC (n = 30) found barely detectable amounts in healthy controls and widely varying values in patient samples (20 pg/mL to 3.6 ng/mL in serum) with no correlation with tumor size [164]. A subsequent larger study of 91 patients with RCC found mean serum CAIX levels to be significantly higher in patients with metastatic versus localized disease ($p = 0.004$) and versus healthy controls ($p = 0.001$). Higher serum levels were correlated with tumor grade, size, and stage and were associated with a higher rate of relapse in patients with localized disease [165]. We subsequently examined serum CAIX prenephrectomy in 216 patients with clear cell RCC including 24% with metastatic disease and found it to be independently prognostic for OS only although prognostic for CSS, DFS, and OS on univariate analysis [166]. The combination of CAIX, CRP, and osteopontin (OPN) outperformed stage however, and CAIX identified a group of stage I/II patients with poor outcome. In a larger study involving 361 RCC patients, serum CAIX was significantly correlated with stage and patients with CAIX concentrations higher than the median tended to have shorter survival but this did not reach significance [167]. Recently higher serum CAIX was shown to be significantly associated (on univariate analysis) with poorer OS in 70 patients with metastatic disease undergoing a variety of treatments but was not associated with other clinical factors including response to treatment [168]. Similar results were obtained for 66 patients with metastatic disease in the placebo arm of the TARGET study as discussed in the previous section [156].

Serum Amyloid A

The acute-phase protein serum amyloid A (SAA) has been reported to be elevated in serum samples from patients with RCC (n = 98) if metastatic disease is present. Independent prognostic value was shown for CSS (HR 2.51; 95% CI, 1.09–5.78; $p = 0.030$) [169]. We subsequently confirmed this with a cohort of 119 RCC patients for CSS with a similar HR, although when CRP (which strongly correlated with SAA) was included in the model, SAA was no longer independently significant [170]. SAA was not independently significant with regard to OS or DFS.

Matrix Metalloproteinase-7

Matrix metalloproteinase-7 (MMP-7) is a member of a family of zinc-containing enzymes capable of degrading various components of the extracellular matrix. At a tissue level, MMP-7 has been shown to be overexpressed in RCC, correlating with higher tumor stage and grade and independently predictive of poor CSS (HR 8.61; 95% CI, 1.10–67.28; $p = 0.04$) [171]. A later study involving 98 RCC patients found a similar relationship of MMP-7 with stage and grade and with MMP-7 having independent prognostic significance for OS [172]. Plasma levels in 97 RCC patients were subsequently evaluated using a commercial assay, detecting both pro- and mature forms of the protein. Levels were significantly elevated in patients with metastatic (but not localized) disease, although they did not correlate with burden or site of metastases. On multivariate analysis, levels were independently prognostic for CSS (HR 2.70; 95% CI, 1.39–5.24; $p = 0.003$) [173].

Osteopontin

OPN is a secreted protein belonging to the small integrin-binding ligand *N*-linked glycoprotein family. Originally discovered in the extracellular matrix of bone, it is expressed by many cell types and has a diverse array of functions including bone remodeling, immune and inflammatory responses, and cell adhesion. OPN is upregulated in many cancers, possibly through hypoxia and implicated in angiogenesis through the HIF/VEGF pathway [174]. Analysis of tissue OPN expression in clear cell RCC ($n = 171$) found a strong association with poor survival at univariate level [175]. Using plasma ($n = 75$) and serum ($n = 116$) samples, OPN concentrations were found to be of independent prognostic significance for patients with RCC with higher concentrations being associated with shorter CSS and plasma being more strongly prognostic [176]. Lower concentrations are measured in serum generally due to proteolytic cleavage during clotting and plasma is the recommended fluid. Interestingly, tumor stage and grade were not independently prognostic in this study. We subsequently analyzed plasma OPN in 216 patients with clear cell RCC where preoperative plasma OPN was found to be significantly prognostic for OS, CSS, and DFS but only at the univariate level [166]. This may have been due to the inclusion of CRP in our model which was closely correlated with OPN and in our study we only included clear cell RCC patients of whom 24% had metastatic disease whereas in the earlier study [176] other subtypes were included and almost 50% of the patients had metastatic disease. Interestingly in our study, high concentrations of OPN in patients with T1 stage disease identified a group of high-risk patients who mostly died from noncancer-related causes [166] suggesting a possible role in identifying patients who may not benefit from surgery for a small renal mass.

Immunosuppressive Acidic Protein

Immunosuppressive acidic protein was first described as a potential serum prognostic marker in RCC almost two decades ago. Since then there have been a number of studies supporting its role in both staging disease and determining prognosis in metastatic patients. In a study of 44 patients with recurrent mRCC, 3-year survival was 42% and 0% in patients below and above 800 μg/mL, respectively. Similar findings were reported among the 40 patients presenting with metastatic disease at diagnosis [177]. Subsequently the doubling time for this protein in patients with recurrent RCC after nephrectomy has been shown to be independently prognostic for survival ($p = 0.0026$), with concentrations being measured longitudinally before detection of recurrent disease ($n = 78$). The 3-year survival rate in patients with a doubling time of greater or less than 200 days was 58.9 and 12.5%, respectively [178].

Other Proteins

Several proteins show promise but require further independent validation. Cathepsin D is a lysosomal protease and increased tissue expression has been correlated with poorer outcome in a number of other cancers, except RCC, where the opposite relationship has been reported [179]. We identified cathepsin D as a candidate biomarker and found high urinary cathepsin D in RCC patients ($n = 239$) to be significantly associated with poorer OS (HR 1.33; 95%CI, 1.09–1.63; $p = 0.005$) on univariate analysis and approaching significance on multivariate analysis using preoperative variables ($p = 0.056$) [180]. Serum concentrations of tumor M2 pyruvate kinase (TuM2-PK) and thymidine kinase 1 (TK1) have been reported to be significantly elevated in patients with RCC ($n = 116$) with 5-year survival rates being significantly lower for patients with elevated concentrations relative to healthy controls (55% vs. 94%, $p < 0.001$ for TuM2-PK and 21% vs. 90%, $p = 0.002$ for TK1). Both were independent predictors of RFS [181]. A smaller ($n = 89$) recent study found significant correlations of plasma TuM2-PK with tumor grade and size but outcome wasn't examined [182].

Clearly, several circulating markers reflecting diverse aspects of RCC biology appear promising but need further systematic evaluation including whether a multiplex panel would be most effective and the optimal combination and whether these could add value to or outperform existing clinic-pathological scoring systems.

Predictive Biomarkers

With the shift toward "personalized or stratified medicine" and tailoring of specific treatments to individual patients, together with various other challenges such as an increase in late phase attrition, the drive within the pharmaceutical industry for companion biomarkers has intensified [183]. The ability to predict prior to or soon after starting treatment, the subgroup of patients destined to respond, carries obvious benefits. Non-responders can be considered for alternative, potentially more effective, therapies, avoiding unnecessary toxicity. Treatments that are often very expensive can be reserved for responders, carrying important economic implications. Indeed, it is likely that in future, novel anticancer agents will require predictive markers to have been developed in parallel, prior to their approval. Such markers may also provide novel insight into mechanisms of drug resistance, further highlighting their importance.

With the introduction of several efficacious yet expensive and toxic biological therapies (illustrated in Fig. 9.1), identifying predictive biomarkers in renal cancer treatment has never been so relevant. The treating oncologist now has a number of therapeutic options for patients with advanced disease and, therefore, defining the optimal sequence of drugs for individual patients is important. Disappointingly, despite intense efforts, few, if any, markers have emerged that are sufficiently robust to be useful in the clinic but several have shown potential utility and warrant further prospective validation in the trial setting [184]. Among the many caveats in interpreting biomarker studies, particular attention must be given to studies examining survival in patients on therapy, as some markers, which may appear to be predictive, may simply be prognostic and their apparent success related to that.

VEGFR-Related TKIs

This class of drug remains the most widely used in patients with mRCC. At a clinical level, it is apparent that non-clear cell RCCs respond less well to TKIs than their clear cell counterparts [185]. In the first-line setting, approximately 80% of patients with clear cell RCC benefit. These agents function through a well-described and common mechanism, targeting the

VEGF receptor, yet the identification of a robust surrogate molecular marker of activity has, to date, proven elusive. This may, in part, reflect the complex relationship between the tumor and the stroma that is likely to govern response and resistance to antiangiogenic drugs [186].

A number of treatment-induced toxicities have emerged as independent biomarkers of response, including hypertension, neutropenia, hand-foot syndrome, and hypothyroidism [187–190]. These data are intriguing and suggest toxicity may serve as a surrogate for increased exposure to TKIs, which in turn may be associated with improved outcomes. In a randomized first-line study of axitinib, mRCC patients randomized to titration of drug (vs. placebo titration) until development of hypertension (or other dose-limiting toxicity), achieved a higher objective response rate (54% vs. 34%) [191]. However, this was not accompanied by an improvement in PFS and, given the associated increased toxicity, it is not clear that this strategy can be routinely delivered in the clinic.

Beyond this, many types of markers have been examined, including at a genetic (germline SNPs and somatic mutations), protein (circulating/tissue based), and radiological (functional imaging) level. Studies examining germline SNPs in genes related to drug metabolism (e.g., CYP3A 4/5) and angiogenesis (e.g., IL-8/FGF2/VEGFR-3) have reported variable results, with a current lack of consistency in their findings [192–196]. VHL mutation status, which might be expected to be relevant to antiangiogenic drug activity, has failed to show predictive value and VEGF may have utility but results are variable with further studies needed [186,197–200]. In many cases numbers are low and study designs are variable which may be contributing factors. In an early pharmacodynamics study involving sorafenib as part of a Phase II trial with 63 patients, plasma VEGF and placenta growth factor (PlGF) levels increased relative to baseline, while soluble VEGFR-2 (sVEGFR-2) and VEGFR-3 (sVEGFR-3) concentrations decreased in response to treatment. Levels of each marker tended to return to near-baseline levels 2 weeks after stopping treatment, suggesting a drug-dependent effect. Baseline levels did not predict for response, although the magnitude of change in levels of VEGF, sVEGFR-2, and sVEGFR-3 was larger in the 25 patients who achieved a radiological PR [198]. An increase in circulating VEGF-A and PlGF and decrease in sVEGFR-3 and VEGF-C has also been demonstrated in patients treated with sunitinib. The study, in 61 patients with bevacizumab-refractory mRCC, also reported that patients with lower baseline levels of sVEGFR-3 and VEGF-C had a higher response rate to sunitinib and a longer PFS [199].

As previously mentioned in the context of prognosis, within the TARGET study comparing sorafenib with placebo, higher baseline plasma VEGF was associated with a worse OS and was independently prognostic [155]. Stratification of patients by median VEGF level (131 pg/mL), demonstrated that both groups in fact benefited equally to sorafenib, in terms of PFS. However, in an exploratory analysis using the 25th and 75th percentiles, patients in the high-VEGF group demonstrated a trend toward higher response [155] and VEGF increased significantly during treatment with sorafenib but not placebo, showing a reciprocal change with sVEGFR-2 as previously reported [156].

Numerous studies have retrospectively examined circulating levels of various cytokines (alone or in combination) as well as routine hematological and biochemical parameters, and circulating proteins including CRP [201], NLR [202], TNF-alpha [203], MMP-9 [203], OPN [204], CAIX [204], and *N*-terminal precursor of brain natriuretic peptide [205]. These studies all report associations with outcome in TKI-treated mRCC patients, although their retrospective nature, involving small numbers of patients, often still as single studies and with no non-TKI treated group, raises questions about the likelihood of subsequent confirmation. A notable exception (although still retrospective in nature) comes from a study examining cytokines and angiogenesis factors (CAFs) at baseline from patients treated with pazopanib within two clinical trials. CAFs were initially assayed in patient samples within a Phase II trial (n = 215), using two independent platforms, before validation in a second, placebo-controlled, Phase III trial (n = 344). Among the 17 CAFs examined, high interleukin-6 was identified as predictive of improved relative PFS benefit from pazopanib compared to placebo [206]. Both groups (high vs. low), however, derived a benefit, making the result difficult to translate to the clinic.

mTOR Inhibitors

The place of the mTOR inhibitors in RCC therapy is likely to change significantly in the near future. Two randomized Phase III trials, involving nivolumab [28] and cabozantinib [207], a PD-1 antibody and c-MET inhibitor, respectively, have very recently shown superiority to everolimus in the second-line setting. In the United Kingdom, everolimus is currently reserved for second-line use for patients in whom axitinib is contraindicated or not tolerated. Although overall response rates are low (around 5%) with a median time to progression of 4–5 months, there are undoubtedly a group of patients who are recognized to do well when treated with this class of drug. Identification of a robust predictive marker is essential if they are to remain central to management.

Serum LDH has been correlated with OS benefit in patients treated with temsirolimus versus interferon alpha, using samples derived from a randomized Phase III trial ($n = 404$) [208]. Patients with an LDH > 1 × upper limit of normal had a significantly improved OS (6.9 vs. 4.2 months; $p < 0.002$). In comparison, patients with an LDH < 1 × upper limit of normal derived no additional benefit in comparison to IFN-alpha (11.7 vs. 10.4 months; $p = 0.51$) [209]. It is apparent that LDH is both prognostic as well as predictive in this setting. Whether these data equally apply to everolimus-treated patients remain unclear.

Interleukin-2

IL-2, when administered at high dose intravenously, can induce complete responses in small numbers of patients that are durable. However, such treatment is associated with significant morbidity and even patient death. Careful patient selection is therefore important. Pathological criteria predicting response have been best defined by Upton and coworkers in a study examining 231 RCC patient tumors treated with IL-2. Clear cell RCC patients with more than 50% alveolar features and no granular or papillary features had a 39% response rate (good-risk) compared to just 3% among patients with >50% granular or any papillary features (poor-risk). Response rates were equally poor in patients with nonclear cell histology [210]. A model based on these pathological criteria plus CAIX tissue expression was subsequently described on a relatively small number of patients ($n = 66$) with mRCC, stratifying patients into a good- and poor-risk group, and reporting highly promising results [211]. Disappointingly, however, a prospective trial (SELECT) failed to validate the model, highlighting the importance of rigorous biomarker validation in well-designed prospective studies [212].

CONCLUSIONS

Cancer biomarkers have revolutionized our approach to patient care in many cancer types. They have major potential benefits for patients, particularly in contributing to "personalized" medicine and improved biomarkers should ultimately lead to improvements in outcomes and more efficient, safe, cost-effective, and evidence-based use of health resources. In RCC, no such markers are in routine clinical use. Much interest is focused on identifying prognostic markers, which can identify high-risk patients that should be targeted for trials of adjuvant therapy or to guide surveillance. Ultimately, large numbers of patients are likely to receive some form of treatment, in either the early and/or metastatic setting. Thus, biomarkers that predict response to these therapies are of equal importance and are also urgently required. The role of neoadjuvant therapy is also now beginning to be explored and markers that determine outcome preoperatively will be required to guide management decisions in this setting.

Currently employed nomograms are limited by their reliance on standard clinicopathologic criteria. The development of accurate prognostic and/or predictive models that are universally applicable will require a concerted effort from the international community. In 2004, an International Kidney Cancer Working Group was established to identify independent, validated predictors of survival, by collecting data on >4000 previously untreated RCC patients. Clearly several potential markers have been identified and more are likely to emerge from the proteomic and genomic initiatives. However, key to their successful exploitation and translation will be the establishment of the necessary infrastructure including high quality annotated sample banks, assays development, and design of multicenter trials to evaluate them with evidence-based progression through this pipeline to the clinic.

ACKNOWLEDGMENTS

This chapter presents an independent review undertaken by the authors within the activities of the NIHR under its Programme Grants for Applied Research Programme (Grant Reference Number RP-PG-0707-10101) (www.biomarkerpipeline.org). The views expressed are those of the authors and not necessarily those of the NHS, the NIHR, or the Department of Health.

REFERENCES

[1] Jonasch E, Gao J, Rathmell WK. Renal cell carcinoma. BMJ 2014;349:g4797.
[2] Davis CF, Ricketts CJ, Wang M, et al. The somatic genomic landscape of chromophobe renal cell carcinoma. Cancer Cell 2014;26(3):319–30.
[3] Slater AA, Alokail M, Gentle D, et al. DNA methylation profiling distinguishes histological subtypes of renal cell carcinoma. Epigenetics 2013;8(3):252–67.
[4] Buttner F, Winter S, Rausch S, et al. Survival prediction of clear cell renal cell carcinoma based on gene expression similarity to the proximal tubule of the nephron. Eur Urol 2015;68:1016–20.
[5] Shuch B, Amin A, Armstrong AJ, et al. Understanding pathologic variants of renal cell carcinoma: distilling therapeutic opportunities from biologic complexity. Eur Urol 2015;67(1):85–97.
[6] Fuhrman SA, Lasky LC, Limas C. Prognostic significance of morphologic parameters in renal cell carcinoma. Am J Surg Pathol 1982;6(7):655–63.
[7] Latif F, Tory K, Gnarra J, et al. Identification of the von Hippel–Lindau disease tumor suppressor gene. Science 1993;260(5112):1317–20.

[8] Scelo G, Riazalhosseini Y, Greger L, et al. Variation in genomic landscape of clear cell renal cell carcinoma across Europe. Nat Commun 2014;5:5135.
[9] Young AC, Craven RA, Cohen D, et al. Analysis of VHL gene alterations and their relationship to clinical parameters in sporadic conventional renal cell carcinoma. Clin Cancer Res 2009;15(24):7582–92.
[10] Nickerson ML, Jaeger E, Shi Y, et al. Improved identification of von Hippel–Lindau gene alterations in clear cell renal tumors. Clin Cancer Res 2008;14(15):4726–34.
[11] Frew IJ, Moch H. A clearer view of the molecular complexity of clear cell renal cell carcinoma. Annu Rev Pathol 2015;10:263–89.
[12] Varela I, Tarpey P, Raine K, et al. Exome sequencing identifies frequent mutation of the SWI/SNF complex gene PBRM1 in renal carcinoma. Nature 2011;469(7331):539–42.
[13] Dalgliesh GL, Furge K, Greenman C, et al. Systematic sequencing of renal carcinoma reveals inactivation of histone modifying genes. Nature 2010;463(7279):360–3.
[14] Durinck S, Stawiski EW, Pavia-Jimenez A, et al. Spectrum of diverse genomic alterations define non-clear cell renal carcinoma subtypes. Nat Genet 2015;47(1):13–21.
[15] Kovac M, Navas C, Horswell S, et al. Recurrent chromosomal gains and heterogeneous driver mutations characterise papillary renal cancer evolution. Nat Commun 2015;6:6336.
[16] Albiges L, Guegan J, Le Formal A, et al. MET is a potential target across all papillary renal cell carcinomas: result from a large molecular study of pRCC with CGH array and matching gene expression array. Clin Cancer Res 2014;20(13):3411–21.
[17] Marsaud A, Dadone B, Ambrosetti D, et al. Dismantling papillary renal cell carcinoma classification: the heterogeneity of genetic profiles suggests several independent diseases. Genes Chromosomes Cancer 2015;54(6):369–82.
[18] Kapur P, Pena-Llopis S, Christie A, et al. Effects on survival of BAP1 and PBRM1 mutations in sporadic clear-cell renal-cell carcinoma: a retrospective analysis with independent validation. Lancet Oncol 2013;14(2):159–67.
[19] Laguna MP, Algaba F, Cadeddu J, et al. Current patterns of presentation and treatment of renal masses: a clinical research office of the endourological society prospective study. J Endourol 2014;28(7):861–70.
[20] Richard PO, Jewett MA, Bhatt JR, et al. Renal tumor biopsy for small renal masses: a single-center 13-year experience. Eur Urol 2015;68:1007–13.
[21] Schachter LR, Cookson MS, Chang SS, et al. Second prize: frequency of benign renal cortical tumors and histologic subtypes based on size in a contemporary series: what to tell our patients. J Endourol 2007;21(8):819–23.
[22] Motzer RJ, Hutson TE, Tomczak P, et al. Overall survival and updated results for sunitinib compared with interferon alfa in patients with metastatic renal cell carcinoma. J Clin Oncol 2009;27(22):3584–90.
[23] Motzer RJ, Hutson TE, Tomczak P, et al. Sunitinib versus interferon alfa in metastatic renal-cell carcinoma. N Engl J Med 2007;356(2):115–24.
[24] Sternberg CN, Davis ID, Mardiak J, et al. Pazopanib in locally advanced or metastatic renal cell carcinoma: results of a randomized phase III trial. J Clin Oncol 2010;28(6):1061–8.
[25] Motzer RJ, McCann L, Deen K. Pazopanib versus sunitinib in renal cancer. N Engl J Med 2013;369(20):1970.
[26] Hudes G, Carducci M, Tomczak P, et al. Temsirolimus, interferon alfa, or both for advanced renal-cell carcinoma. N Engl J Med 2007;356(22):2271–81.
[27] Motzer RJ, Escudier B, Oudard S, et al. Efficacy of everolimus in advanced renal cell carcinoma: a double-blind, randomised, placebo-controlled phase III trial. Lancet 2008;372(9637):449–56.
[28] Motzer RJ, Escudier B, McDermott DF, et al. Nivolumab versus everolimus in advanced renal-cell carcinoma. N Engl J Med 2015;373:1803–13.

[29] Haas NB, Manola J, Uzzo RG, et al. Initial results from ASSURE (E2805): adjuvant sorafenib or sunitinib for unfavourable renal carcinoma, an ECOG-ACRIN-led, NCTN phase III trial. 2015 Genitourinary Cancers Symposium; p. 2915; 2015.
[30] Kim EH, Strope SA. Postoperative surveillance imaging for patients undergoing nephrectomy for renal cell carcinoma. Urol Oncol 2015;33:499–502.
[31] Sobin L, Gospodarowicz M, Wittekind C. TNM classification of malignant tumours. New York, NY: Wiley; 2009.
[32] Edge SB, Byrd DR, Compton CC, Fritz AG, Greene FL, Trotti A. AJCC cancer staging manual. 7th ed. New York, NY: Springer; 2010.
[33] Crispen PL, Boorjian SA, Lohse CM, Leibovich BC, Kwon ED. Predicting disease progression after nephrectomy for localized renal cell carcinoma: the utility of prognostic models and molecular biomarkers. Cancer 2008;113(3):450–60.
[34] Al-Aynati M, Chen V, Salama S, Shuhaibar H, Treleaven D, Vincic L. Interobserver and intraobserver variability using the Fuhrman grading system for renal cell carcinoma. Arch Pathol Lab Med 2003;127(5):593–6.
[35] Delahunt B. Advances and controversies in grading and staging of renal cell carcinoma. Mod Pathol 2009;22:S24–36.
[36] Leibovich BC, Blute ML, Cheville JC, et al. Prediction of progression after radical nephrectomy for patients with clear cell renal cell carcinoma: a stratification tool for prospective clinical trials. Cancer 2003;97(7):1663–71.
[37] Heng DY, Xie W, Regan MM, et al. Prognostic factors for overall survival in patients with metastatic renal cell carcinoma treated with vascular endothelial growth factor-targeted agents: results from a large, multicenter study. J Clin Oncol 2009;27(34):5794–9.
[38] Heng DY, Xie W, Regan MM, et al. External validation and comparison with other models of the International Metastatic Renal-Cell Carcinoma Database Consortium prognostic model: a population-based study. Lancet Oncol 2013;14(2):141–8.
[39] Karakiewicz PI, Suardi N, Capitanio U, et al. A preoperative prognostic model for patients treated with nephrectomy for renal cell carcinoma. Eur Urol 2009;55(2):287–95.
[40] Pavlou MP, Diamandis EP, Blasutig IM. The long journey of cancer biomarkers from the bench to the clinic. Clin Chem 2013;59(1):147–57.
[41] Fuzery AK, Levin J, Chan MM, Chan DW. Translation of proteomic biomarkers into FDA approved cancer diagnostics: issues and challenges. Clin Proteomics 2013;10(1):13.
[42] Diamandis EP. Cancer biomarkers: can we turn recent failures into success? J Natl Cancer Inst 2010;102(19):1462–7.
[43] Heegaard NH, Ostergaard O, Bahl JM, et al. Important options available—from start to finish—for translating proteomics results to clinical chemistry. Proteomics Clin Appl 2015;9(1–2):235–52.
[44] McShane LM, Altman DG, Sauerbrei W, et al. REporting recommendations for tumour MARKer prognostic studies (REMARK). Eur J Cancer 2005;41(12):1690–6.
[45] Bossuyt PM, Reitsma JB, Bruns DE, et al. Toward complete and accurate reporting of studies of diagnostic accuracy. The STARD initiative. Am J Clin Pathol 2003;119(1):18–22.
[46] Barker AD, Compton CC, Poste G. The National Biomarker Development Alliance accelerating the translation of biomarkers to the clinic. Biomark Med 2014;8(6):873–6.
[47] Poste G, Compton CC, Barker AD. The national biomarker development alliance: confronting the poor productivity of biomarker research and development. Expert Rev Mol Diagn 2015;15(2):211–8.
[48] Vasudev NS, Selby PJ, Banks RE. Renal cancer biomarkers: the promise of personalized care. BMC Med 2012;10:112.

[49] Atkins MB, Bukowski RM, Escudier BJ, et al. Innovations and challenges in renal cancer: summary statement from the Third Cambridge Conference. Cancer 2009;115 (10 Suppl.):2247–51.
[50] Oosterwijk E, Rathmell WK, Junker K, et al. Basic research in kidney cancer. Eur Urol 2011;60(4):622–33.
[51] Brooks SA, Brannon AR, Parker JS, et al. ClearCode34: a prognostic risk predictor for localized clear cell renal cell carcinoma. Eur Urol 2014;66(1):77–84.
[52] Gu L, Li H, Chen L, et al. MicroRNAs as prognostic molecular signatures in renal cell carcinoma: a systematic review and meta-analysis. Oncotarget 2015;6(32):32545–60.
[53] Valera VA, Merino MJ. Misdiagnosis of clear cell renal cell carcinoma. Nat Rev Urol 2011;8(6):321–33.
[54] Reuter VE, Tickoo SK. Differential diagnosis of renal tumours with clear cell histology. Pathology 2010;42(4):374–83.
[55] Kummerlin I, ten Kate F, Smedts F, et al. Diagnostic problems in the subtyping of renal tumors encountered by five pathologists. Pathol Res Pract 2009;205(1):27–34.
[56] Skates S, Iliopoulos O. Molecular markers for early detection of renal carcinoma: investigative approach. Clin Cancer Res 2004;10(18 Pt. 2):6296S–301S.
[57] Banks RE, Craven RA, Harnden P, Madaan S, Joyce A, Selby PJ. Key clinical issues in renal cancer: a challenge for proteomics. World J Urol 2007;25(6):537–56.
[58] Gbormittah FO, Lee LY, Taylor K, Hancock WS, Iliopoulos O. Comparative studies of the proteome, glycoproteome, and *N*-glycome of clear cell renal cell carcinoma plasma before and after curative nephrectomy. J Proteome Res 2014;13(11):4889–900.
[59] Kim DS, Choi YP, Kang S, et al. Panel of candidate biomarkers for renal cell carcinoma. J Proteome Res 2010;9(7):3710–9.
[60] Su Kim D, Choi YD, Moon M, et al. Composite three-marker assay for early detection of kidney cancer. Cancer Epidemiol Biomarkers Prev 2013;22(3):390–8.
[61] Thongboonkerd V. Urinary proteomics: towards biomarker discovery, diagnostics and prognostics. Mol Biosyst 2008;4(8):810–5.
[62] Morrissey JJ, London AN, Luo J, Kharasch ED. Urinary biomarkers for the early diagnosis of kidney cancer. Mayo Clin Proc 2010;85(5):413–21.
[63] Morrissey JJ, Kharasch ED. The specificity of urinary aquaporin 1 and perilipin 2 to screen for renal cell carcinoma. J Urol 2013;189(5):1913–20.
[64] Morrissey JJ, Mobley J, Song J, et al. Urinary concentrations of aquaporin-1 and perilipin-2 in patients with renal cell carcinoma correlate with tumor size and stage but not grade. Urology 2014;83(1):256.e9–256.e14.
[65] Morrissey JJ, Mobley J, Figenshau RS, Vetter J, Bhayani S, Kharasch ED. Urine aquaporin 1 and perilipin 2 differentiate renal carcinomas from other imaged renal masses and bladder and prostate cancer. Mayo Clin Proc 2015;90(1):35–42.
[66] Morrissey JJ, Mellnick VM, Luo J, et al. Evaluation of urine aquaporin-1 and perilipin-2 concentrations as biomarkers to screen for renal cell carcinoma: a prospective cohort study. JAMA Oncol 2015;1(2):204–12.
[67] Sreedharan S, Petros JA, Master VA, et al. Aquaporin-1 protein levels elevated in fresh urine of renal cell carcinoma patients: potential use for screening and classification of incidental renal lesions. Dis Markers 2014;2014:135649.
[68] Huang S, Rhee E, Patel H, Park E, Kaswick J. Urinary NMP22 and renal cell carcinoma. Urology 2000;55(2):227–30.
[69] Kaya K, Ayan S, Gokce G, Kilicarslan H, Yildiz E, Gultekin EY. Urinary nuclear matrix protein 22 for diagnosis of renal cell carcinoma. Scand J Urol Nephrol 2005;39(1):25–9.
[70] Han WK, Alinani A, Wu CL, et al. Human kidney injury molecule-1 is a tissue and urinary tumor marker of renal cell carcinoma. J Am Soc Nephrol 2005;16(4):1126–34.

[71] Morrissey JJ, London AN, Lambert MC, Kharasch ED. Sensitivity and specificity of urinary neutrophil gelatinase-associated lipocalin and kidney injury molecule-1 for the diagnosis of renal cell carcinoma. Am J Nephrol 2011;34(5):391–8.
[72] Zhang PL, Mashni JW, Sabbisetti VS, et al. Urine kidney injury molecule-1: a potential non-invasive biomarker for patients with renal cell carcinoma. Int Urol Nephrol 2014;46(2):379–88.
[73] Frantzi M, Metzger J, Banks RE, et al. Discovery and validation of urinary biomarkers for detection of renal cell carcinoma. J Proteomics 2014;98:44–58.
[74] Shuch B, Hofmann JN, Merino MJ, et al. Pathologic validation of renal cell carcinoma histology in the Surveillance, Epidemiology, and End Results Program. Urol Oncol 2014;32(1):23.e9–23.e13.
[75] Delahunt B, Srigley JR, Montironi R, Egevad L. Advances in renal neoplasia: recommendations from the 2012 International Society of Urological Pathology Consensus Conference. Urology 2014;83(5):969–74.
[76] Srigley JR, Delahunt B, Eble JN, et al. The International Society of Urological Pathology (ISUP) Vancouver classification of renal neoplasia. Am J Surg Pathol 2013;37(10):1469–89.
[77] Yap NY, Rajandram R, Ng KL, Pailoor J, Fadzli A, Gobe GC. Genetic and chromosomal aberrations and their clinical significance in renal neoplasms. Biomed Res Int 2015;2015:476508.
[78] Moch H, Srigley J, Delahunt B, Montironi R, Egevad L, Tan PH. Biomarkers in renal cancer. Virchows Arch 2014;464(3):359–65.
[79] Kuroda N, Tanaka A, Ohe C, Nagashima Y. Recent advances of immunohistochemistry for diagnosis of renal tumors. Pathol Int 2013;63(8):381–90.
[80] Thompson RH, Gillett MD, Cheville JC, et al. Costimulatory B7-H1 in renal cell carcinoma patients: indicator of tumor aggressiveness and potential therapeutic target. Proc Natl Acad Sci USA 2004;101(49):17174–9.
[81] Thompson RH, Kuntz SM, Leibovich BC, et al. Tumor B7-H1 is associated with poor prognosis in renal cell carcinoma patients with long-term follow-up. Cancer Res 2006;66(7):3381–5.
[82] Jilaveanu LB, Shuch B, Zito CR, et al. PD-L1 expression in clear cell renal cell carcinoma: an analysis of nephrectomy and sites of metastases. J Cancer 2014;5(3):166–72.
[83] Krambeck AE, Thompson RH, Dong H, et al. B7-H4 expression in renal cell carcinoma and tumor vasculature: associations with cancer progression and survival. Proc Natl Acad Sci USA 2006;103(27):10391–6.
[84] Jung SG, Choi KU, Lee SD, Lee ZZ, Chung MK. The relationship between B7-H4 expression and clinicopathological characteristics in clinical stage T1 conventional renal cell carcinoma. Korean J Urol 2011;52(2):90–5.
[85] Crispen PL, Sheinin Y, Roth TJ, et al. Tumor cell and tumor vasculature expression of B7-H3 predict survival in clear cell renal cell carcinoma. Clin Cancer Res 2008;14(16):5150–7.
[86] Frigola X, Inman BA, Lohse CM, et al. Identification of a soluble form of B7-H1 that retains immunosuppressive activity and is associated with aggressive renal cell carcinoma. Clin Cancer Res 2011;17(7):1915–23.
[87] Jiang Z, Chu PG, Woda BA, et al. Analysis of RNA-binding protein IMP3 to predict metastasis and prognosis of renal-cell carcinoma: a retrospective study. Lancet Oncol 2006;7(7):556–64.
[88] Hoffmann NE, Sheinin Y, Lohse CM, et al. External validation of IMP3 expression as an independent prognostic marker for metastatic progression and death for patients with clear cell renal cell carcinoma. Cancer 2008;112(7):1471–9.

[89] Jiang Z, Chu PG, Woda BA, et al. Combination of quantitative IMP3 and tumor stage: a new system to predict metastasis for patients with localized renal cell carcinomas. Clin Cancer Res 2008;14(17):5579–84.
[90] Pei X, Li M, Zhan J, et al. Enhanced IMP3 expression activates NF-κB pathway and promotes renal cell carcinoma progression. PLoS One 2015;10(4). e0124338.
[91] Jiang Z, Lohse CM, Chu PG, et al. Oncofetal protein IMP3: a novel molecular marker that predicts metastasis of papillary and chromophobe renal cell carcinomas. Cancer 2008;112(12):2676–82.
[92] Oosterwijk E, Ruiter DJ, Hoedemaeker PJ, et al. Monoclonal antibody G 250 recognizes a determinant present in renal-cell carcinoma and absent from normal kidney. Int J Cancer 1986;38(4):489–94.
[93] Grabmaier K, Vissers JL, De Weijert MC, et al. Molecular cloning and immunogenicity of renal cell carcinoma-associated antigen G250. Int J Cancer 2000;85(6):865–70.
[94] Ivanov S, Liao SY, Ivanova A, et al. Expression of hypoxia-inducible cell-surface transmembrane carbonic anhydrases in human cancer. Am J Pathol 2001;158(3):905–19.
[95] Bui MH, Seligson D, Han KR, et al. Carbonic anhydrase IX is an independent predictor of survival in advanced renal clear cell carcinoma: implications for prognosis and therapy. Clin Cancer Res 2003;9(2):802–11.
[96] Bui MH, Seligson DB, Kim HL, et al. Prognostic value of carbonic anhydrase IX and Ki67 as predictors of survival for renal clear cell carcinoma. J Urol 2004;171(4):200.
[97] Leibovich BC, Sheinin Y, Lohse CM, et al. Carbonic anhydrase IX is not an independent predictor of outcome for patients with clear cell renal cell carcinoma. J Clin Oncol 2007;25(30):4757–64.
[98] Zhang BY, Thompson RH, Lohse CM, et al. Carbonic anhydrase IX (CAIX) is not an independent predictor of outcome in patients with clear cell renal cell carcinoma (ccRCC) after long-term follow-up. BJU Int 2013;111(7):1046–53.
[99] Choueiri TK, Cheng S, Qu AQ, Pastorek J, Atkins MB, Signoretti S. Carbonic anhydrase IX as a potential biomarker of efficacy in metastatic clear-cell renal cell carcinoma patients receiving sorafenib or placebo: analysis from the treatment approaches in renal cancer global evaluation trial (TARGET). Urol Oncol 2013;31(8):1788–93.
[100] Patard JJ, Fergelot P, Karakiewicz PI, et al. Low CAIX expression and absence of VHL gene mutation are associated with tumor aggressiveness and poor survival of clear cell renal cell carcinoma. Int J Cancer 2008;123(2):395–400.
[101] Tollefson MK, Thompson RH, Sheinin Y, et al. Ki-67 and coagulative tumor necrosis are independent predictors of poor outcome for patients with clear cell renal cell carcinoma and not surrogates for each other. Cancer 2007;110(4):783–90.
[102] Abel EJ, Bauman TM, Weiker M, et al. Analysis and validation of tissue biomarkers for renal cell carcinoma using automated high-throughput evaluation of protein expression. Hum Pathol 2014;45(5):1092–9.
[103] Parker AS, Kosari F, Lohse CM, et al. High expression levels of survivin protein independently predict a poor outcome for patients who undergo surgery for clear cell renal cell carcinoma. Cancer 2006;107(1):37–45.
[104] Byun SS, Yeo WG, Lee SE, Lee E. Expression of survivin in renal cell carcinomas: association with pathologic features and clinical outcome. Urology 2007;69(1):34–7.
[105] Lei Y, Geng Z, Guo-Jun W, He W, Jian-Lin Y. Prognostic significance of survivin expression in renal cell cancer and its correlation with radioresistance. Mol Cell Biochem 2010;344(1–2):23–31.
[106] Baytekin F, Tuna B, Mungan U, Aslan G, Yorukoglu K. Significance of P-glycoprotein, p53, and survivin expression in renal cell carcinoma. Urol Oncol 2011;29(5):502–7.
[107] Krambeck AE, Dong H, Thompson RH, et al. Survivin and B7-H1 are collaborative predictors of survival and represent potential therapeutic targets for patients with renal cell carcinoma. Clin Cancer Res 2007;13(6):1749–56.

[108] Parker AS, Leibovich BC, Lohse CM, et al. Development and evaluation of BioScore. Cancer 2009;115(10):2092–103.
[109] Kim HL, Seligson D, Liu XL, et al. Using tumor markers to predict the survival of patients with metastatic renal cell carcinoma. J Urol 2005;173(5):1496–501.
[110] Klatte T, Seligson DB, LaRochelle J, et al. Molecular signatures of localized clear cell renal cell carcinoma to predict disease-free survival after nephrectomy. Cancer Epidemiol Biomarkers Prev 2009;18(3):894–900.
[111] Eckel-Passow JE, Lohse CM, Sheinin Y, Crispen PL, Krco CJ, Kwon ED. Tissue microarrays: one size does not fit all. Diagn Pathol 2010;5:48.
[112] Biswas S, Charlesworth PJ, Turner GD, et al. CD31 angiogenesis and combined expression of HIF-1alpha and HIF-2alpha are prognostic in primary clear-cell renal cell carcinoma (CC-RCC), but HIFalpha transcriptional products are not: implications for antiangiogenic trials and HIFalpha biomarker studies in primary CC-RCC. Carcinogenesis 2012;33(9):1717–25.
[113] Rini B, Goddard A, Knezevic D, et al. A 16-gene assay to predict recurrence after surgery in localised renal cell carcinoma: development and validation studies. Lancet Oncol 2015;16(6):676–85.
[114] Donskov F. Immunomonitoring and prognostic relevance of neutrophils in clinical trials. Semin Cancer Biol 2013;23(3):200–7.
[115] Jagdev SP, Gregory W, Vasudev NS, et al. Improving the accuracy of pre-operative survival prediction in renal cell carcinoma with C-reactive protein. Br J Cancer 2010;103(11):1649–56.
[116] Ohno Y, Nakashima J, Ohori M, Hatano T, Tachibana M. Pretreatment neutrophil-to-lymphocyte ratio as an independent predictor of recurrence in patients with non-metastatic renal cell carcinoma. J Urol 2010;184(3):873–8.
[117] de Martino M, Pantuck AJ, Hofbauer S, et al. Prognostic impact of preoperative neutrophil-to-lymphocyte ratio in localized nonclear cell renal cell carcinoma. J Urol 2013;190(6):1999–2004.
[118] Pichler M, Hutterer GC, Stoeckigt C, et al. Validation of the pre-treatment neutrophil-lymphocyte ratio as a prognostic factor in a large European cohort of renal cell carcinoma patients. Br J Cancer 2013;108(4):901–7.
[119] Brookman-May S, May M, Ficarra V, et al. Does preoperative platelet count and thrombocytosis play a prognostic role in patients undergoing nephrectomy for renal cell carcinoma? Results of a comprehensive retrospective series. World J Urol 2013;31(5):1309–16.
[120] Bensalah K, Leray E, Fergelot P, et al. Prognostic value of thrombocytosis in renal cell carcinoma. J Urol 2006;175(3):859–63.
[121] Suppiah R, Shaheen PE, Elson P, et al. Thrombocytosis as a prognostic factor for survival in patients with metastatic renal cell carcinoma. Cancer 2006;107(8):1793–800.
[122] Karakiewicz PI, Trinh QD, Lam JS, et al. Platelet count and preoperative haemoglobin do not significantly increase the performance of established predictors of renal cell carcinoma-specific mortality. Eur Urol 2007;52(5):1428–37.
[123] Erdem S, Amasyali AS, Aytac O, Onem K, Issever H, Sanli O. Increased preoperative levels of plasma fibrinogen and D dimer in patients with renal cell carcinoma is associated with poor survival and adverse tumor characteristics. Urol Oncol 2014;32(7):1031–40.
[124] Pichler M, Hutterer GC, Stojakovic T, Mannweiler S, Pummer K, Zigeuner R. High plasma fibrinogen level represents an independent negative prognostic factor regarding cancer-specific, metastasis-free, as well as overall survival in a European cohort of non-metastatic renal cell carcinoma patients. Br J Cancer 2013;109(5):1123–9.
[125] Du J, Zheng JH, Chen XS, et al. High preoperative plasma fibrinogen is an independent predictor of distant metastasis and poor prognosis in renal cell carcinoma. Int J Clin Oncol 2013;18(3):517–23.

[126] Xiao B, Ma LL, Zhang SD, et al. Correlation between coagulation function, tumor stage and metastasis in patients with renal cell carcinoma: a retrospective study. Chin Med J 2011;124(8):1205–8.
[127] Motzer RJ, Bacik J, Murphy BA, Russo P, Mazumdar M. Interferon-alfa as a comparative treatment for clinical trials of new therapies against advanced renal cell carcinoma. J Clin Oncol 2002;20(1):289–96.
[128] Manola J, Royston P, Elson P, et al. Prognostic model for survival in patients with metastatic renal cell carcinoma: results from the international kidney cancer working group. Clin Cancer Res 2011;17(16):5443–50.
[129] Magera JS Jr, Leibovich BC, Lohse CM, et al. Association of abnormal preoperative laboratory values with survival after radical nephrectomy for clinically confined clear cell renal cell carcinoma. Urology 2008;71(2):278–82.
[130] Yao M, Murakami T, Shioi K, et al. Tumor signatures of PTHLH overexpression, high serum calcium, and poor prognosis were observed exclusively in clear cell but not non clear cell renal carcinomas. Cancer Med 2014;3(4):845–54.
[131] Vasudev NS, Brown JE, Brown SR, et al. Prognostic factors in renal cell carcinoma: association of preoperative sodium concentration with survival. Clin Cancer Res 2008;14(6):1775–81.
[132] Jeppesen AN, Jensen HK, Donskov F, Marcussen N, von der Maase H. Hyponatremia as a prognostic and predictive factor in metastatic renal cell carcinoma. Br J Cancer 2010;102(5):867–72.
[133] Furukawa J, Miyake H, Kusuda Y, Fujisawa M. Hyponatremia as a powerful prognostic predictor for Japanese patients with clear cell renal cell carcinoma treated with a tyrosine kinase inhibitor. Int J Clin Oncol 2015;20(2):351–7.
[134] Schutz FA, Xie W, Donskov F, et al. The impact of low serum sodium on treatment outcome of targeted therapy in metastatic renal cell carcinoma: results from the International Metastatic Renal Cell Cancer Database Consortium. Eur Urol 2014;65(4):723–30.
[135] Hofbauer SL, Stangl KI, de Martino M, et al. Pretherapeutic gamma-glutamyltransferase is an independent prognostic factor for patients with renal cell carcinoma. Br J Cancer 2014;111(8):1526–31.
[136] Dalpiaz O, Pichler M, Mrsic E, et al. Preoperative serum-gamma-glutamyltransferase (GGT) does not represent an independent prognostic factor in a European cohort of patients with non-metastatic renal cell carcinoma. J Clin Pathol 2015;68(7): 547–51.
[137] Ohno Y, Nakashima J, Nakagami Y, et al. Clinical implications of preoperative serum total cholesterol in patients with clear cell renal cell carcinoma. Urology 2014;83(1):154–8.
[138] de Martino M, Leitner CV, Seemann C, et al. Preoperative serum cholesterol is an independent prognostic factor for patients with renal cell carcinoma (RCC). BJU Int 2015;115(3):397–404.
[139] Ramsey S, Lamb GWA, Aitchison M, McMillan DC. Prospective study of the relationship between the systemic inflammatory response, prognostic scoring systems and relapse-free and cancer-specific survival in patients undergoing potentially curative resection for renal cancer. BJU Int 2008;101(8):959–63.
[140] Karakiewicz PI, Hutterer GC, Trinh QD, et al. C-reactive protein is an informative predictor of renal cell carcinoma-specific mortality—a European study of 313 patients. Cancer 2007;110(6):1241–7.
[141] Iimura Y, Saito K, Fujii Y, et al. Development and external validation of a new outcome prediction model for patients with clear cell renal cell carcinoma treated with nephrectomy based on preoperative serum C-reactive protein and TNM classification: the TNM-C score. J Urol 2009;181(3):1004–12. discussion 12.

[142] Nakayama T, Saito K, Ishioka J, et al. External validation of TNM-C score in three community hospital cohorts for clear cell renal cell carcinoma. Anticancer Res 2014;34(2):921–6.
[143] de Martino M, Klatte T, Seemann C, et al. Validation of serum C-reactive protein (CRP) as an independent prognostic factor for disease-free survival in patients with localised renal cell carcinoma (RCC). BJU Int 2013;111(8):E348–53.
[144] Bedke J, Chun FK, Merseburger A, et al. Inflammatory prognostic markers in clear cell renal cell carcinoma—preoperative C-reactive protein does not improve predictive accuracy. BJU Int 2012;110(11 Pt. B):E771–7.
[145] Steffens S, Kohler A, Rudolph R, et al. Validation of CRP as prognostic marker for renal cell carcinoma in a large series of patients. BMC Cancer 2012;12:399.
[146] Ito K, Yoshii H, Sato A, et al. Impact of postoperative C-reactive protein level on recurrence and prognosis in patients with N0M0 clear cell renal cell carcinoma. J Urol 2011;186(2):430–5.
[147] Jabs WJ, Busse M, Kruger S, Jocham D, Steinhoff J, Doehn C. Expression of C-reactive protein by renal cell carcinomas and unaffected surrounding renal tissue. Kidney Int 2005;68(5):2103–10.
[148] Johnson TV, Ali S, Abbasi A, et al. Intratumor C-reactive protein as a biomarker of prognosis in localized renal cell carcinoma. J Urol 2011;186(4):1213–7.
[149] Schips L, Dalpiaz O, Lipsky K, et al. Serum levels of vascular endothelial growth factor (VEGF) and endostatin in renal cell carcinoma patients compared to a control group. Eur Urol 2007;51(1):168–73. discussion 74.
[150] Guethbrandsdottir G, Hjelle KM, Frugard J, Bostad L, Aarstad HJ, Beisland C. Preoperative high levels of serum vascular endothelial growth factor are a prognostic marker for poor outcome after surgical treatment of renal cell carcinoma. Scand J Urol 2015;49:388–94.
[151] Rioux-Leclercq N, Fergelot P, Zerrouki S, et al. Plasma level and tissue expression of vascular endothelial growth factor in renal cell carcinoma: a prospective study of 50 cases. Hum Pathol 2007;38(10):1489–95.
[152] Yang H, Zhao K, Yu Q, Wang X, Song Y, Li R. Evaluation of plasma and tissue S100A4 protein and mRNA levels as potential markers of metastasis and prognosis in clear cell renal cell carcinoma. J Int Med Res 2012;40(2):475–85.
[153] Patard JJ, Rioux-Leclercq N, Masson D, et al. Absence of VHL gene alteration and high VEGF expression are associated with tumour aggressiveness and poor survival of renal-cell carcinoma. Br J Cancer 2009;101(8):1417–24.
[154] Fujita N, Okegawa T, Terado Y, Tambo M, Higashihara E, Nutahara K. Serum level and immunohistochemical expression of vascular endothelial growth factor for the prediction of postoperative recurrence in renal cell carcinoma. BMC Res Notes 2014;7:369.
[155] Escudier B, Eisen T, Stadler WM, et al. Sorafenib for treatment of renal cell carcinoma: final efficacy and safety results of the phase III treatment approaches in renal cancer global evaluation trial. J Clin Oncol 2009;27(20):3312–8.
[156] Pena C, Lathia C, Shan M, Escudier B, Bukowski RM. Biomarkers predicting outcome in patients with advanced renal cell carcinoma: results from sorafenib phase III Treatment Approaches in Renal Cancer Global Evaluation Trial. Clin Cancer Res 2010;16(19):4853–63.
[157] Banks RE, Forbes MA, Kinsey SE, et al. Release of the angiogenic cytokine vascular endothelial growth factor (VEGF) from platelets: significance for VEGF measurements and cancer biology. Br J Cancer 1998;77(6):956–64.
[158] Verheul HM, Hoekman K, Luykx-de Bakker S, et al. Platelet: transporter of vascular endothelial growth factor. Clin Cancer Res 1997;3(12 Pt. 1):2187–90.

[159] Gunsilius E, Petzer A, Stockhammer G, et al. Thrombocytes are the major source for soluble vascular endothelial growth factor in peripheral blood. Oncology 2000;58(2):169–74.
[160] Niers TM, Richel DJ, Meijers JC, Schlingemann RO. Vascular endothelial growth factor in the circulation in cancer patients may not be a relevant biomarker. PLoS One 2011;6(5):e19873.
[161] Webb NJ, Lewis MA, Bruce J, et al. Unilateral multicystic dysplastic kidney: the case for nephrectomy. Arch Dis Child 1997;76(1):31–4.
[162] Salgado R, Vermeulen PB, Benoy I, et al. Platelet number and interleukin-6 correlate with VEGF but not with bFGF serum levels of advanced cancer patients. Br J Cancer 1999;80(5–6):892–7.
[163] Kisand K, Kerna I, Kumm J, Jonsson H, Tamm A. Impact of cryopreservation on serum concentration of matrix metalloproteinases (MMP)-7, TIMP-1, vascular growth factors (VEGF) and VEGF-R2 in biobank samples. Clin Chem Lab Med 2011;49(2):229–35.
[164] Zavada J, Zavadova Z, Zat'ovicova M, Hyrsl L, Kawaciuk I. Soluble form of carbonic anhydrase IX (CA IX) in the serum and urine of renal carcinoma patients. Br J Cancer 2003;89(6):1067–71.
[165] Li G, Feng G, Gentil-Perret A, Genin C, Tostain J. Serum carbonic anhydrase 9 level is associated with postoperative recurrence of conventional renal cell cancer. J Urol 2008;180(2):510–3. discussion 3–4.
[166] Sim SH, Messenger MP, Gregory WM, et al. Prognostic utility of pre-operative circulating osteopontin, carbonic anhydrase IX and CRP in renal cell carcinoma. Br J Cancer 2012;107(7):1131–7.
[167] Papworth K, Sandlund J, Grankvist K, Ljungberg B, Rasmuson T. Soluble carbonic anhydrase IX is not an independent prognostic factor in human renal cell carcinoma. Anticancer Res 2010;30(7):2953–7.
[168] Gigante M, Li G, Ferlay C, et al. Prognostic value of serum CA9 in patients with metastatic clear cell renal cell carcinoma under targeted therapy. Anticancer Res 2012;32(12):5447–51.
[169] Ramankulov A, Lein M, Johannsen M, et al. Serum amyloid A as indicator of distant metastases but not as early tumor marker in patients with renal cell carcinoma. Cancer Lett 2008;269(1):85–92.
[170] Wood SL, Rogers M, Cairns DA, et al. Association of serum amyloid A protein and peptide fragments with prognosis in renal cancer. Br J Cancer 2010;103(1):101–11.
[171] Miyata Y, Iwata T, Ohba K, Kanda S, Nishikido M, Kanetake H. Expression of matrix metalloproteinase-7 on cancer cells and tissue endothelial cells in renal cell carcinoma: prognostic implications and clinical significance for invasion and metastasis. Clin Cancer Res 2006;12(23):6998–7003.
[172] Lu HS, Yang ZH, Zhang H, Gan MF, Zhou T, Wang SL. The expression and clinical significance of matrix metalloproteinase 7 and tissue inhibitor of matrix metalloproteinases 2 in clear cell renal cell carcinoma. Exp Ther Med 2013;5(3):890–6.
[173] Ramankulov A, Lein M, Johannsen M, Schrader M, Miller K, Jung K. Plasma matrix metalloproteinase-7 as a metastatic marker and survival predictor in patients with renal cell carcinomas. Cancer Sci 2008;99(6):1188–94.
[174] Bandopadhyay M, Bulbule A, Butti R, et al. Osteopontin as a therapeutic target for cancer. Expert Opin Ther Targets 2014;18(8):883–95.
[175] Matusan K, Dordevic G, Stipic D, Mozetic V, Lucin K. Osteopontin expression correlates with prognostic variables and survival in clear cell renal cell carcinoma. J Surg Oncol 2006;94(4):325–31.

[176] Papworth K, Bergh A, Grankvist K, Ljungberg B, Sandlund J, Rasmuson T. Osteopontin but not parathyroid hormone-related protein predicts prognosis in human renal cell carcinoma. Acta Oncol 2013;52(1):159–65.
[177] Igarashi T, Tobe T, Kuramochi H, et al. Serum immunosuppressive acidic protein as a potent prognostic factor for patients with metastatic renal cell carcinoma. Jpn J Clin Oncol 2001;31(1):13–7.
[178] Araki K, Igarashi T, Tobe T, et al. Serum immunosuppressive acidic protein doubling time as a prognostic factor for recurrent renal cell carcinoma after nephrectomy. Urology 2006;68(6):1178–82.
[179] Merseburger AS, Hennenlotter J, Simon P, et al. Cathepsin D expression in renal cell cancer-clinical implications. Eur Urol 2005;48(3):519–26.
[180] Vasudev NS, Sim S, Cairns DA, et al. Pre-operative urinary cathepsin D is associated with survival in patients with renal cell carcinoma. Br J Cancer 2009;101(7):1175–82.
[181] Nisman B, Yutkin V, Nechushtan H, et al. Circulating tumor M2 pyruvate kinase and thymidine kinase 1 are potential predictors for disease recurrence in renal cell carcinoma after nephrectomy. Urology 2010;76(2):513.e1–6.
[182] Gayed BA, Gillen J, Christie A, et al. Prospective evaluation of plasma levels of ANGPT2, TuM2PK, and VEGF in patients with renal cell carcinoma. BMC Urol 2015;15(1):24.
[183] Deyati A, Younesi E, Hofmann-Apitius M, Novac N. Challenges and opportunities for oncology biomarker discovery. Drug Discov Today 2013;18(13–14):614–24.
[184] Maroto P, Rini B. Molecular biomarkers in advanced renal cell carcinoma. Clin Cancer Res 2014;20(8):2060–71.
[185] Gore ME, Szczylik C, Porta C, et al. Safety and efficacy of sunitinib for metastatic renal-cell carcinoma: an expanded-access trial. Lancet Oncol 2009;10(8):757–63.
[186] Vasudev NS, Goh V, Juttla JK, et al. Changes in tumour vessel density upon treatment with anti-angiogenic agents: relationship with response and resistance to therapy. Br J Cancer 2013;109(5):1230–42.
[187] Rini BI, Cohen DP, Lu DR, et al. Hypertension as a biomarker of efficacy in patients with metastatic renal cell carcinoma treated with sunitinib. J Natl Cancer Inst 2011;103(9):763–73.
[188] Donskov F, Michaelson MD, Puzanov I, et al. Sunitinib-associated hypertension and neutropenia as efficacy biomarkers in metastatic renal cell carcinoma patients. Br J Cancer 2015;113:1571–80.
[189] Fujita T, Wakatabe Y, Matsumoto K, Tabata K, Yoshida K, Iwamura M. Leukopenia as a biomarker of sunitinib outcome in advanced renal cell carcinoma. Anticancer Res 2014;34(7):3781–7.
[190] Schmidinger M, Vogl UM, Bojic M, et al. Hypothyroidism in patients with renal cell carcinoma: blessing or curse? Cancer 2011;117(3):534–44.
[191] Rini BI, Melichar B, Ueda T, et al. Axitinib with or without dose titration for first-line metastatic renal-cell carcinoma: a randomised double-blind phase 2 trial. Lancet Oncol 2013;14(12):1233–42.
[192] Xu CF, Johnson T, Garcia-Donas J, et al. IL8 polymorphisms and overall survival in pazopanib- or sunitinib-treated patients with renal cell carcinoma. Br J Cancer 2015;112(7):1190–8.
[193] Xu CF, Bing NX, Ball HA, et al. Pazopanib efficacy in renal cell carcinoma: evidence for predictive genetic markers in angiogenesis-related and exposure-related genes. J Clin Oncol 2011;29(18):2557–64.
[194] Garcia-Donas J, Esteban E, Leandro-Garcia LJ, et al. Single nucleotide polymorphism associations with response and toxic effects in patients with advanced renal-cell carcinoma treated with first-line sunitinib: a multicentre, observational, prospective study. Lancet Oncol 2011;12(12):1143–50.

[195] van der Veldt AA, Eechoute K, Gelderblom H, et al. Genetic polymorphisms associated with a prolonged progression-free survival in patients with metastatic renal cell cancer treated with sunitinib. Clin Cancer Res 2011;17(3):620–9.

[196] Choueiri TK, Vaziri SAJ, Jaeger E, et al. von Hippel–Lindau gene status and response to vascular endothelial growth factor targeted therapy for metastatic clear cell renal cell carcinoma. J Urol 2008;180(3):860–5.

[197] Choueiri TK, Fay AP, Gagnon R, et al. The role of aberrant VHL/HIF pathway elements in predicting clinical outcome to pazopanib therapy in patients with metastatic clear-cell renal cell carcinoma. Clin Cancer Res 2013;19(18):5218–26.

[198] Deprimo SE, Bello CL, Smeraglia J, et al. Circulating protein biomarkers of pharmacodynamic activity of sunitinib in patients with metastatic renal cell carcinoma: modulation of VEGF and VEGF-related proteins. J Transl Med 2007;5:32.

[199] Rini BI, Michaelson MD, Rosenberg JE, et al. Antitumor activity and biomarker analysis of sunitinib in patients with bevacizumab-refractory metastatic renal cell carcinoma. J Clin Oncol 2008;26(22):3743–8.

[200] Motzer RJ, Hutson TE, Hudes GR, et al. Investigation of novel circulating proteins, germ line single-nucleotide polymorphisms, and molecular tumor markers as potential efficacy biomarkers of first-line sunitinib therapy for advanced renal cell carcinoma. Cancer Chemother Pharmacol 2014;74(4):739–50.

[201] Fujita T, Iwamura M, Ishii D, et al. C-reactive protein as a prognostic marker for advanced renal cell carcinoma treated with sunitinib. Int J Urol 2012;19(10):908–13.

[202] Keizman D, Ish-Shalom M, Huang P, et al. The association of pre-treatment neutrophil to lymphocyte ratio with response rate, progression free survival and overall survival of patients treated with sunitinib for metastatic renal cell carcinoma. Eur J Cancer 2012;48(2):202–8.

[203] Perez-Gracia JL, Prior C, Guillen-Grima F, et al. Identification of TNF-alpha and MMP-9 as potential baseline predictive serum markers of sunitinib activity in patients with renal cell carcinoma using a human cytokine array. Br J Cancer 2009;101(11):1876–83.

[204] Zurita AJ, Jonasch E, Wang X, et al. A cytokine and angiogenic factor (CAF) analysis in plasma for selection of sorafenib therapy in patients with metastatic renal cell carcinoma. Ann Oncol 2012;23(1):46–52.

[205] Papazisis KT, Kontovinis LF, Papandreou CN, et al. Brain natriuretic peptide precursor (NT-pro-BNP) levels predict for clinical benefit to sunitinib treatment in patients with metastatic renal cell carcinoma. BMC Cancer 2010;10:489.

[206] Tran HT, Liu Y, Zurita AJ, et al. Prognostic or predictive plasma cytokines and angiogenic factors for patients treated with pazopanib for metastatic renal-cell cancer: a retrospective analysis of phase 2 and phase 3 trials. Lancet Oncol 2012;13(8):827–37.

[207] Choueiri TK, Escudier B, Powles T, et al. Cabozantinib versus everolimus in advanced renal-cell carcinoma. N Engl J Med 2015;373:1814–23.

[208] Hudes G, Carducci M, Tomczak P, et al. Temsirolimus, interferon alfa, or both for advanced renal-cell carcinoma. N Engl J Med 2007;356(22):2271–81.

[209] Armstrong AJ, George DJ, Halabi S. Serum lactate dehydrogenase predicts for overall survival benefit in patients with metastatic renal cell carcinoma treated with inhibition of mammalian target of rapamycin. J Clin Oncol 2012;30(27):3402–7.

[210] Upton MP, Parker RA, Youmans A, McDermott DF, Atkins MB. Histologic predictors of renal cell carcinoma response to interleukin-2-based therapy. J Immunother 2005;28(5):488–95.

[211] Atkins M, Regan M, McDermott D, et al. Carbonic anhydrase IX expression predicts outcome of interleukin 2 therapy for renal cancer. Clin Cancer Res 2005;11(10):3714–21.

[212] McDermott DF, Cheng SC, Signoretti S, et al. The high-dose aldesleukin "select" trial: a trial to prospectively validate predictive models of response to treatment in patients with metastatic renal cell carcinoma. Clin Cancer Res 2015;21(3):561–8.

CHAPTER TEN

Proteomics and Advancements in Urinary Biomarkers of Diabetics Kidney Disease

M.L. Merchant, PhD* and J.B. Klein, MD, PhD**
*Division of Nephrology and Hypertension, Department of Medicine, University of Louisville, Louisville, KY, United States
**Robley Rex Veterans Administration Medical Center, Louisville, KY, United States

Contents

INTRODUCTION

Diabetes is a highly prevalent disease estimated by the World Health Organization to affect almost 350 million persons worldwide [1]. In the United States, diabetes affects 29 million individuals with nearly 29% cases undiagnosed [2]. While the overall rates of increase for diabetic kidney disease appear to have plateaued the disease disproportionately affects the old, the poor, and individuals traditionally with reduced access to health care. Moreover the most recent increases in diabetic kidney disease can be attributable to type-2 diabetes (T2D) and obesity [3,4]. These factors present challenges to treating diabetic complication and underscore the need to early detection and prediction of those individuals susceptible to severe progressive loss of renal function as well as strategies to manage the complex network of diabetic complications and comorbidities.

Diabetic nephropathy (DN) is a severe microvascular complication of diabetes and accounts for 44% of all 2011 incident cases of end-stage renal disease (ESRD) [5]. The key tools utilized to identify and monitor progression of DN remains serum creatinine-based estimations of glomerular

Biomarkers of Kidney Disease. http://dx.doi.org/10.1016/B978-0-12-803014-1.00010-8

filtration rate (eGFR) and also urinary albumin based estimations of renal damage [6,7]. While these tools have formed a mainstay for clinical management of DN there are gaps in their application for diagnosis of disease onset and disease progression. While serum creatinine can efficiently be used to estimate GFR values in individuals with chronic kidney disease (CKD) stages 3, 4, and 5, its application to estimate normal filtration rates and hyperfiltration is lost. Our goal in this manuscript is to aggregate and review recent research wherein broad based discovery and targeted proteomic methods have been used to identify and evaluate urinary proteins and peptides for their ability to augment or replace the use of serum creatinine or urinary albumin to diagnose DN and effectively predict renal function decline. These studies address both diabetic kidney disease association with type-1 diabetes (T1D) and T2D.

From the Tissue to the Urine—Application of Discovery Proteomics to Study Type-1 Diabetes and Diabetic Kidney Disease

The urinary proteome is enigmatic in that it represents both the systemic circulating proteome as well as the compartmentalized renal proteome. Understanding a protein's origin is critical to its future use as a candidate biomarker or developing a hypothesis for the progression of a disease. An inability to know where a protein originates can limit its value to that of a surrogate disease biomarker. However, the knowledge of the proteomic origin(s) may provide information into the mechanism of renal dysfunction itself. The directed goal of understanding renal-to-urinary proteomic changes is made difficult by the phenotypic differences between classical animal models of diabetes used to study DN and the clinical samples obtained through patient study enrollment in the clinic. The simplest level of investigation is through the use of an animal model of diabetic kidney disease and seeking to translate renal tissue changes in the laboratory animal into the patient urinary proteome. An intermediate level of investigation is through the directed study of human urinary extracellular vesicles. Lastly the highest level of directed study and the most difficult to achieve is the study of the urinary proteome with contemporaneous, matched renal biopsies.

Four recent proteomic studies explore T1D and early changes in the urinary proteome. Proteomic analysis of T1D, as compared to T2D, is attractive because diabetes onset occurs early, before an individual has had prolonged exposure to environmental factors that might introduce complex epigenetic mechanisms into the disease progression. The first of the studies highlighted in this section is by Suh et al. [8] and addressed the impact of extended hyperglycemic excursions, associated with T1D, on the kidney. Their hypothesis was inflammatory-centric, as prolonged periods of high glucose have been shown to be associated with increased levels of advance glycation end productions and heightened levels of inflammation. The primary goal was not biomarker discovery, but rather a proof in concept that urinary proteome changes reflect proinflammatory processes. The results of these studies would then be used to guide future biomarker research.

Suh et al. utilized a label free, liquid chromatographic-mass spectrometric (LCMS) proteomic approach to study large numbers of patient samples. The authors focused on the urinary proteome of a balanced gender mix (22 male/18 female) of juvenile (n = 40, mean age 12.4 ± 3.4 years) T1D patients with a mean duration of diabetes of 6.3 ± 3.4 years with a HbA1c of 8.4 ± 1.8. The controls (21 male/20 female) were by design healthy siblings (n = 41) whose parents had consented them into the study. Urine samples were collected and stored at −80°C until needed for analysis. The authors used a series of sample handling (e.g., FASP) and proteolytic digest fractionation (off-line neutral pH reversed phase fractionation) steps that have been showed in other settings to enhance low abundant proteome discovery efforts with LCMS workflows. Relative quantitative differences in the proteomes were established using the software tool APEX following database search strategies to make protein assignments and in silico normalization strategies. Qualitative differences in the urinary proteome of T1D versus sib-controls based on gene ontology was determined using the bioinformatics tool PANTHER. Quantitative differences between cases and controls was achieved using a combination of three approaches: (1) consensus ranking by Wilcoxon tests with GeneSelector R package, (2) a stability selection tool (R package *glmnet*), and (3) permutation-based analysis of variance (*PermANOVA*). At a confidence level of 99% and a false discovery rate of ≤1% (based on ProteinProphet modeling) a total of 5,046 proteins were identified with no large-scale differences found between GO terminologies of cases and controls. A combination of the three statistical approaches identified a list of 90 differentially abundant urine proteins with enrichment of GO terminology including proteolysis, cell adhesion, carbohydrate metabolism, organo-nitrogen compound metabolism, and glycoside catabolism. A closer inspection of these proteins included 22 proteins related by GO terms to inflammation or metabolism. Two proteins were decreased and twenty proteins increased in the urine of

T1D; including fourteen proteins of lysosomal origin and eight proteins of platelet origin. A second set of 23 proteins were related by GO terms to biological adhesion and the microvasculature including adhesion molecules and peptidases. For purposes of supporting LCMS based observations and not for biomarker confirmation, antibodies for six highly regulated proteins (plasma α-fucosidase, *N*-acetyl-galactosaminidase, collectin-12, CD166, metalloproteinase inhibitor-1, and apolipoprotein-M) were used for immunoblot of urine proteins samples. Two regulated lysosomal proteins (plasma α-fucosidase and *N*-acetyl-galactosaminidase) were confirmed while immunoblot testing of the remaining four proteins failed due to no differences observed, antibodies lacking sensitivity or the detection of multiple bands (presumably protein fragments).

A general problem with biomarker discovery is a wide dynamic range of protein abundance in biofluids such as serum and urine. One strategy, used in the second study, to address this problem is proteome fractionation prior to LCMS analysis. Caseiro et al. [9] used an approach termed GelC-MS/MS to study the urinary proteome of age/sex matched T1D patients and healthy volunteers. In this approach a first dimension of fractionation is achieved using SDS-PAGE and resolved proteins by molecular weight. The SDS-PAGE gel lanes were digested and analyzed by a LCMS proteomic approach. In this study the patients were balanced between patients with nephropathy + retinopathy (n = 5), retinopathy (n = 5), no discernable microvascular disease (n = 5), and "controls" (n = 5). All T1D patients had a disease duration of at least 15 years and HbA1c of 7.7% or greater. Nephropathy was based on urinary albumin of 300 mg in a 24-h collection. Retinopathy was based on ophthalmologist screening. A total of 219 proteins were identified by GelC-MS/MS analysis including 75 urinary proteins identified across all samples, 8 proteins identified exclusively in diabetic samples, 31 proteins identified exclusive in control samples. Diabetic urine samples could be differentiated by the relative abundances of 79 proteins (59 increased in T1D with retinopathy; 9 increased in T1D with nephropathy and retinopathy; 11 in T1D without microvascular disease). The proteins unique to T1D samples included Ig-kappa chains, gelsolin, proline-rich 4, antithrombin, carboxy-terminal PDZ ligand, contactin-1, dyslexia-associated protein KIAA03119, and hornerin.

In the third and fourth studies from the same group, Zubiri et al. compared the urinary proteome of diabetic patients and controls using a combination of electrophoresis and shotgun LCMS proteomic methods. In the first study Zubiri et al. [10] established an ultracentrifugation-high abundant protein immunodepletion-ultracentrifugation sequence for urinary extracellular vesicles (uEVs) isolation. uEVs are shed into the urine through both regulated (exosome) and unregulated (microparticles or membrane blebs) events. Exosomes have characteristic size (50–80 nm) and composition (e.g., tetraspanins such as TSG101 and endosomal sorting complex required for transport, ESCRTs). These structures are comprised of a delimiting phospholipid bilayer structure as well as membrane bound proteins and luminal proteins and nucleic acids. From a proteomic perspective uEVs represent a unique opportunity to gain increased sensitivity to detection of low abundant proteins as well as proteins whose function may be relevant to the ongoing disease process. Zubiri et al. validated their preparation using a combination of electron microscopy to determine uEV diameter and immunoblot for TSG101 and the ESCRT protein Alix. A label-free, semi-quantitative LCMS method was used to determine relative differences in the isolated proteins. A total of 562 proteins were observed at a 2-peptide, 95% peptide confidence, and 99% protein confidence intervals. Many of these proteins had not previously been reported in the Exocarta database (http://www.exocarta.org/) for exosomal proteins; suggesting DN results in alterations to the urinary exosomal proteome. A large number of identified proteins were observed only in DN ($n = 89$; 15.8%) or control ($n = 75$, 13.3%). These numbers however reflect a frequent problem with low abundant proteins that is the infrequent detection across all samples. For statistical comparisons the authors chose to consider only those proteins identified in at least 7 out of 10 samples or 3 out of 5 if detected only in case or in control samples. Twenty-five proteins were differentially abundant at a p-value < 0.01. Eighteen proteins were increased in controls and seven proteins were increased in DN.

The authors selected three proteins (alpha-1-microglobulin, voltage-dependent anion selective channel protein 1, and isoform 1 of histone-lysine *N*-methyltransferase MLL3) for confirmatory studies using LCMS methods. The most widespread application for mass spectrometry and proteomics is the identification and relative quantification of proteins [11–14]. An emerging method in proteomics is the use of mass spectrometers and peptide specific internal standards for the absolute protein quantification [15,16]. While there are variations in the application the most common is the addition of a synthetic labeled peptide to a protein tryptic digest and quantification of selected fragment ions for both the synthetic labeled peptide and the tryptic peptides from the naturally occurring protein. The fragment ion ratios for the naturally occurring tryptic peptides and the labeled peptides along with an external calibration curve can be used to calculate the absolute quantification of the target peptide. The assumption is that the peptide will be present in the digest at a stoichiometry of 1:1 with the parent pretryptic digest protein. This assumption will be true as long as (1) we are not discussing posttranslational states, (2) if there are no other sequence variants for that protein, and (3) the amino acid sequence cannot be found in any other known protein (referred to as proteospecific or proteotypic). The technique described has been used for small molecule applications in drug chemistry and metabolomics and referred to as stable isotope dilution mass spectrometry. In the field of proteomics this is referred to as multiple reaction monitoring (MRM) or selected reaction monitoring (SRM). The direct quantification of target proteins in biological matrices and affinity enriched protein complexes is comparable to absolute quantification of proteins using standard antibody based assays (e.g., enzyme-linked immunosorbent assay, ELISA). The benefit of the MRM or SRM method is the selectivity of the method allows for accurate quantification of a specific peptide and not quantification of an immunoreactive epitope that is often biased by nonspecific binding. Here the authors chose to use MRM in a relative quantification approach that supported the original label-free results. One limitation of the study was a small number of diabetic patients ($n = 5$ case and control) in the discovery phase and the confirmatory phase ($n = 3$ case and control). Moreover the discovery phase included DN patients with a wide range of albuminuria (3–1325mg/dL).

In the second example by Zubiri et al. [17], the authors used their methods in a refined study that first investigated DN-induced tissue proteome differences in an animal model to translate into a human urinary proteome study. Here early changes in the rat renal proteome induced by intraperitoneal (IP) administration of the pancreatic beta cell toxin streptozotocin (STZ) were studied. Five-week old Wistar-Kyoto rats received a single IP dose of STZ (50 mg/kg in 0.1M citrate pH 4.5) or vehicle. At the time of sacrifice 8 weeks post IP challenge, animals receiving STZ had blood glucose levels of >600 mg/dL while rats receiving vehicle has blood glucose levels <200 mg/dL. Renal immunohistochemical (IHC) analysis classified the DN changes as class IIa with mild mesangial expansion, glomerular basement thickening, and mild extracellular matrix accumulation. The renal proteome was compared using a two-dimensional gel electrophoresis (2DE) method called difference gel electrophoresis or DIGE. With 2DE DIGE the authors were able to chemically derivatize a pair of protein samples with one of two fluorescent Cy-dye labels. After 2DE separation of an admixture of labeled samples, the relative contributions to each 2DE-resolved protein spot can be estimated by the proportion of each Cy-dye fluorescence. The highest differentially regulated protein in the tissue proteomics (2DE) was identified as regucalcin (senescence marker protein 30, SMP30). Regulcalcin is a calcium-binding protein, expressed in multiple tissues (highly expressed in the kidney and liver) including a secreted form and the expression is diminished with age. The contributions of regulcalin toward regulating intracellular calcium homeostasis is believed to be a key element of its role in the kidney. In addition to binding calcium and regulating its levels, regulcalcin either directly or indirectly activates or inhibits multiple intracellular signaling pathways, and has cytoprotective, prosurvival activities [18,19]. In this study, regulcalcin was downregulated 4.55-fold with DN a finding that was validated by immunoblot analysis. Periodic acid-Schiff staining of the rat kidneys suggested STZ resulted in tubular cell injury (brush border loss and widening of tubular lumen but not atrophy) that was supported by IHC analysis of KIM-1 expression. The authors followed this up by examining urinary exosomes for the presence of regulcalcin. To do this the authors needed to use pooled rat urine to achieve significant amounts of urinary exosomes. Using an immunoblot and LC-SRM-MS approach, Zubiri et al. demonstrated that regulcalcin was present in the urinary exosomes of normal rats but undetectable in the urinary exosomes of diabetic animals. The authors confirmed these tissue levels and urinary exosome observations using a small number of available bioreposited biopsies and urinary exosomes from DN patients with an undisclosed stage of CKD. These data identify an association of regulcalcin abundance in the kidney with early stages of DN in an animal model of diabetes as well as changes in the packaging and secretion of regucalcin in exosomes. While the confirmatory work in DN patient samples is promising the lack of firm data on patient CKD status and the small numbers of patients prevents any firm conclusions on a role for regucalcin in the progression of DN.

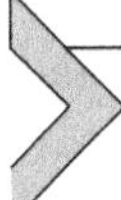

URINARY PEPTIDES AS BIOMARKERS OF DIABETIC KIDNEY DISEASE

Proteolysis of urinary proteins is a constitutive process that can be linked to innocuous processes such as protein catabolism and to biologically significant events such as peptide hormone activation or degradation of extracellular matrices and metastasis. The urinary peptidome at any given time point represents a snapshot of whole proteome metabolism and varies based on pathophysiology. A series of manuscripts has used capillary electrophoresis–mass spectrometry (CE–MS) to establish a large urinary peptide panel (n = 273) termed CKD273 that demonstrates a capacity for diagnosis and prediction of CKD and CKD progression [20–22]. These peptides are comprised of fragments of extracellular matrix, serum, and urinary proteins and were identified during a series of comparisons of healthy control patients and patients with various biopsy-proven renal disease that includes vasculitis, systemic lupus erythematosus, membranous glomerulonephritis, minimal change disease, IgA nephropathy, focal segmental glomerular sclerosis, and DN. Recently a multicenter prospective validation study by Siwy et al. [22], the PRIORITY trial, of the CKD273 classifier was used to test the ability for early diagnosis of T2DN. In this study 165 T2D patients from 9 different centers were prospectively enrolled and clinical characteristics of patients assess at enrollment and in some follow-ups included urinary albumin levels, serum creatinine, eGFR, albumin to creatinine levels or proteinuria. The CKD273 classifier was independently tested for the ability to differentiate T2D patients with known normoalbuminuria and with macroalbuminuria. The CKD273 classification scores of T2D patients were used to construct receiver operating curves (ROC) and the area under the curve (AUC) values determined. The ROC AUC values for the nine centers varied between 0.9 and 1.0 with a grand average of 0.95. The CKD273 classifier correctly identified 147 patients as T2D with normoalbuminuria or macroalbuminuria. Moreover these results were independent of gender and age. Incorrect classification also occurred and included 10 false positive DN and 8 false negative DN. To investigate false positive DN assignments the patient follow-up data from four available patients were studied. These data demonstrated that one false positive control patient developed microalbuminuria and one false positive control patient developed macroalbuminuria at 1-year enrollment. The remaining two false positive control patients remained normoalbuminuric. In addition to classification the authors investigated which CKD273 peptides were most prevalently observed and used in the classification. The most consistent peptides observed across all centers correlating with T2DN were fragments of alpha-1 antitrypsin, collagen IαI, collagen IIIαI, serum albumin, uromodulin, alpha-2-HS glycoprotein, sodium/potassium-transporting ATPase gamma, polymeric-immunoglobulin receptor, and transthyretin. These data are encouraging that the CKD273 classifier is a feasible tool to detect T2DN and predict its progression.

Targeted Urinary Proteomics and Biomarkers of Diabetic Kidney Disease

Urine is considered as the ideal biofluid for clinical testing and the diagnosis of renal disease or the prediction of renal function changes, in that it is abundant and can be obtained in a noninvasive manner. As such a number of studies have tested the ability of specific markers for diagnostic or prognostic sensitivity to identify renal injury and renal hyperfiltration or predict early diabetic changes to the renal proteome, incident CKD and progression of CKD. The majority of these studies utilize urine samples collected as spot urine samples form a cross-sectional cohort of patients enrolled in ongoing studies. In some studies the patients are prospectively enrolled in well-controlled studies.

Type 1 Diabetes

Three targeted studies that followed up target leads from previous OMICs studies examined the association of the urinary proteome with early changes in T1D. In the first study Cherney et al. [23] used quantitative ELISAs to examine whether urinary angiotensin converting enzyme-2 (uACE2) activity and protein levels as well as angiotensinogen (AGT) and urinary angiotensin converting enzyme (ACE) were elevated in early T1D. This question was raised in respect to the earliest stages of T1DN as the renin-angiotensin-aldosterone system (RASS) regulates angiotensin II levels in part through renal ACE2 enzymatic activity. In experimental models of DN ACE2 maybe renoprotective, unlike other components of the RASS. ACE2 has been demonstrated to be increased in patients with CKD and also in T2D patient with an independent association with microalbuminuria. To examine a more uncomplicated model, Cherney et al. studied uACE2 levels in a model of clamped euglycemia in T1D with uncomplicated T1D (n = 58) versus normoglycemic controls (n = 21). In this study patients adhered to a high sodium (>140 mmol/day) moderate protein (<1.5 g/kg per day) diet for 7 days. In T1D patients clamped on two sequential days, first in euglycemia (4–6 mmol/L) and second in hyperglycemia (9–11 mmol/L). The glycemic levels were maintained for 6 h prior to and during experimental studies. After the targeted glycemia conditions were established the glomerular filtration rate was measured using inulin clearance. Healthy controls had a mean glomerular filtration rate of 120 ± 16 (mL/min per 1.73 m^2) and the T1D had a mean glomerular filtration rate of 144 ± 40 (mL/min per 1.73 m^2). In these tightly controlled studies, ACE2 levels but not ACE or AGT levels and ACE2 activity were higher ($p < 0.0001$) in T1D compared to controls and increased in euglycemic clamped T1D compared to hyperglycemic clamped T1D but not AGT or ACE excretion. These studies suggest the relative levels of urinary enzymes in uncomplicated T1D can be responsive to systemic levels of blood glucose. While this study is not a classical proteomics study this study is a proteomic quantification study that follows up on multiple protein biomarkers that have potential as surrogate markers of disease. While antibody assays are quantitative and robust we expect that LC-MRM-MS based assays will at some point replace or augment classical ELISA tests [24].

In the second study by Har et al. [25] of adolescents with T1D, a larger panel of biomarkers was quantified using a bead based antibody capture and antibody reporter system and used to assess a signature of renal hyperfiltration. This large-scale study on targeted candidate biomarkers was guided from prior studies by the authors on hyperglycemia or hyperfiltration in young adults. The authors hypothesized that some or all of the candidate biomarkers known to increase in adults with hyperfiltration would be increased in adolescents. These markers include the following list of cytokines and chemokines: eotaxin, FGF-2, GM-CSF, IFNα2, IL-2, IL-12, MCP-3, MCP-1, MDC, MIP-1α, TNFß, sCD40L, PDGF-AB/BB. The glomerular filtration rate was estimated from serum cystatin-C measurements and urinary albumin to creatinine values were calculated using spot urine samples. In this study urinary cytokines and chemokines were measured using a multiplex, bead-based (Luminex) assay in normotensive, normoalbuminuric, hyperfiltering T1D (n = 31, eGFR >135) versus normofiltering T1D (n = 111, eGFR < 135) and healthy age matched controls (n = 59) to determine an association with CysC based eGFR. Following adjustment for HbA1c and albumin to creatinine ratio, urinary IL-12 significantly (ANOVA p-value < 0.005) increased healthy control < T1D normofiltering < T1D hyperfiltering. Urinary IFNα2 was significantly (ANOVA p-value 0.0019) increased in T1D hyperfiltering versus T1D normofiltering or versus healthy controls. Significant differences in IL-2 (ANOVA p-value 0.0002), sCD40L (ANOVA p-value 0.001), and FGF-2 (ANOVA p-value 0.0038) increased stepwise from healthy control < T1D normofiltering < T1D hyperfiltering. Significant differences for TNFβ and MIP-1α were observed only between healthy control < T1D hyperfiltering. To determine if differences in serum levels of these cytokines and chemokines additional multiplex assays were conducted and only serum IL-2 was significantly different between healthy controls and diabetics. These data are compelling and suggest that urinary and serum proteins can be used to identify differences in patient populations. What is at a loss here is the understanding as to whether these epitopes that were quantified by antibodies in the urine or serum were accessing the mature biologically active form of the cytokine/chemokine or if the data were an aggregate of all epitope bearing protein forms (preproteins, mature proteins, or protein fragments) and pseudoepitope bearing proteins (nonspecific binding).

The third targeted study is an elegant study by Fufaa et al. [26] geared to assess the association of early tissue level morphometric changes in T1D with urinary biomarkers of renal inflammation, MCP-1 and hepcidin. The patient population studied were normotensive, normoalbuminuric T1D from the RASS study with a mean age of 24.6 years, mean duration ofT1D of 11.2 years, a median albumin excretion rate of 6.4 μg/min and mean iohexol-based glomerular filtration rate of 129 mL/min per 1.73m^2. A key aspect to the unique nature of this study in the prospective collection of urine samples prior to a renal biopsy. All baseline urine samples (n = 224) had contemporaneous baseline biopsies and 5-year follow-up urine and biopsy samples which were used to quantify urinary MCP-1 and urinary hepcidin. While no associations were found between urinary hepcidin to creatinine ratios with morphometric values, higher urinary MCP-1 to creatinine ratios were associated with higher interstitial fractional volume at baseline and after 5 years in women but not men. uMCP-1 was not associated with glomerular lesions in either sex. These data highlight the significance of inflammatory processes driving early lesions in female T1D. As demonstrated by these last three targeted studies antibody based assays play a key role in the surrogate biomarker confirmatory study workflow. However due to developments with the fields of mass spectrometry field and the software development, we propose that within 10-years quantitative LC-MRM-MS based tools will be developed as standalone tests or in combination with antibody based enrichment methods to quantified comprehensive biomarker panels.

Type-2 Diabetes

T2D is a complex disease and most targeted or discovery proteomic studies have focused on the goal of quantifying markers that can be used to predict progression of renal function decline. Here two recent studies have been conducted for targeted quantification of renal function decline in T2D. These two studies are complementary in that they both use a large-scale shotgun approach. In the first Pena et al. [27] used an ELISA approach to survey 28 biomarkers that represent a combination of various disease pathways including proinflammatory, profibrotic, and angiogenic processes. These 28 biomarkers were assessed in patients ($n = 82$) seen at a single out-patient center in the Netherlands. The average baseline characteristics of these patients included average age (63.5 ± 9.4 years), body mass index (32.4 ± 6.3), systolic blood pressure (135.2 ± 16.3 mmHg), diastolic pressure (72.7 ± 10.5 mmHg), urine albumin to creatinine ratio (1.2), serum creatinine (88.4 ± 33.5 μmol/L), estimated glomerular filtration rate (77.9 ± 22.6 mL/min per 1.73 m^2), and HbA1c (%) (7.7 ± 1.3). A univariate and multivariate testing of quantified markers determined a significant association (p-value < 0.05) of increased urinary proteins with eGFR decline including two markers associated with inflammation (tumor necrosis factor receptor-1, chitinase 3-like 1), two proteins associated with fibrosis (matrix metallopeptidase 2, matrix metallopeptidase 7) and one protein with angiogenesis (endostatin). Importantly three of these markers (tumor necrosis factor receptor-1, chitinase 3-like 1, and matrix metallopeptidase 7), were independently associated with decline in the glomerular filtration rate even after adjusting for traditional risk markers. A twofold increase in the log concentration of matrix metallopeptidase 7, chitinase 3-like 1, and tumor necrosis factor receptor-1 corresponded to a decrease in the estimated glomerular filtration rate of 0.77 ($p = 0.04$), 0.90 ($p = 0.02$), and 2.1 ($p = 0.03$) respectively. While the authors were able to use statistical modeling to demonstrate the ability of a 19 novel markers and establish risk markers to optimally predict accelerated renal function decline, give the average rate of glomerular filtration loss it is unclear how these markers would be implemented in clinical practice.

In the second, large scale targeted biomarker study, Agarwal et al. [28] assessed a combination of blood and urine biomarkers to predict CKD progression in a population of US veterans with CKD versus age matched controls with no existing CKD, hypertension or cardiovascular disease. Included in this list of 24 biomarkers are seven that overlap with the study of Pena et al.—vascular endothelial growth factor A, matrix metalloproteinase-7, connective tissue growth factor, soluble tumor necrosis factor receptor-1, and soluble tumor necrosis factor receptor-2, monocyte chemotactic protein 1, and fibroblast growth factor-23. The samples for this study were from patient examination at enrollment and from prospective sample collection over a 2–6 year follow-up period. The average baseline characteristics of the patients (controls/cases) included average age (60.3 ± 9.8/67.4 ± 9.4 years), body mass index (27.7 ± 4/32 ± 4.7), systolic blood pressure (114 ± 7.9/121 ± 12.4 mmHg), diastolic pressure

(67.9 ± 5.9/61.6 ± 8.7 mmHg), estimated glomerular filtration rate (86 ± 9/43 ± 13 mL/min per 1.73 m^2), and HbA1c (%) (5.8 ± 0.3/7.8 ± 1.3).

To examine the biomarker panel's ability to assess hard endpoints, CKD progression was defined as (1) the rate of change for the estimated GFR over time (eGFR versus time plots and the slope progression method), (2) end stage renal disease, ESRD, and (3) composition of ESRD and death. Nine urinary markers emerged as significantly different (p-value < 0.05) between cases and controls including nephrin, soluble tumor necrosis factor receptor 1, soluble tumor necrosis factor receptor 2, high molecular weight collagen IV, fibronectin, collagen IV, connective tissue growth factor, tenascin C, and cystatin C. The unadjusted relationship of biomarker concentrations with three end points demonstrated more were correlated with ESRD, then CKD slope progression, and least with ESRD and death. After adjustment for log urine ACR and baseline eGFR only one marker was associated with an end point. Urinary C-terminal fibroblast growth factor-23 (FGF-23) demonstrated the strongest primary association with CKD slope progression (adjusted odds ratio 2.08, $p = 008$). Plasma VEGF had the strongest association with ESRD (adjusted odds ratio 1.44, $p = 0.027$). Plasma C-terminal FGF-23 was associated with composite outcome of ESRD and death (adjusted odds ratio 3.07, $p = 0.008$). Within this study FGF-23 was found to be predictive of two end-points of CKD progression. FGF-23 is a member of the fibroblast growth factor family and significantly contributes to the regulation of phosphate homeostasis. FGF-23 is secreted by osteocytes in response to calcitriol whereupon it acts to activate vitamin D and in the renal proximal tubule to decrease the expression of the sodium-phosphate cotransporter 2a, NPT2a, thus ultimately increasing phosphate excretion. These data are consistent with the findings of several other groups for the association of increased plasma c-terminal FGF-23 and a faster time to doubling for serum creatinine in nondiabetic patients and in T2D who developed ESRD or died before reaching ESRD.

FUTURE DEVELOPMENTS AND APPLICATIONS OF PROTEOMICS FOR BIOMARKER DISCOVERY

Diabetes is a complex disease that presents with a range of attendant complications such as DN. Discovery-based proteomic methods and targeted LCMS based methods have been used and will continue to be used to develop lists of candidate biomarkers associated with diabetic kidney disease onset and progression. As of yet, several protein markers of chronic disease and inflammation are providing statistically significant results in the study of populations. However, no single protein or small group of proteins have been identified to provide sufficient sensitivity and specificity to be used in the clinical management of DN. Nonetheless, as label-free MS methods of protein quantification permeate general proteomic research fields we should begin to see these methods successfully applied toward the identification of diagnostic and prognostic biomarkers of DN.

REFERENCES

[1] Diabetes. Fact Sheet No312. Available from: http://www.who.int/mediacentre/factsheets/fs312/en/
[2] Prevention. CfDCa: Estimates of diabetes and its burden in the United States, 2014. In: Edited by Services DoHaH, vol. National Diabetes Statistics Report: Atlanta, GA: U.S.
[3] de Boer IH, Rue TC, Hall YN, Heagerty PJ, Weiss NS, Himmelfarb J. Temporal trends in the prevalence of diabetic kidney disease in the United States. JAMA 2011;305(24):2532–9.
[4] Geiss LS, Wang J, Cheng YLJ, Thompson TJ, Barker L, Li YF, Albright AL, Gregg EW. Prevalence and incidence trends for diagnosed diabetes among adults aged 20 to 79 years, United States, 1980–2012. JAMA 2014;312(12):1218–26.
[5] Tuttle KR, Bakris GL, Bilous RW, Chiang JL, de Boer IH, Goldstein-Fuchs J, Hirsch IB, Kalantar-Zadeh K, Narva AS, Navaneethan SD, et al. Diabetic kidney disease: a report from an ADA consensus conference. Am J Kidney Dis 2014;64(4):510–33.
[6] Parving HH, Persson F, Rossing P. Microalbuminuria: a parameter that has changed diabetes care. Diabetes Res Clin Pract 2015;107(1):1–8.
[7] Levey AS, Becker C, Inker LA. Glomerular filtration rate and albuminuria for detection and staging of acute and chronic kidney disease in adults: a systematic review. JAMA 2015;313(8):837–46.
[8] Suh MJ, Tovchigrechko A, Thovarai V, Rolfe MA, Torralba MG, Wang J, Adkins JN, Webb-Robertson BJ, Osborne W, Cogen FR, et al. Quantitative differences in the urinary proteome of siblings discordant for type 1 diabetes include lysosomal enzymes. J Proteome Res 2015;14(8):3123–35.
[9] Caseiro A, Barros A, Ferreira R, Padrao A, Aroso M, Quintaneiro C, Pereira A, Marinheiro R, Vitorino R, Amado F. Pursuing type 1 diabetes mellitus and related complications through urinary proteomics. Transl Res 2014;163(3):188–99.
[10] Zubiri I, Posada-Ayala M, Sanz-Maroto A, Calvo E, Martin-Lorenzo M, Gonzalez-Calero L, de la Cuesta F, Lopez JA, Fernandez-Fernandez B, Ortiz A, et al. Diabetic nephropathy induces changes in the proteome of human urinary exosomes as revealed by label-free comparative analysis. J Proteomics 2014;96:92–102.
[11] Zhang G, Annan RS, Carr SA, Neubert TA. Overview of peptide and protein analysis by mass spectrometry. Curr Protoc Mol Biol 2014;108.
[12] Yates JR 3rd, Washburn MP. Quantitative proteomics. Anal Chem 2013;85(19):8881.
[13] Zhang Y, Fonslow BR, Shan B, Baek MC, Yates JR 3rd. Protein analysis by shotgun/bottom-up proteomics. Chem Rev 2013;113(4):2343–94.
[14] Yates JR 3rd. The revolution and evolution of shotgun proteomics for large-scale proteome analysis. J Am Chem Soc 2013;135(5):1629–40.
[15] Carr SA, Abbatiello SE, Ackermann BL, Borchers C, Domon B, Deutsch EW, Grant RP, Hoofnagle AN, Huttenhain R, Koomen JM, et al. Targeted peptide measurements in biology and medicine: best practices for mass spectrometry-based assay development using a fit-for-purpose approach. Mol Cell Proteomics 2014;13(3):907–17.

[16] Gillette MA, Carr SA. Quantitative analysis of peptides and proteins in biomedicine by targeted mass spectrometry. Nat Methods 2013;10(1):28–34.
[17] Zubiri I, Posada-Ayala M, Benito-Martin A, Maroto AS, Martin-Lorenzo M, Cannata-Ortiz P, de la Cuesta F, Gonzalez-Calero L, Barderas MG, Fernandez-Fernandez B, et al. Kidney tissue proteomics reveals regucalcin downregulation in response to diabetic nephropathy with reflection in urinary exosomes. Transl Res 2015;166(5):474–84. e474.
[18] Marques R, Maia CJ, Vaz C, Correia S, Socorro S. The diverse roles of calcium-binding protein regucalcin in cell biology: from tissue expression and signalling to disease. Cell Mol Life Sci 2014;71(1):93–111.
[19] Yamaguchi M. The potential role of regucalcin in kidney cell regulation: involvement in renal failure (review). Int J Mol Med 2015;36(5):1191–9.
[20] Mischak H, Coon JJ, Novak J, Weissinger EM, Schanstra JP, Dominiczak AF. Capillary electrophoresis-mass spectrometry as a powerful tool in biomarker discovery and clinical diagnosis: an update of recent developments. Mass Spectrom Rev 2009;28(5):703–24.
[21] Good DM, Zurbig P, Argiles A, Bauer HW, Behrens G, Coon JJ, Dakna M, Decramer S, Delles C, Dominiczak AF, et al. Naturally occurring human urinary peptides for use in diagnosis of chronic kidney disease. Mol Cell Proteomics 2010;9(11):2424–37.
[22] Siwy J, Schanstra JP, Argiles A, Bakker SJ, Beige J, Boucek P, Brand K, Delles C, Duranton F, Fernandez-Fernandez B, et al. Multicentre prospective validation of a urinary peptidome-based classifier for the diagnosis of type 2 diabetic nephropathy. Nephrol Dial Transplant 2014;29(8):1563–70.
[23] Cherney DZ, Xiao F, Zimpelmann J, Har RL, Lai V, Scholey JW, Reich HN, Burns KD. Urinary ACE2 in healthy adults and patients with uncomplicated type 1 diabetes. Can J Physiol Pharmacol 2014;92(8):703–6.
[24] Liebler DC, Zimmerman LJ. Targeted quantitation of proteins by mass spectrometry. Biochemistry 2013;52(22):3797–806.
[25] Har RL, Reich HN, Scholey JW, Daneman D, Dunger DB, Moineddin R, Dalton RN, Motran L, Elia Y, Deda L, et al. The urinary cytokine/chemokine signature of renal hyperfiltration in adolescents with type 1 diabetes. PLoS One 2014;9(11):e111131.
[26] Fufaa GD, Weil EJ, Nelson RG, Hanson RL, Knowler WC, Rovin BH, Wu H, Klein JB, Mifflin TE, Feldman HI, et al. Urinary monocyte chemoattractant protein-1 and hepcidin and early diabetic nephropathy lesions in type 1 diabetes mellitus. Nephrol Dial Transplant 2015;30(4):599–606.
[27] Pena MJ, Heinzel A, Heinze G, Alkhalaf A, Bakker SJ, Nguyen TQ, Goldschmeding R, Bilo HJ, Perco P, Mayer B, et al. A panel of novel biomarkers representing different disease pathways improves prediction of renal function decline in type 2 diabetes. PLoS One 2015;10(5):e0120995.
[28] Agarwal R, Duffin KL, Laska DA, Voelker JR, Breyer MD, Mitchell PG. A prospective study of multiple protein biomarkers to predict progression in diabetic chronic kidney disease. Nephrol Dial Transplant 2014;29(12):2293–302.

CHAPTER ELEVEN

Biomarkers of Cardiovascular Risk in Chronic Kidney Disease

Z.H. Endre, BScMed, MBBS, PhD, RACP, FASN*,,† and R.J. Walker, MBChB, MD, FRACP, FASN, FAHA‡**

*Department of Nephrology, Prince of Wales Hospital and Clinical School, University of New South Wales, Sydney, NSW, Australia
**Department of Medicine, University of Otago, Christchurch, New Zealand
†School of Medicine, University of Queensland, Brisbane, QLD, Australia
‡Department of Nephrology, Dunedin Hospital and University of Otago, Dunedin, New Zealand

Contents

INTRODUCTION

Chronic kidney disease (CKD) is an established independent risk factor for cardiovascular disease. Likewise cardiovascular disease is the major cause of morbidity and mortality in patients with CKD [1]. With the progression of CKD, there is a change from the traditional atherosclerotic related cardiovascular events, such as stroke and myocardial infarction, to increased arterial stiffness, arteriosclerosis, left ventricular hypertrophy, cardiac fibrosis, and heart failure. This change in pathophysiology is associated with increased risk of arrhythmias and sudden cardiac death. As a consequence

Biomarkers of Kidney Disease. http://dx.doi.org/10.1016/B978-0-12-803014-1.00011-X

of the altered pathophysiology, biomarkers of cardiac injury are more detectable in patients with CKD. These changes in the biomarkers need to be interpreted in the clinical context associated with CKD. This chapter will review traditional markers of kidney disease, traditional markers of cardiovascular disease, and novel markers of kidney damage as markers of cardiovascular risk in subjects with CKD.

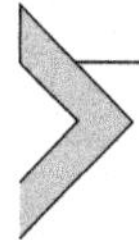

THE DEFINITION OF CKD AND THE RISK OF CARDIOVASCULAR DISEASES: GFR, ALBUMINURIA, AND PROTEINURIA

CKD is common. Defined as a glomerular filtration rate of less than 60 mL/min per 1.73 m^2, or the presence of kidney damage regardless of cause for more than 3 months, the prevalence of CKD in western populations is of the order of 10–16% and increasing [2–4]. In 2013, there were approximately 470 million cases of CKD globally, and, depending on cause, the prevalence had increased from 1995 by 26.8% for CKD from hypertension to 82.5% for CKD due to diabetes mellitus, with intermediate increases for CKD arising from other causes [5].

Most subjects with CKD never reach dialysis. It has long been appreciated that the majority of these subjects with CKD (50%) will die of cardiovascular disease before ever requiring renal replacement therapy [6]. The association between CKD and cardiac pathology (hypertrophy) at post mortem was first reported by Richard Bright in 1836 (cited by Gansevoort [7]). Most recently, it has become accepted that all severity stages of CKD from mild disease to the dialysis-dependent end-stage renal disease (ESRD) are associated with an increase in cardiovascular risk and mortality [1,7–9], with the absolute risk of death increasing exponentially with decreasing kidney function [1,10]. Nevertheless, guidelines on the management of cardiovascular risk have paid limited attention to CKD as a major risk factor. This section will address how the development of the definition of CKD has allowed cardiovascular disease risk to be quantified and examine the robustness of this definition as a risk factor for mortality, particularly cardiovascular mortality.

Nephrologists and others have become increasingly aware of the frequency of cardiovascular disease in CKD, recognizing that this represents the major cause of death among CKD patients [3]. This recognition was assisted by the introduction of a conceptual model in 2002 for definition and classification of CKD by the National Kidney Foundation's Kidney

Disease Outcomes Quality Initiative (KDOQI), which was subsequently endorsed by the Kidney Disease: Improving Global Outcomes (KDIGO) [11]. CKD was defined as kidney damage or a glomerular filtration rate, GFR, <60 mL/min per 1.73 m^2 for 3 months or more, irrespective of cause. Kidney damage was defined by structural or functional abnormalities that could lead to decreased GFR, manifest either by pathologic abnormalities or markers of kidney damage, including abnormalities in the composition of blood or urine, or abnormalities in imaging tests. Albuminuria was accepted as a marker of kidney damage at a threshold albumin-to-creatinine ratio (ACR) of 30 mg/g. Estimates of GFR (eGFR) based on serum creatinine and calculated using estimating equations were also accepted. CKD was subdivided into five stages defined respectively by the eGFR bands 1, ≥90; 60–89; 30–59; 15–29, and <15 mL/min per 1.73 m^2. Early application of the classification to risk assessment revealed that there was an independent, graded association between a reduced eGFR and the risk of death, cardiovascular events and hospitalization [1].

The high prevalence of CKD revealed by this classification has enormous implications for clinical practice, research and public health policy. In the United States, some 12% of the population (25 million) was estimated to have CKD, while <0.2% of the population (>500,000) was treated with dialysis or transplantation [11]. Similar prevalence data have been obtained in other countries. For instance, a cross-sectional survey of Australian adults over 25 years of age revealed a high prevalence of 16% with either kidney damage as shown by proteinuria, or hematuria, and/or a reduced GFR, with over 50% of subjects over 65 years of age having an eGFR <60 mL/min per 1.73 m^2 [4]. While this was revealing, the public health and cost implications of the high prevalence of earlier stages of CKD compared with the incidence of treated kidney failure, and the concern that thresholds of albuminuria and GFR might vary according to age, sex, or race were issues that made the definition and classification of CKD controversial. This led to a reassessment of the CKD definitions at a KDIGO controversies conference in London in 2009 [12]. This meeting provided an opportunity to revisit the definitions after analysis of a large data set for all-cause and cardiovascular mortality, and for kidney outcomes with data pooled from 45 large cohorts totaling 1,555,332 subjects including general population, high risk, and CKD cohorts.

The subdivision of CKD into five stages allowed assessment of the relative risk of five outcomes to declining eGFR and three severity grades of albuminuria (an ACR < 30 mg/g, 30–299 mg/g, and ≥300 mg/g). The

Composite ranking for relative risks by GFR and albuminuria (KDIGO 2009)				Albuminuria stages, description and ragne (mg/g)				
				A1		A2	A3	
				Optimal and high-normal		High	Very high and nephrotic	
				<10	10–29	30–299	300–1999	≥ 2000
GFR stages, description and range (mL/min per 1.73 m^2)	G1	High and optimal	>105					
			90–104					
	G2	Mild	75–89					
			60–74					
	G3a	Mild-moderate	45–59					
	G3b	Moderate-severe	30–44					
	G4	Severe	15–29					
	G5	Kidney failure	<15					

Figure 11.1 ***KDIGO Consensus Classification of Chronic Kidney Disease by Severity.*** The figure shows the composite ranking for relative risks by glomerular filtration rate *(GFR)* and albuminuria (KDIGO, 2009 reported in Ref. [12]) reproduced with permission. Colors reflect the ranking of adjusted relative risk. The ranks assigned in Figure 5 were averaged across all 5 outcomes for the 28 GFR and albuminuria categories. The categories with mean rank numbers 1–8 are *black* (green in the web version), mean rank numbers 9–14 are *light gray* (yellow in the web version), mean rank numbers 15–21 are *gray* (orange in the web version), and mean rank numbers 22–28 are *dark gray* (red in the web version). Color for twelve additional cells with diagonal hash marks is extrapolated based on results from the metaanalysis of chronic kidney disease cohorts [12]. The highest level of albuminuria is termed "nephrotic" to correspond with nephrotic range albuminuria and expressed as ×2000 mg/g.

outcomes assessed were all-cause mortality, cardiovascular mortality, ESRD, acute kidney injury, and progressive kidney disease. Cardiovascular mortality was defined as death due to myocardial infarction, heart failure, sudden cardiac death, or stroke. ESRD was defined as initiation of dialysis or transplantation or death coded as due to kidney disease other than acute kidney injury (AKI). AKI was defined ICD-9 code 584 as primary or additional discharge code. Kidney disease progression was defined as an average annual decline in eGFR during follow-up of at least 2.5 mL/min per 1.73 m^2 per year and a last eGFR value of < 45 mL/min per 1.73 m^2 (Fig. 11.1).

As summarized by Levey et al. [12], the incidence rates were higher for mortality than kidney outcomes, but the risk relationships had similar shape and albuminuria patterns, with higher relative risks associated with lower eGFR and higher levels of albuminuria, suggesting that groups at risk for one outcome were at risk for all outcomes. The graded increase in risk for higher albuminuria categories was independent of eGFR, without an apparent threshold value. The increased relative risk for all outcomes was significant for eGFR of <60 mL/min per 1.73 m^2 in the continuous analysis and in the range 45–59 mL/min per 1.73 m^2 in the categorical analysis, consistent with the threshold value for the definition of CKD (<60 mL/min per 1.73 m^2). The predictive value of albuminuria at all levels of eGFR supported the need to add albuminuria stages to all eGFR stages. Age-stratified analyses showed quantitatively similar patterns in subjects with age <65 and ≥65 years. With the exception of all-cause mortality, the relative hazards were similar above and below 65 years of age, which argued against the use of age-specific GFR thresholds for the definition of CKD.

Multiple analyses support and extend these conclusions. The Chronic Kidney Disease Prognosis Consortium was established by KDIGO to compile and metaanalyze the best available data to provide a more comprehensive assessment of the independent and combined associations of eGFR and albuminuria with mortality and kidney outcomes. Metaanalysis of over 100,000 subjects in 21 general population cohorts in 2009 for all-cause and cardiovascular mortality confirmed that both eGFR and albuminuria were independent predictors of mortality risk in the general population [13]. An exponential increase in mortality risk became significant around an eGFR of 60 mL/min per 1.73 m^2 and was twofold higher around an eGFR of 30–45 mL/min per 1.73 m^2 compared with optimum eGFR levels of 90–104 mL/min per 1.73 m^2 independently of albuminuria. In contrast to the threshold effect observed with eGFR, the association between albuminuria and mortality was linear on the log–log scale with a twofold higher risk at a urinary albumin-creatinine ratio (ACR) of approximately 11.3 mg/mmol (~microalbuminuria) compared with an optimal ACR (0.6 mg/mmol) independently of eGFR and conventional risk factors. Mortality risk was significant even at ACR = 1.1 mg/mmol (~10 mg/g) compared with 0.6 mg/mmol. Further metaanalysis with individual level data from 637,315 adults (>18 years) in 24 cohorts followed for a median of 4 years, comprising 19 general-population cohorts, three high risk cohorts of individuals with diabetes, and two CKD cohorts (Fig. 11.2) [14] demonstrated that the addition of eGFR and ACR significantly improved the

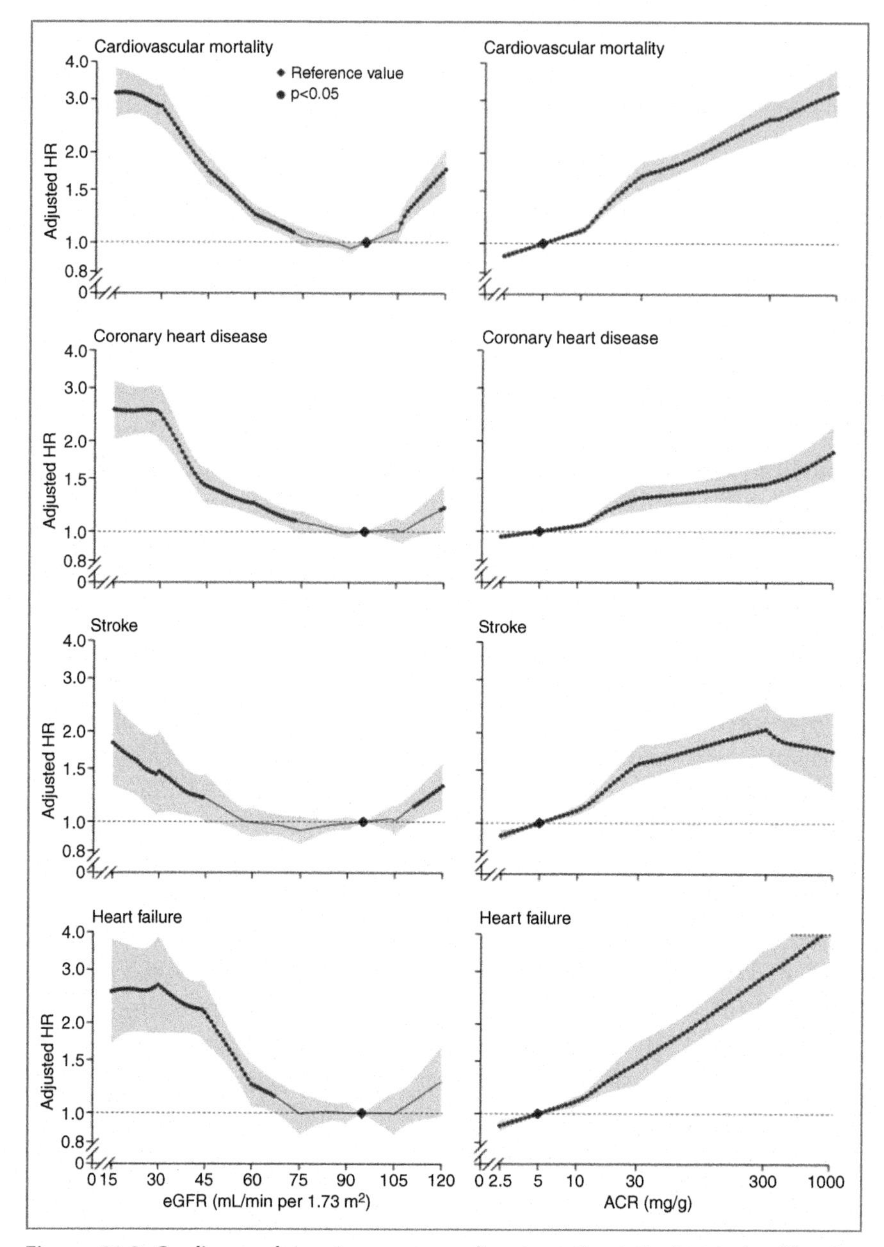

Figure 11.2 ***Cardiovascular outcomes according to estimated glomerular filtration rate*** **(eGFR)** ***and albumin-to-creatinine ratio*** **(ACR)** ***in combined general-population and high-risk cohorts.*** The shaded areas represent 95% CI for cardiovascular mortality, coronary heart disease, stroke and heart failure. *Reproduced with permission from Matsushita K, Coresh J, Sang Y, et al. Estimated glomerular filtration rate and albuminuria for prediction of cardiovascular outcomes: a collaborative meta-analysis of individual participant data. Lancet Diabetes Endocrinol 2015;3:514–525 [14].*

discrimination of cardiovascular outcomes beyond traditional risk factors in general populations, with a greater improvement coming from ACR than eGFR and more evident for cardiovascular mortality and heart failure, than coronary disease and stroke (ibid). Adjusted cardiovascular risk was constant for eGFR values of 75–105 mL/min per 1.73 m^2, and increased steadily below this range (Fig. 11.2). The area under the ROC curve (AUC or C statistic) for cardiovascular outcomes based on traditional risk factors ranged from 0.729 to 0.838 and these values were significantly improved by the addition of either measure of kidney disease. Discrimination improvement was greater with ACR than eGFR and especially evident in individuals with diabetes or hypertension, but still significant in the absence of these disorders [14]. Dipstick proteinuria improved risk prediction to a lesser extent than ACR. As identified previously, these conclusions remain valid regardless of associated comorbidities and the increased cardiovascular risk in patients with CKD is not the result of these underlying diseases [9,15]. This conclusion is strengthened by the observation [10] that 40–50% of patients with low eGFR and albuminuria do not have diabetes or hypertension, and the associations of eGFR and albuminuria with cardiovascular mortality are similar in these patients to individuals with CKD with diabetes or hypertension.

The effect of age was further assessed by the consortium in an individual-level metaanalysis including 2,051,244 participants from 33 general population and high risk (of vascular disease) cohorts and 13 CKD cohorts from Asia, Australasia, Europe, and North and South America followed up for up to 31 years (mean 5.8 years) [16]. Mortality and ESRD risk increased independently with decreasing eGFR below 60 mL/min per 1.73 m^2 regardless of age. Mortality showed a lower relative risk but higher absolute risk differences at older age. In cohorts specifically selected for CKD, age did not modify the mortality associations. The authors further discuss the definition of CKD, highlighting that this could be based on a "normal" distribution in apparently healthy individuals or on kidney markers associated with increased risk of future adverse outcomes. Supporters of the former approach argue for age-specific cut-offs for defining normal kidney function and note an increasing prevalence of nephrosclerosis in renal biopsies with age, even among healthy kidney donors and suggest the concept of "natural inevitable senescence" of the kidneys [17]. However, repeated very large-scale metaanalyses (with over 1 million subjects) conducted by the CKD Consortium document that the risk of mortality and ESRD is increased at a threshold of 60 mL/min per 1.73 m^2 across all age groups. The relative mortality risk associated with low eGFR attenuates with increasing age, as with traditional risk factors and in assessing the interaction with age, the slopes of the hazard ratios versus eGFR at less than 45 mL/min per 1.73 m^2 were largely parallel across the age categories. Similar results were obtained with albuminuria, except, as observed and discussed previously; there was no threshold effect with albuminuria.

Consistent with these observations, CKD confers increased short and long-term cardiovascular mortality risk after surgery. Review of 260,352 evaluable cases in the American College of Surgeons National Surgical Quality Improvement Program (ACS-NSQIP) data sets for 2005–07 demonstrated that increasing severity of CKD stage was associated with progressive increases in short term (30 day) mortality after any Surgery [18]. Recent studies have also demonstrated increased long-term mortality after surgery irrespective of whether acute kidney injury (AKI) was a complication of that surgery [8,19]. While previous studies were focused on all-cause mortality rather than cardiovascular causes, on the assumption that progression to ESRD was the underlying mechanism for observed long-term mortality, these recent studies have examined the association between kidney disease and cause-specific mortality, particularly cardiovascular-specific mortality. In a single-center cohort of 51,457 adult surgical patients, cardiovascular-specific mortality was modeled using a multivariable subdistributional hazards model, treating any other cause of death as a competing risk [8].

This demonstrated that cardiovascular-specific mortality was significantly higher among patients with kidney disease, independently of the progression to ESRD. Preexisting CKD and ESRD and postoperative AKI were the main independent predictors. After a median follow-up of 7 years (IQR: 5–10 years), the two commonest causes of death were cancer (5051/15247, i.e., 33%) and cardiovascular disease (4269/15247, i.e., 28%). Cardiovascular disease accounted for more deaths among patients with any type of kidney disease compared to 18% among patients with no kidney disease ($P < 0.0001$), ranging from 29% for patients with AKI and no CKD, 35% for patients with CKD but no AKI, 45% for those with CKD and AKI during admission, and 55% for ESRD patients). In contrast, cancer accounted for fewer deaths among patients with kidney disease compared to those without. At 10-year follow-up, adjusted cardiovascular-specific mortality estimates were 6, 11, 12, 19, and 27% for patients with no kidney disease, AKI with no CKD, CKD with no AKI, AKI with CKD, and ESRD, respectively ($P < 0.001$). This association remained after excluding 916 patients who progressed to ESRD after discharge although it was significantly amplified among them. Compared to patients with no kidney disease, adjusted hazard ratios for cardiovascular mortality were significantly higher among patients with kidney disease ranging from 1.95 (95% CI: 1.80–2.11) for patients with de novo AKI to 5.70 (CI: 5.00–6.49) for patients with preexisting ESRD. The increasing cardiovascular mortality was proportional to the severity of kidney disease independent of patients' age, gender, comorbidity burden on admission, other postoperative complications or the type of operation in the cohort that included a wide range of major surgical procedures, including noncardiac and specialty surgeries. Thus there is a similar and significant negative impact of both AKI and CKD on long-term survival after surgery due to increased cardiovascular-specific mortality [8].

Similar conclusions were drawn by Huber et al. [19], who examined the association between kidney disease and cardiovascular-specific mortality in a recent large cohort analysis of 3646 subjects undergoing cardiac surgery at a single center from 2000 to 2010. The two commonest causes of the 1577 deaths in the cohort were cardiovascular disease (845 deaths, 56.4%) and cancer (173 deaths, 11.0%). Adjusted cardiovascular mortality estimates at 10 years (after adjusting for demographic variables including age, surgery type, and admission hemoglobin levels) were 17, 31, 30, and 41% respectively for patients with no kidney disease, AKI without CKD, CKD without AKI, and AKI with CKD. The adjusted hazard ratios for cardiovascular mortality were significantly greater for all types of kidney disease than for no kidney disease, even for CKD without AKI (2.01; CI: 1.46–2.78). These were significantly greater than for other risk factors including increasing age, emergent surgery and lower admission hemoglobin levels (less than 10 g/dL versus greater than 12 g/dL).

Biomarkers of Progression of CKD

CKD clearly predicts increased mortality and increasing severity of CKD identifies increasing cardiovascular mortality risk. This raises the question of what predicts progression of CKD. As identified by the earlier large-scale metaanalyses of the CKD Prognosis Consortium, reduced eGFR and severity of albuminuria or proteinuria predict ESRD and AKI but reduced eGFR does not alone predict progression of CKD [12] (Fig. 11.3).

Mortality is associated with the rate of GFR function decline [20] although the mechanism remains unknown [21]. In turn, proteinuria of increasing severity is associated with a faster rate of function decline, regardless of baseline GFR [22]. While past decline in renal function contributes more to the absolute risk of ESRD at lower levels of eGFR, current eGFR was more strongly associated with future ESRD risk than the magnitude of

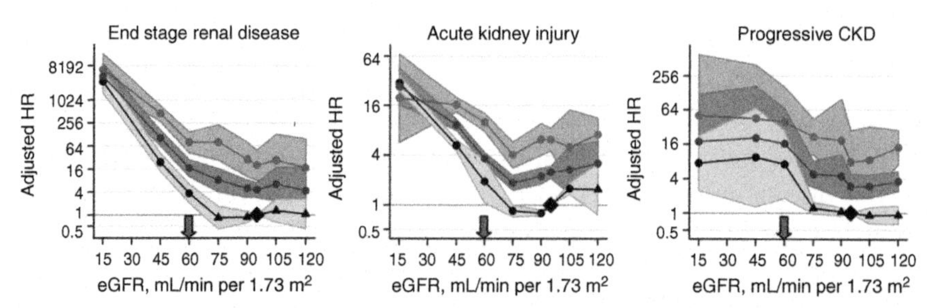

Figure 11.3 ***Kidney outcomes in general population cohorts as a function of eGFR and albuminuria or proteinuria.*** eGFR is expressed as a continuous variable. The three lines represent urine ACR of <30 mg/g or dipstick negative and trace (*light gray*; blue in the web version), urine ACR 30–299 mg/g or dipstick 1+ positive (*gray*; green in the web version), and urine ACR ≥300 mg/g or dipstick ≥2+ positive (*dark gray*; red in the web version). All results are adjusted for covariates and compared with reference point of eGFR of 95 mL/min per 1.73 m² and ACR of 30 mg/g or dipstick negative *(diamond)*. Each point represents the pooled relative risk from a metaanalysis. *Solid circles* indicate statistical significance compared with the reference point ($P < 0.05$); *triangles* indicate nonsignificance. *Arrows* indicate eGFR of 60 mL/min per 1.73 m², threshold value of eGFR for the current definition of CKD. *HR*, hazards ratio; *OR*, odds ratio. *Modified from Levey AS, de Jong PE, Coresh J, et al. The definition, classification, and prognosis of chronic kidney disease: a KDIGO Controversies Conference report. Kidney Int 2011;80:17–28 [12].*

prior decrease, with both factors contributing substantially to ESRD risk when eGFR was <30 mL/min per 1.173 m² [23].

The inability to identify patients at high risk of progression while in early stages of CKD confounds effective management. Several novel biomarkers of tubular injury have also been examined to see if these predict progression of CKD. These include evaluation of the association of liver fatty acid binding protein (L-FABP), kidney injury molecule 1 (KIM-1), *N*-acetyl-β-D-glucosaminidase (NAG), and neutrophil gelatinase–associated lipocalin (NGAL) with incident ESRD in a community-based sample from the Atherosclerosis Risk in Communities Study [24]. This was a matched case-control study of 135 patients with ESRD and 186 controls who were matched on sex, race, kidney function, and diabetes status at baseline. The biomarkers were indexed to urinary creatinine. No significant associations were observed for NAG/Cr, NGAL/Cr, or L-FABP/Cr with ESRD. Those in the highest category for KIM-1/Cr had a higher risk of ESRD compared with those with undetectable biomarker levels (reference group) in unadjusted models (odds ratio, 2.24; 95% confidence interval, 1.97–4.69; $P = 0.03$) or adjustment for age (odds ratio, 2.23; 95% confidence interval,

1.06–4.67; $P = 0.03$). No association between KIM-1/Cr and ESRD was found when KIM-1/Cr was analyzed as a continuous variable.

In a prospective observational Canadian cohort study in 2544 CKD patients with eGFR of 15–45 mL/min per 1.73 m^2, the utility of the newer biomarkers, cystatin C, high sensitivity C-reactive protein (hsCRP), interleukin 6 (IL6), transforming growth factor β1 (TGFβ1), fibroblast growth factor 23 (FGF23), N-terminal probrain natriuretic peptide (NT-proBNP), troponin I and asymmetric dimethylarginine (ADMA), was assessed with respect to prediction of renal and mortality outcomes during the first year of enrolment [25]. The biomarkers did not improve prediction of dialysis, when added to conventional risk factors, such as eGFR, uACR, hemoglobin, phosphate, and albumin. However, in predicting death within 1 year, cystatin C, NT-proBNP, hsCRP and FGF23 values significantly improved model discrimination and reclassification: increasing the c statistic for absolute risk by 4.3% and Net Reclassification Improvement for categories of low, intermediate and high risk by 11.2%.

While CKD has been presumed to follow an unremittingly progressive decline over time, increasing evidence demonstrates that many subjects remain stable over time while a significant minority of 3–9% actually have improving function [25–27]. Several prospective cohort studies of CKD subjects have demonstrated that the majority of subjects do not progress to more severe stages of CKD or end stage kidney disease.

For example, among the recruited Sunnybrook cohort of 1607 subjects referred with Stage 3 CKD, 21% progressed to Stage 4 CKD over a median follow-up time of 3 years, whereas only 3% of patients developed ESRD [28]. Progression to Stage 4 CKD was associated with a nearly twofold higher risk of mortality, AKI and all-cause hospitalizations when compared with patients who did not progress by the end of follow-up, and these risks were similar in subgroups of patients with either Stage 3A or 3B CKD at the origin. The risk of cardiovascular-related hospitalization after progression to Stage 4 was higher than noncardiovascular-related hospitalization. Importantly, the observed risks of mortality and morbidity were due to progression to Stage 4 CKD itself rather than as a result of subsequent ESRD. The risk of subsequent mortality and hospitalization remained significant after adjustment for baseline eGFR, age, and other potential confounders, suggesting that progression to Stage 4 was independently associated with long-term adverse clinical outcomes. Survivor bias was thoughtfully avoided in this analysis (in patients presenting with Stage 4 CKD, who would have had to endure a longer duration of CKD and consequently, a longer period of elevated mortality risk, compared with patients with Stage 3 CKD) by treating the progression to Stage 4 CKD as a time-dependent variable, comparing hospitalization and mortality risks subsequent to the progression, for those who progressed to Stage 4, with otherwise similar patients who did not progress to Stage 4 during the observation period. In the CanPREDDICT study, a prospective multicentre cohort of 2544 adult CKD patients with an eGFR of 15–45 mL/min per 1.73 m^2 enrolled across Canada from 2008 to 2009 over 18 months, were followed for 3 years or until death [25]. After 12 months, a total of 142 patients (5.9%) had progressed to RRT and 137 patients (5.7%) had died [15 (0.6%) after starting RRT]. Twelve months, 10% progressed rapidly (defined as decline of renal function of more than 30%), 13% had slower decline of renal function (20–30%), and 69% were stable, as defined by the change within 20% of baseline, while 9% improved (defined as increase in eGFR > 20%).

Progression from Stage 3 to Stage 4 CKD or to ESRD identifies patients at especially high risk of cardiovascular adverse events who require extra vigilance and some of whom might benefit from early instigation of treatment based on guidelines for CKD 4 subjects [28]. However, there are currently no biomarkers to identify subjects who will progress. This has important clinical and public health implications, some of which have been reviewed (ibid). Consider, for example, the cost-benefit implications of treating all CKD Stage 3 subjects, given the proportion of the population with Stage 3 CKD is 15 times as large as that for Stage 4 CKD (Fig. 11.4).

Summary

In conclusion, no biomarker convincingly predicts progression of CKD. However, in both general and high-risk populations, the risk of cardiovascular disease and cardiovascular-specific mortality increases with stage of kidney disease and loss of GFR. Importantly, from a population perspective, individuals with earlier stages of CKD are more likely to die of cardiovascular disease than to develop ESRD and require dialysis. The definition of CKD utilizing both eGFR and albuminuria appears sound [7]. Incorporation of eGFR or ACR or both into prediction models provides similar or better risk prediction than most of the traditional risk factors including blood pressure, lipids, and smoking [14] and supports the use of both measures to improve assessment of cardiovascular risk in the general population.

The next section examines what is the role of other biomarkers of cardiac disease (both established and novel newer markers) in predicting

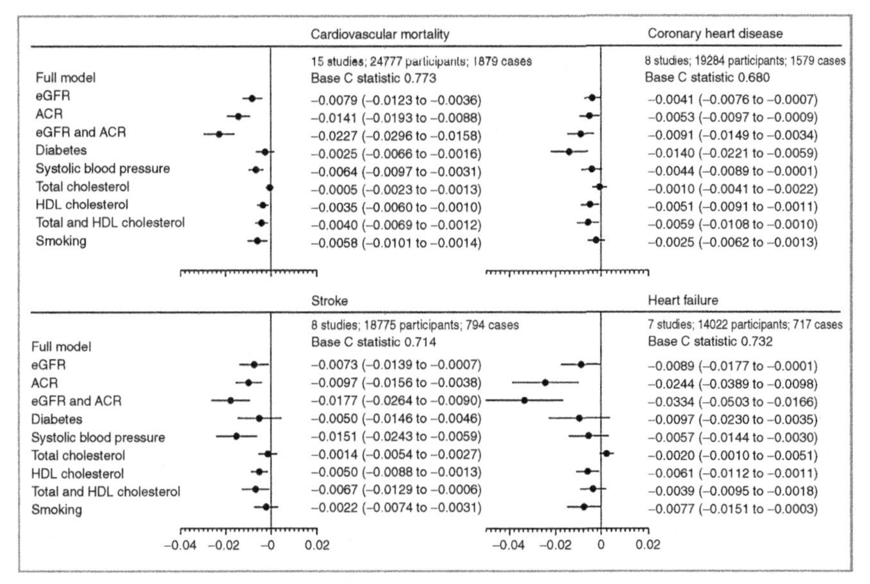

Figure 11.4 ***Cardiovascular outcomes with omission of kidney disease measures and traditional risk factors in a CKD population.*** The figure shows the difference in C statistics and 95% CIs for four cardiovascular outcomes with omission of kidney disease measures and traditional risk factors from a model with all risk factors in a CKD population. Systolic blood pressure includes two different coefficients for those with and without antihypertensive drugs. Total and HDL cholesterol refers to simultaneous omission of these two predictors. *From Matsushita K, Coresh J, Sang Y, et al. Estimated glomerular filtration rate and albuminuria for prediction of cardiovascular outcomes: a collaborative meta-analysis of individual participant data. Lancet Diabetes Endocrinol 2015;3:514–525 [14].*

cardiovascular outcomes in individuals with CKD compared to those with normal kidney function.

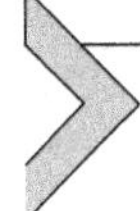

CARDIAC BIOMARKERS IN CKD

Troponins

Elevated troponins are the biomarkers of choice for cardiac injury. The cardiac troponin complex plays a critical role in the regulation of excitation–contraction coupling in the heart. The release of cardiac specific troponins into the circulation are highly specific for cardiac myocyte injury [29] and have become the biomarker of choice for the detection of cardiac injury but do not define the mechanism of injury [30]. With the use of highly sensitive cardiac troponin assays (hs-cTn), low hs-cTn concentrations can be detected in 50–90% of the healthy population, such that an elevation of hs-cTn above the 99th percentile provides a sensitive and diagnostic marker of cardiac injury [31]. Elevated cTn values in patients with acute ischemic presentations, with and without ECG changes, identify patients at higher risk for the development of cardiac events both during short- and long-term follow-up. The short-term outcome is closely associated with the higher acute thrombotic risk of the underlying unstable plaque, where as the long-term prognosis most likely reflects the higher prevalence of a more severe and complex coronary anatomy [29].

CKD is associated with elevated levels of cardiac troponins (both cTnI and cTnT) which in the absence of symptoms of an acute coronary syndrome most likely reflect subtle myocardial injury facilitated by the impact of the CKD [32,33]. Michos EA and coworkers have undertaken a systematic review and metaanalysis of troponin testing in patients with CKD without evidence of an acute coronary syndrome [34]. They analyzed papers that investigated troponin testing in patients on dialysis and for those not on dialysis. For those on dialysis, they identified 11 studies, measuring cTnT where they were able to metaanalyze the pool hazard ratios that had been adjusted for age and coronary artery disease or a risk equivalent. They found that there was a threefold increased risk of all-cause mortality [HR = 3.0 (95% CI: 2.4–4.3)]. For cardiovascular mortality, adjusted for age and coronary artery disease, there was a metaanalysis of the pooled hazards ratio for five studies which demonstrated a threefold increase in cardiovascular mortality [HR = 3.3 (95% CI: 1.8–5.4)] but there was significant heterogeneity between the studies. When they undertook a metaanalysis of the odds risk, there were nine comparable studies that demonstrated an odds risk of cardiovascular mortality of 4.3 (95% CI: 3.0–6.4). Similar results were seen for cTnI assays [34].

A smaller number of studies investigated the predictive value of troponins in patients with CKD not on dialysis [34]. There were two studies measuring TnT levels and all-cause mortality adjusted for age and coronary artery disease and the metaanalysis of the pooled hazards ratio demonstrated a 3.4 fold risk (95% CI: 1.1–11.0) and a metaanalysis of the odds risk from 5 studies demonstrated an OR = 3.0 (95% CI: 1.4–6.7). They concluded that an elevated baseline troponin was associated with a higher risk (approximately two- to threefold) for all-cause mortality, cardiovascular mortality, and major adverse cardiac events, for individuals on dialysis without suspected acute coronary syndrome. There was a similar trend for those with CKD not on dialysis but there were fewer studies available for analysis [34].

A few studies have compared cardiac troponins alone or in conjunction with other biomarkers to better clarify cardiovascular risk in CKD or using cardiac troponins to identify intervention strategies to reduce cardiovascular risk. In the Atherosclerosis Risk in the Community (ARIC) study, the associations of cTnT and NT-proBNP with CVD were independent of kidney function and actually stronger in those with CKD than in those without CKD [35]. Similar findings have reported in the Chronic Renal Insufficiency Cohort (CRIC) study which demonstrated that hsTnT and NT-proBNP were strongly associated with incident heart failure among a diverse cohort of individuals with mild to severe CKD. Elevations in these biomarkers may indicate subclinical changes in volume and myocardial stress that subsequently contribute to the development of clinical heart failure [36].

In CKD patients presenting with symptoms of an acute coronary syndrome, the clinical utility of cardiac troponins is less clear as there is limited data to establish the diagnostic cutoff values for patients with CKD or the required magnitude of change in serial values to define an acute myocardial event [33]. In a subgroup analysis of a prospective multicenter study investigating the utility of different cardiac troponin assays for patients presenting with symptoms of an acute coronary syndrome, for patients with CKD, where the cardiac troponins were greater than the 99th percentile (normal range), these supported the adjudicated diagnosis of an acute myocardial infarction in 45–80% cases [37]. However, the optimal receiver–operator characteristic curve derived cTn cutoff levels in patients with renal dysfunction were significantly higher compared with those in patients with normal renal function by a factor of 1.9–3.4 [37]. In another study investigating weekly variations of cardiac troponins in stable healthy controls and hemodialysis patients, the investigators suggested that a greater than 25% change for hs cTnT concentrations in the hemodialysis group was diagnostic with a low probability for a false positive result. Using change would reduce the need for diagnostic cutoffs for the hs cTnT assay [38,39].

Natriuretic Peptides

Natriuretic peptides play an important role in volume homeostasis via vasodilation, natriuresis, and diuresis to protect the cardiovascular system from volume overload. There are three natriuretic peptides, atrial natriuretic peptide (ANP), brain natriuretic peptide (BNP), and C-type natriuretic peptide (CNP). ANP is produced and secreted from atria, BNP is predominantly synthesized in the cardiac ventricles with myocardial wall stress and stretch the major stimuli for release. CNP is released from the endothelium as well as the myocardium [40]. Natriuretic peptides, in particular BNP, are useful markers for the diagnosis and evaluation of patients with heart failure and volume overload as well as having prognostic value in a number of cardiovascular conditions in both the general population as well as in patients with CKD [40,41].

There have been few studies that have investigated the role of ANP as a biomarker of cardiovascular disease in patients with CKD. Initial studies investigated the role of ANP as a biomarker of volume status in dialysis patients [42–44]. These studies demonstrated that ANP concentrations were closely related to atrial volume and or cardiac filling pressures. However cardiac function was a major confounder in the interpretation of the ANP concentrations documented in patients with CKD. Zoccali and coworkers examined both the predictive value of both BNP and ANP for all-cause mortality and cardiovascular mortality [44]. They demonstrated that BNP was a better indicator of cardiovascular outcomes compared to ANP, and in a Cox's model controlling for left ventricular mass and ejection fraction, only BNP remained an independent predictor of death [44]. Although C-natriuretic peptide has been linked to vascular and cardiac function, there have been no studies examining the predictive role of CNP for cardiovascular outcomes in patients with CKD.

Brain Natriuretic Peptide

Brain natriuretic peptide (BNP) arises from pre-pro BNP, which is synthesized in the cardiomyocytes and is initially cleaved to pro-BNP, which is further cleaved to the active BNP and the inactive N terminal pro BNP (NT proBNP). NT-proBNP is more stable with a longer half-life and is considered a better marker for heart failure and volume overload, as well as for hard outcomes of death and cardiovascular events (Fig. 11.3) [33]. BNP and NT-proBNP plasma concentrations are expressed as pmol/L or pg/mL. The conversion factor for BNP is 1 pg/mL = 0.289 pmol/L. For NT-proBNP the conversion is 1 pg/mL = 0.118 pmol/L [45] (Fig. 11.5).

Although there has been some concern that natriuretic peptides are metabolized and cleared by the kidneys [45,46], a number of studies have demonstrated that left ventricular dysfunction is the major determinant of elevated BNP and NT-proBNP concentrations independent of the severity of kidney dysfunction [35,36,47]. In the CRIC study, elevated NT-proBNP levels were a strong predictor of incipient heart failure. When compared to individuals in the lowest quartile for NT-proBNP (<47.6 ng/mL)

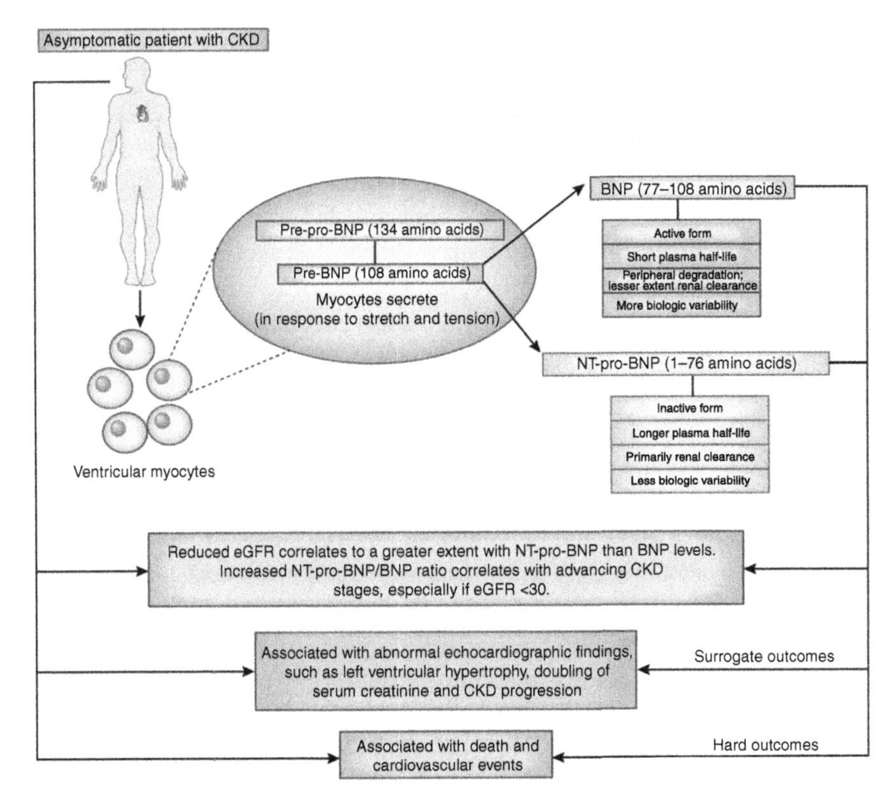

Figure 11.5 ***Brain natriuretic peptides (BNP) in patients with CKD.*** In response to increased stretch or tension, left ventricular myocytes release BNP and N-terminal-pro-BNP (NT-pro-BNP) from precursors. BNP is an active molecule with a short plasma half-life and is degraded in the circulation by enzymatic action. NT-pro-BNP is the inactive form of BNP, with a longer half-life. It is primarily cleared by the kidneys. Reduced eGFR correlates to a greater extent with elevated plasma NT-pro-BNP than to BNP levels. Increased NT-pro-BNP/BNP ratio correlates with advancing CKD stages, especially if the eGFR is 30 mL/min per 1.73 m^2. However, both BNP and NT-pro-BNP are associated with surrogate and hard clinical outcomes in asymptomatic patients with CKD. *From Colbert G, Jain N, de Lemos JA, Hedayati SS. Utility of traditional circulating and imaging-based cardiac biomarkers in patients with predialysis CKD. CJASN 2015;10:515–529 [33].*

participants in the highest quartile (>433.0 ng/mL) had a significantly higher rate of heart failure (HR 9.57: 95% CI: 4.40–20.53) [36].This association remained strong, even after the adjustment for a broad range of traditional and nontraditional cardiovascular risk factors The risk of heart failure with elevated NT-proBNP was similar in CKD patients in both those with incipient heart failure and those with normal cardiac function [36]. Similar findings have been reported in the Atherosclerosis Risk in the Community (ARIC) study, which demonstrated that cardiac markers NT-proBNP and cardiac troponin T were more sensitive for predicting cardiovascular disease in patients with CKD than kidney markers (cystatin C, β_2 microglobulin, β-trace protein) or hsCRP [35]. Of interest, the prediction of cardiovascular disease in individuals with CKD by cTNT and NT-proBNP was considerably higher than that reported for traditional risk factors of total cholesterol, LDL-cholesterol and hsCRP [35,47,48]. This is not surprising in that both elevations in hs cTnT and NT-proBNP reflect subclinical changes in cardiac structure and function which are highly likely in patients with CKD [35]. For further review of the various studies reporting the strong association of NT-proBNP and cardiovascular events, see Table 2 in the well-summarized review by Colbert and coworkers [33].

For patients with end stage kidney disease on dialysis, (hemodialysis or peritoneal dialysis), BNP and NT-proBNP are associated with left ventricular structural and functional abnormalities, which are prevalent in almost all ESKD patients. In a comprehensive review by Wang and Lai [41] they commented that BNP and NT-proBNP have a limited role in assessing acute changes in extracellular volume and the diagnostic utility for BNP is to rule out systolic dysfunction and predict the presence of left ventricular hypertrophy. Whereas BNP and NT-proBNP consistently have been shown to have powerful prognostic value for mortality and cardiovascular death in both hemodialysis patients [49,50] and peritoneal dialysis patients [49], this association is independent of and well beyond that contributed by alterations in left ventricular mass and function [40,41,44,50].

With regard to clinical utility, both BNP and NT-proBNP can be used for prognosis in patients with CKD with respect to cardiovascular events and death. What is less clear is the role of NT-proBNP as a useful diagnostic test for acute congestive heart failure in patients with CKD [33]. In the PRIDE study, the investigators, using a higher cut off value for NT-proBNP of >1200 pg/mL (>142 pmol/L) for patients with an eGFR <60 mL/min per 1.73 m^2, NT-proBNP was found to be a useful diagnostic tool, for acute decompensated heart failure with a sensitivity of 89% and specificity of 72% [51]. Further prospective studies are required to corroborate these findings.

C-Reactive Protein

C-reactive protein (CRP) is an acute phase reactant and is measured as high-sensitivity C-reactive protein (hs-CRP). hs-CRP is an established marker of systemic inflammation as well as a predictor of future cardiac events, in the general population and in patients with CKD [40,52–54]. A number of studies have reported a significant association between hs-CRP and all-cause mortality in ESKD patients. In a study of 399 hemodialysis patients, 46% of patients had an elevated concentration of hs-CRP, which was associated with a 2 year mortality rate of 44% [48]. In the majority of dialysis patients hs-CRP values can be variable related to a number of different inflammatory stimuli, and a number of authors have suggested that serial measurements of hs-CRP may be more meaningful, with only those individuals with a persistently elevated hs-CRP (>10 mg/L) at increased risk of death [52,53].

In CKD patients, similarly hs-CRP levels can be quite variable, but elevated levels of hs-CRP have been associated with a higher risk for cardiovascular events in a community study of elderly patients with CKD [55]. When compared to patients with low hs-CRP and no CKD, the adjusted hazard ratio for those individuals with a high hs-CRP (>2.1 mg/L) and CKD (eGFR < 60 mL/min) was 1.93 (95% CI: 1.45–2.89) [55]. In the Chronic Renal Insufficiency Cohort study, higher levels of hs-CRP were associated with left ventricular hypertrophy and impaired function [53]. Overall, hs-CRP may be useful as an additional assessment of cardiovascular risk, but is of less value if there is already established coronary artery disease. Also on its own, hs-CRP is less predictive of cardiovascular mortality than NT-BNP or cTnT levels [40]. Interestingly, the predictive power of ACR and eGFR either alone or together was substantially higher than that observed in a metaanalysis using hs-CRP as a predictor of cardiovascular outcomes [14,47].

Fibroblast Growth Factor 23

Components of the CKD–mineral bone disorder (CKD–MBD), phosphate, parathyroid hormone, vitamin D, and more recently fibroblast growth factor 23 (FGF-23) have been independently associated with the risk of all-cause mortality in patients with CKD and on dialysis, as well as heart failure, cardiovascular events and death in the general population [40]. FGF-23 levels increase in CKD as a physiological adaptation to maintain neutral phosphate balance and normal serum phosphate concentrations in the setting of reduced phosphate excretion. In the Chronic Renal Insufficiency Cohort (CRIC) study, elevated FGF-23 was independently associated with a greater risk of death and this mortality risk increased by each quartile of measured FGF-23 with the hazard ratio (HR) for the second quartile 1.3 (95% CI: 0.8–2.2), for the third quartile HR 2.0 (95% CI: 1.2–3.3) and for the fourth quartile HR 3.0 (95% CI: 1.85–5.1) [56]. In addition there was a graded and independent association with heart failure and atherosclerotic events [57]. A more recent publication from the CRIC study demonstrated that FGF-23 is associated with left ventricular hypertrophy but does not promote vascular calcification. In contrast, higher serum phosphate levels were associated with the prevalence and severity of coronary artery calcium on CT coronary angiography, even after adjustment for FGF23 [58]. The association of FGF-23 with total mortality was robust across all eGFR strata, supporting the notion that FGF-23 mediates total mortality independently of bone mineral disease and other CKD-mediated comorbidities [59].

In the Cardiovascular Health Study of older people living in the community [60], they examined the strength of the association of FGF-23 with kidney function at the baseline visit and the extent to which associations of FGF-23 with longitudinal outcomes were dependent on the presence of concomitant CKD. Modest decrements in eGFR and elevations in urine Albumin Creatinine Ratios are each independently associated with higher FGF-23. In the CKD group (n = 1,128), the highest FGF-23 quartile had adjusted hazards ratios (HR) of 1.87 (95% confidence interval, CI: 1.47–2.38) for all-cause death, 1.94 (95% CI: 1.32–2.83) for incident HF, and 1.49 (95% CI: 1.02–2.18) for incident CVD events compared with the lowest quartile. These results support the findings in the CRIC study demonstrating an association between, CKD, elevated FGF-23 concentrations and heart failure as well as for all-cause mortality [60]. Similar findings have also been reported in the Homocysteine in Kidney and End Stage Renal Disease (HOST) which also demonstrated in patients with advanced CKD, FGF-23 was strongly and independently associated with all-cause mortality, cardiovascular events, and initiation of chronic dialysis [61].

Thus measurement of FGF-23 may contribute to our understanding of mechanisms linking CKD to cardiovascular disease but the clinical utility for FGF 23 is yet to be established. FGF-23 is not routinely available in clinical practice, nor has the predictive value been evaluated against other well-documented cardiovascular risk assessment tools. Further studies are required to compare FGF-23 to other biomarkers either alone or in combination to identify its' true predictive value [62].

A number of newer markers have been identified that are associated with the increased cardiovascular risk in the setting of CD and possibly relate to the overall increased proinflammatory state of progressive CKD. Most of these newer markers have been identified from prospective longitudinal cohort studies where these has been a single baseline sample taken.

As such, causality cannot be demonstrated merely an association with increased mortality and cardiovascular risk. The markers discussed later require further clinical studies to establish a true predictive role and as such are not yet in routine clinical use.

Placental Growth Factor

Placental Growth Factor (PlGF) is a pleiotropic cytokine, similar to vascular endothelial growth factor (VEGF), which binds to its membrane bound receptor—fms tyrosine kinase 1 (Flt-1) to stimulate angiogenesis. PlGF expression is increased in atherosclerotic plaques with activation of monocytes and macrophages and subsequent increased release of inflammatory angiogenic mediators, increasing the risk of plaque rupture [63]. In the Novel Assessment of Risk management for Atherosclerotic disease in CKD (NARA-CKD) cohort of 1351 individuals followed for a median of 3 years, Matsui and coworkers demonstrated that serum PlGF concentrations are strongly and independently associated with all-cause mortality and cardiovascular events in CKD patients [63]. Using the median of the lowest quartile for serum PlGF (7.4 pg/mL) as the reference, in adjusted analyses the hazard ratio for mortality and cardiovascular risk increased in each successive quartile of serum PlGF level; hazard ratios (HRs) [95% confidence intervals (95% CIs)] for mortality and cardiovascular risk, respectively, were 1.59 (0.83–3.16) and 1.55 (0.92–2.66) for the second quartile, 2.97 (1.67–5.59) and 3.39 (2.20–5.41) for the third quartile, and 3.87 (2.24–7.08) and 8.42 (5.54–13.3) for the fourth quartile (see figure) [63]. With stratification for age, sex, traditional risk factors, and eGFR, serum PlGF still remained an independent risk factor for all-cause mortality and cardiovascular events in this cohort of patients with CKD [63].

There are some limitations to the interpretation of these results. It is implied that all the cardiovascular events are related to ischemic cardiac events and plaque rupture. While this might hold true for acute coronary syndromes, it is not necessarily a valid assumption for sudden cardiac death and heart failure in patients with CKD, where there are other contributing factors. Likewise, it is a single time point measurement at baseline in an observational study. It is not clear what happens to PlGF over time and with progression of CKD (Fig. 11.6).

Galectin 3

Galectin 3 is a member of the multifunctional galectin family, which is ubiquitously expressed in the heart, the kidney, blood vessels, and macrophages

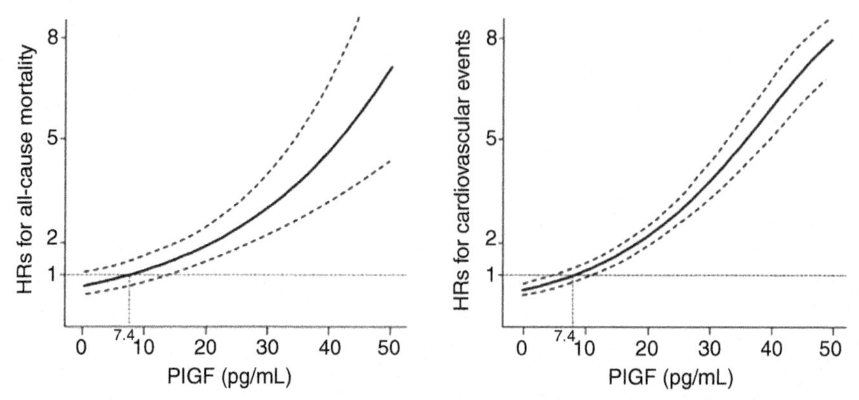

Figure 11.6 ***Multivariable-adjusted hazard ratios*** **(HRs)** ***for all-cause mortality and for cardiovascular events over the distribution of serum placental growth factor*** **(PlGF)** ***concentrations.*** The median of the lowest quartile of PlGF concentration served as the reference group (7.4 pg/mL). The Cox model was adjusted for age, sex, diabetes, hypertension, dyslipidemia, obesity, smoking, previous coronary artery disease, previous stroke, CKD stage, hemoglobin, serum albumin, calcium, phosphorus, C-reactive protein, HbA1c, LDL cholesterol, HDL cholesterol, triglycerides, and use of RAS blockers, calcium channel blockers, β-blockers, mineralo-corticoid receptor antagonists, lipid-lowering agents, diuretics, and antiplatelet agents. The *solid lines* show HRs for mortality and cardiovascular events. *Dashed lines* indicate 95% CIs. *From Matsui M, Uemura S, Takeda Y, et al. Placental growth factor as a predictor of cardiovascular events in patients with CKD from the NARA-CKD study. J Am Soc Nephrol 2015;26:2871–2881 [63].*

and plays a role in tissue fibrosis, immunity, and the inflammatory response. In a large combined cohort of patients from the 4D (German Diabetes Dialysis) and the Ludwigshafen Risk and Cardiovascular Health (LURIC) study, circulating galectin-3 concentrations increase in parallel with decreasing kidney function and were markedly elevated in dialysis patients with type 2 diabetes mellitus [64]. In this study, Galectin-3 was significantly associated with clinical end points, but only in patients with reduced kidney function [64]. It remains unclear if galectin-3 adds to eGFR as a marker of cardiovascular risk.

Neutrophil Gelatinase-Associated Lipocalin

Neutrophil Gelatinase-Associated Lipocalin (NGAL) is a member of the lipocalin iron-carrying protein family expressed in a variety of tissues including neutrophils, kidney, and cardiomyocytes. At steady state plasma and urinary NGAL concentrations are negligible. Plasma and urinary levels are increased in both acute kidney injury and CKD. Elevated plasma NGAL is a strong predictor of acute kidney injury, the subsequent need for acute renal replacement therapy and in-hospital mortality [65]. Urinary NGAL concentrations are increased with CKD [66], however the actual discrimination value of urinary NGAL as a predictor of CKD progression is marginal [67].

The role of NGAL as a marker of cardiovascular risk is much more limited, with only a few studies demonstrating a small predictive benefit of elevated serum NGAL concentrations for cardiac events. In a study of 673 patients presenting with a ST segment elevation myocardial infarction (STEMI), and who underwent percutaneous coronary intervention, a serum NGAL was measured on hospital admission [68]. These patients all had a normal serum creatinine on admission. A serum NGAL > 110 pg/mL was associated with a 1 year mortality of 20% suggesting that serum NGAL measurement can enhance the prognostic score for 1 year mortality rates in patients presenting with STEMI [68]. Liu and coworkers examined the predictive value of urinary NGAL in the Chronic Renal Insufficiency Cohort study. This demonstrated that higher urinary NGAL concentrations adjusted hazard ratio (HR) for the highest (>49 ng/mL) versus the lowest (<6.9 ng/mL) quartile: 1.83 (95% CI: 1.20–2.81), was associated independently with ischemic atherosclerotic events but not heart failure events or death [69].

The different predictive values of serum NGAL or urinary NGAL for cardiovascular events particularly in the setting of CKD where the NGAL values are already elevated, needs clarification as to what they are actually measuring. For example, are the elevated urinary NGAL concentrations reflecting changes in kidney function in the setting of acute or chronic cardiovascular events or actually reflecting the cardiac events themselves. Given that serum NGAL concentrations can be derived from cardiac tissue, neutrophils as well as kidneys, makes it less clear as to the role of NGAL as a predictive tool. At present, studies are based on a single baseline sample. Further studies are required to determine the importance of changes over time, as well as the actual renal clearance of serum NGAL.

SUMMARY

CKD is an established independent risk factor for cardiovascular disease. Likewise cardiovascular disease is the major cause of morbidity and mortality in patients with CKD [1]. There are a number of biomarkers of both cardiac and kidney in origin that have predictive value in identifying the increased risk of cardiovascular events in individuals with CKD. Biomarkers can be useful both to add diagnosis of cardiovascular events and as a prognostic tool with respect to long-term outcomes. To date, the addition of eGFR and urinary albumin/creatinine ratio (ACR) significantly improves the discrimination of cardiovascular outcomes beyond traditional risk factors in general populations, with a greater improvement coming from ACR than eGFR and more evident for cardiovascular mortality and heart failure, than coronary disease and stroke [14].

The associations of eGFR and albuminuria with cardiovascular mortality are even stronger in those patients with CKD [10] and the predictive power of ACR and eGFR either alone or together is substantially higher than that observed in a metaanalysis using hs-CRP as a predictor of cardiovascular outcomes [14,47]. While newer biomarkers have been demonstrated to have predictive value for cardiovascular outcomes, these require larger clinical studies to really identify their diagnostic and prognostic roles in clinical management.

REFERENCES

[1] Go AS, Chertow GM, Fan D, Mcculloch CE, Hsu C-Y. Chronic kidney disease and the risks of death, cardiovascular events, and hospitalization. N Engl J Med 2004;351:1296–305.

[2] Coresh J, Selvin E, Stevens LA, et al. Prevalence of chronic kidney disease in the United States. JAMA 2007;298:2038–47.

[3] Sarnak MJ, Levey AS, Schoolwerth AC, et al. Kidney disease as a risk factor for development of cardiovascular disease: a statement from the American Heart Association Councils on Kidney in Cardiovascular Disease, High Blood Pressure Research, Clinical Cardiology, and Epidemiology and Prevention. Circulation 2003;108:2154–69.

[4] Chadban SJ, Briganti EM, Kerr PG. Prevalence of kidney damage in Australian adults: the AusDiab kidney study. J Am Soc Nephrol 2003;14:S131–8.

[5] Iburg KM, Voss T, Salomon J, Murray CJ. Global, regional, and national incidence, prevalence, and years lived with disability for 301 acute and chronic diseases and injuries in 188 countries, 1990-2013: a systematic analysis for the Global Burden of Disease Study 2013. Lancet 2015;386:743–800.

[6] Levey AS, Beto JA, Coronado BE, Eknoyan G. Controlling the epidemic of cardiovascular disease in chronic renal disease: what do we know? What do we need to learn? Where do we go from here? Am J Kidney Dis 1998;32:853–906.

[7] Gansevoort RT, Correa-Rotter R, Hemmelgarn BR, et al. Chronic kidney disease and cardiovascular risk: epidemiology, mechanisms, and prevention. Lancet 2013;382:339–52.

[8] Ozrazgat-Bislanti T, Matthew H, Kent B, et al. Acute and chronic kidney disease and cardiovascular mortality after major surgery. Ann Surg 2016. (Epub ahead of print).

[9] Fox CS, Matsushita K, Woodward M, Bilo H. Associations of kidney disease measures with mortality and end-stage renal disease in individuals with and without diabetes: a meta-analysis. Lancet 2012;380:1662–73.

[10] Tonelli M, Wiebe N, Culleton B, et al. Chronic kidney disease and mortality risk: a systematic review. J Am Soc Nephrol 2006;17:2034–47.

[11] Levey AS, Eckardt KU, Tsukamoto Y, et al. Definition and classification of chronic kidney disease: a position statement from Kidney Disease: Improving Global Outcomes (KDIGO). Kidney Int 2005;67:2089–100.

[12] Levey AS, de Jong PE, Coresh J, et al. The definition, classification, and prognosis of chronic kidney disease: a KDIGO Controversies Conference report. Kidney Int 2011;80:17–28.

[13] Chronic Kidney Disease Prognosis Consortium. Association of estimated glomerular filtration rate and albuminuria with all-cause and cardiovascular mortality in general population cohorts: a collaborative meta-analysis. Lancet 2010;375:2073–81.
[14] Matsushita K, Coresh J, Sang Y, et al. Estimated glomerular filtration rate and albuminuria for prediction of cardiovascular outcomes: a collaborative meta-analysis of individual participant data. Lancet Diabetes Endocrinol 2015;3:514–25.
[15] Mahmoodi BK, Matsushita K, Woodward M, et al. Associations of kidney disease measures with mortality and end-stage renal disease in individuals with and without hypertension: a meta-analysis. Lancet 2012;380:1649–61.
[16] Hallan SI, Matsushita K, Sang Y, et al. Age and association of kidney measures with mortality and end-stage renal disease. JAMA 2012;308:2349–60.
[17] Glassock RJ, Rule AD. The implications of anatomical and functional changes of the aging kidney: with an emphasis on the glomeruli. Kidney Int 2012;82:270–7.
[18] Gaber AO, Moore LW, Aloia TA, et al. Cross-sectional and case-control analyses of the association of kidney function staging with adverse postoperative outcomes in general and vascular surgery. Ann Surg 2013;258:169–77.
[19] Huber M, Ozrazgat-Baslanti T, Thottakkara P, et al. Mortality and cost of acute and chronic kidney disease after vascular surgery. Ann Vasc Surg 2015;30. 72.e1-2–81.e1-2.
[20] Al-Aly Z, Zeringue A, Fu J, et al. Rate of kidney function decline associates with mortality. J Am Soc Nephrol 2010;21:1961–9.
[21] Kovesdy CP. Rate of kidney function decline associates with increased risk of death. J Am Soc Nephrol 2010;21:1814–6.
[22] Turin TC, James M, Ravani P, et al. Proteinuria and rate of change in kidney function in a community-based population. J Am Soc Nephrol 2013;24:1661–7.
[23] Kovesdy CP, Coresh J, Ballew SH. Past decline versus current eGFR and subsequent ESRD risk. J Am Soc Nephrol 2016;27:2447–55.
[24] Foster MC, Coresh J, Bonventre JV, et al. Urinary biomarkers and risk of ESRD in the atherosclerosis risk in communities study. CJASN 2015;10:1956–63.
[25] Levin A, Rigatto C, Brendan B, et al. Cohort profile: Canadian study of prediction of death, dialysis and interim cardiovascular events (CanPREDDICT). BMC Nephrol 2013;14P 121.
[26] Levin A, Rigatto C, Barrett B, et al. Biomarkers of inflammation, fibrosis, cardiac stretch and injury predict death but not renal replacement therapy at 1 year in a Canadian chronic kidney disease cohort. Nephrol Dial Transplant 2014;29:1037–47.
[27] Hu B, Gadegbeku C, Lipkowitz MS, et al. Kidney function can improve in patients with hypertensive CKD. J Am Soc Nephrol 2012;23:706–13.
[28] Sud M, Tangri N, Pintilie M, Levey AS, Naimark DMJ. Progression to Stage 4 chronic kidney disease and death, acute kidney injury and hospitalization risk: a retrospective cohort study. Nephrol Dial Transplant 2015.
[29] Thygesen K, Mair J, Katus H, et al. Recommendations for the use of cardiac troponin measurement in acute cardiac care. Eur Heart J 2010;31:2197–204.
[30] Babuin L, Jaffe AS. Troponin: the biomarker of choice for the detection of cardiac injury. CMAJ 2005;173:1191–202.
[31] Reichlin T, Hochholzer W, Bassetti S, et al. Early diagnosis of myocardial infarction with sensitive cardiac troponin assays. N Engl J Med 2009;361:858–67.
[32] Mallamaci F, Tripepi G. Value of troponin T as a screening test for left ventricular hypertrophy in CKD. Am J Kidney Dis 2013;61:689–91.
[33]Colbert G, Jain N, de Lemos JA, Hedayati SS. Utility of traditional circulating and imaging-based cardiac biomarkers in patients with predialysis CKD. CJASN 2015;10:515–29.
[34] Michos ED, Wilson LM, Yeh H-C, et al. Prognostic value of cardiac troponin in patients with chronic kidney disease without suspected acute coronary syndrome: a systematic review and meta-analysis. Ann Intern Med 2014;161:491–501.

[35] Matsushita K, Sang Y, Ballew SH, et al. Cardiac and kidney markers for cardiovascular prediction in individuals with chronic kidney disease: the Atherosclerosis Risk in Communities Study. Arterio Thromb Vasc Biol 2014;34:1770–7.
[36] Bansal N, Hyre Anderson A, Yang W, et al. High-sensitivity troponin T and N-terminal pro-B-type natriuretic peptide (NT-proBNP) and risk of incident heart failure in patients with CKD: the Chronic Renal Insufficiency Cohort (CRIC) Study. J Am Soc Nephrol 2015;26:946–56.
[37] Twerenbold R, Wildi K, Jaeger C, et al. Optimal cutoff levels of more sensitive cardiac troponin assays for the early diagnosis of myocardial infarction in patients with renal dysfunction. Circulation 2015;131:2041–50.
[38] Aakre KM, Røraas T, Petersen PH, et al. Weekly and 90-minute biological variations in cardiac troponin T and cardiac troponin I in hemodialysis patients and healthy controls. Clin Chem 2014;60:838–47.
[39] Khalili H, de Lemos JA. What constitutes a relevant change in high-sensitivity troponin values over serial measurement? Clin Chem 2014;60:803–5.
[40] D'Marco L, Bellasi A, Raggi P. Cardiovascular biomarkers in chronic kidney disease: state of current research and clinical applicability. Dis Markers 2015;2015:586569–616.
[41] Wang AY-M, Lai K-N. Use of cardiac biomarkers in end-stage renal disease. J Am Soc Nephrol 2008;19:1643–52.
[42] Zoccali C, Ciccarelli M, Mallamaci F, et al. Effect of ultrafiltration on plasma concentrations of atrial natriuretic peptide in haemodialysis patients. Nephrol Dial Transplant 1986;1:188–91.
[43] Yoshihara F, Horio T, Nakamura S, et al. Adrenomedullin reflects cardiac dysfunction, excessive blood volume, and inflammation in hemodialysis patients. Kidney Int 2005;68:1355–63.
[44] Zoccali C, Mallamaci F, Benedetto FA, et al. Cardiac natriuretic peptides are related to left ventricular mass and function and predict mortality in dialysis patients. J Am Soc Nephrol 2001;12:1508–15.
[45] Almirez R, Protter AA. Clearance of human brain natriuretic peptide in rabbits; effect of the kidney, the natriuretic peptide clearance receptor, and peptidase activity. J Pharmacol Exp Ther 1999;289:976–80.
[46] Takami Y, Horio T, Iwashima Y, Takiuchi S. Diagnostic and prognostic value of plasma brain natriuretic peptide in non-dialysis-dependent CRF. Am J Kidney Dis 2004;44:420–8.
[47] Kaptoge S, Di Angelantonio E, Emerging Risk Factors Collaboration. et al. C-reactive protein, fibrinogen, and cardiovascular disease prediction. N Engl J Med 2012;367:1310–20.
[48] Apple FS, Murakami MM, Pearce LA, Herzog CA. Multi-biomarker risk stratification of N-terminal pro-B-type natriuretic peptide, high-sensitivity C-reactive protein, and cardiac troponin T and I in end-stage renal disease for all-cause death. Clin Chem 2004;50:2279–85.
[49] Wang AY-M, Lam CW-K, Yu C-M, et al. N-terminal pro-brain natriuretic peptide: an independent risk predictor of cardiovascular congestion, mortality, and adverse cardiovascular outcomes in chronic peritoneal dialysis patients. J Am Soc Nephrol 2007;18:321–30.
[50] Satyan S, Light RP, Agarwal R. Relationships of N-terminal pro-B-natriuretic peptide and cardiac troponin T to left ventricular mass and function and mortality in asymptomatic hemodialysis patients. Am J Kidney Dis 2007;50:1009–19.
[51] Anwaruddin S, Lloyd-Jones DM, Baggish A, et al. Renal function, congestive heart failure, and amino-terminal pro-brain natriuretic peptide measurement: results from the ProBNP Investigation of Dyspnea in the Emergency Department (PRIDE) Study. J Amer Coll Cardiol 2006;47:91–7.

[52] Nascimento MM, Pecoits-Filho R, Qureshi AR, et al. The prognostic impact of fluctuating levels of C-reactive protein in Brazilian haemodialysis patients: a prospective study. Nephrol Dial Transplant 2004;19:2803–9.
[53] Snaedal S, Heimbürger O, Qureshi AR, et al. Comorbidity and acute clinical events as determinants of C-reactive protein variation in hemodialysis patients: implications for patient survival. Am J Kidney Dis 2009;53:1024–33.
[54] Gupta J, Dominic EA, Fink JC, et al. Association between Inflammation and Cardiac Geometry in Chronic Kidney Disease: Findings from the CRIC Study. PLoS One 2015;10:e0124772.
[55] Jalal D, Chonchol M, Etgen T, Sander D. C-reactive protein as a predictor of cardiovascular events in elderly patients with chronic kidney disease. J Nephrol 2012;25:719–25.
[56] Isakova T, Xie H, Yang W, et al. Fibroblast growth factor 23 and risks of mortality and end-stage renal disease in patients with chronic kidney disease. JAMA 2011;305:2432–9.
[57] Isakova T, Wahl P, Vargas GS, et al. Fibroblast growth factor 23 is elevated before parathyroid hormone and phosphate in chronic kidney disease. Kidney Int 2011;79:1370–8.
[58] Scialla JJ, Lau WL, Reilly MP, et al. Fibroblast growth factor 23 is not associated with and does not induce arterial calcification. Kidney Int 2013;83:1159–68.
[59] Denker M, Boyle S, Anderson AH, et al. Chronic Renal Insufficiency Cohort Study (CRIC): overview and summary of selected findings. CJASN 2015;10:2073–83.
[60] Ix JH, Katz R, Kestenbaum BR, et al. Fibroblast growth factor-23 and death, heart failure, and cardiovascular events in community-living individuals: CHS (Cardiovascular Health Study). J Amer Coll Cardiol 2012;60:200–7.
[61] Kendrick J, Cheung AK, Kaufman JS, et al. FGF-23 associates with death, cardiovascular events, and initiation of chronic dialysis. J Am Soc Nephrol 2011;22:1913–22.
[62] Larsson TE. The role of FGF-23 in CKD–MBD and cardiovascular disease: friend or foe? Nephrol Dial Transplant 2010;25:1376–81.
[63] Matsui M, Uemura S, Takeda Y, et al. Placental growth factor as a predictor of cardiovascular events in patients with CKD from the NARA-CKD study. J Am Soc Nephrol 2015;26:2871–81.
[64] Drechsler C, Delgado G, Wanner C, et al. Galectin-3, renal function, and clinical outcomes: results from the LURIC and 4D studies. J Am Soc Nephrol 2015;26:2213–21.
[65] Haase M, Bellomo R, Devarajan P, Schlattmann P, Haase-Fielitz A. NGAL Meta-analysis Investigator Group. Accuracy of neutrophil gelatinase-associated lipocalin (NGAL) in diagnosis and prognosis in acute kidney injury: a systematic review and meta-analysis. Am J Kidney Dis 2009;54:1012–24.
[66] Fassett RG, Venuthurupalli SK, Gobe GC, Coombes JS, Cooper MA, Hoy WE. Biomarkers in chronic kidney disease: a review. Kidney Int 2011;80:806–21.
[67] Liu KD, Yang W, Anderson AH, et al. Urine neutrophil gelatinase-associated lipocalin levels do not improve risk prediction of progressive chronic kidney disease. Kidney Int 2013;83:909–14.
[68] Helanova K, Littnerova S, Kubena P, et al. Prognostic impact of neutrophil gelatinase-associated lipocalin and B-type natriuretic in patients with ST-elevation myocardial infarction treated by primary PCI: a prospective observational cohort study. BMJ Open 2015;5:e006872.
[69] Liu KD, Yang W, Go AS, et al. Urine neutrophil gelatinase-associated lipocalin and risk of cardiovascular disease and death in CKD: results from the Chronic Renal Insufficiency Cohort (CRIC) Study. Am J Kidney Dis 2015;65:267–74.

CHAPTER TWELVE

Diagnostic and Prognostic Biomarkers in Autosomal Dominant Polycystic Kidney Disease

G. Fick-Brosnahan, MD and B.Y. Reed, PhD
Division of Renal Diseases and Hypertension, Anschutz Medical Campus, Aurora, CO, United States

Contents

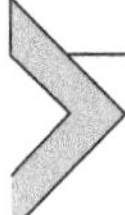

GENETIC TESTING FOR DIAGNOSIS AND PROGNOSIS

Effect of Gene Locus

Autosomal dominant polycystic kidney disease results from a mutation in either the *PKD1* or *PKD2* genes [1,2]. A third genetic locus, GANAB has recently been confirmed in some previously genetically unresolved families with mild ADPKD [3]. Accordingly the gold standard for diagnosis of ADPKD is genetic testing. Determination of the gene affected by mutation has significant prognostic value. On average those patients who harbor a *PKD1* mutation reach ESRD by the mid to late 50's. In contrast those with a *PKD2* mutation have a more slowly progressive disease with onset of ESRD delayed by 10–20 years compared to those with PKD1 [4,5]. Even

Biomarkers of Kidney Disease. http://dx.doi.org/10.1016/B978-0-12-803014-1.00012-1

in children with ADPKD differences in cyst burden associated with the mutated gene have been reported. In a study of 50 PKD1 and 10 PKD2 children Fencl et al. reported that the PKD1 children had larger kidneys and greater cyst counts compared to those with PKD2 [6]. This study confirmed the earlier results largely from the adult Consortium of Radiologic Imaging Studies of Polycystic Kidney Disease (CRISP) observational study regarding cyst burden in PKD1 compared to PKD2 [7]. Thus, genic heterogeneity accounts for a significant proportion of interfamilial variability in disease severity and therefore identification of the mutated gene provides a useful biomarker of disease course. At this stage it is relevant to add a brief description of currently available methods for genetic diagnosis including linkage analysis and direct sequence analysis of the PKD genes. Linkage analysis based on genotyping single nucleotide polymorphisms (SNP's) which are located in close proximity to either the *PKD1* or *PKD2* gene loci allows determination of whether at-risk family members are affected or not, in addition to identifying linkage to *PKD1* or *PKD2* as the disease associated causative gene [8,9]. This method requires prior diagnosis of at least one family member and is therefore limited by availability and consent of family members to DNA testing. In cases where a new mutation is suspected this technique is not applicable [10]. Direct sequence analysis of the *PKD* genes was hindered by the complex structure of *PKD1*. The first 33 exons of *PKD1* are duplicated in 6 pseudo genes which share approximately 98% identity with *PKD1*. More recent development of methods allowing specific amplification of *PKD1* allow diagnosis by direct sequence analysis of the PKD genes [11]. However, it should be stated that definitive mutations are not detected in 5–10% of individuals tested slightly limiting the efficacy of genetic testing.

Effect of Mutation Position and Type on Prognosis

To date the Mayo Clinic PKD Mutation database lists over 1000 pathogenic mutations in *PKD1* and over 200 in *PKD2* (www.pkdb.mayo.edu). Several publications have described an association between mutation position or type on PKD severity. In the Genkyst cohort which comprised 741 individuals from 519 pedigrees an association between *PKD1* mutation type but not mutation position was reported [4]. The median age at onset of ESRD was 55 years for those individuals carrying a truncating mutation and 67 years for those individuals with a nontruncating mutation. An earlier study reported an association between susceptibility to a vascular phenotype and more 5′ location of mutations in *PKD1* [12]. The association between mutation position or type in *PKD2* has been less definitive [13,14]. Therefore, with respect to PKD1 both mutation type (truncating vs. nontruncating) and mutation position may serve as prognostic biomarkers for disease severity. However, the complexity of sequence analysis of the *PKD1* gene, the high costs for mutation identification, in addition to the potential effects of other genes that may modify the severity of ADPKD currently limit the applicability of mutation identification for use as a prognostic biomarker.

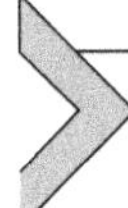

TOTAL KIDNEY VOLUME

Kidney Volume and Renal Function

Recently use of total kidney volume as a biomarker of disease progression in ADPKD or for selection of patients for inclusion in clinical trials is gaining increasing acceptance [15,16]. This fact has been driven by the need to identify better biomarkers in the early stages of disease when serum creatinine remains in the normal range. Interventions designed to slow progression of renal disease in ADPKD are likely to be more effective when administered early, before irreparable kidney damage has occurred [17,18]. Cyst growth with associated renal volume increase is ongoing from birth [19–21]. However, despite the exponential increase in renal volume, renal function is maintained until kidney volume increases to about 5 times normal size, typically occurring between the 4th and 5th decade of life [21]. At that time at least 50% of the renal parenchyma has been destroyed [16]. Compensation by the remaining functioning nephrons in addition to hyperfiltration serves to maintain renal function in the earlier stages of ADPKD [18,22,23]. At present regulatory agency approved end-points for progression of kidney disease center on renal function decline. As a consequence clinical trials must include patients in later stages of the disease, when therapeutic interventions may have limited efficacy, or they must commit to a longer duration when initiated in an earlier stage of ADPKD, thus incurring prohibitively higher cost.

Several studies have evaluated the relationship between renal volume and progression in ADPKD. In a longitudinal study of 108 children with renal imaging performed by ultrasound, children with severe renal enlargement at a younger age were shown to have faster renal growth compared to those with only mild renal enlargement or normal kidney volume [20]. Children with larger kidney volumes also had higher blood pressures and more symptoms, such as gross hematuria, flank and back pain, and proteinuria [19,20]. Several other studies illustrate the utility of renal volume as a marker of renal disease progression. A longitudinal ultrasound imaging study of 229 adults with ADPKD with sequential ultrasound examinations performed over a mean interval of 7.8 years showed a significant negative correlation between estimated glomerular filtration rate (eGFR) and renal volume over time [24]. The CRISP study which included 241 patients with early stage ADPKD (creatinine clearance > 70 mL/min per 1.73m^2) aged 15–45 years demonstrated that age adjusted renal, cyst and percent cyst volume were inversely related to GFR measured by iothalamate clearance and directly related to 24-h urine albumin excretion [18,21]. Magnetic resonance imaging (MRI) was used in CRISP and kidney volume calculated by stereology. Over 3 years follow-up, kidney volume increased exponentially, and baseline total kidney volume predicted the rate of increase in volume independent of age [21]. A baseline total kidney volume > 1500 mL was associated with a significant decline in GFR over 3 years, thus baseline total kidney volume predicts decline in renal function [21]. Continued follow-up of the CRISP cohort for an additional 5 years, using multivariate analysis, demonstrated that for each increment in baseline height corrected total kidney volume (htTKV) of 100 mL/m there was a significant 48% increase in the odds for development of chronic kidney disease (CKD) stage 3 (i.e., GFR below 60 mL/min per 1.73 m^2) within 8 years [25]. In further receiver operator characteristics (ROC) analyses a baseline htTKV of 600 mL/m was the best cut point for predicting the risk of developing stage 3 CKD within 8 years, with 74% sensitivity and 75% specificity [25]. Most notably, baseline htTKV was a better predictor of renal outcome than baseline age. Thus, the CRISP study demonstrates the value of htTKV as a prognostic biomarker for renal disease progression in ADPKD.

Other epidemiological studies have also reported a strong association between total kidney volume and decline in GFR. A prospective observational cohort of 541 Chinese ADPKD patients aged 4–77 years and with eGFR ≥30 mL/min per 1.73m^2 was followed for a median of 14.3 ± 10.6 months [26]. Total kidney volume was determined by sequential MRI assessments at 6-monthly intervals. Despite the wide discrepancy in eGFR and total kidney volumes among the participants, baseline $\log_{10}$ transformed total kidney volume and urinary protein/creatinine ratio emerged as the most significant predictors for a faster decline in eGFR and were associated with a higher total kidney volume growth rate [26]. A similar relationship between MRI-assessed TKV and eGFR decline was reported in a Japanese population with ADPKD [27].

Use of total kidney volume in conjunction with baseline eGFR will significantly facilitate selection of patients for inclusion in clinical trials by allowing inclusion of those patients most likely to reach a prespecified endpoint within a given trial duration [15]. The Polycystic Kidney Disease Outcomes Consortium (PKDOC) which is a collaboration between the PKD Foundation, Critical Path Institute, US Food and Drug Agency, Pharmaceutical Industry representatives, and major ADPKD research groups sought regulatory agency approval for use of total kidney volume as a prognostic biomarker for patient inclusion in clinical trials. To date the PKDOC has obtained qualified approval from both the FDA and the European Medicines Agency. Total kidney volume currently represents one of the most promising prognostic biomarkers in ADPKD.

Kidney Volume and Pain

Pain is the most common complication among patients with ADPKD and is in part related to pressure on the renal capsule from the expanding cysts [18,28,29]. While it may seem apparent that larger kidneys predispose to pain, there have been inconsistent reports linking kidney volume to pain. A pain questionnaire was completed by 555 patients with eGFR $>$ 60 mL/min per 1.73 m^2 who were participants in the HALT PKD clinical trial. There was no association between pain and htTKV determined from MRI assessment except in patients with htTKV $>$ 1000 mL/m [30]. However, a limitation of this study is that patients with eGFR $<$ 60 mL/min per 1.73 m^2 were not evaluated by imaging, thus htTKV is unavailable for those subjects with lower eGFR and accordingly those most likely to have the largest kidneys. With very large kidneys mechanical lower back pain may become prominent due to their weight and strain on the lumbar spine [29].

Measurement of Kidney Volume

In order for a biomarker to be adopted for use in clinical practice it must be generally easy and inexpensive to measure. In clinical trials MRI is the most widely used imaging modality with calculation of kidney volume based on stereology [31,32]. The main limitations associated with stereologic measurement are the time required to calculate renal and cyst volume in addition to the software costs. More rapid methods for estimating kidney and cyst volumes from MRI scans based on the area of a coronal midslice multiplied by the total number of slices have been shown to closely approximate total volume [33]. However, ultrasound is still widely used outside of clinical trials both for diagnosis and monitoring disease progression. In addition to the lower resolution associated with ultrasound imaging, calculation of renal volume based on the formula for a modified ellipsoid $\frac{4}{3}\pi x\left(\frac{1}{4}\text{anterior} - \text{posterior diameter} + \frac{1}{4}\text{width}\right)^2 x\frac{1}{2}\text{length}$ [24] has been shown to underestimate total kidney volume in patients with larger kidneys compared to similarly calculated volume determined by computed tomography (CT)-scan [34,35]. Correlation between MRI and CT assessment with respect to kidney volume measurement is excellent [34]. However, larger errors associated with ultrasound measurements will obscure small changes in TKV over time [34], limiting the usefulness of longitudinal ultrasound imaging for interventional trials. To address the potential value of renal ultrasound measurements for predicting renal function loss in clinical practice, Bhutani et al. reexamined data from the CRISP cohort in which renal ultrasound was performed in addition to MRI at the baseline visit [36]. This study showed that both htTKV and kidney length determined by either modality predicted future CKD stage 3 over a mean follow-up of 9.3 years [36]. Significantly, a baseline cut point of kidney length over 16.5 cm or htTKV > 650 mL/m, both obtained by ultrasound measurements, predicted development of CKD stage 3 with similar sensitivity and specificity as MRI-based cut points of 16.0 cm for kidney length or 550 mL/m for htTKV [36]. This indicates that the parameter of kidney length determined by ultrasound may be sufficient to stratify patient risk for progression to renal insufficiency and thus representative of a measure more amenable to adoption in clinical practice.

RENAL BLOOD FLOW

One of the earliest changes observed in the ADPKD kidney is a reduction in renal blood flow. This may even precede development of hypertension [37]. Many factors contribute to the reduction in renal blood flow including vasoconstriction, endothelial dysfunction, and abnormalities of the microvasculature [38,39]. The utility of renal blood flow as a biomarker of ADPKD progression has been evaluated in several clinical studies. In a cross sectional analysis of 127 patients participating in the CRISP study renal blood flow measured by MRI was strongly correlated with renal anatomical parameters including total kidney volume, total cyst volume and percent cyst volume, and with renal function [40]. In multiple linear regression analysis only age, renal blood flow and renal vascular resistance were identified as separate predictors of GFR. In a 3-year follow-up study of the

CRISP participants with annual assessment of total kidney and cyst volumes and annual renal blood flow measurement performed by MRI, renal blood flow reduction preceded the decline in GFR and was negatively correlated with increase in TKV [41]. Renal blood flow was an independent predictor of renal structural disease progression defined as increase in total kidney volume and cyst volume. Lower renal blood flow and larger total kidney volume both emerged as independent predictors of GFR decline [41].

A more recent European study of 103 ADPKD patients and 103 age- and gender-matched healthy controls evaluated the prevalence of renal function abnormalities in early ADPKD, including measurements of GFR and effective renal plasma flow using ^{125}I-iothalamate and ^{131}I-hippuran clearances [42,43]. Patients and controls were divided into quartiles based on age (median ages 28, 37, 42, and 52 years). Although ADPKD patients in the first quartile of age had normal GFR, effective renal plasma flow [*renal blood flow* = *effective renal plasma flow*/(1 − *hematocrit*)] was significantly lower and filtration fraction significantly higher than in healthy controls [42]. ADPKD patients also had lower 24-h urinary osmolality and higher urinary albumin excretion. This indicates that young adults with ADPKD already have measurable renal abnormalities despite a normal GFR, and that a reduction in renal blood flow precedes the reduction in GFR. In conclusion, renal blood flow is a biomarker for both renal structural and functional disease progression in ADPKD. However, complexity associated with the accurate measurement of renal blood flow limits its applicability to use in clinical trials conducted in centers with expertise in MRI.

SERUM AND URINE BIOMARKERS

The exploration of serum and urinary biomarkers for ADPKD is still in its infancy. Serum and/or urinary biomarkers may be useful for diagnosis in cases where renal imaging is negative for cysts, such as in very young subjects, or where they are equivocal or inconclusive, such as in very mild or in atypical cases with no family history of ADPKD. Although genetic testing is feasible, it is very costly due to the complexity of the genes involved, and no mutation is detected in 10–15% of patients [44]. Therefore a biomarker in serum or urine that is easily measurable and can distinguish affected from healthy subjects is desirable.

The extreme variability in the age at development of ESRD due to ADPKD underlines the utility of a prognostic biomarker for patient counseling and management, especially in early disease. In addition, a biomarker that could serve as a surrogate endpoint for ESRD in clinical trials would reduce necessary trial duration and associated costs. Unfortunately, no such biomarkers exist today. Moreover it is very unlikely that any single laboratory biomarker can distinguish ADPKD from other renal diseases or predict the course. More likely a specific set of markers in combination will be required to identify patients with ADPKD and predict disease severity.

Here we summarize the various studies aimed at finding diagnostic and prognostic serum and urinary biomarkers for ADPKD.

Diagnostic Biomarker Studies (Urine)

Proteomic and metabolomic studies of urine have been performed in small cohorts of ADPKD patients and compared with healthy controls as well as patients with other kidney diseases. An initial analysis of the urine proteome of 41 young (mean age 31 years) patients with ADPKD and 812 controls used capillary electrophoresis coupled online to mass spectrometry (CE–MS) and identified 197 proteins and peptides whose urinary excretion was significantly different in ADPKD subjects from healthy controls and patients with other renal diseases [45]. Considering only the most consistent of these polypeptides led to a reduction to 38 potential biomarkers, which were then tested in a training set of 17 ADPKD subjects and 86 controls, using a support vector machine-(SVM) based model. A score derived from this biomarker model distinguished subjects with ADPKD from controls with high sensitivity and specificity, and this was confirmed in an independent validation set of 24 ADPKD patients and 35 age- and gender-matched healthy controls. The authors were able to obtain amino acid sequences for 38 of the differentially excreted peptides, which revealed a high number of breakdown products of collagen and uromodulin, suggesting an important role for extracellular matrix remodeling in the pathogenesis of ADPKD [45].

This group of researchers greatly expanded their analyses in a second report comprising 1048 subjects, 292 of whom had ADPKD [46]. Using CE–MS in a training set of 41 young adults from the SUISSE ADPKD study and 189 healthy controls, they discovered 657 peptides that were differentially excreted in the urine of patients and controls. Of those, 142 peptides were most consistently altered in ADPKD and were used in a SVM-based diagnostic score, which was then tested in a validation set of 224 young ADPKD adults from the CRISP cohort, an additional 27 subjects from the SUISSE ADPKD cohort, an additional 86 healthy controls, and 481 controls with other renal or systemic diseases. In this validation cohort the urinary peptide score had a diagnostic sensitivity for ADPKD of 84.5% and a specificity of

90.8%. However, similar to renal ultrasound, the sensitivity of the biomarker score was somewhat lower in younger (<30 years) patients and in those with PKD2. In patients with PKD1 who were at least 20 years old the score performed well for the diagnosis of ADPKD, with a sensitivity of 92% and specificity of 93%. Similar to the earlier report, most of the identified urinary peptides (209 of 657 could be sequenced using high-resolution tandem mass spectrometry) were fragments of collagen I and III [46].

As this urinary biomarker score was not tested in any ADPKD subject who had not yet developed renal cysts detectable by ultrasound imaging, its performance in these patients is unknown and it cannot be used for diagnosing or ruling out ADPKD in young at-risk individuals [46]. Imaging and genetic testing will remain the main diagnostic tools at present.

The authors also explored whether their diagnostic score correlated with any measures of disease severity and found a positive correlation with height-adjusted total kidney volume (htTKV) measured by MRI. There also was a positive correlation with the annual absolute increase in TKV and a negative correlation with glomerular filtration rate (GFR). However, these correlations were weak and there was no correlation with albuminuria or proteinuria [46]. The authors then examined all detectable urinary peptides for a correlation with htTKV in 134 randomly selected ADPKD subjects, identified 99 such peptides and combined them in a linear model. This resulted in a stronger correlation with htTKV than the original diagnostic score in the training set (n = 134) as well as in a validation set of 158 ADPKD patients. As data on disease progression were not available, a true prognostic biomarker profile cannot be determined yet but may be developed in the future when long-term follow-up of these ADPKD patient cohorts is available [46].

A different group of researchers took a metabolomics approach, analyzing urine specimens with multidimensional-multinuclear nuclear magnetic resonance (NMR) spectroscopy with SVM-based classification [47]. In this cross-sectional study the authors compared the urinary metabolite pattern of 54 ADPKD patients who had mostly normal GFR with 46 healthy controls, 10 ADPKD patients on hemodialysis, 16 kidney transplant recipients, and 52 diabetic patients with chronic kidney disease. The urinary NMR pattern of the 54 ADPKD patients was distinctive and allowed prediction of ADPKD with 85% accuracy, however this was not validated in an independent group of patients. As expected, the urine metabolite pattern of the ADPKD hemodialysis patients was very different from the pattern of subjects with preserved renal function. Much larger studies with longer follow-up will need to be conducted before a reliable diagnostic or prognostic biomarker profile can be utilized in practice.

Another promising urinary biomarker test is being developed by Hogan et al. [48]. These authors studied differential protein abundance on urinary exosome-like vesicles (ELVs) from subjects with ADPKD compared to controls, and found that among more than 2000 ELV proteins 8 were reduced in ADPKD patients and 1 was consistently increased. Specifically, polycystin 1 (PC1) and polycystin 2 (PC2) were reduced to 54 and 53% of normal, whereas Transmembrane protein 2 (TMEM2), a fibrocystin homolog, was increased by 206%. These results were obtained from a discovery cohort of 13 subjects with confirmed *PKD1* mutations, all younger than 40 years (mean age 29.5 years) and with normal renal function, compared to 18 age-matched healthy controls. Using tandem mass spectrometry data in this cohort, the authors found an inverse correlation between the PC1/TMEM2 ratio in urinary ELVs and htTKV (determined by MRI in 12 of the 13 patients), suggesting that this ratio might be useful as a prognostic biomarker.

Further, the authors also developed a quantitative western blot to determine the PC1/TMEM2 and PC2/TMEM2 ratios and tested their diagnostic value in a different confirmation cohort of 12 patients with *PKD1* mutations, 3 with *PKD2* mutations and one with undetectable mutation, as compared to 23 normal controls [48]. Both the PC1/TMEM2 and PC2/TMEM2 ratios, determined by western blot, were significantly lower in the patients with *PKD1* mutations than in the controls and could discriminate between affected and nonaffected status. There was no significant difference in these ratios between the subjects with *PKD2* mutations and healthy controls, which may have been due to the low number of subjects and the relative inaccuracy of the western blot [48].

These initial studies will need to be extended to larger cohorts and to patients with chronic kidney diseases other than ADPKD, as well as to children with confirmed PKD mutations but as yet undetectable cysts, to further define their discriminatory value. Further, it is not known whether the reduced PC1/TMEM2 ratio remains constant in an individual over time or changes as the disease progresses. Only long-term follow-up of well-characterized patient populations will allow an evaluation of their prognostic value.

Prognostic Biomarkers

To find prognostic biomarkers to predict the course of ADPKD in an individual at an early age has been of high interest for patients and researchers for some time because treatments that start late in the course are unlikely to be of much benefit. Yet, subjecting people who will have a mild course to treatments that potentially have substantial adverse effects is not desirable either. Until recently the only established marker to predict the development of ESRD has been an increase in serum creatinine concentration. However, because of compensatory hyperfiltration by noncystic nephrons, serum creatinine remains normal until the majority of renal parenchyma has been destroyed by the cystic process. As discussed earlier in this chapter, increased total kidney volume at a given age is a potential biomarker for more severe disease, but determination of TKV is laborious and expensive. Therefore, an easily measurable laboratory biomarker would be very helpful in this respect. Several potential prognostic serum and urine biomarkers have been explored, mostly in cross-sectional and limited prospective observations, but none of them have been rigorously studied or validated.

Urinary Biomarker Studies

An early biomarker study examined urinary excretion of monocyte chemoattractant protein-1 (MCP-1) in 55 patients with ADPKD and 19 healthy controls [49], based on the finding of increased expression of MCP-1 in a rat model of ADPKD. In the human study, very high levels of MCP-1 were found in cyst fluids from nephrectomy specimens, and urinary excretion of MCP-1 was increased compared to controls in 42 of the 55 patients, 27 of whom had normal serum creatinine concentrations and normal urine protein excretion, suggesting that elevated urinary MCP-1 excretion may predict a decline in GFR [49].

Another cross-sectional analysis of selected urinary biomarkers was performed by Meijer et al. [50]. This group studied 102 patients with ADPKD, mean age 40 ± 11 years, with measured GFR 77 ± 31 mL/min per 1.73 m^2, and compared them to 102 age- and sex-matched healthy controls. 24-h urinary excretion of immunoglobulin G, albumin, kidney injury molecule 1 (KIM-1), *N*-acetyl-β-D-glucosaminidase, neutrophil gelatinase-associated lipocalin (NGAL), β2-microglobulin, heart-type fatty acid binding protein (H-FABP), and MCP-1 were all significantly higher in ADPKD patients compared with controls. Excretion of NGAL, β2-microglobulin and H-FABP correlated with measured GFR and renal blood flow independent of urinary albumin excretion [50].

Piazzon et al. evaluated urinary excretion of fetuin A as a potential biomarker, based on the finding of increased expression levels of fetuin A in the kidneys of two mouse models of PKD [51]. The investigators found an increased urinary excretion of fetuin A per mmol creatinine in 66 patients with ADPKD (17.5 ± 12.5 μg/mmol) compared with

17 healthy controls (8.5 ± 3.8 μg/mmol) and 50 subjects with renal diseases other than ADPKD (6.2 ± 2.9 μg/mmol). Moreover, urinary fetuin was already elevated at 13.8 ± 9.6 μg/mmol in ADPKD patients with normal GFR (mean 102 mL/min per 1.73 m^2), and increased further to 21.8 ± 14.3 μg/mmol in patients with decreased GFR (mean 35 mL/min per 1.73 m^2), in contrast to the 50 individuals with other renal diseases who had a mean GFR of 30 mL/min per 1.73 m^2. Therefore, increased urinary excretion of fetuin A appears to be specific for ADPKD. A subset of 19 ADPKD patients with early disease was followed for 2 years, with urinary fetuin A measurements every 6 months. Urinary fetuin A excretion increased progressively and significantly over the 2 years, whereas estimated GFR remained unchanged [51]. Further exploration of this biomarker may be warranted.

Serum Biomarker Studies

Of potential serum biomarkers the plasma copeptin concentration has been studied the most because of its pathophysiological significance. Copeptin is also known as AVP-associated glycopeptide and is part of preprovasopressin in the posterior pituitary gland [52]. After cleavage it is secreted together with vasopressin in a 1:1 ratio and thus a marker of endogenous vasopressin levels. Copeptin is much more stable in the circulation and can be measured more reliably than vasopressin itself, using a sandwich immunoassay [52]. Arginine vasopressin (AVP) also functions as antidiuretic hormone. It appears to be required for cyst formation, as rats with a polycystic gene mutation but genetic absence of vasopressin do not develop significant cystic renal disease [53]. Vasopressin binding to its receptor causes cAMP generation which in turn stimulates cyst cell proliferation and fluid secretion [54–57]. Vasopressin is constantly present in the human circulation and is increased in subjects with ADPKD, in order to retain water as a compensation for the early urine concentration defect caused by medullary disruption by the cysts [56–58].

In a cross-sectional study of 102 consecutive ADPKD patients with mildly reduced renal function (measured GFR 77 ± 31 mL/min per 1.73 m^2), plasma copeptin concentration correlated positively with TKV assessed by MRI and negatively with measured GFR and effective renal blood flow, independent of age, gender, and use of diuretics [59]. However, there was no follow-up of these patients to assess prediction of progression by copeptin level. For this purpose baseline plasma copeptin concentration was measured in 225 participants of the CRISP Cohort [60]. In these individuals, baseline plasma copeptin was associated with baseline TKV and correlated negatively with baseline GFR. More importantly, over a follow-up period of 8–9 years, higher plasma copeptin levels were associated with a greater increase in TKV, even after adjustment for age, sex, cardiovascular risk factors, and diuretic use [60]. There was no significant association with GFR decline, and hydration status was not standardized at the time of copeptin measurement.

Another study examined associations with disease severity and progression not only of plasma copeptin but also of its counteracting peptide apelin [61]. Apelin is produced by hypothalamic neurons, it inhibits vasopressin release and thus contributes to fine-tuning of water balance. Water deprivation leads to an increase in serum vasopressin and a decrease in serum apelin levels, which can be measured using a commercially available ELISA test [61]. In a group of 52 ADPKD patients, of whom 42 had decreased GFR, and 50 age-and gender-matched healthy controls, plasma copeptin levels were significantly higher and apelin levels significantly lower in patients than in controls, even in the 10 subjects with ADPKD and normal GFR [61]. Apelin was positively correlated with measured GFR and negatively with TKV and albuminuria at baseline. Over a 2-year follow-up period, a lower apelin level at baseline was associated with a larger decrease in GFR and a greater increase in TKV, whereas for copeptin a higher level was associated with faster progression. However, most subjects with ADPKD in this study already had advanced chronic kidney disease (mean serum creatinine 2.83 ± 2.3 mg/dL), therefore this was not a study of an early biomarker. These preliminary findings need to be confirmed in larger prospective studies with longer follow-up.

Other Studies

A very preliminary investigation of urine microRNA profiles from 20 ADPKD patients, age 25–70 years, with estimated GFR ranging from 20–130 mL/min per 1.73 m^2 was reported by Ben-Dov et al. [62]. Twenty patients with other renal diseases, matched for age, gender, race, and renal function, served as controls. There were significant differences in urine microRNA profiles between the two groups which were partly mirrored by differences in expression levels by primary cell cultures derived from ADPKD cysts and normal adult kidney cells. Urine from ADPKD patients contained higher levels of microRNAs from inflammatory cell and fibroblast origin and lower levels of kidney tumor suppressor-associated microRNAs than urine from controls [62].

There are several other exploratory studies examining associations of various urine and serum biomarkers with disease severity in ADPKD, but these studies are limited by small numbers of patients, mostly cross-sectional design, a wide range of GFR of included subjects with ADPKD, short (if any) follow-up, and often absence of control groups, either healthy volunteers or patients with other chronic kidney diseases [63–72]. Moreover, almost all correlational studies were done in patients with moderate to advanced ADPKD. The study by Reed et al. is the only one undertaken in adolescents (mean age 16 ± 4 years, $n = 71$), when therapies aimed at slowing cyst growth might be most efficient [71]. These investigators found a significant correlation between serum levels of vascular endothelial growth factor (VEGF) and total renal volume, total cyst volume, and left ventricular mass index. A similar positive correlation was also found for serum angiopoietin 1. However, the study was cross-sectional and did not contain a control group.

In conclusion, we have presented the rationale for the quest for a reliable biomarker for diagnosis and prognostication in early ADPKD, as well as an overview of potential genetic, imaging, and laboratory biomarkers that have been investigated so far. While genetic and imaging biomarkers may already be applicable for clinical trial design, much work remains to be done before any serum or urine biomarker can be used in the clinic, either for diagnosis or prediction of disease course in ADPKD.

REFERENCES

[1] Hughes J, Ward CJ, Peral B, et al. The polycystic kidney disease 1 (PKD1) gene encodes a novel protein with multiple cell recognition domains. Nat Genet 1995;10:151–60.
[2] Kimberling WJ, Kumar S, Gabow PA, Kenyon JB, Connolly CJ, Somlo S. Autosomal dominant polycystic kidney disease: localization of the second gene to chromosome 4q13-q23. Genomics 1993;18:467–72.
[3] Porath B, Gainullin VG, Cornec-Le Gall E, et al. Mutations in GANAB, encoding the glucosidase IIα subunit, cause autosomal-dominant polycystic kidney and liver disease. Am J Hum Genet 2016;98:1193–207.
[4] Cornec-Le Gall E, Audrézet MP, Chen JM, et al. Type of PKD1 mutation influences renal outcome in ADPKD. J Am Soc Nephrol 2013;24:1006–13.
[5] Hateboer N, vanDijk MA, Bogdanova N, et al. European PKD1-PKD2 study group: Comparison of phenotypes of polycystic kidney disease types 1 and 2. Lancet 1999;352:103–7.
[6] Fencl F, Janda J, Bláhová K, et al. Genotype-phenotype correlation in children with autosomal dominant polycystic kidney disease. Pediatr Nephrol 2009;24:983–9.
[7] Harris PC, Bae KT, Rossetti S, et al. Cyst number but not the rate of cystic growth is associated with the mutated gene in autosomal dominant polycystic kidney disease. J Am Soc Nephrol 2006;17:3013–9.

[8] Coto E, Sanz de Castro S, Aguado S, et al. DNA microsatellite analysis of families with autosomal dominant polycystic kidney diseasetypes 1 and 2: evaluation of clinical heterogeneity between both forms of the disease. J Med Genet 1995;32:442–5.
[9] Torra R, Badenas C, Darnell A, et al. Linkage, clinical features, and prognosis of autosomal dominant polycystic kidney disease types 1 and 2. J Am Soc Nephrol 1996;7:2142–51.
[10] Pei Y. Diagnostic approach in autosomal dominant polycystic kidney disease. Clin J Am Soc Nephrol 2006;1:1108–14.
[11] Rossetti S, Chauveau D, Walker D, et al. A complete mutation screen of the ADPKD genes by DHPLC. Kidney Int 2002;61:1588–99.
[12] Rossetti S, Chauveau D, Kubly V, et al. Association of mutation position in polycystic kidney disease 1 (PKD1) gene and development of a vascular phenotype. Lancet 2003;361:2196–201.
[13] Hateboer N, Veldhuisen B, Peters D, et al. Location of mutations within the PKD2 gene influences clinical outcome. Kidney Int 2000;57:1444–51.
[14] Magistroni R, He N, Wang K, et al. Genotype-renal function correlation in type 2 autosomal dominant polycystic kidney disease. J Am Soc Nephrol 2003;14:1164–74.
[15] Irazabal MV, Rangel LJ, Bergstralh EJ, et al. Imaging classification of autosomal dominant polycystic kidney disease: A simple model for selecting patients for clinical trials. J Am Soc Nephrol 2015;26:160–72.
[16] Alam A, Dahl NK, Lipschutz JH, et al. Total kidney volume in autosomal dominant polycystic kidney disease: a biomarker of disease progression and therapeutic efficacy. Am J Kid Dis 2015;66(4):564–76.
[17] Cadnapaphornchai MA, George DM, McFann K, et al. Effect of pravastatin on total kidney volume, left ventricular mass index, and microalbuminuria in pediatric autosomal dominant polycystic kidney disease. Clin J Am Soc Nephrol 2014;9:889–96.
[18] Grantham JJ, Chapman AB, Torres VE. Volume progression in autosomal dominant polycystic kidney disease: the major factor determining clinical outcomes. Clin J Am Soc Nephrol 2006;1:148–57.
[19] Fick GM, Duley IT, Johnson AM, Strain JD, Manco-Johnson ML, Gabow PA. The spectrum of autosomal dominant polycystic kidney disease in children. J Am Soc Nephrol 1994;4:1654–60.
[20] Fick-Brosnahan GM, Tran ZV, Johnson AM, Strain JD, Gabow PA. Progression of autosomal-dominant polycystic kidney disease in children. Kidney Int 2001;59:1654–62.
[21] Grantham JJ, Torres VE, Chapman AB, et al. Volume progression in polycystic kidney disease. New Engl J Med 2006;354:2122–30.
[22] Helal I, Fick-Brosnahan GM, Reed-Gitomer B, Schrier RW. Glomerular hyperfiltration: definitions, mechanisms and clinical applications. Nat Rev Nephrol 2012;18:293–300.
[23] Helal I, Reed B, McFann K, Yan XD, Fick-Brosnahan GM, Cadnapaphornchai MA, Schrier RW. Glomerular hyperfiltration and renal progression in children with autosomal dominant polycystic kidney disease. Clin J Am Soc Nephrol 2011;6: 2439–43.
[24] Fick-Brosnahan GM, Belz MM, McFann KK, Johnson AM, Schrier RW. Relationship between renal volume growth and renal function in autosomal dominant polycystic kidney disease: a longitudinal study. Am J Kidney Dis 2002;39:1127–34.
[25] Chapman AB, Bost JE, Torres VE, et al. Kidney volume and functional outcomes in autosomal dominant polycystic kidney disease. Clin J Am Soc Nephrol 2012;7:470–86.
[26] Chen D, Ma Y, Wang X, et al. Clinical characteristics and disease predictors of a large Chinese cohort of patients with autosomal dominant polycystic kidney disease. PloS One 2014;9:e92232.
[27] Tokiwa S, Muto S, China T, Horie S. The relationship between renal volume and renal function in autosomal dominant polycystic kidney disase. Clin Exp Nephrol 2011;15:539–45.

[28] Masoumi A, Elhassan E, Schrier RW. Interpretation of renal volume in autosomal dominant polycystic kidney disease and relevant clinical interpretations. Iran J Kidney Dis 2011;5:1–8.
[29] Casteleijn NF, Visser FW, Drenth JP, et al. A stepwise approach for effective management of chronic pain in autosomal-dominant polycystic kidney disease. Nephrol Dial Transplant 2014;29(Suppl 4):iv142–53.
[30] Miskulin DC, Abebe KZ, Chapman AB, et al. Health-related quality of life in patients with autosomal dominant polycystic kidney disease and CKD stages 1–4: a cross sectional study. Am J Kidney Dis 2014;63:214–26.
[31] Chapman AB, Guay-Woodford LM, Grantham JJ, et al. Renal structure in early autosomal dominant polycystic kidney disease (ADPKD): the Consortium for Radiologic Imaging Studies of Polycystic Kidney Disease (CRISP) cohort. Kidney Int 2003;64:2214–21.
[32] Bae KT, Commean PK, Lee J. Volumetric measurement of renal cysts and parenchyma using MRI: phantoms and patients with polycystic kidney disease. J Comput Assist Tomogr 2000;24:614–9.
[33] Bae KT, Tao C, Wang J, et al. Novel approach to estimate kidney and cyst volumes using mid-slice magnetic resonance images in polycystic kidney disease. Am J Nephrol 2013;38:333–41.
[34] Chapman AB, Wei W. Imaging approaches to patients with polycystic kidney disease. Semin Nephrol 2011;31:237–44.
[35] Hammoud S, Tissier AM, Elie C, et al. Ultrasonographic renal volume measurements in early autosomal dominant polycystic kidney disease. Diagn Interv Imaging 2015;96:65–71.
[36] Bhutani H, Smith V, Rahbari-Oskoui F, et al. A comparison of ultrasound and magnetic resonance imaging shows that kidney length predicts chronic kidney disease in autosomal dominant polycystic kidney disease. Kidney Int 2015;88(1):146–51.
[37] Barrett BJ, Foley R, Morgan J, Hefferton D, Parfrey P. Differences in hormonal and renal vascular responses between normotensive patients with autosomal dominant polycystic kidney disease and unaffected family members. Kidney Int 1994;46:1118–23.
[38] Helal I, Reed B, Schrier RW. Emergent early markers of renal progression in autosomal dominant polycystic kidney disease: implications for treatment and prevention. Am J Nephrol 2012;36:162–7.
[39] Xu R, Franchi F, Miller B, et al. Polycystic kidneys have decreased vascular density: a micro-CT study. Microcirculation 2013;20:183–9.
[40] King BF, Torres VE, Brummer ME, et al. Magnetic resonance measurements of renal blood flow as a marker of disease severity in autosomal dominant polycystic kidney disease. Kidney Int 2003;64:2214–21.
[41] Torres VE, King BF, Chapman AB, et al. Magnetic resonance measurements of renal blood flow and disease progression in autosomal dominant polycystic kidney disease. Clin J Am Soc Nephrol 2007;2:112–20.
[42] Meijer E, Rook M, Tent H, et al. Early renal abnormalities in autosomal dominant polycystic kidney disease. Clin J Am Soc Nephrol 2010;5:1091–8.
[43] Donker AJ, van der Hem GK, Sluiter WJ, Beekhuis H. A radioisotope method for simultaneous determination of the glomerular filtration rate and the effective renal plasma flow. Neth J Med 1977;20:97–103.
[44] Schrier RW, Abebe KZ, Perrone RD, et al. Blood pressure in early autosomal dominant polycystic kidney disease. N Engl J Med 2014;371:2255–66.
[45] Kistler AD, Mischak H, Poster D, Dakna M, Wüthrich RP, Serra AL. Identification of a unique urinary biomarker profile in patients with autosomal dominant polycystic kidney disease. Kidney Int 2009;76:89–96.
[46] Kistler AD, Serra AL, Siwy J, et al. Urinary proteomic biomarkers for diagnosis and risk stratification of autosomal dominant polycystic kidney disease: a multicentric study. PLoS One 2013;8:e53016.

[47] Gronwald W, Klein MS, Zeltner R, et al. Detection of autosomal dominant polycystic kidney disease by NMR spectroscopic fingerprinting of urine. Kidney Int 2011;79:1244–53.
[48] Hogan MC, Bakeberg JL, Gainullin VG, et al. Identification of biomarkers for PKD1 using urinary exosomes. J Am Soc Nephrol 2015;27:1661–70.
[49] Zheng D, Wolfe M, Cowley BD, Wallace DP, Yamaguchi T, Grantham JJ. Urinary excretion of monocyte chemoattractant protein-1 in autosomal dominant polycystic kidney disease. J Am Soc Nephrol 2003;14:2588–95.
[50] Meijer E, Boertien WE, Nauta FL, et al. Association of urinary biomarkers with disease severity in patients with autosomal dominant polycystic kidney disease: a cross-sectional analysis. Am J Kidney Dis 2010;56:883–95.
[51] Piazzon N, Bernet F, Guihard L, et al. Urine Fetuin-A is a biomarker of autosomal dominant polycystic kidney disease progression. J Transl Med 2015;13:103.
[52] Morgenthaler NG, Struck J, Jochberger S, Dünser MW. Copeptin: clinical use of a new biomarker. Trends Endocrinol Metab 2008;19:43–9.
[53] Wang X, Wu Y, Ward CJ, Harris PC, Torres VE. Vasopressin directly regulates cyst growth in polycystic kidney disease. J Am Soc Nephrol 2008;19:102–8.
[54] Mangoo-Karim R, Uchic M, Lechene C, Grantham JJ. Renal epithelial cyst formation and enlargement in vitro: dependence on cAMP. Proc Natl Acad Sci USA 1989;86:6007–11.
[55] Sullivan LP, Wallace DP, Grantham JJ. Chloride and fluid secretion in polycystic kidney disease. J Am Soc Nephrol 1998;9:903–16.
[56] Torres VE, Harris PC. Strategies targeting cAMP signaling in the treatment of polycystic kidney disease. J Am Soc Nephrol 2014;25:18–32.
[57] Zittema D, Boertien WE, van Beek AP, et al. Vasopressin, copeptin, and renal concentrating capacity in patients with autosomal dominant polycystic kidney disease without renal impairment. Clin J Am Soc Nephrol 2012;7:906–13.
[58] Zittema D, van den Berg E, Meijer E, et al. Kidney function and plasma copeptin levels in healthy kidney donors and autosomal dominant polycystic kidney disease patients. Clin J Am Soc Nephrol 2014;9:1553–62.
[59] Meijer E, Bakker SJ, van der Jagt EJ, et al. Copeptin, a surrogate marker of vasopressin, is associated with disease severity in autosomal dominant polycystic kidney disease. Clin J Am Soc Nephrol 2011;6:361–8.
[60] Boertien WE, Meijer E, Li J, et al. Relationship of copeptin, a surrogate marker for arginine vasopressin, with change in total kidney volume and GFR decline in autosomal dominant polycystic kidney disease: results from the CRISP cohort. Am J Kidney Dis 2013;61:420–9.
[61] Lacquaniti A, Chirico V, Lupica R, et al. Apelin and copeptin: two opposite biomarkers associated with kidney function decline and cyst growth in autosomal dominant polycystic kidney disease. Peptides 2013;49:1–8.
[62] Ben-Dov IZ, Tan YC, Morozov P, et al. Urine microRNA as potential biomarkers of autosomal dominant polycystic kidney disease progression: description of miRNA profiles at baseline. PLoS One 2014;9:e86856.
[63] Parikh CR, Dahl NK, Chapman AB, et al. Evaluation of urine biomarkers of kidney injury in polycystic kidney disease. Kidney Int 2012;81:784–90.
[64] Park HC, Hwang JH, Kang AY, et al. Urinary N-acetyl-β-D glucosaminidase as a surrogate marker for renal function in autosomal dominant polycystic kidney disease: 1 year prospective cohort study. BMC Nephrol 2012;13:93.
[65] Romaker D, Puetz M, Teschner S, et al. Increased expression of secreted frizzled-related protein 4 in polycystic kidneys. J Am Soc Nephrol 2009;20:48–56.
[66] Zschiedrich S, Budde K, Nürnberger J, et al. Secreted frizzled-related protein 4 predicts progression of autosomal dominant polycystic kidney disease. Nephrol Dial Transplant 2016;31:284–9.

[67] Kurultak I, Sengul S, Kocak S, et al. Urinary angiotensinogen, related factors and clinical implications in normotensive autosomal dominant polycystic kidney disease patients. Ren Fail 2014;36:717–21.
[68] Kawano H, Muto S, Ohmoto Y, et al. Exploring urinary biomarkers in autosomal dominant polycystic kidney disease. Clin Exp Nephrol 2015;19:968–73.
[69] Petzold K, Poster D, Krauer F, et al. Urinary biomarkers at early ADPKD disease stage. PLoS One 2015;10:e0123555.
[70] Zhou J, Ouyang X, Cui X, et al. Renal CD14 expression correlates with the progression of cystic kidney disease. Kidney Int 2010;78:550–60.
[71] Reed BY, Masoumi A, Elhassan E, et al. Angiogenic growth factors correlate with disease severity in young patients with autosomal dominant polycystic kidney disease. Kidney Int 2011;79:128–34.
[72] Park HC, Kang AY, Jang JY, et al. Increased urinary angiotensinogen/creatinine (AGT/Cr) ratio may be associated with reduced renal function in autosomal dominant polycystic kidney disease patients. BMC Nephrol 2015;16:86.

Biomarkers in Glomerular Disease

J.M. Arthur, MD, PhD, E. Elnagar, MBBS, MPH and N. Karakala, MD
Division of Nephrology, University of Arkansas for Medical Sciences and Central Arkansas Veterans Healthcare System, Little Rock, AR, United States

Contents

Glomerular diseases are a major cause of end-stage renal disease. Many glomerular diseases are treatable if the specific diagnosis is known. A renal biopsy is required to make a definitive diagnosis of the cause of the disease in most cases. Although renal biopsy is the gold standard for diagnosis, it is invasive and has potential complications, such as bleeding, infection, and death [1]. Although serial renal biopsies might improve treatment, they are often difficult to justify because of the risk, discomfort, and expense. New methods are needed to identify the cause of renal disease and prognosis without a biopsy. Urine or blood testing for biomarkers could replace renal biopsy as a simple, safe, and accurate test that could be repeated to follow progression of the disease, monitor response to therapy, and potentially guide the choice of therapy. Immunohistochemistry of biomarker proteins within the glomerulus could help guide therapy and aid in prediction of prognosis. The clinical course of many glomerular diseases is highly variable so it is difficult to predict if any given patient will lose renal function or suffer the associated complications of renal disease. The major impediment to the development of new treatments is the inability to identify patients that

Biomarkers of Kidney Disease. http://dx.doi.org/10.1016/B978-0-12-803014-1.00013-3

would benefit from them. Biomarkers provide the opportunity to identify the disease, predict prognosis, and predict the response to therapy in a noninvasive way.

BIOMARKERS IN GLOMERULAR DISEASES

A number of biomarkers have been proposed to predict the rate of progression or the underlying cause of glomerular diseases. The best characterized is urinary albumin as a predictor of progression in diabetic nephropathy. Microalbuminuria is associated with an increased rate of progression of diabetic renal disease in patients with both IDDM [2] and NIDDM [3]. Unfortunately, microalbuminuria is not an ideal marker for progression of diabetic nephropathy. A large study of patients with insulin-dependent diabetes points out the problems with microalbuminuria as a marker for progression of diabetic nephropathy [4]. After 5 years, 33% of patients with microalbuminuria had returned to normoalbuminuria while only 19% had diabetic nephropathy. In another study of 386 patients with type I diabetes and persistent microalbuminuria, 58% experienced regression of the rate of albumin excretion by 50% or greater over at least one 2-year period [5]. The role of biomarkers in diabetic nephropathy is the subject of another chapter in this book. In this chapter, we will focus on biomarkers in other glomerular diseases, such as lupus nephritis, membranous nephropathy (MN), FSGS, minimal change disease, and IgA nephropathy. In addition, we will highlight studies using discovery techniques, such as proteomics to identify novel biomarkers.

PREDICTORS OF OUTCOME IN GLOMERULAR DISEASES

Glomerular diseases can have a range of outcomes from complete spontaneous remission to progression to ESRD. While renal biopsy can help identify the disease, it is not as good at predicting the outcome. Biomarkers which can predict outcome would be tremendously helpful to guide treatment decisions. *N*-acetyl-β-glucosamynidase (NAG) is a tubular enzyme which is increased in the urine during renal injury. In patients with nephrotic syndrome, NAG concentrations above a cutoff value predicted progression to chronic renal failure better than did the level of proteinuria [6]. While this finding needs further verification, it is a promising approach to prediction of outcomes in patients with glomerular diseases. Tubular injury markers may predict long-term outcomes in patients with glomerular diseases because they reflect more tubulointerstitial injury. For instance urinary NGAL and KIM-1 are increased in children with both steroid resistant and steroid sensitive nephrotic syndrome [7]. A number of other studies have been done in patients with specific glomerular diseases.

BIOMARKERS IN LUPUS NEPHRITIS

Treatment of lupus nephritis has met with limited success for a number of reasons. Current therapeutic regimens typically produce complete and partial remission rates of less than 50% [8,9]. Furthermore many therapies are highly toxic. Flares of nephritis can be difficult to predict ahead of time which makes timely treatment more difficult. Finally, renal biopsy is often not repeated which makes tailoring treatment more difficult. Biomarkers which can predict flares early, which can predict the class of nephritis present, and which can help guide treatment would be extremely useful and help improve outcomes in lupus nephritis.

PREDICTORS OF LUPUS NEPHRITIS CLASS

One of the key areas of interest in the field is biomarkers that can predict the type of nephritis present. Currently, examination of urine sediment and proteinuria can be helpful but renal biopsy is required to differentiate specific classes of lupus nephritis. A number of potential candidate markers have been evaluated to differentiate between classes of nephritis. Messenger RNA levels in urine of IP-10, its receptor (CXCR3), TGF-β, and VEGF were evaluated for their ability to discriminate between class IV lupus nephritis and other classes [10]. All four analytes had significantly different levels between class IV lupus nephritis and other classes. ROC curve analysis showed that IP-10 had the best discriminative power with an AUC value of 0.89. When individual values were plotted; however, there was a significant amount of overlap between the number of copies of mRNA in patients with class IV lupus and other classes, demonstrating that better discriminating power is needed before a test based on IP-10 message levels could be used to base treatment on with toxic agents in patients with lupus nephritis. The levels of β1 integrin (CD29) expression on peripheral blood T cells has been compared between patients with class IV lupus nephritis and other classes of nephritis [11]. Patients with class IV lupus nephritis showed a significantly higher level of β1 integrin expression on T cells than patients with other classes of lupus nephritis or healthy volunteers. Furthermore, there was an inverse correlation between β1 integrin expression and serum complement levels which is an indicator of the severity of the lupus flare. In this study all patients with increased β1 integrin expression had class IV lupus although not all patients with class IV lupus had increased β1 integrin expression. These data are exciting in that they suggest that β1 integrin may be a highly specific marker for class IV nephritis but will need to be confirmed in other populations. Proteomic signatures have also been used to predict the class of nephritis present [12]. Two-dimensional gel electrophoresis (2DE) of urine proteins was used to develop an algorithm which could differentiate between classes of nephritis identified by renal biopsy. A combination of spots was identified using a machine learning algorithm that could predict the cause of disease with AUC values between 0.85 and 0.95. In the analysis 213 proteins were used although most of the accuracy of the analysis was contributed by 10 spots. Although this represents an improvement in accuracy compared to single analytes, measurement of larger numbers of proteins is correspondingly more difficult. Furthermore, the results have not yet been independently confirmed. Overall, the existing studies do not support any currently available tests to determine lupus class without a renal biopsy. Studies that identify novel biomarkers will be useful and successful methods will likely use a combination of markers to predict the class of lupus nephritis that is present.

BIOMARKERS THAT PREDICT RENAL LUPUS FLARES

A potential stumbling block to efficacious treatment of lupus nephritis is the timing of treatment. Typically, nephritis is treated after structural changes to the glomerulus have already occurred. These structural changes are indicated by increases in proteinuria, hematuria, and serum creatinine. If treatment could begin earlier, it may be more successful. A number of studies have attempted to identify biomarkers that can predict flares in renal lupus activity. Urinary FOXP3 mRNA has been proposed as a candidate marker for lupus renal disease activity [13]. FoxP3 is a regulator of the development and function of regulatory T cells. FoxP3 mRNA levels were higher in patients with active lupus nephritis than those without. Furthermore, among patients with active renal disease, levels were higher in patients with proliferative disease. Finally, levels were higher in the group of patients that did not respond to therapy (57.6 ± 69.8 copies) than in the group that did (2.4 ± 1.9 copies). This finding suggests that FOXP3 mRNA levels may be useful in predicting renal activity and response to treatment but will need to be further evaluated. The association of urinary TWEAK with lupus renal activity has been investigated in a multicenter cohort study [14]. Urinary TWEAK levels were higher in patients with lupus nephritis than in patients with lupus without nephritis. High levels of urinary TWEAK predicted lupus renal activity with an odds ratio of 7.36. Urinary TWEAK levels peaked during a renal flare and were significantly higher during the flare than before or after the flare. Whether TWEAK can be used to anticipate renal flares has not been shown.

Neutrophil gelatinase associated lipocalin (NGAL, also known as lipocalin-2) has been shown to be a marker of acute kidney injury (AKI) and has been proposed as a candidate marker for lupus flares. In a prospective study of emergency department patients, Nickolas et al studied the ability of a single measurement of urinary NGAL to identify patients presenting with AKI [15]. Urine NGAL was measured in 635 patients admitted through the emergency department. Mean NGAL concentration was significantly higher in patients with AKI. Based on this and other studies showing its predictive value for kidney injury in AKI, NGAL has been evaluated as a predictive marker in lupus nephritis. Antidouble stranded DNA antibodies can up regulate the expression of NGAL in mesangial cells in the glomerulus. The levels of urinary NGAL have been compared between patients with lupus nephritis and those with lupus without renal disease activity [16]. Urinary levels were significantly higher in patients with lupus nephritis. Furthermore, the urinary levels of NGAL correlated significantly ($r = 0.452, p = 0.009$) with renal disease activity scores but not with extrarenal disease scores. NGAL has also been associated with renal flares in children with lupus nephritis [17]. In this study, urinary NGAL levels greater than 0.6 mg/dL were 90% sensitive and 100% specific for identifying childhood lupus nephritis relative to those with juvenile rheumatoid arthritis. Similarly, in an adult population higher NGAL levels were found in patients with active lupus nephritis [16]. These studies show an association with disease activity but do not necessarily demonstrate that NGAL can be used as an early marker to identify patients who have an imminent flare of lupus nephritis. However, several studies have looked at the predictive ability of NGAL. Hinze looked at urine from a group of 111 pediatric patients with SLE to determine if NGAL levels increased before a clinical flare of lupus nephritis [18]. Urinary NGAL levels increased by as much as 104% 3 months prior to worsening of lupus nephritis. Plasma levels also increased prior to a renal flare but to a lesser extent. Rubinstein and coworkers have recently looked at a similar question [19]. Urinary NGAL levels from the previous visit were used to determine if they could predict renal disease activity at each visit. The AUC of urinary NGAL as a predictor of a renal flare was 0.76. This was greater than the AUC for antidouble stranded DNA antibody titers and similar to the AUC values obtained for C3 and C4 levels. These studies demonstrate that urinary NGAL may be an early predictor for flares of lupus nephritis. New tests with better predictive value would be useful, however.

Urinary MCP-1 has also been proposed as a potential predictor of renal lupus flares [20]. Urinary MCP-1 concentration was higher at the time of a renal flare than it was at nonrenal flare and in healthy and nonlupus renal controls. Furthermore, the increase in MCP-1 was seen 2–4 months prior to the clinical flare indicating that it may be an early predictor. Levels of urine IL-6 have also been proposed as markers of lupus renal activity [21]. Urine IL-6 levels correlated with disease activity as well as with active urine sediment. This study did not evaluate whether IL-6 levels could be used as early predictors of renal lupus flare. Similar results were found for urinary VCAM-1 [22] and urine osteoprotegerin [23]. From these studies of early predictors of lupus renal disease activity, NGAL and MCP-1 appear to be the most promising. Further evaluation of these candidates as well as others will help to better guide treatment in patients with lupus nephritis. Recent studies have focused on predicting decline of renal function in patients with lupus where multianalyte panels appear promising [24,25].

MEMBRANOUS NEPHROPATHY

MN is a frequent cause of nephrotic syndrome. The spontaneous outcome of MN can be difficult to predict which make decisions about treatment challenging. Without treatment, only about a third of patients with MN will have their renal function decline to the point that they need dialysis over 10 years [26]. The remaining two-thirds of patients will not receive any benefit from treatment since the natural history of their disease is relatively benign. Better methods are needed to predict which patients will have progressive worsening of renal function so that treatment can be targeted to these patients while patients that will not progress are spared from cytotoxic therapies. Protein expression in renal biopsy tissue is one potential method to predict the patient outcome regarding progression of the disease. Interstitial smooth muscle actin (SMA) was stained for on the renal biopsy to determine if it could be used as a predictor of progression [27]. SMA staining in the myofibroblasts strongly correlated with GFR after 7 years of follow-up. In a similar study, interstitial alpha SMA staining on the renal biopsy was strongly associated with progression to ESRD among patients with MN. In addition MCP-1 positive infiltrating mononuclear cells were strongly associated with progression to ESRD [28]. These studies show that findings on biopsy may be useful to predict progression of idiopathic MN but the findings will need to be further defined in larger and more diverse populations. Staining of biopsies for specific proteins that can predict outcome is promising but it is less likely to provide the ability to longitudinally follow the disease course in patients. Urinary levels of a number of proteins have also been used to predict the progression of renal disease in idiopathic MN. In these studies beta-2 microglobulin (β2M), IgG, urinary complement levels, NAG and L-fatty acid binding protein (L-FABP) have been used with varying success to predict progression of renal disease. Unfortunately, a number of definitions of progression of disease have been used and none have used hard outcomes, such as progression to ESRD. Urinary β2M levels and IgG levels were used to attempt to predict outcome of MN [29]. All patients in this study had a baseline serum creatinine which was less than 1.5 mg/dL. The endpoint was serum creatinine greater than 1.5 mg/dL or 50% rise in serum creatinine. After 53 months, 44% of the patients had met this endpoint. After multivariate analysis, urinary β2M was the strongest predictor for progression. Sensitivity and specificity to predict progression were 88 and 91%. Urinary IgG performed slightly less well. Although the sensitivity and specificity are not high enough yet to base the use of cytotoxic therapy on these markers, the results are encouraging. Activation of the complement system may play a role in the pathogenesis and progression of MN. Urinary C3dg and C5b-9 levels have been measured in patients with IgA and MN [30]. High urinary concentration of these two complement activation markers were found to correlate with MN and not IgA nephropathy. Furthermore, 66% of patients with MN who had high levels of urinary C5b-9 showed an unstable clinical course with deteriorating renal function compared to only 18% of those with a low level. This study demonstrates that urinary complement levels may be beneficial in predicting which patients should be treated with MN but markers that would provide better separation of progressors from nonprogressors are needed. Another urinary marker that has been used to predict outcome in MN is L-FABP [31]. In a study of 40 patients, urinary levels of L-FABP predicted worsening of renal function with a sensitivity of 81% and a specificity of 83%. Renal failure (defined as an increase in serum creatinine greater than 25% and exceeding 1.5 mg/dL) occurred in approximately 15% of patients with L-FABP levels lower than 5.7 μg/mmol while approximately 82% of patients with higher levels of L-FABP had renal failure. This separation of progressors from non-progressors is slightly better than that seen with urinary complement levels but still not adequately predictive to be used to change treatment decisions. The same authors have examined the role of urinary β2M in predicting prognosis in MN [32]. In this study, the end point was slightly different than the study of L-FABP. Renal failure was defined as an increase in serum creatinine of 50% or a serum creatinine greater than 1.5 mg/dL. Renal survival was 32% at 1 year in the group with high β2M and 93% in the group with low β2M. β2M was superior at predicting progression of renal disease to NAG but similar to L-FABP. In summary, a number of potential urinary biomarkers to predict progression of idiopathic MN have been proposed. The most promising are urinary β2M, L-FABP and the complement activation marker C5b-9. These urinary markers have added advantage of being able to follow them sequentially during treatment, unlike findings on renal biopsy. These biomarkers have the potential to greatly improve our ability to select patients who should be treated for MN. The majority of the studies however have used small increases in serum creatinine as an endpoint. While this is an important first step since these patients are more likely to progress further, it does not definitively identify patients who will develop ESRD. Future studies should focus on identifying patients with hard outcomes, such as progression to ESRD, including larger numbers of patients, using combinations of markers to predict outcome and identifying novel markers. The advances related to measurement of antibodies against the M-type phospholipase A_2 receptor described later are exciting but do not yet answer the important questions about prediction of progression of MN.

MN is caused by deposition of circulating antibodies in the basement membrane of the glomerulus. The epitope for these antibodies has not been clear. In the experimental rat model of Heymann nephritis, circulating autoantibodies bind intrinsic podocyte antigens. The antigen was identified as megalin, a 516 kDa glycoprotein, member of the LDL receptor family, and is expressed with clathrin on the cell membrane of the podocyte foot processes. These antibodies activate complement, resulting in the generation of the membrane attack complex and leading to injury. These immune complexes are subsequently degraded to form discontinuous subepithelial deposits. Maternal antineutral endopeptidase antibodies were also described to cross the placenta, resulting in neonatal membranous nephritis. Neither megalin, nor NEP causes adult MN however. Nevertheless, they provided the framework leading to the discovery of antibodies against the M-type phospholipase A_2 receptor (PLA_2R) which appear to be causative for many cases of idiopathic MN [33] and are one of the most exciting biomarkers identified in the glomerular disease area.

In order to identify the pathogenic antibodies in MN, investigators ran extracted proteins from glomeruli isolated from deceased donor kidneys on a gel [34]. Western blotting of the human glomerular extract (the antigen) with serum samples (the antibody) from patients with idiopathic MN was performed. A 185 kDa protein band was identified in 70% of the serum samples from patients with idiopathic MN but not when serum from patients with other glomerular diseases, including secondary MN or autoimmune disorders and serum from normal (volunteer) controls was used. The band was excised and analyzed by liquid chromatography/tandem mass spectrometry and a list of candidate proteins was generated. One of the proteins in the list was PLA_2R. Antibodies against PLA_2R identified a glomerular band of the same size, recombinant PLA_2R migrated to the same molecular weight and the 70% of idiopathic MN serum samples that were reactive to glomerular extracts also recognized recombinant PLA_2R. Moreover, PLA_2R was identified in podocytes by immunostaining of tissue sections and PCR and western blotting of cultured podocytes. Additional evidence that PLA_2R antibodies were associated with MN was that IgG eluted from kidneys of patients with idiopathic MN recognized a band at the size of PLA_2R from glomeruli and recombinant PLA_2R on western blots. This landmark study has led to the use of assays for PLA_2R as a clinical test in idiopathic MN.

Serum assays of anti-PLA2R antibodies are currently in clinical use to aid in the diagnosis of primary MN and may prove valuable in monitoring of disease activity. Ant-PLA_2R are occasionally described in patients with lupus nephritis, cancer, and hepatitis B virus infection, but has not been found in any proteinuric kidney disease other than MN. In the future serum assays may be used to diagnose MN, whether primary or secondary, when kidney biopsy poses significant risk or is contraindicated. It is worth noting that seropositivity develops after the development of clinical MN, and that spontaneous or treatment-induced decline or disappearance of circulating anti-PLA_2R antibodies precedes clinical remission by months. Circulating anti-PLA_2R antibody titers are also closely correlated with the degree of proteinuria, and high titers are usually associated with the nephrotic syndrome rather than asymptomatic proteinuria [35]. Further studies are needed to determine if high titers are associated with decline in kidney function. The presence of anti-PLA_2R IgG4 antibodies at the time of kidney transplantation highly predicts recurrence of primary MN on the kidney allograft, distinguishing it from de novo MN after transplantation. de novo MN after transplantation is mostly associated with IgG1 subclass, and usually tests negative for anti-PLA_2R antibodies and renal biopsy tissue stains negative for PLA_2R.

In addition to serum antibody testing, immunostaining for PLA_2R on kidney biopsy specimens is also a sensitive and specific test to diagnose PLA_2R-associated MN. Kidneys may stain positive for PLA_2R despite absence of circulating anti-PLA_2R antibodies. This may have to do with the kidney functioning as a sponge trapping anti-PLA_2R antibodies, or the methods of detecting such antibodies. On the other hand, a minority of patients have negative staining for PLA_2R despite the presence of circulating anti-PLA_2R antibodies. Just as circulating anti-PLA_2R IgG4 antibodies are occasionally described in patients with lupus nephritis, cancer, and hepatitis B virus infection; staining for PLA_2R was also described in patients with secondary MN due to lupus, hepatitis, sarcoidosis, and cancer [35]. Whether this is a concomitant primary MN or secondary MN will need to be proven.

A similar method to the discovery of anti-PLA_2R antibodies was used to identify a second podocyte antigen, the thrombospondin type 1 domain-containing 7A (THSD7A) [36]. Like the PLA_2R, THSD7A is expressed on podocytes, and is involved in subepithelial deposits. It accounts for up to 5% of cases of idiopathic MN in Western countries. These studies demonstrate a novel approach to the identification of glomerular disease biomarkers which has rapidly changed practice related to diagnosis and treatment of patients with idiopathic MN.

FOCAL SEGMENTAL GLOMERULOSCLEROSIS

Primary Focal Segmental Glomerulosclerosis (FSGS) has become one of the most common causes of idiopathic glomerular diseases in adults. There has been a steady increase in the incidence of FSGS in the last two decades [37,38]. FSGS is a progressive disease and is one of the leading cause of end-stage renal disease. Initial treatment is typically done with corticosteroids for a prolonged period but many patients do not respond. Markers which could predict which patients would respond could help avoid the toxicity of long-term treatment in patients that ultimately will not benefit from it. Mastroianni and coworkers examined the value of urinary retinol-binding protein as a prognostic marker in the treatment of nephrotic syndrome [39]. Urine levels of plasma retinol binding protein were measured in patients with FSGS, minimal change disease or mesangial proliferative glomerulonephritis. Patients with pretreatment levels less than 1 mg/mL were 30 times more likely to respond to treatment than those with higher levels. Furthermore patients with higher baseline levels of urinary plasma retinol binding protein that normalized during treatment were more likely to respond to treatment than those patients that did not. This finding is promising but will need to be validated in additional patients.

It has long been suspected that a circulating factor is responsible for causing primary FSGS. Reports of recurrent biopsy proven FSGS after transplant and successful transplant of a kidney from a recurrent FSGS patient into a non-FSGS patient supported the role for a circulating factor [40]. In vitro experiments demonstrated that this circulating factor of FSGS could be 30–50 kDa fraction of serum and has been called the permeability factor [41].

Serum soluble urokinase receptor (suPAR) has emerged as a potential biomarker in the diagnosis of FSGS. The urokinase receptor (uPAR) is a three domain (D_I, D_{II} and D_{III}) glycosylphosphatidylinositol (GPI) anchored membrane protein with a molecular weight of 35–60 kDa depending on the degree of glycosylation. It is expressed by multiple cells including osteoclasts, vascular smooth muscle cells, and endothelial cells. uPAR can bind to serum and extracellular–matrix protein vitronectin, which is a ligand of $\alpha_5\beta_3$ integrin activation. uPAR influences integrin dependent cell adhesion, migration, and proliferation. Signaling of uPAR in the kidney induces podocyte motility, foot process effacement, cellular proliferation, increases in glomerular fibrin deposits [42], and increased urinary protein loss. Membrane anchored uPAR is released into the serum as a soluble molecule (suPAR) following cleavage of the GPI anchor. suPAR can be further cleaved between the D_I and D_{II} domain forming a D_I and $D_{II,III}$ fragments. It has been hypothesized that these circulating suPAR fragments could be the presumed permeability factors.

Wei et al. [43] demonstrated that serum suPAR levels were significantly higher in patients with FSGS compared to healthy subjects. Immunoprecipitaion studies of FSGS serum found a 22 kDa fragment that was increased relative to non-FSGS serum. Serum suPAR concentration was able to differentiate FSGS patients from other glomerular disease with proteinuria (minimal change disease, MN and preeclampsia). Serum suPAR concentrations of >3000 pg/mL were reported to be an optimal cutoff for diagnosing FSGS. The most important clinical implication of serum suPAR appears to be in renal transplant patients. Pretransplant serum suPAR concentration have been shown to predict recurrence of posttransplant FSGS. This could play an important role in modifying the management of patients with high risk of disease recurrence. Serum suPAR levels could be helpful in monitoring treatment as the levels decrease following plasmapheresis [43]. Serum suPAR concentrations correlated with the presence but not with the degree of proteinuria. However, there are a number of factors that could confound interpretation. Serum suPAR concentration is effected by many other independent factors. For instance, higher levels are seen in females and elderly patients. Serum suPAR levels strongly correlate with C-reactive protein levels, indicating that higher levels are a sign of inflammation. Renal function was the strongest determinant of suPAR concentration. Estimated glomerular filtration rate is inversely related to serum suPAR concentrations both in FSGS and non-FSGS patients. In one study, serum suPAR concentration of >3000 pg/mL were observed in 39, 88, and 95% respectively of both FSGS and non-FSGS patients with eGFR of 45–60, 30–45 and, <30 mL/min per 1.73 m^2 [44].

Though the in vitro, animal studies and the initial human study appeared promising other investigators could not reproduce similar results. In recent follow-up trials serum suPAR levels were not helpful in differentiating FSGS from other proteinuria glomerular diseases. In one study, serum suPAR concentrations of >3000 pg/mL had a sensitivity of only 31% for identifying FSGS but improved to 75% when the cutoff was lowered to 2442.5 pg/mL. Even with this cutoff, the area under the ROC curve was 0.62 for differentiating FSGS from other glomerular diseases [45]. There is also conflicting evidence if serum suPAR levels have any prognostic value in FSGS patients treated with immunosuppressive therapy. Other molecules like CD80 (B7-1) [46] and Cardiotrophin like cytokine-1 (CLC-1) have been proposed as potential biomarkers in differentiating FSGS from other glomerular diseases. In summary, while suPAR may be useful in some populations, data do not yet support its widespread use as a biomarker of FSGS.

MINIMAL CHANGE DISEASE

Minimal change disease is the most common glomerular disease in children and also a common cause of nephrotic syndrome in adults. Renal biopsy is typically not performed in children presenting with nephrotic syndrome until after a treatment attempt with steroids has failed. A method to differentiate minimal change disease from other diseases would be very helpful in guiding treatment. Garin and coworkers measured the urine levels of soluble CD80 (sCD80) in patients with relapsed minimal change disease as well as several other glomerular diseases [47]. Urinary concentrations of sCD80 were significantly higher in patients with relapsed minimal change disease compared to patients in remission, patients with other glomerular diseases and normal controls. In a follow-up study, the authors compared urinary concentrations of sCD80 between patients with relapsed minimal change disease and patients with FSGS [48]. Levels were significantly higher in patients with relapsed minimal change disease. An ROC curve was made which showed that the AUC value for differentiating relapsed minimal change disease and FSGS was 0.99 and the AUC value for relapsed minimal change versus remission was 1.0. Furthermore, comparison of protein expression in renal biopsy tissue showed CD80 in 7/7 biopsies from patients with relapsed MCD while no staining was seen in biopsy tissue from two patients with FSGS and one patient with MCD in remission. These data are exciting in that they are the first to show that a single marker may be able to differentiate between two diseases that may be otherwise difficult to differentiate without a renal biopsy. Furthermore they have been replicated in a separate, chronologically distinct population and have biopsy data to support it. They suggest that urinary sCD80 may be a biomarker to differentiate FSGS from minimal change disease.

CD80/B7-1 IN MINIMAL CHANGE AND FSGS

Recently a number of interesting studies have expanded on these studies for patients with glomerular diseases including minimal change and FSGS. CD80, also called B7-1, is a costimulatory ligand that facilitates binding of antigen presenting cells to the T-cell receptors CD28 and CTLA-4. This binding is necessary for activating and regulating T-cell immunity [49]. B7-1 was identified in podocytes in a number of animal models of glomerular disease [50]. Moreover, the same study showed that podocyte expression of B7-1 correlated with the severity of lupus nephritis in humans and that that mice lacking B7-1 are resistant to LPS-induced nephrotic syndrome. Further studies showed that B7-1 was also induced in podocytes from patients with diabetic kidney disease and murine models of diabetic kidney disease, high glucose concentrations in vitro led to upregulation of B7-1 in cultured podocytes and that the expression of B7-1 was associated with apoptosis and cytoskeletal disruption that was reversed by antibody against CTLA-4 [51]. Excitement peaked when abatacept, a clinically available

CTLA-4 fusion protein which blocks stimulation by B7-1, induced partial or complete remission in five patients with FSGS who had B7-1 immunostaining on podocytes [52]. Moreover, CD80/B7-1 have been found in urine where it has primarily been associated with minimal change disease [46,53,54]. In combination, the results suggest that CD80/B7-1 in urine or in podocytes on renal biopsy may serve as a biomarker for glomerular disease and that these biomarkers may be able to guide the successful use of costimulatory blocking therapies like abatacept. Recently however, a number of studies have called the presence of B7-1 on podocytes into doubt [55–57]. At this point, it is not clear if podocyte CD80/B7-1 expression is an important biomarker although the studies regarding CD80 in urine of subjects with minimal change disease are compelling.

Other studies of biomarkers in minimal change disease have looked at the association of proteins with relapse or response to treatment. NAG levels have been evaluated to determine their association with remission and relapse in children with minimal change disease. [58]. Levels of urinary NAG (normalized for urine creatinine) were the same as normal controls. In patients with relapse, urinary levels were elevated demonstrating that NAG may be an indicator of relapse. Levels of urinary NAG have not been measured prior to clinical relapse to determine if they may be used as an early marker to predict relapse. Woroniecki evaluated the ability of a set of urinary cytokines to distinguish between MCD and FSGS as well as between steroid responsive and steroid-resistant nephrotic syndrome [59]. There were no significant differences in the levels of ICAM-1 and TGF-beta1 related to steroid responsiveness. Urinary concentrations of TGF-beta1 were significantly higher in patients with FSGS compared to those with minimal change disease.

IgA NEPHROPATHY

IgA nephropathy is the most common primary glomerular disease worldwide. The clinical course of IgA nephropathy can be highly variable which makes decisions about the treatment difficult. Many patients have a long-term indolent course of their disease. A number of studies have evaluated the ability of protein markers in biopsied tissue or in urine or serum to predict the progression of the disease. Renal biopsy is usually performed so evaluation of tissue can be an informative method to obtain information about prognosis. The crescentic variant of IgA nephropathy has a worse prognosis than variants without crescents. Bazzi and coworkers evaluated the ability of several different predictors of progression in crescentic IgA nephropathy after biopsy [60]. Serum creatinine alone was the best single predictor of progression with an AUC of 0.92. The fractional excretion of IgG normalized for the percent of glomeruli that were not globally sclerotic on the renal biopsy was the second best with an AUC value of 0.90. Interestingly, the combination of serum creatinine and normalized fractional excretion of IgG was able to stratify the patients into a high and low risk group in which 100% of the patients in the high risk group progressed and none of the patients in the low risk group did. Although this discrimination requires renal biopsy, it indicates that urinary excretion of IgG can be useful in helping determine the prognosis of patients with crescentic IgA nephropathy. Similarly, Van Es did a retrospective analysis of renal biopsies from 50 patients with IgA nephropathy to determine if GMP-17-positive T lymphocytes in renal tubules predict progression in early stages of IgA nephropathy [61]. They found that there was a positive association between GMP-17-positive cytotoxic T lymphocytes in intact renal tubules and progression of IgA nephropathy indicating this may be another marker of progression that can be determined using renal biopsy. A third study used findings on biopsy to predict which patients with IgA nephropathy would have a positive response to treatment with steroids. This study used the number of cells in the biopsy that were positive for fibroblast-specific protein 1 (FSP1) in patients with IgA nephropathy [62]. The investigators compared the ability of serum creatinine, estimated GFR, severity of mesangial proliferation, percent of sclerotic glomeruli, extent of interstitial damage, and FSP1 positive cell number to predict the response to treatment with corticosteroids. The number of FSP1 positive cells was the strongest predictor of response. When patient biopsies had more than 32.6 positive cells/high powered field, they were more likely to show steroid resistance. A similar study of FSP-1 in renal biopsy also showed correlation with prognosis [63]. There have not been any follow up studies in this area since 2008. These studies demonstrate that findings on renal biopsy may be able to predict response (or lack of response).

Differentiating between IgA nephropathy and a host of other glomerular diseases can be a difficult problem. Boor compared the serum levels of PDGF-DD levels in patients with IgA nephropathy to patients with lupus nephritis, FSGS, membranous glomerulonephritis, and ANCA-associated vasculitis as well as healthy controls [64]. Only patients with IgA nephropathy had levels which were significantly higher than control patients, suggesting that serum PDGF-DD may be useful as part of a panel of biomarkers to distinguish glomerular diseases.

Tubulointerstitial injury occurs with IgA nephropathy and can be an important part of the injury process. The injury can be detected by renal biopsy but biomarkers can help with diagnosis. Urinary concentrations of NGAL have been compared to evidence of tubulointerstitial injury on biopsy [14,65]. Both NGAL and NAG were elevated in patients with IgA nephropathy and tubulointerstitial injury but increases in NGAL were seen with earlier (Lee grade II) lesions than were increases in NAG. This suggests that urinary NGAL levels could potentially be used to identify and follow tubulointerstitial injury. The prognostic value of this association has not been evaluated.

A very interesting study looked at the prognostic value of urinary interleukin 6 in patients with IgA nephropathy [17,66]. Urinary IL-6 levels were measured in 59 patients with IgA nephropathy who were followed for a mean of 8 years. IL-6 levels were significantly higher in progressors. Furthermore, patients with urinary IL-6 levels greater than 2.5 ng/day at diagnosis had a 7.8-fold higher risk of progression than patients with lower levels. The finding was not confirmed however in a subsequent study [67].

Serum IgA/C3 ratio has also been measured to attempt to correlate the levels with renal prognosis. Increasing values of the ratio were associated with worsening prognosis [68,69]. While the association was not as strong as that seen in the initial studies of urinary IL-6, serum IgA/C3 ratio may provide important additional information about prognosis. The ratio of epidermal growth factor (EGF) to monocyte chemotactic peptide-1 (MCP-1) in the urine has also been used to predict renal prognosis in IgA nephropathy [70]. This study was based on previous findings that EGF may modulate renal response to injury whereas MCP-1 plays a role in progression of renal disease possibly by recruiting inflammatory cells into the interstitium. Patients were divided into tertiles based on the ratio of EGF to MCP-1. Patients in the lowest tertile had a significant decline in renal function while those in the highest tertile had a 100% renal survival at 48 and 84 months of follow-up. The area under the receiver operator characteristics curve was used to determine the quality of the tests as a predictor of adverse outcomes. MCP-1 had an AUC of 0.57. EGF alone had an AUC value of 0.83 while the ratio had an AUC of 0.91. These data indicate that the ratio of EGF to MCP-1 is another compelling candidate to predict progression of IgA nephropathy. Complement factor H plays a role in regulating complement activation. Based on these findings, the ability of urinary complement factor H as a biomarker of IgA nephropathy disease activity has been investigated [71]. The urinary levels of complement factor H were associated with disease activity as measured by serum creatinine and the amount of proteinuria. Whether this analyte can be used to predict prognosis in patients with IgA nephropathy has not been evaluated. Serum levels of TNF receptor 1 have also been proposed as a candidate biomarker with a hazard ratio of 7.48 for progression for patients in the highest quartile [72].

In summary, a number of promising candidates to predict the risk of worsening renal function in IgA nephropathy have been proposed. The numbers of GMP-17 positive T lymphocytes and FSP-1 positive cells look promising as potential markers in renal biopsy. This is an area which could be amenable to proteomic analysis of isolated glomeruli which identify candidate proteins for further evaluation. Studies of urinary ratio of EGF to MCP-1 look particularly promising but will need to be further evaluated. Studies which combine measurement of urinary IL-6, EGF and MCP-1 may be particularly enlightening. Measurement of urinary values has the added benefit that the measurements can be done serially to determine if they can be used to assess response to treatment or to time the initiation of treatment.

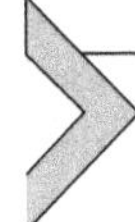

DISCOVERY OF NEW BIOMARKERS USING PROTEOMICS

Proteomic analysis provides an additional tool to identify novel biomarkers. These novel markers can then be tested to determine their validity. In addition, combinations of biomarkers can be analyzed using statistical or informatic tools. These approaches have been used to attempt to diagnose glomerular diseases and to provide prognostic information.

One of the most powerful proteomic techniques is liquid chromatography/tandem mass spectrometry (LC/MS-MS) in which proteins from urine, plasma, or tissue are digested with trypsin, peptides separated by chromatography, and the peptides are identified by tandem mass spectrometry. Proteins can be identified and information about relative abundance between samples can be obtained. An example of the use of this technique to identify urine biomarkers was the separation of urine from patients with diabetes who either did (progressors) or did not (non-progressors) develop diabetic nephropathy over time [73]. A number of candidate urine biomarkers were identified and urinary haptoglobin was validated as a biomarker for progression to diabetic nephropathy. Rood and coworkers used a similar technique to identify biomarkers that can distinguish idiopathic MN from other glomerular diseases [74]. LC/MS-MS was done on urinary microvessels obtained from patients with idiopathic MN and FSGS. Lysosome membrane protein-2 was increased in patients with idiopathic membranous nephropathy. Immunofluorescence microscopy confirmed that the proteins were expressed in glomeruli of patients with idiopathic membranous nephropathy at higher levels than in patients with FSGS, minimal change nephropathy, IgA nephropathy, or membranoproliferative glomerulonephritis.

A promising technique is capillary electrophoresis coupled to mass spectrometry (CE/MS). This approach was used to identify patterns of polypeptides that can differentiate between glomerular diseases including differentiating minimal change disease from FSGS [75]. While the approach needs to be further validated, it provides an interesting insight into the potential of such techniques. Julian and coworkers used a similar approach to characterize IgA related renal diseases [24,76]. In this study urine from patients with IgA nephropathy, Henoch–Schonlein purpura, and IgA associated nephropathy secondary to hepatitis C virus was analyzed by CE/MS. A pattern of polypeptides was identified which was seen in 90% of patients with IgA nephropathy and Henoch–Schonlein purpura but only 1% of patients with hepatitis C related disease and no normal control patients. While these findings using CE/MS are interesting, the technology is not currently available to measure the multiple polypeptides necessary on a clinical basis. Another proteomic technique with the potential to identify novel biomarkers is 2DE. We used 2DE to identify a set of proteins that can predict the cause of glomerular disease from among diabetic nephropathy, FSGS, membranous nephropathy, and lupus nephritis [77]. Interestingly, the urinary proteins that allowed the differentiation were glycosylated charge forms of proteins. This finding implies that different glomerular diseases have different glomerular permeability to charged proteins. This finding has not yet been replicated in a different set of patients. 2DE has also been used to identify urine proteins which are present in patients with IgA nephropathy but not in the urine of normal controls [78]. The investigators found 82 protein spots that were increased in urine from patients with IgA nephropathy relative to controls and 134 that were decreased. 84 of the proteins were identified by peptide mass fingerprinting. Although it is not known if these proteins are also differentially present in other renal diseases, it provides a list of protein candidates known to be present in the urine of patients with IgA nephropathy. An interesting approach to determine the response to treatment with ACE inhibitors of patients with IgA nephropathy used 2DE to identify urine proteins that were different between responders and nonresponders [79]. Kininogen, interalpha-trypsin-inhibitor heavy chain, and transthyretin were identified as proteins that differed between patients with IgA nephropathy that did and did not improve with ACE inhibitor treatment. Immunoblotting was then used to confirm that patients with low levels of kininogen at baseline were less likely to respond to treatment with ACE inhibitors. Another approach to identify candidate biomarkers is to analyze proteins and peptides using surface enhanced laser desorption and ionization mass spectrometry (SELDI-MS). In this technique proteins are adsorbed to a surface based on specific chemistries, unbound proteins are washed off and the remaining proteins are analyzed by mass spectrometry. Using this technique, the urine proteome of children with steroid-sensitive and steroid resistant nephrotic syndrome has been evaluated [80]. The authors identified a pattern of polypeptides which could predict steroid responsiveness 100% of the time. A protein of mass 4144 Daltons was identified as the most important classifier in the group of polypeptides but the identity of the peptide was not determined. These results are promising in that they show the potential of combinations of proteins to predict outcome but will need to be validated in independent sets. Another study that used SELDI-MS serially examined urine from patients with lupus nephritis [81]. Urine from 19 patients was obtained at baseline, preflare, flare, and postflare. SELDI-MS analysis followed by tandem mass spectrometry sequencing of relevant peptides was done. Several proteins were identified which had concentrations that peaked during or before renal flares. The 20 amino acid isoform of hepcidin was found to increase 4 months before renal flare and returned to baseline at the time of the flare. In contrast, the 25 amino acid isoform of hepcidin decreased at flare and returned to baseline 4 months after. These data are promising about the ability to identify and use early markers for renal flare in lupus but will need to be further validated.

A number of candidate markers have been identified in glomerular diseases. The most promising are markers for early prediction of flares in patients with lupus nephritis, prediction of outcome in patients with membranous nephropathy, differentiation between children with minimal change disease and FSGS and progression of IgA nephropathy. Others still require large amounts of validation and may ultimately not be sufficiently discriminatory to serve as useful markers. The use of discovery techniques, such as proteomics may provide additional new candidate markers. Ultimately, some diseases are likely to require the use of combinations of markers in useful clinical assays.

REFERENCES

[1] Whittier WL, Korbet SM. Timing of complications in percutaneous renal biopsy. J Am Soc Nephrol 2004;15:142–7.
[2] Mogensen CE, Christensen CK. Predicting diabetic nephropathy in insulin-dependent patients. N Engl J Med 1984;311:89–93.
[3] Mogensen CE. Microalbuminuria predicts clinical proteinuria and early mortality in maturity-onset diabetes. N Engl J Med 1984;310:356–60.
[4] Almdal T, Norgaard K, Feldt-Rasmussen B, Deckert T. The predictive value of microalbuminuria in IDDM. A five-year follow-up study. Diabetes Care 1994;17:120–5.
[5] Perkins BA, Ficociello LH, Silva KH, Finkelstein DM, Warram JH, Krolewski AS. Regression of microalbuminuria in type 1 diabetes. N Engl J Med 2003;348:2285–93.
[6] Bazzi C, Petrini C, Rizza V, Arrigo G, Napodano P, Paparella M, D'Amico G. Urinary N-acetyl-beta-glucosaminidase excretion is a marker of tubular cell dysfunction and a predictor of outcome in primary glomerulonephritis. Nephrol Dial Transplant 2002;17:1890–6.
[7] Bienias B, Zajaczkowska M, Borzecka H, Sikora P, Wieczorkiewicz-Plaza A, Wilczynska B. Early markers of tubulointerstitial fibrosis in children with idiopathic nephrotic syndrome: preliminary report. Medicine 2015;94:e1746.
[8] Appel GB, Contreras G, Dooley MA, Ginzler EM, Isenberg D, Jayne D, Li LS, Mysler E, Sanchez-Guerrero J, Solomons N, Wofsy D. Aspreva Lupus Management Study Group: mycophenolate mofetil versus cyclophosphamide for induction treatment of lupus nephritis. J Am Soc Nephrol 2009;20:1103–12.
[9] Bao H, Liu ZH, Xie HL, Hu WX, Zhang HT, Li LS. Successful treatment of class V + IV lupus nephritis with multitarget therapy. J Am Soc Nephrol 2008;19:2001–10.
[10] Avihingsanon Y, Phumesin P, Benjachat T, Akkasilpa S, Kittikowit V, Praditpornsilpa K, Wongpiyabavorn J, Eiam-Ong S, Hemachudha T, Tungsanga K, Hirankarn N. Measurement of urinary chemokine and growth factor messenger RNAs: a noninvasive monitoring in lupus nephritis. Kidney Int 2006;69:747–53.
[11] Nakayamada S, Saito K, Nakano K, Tanaka Y. Activation signal transduction by beta1 integrin in T cells from patients with systemic lupus erythematosus. Arthritis Rheum 2007;56:1559–68.
[12] Oates JC, Varghese S, Bland AM, Taylor TP, Self SE, Stanislaus R, Almeida JS, Arthur JM. Prediction of urinary protein markers in lupus nephritis. Kidney Int 2005;68:2588–92.
[13] Wang G, Lai FM, Tam LS, Li EK, Kwan BC, Chow KM, Li PK, Szeto CC. Urinary FOXP3 mRNA in patients with lupus nephritis—relation with disease activity and treatment response. Rheumatology 2009;48:755–60.
[14] Schwartz N, Rubinstein T, Burkly LC, Collins CE, Blanco I, Su L, Hojaili B, Mackay M, Aranow C, Stohl W, Rovin BH, Michaelson JS, Putterman C. Urinary TWEAK as a biomarker of lupus nephritis: a multicenter cohort study. Arthritis Res Ther 2009;11:R143.
[15] Nickolas TL, O'Rourke MJ, Yang J, Sise ME, Canetta PA, Barasch N, Buchen C, Khan F, Mori K, Giglio J, Devarajan P, Barasch J. Sensitivity and specificity of a single emergency department measurement of urinary neutrophil gelatinase-associated lipocalin for diagnosing acute kidney injury. Ann Intern Med 2008;148:810–9.
[16] Pitashny M, Schwartz N, Qing X, Hojaili B, Aranow C, Mackay M, Putterman C. Urinary lipocalin-2 is associated with renal disease activity in human lupus nephritis. Arthritis Rheum 2007;56:1894–903.
[17] Brunner HI, Mueller M, Rutherford C, Passo MH, Witte D, Grom A, Mishra J, Devarajan P. Urinary neutrophil gelatinase-associated lipocalin as a biomarker of nephritis in childhood-onset systemic lupus erythematosus. Arthritis Rheum 2006;54:2577–84.
[18] Hinze CH, Suzuki M, Klein-Gitelman M, Passo MH, Olson J, Singer NG, Haines KA, Onel K, O'Neil K, Silverman ED, Tucker L, Ying J, Devarajan P, Brunner HI. Neutrophil gelatinase-associated lipocalin is a predictor of the course of global and

renal childhood-onset systemic lupus erythematosus disease activity. Arthritis Rheum 2009;60:2772–81.
[19] Rubinstein T, Pitashny M, Levine B, Schwartz N, Schwartzman J, Weinstein E, Pego-Reigosa JM, Lu TY, Isenberg D, Rahman A, Putterman C. Urinary neutrophil gelatinase-associated lipocalin as a novel biomarker for disease activity in lupus nephritis. Rheumatology 2010;49:960–71.
[20] Rovin BH, Song H, Birmingham DJ, Hebert LA, Yu CY, Nagaraja HN. Urine chemokines as biomarkers of human systemic lupus erythematosus activity. J Am Soc Nephrol 2005;16:467–73.
[21] Peterson E, Robertson AD, Emlen W. Serum and urinary interleukin-6 in systemic lupus erythematosus. Lupus 1996;5:571–5.
[22] Molad Y, Miroshnik E, Sulkes J, Pitlik S, Weinberger A, Monselise Y. Urinary soluble VCAM-1 in systemic lupus erythematosus: a clinical marker for monitoring disease activity and damage. Clin Exp Rheumatol 2002;20:403–6.
[23] Kiani AN, Johnson K, Chen C, Diehl E, Hu H, Vasudevan G, Singh S, Magder LS, Knechtle SJ, Petri M. Urine osteoprotegerin and monocyte chemoattractant protein-1 in lupus nephritis. J Rheumatol 2009;36:2224–30.
[24] Abulaban KM, Song H, Zhang X, Kimmel PL, Kusek JW, Nelson RG, Feldman HI, Vasan RS, Ying J, Mauer M, Nelsestuen GL, Bennett M, Brunner HI, Rovin BH. Predicting decline of kidney function in lupus nephritis using urine biomarkers. Lupus 2016;25:1012–8.
[25] Wolf BJ, Spainhour JC, Arthur JM, Janech MG, Petri M, Oates JC. Development of biomarker models to predict outcomes in lupus nephritis. Arthritis Rheumatol 2016;68:1955–63.
[26] Jha V, Ganguli A, Saha TK, Kohli HS, Sud K, Gupta KL, Joshi K, Sakhuja V. A randomized, controlled trial of steroids and cyclophosphamide in adults with nephrotic syndrome caused by idiopathic membranous nephropathy. J Am Soc Nephrol 2007;18:1899–904.
[27] Badid C, Desmouliere A, McGregor B, Costa AM, Fouque D, Hadj Aissa A, Laville M. Interstitial alpha-smooth muscle actin: a prognostic marker in membranous nephropathy. Clin Nephrol 1999;52:210–7.
[28] Yoshimoto K, Wada T, Furuichi K, Sakai N, Iwata Y, Yokoyama H. CD68 and MCP-1/CCR2 expression of initial biopsies reflect the outcomes of membranous nephropathy. Nephron Clin Pract 2004;98:c25–34.
[29] Branten AJ, du Buf-Vereijken PW, Klasen IS, Bosch FH, Feith GW, Hollander DA, Wetzels JF. Urinary excretion of beta2-microglobulin and IgG predict prognosis in idiopathic membranous nephropathy: a validation study. J Am Soc Nephrol 2005;16:169–74.
[30] Brenchley PE, Coupes B, Short CD, O'Donoghue DJ, Ballardie FW, Mallick NP. Urinary C3dg and C5b-9 indicate active immune disease in human membranous nephropathy. Kidney Int 1992;41:933–7.
[31] Hofstra JM, Deegens JK, Steenbergen EJ, Wetzels JF. Urinary excretion of fatty acid-binding proteins in idiopathic membranous nephropathy. Nephrol Dial Transplant 2008;23:3160–5.
[32] Hofstra JM, Deegens JK, Willems HL, Wetzels JF. Beta-2-microglobulin is superior to N-acetyl-beta-glucosaminidase in predicting prognosis in idiopathic membranous nephropathy. Nephrol Dial Transplant 2008;23:2546–51.
[33] Sinico RA, Mezzina N, Trezzi B, Ghiggeri G, Radice A. Immunology of membranous nephropathy: from animal models to humans. Clin Exp Immunol 2016;183:157–65.
[34] Beck LH Jr, Bonegio RG, Lambeau G, Beck DM, Powell DW, Cummins TD, Klein JB, Salant DJ. M-type phospholipase A2 receptor as target antigen in idiopathic membranous nephropathy. N Engl J Med 2009;361:11–21.
[35] Francis JM, Beck LH Jr, Salant DJ. Membranous nephropathy: a journey from bench to bedside. Am J Kidney Dis 2016;68(1):138–47.

[36] Tomas NM, Beck LH Jr, Meyer-Schwesinger C, Seitz-Polski B, Ma H, Zahner G, Dolla G, Hoxha E, Helmchen U, Dabert-Gay AS, Debayle D, Merchant M, Klein J, Salant DJ, Stahl RA, Lambeau G. Thrombospondin type-1 domain-containing 7A in idiopathic membranous nephropathy. N Engl J Med 2014;371:2277–87.
[37] Haas M, Spargo BH, Coventry S. Increasing incidence of focal-segmental glomerulosclerosis among adult nephropathies: a 20-year renal biopsy study. Am J Kidney Dis 1995;26:740–50.
[38] Swaminathan S, Leung N, Lager DJ, Melton LJ 3rd, Bergstralh EJ, Rohlinger A, Fervenza FC. Changing incidence of glomerular disease in olmsted county, minnesota: a 30-year renal biopsy study. Clin J Am Soc Nephrol 2006;1:483–7.
[39] Mastroianni Kirsztajn G, Nishida SK, Silva MS, Ajzen H, Pereira AB. Urinary retinol-binding protein as a prognostic marker in the treatment of nephrotic syndrome. Nephron 2000;86:109–14.
[40] Gallon L, Leventhal J, Skaro A, Kanwar Y, Alvarado A. Resolution of recurrent focal segmental glomerulosclerosis after retransplantation. N Engl J Med 2012;366:1648–9.
[41] Savin VJ, Sharma R, Sharma M, McCarthy ET, Swan SK, Ellis E, Lovell H, Warady B, Gunwar S, Chonko AM, Artero M, Vincenti F. Circulating factor associated with increased glomerular permeability to albumin in recurrent focal segmental glomerulosclerosis. N Engl J Med 1996;334:878–83.
[42] Xu Y, Berrou J, Chen X, Fouqueray B, Callard P, Sraer JD, Rondeau E. Induction of urokinase receptor expression in nephrotoxic nephritis. Exp Nephrol 2001;9:397–404.
[43] Wei C, El Hindi S, Li J, Fornoni A, Goes N, Sageshima J, Maiguel D, Karumanchi SA, Yap HK, Saleem M, Zhang Q, Nikolic B, Chaudhuri A, Daftarian P, Salido E, Torres A, Salifu M, Sarwal MM, Schaefer F, Morath C, Schwenger V, Zeier M, Gupta V, Roth D, Rastaldi MP, Burke G, Ruiz P, Reiser J. Circulating urokinase receptor as a cause of focal segmental glomerulosclerosis. Nat Med 2011;17:952–60.
[44] Meijers B, Maas RJ, Sprangers B, Claes K, Poesen R, Bammens B, Naesens M, Deegens JK, Dietrich R, Storr M, Wetzels JF, Evenepoel P, Kuypers D. The soluble urokinase receptor is not a clinical marker for focal segmental glomerulosclerosis. Kidney Int 2014;85:636–40.
[45] Wada T, Nangaku M, Maruyama S, Imai E, Shoji K, Kato S, Endo T, Muso E, Kamata K, Yokoyama H, Fujimoto K, Obata Y, Nishino T, Kato H, Uchida S, Sasatomi Y, Saito T, Matsuo S. A multicenter cross-sectional study of circulating soluble urokinase receptor in Japanese patients with glomerular disease. Kidney Int 2014;85:641–8.
[46] Garin EH, Mu W, Arthur JM, Rivard CJ, Araya CE, Shimada M, Johnson RJ. Urinary CD80 is elevated in minimal change disease but not in focal segmental glomerulosclerosis. Kidney Int 2010;78:296–302.
[47] Garin EH, Diaz LN, Mu W, Wasserfall C, Araya C, Segal M, Johnson RJ. Urinary CD80 excretion increases in idiopathic minimal-change disease. J Am Soc Nephrol 2009;20:260–6.
[48] Garin EH, Mu W, Arthur JM, Rivard CJ, Araya CE, Shimada M, Johnson RJ. Urinary CD80 is elevated in minimal change disease but not in focal segmental glomerulosclerosis. Kidney Int 2010;78:296–302.
[49] Bhatia S, Edidin M, Almo SC, Nathenson SG. B7-1 and B7-2: similar costimulatory ligands with different biochemical, oligomeric and signaling properties. Immunol Lett 2006;104:70–5.
[50] Reiser J, von Gersdorff G, Loos M, Oh J, Asanuma K, Giardino L, Rastaldi MP, Calvaresi N, Watanabe H, Schwarz K, Faul C, Kretzler M, Davidson A, Sugimoto H, Kalluri R, Sharpe AH, Kreidberg JA, Mundel P. Induction of B7-1 in podocytes is associated with nephrotic syndrome. J Clin Invest 2004;113:1390–7.
[51] Fiorina P, Vergani A, Bassi R, Niewczas MA, Altintas MM, Pezzolesi MG, D'Addio F, Chin M, Tezza S, Ben Nasr M, Mattinzoli D, Ikehata M, Corradi D, Schumacher

V, Buvall L, Yu CC, Chang JM, La Rosa S, Finzi G, Solini A, Vincenti F, Rastaldi MP, Reiser J, Krolewski AS, Mundel PH, Sayegh MH. Role of podocyte B7-1 in diabetic nephropathy. J Am Soc Nephrol 2014;25:1415–29.
[52] Yu CC, Fornoni A, Weins A, Hakroush S, Maiguel D, Sageshima J, Chen L, Ciancio G, Faridi MH, Behr D, Campbell KN, Chang JM, Chen HC, Oh J, Faul C, Arnaout MA, Fiorina P, Gupta V, Greka A, Burke GW 3rd, Mundel P. Abatacept in B7-1-positive proteinuric kidney disease. N Engl J Med 2013;369:2416–23.
[53] Cara-Fuentes G, Araya C, Wei C, Rivard C, Ishimoto T, Reiser J, Johnson RJ, Garin EH. CD80, suPAR and nephrotic syndrome in a case of NPHS2 mutation. Nefrologia 2013;33:727–31.
[54] Ling C, Liu X, Shen Y, Chen Z, Fan J, Jiang Y, Meng Q. Urinary CD80 levels as a diagnostic biomarker of minimal change disease. Pediatr Nephrol 2015;30:309–16.
[55] Gagliardini E, Novelli R, Corna D, Zoja C, Ruggiero B, Benigni A, Remuzzi G. B7-1 is not induced in podocytes of human and experimental diabetic nephropathy. J Am Soc Nephrol 2016;27:999–1005.
[56] Novelli R, Gagliardini E, Ruggiero B, Benigni A, Remuzzi G. Any value of podocyte B7-1 as a biomarker in human MCD and FSGS? Am J Physiol Renal Physiol 2016;310:F335–41.
[57] Larsen CP, Messias NC, Walker PD. B7-1 immunostaining in proteinuric kidney disease. Am J Kidney Dis 2014;64:1001–3.
[58] Dillon SC, Taylor GM, Shah V. Diagnostic value of urinary retinol-binding protein in childhood nephrotic syndrome. Pediatr Nephrol 1998;12:643–7.
[59] Woroniecki RP, Shatat IF, Supe K, Du Z, Kaskel FJ. Urinary cytokines and steroid responsiveness in idiopathic nephrotic syndrome of childhood. Am J Nephrol 2008;28:83–90.
[60] Bazzi C, Rizza V, Raimondi S, Casellato D, Napodano P, D'Amico G. In crescentic IgA nephropathy, fractional excretion of IgG in combination with nephron loss is the best predictor of progression and responsiveness to immunosuppression. Clin J Am Soc Nephrol 2009;4:929–35.
[61] van Es LA, de Heer E, Vleming LJ, van der Wal A, Mallat M, Bajema I, Bruijn JA, de Fijter JW. GMP-17-positive T-lymphocytes in renal tubules predict progression in early stages of IgA nephropathy. Kidney Int 2008;73:1426–33.
[62] Harada K, Akai Y, Yamaguchi Y, Kimura K, Nishitani Y, Nakatani K, Iwano M, Saito Y. Prediction of corticosteroid responsiveness based on fibroblast-specific protein 1 (FSP1) in patients with IgA nephropathy. Nephrol Dial Transplant 2008;23:3152–9.
[63] Nishitani Y, Iwano M, Yamaguchi Y, Harada K, Nakatani K, Akai Y, Nishino T, Shiiki H, Kanauchi M, Saito Y, Neilson EG. Fibroblast-specific protein 1 is a specific prognostic marker for renal survival in patients with IgAN. Kidney Int 2005;68:1078–85.
[64] Boor P, Eitner F, Cohen CD, Lindenmeyer MT, ERCB-Consortium, Mertens PR, Ostendorf T, Floege J. Patients with IgA nephropathy exhibit high systemic PDGF-DD levels. Nephrol Dial Transplant 2009;24:2755–62.
[65] Ding H, He Y, Li K, Yang J, Li X, Lu R, Gao W. Urinary neutrophil gelatinase-associated lipocalin (NGAL) is an early biomarker for renal tubulointerstitial injury in IgA nephropathy. Clin Immunol 2007;123:227–34.
[66] Harada K, Akai Y, Kurumatani N, Iwano M, Saito Y. Prognostic value of urinary interleukin 6 in patients with IgA nephropathy: an 8-year follow-up study. Nephron 2002;92:824–6.
[67] Stangou M, Alexopoulos E, Papagianni A, Pantzaki A, Bantis C, Dovas S, Economidou D, Leontsini M, Memmos D. Urinary levels of epidermal growth factor, interleukin-6 and monocyte chemoattractant protein-1 may act as predictor markers of renal function outcome in immunoglobulin A nephropathy. Nephrology 2009;14:613–20.

[68] Ishiguro C, Yaguchi Y, Funabiki K, Horikoshi S, Shirato I, Tomino Y. Serum IgA/C3 ratio may predict diagnosis and prognostic grading in patients with IgA nephropathy. Nephron 2002;91:755–8.
[69] Maeda A, Gohda T, Funabiki K, Horikoshi S, Shirato I, Tomino Y. Significance of serum IgA levels and serum IgA/C3 ratio in diagnostic analysis of patients with IgA nephropathy. J Clin Lab Anal 2003;17:73–6.
[70] Torres DD, Rossini M, Manno C, Mattace-Raso F, D'Altri C, Ranieri E, Pontrelli P, Grandaliano G, Gesualdo L, Schena FP. The ratio of epidermal growth factor to monocyte chemotactic peptide-1 in the urine predicts renal prognosis in IgA nephropathy. Kidney Int 2008;73:327–33.
[71] Zhang JJ, Jiang L, Liu G, Wang SX, Zou WZ, Zhang H, Zhao MH. Levels of urinary complement factor H in patients with IgA nephropathy are closely associated with disease activity. Scand J Immunol 2009;69:457–64.
[72] Oh YJ, An JN, Kim CT, Yang SH, Lee H, Kim DK, Joo KW, Paik JH, Kang SW, Park JT, Lim CS, Kim YS, Lee JP. Circulating tumor necrosis factor alpha receptors predict the outcomes of human IgA nephropathy: a prospective cohort study. PLoS One 2015;10:e0132826.
[73] Bhensdadia NM, Hunt KJ, Lopes-Virella MF, Michael Tucker J, Mataria MR, Alge JL, Neely BA, Janech MG, Arthur JM. Urine haptoglobin levels predict early renal functional decline in patients with type 2 diabetes. Kidney Int 2013;83:1136–43.
[74] Rood IM, Merchant ML, Wilkey DW, Zhang T, Zabrouskov V, van der Vlag J, Dijkman HB, Willemsen BK, Wetzels JF, Klein JB, Deegens JK. Increased expression of lysosome membrane protein 2 in glomeruli of patients with idiopathic membranous nephropathy. Proteomics 2015;15:3722–30.
[75] Wittke S, Mischak H, Walden M, Kolch W, Radler T, Wiedemann K. Discovery of biomarkers in human urine and cerebrospinal fluid by capillary electrophoresis coupled to mass spectrometry: towards new diagnostic and therapeutic approaches. Electrophoresis 2005;26:1476–87.
[76] Julian BA, Wittke S, Novak J, Good DM, Coon JJ, Kellmann M, Zurbig P, Schiffer E, Haubitz M, Moldoveanu Z, Calcatera SM, Wyatt RJ, Sykora J, Sladkova E, Hes O, Mischak H, McGuire BM. Electrophoretic methods for analysis of urinary polypeptides in IgA-associated renal diseases. Electrophoresis 2007;28:4469–83.
[77] Varghese SA, Powell TB, Budisavljevic MN, Oates JC, Raymond JR, Almeida JS, Arthur JM. Urine biomarkers predict the cause of glomerular disease. J Am Soc Nephrol 2007;18:913–22.
[78] Park MR, Wang EH, Jin DC, Cha JH, Lee KH, Yang CW, Kang CS, Choi YJ. Establishment of a 2-D human urinary proteomic map in IgA nephropathy. Proteomics 2006;6:1066–76.
[79] Rocchetti MT, Centra M, Papale M, Bortone G, Palermo C, Centonze D, Ranieri E, Di Paolo S, Gesualdo L. Urine protein profile of IgA nephropathy patients may predict the response to ACE-inhibitor therapy. Proteomics 2008;8:206–16.
[80] Woroniecki RP, Orlova TN, Mendelev N, Shatat IF, Hailpern SM, Kaskel FJ, Goligorsky MS, O'Riordan E. Urinary proteome of steroid-sensitive and steroid-resistant idiopathic nephrotic syndrome of childhood. Am J Nephrol 2006;26:258–67.
[81] Zhang X, Jin M, Wu H, Nadasdy T, Nadasdy G, Harris N, Green-Church K, Nagaraja H, Birmingham DJ, Yu CY, Hebert LA, Rovin BH. Biomarkers of lupus nephritis determined by serial urine proteomics. Kidney Int 2008;74:799–807.

CHAPTER FOURTEEN

Biomarkers in Preeclampsia

S.A. Karumanchi, MD
Department of Medicine, Obstetrics and Gynecology, Beth Israel Deaconess Medical Center and Harvard Medical School, Boston, MA, United States

Contents

DEFINITION AND PREVALENCE OF THE DISEASE

Preeclampsia, the most frequently encountered renal complication of pregnancy, is a leading cause of maternal and perinatal morbidity and mortality worldwide. It is a multisystemic disease that complicates 3–8% of pregnancies, and that is characterized by the new onset hypertension and proteinuria after 20 weeks of gestation [1,2]. In the mother, the disease can progress to widespread endothelial dysfunction affecting mainly the liver, brain, and kidney. In the fetus it is associated with intrauterine growth restriction (IUGR) and prematurity [3].

Biomarkers of Kidney Disease. http://dx.doi.org/10.1016/B978-0-12-803014-1.00014-5

In developing countries where access to healthcare is limited, preeclampsia is a leading cause of maternal mortality. Of the estimated 60,000 or more deaths from preeclampsia worldwide each year, >90% of the deaths are in low- and middle-income countries [1]. In the developed world, the burden falls on the neonate, as premature deliveries are performed to preserve the health of the mother. Although technological advances in perinatal and neonatal care have reduced infant mortality due to preterm birth, morbidity remains a serious problem. These babies are at an increased risk of neurodevelopment disabilities, such as cerebral palsy, mental retardation, sensory deficits, and behavioral impairments [4], and are also more vulnerable to metabolic disorders and cardiovascular disease (CVD) later in life [5–7].

Preeclampsia is not only responsible for adverse pregnancy outcomes, but also predisposes women to long-term health complications entailing a major economic and familial burden in the society. Many factors have been associated with an increased risk of developing preeclampsia, including familial obstetric history, preexisting medical conditions, obesity, advanced maternal age, and characteristics of pregnancy, such as parity. In a systematic review of controlled studies, Duckitt and Harrington reported that nulliparity, multiple pregnancy, family history of preeclampsia, history of preeclampsia in a previous pregnancy, a time span of more than 10 years since the last pregnancy, maternal age above 40, raised body mass index, raised blood pressure at booking, and preexisting medical conditions, such as antiphospholipid antibodies (APL), diabetes, hypertension, and renal disease, were associated with an increased risk of developing preeclampsia [8]. Nulliparity and multifetal gestations increased the risk almost threefold, and APL over ninefold [8]. Paternal factors have also been implicated. Reproductive practices that minimize exposure to sperm, such as barrier contraception, nonpartner donor insemination, and short duration of sexual cohabitation with the father before conception are associated with an increased risk of preeclampsia. Indeed, multiparous women pregnant with a new partner have a risk similar to nulliparous women. It is still not clear if this effect is due to the change in paternity per se or the greater risk associated with increased interpregnancy interval [9]. Bartsch et al. recently performed a systematic review and metaanalyses of all the large cohort studies that evaluated risk factors for preeclampsia and noted that women with APL syndrome had the highest pooled rate of preeclampsia (17.3%; 95% CI: 6.8–31.4%) and those with chronic hypertension had the second highest pooled rate (16.0%; CI: 12.6–19.7%) [10]. Pregestational diabetes [pooled rate: 11.0% (8.4–13.8%); pooled relative risk: 3.7 (3.1–4.3)], prepregnancy body mass index > 30 [7.1% (6.1–8.2%); 2.8 (2.6 to 3.1)], and the use of assisted reproductive technology [6.2% (4.7–7.9%); 1.8 (1.6–2.1)] were other prominent risk factors.

PATHOPHYSIOLOGY AND MECHANISMS

It has long been recognized that preeclampsia will not resolve until after complete placental delivery. Further, as illustrated by cases of molar and extrauterine pregnancies, while the placenta is required for developing preeclampsia, the fetus is not [11,12]. In addition, cases of postpartum eclampsia have been associated with retained placental fragments, with rapid improvement after uterine curettage [13]. Taken together, these observations suggest that the placenta is both necessary and sufficient for the development of preeclampsia. First proposed by Page in 1939 [14], it is now widely recognized that the placenta is the central culprit in the pathogenesis of the disease.

Research has focused on the placenta as the source of the disease, and has tried to unravel the mechanisms that ultimately lead to generalized maternal endothelial dysfunction. Many hypotheses have emerged that attempt to gather a causal framework for the disease, causing preeclampsia to be named the "disease of theories." Despite intensive investigation, its etiology and pathogenesis is not completely understood, and as of 2016 there is no cure other than delivery of the placenta. Nevertheless, knowledge in the field is progressing substantially with recent findings, opening new perspectives for the near future, specifically key discoveries about alterations in placental antiangiogenic factors in the pathogenesis of the clinical syndrome.

It has been suggested that preeclampsia is caused by placental dysfunction followed by the release of factors by the diseased placenta into the maternal circulation, inducing widespread endothelial dysfunction that heralds the classic manifestations of the disease [15] (Fig. 14.1). In this regard, two antiangiogenic proteins overproduced by the placenta that gain access to the maternal circulation have become candidate molecules responsible for the phenotype of preeclampsia [16]. Soluble Fms-like tyrosine kinase 1 (sFlt-1), an endogenous inhibitor of vascular endothelial growth factor (VEGF), placental growth factor (PlGF), and soluble endoglin (sEng), a circulating coreceptor of transforming growth factor beta (TGF-β), have been shown to be at increased levels in the serum of preeclamptic women, as compared to normal pregnancy, weeks before the appearance of overt clinical manifestations of the disease [17–19]. In addition, when injected into pregnant rodents, these molecules produce systemic endothelial dysfunction resulting in a preeclampsia-like phenotype, including severe hypertension, proteinuria, glomerular endotheliosis, and features resembling the hemolysis, elevated liver enzymes, low platelets (HELLP) syndrome [20].

An array of insults may contribute to placental damage that is proximally linked to the production of soluble pathogenic factors by this organ. Various pathways have been proposed to have key roles in inducing placental disease, including deficient heme oxygenase expression, impaired corin expression, placental hypoxia, genetic factors, autoantibodies against the angiotensin receptor, oxidative stress, inflammation, altered natural killer cell signaling, and deficient catechol-O-methyl transferase [19,21,22]. Interestingly, most of these were shown to increase placental production of the antiangiogenic factors. Still, the underlying events that induce placental disease activating the cascade of placental damage and antiangiogenic factor production remain unknown.

CLINICAL MANIFESTATIONS

Besides hypertension and proteinuria de novo after 20 weeks of gestation (the hallmarks of the disease), preeclampsia may also be accompanied by other manifestations. Additional signs and symptoms that can occur include edema, acute renal failure, liver abnormalities, thrombocytopenia, microangiopathic hemolytic anemia, placental abruption, visual disturbances, stroke, seizures, and death. These manifestations are the result of widespread endothelial dysfunction affecting mainly the liver, kidney, and brain (Fig. 14.1).

Hemolysis, abnormal elevation of liver enzymes levels, and low platelet count occur together as the HELLP syndrome [23]. Considered by many to be a severe variant of preeclampsia, HELLP syndrome occurs in 5% of cases and can progress rapidly to a life-threatening condition [24]. Another serious complication of preeclampsia is eclampsia, differentiated from the former by the presence of seizures [25]. The onset of eclampsia is often heralded by headache, visual disturbances, and epigastric pain. However, the eclamptic convulsion can occur suddenly and without warning. Once associated with a high mortality rate, improved and aggressive management have decreased the occurrence of convulsions, and nowadays maternal deaths are unusual [26].

The clinical spectrum varies widely from preeclampsia accompanied by mild hypertension and without demonstrable fetal involvement, to preeclampsia with various organ dysfunction, HELLP syndrome, eclampsia,

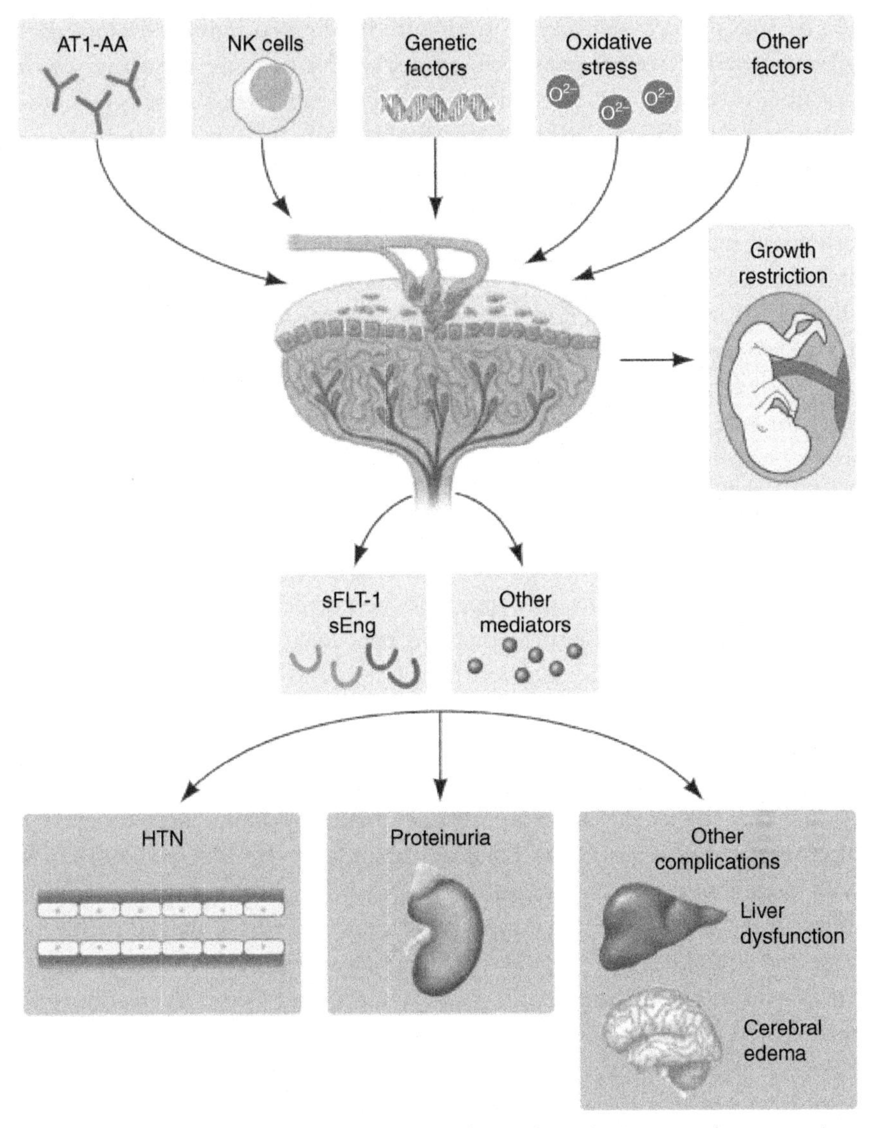

Figure 14.1 ***Summary of the pathogenesis of preeclampsia.*** Immune factors, such as autoantibodies against the angiotensin receptor *(AT1-AA)*, *oxidative stress*, natural killer *(NK)* cell abnormalities, and *other factors* may cause placental dysfunction, which in turn leads to the release of antiangiogenic factors [such as soluble Fms-like tyrosine kinase 1 *(sFlt-1)* and soluble endoglin *(sEng)*] and other inflammatory mediators to induce hypertension *(HTN)*, *proteinuria*, and *other complications* of preeclampsia. *Reproduced with permission from Wang A, Rana S, Karumanchi SA. Preeclampsia: the role of angiogenic factors in its pathogenesis. Physiology 2009;24(3):147–58.*

IUGR, and preterm delivery. Women with mild preeclampsia developing at term generally have pregnancy outcomes similar to those of women with normotensive pregnancies. Some women will experience an atypical presentation of the disease, for example, with the absence of hypertension or proteinuria, or that manifests outside the established gestational time period [27]. It has been reported that 10% of women with other clinical and/or histologic manifestations of preeclampsia have minimal or no proteinuira, and that 14% of women who develop eclampsia have no proteinuria at all [26]. Moreover, either hypertension or proteinuria may be absent in 10–15% of women who develop HELLP syndrome [24].

The kidney: In preeclampsia, renal plasma flow (RPF) and glomerular filtration rate (GFR) decrease by approximately 25%. It is important to note that in normal pregnancy RPF and GFR generally increase from 30% to 50%. Thus, blood urea nitrogen and creatinine in preeclampsia may actually be very approximate to or slightly above those seen in the normal range for nonpregnant woman. The degree of proteinuria varies from minimal to the nephrotic range, and does not appear to be correlated with maternal and fetal outcomes [28]. Hyperuricemia and hypocalciuria also occur. Urinary sediment is usually bland, and red blood cells and cellular casts are rarely seen. The decrement in RPF is attributable to vasoconstriction, whereas the fall in GFR relates both to the decrement of RPF and the development of a particular glomerular lesion termed glomerular endotheliosis. Glomerular endotheliosis is a unique pathologic condition characterized by ultrastructural changes in the renal glomeruli that shares some similarities with thrombotic microangiopathies, but that also shows some intriguing differences [29]. Under light microscopy, the glomeruli appear relatively large and the glomerular capillary lumen appears "bloodless" due to endothelial and mesangial cell swelling and hypertrophy. Fibrin deposition can be detected by immunofluorescence, but thrombosis is definitely unusual. Electron microscopy reveals subendothelial electron dense fibrinoid and granular deposits, as well as loss of endothelial fenestrae [29,30].

Despite marked proteinuria, the epithelial foot processes appear relatively intact. Nonetheless, recent evidence suggests that these podocytes are also affected, as considerable podocyturia accompanies the proteinuria [31,32]. In rare cases, preeclampsia can lead to acute renal failure in pregnancy. After delivery, the glomerular changes usually reverse rapidly, coinciding with the resolution of hypertension and proteinuria. Focal segmental glomerulosclerosis is seen in some cases. Clinically, even these women regain normal renal function along with the resolution of proteinuria [30].

Liver and coagulation abnormalities: Normal pregnancy is associated with a relatively hypercoagulable state, which is of teleologic advantage in avoiding hemorrhage after delivery. In preeclampsia, this hypercoagulability is accentuated (e.g., reduced antithrombin III, protein S, and protein C) and is usually associated with platelet activation and thrombocytopenia. These occur most likely as a result of endothelial cell damage/thrombotic microangiopathy [30].

The extent of liver damage in preeclampsia depends on disease severity. Increases in transaminases and lactic acid dehydrogenase levels are generally seen. These are usually mild except in the setting of HELLP syndrome. Indeed, transaminases levels are a clinical marker for severity of disease. Examination of the liver can show periportal hemorrhage, ischemic lesions, and fibrin deposition. Bleeding from periportal lesions, or hemorrhage into infarcts, can cause an intrahepatic hematoma. Although rare, subscapsular bleeding leading to hepatic rupture can occur as a catastrophic complication of pregnancy [30].

The brain: Besides convulsions (eclampsia), headache, and blurred vision, scotoma and blindness occur as central nervous system manifestations of preeclampsia. It can be due to retinal detachment or vascular occlusion, but more frequently is of cortical origin. Cortical blindness almost always resolves spontaneously after control of blood pressure, but blindness related to other causes, such as progression of underlying diabetic retinopathy may be permanent. Stroke is a serious but rare complication. Fatal cases of preeclampsia demonstrate various degrees of cerebral bleeding from microscopic petechiae to gross hemorrhage, ischemic brain damage, microinfarcts, and fibrinoid necrosis. The cerebrovascular manifestations of severe preeclampsia are poorly understood, but may represent a form of reversible posterior leukoencephalopathy syndrome. Neuroradiography findings on computed tomography and magnetic resonance imaging (MRI) show vasogenic cerebral edema and infarctions in the subcortical white matter and in adjacent gray matter, predominantly in the parietal and occipital lobes. Interestingly, these same characteristic MRI changes have also been associated with the use of antiangiogenic agents in cancer therapy [30].

Fetal complications: Although the acute manifestation of the disease is more ominous for the mother, the condition affects the fetus as well, imposing an increased risk of iatrogenic and spontaneous prematurity, intrauterine fetal growth restriction, oligohydramnios, and an increased risk of perinatal death. Although the exact pathogenesis of these complications is unknown, impaired uteroplacental blood flow or placental infarction are likely contributors. Many studies are now addressing the effects of preeclampsia per se on the health of these children, and whether or not these children (premature or term) are different from other premature and term children.

Children born to mothers who had preeclampsia, especially those born at term, had an increased risk of being hospitalized for a number of diseases, such as endocrine, nutritional, metabolic diseases, and diseases of the blood and blood-forming organs [33]. In preterm IUGR children with signs of cardiac dysfunction, being born to mothers who had preeclampsia did not influence cardiac performance [34]. In addition, for those exposed to preeclampsia during pregnancy, increased blood pressure in childhood [35,36] and stroke in adult life have been reported [37].

Long-term complications: Traditionally, women have been reassured that the syndrome would resolve postpartum without long-term consequences for the mother other than a higher risk of recurrence of the disease in subsequent pregnancies. However, there is increasing evidence suggesting that these women are at an increased risk for developing end-stage renal disease [38] and CVD [39–42] later in life. Approximately 20% of women with preeclampsia develop hypertension or microalbuminuria within 7 years of a preeclamptic pregnancy, as compared with only 2% among women with uncomplicated pregnancies [43]. In a recent meta-analysis, Bellamy et al. showed that after a pregnancy complicated by preeclampsia, women had an increased risk for hypertension (3.7; 95% CI: 2.70–5.05), ischemic heart disease (2.16; 95% CI: 1.86–2.52), stroke (1.81; 95% CI: 1.45–2.27), and venous thromboembolism (1.19; 95% CI: 1.37–2.33) [39]. The risk of death from cardiovascular and other causes is also increased in these women. Women with preeclampsia at term had a 1.65-fold higher long-term risk of death (95% CI: 1.01–2.70) from cardiovascular causes than women who did not have preeclampsia [40]. In particular, women with a history of preeclampsia and preterm birth had a dramatic 8.12 (95% CI: 4.31–15.33) higher risk when compared to women with normal pregnancies. Mortality from all causes was also increased: 1.04 (95% CI: 0.88–1.23) and 2.71 (95% CI: 1.99–3.68), respectively [40]. Women with early onset, severe preeclampsia appear to be at the highest risk [41,42].

Whether preeclampsia is another manifestation of a shared pathophysiology or is an independent risk factor for CVD is still a matter of debate [44]. Indeed, preeclampsia and CVD share many pathophysiologic mechanisms and risk factors, such as obesity, diabetes, and hypertension that can lead to both preeclampsia and CVD at different times during a woman's life. On the other hand, preeclampsia may itself induce vascular and metabolic changes that increase the risk for developing CVD. Regardless of this debate, preeclampsia raises a red flag concerning the risk of CVD in later life. Proper management targeting lifestyle and risk-factor modification should be implemented.

DIAGNOSIS

Historically, there has been difficulty in reaching a consensus on the diagnostic criteria for preeclampsia, and the definitions have changed over time. In 2013, the American College of Obstetrics and Gynecology recognizing the syndromic nature of the disease, changed the clinical definitions of preeclampsia to include aytpical presentations of the disease. Preeclampsia is now diagnosed with appearance of hypertension (systolic or diastolic at or above 140 and 90 mmHg, respectively) and new onset proteinuria (at or greater than 300 mg/day of protein), or a protein to creatinine ratio in a spot urine of 0.3 mg/dl after 20 weeks of gestation [45]. In the absence of proteinuria the diagnosis can be made when any other of the multisystem aspects were present, such as new onset thrombocytopenia, liver function abnormalities, pulmonary edema, and cerebral and visual system abnormalities [45]. For clinical purposes, preeclampsia can be classified as early (<34 weeks) or late onset (≥34 weeks) of the disease. In the developed countries, early onset of the disease (<34 weeks) is the form that is associated with greater morbidity than when the disorder presents closer to term. However, in developing countries term disease can also be associated with severe morbidity that is largely related to late presentation and failure to recognize and deliver.

Although the criteria presented are clear, accurate diagnosis of preeclampsia may not be straightforward. Accurate diagnosis relies on precise blood pressure and proteinuria measurements, and, of course, general agreement on the criteria. It is well recognized that blood pressure measurement is prone to inaccuracy due to observer and device error [46]. In addition, the 24-h urine specimen is not always available, and studies have shown that urinary dipstick determinations correlate poorly with the amount of protein found in 24-h urine samples [47,48] More recently, the urinary protein to creatinine (P:C) ratio has become the preferred method for quantification of proteinuria in the nonpregnant population. However, its use to estimate 24-h protein excretion for the diagnosis of preeclampsia has been controversial. Several studies have compared P:C ratio with 24-h urine collection in this setting, with discordant conclusions. A metaanalysis showed a pooled sensitivity of 84% and specificity of 76% using P:C ratio cutoff of greater than 30 mg/mmol, as compared with the gold standard of 24-h urine protein excretion > 300 mg/day [49].

Several clinical and laboratory findings suggest severe disease, and should prompt consideration of immediate delivery. Oliguria (less than 500 mL urine in 24 h) is usually transient; acute renal failure, though uncommon, can occur. Persistent headache or visual disturbances can be a prodrome to seizures. Pulmonary edema complicates 2–3% of severe preeclampsia and can lead to respiratory failure. Epigastric or right upper quadrant pain may be associated with liver injury. Elevated liver enzymes can occur alone or as part of the HELLP syndrome.

Preeclampsia should be distinguished from other disorders that can occur with increased blood pressure in pregnancy, including chronic hypertension and gestational hypertension. Chronic hypertension is defined as hypertension that is present and observable before pregnancy, or that is diagnosed before the 20th week of gestation. Gestational hypertension is defined as transient hypertension during pregnancy if preeclampsia is not present at the time of delivery, and blood pressure returns to normal by 12 weeks postpartum [25]. It is important to discriminate between preeclampsia and these conditions, as pregnancy management and outcome differ substantially. Most women with chronic hypertension have uneventful gestations as long as their blood pressure remains at (or is controlled to) levels considered "mild to moderate." In contrast, preeclampsia is associated with many adverse maternal and fetal complications.

The wide clinical variability and presentation of the disease, including atypical presentations, make diagnosis more challenging. Preeclampsia can also appear as early as 12 weeks of gestation in the case of trophoblastic diseases, such as hydatidiform mole [25]. Predicting preeclampsia weeks before its clinical presentation is crucial, and involves close monitoring, earlier recognition of the syndrome, and proper and timely intervention before life-threatening complications develop. Hence, there is an urgent need for biomarkers that can identify women at an increased risk of developing the disease, can accurately rule out diagnoses in suspected cases, and thus identify women at increased risk of adverse outcome. Such a biomarker would also assist the investigation of targeted strategies for prevention and treatment of preeclampsia.

BIOMARKERS

It is important to note that the utility of a predictive test will depend on the overall prevalence of the disease. As the incidence of preeclampsia is relatively low, screening tests with positive test results require high likelihood ratios (LR) to adequately predict the diseasés probability, and tests with negative results require very low LRs to confidently exclude the disorder. Thus, useful prediction for preeclampsia would require a very high LR (>15) for a positive test, as well as a very low LR for a negative result (<0.1) [50,51].

Angiogenic Markers

As alterations in absolute levels of sFlt-1, VEGF, PlGF, and sEng, in the maternal circulation precede the clinical onset of preeclampsia by several weeks to months, they have been proposed as a potential predictive test [17,18,52]. Levine et al. performed a nested case–control study within the Calcium for Preeclampsia Prevention (CPEP) trial, that included 120 preeclamptic women and 120 normotensive pregnancies, and measured serum concentrations of angiogenic factors (total sFtl-1, free VEGF, and free PlGF) measured throughout pregnancy [17]. In normotensive pregnancies, sFlt-1 levels were stable during the early and middle stages of gestation, and started to rise at 33–36 weeks. PlGF concentrations increased during the first two trimesters, peaked at 29–32 weeks, and decreased thereafter. Consistent with what was previously observed, sFlt-1 levels were significantly higher in the preeclamptic group during clinical disease [53,54]. Furthermore, Levine et al. observed that circulating levels of sFlt-1 began to increase 5 weeks before the clinical onset of preeclampsia and correlated with disease severity. In parallel with the increase in sFlt-1 levels, free PlGF and free VEGF levels decreased, suggesting that those levels were the result of binding by sFlt-1. Low levels of PlGF at both 13–20 and at 21–32 weeks were predictive of preterm preeclampsia, and low levels at 33–41 weeks were predictive of term preeclampsia. Associations between sFlt-1 levels and preeclampsia were not observed until closer to the onset of disease. High sFlt-1 levels no earlier than 21–32 weeks were predictive of preterm preeclampsia, while high sFlt-1 levels at 33–41 weeks predicted term preeclampsia. In general, women who developed severe and/or early onset preeclampsia had higher sFlt-1 and lower PlGF levels at each of the time intervals [17].

Later, prompted by the finding that sEng acts together with sFlt-1 in the pathogenesis of preeclampsia, the same group performed another nested case–control study, this time with an additional focus on sEng [18]. In normal pregnancies, serum levels of sEng started to rise at 33–36 weeks of gestation, but rose earlier and more steeply in women who developed preeclampsia, and reached a peak at the onset of clinical disease. sEng levels began to rise 9–11 weeks before the clinical onset of preeclampsia, and began to rise 12–14 weeks before the clinical onset of term preeclampsia. High levels of sEng at 13–20 weeks and at 21–32 weeks were predictive of preterm preeclampsia, and high levels at 21–32 weeks and at 33–41 weeks were predictive of term preeclampsia. Of note, levels were not markedly elevated prior to disease presentation either in women who developed gestational hypertension, or in normotensive women delivering small for gestational age (SGA) babies [18].

The sFlt-1:PlGF ratio (an index of antiangiogenic activity that reflects both increased sFlt-1 and decreased PlGF) was also evaluated, and found to parallel sEng levels. However, multivariate analysis showed that each was independently associated with preeclampsia, and that the sFlt-1:PlGF ratio predicted preeclampsia more reliably than either protein alone. Adding sEng to the equation, (sFlt-1 + sEng):PlGF ratio was more strongly predictive of preeclampsia than were individual biomarkers. Finally, when analyzed, the risk among women with high or low levels of sEng, of sFlt-1:PlGF ratio or of both, the risk of developing preeclampsia was greatest [odds ratio (OR) > 30] among women with higher levels of both sEng and sFlt-1:PlGF ratio, while women with high levels of a single biomarker had only small elevations in the risk of developing preeclampsia (OR: 2.3–7.4).

These two studies reported exciting findings with promising OR, especially for preterm preeclampsia, based on biomarker levels in the second trimester. Various other independent studies analyzed the predictive accuracy of angiogenic factors reporting significant changes in PlGF, sFlt-1, or sEng before the onset of preeclampsia [52,55–67]. Overall, the sensitivity, specificity, and positive and negative LRs of PlGF, sFlt-1, and sEng for all cases of preeclampsia ranged between 40% and 100%, 43% and 100%, 1.4% to infinity, and 0.1% and 0.8%, respectively. For early onset preeclampsia, the sensitivity, specificity, and positive and negative LRs varied between 17% and 100%, 51% and 97%, 1.7% and 24%, and 0.0% and 0.9%, respectively. This wide range in diagnostic performance might be explained by differences in study design, populations included, gestational age at sampling, etc.

Changes in PlGF are also seen as early as the first trimester whereas reproducible alterations in sFlt-1 and sEng are observed only in the mid-to-late second trimester onward. Predictive accuracy appears higher for early onset preeclampsia. The incorporation of these angiogenic factors into a single angiogenic index has improved accuracy in the prediction of preeclampsia. The index that has been extensively studied is sFlt-1:PlGF ratio, but many others have been proposed, such as PlGF:(sEng × VEGFR-1), PlGF:sEng, and PlGF:sFlt-1 [67].

Sequential changes: As circulating concentrations of angiogenic factors change with gestational age, it has been proposed that sequential changes in levels of sFlt-1, PlGF, and sEng could be more informative in assessing the risk for preeclampsia than time point measurements. Rana et al. [68] and Vatten et al. [69] described that sequential changes in angiogenic factors from first to second trimester differ in women destined to develop preeclampsia. A small increase in PlGF and a high increase in sFtl-1 were strong predictors of preeclampsia. The ORs were higher for sequential change than for each measurement alone. Interestingly, the combination of the lowest quartile of PlGF change and the highest quartile of sFlt-1 change was associated with an OR of 35.3 (95% CI: 7.6–164.2) for preterm preeclampsia, and a 3.2 (95% CI: 1.4–7.0) OR for term preeclampsia [69]. Sequential changes of sEng were also predictive of preeclampsia [68]. Consistent with these results, Erez et al. reported that differences in concentration of sEng, sFlt-1, and PlGF between first and second trimesters were associated with an increased risk of preterm preeclampsia and OR of 14.9 (95% CI: 4.9–45.0), 3.9 (95% CI: 1.2–12.6), and 4.3 (95% CI: 1.2–15.5), respectively. A small change in the PlGF:sEng ratio conferred an increased risk for preterm and term preeclampsia, and OR of 7.68 (95% CI: 1.7–34.74) and 2.46 (95% CI: 1.15–5.26), respectively. Differences in concentrations from second to first trimester had a higher OR then isolated measurements [70]. Sequential changes were also measured at later gestational ages. The rate of rise in sFlt-1 and sFlt-1:PlGF ratio accessed from 22–36 weeks was also predictive of overall preeclampsia risk, with areas under the receiver operating characteristic (ROC) curve of 92.4% (95% CI: 86.3–98.5) and 93.8% (95% CI: 88.2–99.4), respectively [71].

Kusanovic et al. reported a remarkable performance of delta and slope of PlGF:sEng ratio, (from early pregnancy and midtrimester) with a positive LR of 55.6 (95% CI: 36.4–55.6) and 89.6 (95% CI: 56.4–89.6), respectively for predicting early onset preeclampsia. Overall, their accuracy was better than that of individual factors. Indeed, the slope PlGF:sEng ratio performed better than any other test [67]. Recently, Myatt et al. also studied low-risk nulliparous women and reported that changes in plasma sFlt-1, sEng, and PlGF between the first and second trimesters had strong utility for predicting early onset preeclampsia with areas under ROC curve of 0.86, 0.84, and 0.86, respectively [72].

High-risk populations: Many of the studies reported are limited to healthy, nulliparous populations in Caucasians [18]. How do angiogenic factors perform in other settings, namely in high-risk populations and across other populations?

Moore-Simas et al. analyzed the performance of angiogenic biomarkers in 94 women with at least one of the following risk factors for preeclampsia: pregestational diabetes mellitus, chronic hypertension, chronic kidney disease, obesity, systemic lupus erythematosus (SLE), APL syndrome, or prior history of preeclampsia [71]. Samples were collected between 22 and 36 weeks, at a 4-week interval. In this high-risk population, they showed that maternal serum sFlt-1 is significantly increased, and PlGF significantly decreased prior to disease onset in women who go on to develop preeclampsia, as compared to women who do not develop preeclampsia [71]. This suggests that these biomarkers are likely to be clinically useful in this population as well. Similar to what was shown in healthy nulliparous women [17], levels increased earlier in women destined to develop preterm preeclampsia, and the sFlt-1:PlGF ratio was more predictive of the development of preeclampsia than sFlt-1 alone [71]. In samples taken at 22–26 weeks, the area under the ROC curve calculated for isolated sFlt-1 and sFlt-1:PlGF ratio for development of preterm preeclampsia was 90.1 (95% CI: 78–100) and 97 (95% CI: 90.8–100.0), respectively.

Sibai et al. evaluated the performance of sFlt-1 and PlGF in 704 women with previous preeclampsia and/or chronic hypertension enrolled in a randomized, placebo-controlled trial of vitamins C and E [73]. Samples were collected at 12–19.9 weeks and at 24–28 weeks of gestation. Angiogenic factor levels at 12–19.9 weeks were not associated with term preeclampsia, but had significant associations with onset of the disease prior to 27 weeks, as did levels obtained at 24–28 weeks with onset of preeclampsia prior to 37 weeks. Although there was a significant association between these markers and the subsequent development of preeclampsia, with very good sensitivity and negative predictive value (NPV) for preeclampsia developing prior to 27 weeks, the corresponding positive predictive values (PPV) were poor (6–8% at a specificity of at least 90%). Thus, the authors concluded that these markers might not be clinically useful for predicting preeclampsia in this high-risk population [73]. However, one of the problems with this study was the wide gestational windows used in the analyses of the data and the heterogeneity of the preeclampsia phenotypes studied.

Circulating angiogenic factors measured during early gestation have a high NPV in ruling out the development of severe adverse maternal and perinatal outcomes among patients with SLE and/or APL syndrome [74]. In a prospective multicenter study, we demonstrated that among high-risk subjects with SLE and/or APL syndrome, the combination of the combination of sFlt-1 and PlGF was most predictive of severe adverse pregnancy outcomes when measured early in pregnancy (16–19 weeks), with risk greatest for subjects with both PlGF in lowest quartile (<70.3 pg/mL) and sFlt-1 in highest quartile (>1872 pg/mL; OR = 31.1; 95% CI: 8.0–121.9; PPV = 58%; NPV = 95%). Severe adverse outcome rate in this high-risk subgroup was 94% (95%CI: 70%–99.8%), if lupus anticoagulant or history of high blood pressure is additionally present [74].

In summary, extensive work clearly identifies angiogenic factors, especially sFlt-1, PlGF, and sEng during early pregnancy, as powerful tools in prediction of early onset preeclampsia; however, the corresponding PPVs are low mainly due to the relatively low prevalence of the disease [75,76]. Current evidence suggests that combination of these biomarkers along with uterine artery Doppler may provide the best predictive accuracy for the identification of early onset preeclampsia. However, there is lack of evidence that interventions or close follow-up may improve maternal and/or fetal outcome even if one were to predict early onset preeclampsia.

Other antiangiogenic states: An antiangiogenic profile may not be unique to preeclampsia, but may underlie other pregnancy complications, such as mirror syndrome [77], unexplained fetal death [78], massive perivillous fibrin deposition [79], placental abruption [80], and delivery of SGA babies, without the presence of preeclampsia [57,81,82].

Low PlGF and high sEng have been associated with an increased risk for delivery of a SGA neonate [57,81,82], as well as with changes in maternal plasma concentrations of sEng, PlGF, or in their ratios between the first and second trimesters of pregnancy [70]. Romero et al. performed a longitudinal nested case–control study to evaluate whether maternal concentrations of angiogenic factors differ prior to development of the disease between women with normal pregnancies and women destined either to develop preeclampsia or to deliver a SGA neonate [83]. Patients destined to deliver SGA neonates showed changes in maternal plasma concentration of sEng and PlGF, but not sFlt-1. These changes differed in timing and magnitude from those in patients destined to develop preterm or term preeclampsia [83]. The difference in the pattern of change reflected two distinct phenotypes of an antiangiogenic state. Patients destined to develop SGA neonates had a higher plasma concentration of sEng from as early as 10-week gestation onward, while patients who developed preeclampsia had higher plasma concentrations only after a 24-week gestation. sFtl-1 levels did not change between patients destined to develop SGA pregnancies versus the controls, but did increase in women who later developed preeclampsia. When compared to controls, PlGF levels were decreased both in women who delivered SGA neonates and in women who developed preeclampsia. In both groups, this decrease was already evident at 10 weeks [83].

In conclusion, pregnancies destined to deliver SGA neonates and those destined to develop preeclampsia presented different antiangiogenic profiles. Alterations in PlGF and sEng levels were already evident as early as 10 weeks of gestation in those pregnancies destined to deliver SGA neonates, while sFlt-1 levels were predictive only for preeclampsia [83]. Also, the profile of maternal plasma concentrations of angiogenic (PlGF) and antiangiogenic factors (sEng and sFlt-1) between the first and second trimesters is significantly different among patients who subsequently had a normal pregnancy versus those destined to develop preeclampsia or to deliver SGA neonates [70].

PlGF in the urine: Free PlGF is freely filtered into urine and therefore has also been accessed as a predictive factor of preeclampsia. Levine et al. evaluated the urinary PlGF levels at 13 weeks of gestation onward [84]. In normal pregnancies, urinary PlGF increased during the first two trimesters, peaked at 29–32 weeks, and then decreased. In preeclamptic pregnancies, the pattern of urinary PlGF was similar to that of normal pregnancies before the onset of preeclampsia, but beginning at 25–28 weeks, and not before, levels were significantly reduced. There were particularly large differences between the controls and the cases with subsequent early onset preeclampsia. For samples collected after 21–31 weeks, the adjusted OR was 22.5 (95% CI: 7.4–67.8). The investigators concluded that decreased urinary PlGF concentrations at midgestation are strongly associated with subsequent early development of preeclampsia [84]. These findings were confirmed by others [85,86]. Recently, Savvidou et al. measured urinary PlGF at 11–14 weeks. They found that in the first trimester, development of preeclampsia was not preceded by altered urinary PlGF [87], confirming that first trimester urinary PlGF levels are not useful for predicting preeclampsia.

Diagnosis and prognosis studies: In addition to being useful in the prediction of preeclampsia before the onset of clinical symptoms, angiogenic factors may also prove useful in diagnosing the disease and in distinguishing it from other hypertensive disorders of pregnancy, such as gestational hypertension and chronic hypertension. The clinical utility of sFlt-1, sEng, and PlGF serum levels in differentiating among hypertensive disorders of pregnancy has been evaluated. The sensitivity and specificity in differentiating preeclampsia from chronic hypertension were 84% and 95% for sFlt-1 and 84% and 79% for sEng, respectively [88]. sFlt-1, PlGF, and sEng also differentiated women with superimposed preeclampsia (i.e., chronic hypertension plus preeclampsia) from those with chronic hypertension without preeclampsia [89,90]. Dramatic changes in angiogenic factors were also reported in patients with the most severe form of the disorder, such as eclampsia or early onset preeclampsia [52,91]. Several groups have demonstrated that measurement for sFlt-1 and PlGF can be used to differentiate preeclampsia from other diseases that mimic preeclampsia, such as chronic kidney disease and gestational thrombocytopenia [92,93]. Circulating anti-angiogenic factors have also been used to differentiate between preeclampsia and other causes of escalating hypertension in pregnant women undergoing hemodialysis [94].

Recently, with the availability of automated assays for the measurement of sFlt-1 and PlGF, several groups have evaluated the role of angiogenic factor analytes in plasma and serum and found them to be useful for routine diagnosis of preeclampsia, particularly in subjects who presented preterm (<37 weeks) [95–99]. To demonstrate the clinical utility of these biomarkers, Rana et al. evaluated the role of angiogenic biomarkers in the prediction of preeclampsia related adverse outcomes among women evaluated at our institution for suspected preeclampsia in a large clinical study. Plasma sFlt1:PlGF ratio on presentation predicts adverse maternal and perinatal outcomes (occurring within 2 weeks) in the preterm setting [100]. This ratio alone performed better than the standard battery of clinical diagnostic measures including blood pressure, proteinuria, uric acid, and other laboratory assays. Several recent studies have confirmed that levels of sFlt1 and PlGF in the triage setting can be used as a robust prognostic test and these levels correlate with the duration of pregnancy [101–103]. Recently, Zeisler et al. demonstrated in a prospective multicenter clinical trial that serum sFlt:PlGF can be used to rule out preeclampsia among patients with suspected disease [104]. The authors performed an elegantly designed study first identifying a ratio cutoff of 38 for a clinically significant predictive value in 500 women.

In a subsequent validation study among 550 additional participants, the NPV of the sFlt1:PlGF ratio was robust at 99.3% (greatly reducing the likelihood of a true diagnosis of preeclampsia in the next 7 days). Despite the excellent NPV in the Zeisler et al. study, a high sFlt-1:PlGF ratio had a PPV of only 38% (not all patients with a positive test developed preeclampsia). This is not surprising as many patients with even subclinical preeclampsia are delivered at 37 weeks and we do not know if these patients with high sFlt1:PlGF ratio would have developed full-blown disease had the obstetrician not intervened. Consistent with this hypothesis, in a recent prospective study of 100 patients admitted for suspected preeclampsia, Baltajian et al. demonstrated that the PPV for sFlt1:PlGF for indicated delivery within 2 weeks of presentation of suspected preeclampsia was 91% [105].

In an attempt to calculate patient specific risk of adverse outcomes based on angiogenic profile, Palomaki et al. published a validated model that assigns patient-specific risks of any severe outcome among women being evaluated in triage for preeclampsia. They concluded that this risk stratification will allow women at high risk to receive close surveillance and will have the benefit for reducing unnecessary preterm deliveries among women at an otherwise low risk of adverse outcomes [106]. Tsiakkas et al. reported that a combination of maternal factors and angiogenic biomarkers during third trimester could predict nearly all cases of preterm preeclampsia and half of those with term preeclampsia at 5% false-positive rate [107].

In summary, angiogenic markers may be useful for the diagnosis and prognosis in patients with suspected preeclampsia in the preterm setting and the recent study conducted by Zeisler et al. is a major step forward in demonstrating clinical utility for these biomarkers [104]. To evaluate the influence of sFlt-1:PlGF ratio in clinician's decision making in women presenting with suspected preeclampsia, Klein et al. performed an open-label study and demonstrated that this diagnostic test influenced initial clinical decision toward appropriate hospitalization in 17% of women [108]. Additional prospective clinical trials are needed to demonstrate that maternal and neonatal outcomes improve when physicians have the knowledge of the sFlt-1:PlGF diagnostic test. In addition to angiogenic proteins, other molecules, such as PP-13 and PAPP-A have also been evaluated as predictive markers (see further).

Placental Protein-13

Placental protein-13 (PP-13) [109] is a member of the galectin family [110], predominantly expressed by the syncytiotrophoblasts, that is involved in normal implantation and placental vascular development [111]. First trimester–circulating levels of PP-13 are significantly lower in women who go on to develop preeclampsia, IUGR, and preterm birth [112,113]. In a prospective nested case–control study involving 290 controls and 47 preeclamptic women, Chafetz et al. observed that preeclamptic pregnancies had lower levels of PP-13 in the first trimester (9–12 weeks) when compared with controls. Results were expressed as multiples of the gestation-specific median (MoM) in controls. Using a cut-off of 0.38 MoM, the OR was 32.1 (95% CI: 14.5–71.0), the sensitivity was 79%, and the specificity 90%. ROC analysis yielded areas under the curve of 0.91 (95% CI: 0.86–0.95) [114].

Romero et al. reported a sensitivity of 100% for early onset preeclampsia and 85% for preterm preeclampsia at 80% specificity and a cutoff of 0.39 MoM. PP-13 did not perform well for prediction of severe preeclampsia and mild preeclampsia at term [115]. Spencer et al. conducted a nested case control study of 446 cases and 88 controls. At a specificity set at 80%, the sensitivity of first trimester PP-13 for all cases of preeclampsia was 40%, and was 50% for early onset preeclampsia [116]. Furthermore, a recent study has shown that there is a benefit in sequential testing with PP-13. Gonen et al. measured levels of PP-13 at 6–10 weeks, 16–20 weeks, and 24–28 weeks of gestation in 1366 women, and reported that PP-13 in the first trimester alone or in combination with the slope between the first and the second trimesters may be a promising marker for assessing the risk of preeclampsia. Combining MoM at 6–10 weeks and a slope between 6–10 and 16–20 weeks, the OR was 55.5 (95% CI: 18.2–169.2), the sensitivity was 78%, and the specificity 94% [117].

PP-13 was also predictive of early onset preeclampsia in a high-risk population. At a MoM cutoff of 0.53, for a false-positive rate of 10%, sensitivity was 71%. Again, it predicted early onset disease is better than disease at term [118]. Second trimester levels of PP-13 (22–32 weeks) taken at a single time point are not useful in predicting preeclampsia, and its prediction did not increase when coupled with Doppler velocimetry [119].

Combining first trimester PP-13 with other parameters may further improve predictive performance. Larger prospective studies are needed to determine whether PP-13 will be a valuable clinical marker for the early prediction of preeclampsia; however, lack of an automated assay for PP-13 has been a significant barrier in testing this marker in large clinical studies.

Pregnancy-Associated Plasma Protein A

Pregnancy-associated plasma protein A (PAPP-A) is a peptidase produced by syncitiotrophoblast with hydrolytic activity for insulin-like growth factor–binding proteins [120,121]. These regulate insulin growth factors known to be important for implantation, for trophoblast invasion of maternal decidua, and for placental growth [122]. It is released into the maternal circulation where it binds the eosinophil major binding protein, an inhibitor of its proteolytic activity [123].

Decreased levels of PAPP-A in the first trimester have been associated with increased risk of adverse pregnancy outcomes, including preeclampsia [124]. It is in fact an established biomarker for trisomy 21. Spencer et al. described an association of a modest increase in the LR of developing preeclampsia with decreasing levels of PAPP-A. At the 5th centile of normal PAPP-A (MoM 0.415), the OR was increased 3.7-fold (95% CI: 2.3–4.8), and at this cut-off, 15% of cases of preeclampsia were identified [125].

A recent systematic review and metaanalysis determined the accuracy in predicting preeclampsia of five serum analytes, used in Down's serum screening. At the 5th centile of normal PAPP-A, the positive LR was 2.10 (95% CI: 1.57–2.81), and the negative LR was 0.95 (95% CI: 0.93–0.98) [126]. First trimester serum PAPP-A was not a good predictor of late onset preeclampsia [127].

Renal Dysfunction–Related Tests

As the kidney is a major target organ of preeclampsia, renal dysfunction–related tests were proposed as possible predictors of preeclampsia.

Serum uric acid: Hyperuricemia observed in preeclampsia led to studies to determine if measuring serum uric acid levels could be used to predict preeclampsia. Unfortunately, uric acid is of limited clinical utility in either distinguishing preeclampsia from other hypertensive disorders of pregnancy or as a clinical predictor of adverse outcomes [128,129].

Proteinuria: Proteinuria is routinely assessed in antenatal care visits from first booking. Proteinuria measurement includes total protein or total albumin excretion during 24 h, microalbuminuria, albumin:creatinine ratio, and dipsticks for spot proteinuria or albuminuria. Pooled estimates of sensitivity and specificity for total proteinuria were 35% (95% CI: 13–68%) and 89% (95% CI: 79–94%); for total albuminuria were 70% (95% CI: 45–87%) and 89% (95% CI: 79–94%); for microalbuminuria were 62% (95% CI: 23–90%) and 68% (95% CI: 57–77%); and for albumin:creatinine ratio were 19% (95% CI: 12–28%) and 75% (95% CI: 73–77%), respectively [130]. Morris et al. performed a systematic review and metaanalysis of urinary spot–protein creatinine ratio in preeclampsia and confirmed that while spot-protein creatinine ratio has a role in diagnosing significant proteinuria, it did not correlate with adverse outcomes related to preeclampsia [131].

Kallikreins: The kallikrein–kinin system is an important paracrine regulator of vessel dilatation, and consequently of blood flow. Millar et al. [132] and Kyle et al. [133] assessed the levels of urinary kallikrein as a predictor of preeclampsia. Kallikreinuria has been shown to be decreased in patients with preeclampsia as compared to uncomplicated pregnancies; however, it is unlikely to be useful as a screening test as the reported specificity values are quite poor.

Free Fetal Nucleic Acids

The first description of the presence of fetal cells in the mother stems is from the 19th century when a German pathologist detected trophoblast cells in the lungs of women who died of eclampsia. In 1969, male fetal cells were found in the blood of healthy pregnant women [134]. Several studies confirmed these results and led to the description of fetal–maternal cell trafficking. Several investigators have described circulating nucleic acids of fetal origin in maternal blood and in relatively more abundance than fetal cells [135,136]. It is now widely recognized that there is transfer of allogeneic fetal cells into the maternal circulation and vice versa [137] and that cell-free fetal nucleic acids (DNA and mRNA) circulate in the maternal blood [138]. Except for the migration of fetal cytotrophoblasts, the exact mechanism leading to bidirectional transplacental migration of cells is largely unknown. Also, the mechanism of release of free extracellular nucleic acids into the circulation is not yet clear. Multiple lines of evidence suggest that the vast majority of the cell-free fetal DNA in the maternal plasma is probably derived from the placenta through apoptosis and necrosis of cytotrophoblasts, although some could be derived from circulating cells.

The examination of fetal cells, specifically erythroblasts, and of cell-free fetal DNA from the blood of pregnant women is the subject of intense research, with the aim of developing new risk-free methods for prenatal diagnosis [139,140]. Cell-free fetal DNA is already in use in determining fetal sex and fetal Rhesus status [140].

In preeclampsia, fetal–maternal cell trafficking is significantly altered with elevated numbers of fetal cells detected in the maternal circulation during those pregnancies [141]. Prospective studies further indicated that this perturbation occurs early in preeclamptic pregnancies [142,143]. In a similar manner, it has been shown that in preeclamptic pregnancies [144] cell-free fetal DNA is elevated long before the clinical onset of the disease [145,146]. These results were confirmed in a large case–control study within the CPEP Trial. Levine et al. reported a two-stage increase in cell-free fetal DNA in maternal sera before the onset of preeclampsia, with an initial elevation starting at 17–28 weeks (36 vs. 16 genomic equivalents/mL, $p < 0.001$), and a secondary elevation beginning about 3 weeks before the onset of clinical syndrome (176 vs. 75 genomic equivalents/mL, $p < 0.001$) [147]. Of note, the fetal DNA was greater from 17 to 20 weeks onward than in the controls, but was not different statistically until 25–28 weeks. In early pregnancy (between 13 and 16 weeks) there was no demonstrable difference. If preeclampsia was severe, the reported differences were greater and at an earlier gestational age, or they were associated with a SGA infant [147].

Crowley et al. performed a nested case–control study to quantify plasma fetal DNA before 20 weeks of gestation in pregnancies subsequently complicated by preeclampsia, in comparison to normal pregnancies. The median gestational age at sample collection was 13 weeks. The sex determining region Y (SRY) gene, which is specific to the Y chromosome, was used as a fetal marker. This gene was detected in 94% of preeclamptic women, and in 78% of normal pregnancies. However, its median levels were similar between cases and controls. The authors concluded that free fetal DNA quantification in maternal plasma before 20 weeks is not a useful predictor of preeclampsia [148]. Using whole blood samples collected at the first antenatal visit (mean 15.7 ± 3.6 weeks), Cotter et al. reported a sensitivity of 39%, and a specificity of 90% for predicting preeclampsia at a cutoff of 50,000 SRY copies/mL [149].

A downside of using these tests as screening tools for predicting preeclampsia is that, as of 2010, the analysis of fetal cells or cell-free fetal DNA is still complex and costly. Owing to high maternal DNA background, detection of fetal DNA from maternal plasma is difficult. In addition, the quantification of fetal DNA is typically based on Y chromosome–specific sequences, that is SYR and DNA Y-chromosome segment (DYS), limits the technique to pregnancies carrying a male fetus. Thus, other approaches have been used to overcome this limitation, such as the utilization of different epigenetic markers between maternal and fetal DNA [150] and fetal RhD gene [151] as universal, gender-independent fetal markers. Total free DNA has also been used, and has been reported to be increased in women who subsequently develop preeclampsia, thus overcoming the gender issue [152]. Interestingly, mRNA of placental origin has also been identified in pregnant women, and research in the area has turned the focus on fetal-free mRNA to produce new biomarkers. The advantage of using mRNAs is that it is of placental/fetal origin, is specific to pregnancy, and is independent of fetal gender. Ng et al. detected and quantified mRNA expression in maternal plasma of human chorionic gonadotropin and human placental lactogen, proteins that are produced exclusively by the placenta. mRNA expression in these proteins was found to be pregnancy-specific and to reflect relative placental gene expression [153].

Purwosunu et al. demonstrated that in preeclamptic pregnancies a panel of free mRNA of placental origin is increased in maternal plasma at gestational weeks 15–20. At a 5% false-positive rate, the detection rate was 84% (95% CI: 71.8–91.5), with area under the ROC curve of 0.927 ($p < 0.001$) [154]. Circulating fetal cells in maternal blood and cell-free fetal nucleic acids are a promising field of research. Future studies will determine whether they will turn out to be good predictive biomarkers for preeclampsia.

Uterine Doppler Velocimetry

Preeclampsia is characterized by an abnormal placenta and a decreased invasion of maternal uterine arteries by cytotrophoblast cells. As a result, the normal vascular remodeling of maternal uterine spiral arteries converting them in high-flow and low-resistance vessels does not occur. Therefore, Doppler ultrasonography has been evaluated as a potential predictive test for preeclampsia.

The uterine artery is identified using color Doppler ultrasonography, and then pulsed-wave Doppler is applied to obtain waveforms. The increased flow resistance within uterine arteries results in an abnormal waveform pattern, which is represented by either an increased resistance index, or pulsatility index (PI), or by the persistence of a unilateral or bilateral diastolic notch. Several indices are then calculated and assessed from flow velocity waveforms, and either alone or combined have been investigated as predictive of preeclampsia. This has revealed varied results. In this regard Cnossen et al. conducted a recent systematic review and metaanalysis to access the use of uterine Doppler ultrasonography to predict preeclampsia [155]. The authors concluded that an increased PI in the second trimester, alone or combined with notching, is the best Doppler index predictor of preeclampsia. In high-risk patients, an increased PI with notching had a positive LR of 21.0 (95% CI: 5.5–80.5), and a negative LR of 0.82 (95% CI: 0.72–0.93). In low-risk patients it had a positive LR of 7.5 (95% CI: 5.4–10.2), and a negative LR of 0.56 (95% CI: 0.47–0.71). Other Doppler indices showed low-to-moderate predictive value, for example, when assessed in the first trimester [155].

Several studies have assessed the predictive accuracy of uterine Doppler velocimetry for early onset preeclampsia. Positive LR ranged from 5 to 20 and negative LR ranged from 0.1 to 0.8 [51]. It appears that irrespective of the index or combinations of index used, uterine artery Doppler velocimetry (UADV) maybe a moderate-to-good predictor for the development of early onset preeclampsia.

Combination of Tests

Angiogenic factors, along with other modalities, may be combined for predicting preeclampsia. To identify patients at risk for severe and/or early onset preeclampsia, Espinoza et al. conducted a prospective study of 3296 women to determine the role of UADV, maternal plasma PlGF and sFlt-1 concentrations in the second trimester [58]. Sample collection and UADV were performed between 22 and 26 weeks, and showed that the combination of abnormal UADV and low-serum PlGF was strongly associated with both early onset and severe preeclampsia, with OR of 35–45. sFlt-1 did not improve prediction of Doppler combined with PlGF. For all of the cases of preeclampsia, the prediction sensitivities of maternal plasma PlGF concentration, abnormal UADV, and the combination of these tests were 61, 35, and 27%, respectively. The corresponding specificities were 51, 90, and 96% and positive LR 1.42 (95% CI: 1.25–1.62), 3.42 (95% CI: 2.60–4.49), and 7.53 (95% CI: 5.27–10.75), respectively. Combination testing improved the specificity, PPV, and positive LR over each test alone for the prediction of early onset preeclampsia, although with a slight reduction in sensitivity [58].

Stepan et al. performed a prospective study of 63 second trimester pregnant women with abnormal uterine perfusion. When combining the measurements of uterine Doppler with sFlt-1 and PlGF levels in the second trimester, the sensitivity and specificity of Doppler alone to predict early onset preeclampsia increased from 67% to 83% and from 76% to 95%, respectively. The combination of parameters performed better than any parameter alone [63]. Later on, the same group [64] also demonstrated that in pregnancies with abnormal uterine perfusions that resulted in the development of preeclampsia, second trimester levels of sEng were also increased. Combined analysis of sEng and sFlt-1 in this population with abnormal uterine Doppler was able to predict early onset preeclampsia with a sensitivity of 100% and a specificity of 93.3% [64].

Combination of markers in the first trimester was also evaluated. Patients who developed preeclampsia requiring delivery before 34 weeks of gestation had lower PP13 serum concentration than did normotensive controls at 11–14 weeks gestation. For a 90% detection rate, the false-positive rate for PP-13 was 12%, and for Doppler analysis alone, performed at the same gestational age, it was 31%. For a 10% false-positive rate, the detection rates were 80% for PP-13 alone and 90% for PP-13 combined with Doppler [156]. In the above study by Spencer et al. [116], the sensitivity of first trimester PP-13 for all cases of preeclampsia increased from 40% to 74%, and for early onset increased from 50% to 74% when combined with UADV. However, serum PP-13 does not improve the prediction of early preeclampsia significantly that is provided by a combination of maternal factors, uterine artery PI, and PAPP-A [157]. Interestingly, PAPP-A does not improve the prediction of early preeclampsia when first trimester PP-13 and second trimester PI are used together [116]. Second trimester levels of PP-13 (22–32 weeks) are not useful in predicting preeclampsia, and prediction did not improve when coupled with Doppler velocimetry [119].

An increase in cell-free fetal DNA has been described in women with an abnormal uterine Doppler and who developed preeclampsia [158]. Studies assessing the combination of Doppler with cell-free fetal DNA for predicting preeclampsia are not available. However, assessment of this metabolite is unlikely to improve the performance of Doppler, because there is a high association between uterine artery PI and plasma cell-free fetal DNA [159].

Poon et al. evaluated 7797 women with singleton pregnancies, during gestational weeks 11–13. This yielded very good results using an algorithm developed by logistic regression that combined the logs of uterine PI, mean arterial pressure, PAPP-A, serum-free PlGF, body mass index, and the presence of nulliparity or previous preeclampsia. At a 5% false-positive rate, the detection rate for early preeclampsia was 93.1% [160]. The calculated positive LR was 16.5 and negative LR was 0.06. O'Gorman more recently confirmed that maternal factors, when combined with PlGF levels and uterine artery PI, can detect 75% of cases of preterm preeclampsia in a large study of over 35,000 patients [161].

In summary, the combination of angiogenic factors, placental proteins, and other parameters, such as Doppler studies increased the sensitivity without losing specificity. Still, more studies are needed to confirm these results and assess the cost effectiveness of this approach.

NOVEL BIOMARKERS AND FUTURE PERSPECTIVES

With the completion of the human genome project, various high-throughput techniques have evolved, allowing the simultaneous examination of thousands of genes (genomics), gene transcripts (transcriptomics), proteins (proteomics), metabolites (metabolomics), protein interaction (interactomics), and chromatin modifications (epigenomics) in single experiments. These novel technologies have greatly increased the number of potential DNA, RNA, and protein biomarkers, leading to a renewed interest in the field. Analysis of a single biomarker, or a combination of only a few, is being replaced by a multiparametric analysis yielding a signature of genes, RNA, or proteins. These promising new methodologies are currently being reported in almost all fields of medicine, such as oncology [162–164], nephrology [165], cardiology [166–168], and many others, to distinguish patterns that help in early diagnosis, classification, prognosis, and in the prediction of response to therapies.

Transcriptomics

The transcriptome is a description of all DNA that is transcribed into RNA (messenger RNA, transfer RNA, microRNA, and other RNA species) at any given moment. It forms the template for protein synthesis, resulting in the corresponding proteome. Transcriptomics refers to a global RNA assessment.

Farina et al. measured a panel of 7 circulating mRNAs in maternal blood from 6 women with preeclampsia, and from 30 controls. A different expression pattern between cases and controls was reported. Inhibin A, p-selectin, and VEGF receptor mRNA values were higher in preeclampsia, whereas human placental lactogen, KISS-1, and plasminogen activator type 1 were lower, both as compared to normotensive controls [169]. The authors suggested that aberrant quantitative expression of this circulating placenta-specific mRNA in serum from preeclamptic women might prove useful for the prediction of this disorder [169]. This was not a large-scale approach. Recently, Tsui et al. described the use of microarray technology for identification of new placenta-specific mRNA markers in maternal plasma [170]. Circulating cells of fetal/placental origin are also a source of mRNA that can be assessed as a potential biomarker. Okazaki et al. performed gene expression profiling and real-time quantitative reverse-transcription polymerase chain reaction (RT-PCR) in the cellular component of maternal blood to identify potential biomarkers of preeclampsia. Microarray analysis was performed in five samples from women with preeclampsia, and in five matched control subjects. This was followed by RT-PCR analysis in 28 blood samples from women affected with preeclampsia and 29 controls. Trophoblast glycoprotein (a trophoblast membrane protein) and pregnancy-specific β1 glycoprotein (protein produced by the syncytiotrophoblasts) mRNA were increased in women with preeclampsia, and there was a direct correlation between pregnancy-specific β1 expression levels and the severity of the disease [171].

Chorionic villous sampling (CVS) is a biopsy of placenta chorionic villous performed under ultrasonic guidance around 10–13 weeks of gestation for prenatal diagnosis. Founds et al. followed 160 pregnant women on whom CVS had been performed. Of these, four developed preeclampsia,, and their banked CVS was matched to eight control CVS of unaffected pregnancies. Microarray analysis was conducted on these samples revealing 36 differentially expressed genes between normal pregnancies and those who went on to develop preeclampsia, 6 months before the onset of clinical symptoms [172]. Consistent with these results, Farina et al. also reported a different CVS gene expression profile in women who went on to develop preeclampsia, as compared to normal pregnancies. Altered expression was found among several genes, including those involved in the invasion of human trophoblasts, inflammatory stress, endothelial aberration, angiogenesis, and blood pressure control. Furthermore, RT-PCR analysis of peripheral blood at term showed significant differences for all the genes studied [173].

In addition to mRNA, small RNA molecules, such as microRNAs (miRNAs) are now being investigated as novel circulating markers. miRNAs are short (19–25 nucleotides), single-stranded nonprotein-coding RNAs that regulate gene expression by binding to the 3′ untranslated region of the target miRNAs. miRNAs are involved in diverse genetic pathways across human tissues including fertility regulation [174]. Pineles et al. [175] and Zhu et al. [176] studied the expression of miRNAs in preeclamptic placentas obtained at delivery, as compared to those from placentas of normal pregnancies. They reported a different expression profile between the two groups. A clinical useful test for risk assessment in preeclampsia should be minimally invasive [50]. In this regard, placenta-specific miRNA have been shown to be secreted into maternal circulation [177] and detectable in maternal plasma samples [178].

Proteomics

The proteome is the total complement of proteins present in any defined biologic compartment, such as a whole organism, a cell, an organelle, or a fluid, such as blood, amniotic fluid, or urine. Proteomics has the advantage over transcriptomics of measuring the protein itself, that is, the functional product of gene expression.

Blumenstein et al. compared the plasma proteome at 20 weeks gestation in women who subsequently developed preeclampsia, to that of healthy women with uncomplicated pregnancies, and reported a different pattern of proteins between the two groups. The differently expressed proteins are involved in lipid metabolism, coagulation, complement regulation, extracellular matrix remodeling, protease inhibitor activity, and acute phase responses [179].

Recently, Buhimschi et al. performed a proteomic profiling of urine from pregnant women, and reported that women with severe preeclampsia requiring mandated delivery presented a unique urine proteomic fingerprint [180]. Furthermore, this characteristic proteomic profile appeared more than 10 weeks before the clinical manifestations, and distinguished preeclampsia from other hypertensive or proteinuric disorders in pregnancy. Proteomic profiling of urine performed better than P:C ratio and sFlt-1:PlGF ratio for the prediction of preeclampsia requiring mandated delivery. Tandem mass spectrometry and de novo sequencing identified the biomarkers as nonrandom cleavage products of SERPINA1 and albumin. Of these, the 21 amino acid C-terminus fragment of SERPINA1 was highly associated with severe forms of preeclampsia requiring early delivery [180]. Follow-up studies from the same group have suggested that preeclampsia shares pathophysiologic features with recognized protein misfolding disorders and have suggested that assessment of global protein misfolding load in pregnancy based on urine congophilia (Congo red dot test) carries diagnostic and prognostic potential for preeclampsia [181].

Metabolomics

Metabolomics is defined as the global analysis of endogenous and secreted metabolites in a biologic system. As with proteomics, studies of the human metabolome can be carried out on routine samples of urine, plasma, or serum, and requires minimal specialist preparation of samples. An advantage of metabolomics is that it involves a smaller and more tractable group of compounds compared to the proteome.

A preliminary study revealed that metabolomic strategies might be appropriate for investigating the metabolic function of trophoblast or placental tissue, and to assess changes in response to altered environmental conditions. Heazell et al. examined the placental metabolome under different oxygen tensions. Placental villous explants were cultured in 1, 6, and 20% oxygen for 96 h, revealing new redox biomarkers [182]. The same group showed that conditioned media from preeclamptic explants has a different metabolic footprint when compared to conditioned media from uncomplicated pregnancies [183]. Metabolomic strategies were also applied to the plasma, and it was found that preeclamptic pregnancies have a different metabolomic profile when compared to normal pregnancies [184]. Using three of the metabolite peak variables, preeclampsia could be distinguished from normal pregnant controls with a sensitivity of 100% and a specificity of 98% [185]. To use metabolomics signature to predict preeclampsia in early pregnancy, Kenny et al. used a two-phase discovery/validation metabolic profile study and reported a 14 metabolite signature that predicted all preeclampsia with an area under ROC curve of 0.92 [186]. This work is very promising and may lead to a robust screening test in the future.

In conclusion, similar to what is occurring in other fields of medicine, the use of these novel technologies in preeclampsia appears quite promising. Although the number of studies is still scarce, they suggest that an aberrant transcriptomic, proteomic, and metabolomic profile may be predictive of the disease, opening a new and exciting avenue in biomarker discovery for preeclampsia. Future studies are warranted, with the collaborative efforts of bioinformatics, biostatistics, researchers, and clinicians.

In addition to predicting the presence of disease, a biomarker can also be used as an indicator of disease severity, prognosis and response to therapeutics. In this regard, gene expression profile of placentas has been shown to be different between preeclampsia and uncomplicated pregnancies and early and late onset preeclampsia [187,188]. Accordingly, proteomic analysis of placentas [189], amniotic fluid [190], plasma [191], and urine [180] of women with established preeclampsia revealed different proteomic profiles from those of normal controls. Similar to what has been reported for other biomarkers, these results highlight the future possibility of applying genomic and proteomic strategies to rule out preeclampsia in complicated cases, to classify the disease in terms of severity, and to assess its prognosis. In addition, this technology could generate a very large data base that could be mined by computational biologists, and could yield new pathways and molecules that may bring new insights into the mechanism of the disease. It would also stimulate hypothesis-driven research, accelerating the efforts to unravel the biology of preeclampsia, and would ultimately lead to new therapies.

CONCLUSIONS

Preeclampsia can be a devastating disease. It remains a major cause of maternal and neonatal mortality. The ability to predict preeclampsia would be a major advance in maternal–fetal medicine. In this regard, several biomarkers have been proposed. Angiogenic factors, PP-13, and combinations of these and other parameters with Doppler analysis hold promise for future predictive testing for preeclampsia. Newer genomic and proteomic technologies are a rapidly emerging field that has enabled biologic samples to be surveyed for biomarkers in ways never before possible, and promises the development of exciting new applications over the next few years.

Several clinical studies have demonstrated a potential role for use of angiogenic biomarkers for aid in the diagnosis and prognosis of preterm preeclampsia. We now need clinical trials demonstrating the utility of these biomarkers in helping obstetrician's management decisions, improve health outcomes, and/or reduce costs to the healthcare system [192]. It is exciting to envision the tremendous impact that an accurate biomarker for preeclampsia would have, that is, reduction of fetal and maternal deaths, improvement in acute and long-term outcomes, reduced health costs, and in addition, the acceleration of drug discovery leading to the ultimate goal: the effective treatment of preeclampsia. More prospective studies are needed to better evaluate the clinical utility of preeclampsia biomarkers.

ABBREVIATIONS

CI Confidence interval
CVD Cardiovascular disease
CVS Chorionic villous sampling
DYS DNA Y-chromosome segment
GFR Glomerular filtration rate
HELLP Hemolysis, elevated liver enzymes, low platelets
IUGR Intrauterine growth restriction
LR Likelihood ratio
miRNA Micro ribonucleic acid
MoM Multiples of the median
MRI Magnetic resonance imaging
mRNA Messenger ribonucleic acid
OR Odds ratio
P:C Protein:creatinine
PAPP-A Pregnancy-associated plasma protein A
PI Pulsatility index
PlGF Placental growth factor
PP-13 Placental protein-13
PPV Positive predictive value
RhD Rhesus D antigen
ROC Receiver operating characteristic
RPF Renal plasma flow
RT-PCR Reverse-transcription polymerase chain reaction

sEng Soluble endoglin
sFlt-1 Soluble Fms-like tyrosine kinase 1
SGA Small for gestational age
SRY Sex determining region Y
TGF-β Transforming growth factor beta
UADV Uterine artery Doppler velocimetry
VEGF Vascular endothelial growth factor

REFERENCES

[1] Duley L. The global impact of pre-eclampsia and eclampsia. Semin Perinatol 2009;33(3):130–7.
[2] Abalos E, Cuesta C, Grosso AL, Chou D, Say L. Global and regional estimates of preeclampsia and eclampsia: a systematic review. Eur J Obstet Gynecol Reprod Biol 2013;170(1):1–7.
[3] Sibai B, Dekker G, Kupferminc M. Pre-eclampsia. Lancet 2005;365(9461):785–99.
[4] Saigal S, Doyle LW. An overview of mortality and sequelae of preterm birth from infancy to adulthood. Lancet 2008;371(9608):261–9.
[5] Hofman PL, Regan F, Jackson WE, et al. Premature birth and later insulin resistance. N Engl J Med 2004;351(21):2179–86.
[6] Irving RJ, Belton NR, Elton RA, Walker BR. Adult cardiovascular risk factors in premature babies. Lancet 2000;355(9221):2135–6.
[7] Hovi P, Andersson S, Eriksson JG, et al. Glucose regulation in young adults with very low birth weight. N Engl J Med 2007;356(20):2053–63.
[8] Duckitt K, Harrington D. Risk factors for pre-eclampsia at antenatal booking: systematic review of controlled studies. BMJ 2005;330(7491):565.
[9] Skjaerven R, Wilcox AJ, Lie RT. The interval between pregnancies and the risk of preeclampsia. N Engl J Med 2002;346(1):33–8.
[10] Bartsch E, Medcalf KE, Park AL, Ray JG. High risk of pre-eclampsia identification G. Clinical risk factors for pre-eclampsia determined in early pregnancy: systematic review and meta-analysis of large cohort studies. BMJ 2016;353:i1753.
[11] Shembrey MA, Noble AD. An instructive case of abdominal pregnancy. Aust N Z J Obstet Gynaecol 1995;35(2):220–1.
[12] Soto-Wright V, Bernstein M, Goldstein DP, Berkowitz RS. The changing clinical presentation of complete molar pregnancy. Obstet Gynecol 1995;86(5):775–9.
[13] Matsuo K, Kooshesh S, Dinc M, Sun CC, Kimura T, Baschat AA. Late postpartum eclampsia: report of two cases managed by uterine curettage and review of the literature. Am J Perinatol 2007;24(4):257–66.
[14] Page E. The relation between hydatid moles, relative ischemia of the gravid uterus and the placental origin of eclampsia. Am J Obstet Gynecol 1939;37:291–3.
[15] Roberts JM, Taylor RN, Musci TJ, Rodgers GM, Hubel CA, McLaughlin MK. Preeclampsia: an endothelial cell disorder. Am J Obstet Gynecol 1989;161(5):1200–4.
[16] Romero R, Chaiworapongsa T. Preeclampsia: a link between trophoblast dysregulation and an antiangiogenic state. J Clin Invest 2013;123(7):2775–7.
[17] Levine RJ, Maynard SE, Qian C, et al. Circulating angiogenic factors and the risk of preeclampsia. N Engl J Med 2004;350(7):672–83.
[18] Levine RJ, Lam C, Qian C, et al. Soluble endoglin and other circulating antiangiogenic factors in preeclampsia. N Engl J Med 2006;355(10):992–1005.
[19] Powe CE, Levine RJ, Karumanchi SA. Preeclampsia, a disease of the maternal endothelium: the role of antiangiogenic factors and implications for later cardiovascular disease. Circulation 2011;123(24):2856–69.

[20] Venkatesha S, Toporsian M, Lam C, et al. Soluble endoglin contributes to the pathogenesis of preeclampsia. Nat Med 2006;12(6):642–9.
[21] Parikh SM, Karumanchi SA. Putting pressure on pre-eclampsia. Nat Med 2008;14(8):810–2.
[22] Cui Y, Wang W, Dong N, et al. Role of corin in trophoblast invasion and uterine spiral artery remodelling in pregnancy. Nature 2012;484(7393):246–350.
[23] Weinstein L. Syndrome of hemolysis, elevated liver enzymes, and low platelet count: a severe consequence of hypertension in pregnancy. Am J Obstet Gynecol 1982;142(2):159–67.
[24] Sibai BM. Diagnosis, controversies, and management of the syndrome of hemolysis, elevated liver enzymes, and low platelet count. Obstet Gynecol 2004;103(5 Pt. 1):981–91.
[25] Report of the National High Blood Pressure Education Program Working Group on High Blood Pressure in Pregnancy. Am J Obstet Gynecol 2000;183(1):S1–S22.
[26] Sibai BM. Diagnosis, prevention, and management of eclampsia. Obstet Gynecol 2005;105(2):402–10.
[27] Sibai BM, Stella CL. Diagnosis and management of atypical preeclampsia-eclampsia. Am J Obstet Gynecol 2009;200(5):481e1–7.
[28] Thangaratinam S, Coomarasamy A, O'Mahony F, et al. Estimation of proteinuria as a predictor of complications of pre-eclampsia: a systematic review. BMC Med 2009;7:10.
[29] Stillman IE, Karumanchi SA. The glomerular injury of preeclampsia. J Am Soc Nephrol 2007;18(8):2281–4.
[30] Baumwell S, Karumanchi SA. Pre-eclampsia: clinical manifestations and molecular mechanisms. Nephron Clin Pract 2007;106(2):c72–e81.
[31] Garovic VD, Wagner SJ, Turner ST, et al. Urinary podocyte excretion as a marker for preeclampsia. Am J Obstet Gynecol 2007;196(4):320e1–7.
[32] Weissgerber TL, Craici IM, Wagner SJ, Grande JP, Garovic VD. Advances in the pathophysiology of preeclampsia and related podocyte injury. Kidney Int 2014;86(2):445.
[33] Wu CS, Nohr EA, Bech BH, Vestergaard M, Catov JM, Olsen J. Health of children born to mothers who had preeclampsia: a population-based cohort study. Am J Obstet Gynecol 2009;201(3):269e1–269e10.
[34] Crispi F, Comas M, Hernandez-Andrade E, et al. Does pre-eclampsia influence fetal cardiovascular function in early-onset intrauterine growth restriction? Ultrasound Obstet Gynecol 2009;34(6):660–5.
[35] Tenhola S, Rahiala E, Halonen P, Vanninen E, Voutilainen R. Maternal preeclampsia predicts elevated blood pressure in 12-year-old children: evaluation by ambulatory blood pressure monitoring. Pediatr Res 2006;59(2):320–4.
[36] Seidman DS, Laor A, Gale R, Stevenson DK, Mashiach S, Danon YL. Pre-eclampsia and offspring's blood pressure, cognitive ability and physical development at 17-years-of-age. Br J Obstet Gynaecol 1991;98(10):1009–14.
[37] Kajantie E, Eriksson JG, Osmond C, Thornburg K, Barker DJ. Pre-eclampsia is associated with increased risk of stroke in the adult offspring: the Helsinki birth cohort study. Stroke 2009;40(4):1176–80.
[38] Vikse BE, Irgens LM, Leivestad T, Skjaerven R, Iversen BM. Preeclampsia and the risk of end-stage renal disease. N Engl J Med 2008;359(8):800–9.
[39] Bellamy L, Casas JP, Hingorani AD, Williams DJ. Pre-eclampsia and risk of cardiovascular disease and cancer in later life: systematic review and meta-analysis. BMJ 2007;335(7627):974.
[40] Irgens HU, Reisaeter L, Irgens LM, Lie RT. Long term mortality of mothers and fathers after pre-eclampsia: population based cohort study. BMJ 2001;323(7323):1213–7.
[41] Gaugler-Senden IP, Berends AL, de Groot CJ, Steegers EA. Severe, very early onset preeclampsia: subsequent pregnancies and future parental cardiovascular health. Eur J Obstet Gynecol Reprod Biol 2008;140(2):171–7.

[42] Habli M, Eftekhari N, Wiebracht E, et al. Long-term maternal and subsequent pregnancy outcomes 5 years after hemolysis, elevated liver enzymes, and low platelets (HELLP) syndrome. Am J Obstet Gynecol 2009;201(4):385e1–5.
[43] Nisell H, Lintu H, Lunell NO, Mollerstrom G, Pettersson E. Blood pressure and renal function seven years after pregnancy complicated by hypertension. Br J Obstet Gynaecol 1995;102(11):876–81.
[44] Chen CW, Jaffe IZ, Karumanchi SA. Pre-eclampsia and cardiovascular disease. Cardiovasc Res 2014;101(4):579–86.
[45] American College of Obstetricians and Gynecologists, Task Force on Hypertension in Pregnancy. Report of the American College of Obstetricians and Gynecologists' Task Force on Hypertension in Pregnancy. Obstet Gynecol 2013;122(5):1122–31.
[46] Higgins JR, de Swiet M. Blood-pressure measurement and classification in pregnancy. Lancet 2001;357(9250):131–5.
[47] Gangaram R, Ojwang PJ, Moodley J, Maharaj D. The accuracy of urine dipsticks as a screening test for proteinuria in hypertensive disorders of pregnancy. Hypertens Pregnancy 2005;24(2):117–23.
[48] Al RA, Baykal C, Karacay O, Geyik PO, Altun S, Dolen I. Random urine protein-creatinine ratio to predict proteinuria in new-onset mild hypertension in late pregnancy. Obstet Gynecol 2004;104(2):367–71.
[49] Cote AM, Brown MA, Lam E, et al. Diagnostic accuracy of urinary spot protein:creatinine ratio for proteinuria in hypertensive pregnant women: systematic review. BMJ 2008;336(7651):1003–6.
[50] Conde-Agudelo A, Villar J, Lindheimer M. World Health Organization systematic review of screening tests for preeclampsia. Obstet Gynecol 2004;104(6):1367–91.
[51] Conde-Agudelo A, Romero R, Lindheimer M. Tests to predict preeclampsia. In: Lindheimer MD, Roberts JM, Cunningham FG, Chesley LC, editors. Chesley's hypertensive disorders in pregnancy. 3rd ed. Amsterdam, Boston: Academic Press; 2009. p. 189–211.
[52] Noori M, Donald AE, Angelakopoulou A, Hingorani AD, Williams DJ. Prospective study of placental angiogenic factors and maternal vascular function before and after preeclampsia and gestational hypertension. Circulation 2010;122(5):478–87.
[53] Koga K, Osuga Y, Yoshino O, et al. Elevated serum soluble vascular endothelial growth factor receptor 1 (sVEGFR-1) levels in women with preeclampsia. J Clin Endocrinol Metab 2003;88(5):2348–51.
[54] Maynard SE, Min JY, Merchan J, et al. Excess placental soluble fms-like tyrosine kinase 1 (sFlt1) may contribute to endothelial dysfunction, hypertension, and proteinuria in preeclampsia. J Clin Invest 2003;111(5):649–58.
[55] Tidwell SC, Ho HN, Chiu WH, Torry RJ, Torry DS. Low maternal serum levels of placenta growth factor as an antecedent of clinical preeclampsia. Am J Obstet Gynecol 2001;184(6):1267–72.
[56] Polliotti BM, Fry AG, Saller DN, Mooney RA, Cox C, Miller RK. Second-trimester maternal serum placental growth factor and vascular endothelial growth factor for predicting severe, early-onset preeclampsia. Obstet Gynecol 2003;101(6): 1266–74.
[57] Thadhani R, Mutter WP, Wolf M, et al. First trimester placental growth factor and soluble fms-like tyrosine kinase 1 and risk for preeclampsia. J Clin Endocrinol Metab 2004;89(2):770–5.
[58] Espinoza J, Romero R, Nien JK, et al. Identification of patients at risk for early onset and/or severe preeclampsia with the use of uterine artery Doppler velocimetry and placental growth factor. Am J Obstet Gynecol 2007;196(4):326e1–326e13.
[59] Hertig A, Berkane N, Lefevre G, et al. Maternal serum sFlt1 concentration is an early and reliable predictive marker of preeclampsia. Clin Chem 2004;50(9):1702–3.

[60] Chaiworapongsa T, Romero R, Kim YM, et al. Plasma soluble vascular endothelial growth factor receptor-1 concentration is elevated prior to the clinical diagnosis of pre-eclampsia. J Matern Fetal Neonatal Med 2005;17(1):3–18.
[61] Lim JH, Kim SY, Park SY, et al. Soluble endoglin and transforming growth factor-beta1 in women who subsequently developed preeclampsia. Prenat Diagn 2009;29(5):471–6.
[62] Lim JH, Kim SY, Park SY, Yang JH, Kim MY, Ryu HM. Effective prediction of preeclampsia by a combined ratio of angiogenesis-related factors. Obstet Gynecol 2008;111(6):1403–9.
[63] Stepan H, Unversucht A, Wessel N, Faber R. Predictive value of maternal angiogenic factors in second trimester pregnancies with abnormal uterine perfusion. Hypertension 2007;49(4):818–24.
[64] Stepan H, Geipel A, Schwarz F, Kramer T, Wessel N, Faber R. Circulatory soluble endoglin and its predictive value for preeclampsia in second-trimester pregnancies with abnormal uterine perfusion. Am J Obstet Gynecol 2008;198(2):175e1–6.
[65] Diab AE, El-Behery MM, Ebrahiem MA, Shehata AE. Angiogenic factors for the prediction of pre-eclampsia in women with abnormal midtrimester uterine artery Doppler velocimetry. Int J Gynaecol Obstet 2008;102(2):146–51.
[66] De Vivo A, Baviera G, Giordano D, Todarello G, Corrado F, D'Anna R. Endoglin, PlGF and sFlt-1 as markers for predicting pre-eclampsia. Acta Obstet Gynecol Scand 2008;87(8):837–42.
[67] Kusanovic JP, Romero R, Chaiworapongsa T, et al. A prospective cohort study of the value of maternal plasma concentrations of angiogenic and anti-angiogenic factors in early pregnancy and midtrimester in the identification of patients destined to develop preeclampsia. J Matern Fetal Neonatal Med 2009;22(11):1021–38.
[68] Rana S, Karumanchi SA, Levine RJ, et al. Sequential changes in antiangiogenic factors in early pregnancy and risk of developing preeclampsia. Hypertension 2007;50(1):137–42.
[69] Vatten LJ, Eskild A, Nilsen TI, Jeansson S, Jenum PA, Staff AC. Changes in circulating level of angiogenic factors from the first to second trimester as predictors of pre-eclampsia. Am J Obstet Gynecol 2007;196(3):239e1–6.
[70] Erez O, Romero R, Espinoza J, et al. The change in concentrations of angiogenic and anti-angiogenic factors in maternal plasma between the first and second trimesters in risk assessment for the subsequent development of preeclampsia and small-for-gestational age. J Matern Fetal Neonatal Med 2008;21(5):279–87.
[71] Moore-Simas TA, Crawford SL, Solitro MJ, Frost SC, Meyer BA, Maynard SE. Angiogenic factors for the prediction of preeclampsia in high-risk women. Am J Obstet Gynecol 2007;197(3):244e1–8.
[72] Myatt L, Clifton RG, Roberts JM, et al. Can changes in angiogenic biomarkers between the first and second trimesters of pregnancy predict development of pre-eclampsia in a low-risk nulliparous patient population? BJOG 2013;120(10):1183–91.
[73] Sibai BM, Koch MA, Freire S, et al. Serum inhibin A and angiogenic factor levels in pregnancies with previous preeclampsia and/or chronic hypertension: are they useful markers for prediction of subsequent preeclampsia? Am J Obstet Gynecol 2008;199(3):268e1–9.
[74] Kim MY, Buyon JP, Guerra MM, et al. Angiogenic factor imbalance early in pregnancy predicts adverse outcomes in patients with lupus and antiphospholipid antibodies: results of the PROMISSE study. Am J Obstet Gynecol 2016;214(1):108e1–108e14.
[75] McElrath TF, Lim KH, Pare E, et al. Longitudinal evaluation of predictive value for preeclampsia of circulating angiogenic factors through pregnancy. Am J Obstet Gynecol 2012;207(5):407e1–7.
[76] Widmer M, Cuesta C, Khan KS, et al. Accuracy of angiogenic biomarkers at 20 weeks' gestation in predicting the risk of pre-eclampsia: a WHO multicentre study. Pregnancy Hypertens 2015;5(4):330–8.

[77] Rana S, Venkatesha S, DePaepe M, Chien EK, Paglia M, Karumanchi SA. Cytomegalovirus-induced mirror syndrome associated with elevated levels of circulating antiangiogenic factors. Obstet Gynecol 2007;109(2 Pt2):549–52.
[78] Espinoza J, Chaiworapongsa T, Romero R, et al. Unexplained fetal death: another anti-angiogenic state. J Matern Fetal Neonatal Med 2007;20(7):495–507.
[79] Whitten AE, Romero R, Korzeniewski SJ, et al. Evidence of an imbalance of angiogenic/antiangiogenic factors in massive perivillous fibrin deposition (maternal floor infarction): a placental lesion associated with recurrent miscarriage and fetal death. Am J Obstet Gynecol 2013;208(4):310e1–310e11.
[80] Signore C, Mills JL, Qian C, et al. Circulating soluble endoglin and placental abruption. Prenat Diagn 2008;28(9):852–8.
[81] Chaiworapongsa T, Espinoza J, Gotsch F, et al. The maternal plasma soluble vascular endothelial growth factor receptor-1 concentration is elevated in SGA and the magnitude of the increase relates to Doppler abnormalities in the maternal and fetal circulation. J Matern Fetal Neonatal Med 2008;21(1):25–40.
[82] Taylor RN, Grimwood J, Taylor RS, McMaster MT, Fisher SJ, North RA. Longitudinal serum concentrations of placental growth factor: evidence for abnormal placental angiogenesis in pathologic pregnancies. Am J Obstet Gynecol 2003;188(1):177–82.
[83] Romero R, Nien JK, Espinoza J, et al. A longitudinal study of angiogenic (placental growth factor) and anti-angiogenic (soluble endoglin and soluble vascular endothelial growth factor receptor-1) factors in normal pregnancy and patients destined to develop preeclampsia and deliver a small for gestational age neonate. J Matern Fetal Neonatal Med 2008;21(1):9–23.
[84] Levine RJ, Thadhani R, Qian C, et al. Urinary placental growth factor and risk of preeclampsia. JAMA 2005;293(1):77–85.
[85] Aggarwal PK, Jain V, Sakhuja V, Karumanchi SA, Jha V. Low urinary placental growth factor is a marker of pre-eclampsia. Kidney Int 2006;69(3):621–4.
[86] Buhimschi CS, Norwitz ER, Funai E, et al. Urinary angiogenic factors cluster hypertensive disorders and identify women with severe preeclampsia. Am J Obstet Gynecol 2005;192(3):734–41.
[87] Savvidou MD, Akolekar R, Zaragoza E, Poon LC, Nicolaides KH. First trimester urinary placental growth factor and development of pre-eclampsia. BJOG 2009;116(5):643–7.
[88] Salahuddin S, Lee Y, Vadnais M, Sachs BP, Karumanchi SA, Lim KH. Diagnostic utility of soluble fms-like tyrosine kinase 1 and soluble endoglin in hypertensive diseases of pregnancy. Am J Obstet Gynecol 2007;197(1):28e1–6.
[89] Perni U, Sison C, Sharma V, et al. Angiogenic factors in superimposed preeclampsia: a longitudinal study of women with chronic hypertension during pregnancy. Hypertension 2012;59(3):740–6.
[90] Verlohren S, Herraiz I, Lapaire O, et al. The sFlt-1/PlGF ratio in different types of hypertensive pregnancy disorders and its prognostic potential in preeclamptic patients. Am J Obstet Gynecol 2012;206(1):58e1–8.
[91] Vaisbuch E, Whitty JE, Hassan SS, et al. Circulating angiogenic and antiangiogenic factors in women with eclampsia. Am J Obstet Gynecol 2011;204(2):152e1–9.
[92] Rolfo A, Attini R, Nuzzo AM, et al. Chronic kidney disease may be differentially diagnosed from preeclampsia by serum biomarkers. Kidney Int 2013;83(1):177–81.
[93] Young B, Levine RJ, Salahuddin S, et al. The use of angiogenic biomarkers to differentiate non-HELLP related thrombocytopenia from HELLP syndrome. J Matern Fetal Neonatal Med 2010;23(5):366–70.
[94] Shan HY, Rana S, Epstein FH, Stillman IE, Karumanchi SA, Williams ME. Use of circulating antiangiogenic factors to differentiate other hypertensive disorders from preeclampsia in a pregnant woman on dialysis. Am J Kidney Dis 2008;51(6):1029–32.

[95] Ohkuchi A, Hirashima C, Suzuki H, et al. Evaluation of a new and automated electrochemiluminescence immunoassay for plasma sFlt-1 and PlGF levels in women with preeclampsia. Hypertens Res 2010;33(5):422–7.
[96] Schiettecatte J, Russcher H, Anckaert E, et al. Multicenter evaluation of the first automated Elecsys sFlt-1 and PlGF assays in normal pregnancies and preeclampsia. Clin Biochem 2010;43(9):768–70.
[97] Sunderji S, Gaziano E, Wothe D, et al. Automated assays for sVEGF R1 and PlGF as an aid in the diagnosis of preterm preeclampsia: a prospective clinical study. Am J Obstet Gynecol 2010;202(1):40e1–7.
[98] Verlohren S, Galindo A, Schlembach D, et al. An automated method for the determination of the sFlt-1/PIGF ratio in the assessment of preeclampsia. Am J Obstet Gynecol 2010;202(2):161e1–161e11.
[99] Verlohren S, Herraiz I, Lapaire O, et al. New gestational phase-specific cutoff values for the use of the soluble fms-like tyrosine kinase-1/placental growth factor ratio as a diagnostic test for preeclampsia. Hypertension 2014;63(2):346–52.
[100] Rana S, Powe CE, Salahuddin S, et al. Angiogenic factors and the risk of adverse outcomes in women with suspected preeclampsia. Circulation 2012;125(7):911–9.
[101] Chaiworapongsa T, Romero R, Korzeniewski SJ, et al. Plasma concentrations of angiogenic/anti-angiogenic factors have prognostic value in women presenting with suspected preeclampsia to the obstetrical triage area: a prospective study. J Matern Fetal Neonatal Med 2014;27(2):132–44.
[102] Chappell LC, Duckworth S, Seed PT, et al. Diagnostic accuracy of placental growth factor in women with suspected preeclampsia: a prospective multicenter study. Circulation 2013;128(19):2121–31.
[103] Moore AG, Young H, Keller JM, et al. Angiogenic biomarkers for prediction of maternal and neonatal complications in suspected preeclampsia. J Matern Fetal Neonatal Med 2012;25(12):2651–7.
[104] Zeisler H, Llurba E, Chantraine F, et al. Predictive value of the sFlt-1:PLGF ratio in women with suspected preeclampsia. N Engl J Med 2016;374(1):13–22.
[105] Baltajian K, Bajracharya S, Salahuddin S, et al. Sequential plasma angiogenic factors levels in women with suspected preeclampsia. Am J Obstet Gynecol 2016;215(1):89e1–89e10.
[106] Palomaki GE, Haddow JE, Haddow HR, et al. Modeling risk for severe adverse outcomes using angiogenic factor measurements in women with suspected preterm preeclampsia. Prenat Diagn 2015;35(4):386–93.
[107] Tsiakkas A, Saiid Y, Wright A, Wright D, Nicolaides KH. Competing risks model in screening for preeclampsia by maternal factors and biomarkers at 30-34 weeks' gestation. Am J Obstet Gynecol 2016;215(1):87e1–87e17.
[108] Klein E, Schlembach D, Ramoni A, et al. Influence of the sFlt-1/PlGF ratio on clinical decision-making in women with suspected preeclampsia. PloS One 2016;11(5):e0156013.
[109] Bohn H, Kraus W, Winckler W. Purification and characterization of two new soluble placental tissue proteins (PP13 and PP17). Oncodev Biol Med 1983;4(5):343–50.
[110] Visegrady B, Than NG, Kilar F, Sumegi B, Than GN, Bohn H. Homology modelling and molecular dynamics studies of human placental tissue protein 13 (galectin-13). Protein Eng 2001;14(11):875–80.
[111] Than NG, Pick E, Bellyei S, et al. Functional analyses of placental protein 13/galectin-13. Eur J Biochem 2004;271(6):1065–78.
[112] Huppertz B, Sammar M, Chefetz I, Neumaier-Wagner P, Bartz C, Meiri H. Longitudinal determination of serum placental protein 13 during development of preeclampsia. Fetal Diagn Ther 2008;24(3):230–6.

[113] Burger O, Pick E, Zwickel J, et al. Placental protein 13 (PP-13): effects on cultured trophoblasts, and its detection in human body fluids in normal and pathological pregnancies. Placenta 2004;25(7):608–22.
[114] Chafetz I, Kuhnreich I, Sammar M, et al. First-trimester placental protein 13 screening for preeclampsia and intrauterine growth restriction. Am J Obstet Gynecol 2007;197(1):35e1–7.
[115] Romero R, Kusanovic JP, Than NG, et al. First-trimester maternal serum PP13 in the risk assessment for preeclampsia. Am J Obstet Gynecol 2008;199(2):122e1–122e11.
[116] Spencer K, Cowans NJ, Chefetz I, Tal J, Meiri H. First-trimester maternal serum PP-13, PAPP-A and second-trimester uterine artery Doppler pulsatility index as markers of pre-eclampsia. Ultrasound Obstet Gynecol 2007;29(2):128–34.
[117] Gonen R, Shahar R, Grimpel YI, et al. Placental protein 13 as an early marker for preeclampsia: a prospective longitudinal study. BJOG 2008;115(12):1465–72.
[118] Khalil A, Cowans NJ, Spencer K, Goichman S, Meiri H, Harrington K. First trimester maternal serum placental protein 13 for the prediction of pre-eclampsia in women with a priori high risk. Prenat Diagn 2009;29(8):781–9.
[119] Spencer K, Cowans NJ, Chefetz I, Tal J, Kuhnreich I, Meiri H. Second-trimester uterine artery Doppler pulsatility index and maternal serum PP13 as markers of preeclampsia. Prenat Diagn 2007;27(3):258–63.
[120] Lawrence JB, Oxvig C, Overgaard MT, et al. The insulin-like growth factor (IGF)-dependent IGF binding protein-4 protease secreted by human fibroblasts is pregnancy-associated plasma protein-A. Proc Natl Acad Sci USA 1999;96(6):3149–53.
[121] Giudice LC, Conover CA, Bale L, et al. Identification and regulation of the IGFBP-4 protease and its physiological inhibitor in human trophoblasts and endometrial stroma: evidence for paracrine regulation of IGF-II bioavailability in the placental bed during human implantation. J Clin Endocrinol Metab 2002;87(5):2359–66.
[122] Hamilton GS, Lysiak JJ, Han VK, Lala PK. Autocrine-paracrine regulation of human trophoblast invasiveness by insulin-like growth factor (IGF)-II and IGF-binding protein (IGFBP)-1. Exp Cell Res 1998;244(1):147–56.
[123] Oxvig C, Sand O, Kristensen T, Gleich GJ, Sottrup-Jensen L. Circulating human pregnancy-associated plasma protein-A is disulfide-bridged to the proform of eosinophil major basic protein. J Biol Chem 1993;268(17):12243–6.
[124] Spencer CA, Allen VM, Flowerdew G, Dooley K, Dodds L. Low levels of maternal serum PAPP-A in early pregnancy and the risk of adverse outcomes. Prenat Diagn 2008;28(11):1029–36.
[125] Spencer K, Cowans NJ, Nicolaides KH. Low levels of maternal serum PAPP-A in the first trimester and the risk of pre-eclampsia. Prenat Diagn 2008;28(1):7–10.
[126] Morris RK, Cnossen JS, Langejans M, et al. Serum screening with Down's syndrome markers to predict pre-eclampsia and small for gestational age: systematic review and meta-analysis. BMC Pregnancy Childbirth 2008;8:33.
[127] D'Anna R, Baviera G, Giordano D, et al. First trimester serum PAPP-A and NGAL in the prediction of late-onset pre-eclampsia. Prenat Diagn 2009;29(11):1066–8.
[128] Lim KH, Friedman SA, Ecker JL, Kao L, Kilpatrick SJ. The clinical utility of serum uric acid measurements in hypertensive diseases of pregnancy. Am J Obstet Gynecol 1998;178(5):1067–71.
[129] Thangaratinam S, Ismail KM, Sharp S, Coomarasamy A, Khan KS. Accuracy of serum uric acid in predicting complications of pre-eclampsia: a systematic review. BJOG 2006;113(4):369–78.
[130] Meads CA, Cnossen JS, Meher S, et al. Methods of prediction and prevention of preeclampsia: systematic reviews of accuracy and effectiveness literature with economic modelling. Health Technol Assess 2008;12(6):iii–iv. 1–270.

[131] Morris RK, Riley RD, Doug M, Deeks JJ, Kilby MD. Diagnostic accuracy of spot urinary protein and albumin to creatinine ratios for detection of significant proteinuria or adverse pregnancy outcome in patients with suspected pre-eclampsia: systematic review and meta-analysis. BMJ 2012;345:e4342.
[132] Millar JG, Campbell SK, Albano JD, Higgins BR, Clark AD. Early prediction of pre-eclampsia by measurement of kallikrein and creatinine on a random urine sample. Br J Obstet Gynaecol 1996;103(5):421–6.
[133] Kyle PM, Campbell S, Buckley D, et al. A comparison of the inactive urinary kallikrein:creatinine ratio and the angiotensin sensitivity test for the prediction of pre-eclampsia. Br J Obstet Gynaecol 1996;103(10):981–7.
[134] Walknowska J, Conte FA, Grumbach MM. Practical and theoretical implications of fetal-maternal lymphocyte transfer. Lancet 1969;1(7606):1119–22.
[135] Lo YM, Corbetta N, Chamberlain PF, et al. Presence of fetal DNA in maternal plasma and serum. Lancet 1997;350(9076):485–7.
[136] Poon LL, Leung TN, Lau TK, Lo YM. Circulating fetal RNA in maternal plasma. Ann NY Acad Sci 2001;945:207–10.
[137] Klonisch T, Drouin R. Fetal-maternal exchange of multipotent stem/progenitor cells: microchimerism in diagnosis and disease. Trends Mol Med 2009;15(11):510–8.
[138] Chiu RW, Lo YM. The biology and diagnostic applications of fetal DNA and RNA in maternal plasma. Curr Top Dev Biol 2004;61:81–111.
[139] Dennis Lo YM, Chiu RW. Prenatal diagnosis: progress through plasma nucleic acids. Nat Rev Genet 2007;8(1):71–7.
[140] Maddocks DG, Alberry MS, Attilakos G, et al. The SAFE project: towards non-invasive prenatal diagnosis. Biochem Soc Trans 2009;37(Pt. 2):460–5.
[141] Holzgreve W, Ghezzi F, Di Naro E, Ganshirt D, Maymon E, Hahn S. Disturbed feto-maternal cell traffic in preeclampsia. Obstet Gynecol 1998;91(5 Pt. 1):669–72.
[142] Holzgreve W, Li JJ, Steinborn A, et al. Elevation in erythroblast count in maternal blood before the onset of preeclampsia. Am J Obstet Gynecol 2001;184(2):165–8.
[143] Al-Mufti R, Hambley H, Albaiges G, Lees C, Nicolaides KH. Increased fetal erythroblasts in women who subsequently develop pre-eclampsia. Hum Reprod 2000;15(7):1624–8.
[144] Lo YM, Leung TN, Tein MS, et al. Quantitative abnormalities of fetal DNA in maternal serum in preeclampsia. Clin Chem 1999;45(2):184–8.
[145] Zhong XY, Holzgreve W, Hahn S. The levels of circulatory cell free fetal DNA in maternal plasma are elevated prior to the onset of preeclampsia. Hypertens Pregnancy 2002;21(1):77–83.
[146] Leung TN, Zhang J, Lau TK, Chan LY, Lo YM. Increased maternal plasma fetal DNA concentrations in women who eventually develop preeclampsia. Clin Chem 2001;47(1):137–9.
[147] Levine RJ, Qian C, Leshane ES, et al. Two-stage elevation of cell-free fetal DNA in maternal sera before onset of preeclampsia. Am J Obstet Gynecol 2004;190(3):707–13.
[148] Crowley A, Martin C, Fitzpatrick P, et al. Free fetal DNA is not increased before 20 weeks in intrauterine growth restriction or pre-eclampsia. Prenat Diagn 2007;27(2):174–9.
[149] Cotter AM, Martin CM, O'Leary JJ, Daly SF. Increased fetal DNA in the maternal circulation in early pregnancy is associated with an increased risk of preeclampsia. Am J Obstet Gynecol 2004;191(2):515–20.
[150] Tsui DW, Chan KC, Chim SS, et al. Quantitative aberrations of hypermethylated RASSF1A gene sequences in maternal plasma in pre-eclampsia. Prenat Diagn 2007;27(13):1212–8.
[151] Cotter AM, Martin CM, O'Leary JJ, Daly SF. Increased fetal RhD gene in the maternal circulation in early pregnancy is associated with an increased risk of pre-eclampsia. BJOG 2005;112(5):584–7.

[152] Farina A, Sekizawa A, Iwasaki M, Matsuoka R, Ichizuka K, Okai T. Total cell-free DNA (beta-globin gene) distribution in maternal plasma at the second trimester: a new prospective for preeclampsia screening. Prenat Diagn 2004;24(9):722–6.
[153] Ng EK, Tsui NB, Lau TK, et al. mRNA of placental origin is readily detectable in maternal plasma. Proc Natl Acad Sci USA 2003;100(8):4748–53.
[154] Purwosunu Y, Sekizawa A, Okazaki S, et al. Prediction of preeclampsia by analysis of cell-free messenger RNA in maternal plasma. Am J Obstet Gynecol 2009;200(4):386e1–7.
[155] Cnossen JS, Morris RK, ter Riet G, et al. Use of uterine artery Doppler ultrasonography to predict pre-eclampsia and intrauterine growth restriction: a systematic review and bivariable meta-analysis. CMAJ 2008;178(6):701–11.
[156] Nicolaides KH, Bindra R, Turan OM, et al. A novel approach to first-trimester screening for early pre-eclampsia combining serum PP-13 and Doppler ultrasound. Ultrasound Obstet Gynecol 2006;27(1):13–7.
[157] Akolekar R, Syngelaki A, Beta J, Kocylowski R, Nicolaides KH. Maternal serum placental protein 13 at 11-13 weeks of gestation in preeclampsia. Prenat Diagn 2009;29(12):1103–8.
[158] Diesch CH, Holzgreve W, Hahn S, Zhong XY. Comparison of activin A and cell-free fetal DNA levels in maternal plasma from patients at high risk for preeclampsia. Prenat Diagn 2006;26(13):1267–70.
[159] Sifakis S, Zaravinos A, Maiz N, Spandidos DA, Nicolaides KH. First-trimester maternal plasma cell-free fetal DNA and preeclampsia. Am J Obstet Gynecol 2009;201(5): 472e1–7.
[160] Poon LC, Kametas NA, Maiz N, Akolekar R, Nicolaides KH. First-trimester prediction of hypertensive disorders in pregnancy. Hypertension 2009;53(5):812–8.
[161] O'Gorman N, Wright D, Syngelaki A, et al. Competing risks model in screening for preeclampsia by maternal factors and biomarkers at 11-13 weeks gestation. Am J Obstet Gynecol 2016;214(1):103e1–103e12.
[162] Fan AC, Deb-Basu D, Orban MW, et al. Nanofluidic proteomic assay for serial analysis of oncoprotein activation in clinical specimens. Nat Med 2009;15(5):566–71.
[163] Cho WC, Cheng CH. Oncoproteomics: current trends and future perspectives. Expert Rev Proteomics 2007;4(3):401–10.
[164] Quackenbush J. Microarray analysis and tumor classification. N Engl J Med 2006;354(23):2463–72.
[165] Magnussen EB, Vatten LJ, Smith GD, Romundstad PR. Hypertensive disorders in pregnancy and subsequently measured cardiovascular risk factors. Obstet Gynecol 2009;114(5):961–70.
[166] Ping P. Getting to the heart of proteomics. N Engl J Med 2009;360(5):532–4.
[167] Chen CH, Budas GR, Churchill EN, Disatnik MH, Hurley TD, Mochly-Rosen D. Activation of aldehyde dehydrogenase-2 reduces ischemic damage to the heart. Science 2008;321(5895):1493–5.
[168] Arab S, Gramolini AO, Ping P, et al. Cardiovascular proteomics: tools to develop novel biomarkers and potential applications. J Am Coll Cardiol 2006;48(9):1733–41.
[169] Farina A, Sekizawa A, Purwosunu Y, et al. Quantitative distribution of a panel of circulating mRNA in preeclampsia versus controls. Prenat Diagn 2006;26(12): 1115–20.
[170] Tsui NB, Lo YM. A microarray approach for systematic identification of placental-derived RNA markers in maternal plasma. Methods Mol Biol 2008;444:275–89.
[171] Okazaki S, Sekizawa A, Purwosunu Y, Farina A, Wibowo N, Okai T. Placenta-derived, cellular messenger RNA expression in the maternal blood of preeclamptic women. Obstet Gynecol 2007;110(5):1130–6.
[172] Founds SA, Conley YP, Lyons-Weiler JF, Jeyabalan A, Hogge WA, Conrad KP. Altered global gene expression in first trimester placentas of women destined to develop preeclampsia. Placenta 2009;30(1):15–24.

[173] Farina A, Morano D, Arcelli D, et al. Gene expression in chorionic villous samples at 11 weeks of gestation in women who develop preeclampsia later in pregnancy: implications for screening. Prenat Diagn 2009;29(11):1038–44.
[174] Luense LJ, Carletti MZ, Christenson LK. Role of Dicer in female fertility. Trends Endocrinol Metab 2009;20(6):265–72.
[175] Pineles BL, Romero R, Montenegro D, et al. Distinct subsets of microRNAs are expressed differentially in the human placentas of patients with preeclampsia. Am J Obstet Gynecol 2007;196(3):261e1–6.
[176] Zhu XM, Han T, Sargent IL, Yin GW, Yao YQ. Differential expression profile of microRNAs in human placentas from preeclamptic pregnancies vs normal pregnancies. Am J Obstet Gynecol 2009;200(6):661e1–7.
[177] Luo SS, Ishibashi O, Ishikawa G, et al. Human villous trophoblasts express and secrete placenta-specific microRNAs into maternal circulation via exosomes. Biol Reprod 2009;81(4):717–29.
[178] Chim SS, Shing TK, Hung EC, et al. Detection and characterization of placental microRNAs in maternal plasma. Clin Chem 2008;54(3):482–90.
[179] Blumenstein M, McMaster MT, Black MA, et al. A proteomic approach identifies early pregnancy biomarkers for preeclampsia: novel linkages between a predisposition to preeclampsia and cardiovascular disease. Proteomics 2009;9(11):2929–45.
[180] Buhimschi IA, Zhao G, Funai EF, et al. Proteomic profiling of urine identifies specific fragments of SERPINA1 and albumin as biomarkers of preeclampsia. Am J Obstet Gynecol 2008;199(5):551e1–551e16.
[181] Buhimschi IA, Nayeri UA, Zhao G, et al. Protein misfolding, congophilia, oligomerization, and defective amyloid processing in preeclampsia. Sci Transl Med 2014;6(245):245ra92.
[182] Heazell AE, Brown M, Dunn WB, et al. Analysis of the metabolic footprint and tissue metabolome of placental villous explants cultured at different oxygen tensions reveals novel redox biomarkers. Placenta 2008;29(8):691–8.
[183] Dunn WB, Brown M, Worton SA, et al. Changes in the metabolic footprint of placental explant-conditioned culture medium identifies metabolic disturbances related to hypoxia and pre-eclampsia. Placenta 2009;30(11):974–80.
[184] Kenny LC, Broadhurst D, Brown M, et al. Detection and identification of novel metabolomic biomarkers in preeclampsia. Reprod Sci 2008;15(6):591–7.
[185] Kenny LC, Dunn WB, Ellis DI, Baker PN, GOPEC Consortium, Kell DB. Novel biomarkers for pre-eclampsia detected using metabolomics and machine learning. Metabolomics 2005;1(3):227–34.
[186] Kenny LC, Broadhurst DI, Dunn W, et al. Robust early pregnancy prediction of later preeclampsia using metabolomic biomarkers. Hypertension 2010;56(4):741–9.
[187] Sitras V, Paulssen RH, Gronaas H, et al. Differential placental gene expression in severe preeclampsia. Placenta 2009;30(5):424–33.
[188] Nishizawa H, Pryor-Koishi K, Kato T, Kowa H, Kurahashi H, Udagawa Y. Microarray analysis of differentially expressed fetal genes in placental tissue derived from early and late onset severe pre-eclampsia. Placenta 2007;28(5–6):487–97.
[189] Jin H, Ma KD, Hu R, et al. Analysis of expression and comparative profile of normal placental tissue proteins and those in preeclampsia patients using proteomic approaches. Anal Chim Acta 2008;629(1–2):158–64.
[190] Park JS, Oh KJ, Norwitz ER, et al. Identification of proteomic biomarkers of preeclampsia in amniotic fluid using SELDI-TOF mass spectrometry. Reprod Sci 2008;15(5):457–68.
[191] Watanabe H, Hamada H, Yamada N, et al. Proteome analysis reveals elevated serum levels of clusterin in patients with preeclampsia. Proteomics 2004;4(2):537–43.
[192] Karumanchi SA. Angiogenic factors in preeclampsia: from diagnosis to therapy. Hypertension 2016;67(6):1072–9.

INDEX

B

D

J

K

N

O

P

Q

R

S

T

U

Printed in the United States
By Bookmasters